MILL OPERATORS CONFERENCE 2024

21–23 OCTOBER 2024
PERTH, AUSTRALIA

The Australasian Institute of Mining and Metallurgy
Publication Series No 9/2024

AusIMM

Published by:
The Australasian Institute of Mining and Metallurgy
Ground Floor, 204 Lygon Street, Carlton Victoria 3053, Australia

ISBN 978-1-922395-41-2

ADVISORY COMMITTEE

David Seaman
MAusIMM
Conference Advisory Committee Co-Chair

Rob Coleman
MAusIMM
Conference Advisory Committee Co-Chair

Katie Barns
MAusIMM

Stuart Emery
MAusIMM

Bianca Newcombe
MAusIMM

Paul Petrucci
MAusIMM

Reg Radford
MAusIMM

Jacqueline Reinhold
AAusIMM

AusIMM

Julie Allen
Head of Events

Joelle Glenister
Senior Manager, Events

Vicki Melhuish
Manager, Events

REVIEWERS

We would like to thank the following people for their contribution towards enhancing the quality of the papers included in this volume:

Peter Allen
MAusIMM(CP)

Katie Barns
MAusIMM

Michael Becker
MAusIMM(CP)

Duncan Bennett
FAusIMM

Ben Bonfils

Fraser Burns

Rob Coleman
MAusIMM

Adrian Dance
FAusIMM

Diana Drinkwater
FAusIMM

Stuart Emery
MAusIMM

Karyn Gardner
MAusIMM

Aidan Giblett
FAusIMM(CP)

Steven Hart
FAusIMM

David Hunter
MAusIMM

Carla Kaboth
FAusIMM(CP)

Fred Kock
FAusIMM

Mike Larson

Virginia Lawson
FAusIMM(CP)

Bianca Newcombe
MAusIMM

Andrew Newell
FAusIMM(CP)

Joe Pease
FAusIMM

Paul Petrucci
MAusIMM

Reg Radford
MAusIMM

Laurie Reemeyer
FAusIMM(CP)

Jacqueline Reinhold
AAusIMM

Brigitte Seaman
MAusIMM

David Seaman
MAusIMM

Joe Seppelt
MAusIMM

Berge Simonian
AAusIMM

Steven Sinclair
MAusIMM(CP)

Christopher Smith
MAusIMM

Luke Vollert

Tom Waters
MAusIMM

FOREWORD

On behalf of AusIMM, the AusIMM Metallurgical Society (MetSoc), and this year's Advisory Committee, we are delighted to welcome you to the Mill Operators Conference 2024.

Since its inception in Mt Isa in 1978, the Mill Operators Conference has consistently promoted the sharing of knowledge in mineral processing plant operations, covering areas such as extractive metallurgy, process control and environmental issues. This year is the 16th iteration of the conference, and we are particularly excited to return to a wholly in-person format in Perth after two conferences in Brisbane. This marks a return to normalcy and re-establishes our biennial routine.

'MillOps' remains the premier mineral processing conference in Australia and is AusIMM's largest conference in this field. The theme of sharing best practices, plant updates, optimisation outcomes and showcasing the latest technologies continues. The term 'Mill Operators' includes the many contributors to our industry's success, such as vendors, technology providers, universities and engineering firms.

This conference has evolved with our industry. We hope it provides opportunities for plant operators, metallurgists, engineers and suppliers to reconnect, learn and network. Advances in technology have transformed our industry, allowing us to interact in new ways. We challenge you to seize these opportunities, whether through this conference or other professional development avenues.

We deeply appreciate our sponsors and exhibitors for their unwavering support. Your contributions make this event possible. Collaboration is fundamental to our industry, and we encourage you to engage with exhibitors and learn about their latest advancements.

In preparing the conference program we received over 100 abstract submissions. Authors, committee members and technical reviewers worked tirelessly to select and fine-tune the most relevant content, producing high-quality proceedings. We have chosen to maintain a single stream of presentations at the conference to provide a more immersive experience for our delegates with a focus on sharing and collaboration.

It is not only the formal speakers and authors who contribute to the sharing of knowledge in the technical sessions, but also our delegates in the Q&As. We encourage you all to participate in the discussions and to share your experiences while learning from others.

We extend our gratitude to the committee members and AusIMM team for their hard work and professional expertise in bringing this conference together. Their efforts are crucial to the success of the Mill Operators Conference.

Yours faithfully,

David Seaman *MAusIMM* and Rob Coleman *FAusIMM*
Mill Operators Conference 2024 Advisory Committee Co-Chairs

SPONSORS

Major Conference Sponsor

MOLYCOP

Platinum Sponsors

Metso

RME RUSSELL MINERAL EQUIPMENT

Gold Sponsors

ME elecmetal

stm MINERALS

SYENSQO

Silver Sponsors

CLARIANT

PiONERA
Pioneering solutions in mineral processing

SCIDEV

xylem

Conference Proceedings Sponsor

MAGOTTEAUX

Stand-up Conference Dinner Sponsor

SEDGMAN

Coffee Cart Sponsors

Strategy Innovation Sponsor

Name Badge and Lanyard Sponsor

Technical Session Sponsors

Conference App Sponsor

Lunch Sponsor

Notepad and Pen Sponsor

Destination Partner

BUSINESS
EVENTS
PERTH

Supporting Partner

PERTH CONVENTION AND
EXHIBITION CENTRE
WESTERN AUSTRALIA

CONTENTS

Best practice

Practical geometallurgy – and let there be light 3
D W Bennett and P D Munro

Bean counting, how hard can it be? – common mistakes in metallurgical accounting 17
D Felipe and P Guerney

Design and delivery of a Professional Certificate in Metal Accounting 27
G B Gnoinski, K M McCaffery and J Jessop

Mill relining automation – applying lessons learned 39
S Gwynn-Jones, T A Ogden, J Bohorquez, D Sims, M Turner, C Kramer and S Smith

Optimising grinding circuit performance – a success story at Harmony Gold's Hidden Valley 59
Mine with Rockwell Automation's Pavilion8™ MPC process optimisation technology
A Jain, T Tlhobo, L Silva and F Anis

The effect of primary grind size on flotation recovery and the theoretical grade-recovery curve 69
at Olympic Dam
Y Li, K Ehrig, V Liebezeit and S Barsby

The journey of process improvement – Ok Tedi's vision 81
H Liang, L Brown and B Wong

Applying geometallurgical principles for metallurgical sample selection 95
I T Lipton, P G Greenhill and L Torres

The evolution of level control and its role in advanced control in the Northparkes 107
flotation circuit
C P Lowes, M D Curtis and M J Garside

Optimisation from mine to mill to mine at Newmont Lihir gold mine 121
J Moilanen, A Remes, G Peachey and J Kaartinen

Strategic ore planning – using metallurgical performance data to drive mine planning for 133
operational optimisation: case study 1 (oxides versus sulfides)
B Newcombe and G Newcombe

Metallurgical test work – the beginning of the end or the end of the beginning? 147
A J H Newell, P D Munro and K Fiedler

Practical considerations in the design of flotation process twins for set point decision 159
automation at MMG Dugald River Mine
W Smith, A Malcolm, I Goode, G Bai and C Martin

Operation updates

A review of operating data over 50 years of secondary crushing prior to AG/SAG mills 177
R Chandramohan and G Lane

Pioneering coarse particle flotation – transformative insights from five years of operation   195
    C Haines, L Vollert, W Downie, I Heath and R Ghattas

Gruyere Gold Mine Western Australia – Part 2 – ramp up to operations, a journey of   211
comminution, optimisation and challenges
    G Heard, J Stevens, M Becker and R Radford

Mastering clay and oxide ores at St Ives Gold Mine – easing circuit constraints and   231
bottlenecks with time-honoured comminution, leach CIP, and tailings circuit wizardry!
    E Mort, M Simpson, B van Saarloos and R Radford

Navigating the operational challenges at Campo Morado – a success story   249
    R Whittering, A Bill, C Meinke, J Mendoza, A Villalvazo and R Chandramohan

Optimisation and improvement

Managing froth tenacity at BHP Carrapateena   271
    F Burns, J Reinhold, J Seppelt, S Assmann, and G Tsatouhas

CSA mine new SAG mills optimisation using grindcurves   299
    G D Figueroa Salguero, B Palmer, S Ntamuhanga and J Buckman

Regrinding – a subtle mix of liberation and chemistry   313
    C J Greet and B Shean

Improved classification with the Cavex® DE hydrocyclone for mill circuit, coarse particle   323
flotation, and tailings applications
    J J Hanhiniemi, J Heo, N Weerasekara, E Wang, H Thanasekaran and A Kilcullen

Improved cone crusher chamber design and operation using design and simulation tools   343
    J J Hanhiniemi, N S Weerasekara and S Buaseng

Grinding circuit expansion and optimisation at the B2Gold Masbate Project, Philippines   359
    A Insalada, K Bartholomew, A J Marcera, E Occena, M Anghag, D Torres, N Avenido,
    J Rajala and R E McIvor

Red Chris flotation circuit expansion – from piloting to full scale   385
    K Li, D R Seaman, B A Seaman and J Baldock

The optimisation of a vertical regrind mill at Renaissance Minerals, Okvau Mine   401
    A Paz, A Siphanya and K Vansana

Whiskey, puddings, shampoo and insanity – a taxonomy of circulating loads   413
    J D Pease and P D Munro

Rockmedia   427
    M S Powell, A N Mainza, L M Tavares and S Kanchibotla

Tooling up for future innovation – turning overlooked process opportunities to immediate and   445
long-term benefit
    M S Powell

Modelling to prepare for and mitigate the impact of increasing ore hardness at Lihir   463
    L Pyle, A Rice, P Griffin, B Seaman, W Valery, K Duffy and A Jankovic

Optimisation of Jameson Cells to improve concentrate quality at BHP Carrapateena   475
    J Reinhold, J Van Sliedregt, F Burns, A Price and J Seppelt

Optimising flotation pulp conditions to improve recovery at Newmont Red Chris Mine   495
    D R Seaman, J Johannson, J Baldock and B A Seaman

Regrind test work review – analysis of procedures and results   509
    J Thomson, B Foggiatto, G Ballantyne and G Lane

Continuous improvement of the Carrapateena grinding circuit   525
    P Toor, J Seppelt, F Burns, J Reinhold, W Valery, L Brennan and K Duffy

BRC copper processing – recovery improvements   541
    I A Torok and J Begelhole

Optimising ball mill grinding circuits – it's not just the mill   551
    J Zela and B Cornish

Sustainability

Plant design comparison of dry VRM milling plus magnetic separation with AG and ball milling   573
plus magnetic separation for Grange Resources' Southdown ore
    D David, D Olwagen, C Gerold, C Schmitz, S Baaken, C Stanton and M Everitt

Redefining the battery limits of processing plants – improving sustainability through the   587
deployment of sensing technologies
    W Futcher, D R Seaman and B Klein

Scats – what are they good for?   605
    T McCredden and C Fitzmaurice

Author index   615

Best practice

Practical geometallurgy – and let there be light

D W Bennett[1] and P D Munro[2]

1. FAusIMM, Principal Consultant, Mineralis Consultants Pty Ltd, Taringa Qld 4068.
   Email: dbennett@mineralis.com.au
2. FAusIMM, Principal Consultant, Mineralis Consultants Pty Ltd, Taringa Qld 4068.
   Email: pmunro@mineralis.com.au

ABSTRACT

Mining and processing operations that do not understand the characteristics of their deposits survive as did J R R Tolkien's Gollum; in the dark until a metallurgical crisis drags them out to search for a precious solution. Geometallurgy is about knowing metallurgical and production outcomes <u>before</u> ore is mined and processed and requires that the key drivers of these outcomes are attributes in the mine block model. Practical geometallurgy is the use of deposit geological 'style' characteristics and the common drivers of performance for that deposit style to generate these attributes. This allows the common features of the deposit with others of the same style with operating history to be 'banked' while focusing attention on differences discovered during a geometallurgical program. Mineralogy controls metallurgy, so a practical geometallurgy program is about measuring the important characteristics of the basis lithology, alteration, and weathering units in the deposit such as mineralogy, mineral associations, mineral liberation, and mineral texture before embarking on extensive and expensive metallurgical test programs. This paper describes the general outline of a practical geometallurgy program from sampling and retaining the characteristics of ore in 3D mineralised space to typical analysis and test programs and using the results of these programs to develop geometallurgical models to populate the block model. Geometallurgy case studies for some common deposit styles are included to give examples of the consistent drivers of performance inherent in each style.

INTRODUCTION

Geometallurgy is about knowing metallurgical and production outcomes <u>before</u> ore is mined and processed. To know these outcomes, the engineering block model (the mining version of the geology block model) must be populated with measured parameters or proxies (attributes) that allow prediction of metallurgical performance. Populating the engineering block model also eliminates time as a variable, allowing changes to the mine production schedule without impacting the predicted value of outcomes from processing an individual ore block.

Geometallurgy practice includes words like 'geology', and 'mining'. Hence there cannot be robust and practical geometallurgy outcomes without engaging the other two key disciplines for mineral processing operations – geology and mining.

Standardising geological and geotechnical measures and practices combined with setting up of a strategic geometallurgy program outline should commence before any holes are drilled, and certainly before samples are submitted for assay. Although it is better to be late starting a geometallurgy program than never, having an early outline in place will prevent re-work and missed key measures, and more importantly mitigate the fundamental business risk of not accurately predicting and achieving planned future production and costs. A common problem is that different disciplines will store their data in different software packages and in different locations – a geometallurgy program should have all data on a 'single tab' so that the assay, geology, geotechnical, mineralogy and metallurgy measures can all be provenanced back to a location in 3D mineralised space.

The geological style of deposit will focus the activities in a geometallurgy program plan and provide a reference case. A geometallurgy axiom is to 'bank the similarities, master the differences' as it is rare that a deposit style is unique. Geological differences or uncertainties of style may also not be geometallurgically important, for example Irish-type lead-zinc deposits and Mississippi Valley type lead-zinc deposits can be considered identical from a metallurgist's perspective.

Geotechnical measures of rock hardness taken during routine core logging can provide important information for geometallurgical ore hardness and comminution modelling. For example, measures such as rock quality designation (RQD) and Point Load Index (PLi) and other measures of rock competence are indicators of the rock's resistance to comminution (Morrell, personal communications, 2019).

Another geometallurgy axiom is to 'measure more, test less'. Measuring the geotechnical characteristics, mineralogy, mineral association, mineral liberation, and mineral texture <u>up front</u> allows benchmarking against other deposits of the same style and will highlight any potential differences or operating or metallurgical challenges. Testing then becomes confirmatory rather than exploratory, minimising the overall costs and time of a program. An analogy for proceeding directly to testing lacking these vital data is that it can create a 'results crime scene'; equivalent to having a body lying on the floor and no idea how it got there. You are then forced to put on a Sherlock Holmes hat and try to gather evidence which takes time and money, delays the program, and creates anxiety for project stakeholders because a metallurgical criminal is on the loose, and may never be caught. If the full characterisation measures are done upfront, to quote Captain Louis Renault in *Casablanca*, you can 'round up the usual suspects' preventing the 'results crime' from being committed.

Depending upon how common the deposit style, a suitable geometallurgy program plan can be developed well before a hole is drilled. For example, a geometallurgy plan including all the important assays, tests, and mineralogy measures for porphyry copper deposits, orogenic gold deposits, and sedimentary exhalative (SEDEX) lead-zinc deposits can be developed to a very high level of detail to capture the key drivers of metallurgical performance without even having a deposit.

From a processing perspective, deposits normally fit into one of two categories; throughput dominated, or metallurgy dominated, and a geometallurgy plan should recognise which of these categories the deposit under investigation is in. Porphyry copper deposits are a good example of throughput dominated – large and low-grade, achieving throughput is critical to the project economics, while volcanogenic massive sulfide (VMS) Cu-Au-Ag-Pb-Zn deposits are a good example of metallurgy dominated – achieving good recoveries into multiple concentrates of acceptable quality is critical. So the geometallurgy plan for a porphyry copper deposit must have comprehensive ore hardness characterisation, while the geometallurgy plan for a VMS will strongly focus on the variable mineralogy, mineral associations, and liberation size of the key minerals.

Including all the key geometallurgical measures and drivers as algorithm-based models of throughput and metallurgical performance into the block model is rarely an experience in achieving outstanding accuracy. Geometallurgy models are often performance trends due to compounding errors in sampling, measuring, and testing combined with the 'noisy' environment of process operations treating variable material. Therefore, care should be taken to avoid over-complicated models and making predictions over short time periods. The purpose of the geometallurgical models is important to understand – are they for life-of-mine, annual, quarterly, monthly production planning, or for short-term performance prediction? Models that don't capture the key drivers of performance or require constant 'tweaking' or that include variables that can't be attributes of the block model are not useful.

THE REFERENCE CASE

Most deposits are members of a geological style. It is rare that a deposit has truly unique geochemistry, rock types, and mineralogy. Some styles are very common, and this allows new deposits of the same style to be easily benchmarked against the large operating data set, with just the measures of mineralogy, mineral association, mineral liberation and texture able to provide a supportable estimate of likely metallurgical performance.

An example of deposit style benchmarking is presented in Figure 1, with a SEDEX Pb-Zn-Ag deposit under study compared with the Broken Hill, Mount Isa, and McArthur River SEDEX Pb-Zn-Ag deposits. The metallurgical performance of this deposit style is heavily influenced by the degree of metamorphism and the resultant textural complexity. Sphalerite, galena and pyrite grain sizes increase, the naturally hydrophilic carbonaceous gangue content decreases, and the spherical/framboidal form of pyrite which is finally intergrown with sphalerite is depleted with increasing degree of metamorphism. Figure 1 shows that the SEDEX under study is 'better' than McArthur River, but 'worse' than Mount Isa, and metallurgical performance (lead and zinc recoveries and concentrate grades) will be between these two.

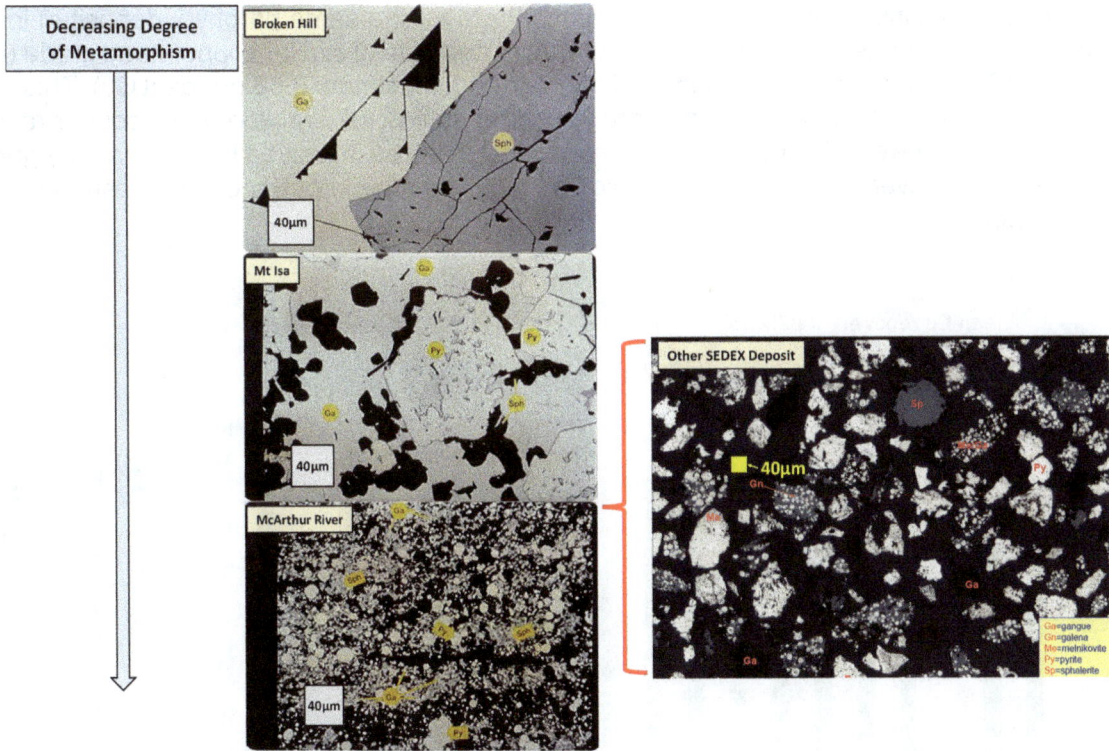

FIG 1 – Benchmarking of a SEDEX deposit against some well-established SEDEX operations.

A second example of deposit style benchmarking is shown in Figure 2 with copper sulfide mineral liberation at the optimum P_{80} primary grind size for a porphyry deposit of interest in comparison with 77 other porphyry deposits. The mean optimum P_{80} primary grind size for the 77 other deposits is 175 µm, while the deposit of interest is 150 µm. Even with the finer grind, copper sulfide mineral liberation is only 42 per cent against a mean of 52 per cent for the other 77 deposits.

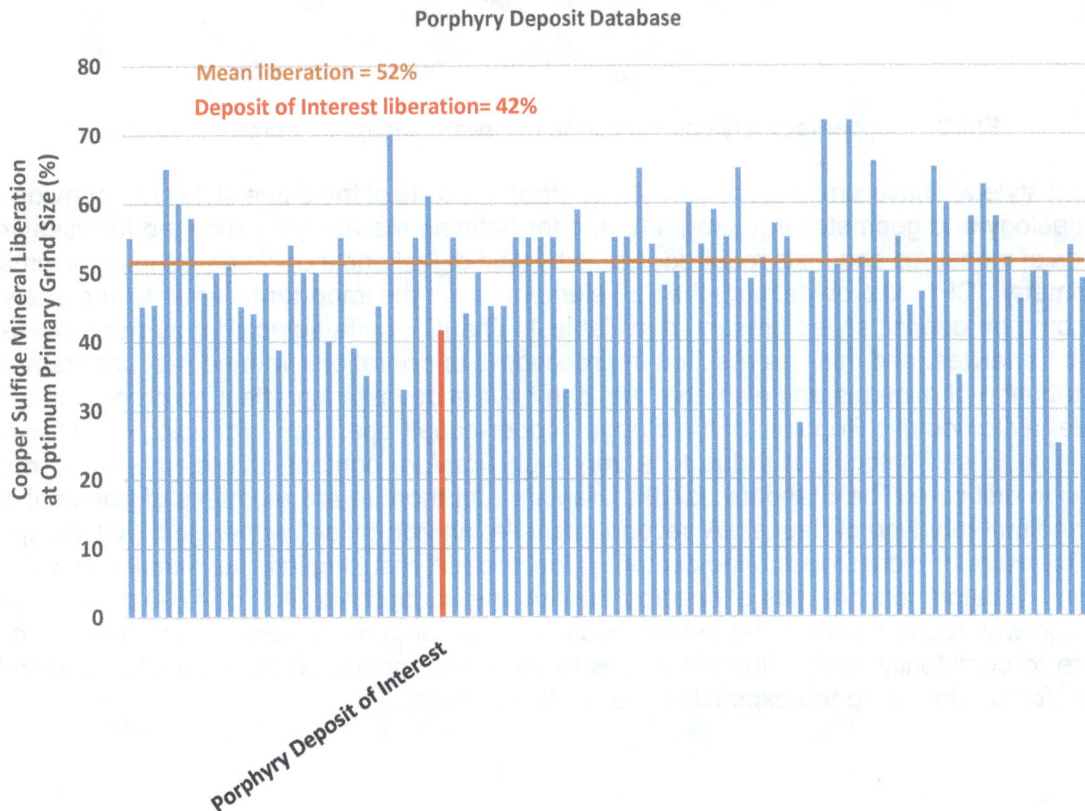

FIG 2 – Porphyry deposits copper sulfide mineral liberation at optimum primary grind size.

Figure 3 compares copper recovery into final concentrate for the same porphyry deposit of interest (head grade of 0.48 per cent Cu) and 25 other porphyry deposits of similar copper head grade (mean head grade of 0.48 per cent Cu, minimum 0.37 per cent Cu, maximum 0.66 per cent Cu). This shows the negative influence of the lower-than-average copper sulfide mineral liberation; copper recovery is almost 5 per cent lower. This copper recovery of course can be improved by finer primary grinding and regrinding. However, this is less economic and highlights the importance of measuring the key drivers of performance.

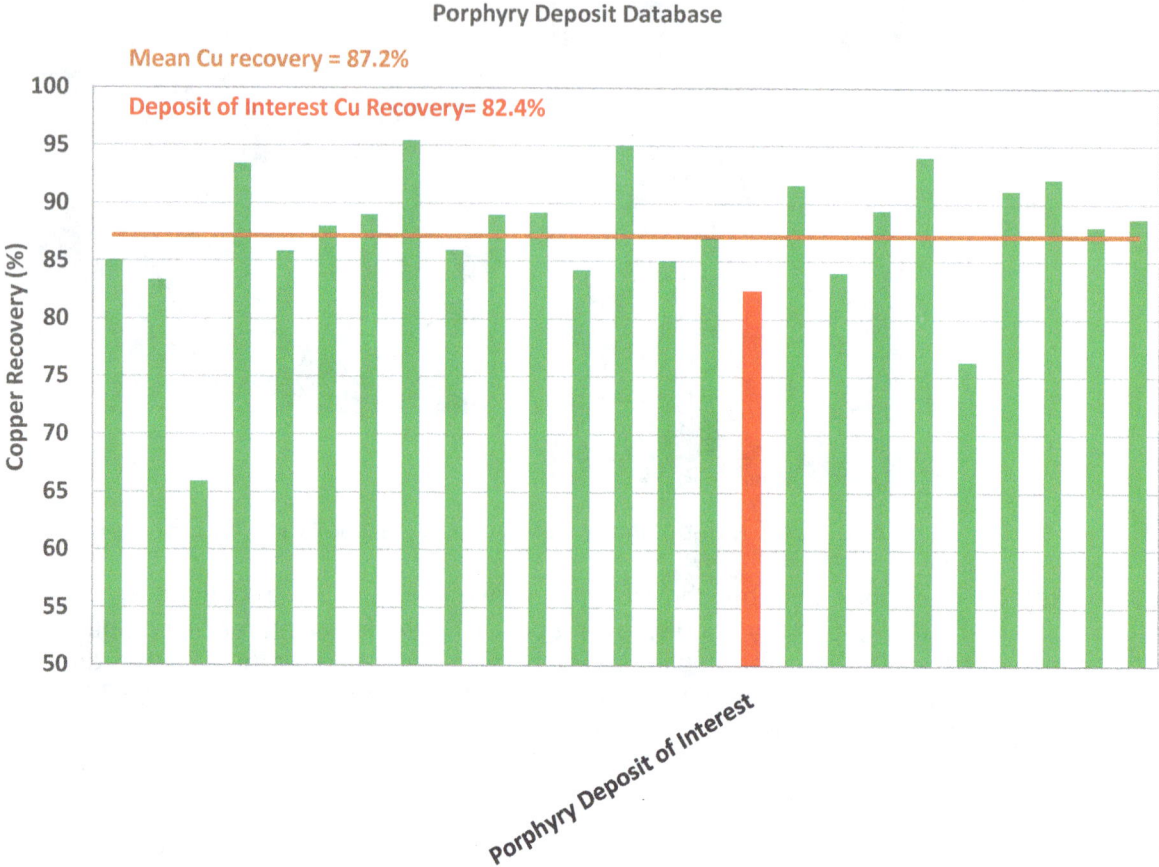

FIG 3 – Copper recovery comparison with similar head grade porphyry deposits.

A deposit style will have similar geochemistry to other deposits of the same style. This provides focus to the geology and geometallurgy program, and for defining the analytes required for assaying and the types of measures and tests required. Assay by acid digest and inductively coupled plasma mass spectrometry (ICP) will provide more than 30 elements, with the important elements and analytes for some common deposit styles presented in Table 1. The ICP suite selected may not provide some important analytes, and this has resulted in proceeding to operations with no real understanding of the distribution of minerals and elements that can have a devastating effect on product quality. For example, a Democratic Republic of the Congo copper-cobalt operation producing cobalt hydroxide was unable to sell batches of it due to a very high uranium content; this element had not been assayed in drill core (The Standard, 2019). Uranium and thorium are elements of potential interest in any sedimentary type of deposit. A second example; an iron-oxide copper-gold (IOCG) operation did not measure or model the fluorine content in the deposit, resulting in initial metallurgical test work producing flotation concentrates exceeding the fluorine acceptance limit of 1200 ppm. Although a correlation was found between the barium assay and the fluorine content, this was not sufficiently accurate to confidently assign fluorine values to the block model, so core samples had to be re-assayed for fluorine using the expensive specific ion method.

TABLE 1

Important analytes for some major deposit styles – analytes in **bold** are not measured in a 47 element 4-acid digest and ICP assay.

Deposit style	Important analytes	Important penalty/toxic analytes	Processing category
Porphyry Copper Deposit – Island Arc	Cu, Fe, S, **Au**, Ag, **Acid Soluble Cu**, **Cyanide Soluble Cu**,	As, Pb, Zn	Throughput dominated
Porphyry Copper Deposit – Cordilleran	Cu, Mo, Fe, S, **Au**, Ag, **Acid Soluble Cu**, **Cyanide Soluble Cu**	As, Pb, Zn	Throughput dominated
Iron Oxide Copper – Gold Deposit	Cu, Fe, S, Au, Ag	Bi, Pb, Zn, **F**, U	Throughput dominated
Volcanogenic Massive Sulfide	Cu, Pb, Zn, Au, Ag, Fe, S	As, Sb, **Hg**	Metallurgy dominated
SEDEX	Pb, Zn, Ag, Fe, S	**C(organic)**, **SiO₂**	Metallurgy dominated
Low – Intermediate Sulfidation Epithermal Gold-Silver	**Au**, Ag, S, **Cyanide Soluble Au and Ag**, **C(organic)**	**Hg**, Se	Throughput dominated
Orogenic/Archean Gold (Newmont Ltd, 2021)	**Au – fire assay, screen fire assay, NaCN soluble Au**, Ag	**S – total, sulfide, sulfate. C – carbonate, organic, preg-robbing capacity**, **Hg**	Throughput dominated

THE 'LAW' OF GEOMETALLURGICAL SAMPLING

Populating the block model with performance-predicting attributes or proxies requires that the metallurgically important characteristics (key drivers) of each block are first determined. Samples need to be provenanced to a location in 3D mineralised space, and the characteristics of that location applied to an ore block in the model. When starting a geometallurgy program, the key drivers are not defined and quantified, even if they may be recognised as likely to be key drivers. Samples are therefore required to retain these key drivers so that they may be measured.

Besides element grades which are always measured, these drivers may be included in the 'LAW' (lithology, alteration and weathering) characteristics of the material so these should form the boundary of geological discontinuities <u>and</u> of each sample. Material with identical lithology, alteration, and weathering characteristics and therefore similar geochemistry can be reasonably expected to behave the same regardless of its location, while crossing boundaries risks blending metallurgically different features. An example of a visually obvious geological boundary is presented in Figure 4, this boundary forms the sampling limit. Sampling across the boundary means that the characteristics of both become mixed, inseparable, and the key drivers of performance cannot be determined.

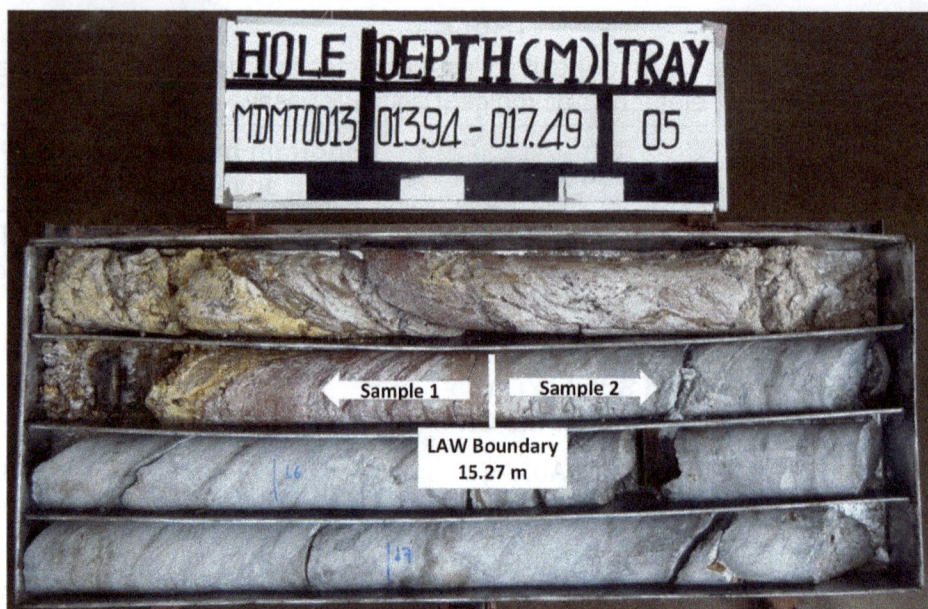

FIG 4 – Core photo showing a geological and sampling boundary.

Mixing samples to form composites of sufficient mass for testing must be done with <u>extreme</u> care, because as the old saying goes – one bad apple can spoil the whole barrel, and it is impossible to 'deconstruct' a composite back into its individual components. Compositing for metallurgical testing introduces time as a variable into 3D mineralised space, ie it assumes that this is how the material will be mined, blended, and processed. A credible mine production schedule is normally the last thing to evolve in the general sequence of study activities for the mining and processing of a deposit.

An example of the consequences of compositing occurred during a test program on a copper-lead-zinc VMS deposit when a sample with a significant talc content was included with other samples to make a master composite for flotation testing. The hydrophobic talc floated into the copper concentrate, making it impossible to produce a saleable copper product due to the low copper grade and high fluorine content. Ironically, the sample with talc came from outside the mine wireframe, thus creating a metallurgical 'problem' that did not even exist. An example of compositing disguising variability was from a low sulfidation epithermal gold-silver deposit, when approximately 80 samples with a 'similar' oxidation level were combined into a bulk composite for cyanide leach testing. The results disguised that the gold in some of the samples was very fine and poorly liberated at the primary grind size. During cyanide leaching operations, all was tragically revealed with long periods of gold recoveries less than 10 per cent when treating this material.

Samples are selected based on LAW geology logging, and even though automated scanning is becoming more common, most logging is still the result of visual inspection and manual measurements. These logging results are what will end up in the block model, so samples must align both spatially and with the LAW characteristics of each block within the block model.

COMMINUTION GEOMETALLURGY

Geotechnical measures of rock competency and hardness taken during core logging can provide important information for geometallurgical ore hardness and comminution modelling. For example, rock quality designation (RQD) and Point Load Index (PLi) and other measures of competence such as core recovery are all potential indicators of the rock's resistance to comminution (Morrell, personal communications, 2019). Some of these measure correlations are consistent across all deposits, for example PLi from testing of lump samples corrected to the 50 mm drill core reference dimension by multiplying the lump result by 1.54 (Morrell, personal communications, 2013) with the SMC test Drop Weight Index (DWi) is presented in Figure 5. Note that the correlation R^2 values range between 0.59 and 0.72 showing that the correlation is fairly weak, but with hundreds or thousands of PLi measures available compared with typically less than 100 SMC tests for a large deposit, ore hardness and resulting semi-autogenous grinding (SAG) mill throughput can be modelled with reasonable confidence. Having a PLi-based geometallurgical block model can also provide life-of-mine planning capability if core throughout the deposit has been routinely tested during logging; zones of high hardness can be delineated and tested in greater detail to help investment case development for future comminution equipment requirements.

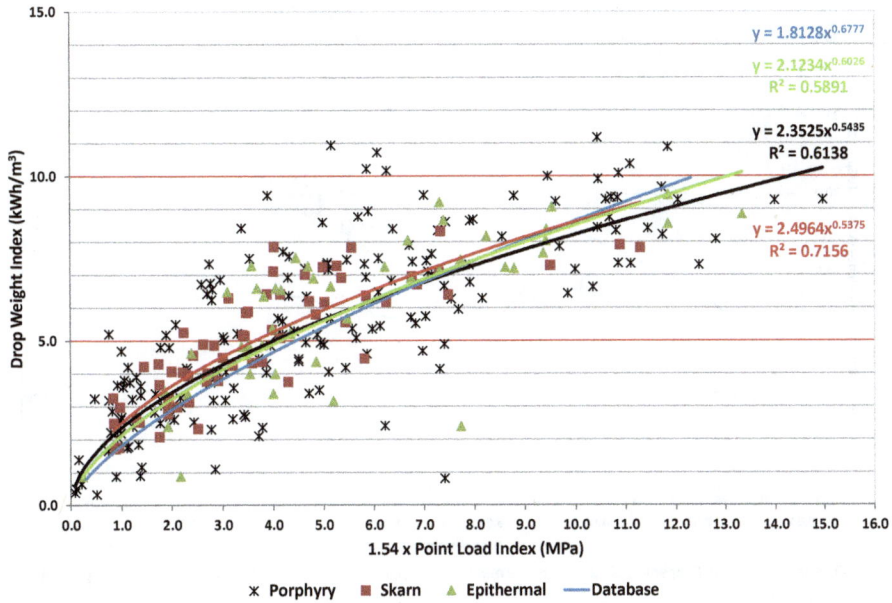

FIG 5 – PLi measures of lump material corrected to 50 mm reference dimension and correlation with SMC test DWi.

Other common and simple measures can also be used to give an indication of rock competency. Porphyry deposits form at depth with a vertical pressure and temperature gradient, which can lead to higher rock competency, hardness, and specific gravity at greater depths. For example, the mean measured DWi for each 25 m of depth interval below surface is presented in Figure 6 for a large copper-gold Island Arc porphyry deposit.

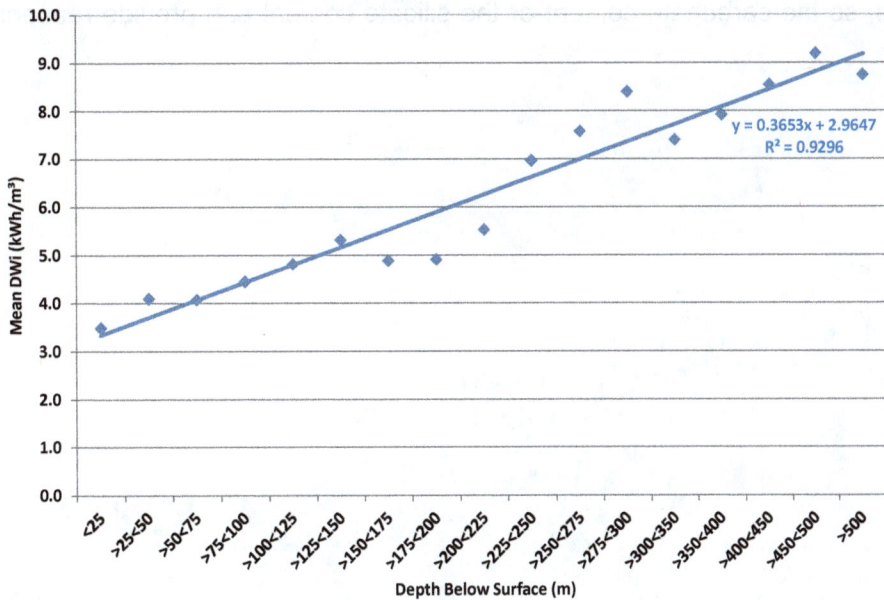

FIG 6 – SMC test DWi for 25 m depth intervals for a porphyry deposit.

Even simpler methods can be used successfully, for example a porphyry-skarn deposit treating three main rock types applied mill throughput values to each rock type with a self-learning empirical modelling method using the SAG mill as the analytical instrument, resulting in monthly throughput prediction errors falling to within ±1.5 per cent of actual (Carpenter and Saunders, 2017).

Bond Ball Mill Work Index (BWi) is often more closely related to individual mineral grain hardness than rock competency as shown in Figure 7 by the inconsistent association between mean DWi and BWi (normalised to a 106 μm closing screen size) values across multiple deposit styles. The marker size indicates the number of test results for each deposit.

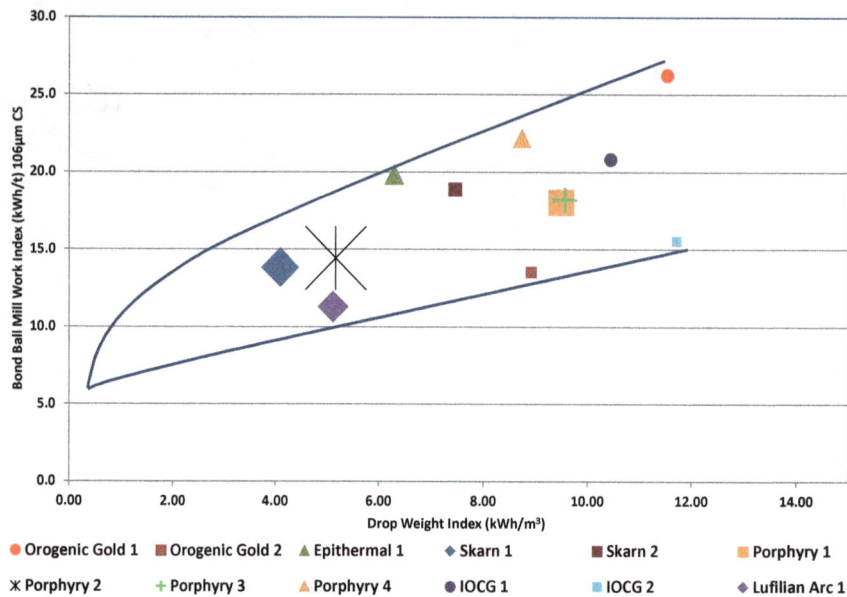

FIG 7 – BWi versus SMC test DWi for different deposit styles.

This mineral grain hardness relationship to BWi can be exploited for geometallurgical modelling. Intelligent use of multi-element data in the block model supported by Quantitative X-ray diffraction (QXRD) can predict rock type, and with other measures such as specific gravity can be used to estimate hardness (Li *et al*, 2021). An example of this is presented in Figure 8, with calcium and magnesium assays proxies for the carbonate minerals content of a Lufilian Arc sediment hosted stratabound copper-cobalt deposit. The majority of non-carbonate minerals in these deposits are harder silicates, so the carbonate content or the silicate content can provide reasonable hardness models.

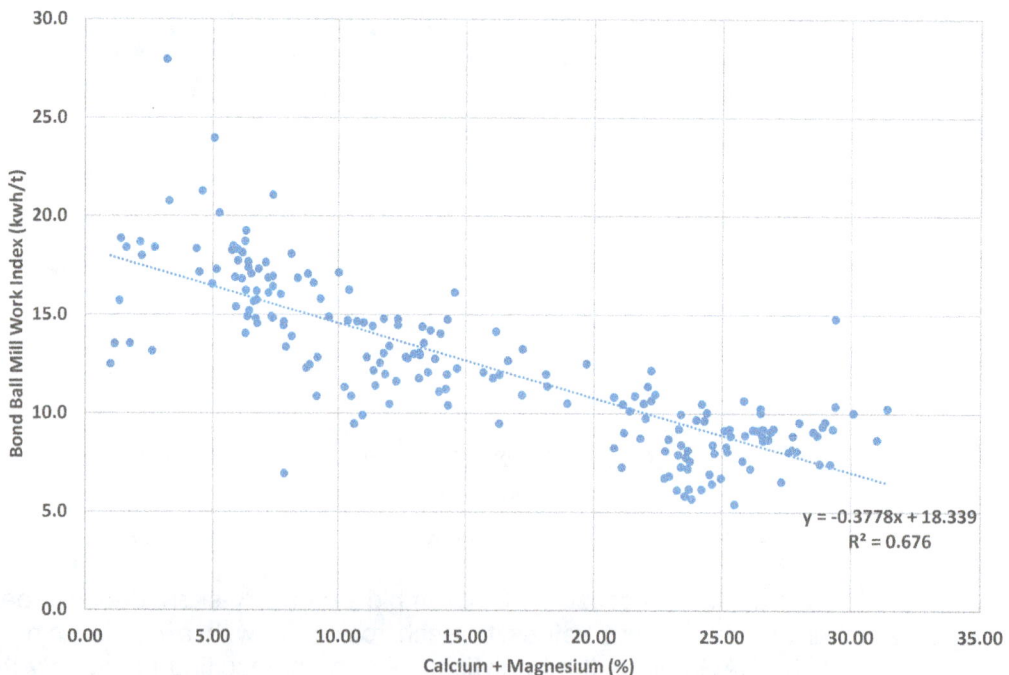

FIG 8 – BWi versus Calcium + Magnesium % for a Lufilian Arc copper-cobalt deposit.

Table 2 gives some typical measures that may be useful for ore hardness modelling and subsequent throughput modelling in combination with the grinding circuit design.

TABLE 2

Some typical ore hardness proxy measures.

Attribute	Model output	Comments
PLi	Axb, DWi	The relationship between PLi and DWi (kWh/m³) is consistent for all deposit styles and can provide a lot of data points at low cost compared with comminution testing
RQD	Axb, DWi	RQD is a measured indicator of rock competency and may have a significant correlation with Axb and DWi
Specific gravity, depth from surface	Axb, DWi	Specific gravity can provide a correlation with mineralogy and usually increases with depth and can have a correlation with comminution parameters such as Axb, DWi, and BWi.
Lithology	Axb, DWi	A lithology (rock type) may be of consistent hardness and give an Axb, DWi, BWi or simple throughput value of acceptable accuracy for modelling
Silicate/carbonate grade	BWi	The combined silicate minerals content has been observed to be positively correlated with BWi at numerous sites, while carbonate content may be negatively correlated.
Sulfur grade	Axb, BWi	Massive sulfide deposits have been observed to have a strong correlation between assayed sulfur grade and BWi and Axb, with increasing sulfur grade resulting in decreasing BWi and Axb.

PROCESSING GEOMETALLURGY

The geometallurgy axioms of 'bank the similarities, master the differences' and 'measure more, test less' generally apply, but especially to processing geometallurgy. A failure to thoroughly understand the deposit mineralogy including the mineral associations, the minerals particle grain size and liberation size, and the minerals texture will risk getting poor metallurgical test results that just cannot be explained after test products are pulverised for assay. Pressure then comes on to 'fix the problem', impacting on study schedule and cost as repeating the sampling and analysis work is often the only possible solution.

A useful analogy for testing before characterising is going straight into surgery before having a check-up.

All samples should therefore be analysed and characterised before any testing is undertaken, with QXRD and QEMSCAN (the chosen discrimination mode is dependent upon mineralogy, grades, and processing requirements) and techniques such as semi-quantitative mineragraphy or Fourier Transform Infra-Red (FTIR) providing important measures. In combination with benchmarking results for the deposit style, the objective is to have good estimates of liberation for primary grind and regrind product sizes, and preliminary test conditions (chemistry, pH, redox, residence time, slurry density etc) available prior to the first test.

A list of mineralogy measuring techniques is provided in Table 3.

TABLE 3

Mineral analysis methods suitable for geometallurgy.

Method	Description	Detail
QEMSCAN®, MLA	Quantitative electron microscopy, mineral liberation analyser	Automated scanning electron microscopy for characterising minerals by type, association, texture, and size. Typically five modes with varying analysis detail: 1. BMA (bulk mineral analysis) which provides fast mineral identification and distribution and quantity. 2. BMAL (bulk mineral analysis with liberation) for mineral composition and estimate of the mineral liberation size. 3. Field image or MLA-XBSE which provides mineralogical and textural information from rock samples. 4. PMA (particle mineral analysis) which characterises the composition of discrete particles. 5. SMS/TMS (specific/trace mineral analysis) or MLA-SPL (sparse phase liberation) which looks in detail for selected minerals.
QXRD	Quantitative X-Ray Diffraction	Used for non-destructive mineral assaying and distribution. Limited capacity for low-level mineral detection and oxide gangue discrimination.
Sequential assay	Multiple wet chemistry methods	Sequential assaying uses a series of discrete assay steps to estimate quantities of different minerals of the same element, and weathering (oxidation) of minerals. Limitations are in discrimination in complex systems such as copper which may have over five mineral species present with two or more species assayed or partly assayed within one step.
Semi-Quantitative Mineragraphy		Optical microscopy and point counting supported by QXRD mineralogy, particularly useful for sulfide mineral characterisation.
FTIR, SWIR	Fourier-Transform and Short Wave Infrared Spectroscopy	Uses reflected light spectra collected across the near, mid and far infra-red spectral ranges to detect and quantify oxide mineral content, particularly useful for certain mineral species such as swelling clays that are undefined by QXRD or QEMSCAN®.

Liberation is critical to economically optimise, but it is the most difficult deposit attribute to model (Preece, Robles and Salazar, 2023). Fortunately most deposits have a reasonably consistent mineral particle size range, but a site geometallurgy program must include measure of liberation of key minerals to ensure mineral particle size won't unexpectedly change and have either a detrimental effect on recovery or present an opportunity to relax grind size if liberation improves.

Once the characterisation is complete, metallurgical testing can commence as confirmation rather than as exploration, minimising the likelihood of surprises in test results. Tests have produced successful results at the first attempt after effective sample characterisation and alignment of test conditions with benchmarked operations treating the same deposit style.

The results of tests are then used to derive geometallurgical models, using the characteristics of the tested samples to determine what the key drivers of performance were. Some typical characteristics that are used for process geometallurgical modelling are presented in Table 4.

TABLE 4

Some typical process geometallurgy modelling measures.

Attribute	Model output	Comments
Head grade	Recovery	Very common driver of process recovery and always available as a block model attribute.
S/Me or Me/S ratio	Recovery/grade	Common in copper, lead and zinc (Me) sulfide flotation process modelling, where the S/Me or Me/S ratio is a proxy for the gangue sulfide/valuable sulfide ratio and gives an indication of position on the final grade-recovery curve.
Metal ratios	Recovery/grade	Certain metal ratios may be used as proxies for degree of weathering and metallurgical response, as some minerals are oxidised before others. For example, the Ag/Au ratio may be a proxy for weathering as silver is depleted while gold is not.
Specific mineral content	Recovery/grade	The presence of certain minerals can have a significant influence on metallurgical performance. If the mineral is associated with a certain lithology or alteration mode or if an element proxy can be found for the mineral (for example, fluorine in talc) it can be an attribute in the block model,
Alteration mode	Recovery/grade	Alteration mode describes a group of mineral types, some of which may be drivers of metallurgical performance. For example, advanced argillic altered material contains many minerals including phyllosilicates and clays that are detrimental to metallurgical and materials handling performance.
Ca, Mg, others	Recovery and acid consumption	Ca, Mg or other elements which are entirely contained in carbonate minerals can provide a proxy for metallurgical performance (carbonate minerals can 'protect' valuable sulfide minerals) and acid consumption in leaching operations.
CO_3, S	Acid generating/ neutralising capacity	Sulfur or sulfide sulfur assays and carbonate mineral assays can be used to classify waste rock and tailings as potentially acid forming or non-acid forming, and provide the likelihood of oxidation during stockpiling and storage.
Conductivity, natural pH, EDTA	Degree of oxidation	Not easily directly measurable on core during logging but can be measured during laboratory tests and on grade control samples to confirm oxidation/weathering state.

GEOMETALLURGICAL MODELLING

The modelling process begins with confirmation of the drivers of performance, and quantification of their influence on processing outcomes. Modelling is an iterative process, the preliminary models developed from the deposit studies must be updated as more data becomes available during operations. Every plant upgrade or process change will require a model review. Best practice is formal detailed annual models review and validation as part of life-of-mine planning and monthly geometallurgy models review as part of production reporting. Trends in performance can then be monitored and models adjusted before becoming unacceptably inaccurate and driving wrong operations behaviours, for example developing to and mining uneconomic material.

Some notes on <u>geometallurgical modelling</u>:

- Model variables must be attributes or attributes by proxy in the block model. Operating conditions such as throughput and grind size cannot be included in models unless they are attributes in the block model.

- Model variables and sign should be intuitive. For example, increasing head grade should give increasing recovery, and increasing acid soluble copper content as a percentage of total copper content should reduce sulfide flotation copper recovery.

- Model variables should be chased back to the mineral that is the driver to confirm and validate their value contribution. For example, if the aluminium assay value provides a good correlation with ore hardness, the aluminium containing mineral that contributes to the ore hardness value should be determined in case other aluminium containing minerals appear and confound the model hardness output.

- Minor element and minor mineral content correlation with model outputs need to be validated. For example, it is unlikely a mineral that is 1 per cent of the ore can have a significant effect on ore hardness.

- Having all characterisation and test data on a 'single tab' allows for fast and simple determination of likely drivers and modelling.

- Models are developed from test work which include drilling and sampling errors, test method errors (Angove and Dunne, 1997), and assay errors. Do not expect model accuracy (on a monthly basis) of better than ±5 per cent relative without a lot of work to characterise ore material and quantify operational variability. Very extensive work on throughput modelling at Minera Los Pelambres in Chile was required to improve monthly throughput estimate accuracy to a mean relative error of 3.0 per cent (Rodriguez *et al*, 2023).

CONCLUSIONS

Geometallurgy is about knowing metallurgical and production outcomes <u>before</u> ore is mined and processed. To know these outcomes, the block model must be populated with measured parameters or proxies (attributes) that allow prediction of throughput and metallurgical performance. Populating the block model also eliminates time as a variable, allowing changes to the mine schedule without impacting the value of outcomes from processing an ore block.

The geological style of a deposit will define the important activities required in a geometallurgy program plan. A geometallurgy axiom is to 'bank the similarities, master the differences' as it is rare that a deposit style is truly unique. Geological differences or uncertainties of style may also not be geometallurgically important.

The geotechnical measures of rock hardness taken during core logging can provide important information for geometallurgical ore hardness and comminution modelling. For example, measures such as rock quality designation (RQD) and Point Load Index (PLi) and other measures of rock competence are all indicators of the rock's resistance to comminution.

The geometallurgy axiom of 'measure more, test less', by measuring the characteristics of the deposit including geotechnical, mineralogy, mineral association, mineral liberation, and mineral texture <u>up front</u> allows benchmarking against other deposits of the same style. Testing then becomes confirmatory rather than exploratory, and the overall costs and time of a program are minimised.

Including all the key geometallurgical measures and drivers as algorithm models of throughput and metallurgical performance into the block model is rarely an experience in achieving outstanding accuracy. Geometallurgy models are often <u>performance trends</u> due to compounding errors in sampling, measuring, and testing combined with the 'noisy' environment of process operations treating variable material. Therefore care should be taken to avoid producing over-complicated models and attempting predictions over short time periods.

Geometallurgical practice and modelling is an iterative ongoing process, the preliminary models developed from the deposit studies must be updated as more data becomes available during operations, and every plant upgrade or process change will require regular model review. Best practice is formal detailed annual models review and validation as part of life-of-mine planning and monthly geometallurgy models review as part of production reporting so trends in performance can be monitored and adjusted before becoming unacceptably inaccurate.

REFERENCES

Angove, J E and Dunne, R C, 1997. A review of standard physical ore property determinations, in *Proceedings World Gold '97 Conference,* pp 139–144 (The Australasian Institute of Mining and Metallurgy: Melbourne).

Carpenter, J and Saunders, B, 2017. Empirical mill throughput modelling and linear programming for blend optimisation at the Phu Kham Copper-Gold Operation, Laos, in *Proceedings Tenth International Mining Geology Conference*, pp 291–296 (The Australasian Institute of Mining and Metallurgy: Melbourne).

Li, Y, Liebezeit, V, Ehrig, K, Smith, M, Pewkliang, B and Macmillan, E, 2021. Predicting mill feed specific energy from Bond ball mill Work index tests at Olympic Dam, in *Proceedings Mill Operators Conference* 2021*, pp 468–479 (The Australasian Institute of Mining and Metallurgy: Melbourne).

Newmont Ltd, 2021. Nevada Operations Nevada, USA Technical Report Summary, p 18. Available from: https://www.sec.gov/Archives/edgar/data/1164727/000116472722000007/exhibit964-nevadagoldmines.htm [Accessed: 25 March 2024].

Preece, R K, Robles, C D and Salazar, A, 2023. Geometallurgy Modelling of the Escondida Deposit, *Mining Metallurgy and Exploration*, 40:1585–1619;1595.

Rodriguez, L, Morales, M, Valery, W, Valle, R, Hayashida, R, Bonfils, B and Plasencia, C, 2023. Developing an Advanced Throughput Forecast Model for Minera Los Pelambres, in Proceedings SAG Conference 2023 (Canadian Institute of Mining, Metallurgy and Petroleum).

The Standard, 1/02/2019. Glencore faces delays in DRC over building of vital cobalt plant. Available from: https://www.standard.co.uk/business/glencore-faces-delays-in-drc-over-building-of-vital-cobalt-plant-a4055186.html [Accessed: 26 March 2024].

Bean counting, how hard can it be? – common mistakes in metallurgical accounting

D Felipe[1] and P Guerney[2]

1. MAusIMM, Senior Process Specialist, JKTech, Indooroopilly Qld 4068.
   Email: d.felipe@jktech.com.au
2. Principal Consultant, MetData, Bellthorpe Qld 4514. Email: phil@metdata.com.au

ABSTRACT

Every production metallurgist will inevitably face metallurgical accounting in their career. It is a vital process to ensure accurate tracking of production through the mine value chain. Despite this, it is commonly perceived as 'bean counting' and relegated to junior members of the metallurgy team, often using a complex web of spreadsheets with minimal documentation.

In recent years, metallurgical accounting has received more attention which has caused many sites to request audits against the recommendations outlined by AMIRA P754 *Principles of Metal Accounting* (Code 1.2 Release 1.3). Typically, sites want the auditing efforts to focus on their software systems, believing that they are the largest source of error in the metallurgical accounting process. While many are keen to highlight the flaws and non-compliance of spreadsheet-based systems, this distracts from the actual issues found to be plaguing accuracy and precision.

In this paper, case studies are used to highlight examples of where the main metallurgical accounting errors observed during audits conducted by JKTech were found to be; weightometer-derived mass flow rates, in-plant samplers and mass balancing techniques. Addressing these areas are essential for improving accuracy of production reporting and therefore decision-making.

INTRODUCTION

Engineers continuously seek to optimise and improve their systems. Mineral process engineers and metallurgists are no different, perpetually striving for that extra 1 per cent from their plant. Nowadays, they have access to vast amounts of data to analyse, but is the abundance of data alone enough to generate precise and accurate results?

Much like accurate bookkeeping and accounting are critical for informed financial decisions, accurate measurements and metallurgical accounting are essential for informed metallurgical decisions.

Commonly referred to as metal accounting or met accounting, Morrison (2008, p 5) defines metallurgical accounting as '…the estimation of (saleable) metal produced by the mine and carried into subsequent process streams over a defined period of time'. It is a fundamental process ensuring accurate tracking and reporting of valuable minerals through the value chain. Common issues can lead to significant discrepancies and financial losses. Despite its significance, metallurgical accounting is commonly perceived as a simple task; merely 'bean counting'.

In the early 2000s, the AMIRA P754 project was initiated to develop ways to improve the metallurgical accounting process. One of the main objectives was to develop an industry standard at corporate governance level which was called the Code of Practice (also referred to as the 'Principles of Metal Accounting'). An accompanying monograph was also published to elaborate on specific aspects of a metallurgical accounting system and the requirements to conform to the Code of Practice. When auditing metallurgical accounting systems, the Code of Practice and accompanying monograph are used as a benchmark for best practice.

The first point in the Code of Practice states:

> The metallurgical accounting system must be based on accurate measurements of mass and metal content. It must be based on a full Check in-Check out system using the Best Practices as defined in this Code, to produce an on-going metal/commodity balance for the operation. The system must be integrated with management information systems, providing a one-way transfer of information to these systems as required. Morrison (2008, p 18)

When an audit has been requested, the focus of many operations' concerns is around the data management and security of their metallurgical accounting system. Whilst good data management and security are core tenets of an ideal metallurgical accounting system, the concerns of data manipulation (intentional or otherwise) are essentially rendered moot if the initial values being input are inaccurate.

Mineral processing plants contain an extensive array of measuring devices. Metallurgists are painfully aware that each value collected will have some degree of error, and some may even have a bias. Formally speaking, error is the difference between a measured value and the true value while bias is a systematic error. It occurs when the method/s used to gather the data present an inaccurate or skewed depiction of the true value/s. Unfortunately, the true value is seldom known in mineral processing; assays are taken on samples from a lot, it is extremely unlikely the entire lot (ie population) will be assayed. Therefore, for metallurgical accounting, an abundance of measurements and derived data alone will not automatically lead to accurate key performance indicators (KPIs). Despite this, they are used to derive long-reaching financial decisions.

This paper aims to highlight the importance of evaluating the accuracy and precision of several key inputs, how to identify common sources of error and what the common causes are. The focus will be the primary metallurgical accounting balance, which is defined as the elemental balance across the entire plant, concerning custody transfer points for plant feed(s) and the plant products. By understanding these mistakes, professionals in the industry can implement preventive measures and improve the integrity of their metallurgical accounting system.

WEIGHTOMETER DERIVED MASS FLOWS

The first point in the Code of Practice highlights the Check-in Check-out (CICO) system. This indicates that all streams entering and exiting the process for which the balance is being performed, are measured for flow, and sampled for analysis. It is uncommon for plants to comply with this requirement, particularly for tailing streams. For the measurements that do occur, there is always some degree of error dependent on the quality of the measurement.

For plants that use 'n-product' calculations for primary accounting, the only mass flow rate measurement used when calculating 'balanced' values is from the mill feed weightometer. Therefore, maintaining accuracy of feed mass flow rate measurement is of the upmost importance as it has a significant influence in the calculated product mass flow rate.

The Code states a minimum installed precision for weighers used for primary accounting. They must have a maximum permissible error rating of OIML (Organisation Internationale de Métrologie Légale) Class 0.5. This is a ± 0.5 per cent in service precision, as per OIML R-50 (or equivalent local standard).

To achieve a high level of accuracy, the weightometer must be installed in a suitable location (as per manufacturer guidelines). Basic installation considerations include, the weightometer's distance from the feed point, consistent belt tension and correctly aligned idlers in the belt scale area which is sheltered from wind. To maintain a high level of accuracy, quality calibrations should be carried out routinely including after any work on the weighframe, belt scale or belt.

In a previous JKTech study, for over 12 months, an operation observed a consistent positive adjustment (or 'write-up') for tonnes of concentrate during the end of month (EOM) reconciliation. The root cause was unknown. An audit was undertaken of the key metallurgical accounting files. The circuit and on-site laboratory physically inspected. Focus areas were on the overall strategy of reconciliation procedures, as well as the source and direction of bias in measurements that could contribute to the recent production write ups.

The 'mass balance' method used on this site required feed tonnage, feed assays, concentrate assays and final tail assays to calculate shiftly and daily concentrate tonnages. Sources of potential biases included the feed weightometer, feed/product samplers and laboratory sample preparation and assay procedures. For reconciliation, additional input measurements were, the concentrate weighbridge, and the stockpile survey.

Following the site visit, it was determined that the samplers, laboratory procedures, concentrate weighbridge and stockpile surveys were unlikely to be the sources of the observed bias. This left the feed weightometer as the remaining suspect instrument, which otherwise had no obvious issues upon visual inspection. Maintenance logbooks indicated that the weightometer typically underwent a calibration during scheduled maintenance shutdowns using static weights. The use of this calibration method (though common) is undesirable as it will not correct for anything that alters the transfer of belt load through the weighing platform to the load cells. This is particularly of concern after any adjustments to the weighing platform and any belt replacements which affect belt tension and which had occurred multiple times in the previous year.

To confirm this theory, a CuSum chart of the difference between the calculated concentrate production (or 'book stock') against the measured concentrate entering the stockpile storage on-site, was plotted. The objective was to assess whether the changes in gradient aligned with the dates of the calibrations. The quality and frequency of the calibrations of the concentrate conveyor's weightometer allowed it to be deemed sufficiently accurate to be used as the reference value (ie the best estimate of the 'truth'). It was also assumed that the biases that may be present due to assays were negligible and/or consistent from month-to-month.

Figure 1 shows the difference between the measured and calculated metal in concentrate produced each shift (which theoretically should equal 0), overlaid with a CuSum transformation of this same data. The values are shown as the equivalent number of shifts to produce the same metal in concentrate. While the raw differences were noisy and without any immediately apparent trends, the CuSum showed multiple inflection points corresponding to changes in the prevailing mean difference of each period. Critically, the inflections in the CuSum trend aligned consistently with the dates of the calibration events. Therefore, the calibration events were established to be the most likely source of bias that ultimately caused the concentrate stock write-ups.

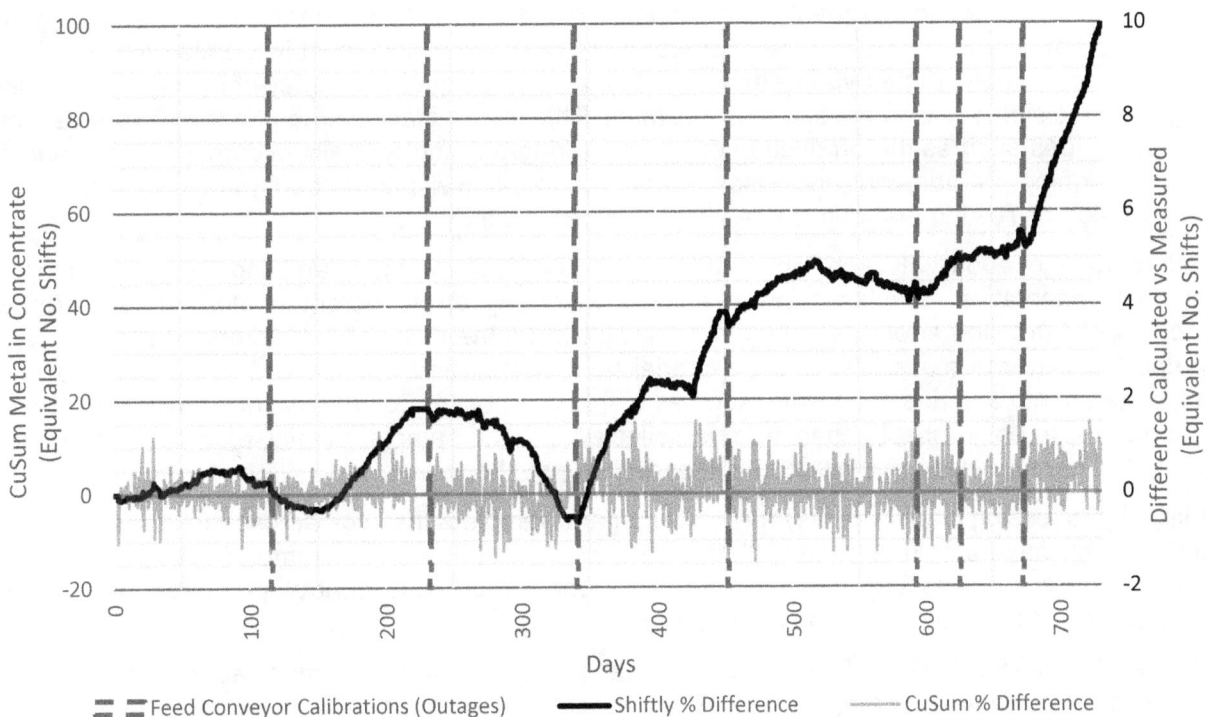

FIG 1 – Difference between measured (final concentrate weightometer) versus calculated concentrate tonnage, with the corresponding CuSum transformation overlaid. The feed weightometer calibration dates (based on information provided) are also shown with vertical reference lines in the chart.

The results from this analysis allowed the site to target their efforts towards the most significant source of systematic bias in their mass balancing process. The site was able to justify the additional capital expenditure to improve calibration quality and reduce the required adjustments during EOM reconciliation.

IN-PLANT SAMPLERS

A CICO system requires all primary accounting streams to be sampled for assay. These samples form a basis for determining the grades for each respective stream. As with any other measurement taken in the plant, they will have a certain sampling accuracy and precision. Error in sampling is not just limited to the physical extraction of material but includes things such as increment frequency. Using the nomenclature by Morrison (2008, pg. 142), total sampling error (TE_i) for sampling stage 'i' is comprised of individual components, summarised in the following equation:

$$TE_i = FE_i + GE_i + QE2_i + QE3_i + WE_i + DE_i + EE_i + PE_i$$

where:

FE_i	Fundamental error
GE_i	Grouping and segregation error
$QE2_i$	Long-range quality fluctuation error
$QE3_i$	Periodic quality fluctuation error
WE_i	Weighting error
DE_i	Increment delimitation error
EE_i	Increment extraction error
PE_i	Preparation error

These components can be broadly grouped into error due to, the ore characteristics, sampling error and preparation and assay. The ore characteristics will have a significant influence on the variability of the stream assay. Higher variability with time requires a higher frequency of increment to achieve a given precision of the composite measurement. This requires assessment of the stream variability by techniques such as the production of a semi-variogram and typically occurs as part of sampler commissioning. Preparation (and assay error) is generally well understood, particularly for National Association of Testing Authorities (NATA) accredited laboratories.

In a majority of audits, JKTech finds that the main contributor of total sampling error is from the use of unrepresentative samplers. This results in weighting error, increment delimitation error and increment extraction error. Weighting error occurs when the mass of the increment collected is not proportional to the mass flow of the material represented and is observed when increments are aggregated into a period composite, such as a shift composite. Delimitation error results in particles not having an equiprobable chance to be collected into the sample. Finally, extraction error occurs when the sampler does not extract the intended sample boundary.

Mineral processing plants typically rely on automatic samplers that 'cut' across the stream at regular intervals to make a shift composite that is sent for assay. It is also common for all or a portion of each sample to pass through a stream analyser for process control purposes before being aggregated at the shift composite collection point.

The AMIRA P754 Code of Practice states that samplers used in primary accounting must take representative samples. Samplers that do not meet this requirement will have some degree of bias that cannot, by definition, be quantified (bias is not uniformly distributed around a central value) and it will be dependent on variables such as flow rate and density.

A representative sample whether from a stream or static lot (stockpile, tank etc) can only be obtained if:

- Every particle in the stream or lot has an equal chance of getting into the sample.

- The sample preserves the size and density distributions of the stream.

In recent years, a number of plants reviewed did not have code compliant samplers installed on all their primary accounting streams; these streams did not have representative samplers installed and therefore collecting biased samples for assay. Taking a representative sample reduces the total sampling error and hence the error propagated to subsequent calculated values.

In another JKTech study, an operation planned to treat third party ore to maintain high utilisation of the plant and required an evaluation on its accounting system for provisions of the toll treatment. The operation's ore was to be fed as per usual from the main comminution and flotation line (Line 1), while the third-party ore was to be fed to a secondary line (Line 2). Therefore, the audit had a particular focus on the samplers on both lines to track the payable metal recovered in each respective concentrate. Figure 2 shows that at the time of the audit, a CICO system was not in place to separately balance the two different ore sources as there were insufficient samplers installed.

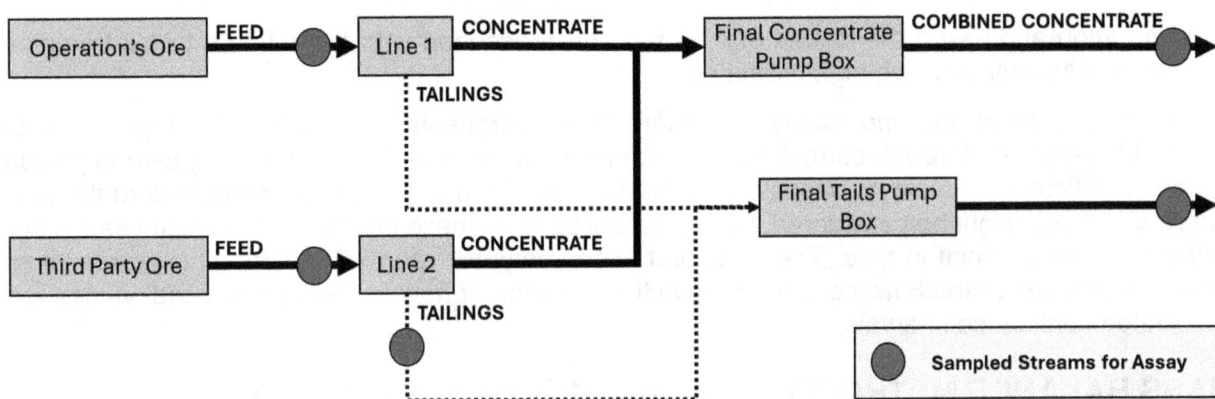

FIG 2 – Simplified diagram of the planned flow sheet for toll treatment campaigns with the streams with the ability to sample for assay (at the time of the audit).

The types of samplers in use, at the time of the audit were:

- Full stream cross-flow cutter.
- Samstat (a three-stage sampler).
- Pressure pipe.
- T-piece off-take (particularly for streams sent to the stream analyser).

A full stream cross flow cutter was located on the Line 1 feed, combined concentrate and combined tailings streams. The delimitation error was minimised as the sampler cut across the stream at (an observed) constant speed, providing all particles an equal probability of being sampled. It also appeared to 'park' outside of the stream and was free of any backsplash. The main concern was the depth and volume that the sampler could hold for the given flow rate. If the shape and/or volume allows for the collected increment (correct delimitation) to be lost due to backsplash or overflow (incorrect extraction), this adds a weighting error to the list. Once the sampler body is full and overflows, heavier and larger particles will typically be favoured to be collected. This can be addressed by modifying the shape to increase the volume of the sampler body.

The combined tail sample was from a Samstat three-stage sampler. This type of sampler was also planned for the Line 2 concentrate and the tailing stream, but was not installed at the time of the audit. As the entire stream is not sampled by the first two stages of static cutters, this introduces some degree of delimitation error which is minimised by an agitator prior to the static cutters to 'homogenise' the slurry and reduce this error. The error is typically not an issue if the cutter 'blades' do not become blocked or excessively worn. The third stage, a full stream cross-flow cutter that empties its contents into a container, should be representative if there are no blockages, the cutter moves across the stream at a constant speed and it 'parks' outside the stream. Although not a perfectly 'correct' sampler, it is generally regarded as a practical and cheaper alternative to a three-stage full-stream cross-cut sampling installation.

There was a pressure pipe sampler (temporarily) installed on the Line 2 concentrate stream. This is commonly used for process control purposes but should not be used for metallurgical accounting purposes as it does not take a representative sample. Delimitation error is introduced as only a portion of the stream is sampled. A weighting error to the shift composite is also introduced as the increment masses collected are not mass-proportional to the main stream. The error can further increase if the collection tube and/or turbulence bars are installed incorrectly or excessively worn. For example, a worn collection pipe would extract sample at a higher location which is less agitated allowing settling of coarser and denser particles away from the sample extraction point.

Line 2 feed and tailings were sampled with a T-piece off-take. Therefore, only a portion of the stream is sampled which will cause a bias due to delimitation and extraction errors. It was common for Line 1 to use a T-piece off-take as a secondary cutter. This was the case for the feed sample, following the full stream cross flow cutter. Therefore, even if the primary cutter sampled representatively, it was undone by the secondary cutter taking an unrepresentative subsample.

The operation addressed the issues highlighted and proceeded with the third-party toll treatment campaigns while maintaining high utilisation of the plant.

The error (accuracy) and uncertainty (precision) for unrepresentative automated samples may be able to be estimated through correct manual sampling campaigns. These sampling campaigns are generally difficult and costly to execute effectively. Even if the error was characterised to the point where an 'offset' might be considered, this is discouraged as these errors and uncertainties will only represent a single point in time. The error and uncertainty will vary with differing flow rates, slurry densities and ore characteristics. So, the question remains of how to report 'balanced' values with associated confidence intervals.

MASS BALANCE METHODOLOGY (AND ERROR WEIGHTING)

During audits, raising issues with the quality of the measurements are often met with resistance to address them due to the capital costs associated with a remedy. It is common for site to question if the capital cost of upgrading a sampler to be representative is worth it, when the current one is producing a sample that results in plausible assays. As single values are typically reported in the daily balance, it gives the impression that there is no uncertainty associated with the value. It is often a mystery at the end of the month where the reported tonnes in the concentrate shed have gone.

As discussed in the previous section, each measurement has a degree of uncertainty, typically expressed as a confidence interval or confidence limit. This uncertainty can be seen as risk when they are being used to make decisions. The use of an appropriate mass balance method and measuring the uncertainty of each measurement, not only produces a balanced data set, but also quantifies the uncertainty of each reported value.

Using a data set collected from a copper concentrator (Table 1), this section demonstrates the effect of measurement quality on the precision of the calculated production values.

TABLE 1

Raw (unbalanced) measured dry metric tonnes (DMT) and copper assay. Also shown are their associated relative standard deviations (SDs).

	Feed DMT	Con DMT	Tail DMT	Feed Cu%	Con Cu%	Tail Cu%
Denoted as	F	C	T	f	c	t
Mean	10627	106	10449	0.36	26.36	0.08
CV	3%	10%	10%	2.68	2.57	7.37

For many metallurgists, 'mass balancing' in metallurgical accounting is obtaining a 'reasonable' set of mass flow rates and assays around a node to allow for the calculation of the flow rate splits and recoveries. It is common to use the 2-product formula (or a 'n-product' formula variation) for recovery as it enables the calculation of recovery of a component (eg copper) in the absence of complete flow rate data:

$$Recovery = \frac{c\,(f-t)}{f\,(c-t)} = \frac{26.36\%\,(0.36\%-0.08\%)}{0.36\%\,(26.36\%-0.08\%)} = 78.0\%\,Cu \qquad (1)$$

At this site, the recovery can also be calculated using the feed and the measured concentrate flow rates and assays:

$$Recovery = \frac{Cc}{Ff} = \frac{Metal\ Units\ OUT}{Metal\ Units\ IN} = \frac{(106)\ (26.36\%)}{(10627)\ (0.36\%)} = 73.0\%\ Cu \qquad (2)$$

These equations will generally calculate a plausible result for the species of interest, but it does not actually balance the data. The data for this reporting period is not balanced in the true sense unless the sum of mass inputs is equal to the mass outputs when calculated using each analyte of interest. The data must be adjusted so that it is balanced, this becomes apparent when there is data redundancy (ie additional mass flow rate measurements) and/or 'balancing' multiple species using this method as each will produce a different mass split. Unbalanced data can lead to multiple values being calculated for the same metric, depending on the equation or method used.

For example, using the data from Table 1 and the 2-product formula for recovery, the recovery was calculated as 78.0 per cent Cu while the second equation calculated the recovery as 73.0 per cent Cu.

The options are compounded when calculating metal units in concentrate:

$$Cu\ DMT = Ff \times Rec\ from\ Eqn.\ 1 = (10627)(0.36\%)(78.0\%) = 29.8\ Cu\ DMT \qquad (3)$$

$$Cu\ DMT = Ff \times Rec\ from\ Eqn.\ 2 = (10627)(0.36\%)(73.0) = 20.4\ Cu\ DMT \qquad (4)$$

Using the values from Table 1, the equations above calculated the metal in concentrate as 29.8 Cu DMT and 27.9 Cu DMT, respectively.

There are multiple recovery and metal units in concentrate values that can be calculated from the same data set depending on the measurements used if the data is not balanced. This can lead to confusion as to which is the 'correct' value. A balanced data set ensures that the same value (eg recovery or metal units in concentrate) will be calculated and reported, regardless of the variant of the equation used. This raises the question, which values should be adjusted, and by how much, when mass balancing?

One method that fulfils these requirements is a 'statistical mass balance' which incorporates an 'error weighting' (represented by an SD) for each measurement. In this method, the measured values are adjusted to minimise the weighted sum of squared errors (WSSQ). This process is typically iterative and continues until it converges on a solution. The SD assigned to each measurement will generally influence how readily each unbalanced value will be adjusted to achieve an overall balance. A larger SD indicates a less precise measurement and will indicate possibly greater adjustments while a lower SD indicates a more precise measurement and will constrain the adjustments. A data set with accurate and precise measurements for all inputs will converge on a balance with a low WSSQ indicating confidence that the balanced values reflect the true values in the plant. Conversely, a data set with one or more inaccurate measurements can still converge on a balance, but it will have a larger WSSQ due to the larger adjustment(s) required. Adjustments made to each input will indicate possible errors in the inputs. Routine balances will also have expected magnitudes for each adjustment which can also be used as a high-level check of the quality of the balance.

The statistical mass balance with error weightings utilises all available measurements and associated SDs, around the node/s to converge on a balance. In the case of the Table 1 data, this includes not only assay data, but also the mass flow rates. This method creates a balanced data set where production values can be calculated from the balanced values (refer to Figure 3). Using this method, the recovery and copper metal in concentrate were calculated to be 76.6 per cent Cu and 29.0 Cu DMT, respectively.

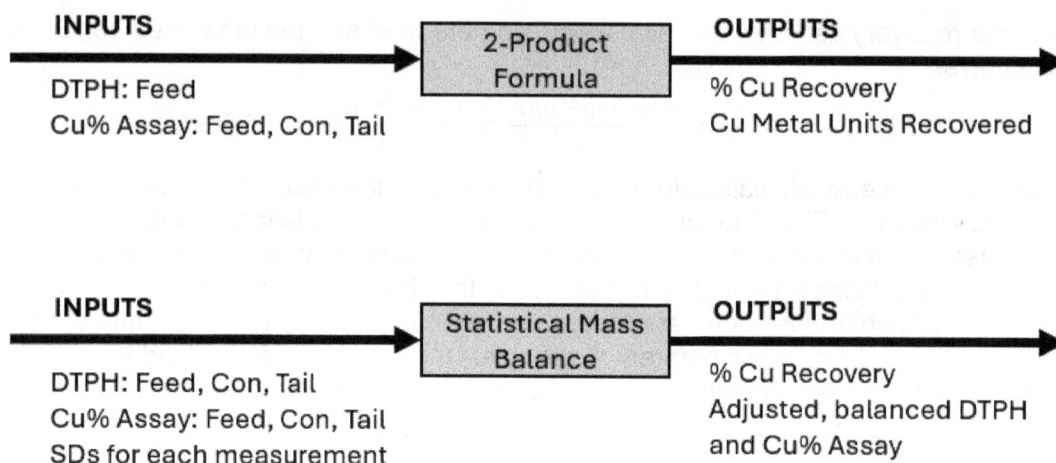

FIG 3 – Inputs and outputs of: (a) 2-product formula; and (b) statistical mass balance using data from Table 1.

Once the data is balanced, the Code also mandates that the quality of the balance must be understood and measured. This is because these numbers are not reported in a vacuum – they are generally used to make financial decisions. The uncertainty and associated risks for key values should be clearly identified and reported to enable informed decision-making and enhance transparency.

Current metallurgical accounting practices seldom consider the uncertainties associated with the measurements used in obtaining the metal balance and the propagated uncertainties in the reported results. As each measurement in Table 1 has an associated SD, the SDs for the calculated values can also be quantified and reported. This can be done by propagation of error calculations or a Monte Carlo simulation. A Monte Carlo simulation is a computation technique typically used to analyse complex processing subject to uncertainty. In this example, the uncertainty was simulated by generating 10 000 iterations (for each method) using combinations of input measurements randomised by a normal distribution using the mean and SD values. To produce the Monte Carlo simulation for the statistical balance method, the mean values for the throughput and assays were randomly varied based on their associated SD. These were then 'balanced' using the statistical mass balance described earlier. This was repeated 10 000 times. The statistical mass balance was automated using VBA to allow the solver routine to run for all iterations. The results are shown in the following histograms (refer to Figure 4) for the calculated Cu recoveries and Cu metal in concentrate.

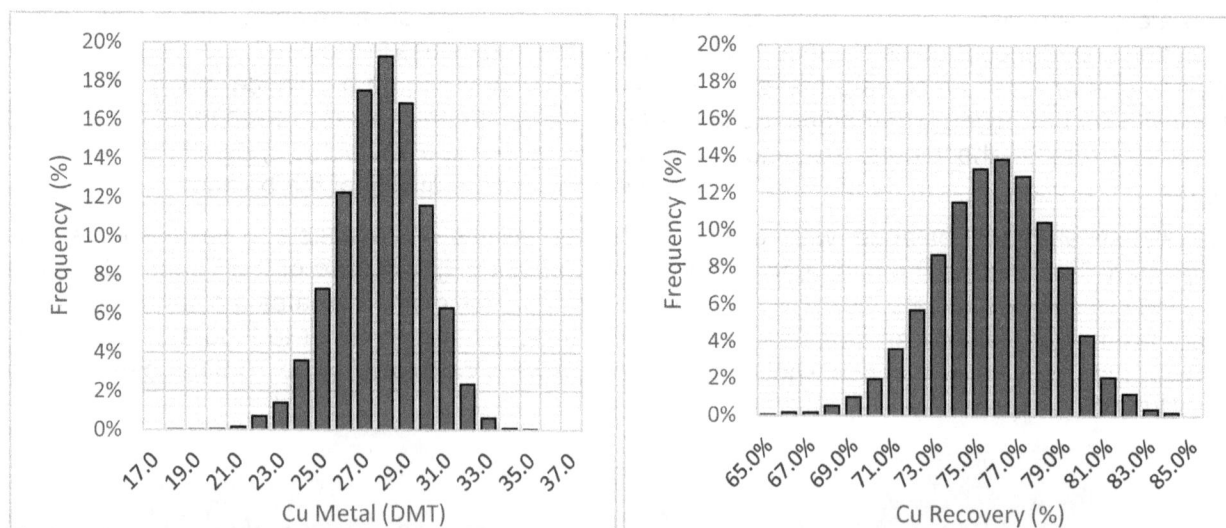

FIG 4 – The histogram of results the Monte Carlo simulation (10 000 iterations) using the statistical mass balance method and values from Table 1.

Given a mean and SD for each measurement, this technique effectively 'rolls the dice' such that multiple sets of calculated results (such as recovery) are produced. If the plant had identical feed every day and identical process performance, the Monte Carlo method produces a distribution of calculated results with the probability of them occurring. These are just a result of calculated results from the random fluctuations in the measurements. Variations in the calculated results each day are not necessarily due to the ore changing or operators not doing their job properly.

Given 10 000 instances (equivalent to 27 years of daily results) the Cu per cent recovery ranged from ~65–85 per cent and the Cu metal in concentrate ranged from ~18–36 DMT using the statistical method. Therefore, Cu recovery and the Cu metal in concentrate reported (at 95 per cent confidence) as 76.6 ± 3.5 per cent Cu and 29.0 ± 2.6 Cu DMT, respectively. The spread of the results will depend on the assigned SDs to each of the six inputs. Is this acceptable or is the uncertainty too great?

It is important to highlight that self-consistent (or balanced) results are not necessarily 'better' than the original values, particularly if one or more measurements are strongly biased as these will effectively bias the balanced results away from the true values. Even accurate and representative measurements will have uncertainty associated with the value measured. These errors will propagate through any calculations, as highlighted by this section. Whether the uncertainties, reported as confidence intervals, are acceptable will depend on the risk tolerance of those using these numbers. Reporting the uncertainty, in the form of confidence intervals, alongside the mean value of each KPI is crucial as it provides decision-makers relying on the data with a better an understanding of the uncertainty (and therefore risk) associated with each number. It also facilitates the establishment of benchmarks that can be monitored over time or compared between operations, enabling timely identification of changes in measurement quality. If the uncertainty is too high, there is no shortcut or offset that can remedy the situation. The only way to reduce it is to improve the quality of the measurements.

CONCLUSIONS

An abundance of measurements alone does not satisfy the requirements for reliable metallurgical accounting. As highlighted in the previous section, depending on the equation or method used, the calculated values can differ from the same set of data. Good instrumentation and correct sampling procedures are required to achieve useful accuracy and precision. A mass balancing method that assigns and utilises an error weighting for each measurement reduces distortion due to lower accuracy values. Target values for measurement error are required to monitor if measurements continue to achieve the target precision. High accuracy and precision of measurement are the pillars of a sound metallurgical accounting system. Without it, no matter how sophisticated the system, poor data cannot be transformed into 'good data'. It is akin to the adage of garbage in, garbage out.

Without reliable data, it becomes impossible to make sound metallurgical or financial decisions.

It is important to have:

- correctly installed, maintained and calibrated weightometers to provide accurate mass flow rates,
- representative in-plant samplers to provide accurate samples for assay,
- a mass balancing technique that assigns and utilises an error weighting for each measurement.

Manufacturers typically have clear guidelines for correct installation and calibrations should occur whenever there are any adjustments carried out on or around the weightometer. A roller/chain calibration is strongly preferred over a static weight on the belt scale or electronic ('zero check') calibration.

Measurement accuracy and precision can only be improved using appropriate instrumentation and taking representative samples followed by correct sample preparation and assay procedures. Any biases will fluctuate inconsistently with changing throughput and ore characteristics. Therefore an 'offset' is not a suitable remedy.

A statistical mass balance technique is quite simple to set-up around a few nodes or less. Textbooks such as Statistical Methods for Mining Engineers (Napier-Munn, 2015) explain how to do this. There is also commercial mass balancing software that utilises these principles that offer a graphical UI.

In conclusion, it is important to remember that the plant, and every part of it, is perfectly balanced. Ignoring inventory changes for this discussion, everything that goes in, goes out again. If the measurements are not in balance within the desired precision, then one or more of the flow rates and/or assays used in the balance must be inaccurate, or there are calculation errors or circuit issues. From JKTech's experience the usual suspects are weightometer-derived mass flows, in-plant samplers and mass balancing techniques.

REFERENCES

Morrison, R D (ed), 2008. *An Introduction to Metal Balancing and Reconciliation*, Julius Kruttschnitt Mineral Research Centre.

Napier-Munn, T J, 2015. *Statistical Methods for Mineral Engineers – How to Design Experiments and Analyse Data*, Julius Kruttschnitt Mineral Research Centre.

Design and delivery of a Professional Certificate in Metal Accounting

G B Gnoinski[1], K M McCaffery[2] and J Jessop[3]

1. MAusIMM, Principal Consultant, VinOre, Perth WA 6000. Email: gail.gnoinski@bigpond.com
2. Principal Consultant, Tastufo Consulting, Deloraine Tas 7304.
   Email: karenmmccaffery@gmail.com
3. Managing Director, Think Advisory Pty Ltd, Deloraine Tas 7304. Email: john@metopto.com.au

ABSTRACT

The Professional Certificate in Metal Accounting is a course requested by and designed for professionals and stakeholders in the minerals industry. Professional development is often informal in the minerals industry and there is a consensus that metal accounting is often not performed well.

Some reasons for this include lack of knowledge of methodologies, site size and commodity differences, and turnover of people resulting in this essential work being delegated to the most junior and least experienced personnel. These shortcomings expose companies to risk associated with lack of transparency and accuracy in the reported metal accounts.

To address this, a Professional Certificate in Metal Accounting was designed and developed by the authors and the Australasian Institute of Mining and Metallurgy (AusIMM), with the intent to introduce best practice guidelines aimed at improving transparency, reliability and accuracy in metal accounting. The course draws on subject matter experts to provide teaching in a collegial environment, supporting learners to develop a rich and integrated understanding of the discipline.

The course framework is based on the AMIRA P754 Code of Practice for Metal Accounting and associated guidelines (AMIRA Code). The fully online course comprises six modules and introduces learners to the principles of the AMIRA Code while gaining an appreciation of the underlying techniques and methodologies required to track metal production accurately and efficiently from mine to process plant and saleable products. The course has been delivered on three occasions and has been well-received to date, across a range of geographies, commodities, organisation sizes and professional roles, with increased enrolments for each delivery.

Building capacity in metal accounting is vital to ongoing success and growth in the minerals industry. Given the time constraints of industry professionals in the modern era, considerable thought was put into learning design to foster engagement across disciplines, first principles problem solving and a continuous improvement mindset in order to develop this capacity as rapidly as possible. Here we present the key principles of learning design used in the course design and reflections following course delivery to a global cohort of minerals industry professionals.

INTRODUCTION

This paper is based on the design and delivery of the Professional Certificate in Metal Accounting, a course requested by and designed for professionals and stakeholders in the minerals industry, and currently being delivered by the Australasian Institute of Mining and Metallurgy (AusIMM).

Metal accounting is a multidisciplinary process requiring inputs from across the mining value chain. The primary purpose of data generated by a metal accounting system is to provide information to operational management. Outputs from the metal accounting system are used in a company's financial reports amongst other purposes. There is a requirement in the minerals industry for transparent and auditable metal accounting systems. Hence, the metal accounting system and its design should be subject to approval by a Competent Person(s), who has sufficient relevant experience and expertise in metal accounting.

Operating, management and laboratory personnel are often not fully trained in all aspects of the metal accounting system perhaps due to cost saving or lack of knowledge of the importance of transparent metal accounting reporting. There is a consensus that metal accounting activities are often not performed well and is far from optimal on most sites and within most companies, with some honourable exceptions. This is partly due to the inevitable difference between sites and commodities, size of the site, and the turnover of people undertaking the metal accounting activities. On a small concentrator, for example, metal accounting responsibilities often falls on the most junior and least experienced.

Retained experience of those who have worked in this space has diminished in recent years, emphasising an already widening skills gap. There is a lack of knowledge of the methodologies involved and a shared frustration at the way the metallurgical accounting function can be delegated to the IT department (who are often unskilled in both the requirements and practice of metal accounting) or left to unskilled juniors in the plant. This is compounded by substandard design and provision of appropriate mass measurement and sampling functionality in both new and existing operating plants. This functionality is frequently compromised by either removing or downgrading equipment installed in new plants in the inevitable quest for reduction of construction capital expenditure. The resulting plant footprint can mean that essential equipment excluded cannot be retrofit at all or without significant expenditure. Both aspects can lead to some real damage to the value of metal accounting within the industry.

Knowledge and skills in metal accounting are acknowledged as essential for graduates and professionals if they are to provide the necessary support to maintain competitiveness in the minerals industry. Building capacity and lifting the level of confidence and competence of professionals in metal accounting is fundamental to the ongoing success and growth of the minerals industry.

Professional development is often informal, taking the form of operational on-site application or other skills-based activities that have the basis in experiential learning, or learning by doing (Kolb, 1984) that is suited to adult learners and builds on their previous work experience. Formal training specifically suited to professionals already working in the minerals industry would assist to develop and expand knowledge and capability in metal accounting practices and improve understanding of non-negotiable plant functionality needed for successful metal accounting capability in new plants.

To address these issues and focus on engaging the next generation of graduates and providing opportunities for professional development of the current workforce, a Professional Certificate in Metal Accounting was designed and developed with the intent to introduce best practice guidelines aimed at improving transparency, reliability and accuracy in metal accounting reporting. This certificate course drew on subject matter experts to provide teaching in a collegial environment, supporting learners to develop a rich and integrated understanding of the discipline, and is an example of authentic teaching (Oliver, 2010) or the 'participation model'; where students participate in the actual work of a professional community, engaging directly in the target community itself (Radinsky *et al*, 1998).

Stakeholders other than metallurgists typically wouldn't have formal training in metal accounting – it is not a part of their tertiary education. To acquire sufficient knowledge to understand metal accounting related issues, collaboration between universities, technical colleges, subject matter experts and minerals industry organisations facilitates delivery of focused training. Oliver (2010) states that collegial conversations about assuring achievement of capabilities inevitably turn to standards. Consequently, this course framework is based on the AMIRA P754 Code of Practice for Metal Accounting and associated guidelines (AMIRA code), that was collaboratively developed by minerals industry organisations as a voluntary industry standard to build capabilities in undertaking metal accounting activities (AMIRA, 2007; Morrison, 2008; Gaylard *et al*, 2006; Morrison and Gaylard, 2011).

Further to this, Oliver (2010) stated that learners need ready access to learning tools that enable them to access information focal to their employability aspirations. Yorke (2006) elaborated on graduate employability stating that students are '*successful in their chosen occupations to the benefit of themselves, the workforce, the community and the economy*'. According to Yorke (2004) '*employability is not merely an attribute of the new graduate. It needs to be continuously refreshed throughout a person's working life*'.

Building capability and capacity in metal accounting is vital to ongoing success and growth in the minerals industry. Given the time constraints of industry professionals in the modern era, considerable thought was put into learning design to foster engagement across disciplines, first principles problem solving and a continuous improvement mindset to develop this capacity and capability as rapidly as possible. According to Stephenson (1998):

> Capable people, have confidence in their ability to take effective and appropriate action, explain what they are seeking to achieve, live and work effectively with others, and continue to learn from their experiences, both as individuals and in association with others, in a diverse and changing society.

Based on the objectives described above, a fully online Professional Certificate in Metal Accounting course comprising six modules was developed to introduce learners to the principles of the AMIRA Code while gaining an appreciation of the underlying techniques and methodologies required to track metal production accurately and efficiently from mine to process plant and saleable products. The Metal Accounting Professional Certificate establishes fundamental guidelines for governance and best practice standards in the minerals industry aimed at transparency, reliability and accuracy of Public Reporting.

Here we present the key principles of learning design used in the course design and reflections following course delivery to a global cohort of minerals industry professionals.

APPROACH

There is overwhelming pressure to meet increasing industry requirements for work-ready and well-prepared professionals (Peach and Matthews, 2011; Patrick *et al*, 2008) and demands by professionals for employable knowledge and skills. An initial scan and industry consultation was undertaken via the AusIMM mineral industry advisory board, to establish the role of metal accounting best practice in the minerals industry. This helped start the course design and development process from an evidence-based approach. Further discussions with industry professionals who managed the AMIRA Code development project identified key reasons and the need for development and delivery of a course in metal accounting, such as the inevitable differences between sites and commodities, the turnover of people doing metal accounting and the lack of knowledge of the methodologies and plant functionality involved in undertaking best practice metal accounting activities.

For reasons explained above, an initial 'map of the terrain' was formed, to establish the intention to design, structure and deliver the Professional Certificate in Metal Accounting as a two-stage project. Subject matter experts examined the course philosophy and developed learning outcomes, storyboards, course content, modules, activities and assessments, based on an appreciation of their knowledge, understanding and observations of metal accounting practices at operations in the minerals industry. Significant thought was given to what inclusions were required as introductory and what should be considered for potential in-depth future masterclasses.

The first stage of the project encompassed design, determining the structure and learning outcomes and development of the course modules. The second stage involved delivery of the course for the first time in October 2022, with subsequent courses held in March 2023, September 2023 and March 2024.

Student cohorts were invited to participate in surveys pre-course, give feedback post-modules completion and at course exit to obtain reflections on completion of course deliveries. In the most recent delivery of the course, the survey response rate was 98 per cent of total enrolments that gave the course a positive recommendation. Information was collected from all participants, for example, location, company affiliation and their role at their organisation, educational achievement and overall perception from their learnings on the course. This has resulted in continuous improvement of course materials and delivery.

Stage 1 – Design and development of the Professional Certificate in Metal Accounting

The Professional Certificate in Metal Accounting fulfils two main roles. First, to ensure every metallurgist has foundational knowledge of metal accounting, the course is a primer to introduce the key topics. Second, to provide broad understanding amongst other mining sector professionals of core metal accounting requirements.

Course learning outcomes

Metal Accounting involves a broad range of technical areas, encompassing plant instrumentation for mass and volume measurement, sampling theory and practice, laboratory procedures, quality assurance, statistics (for example, estimation of precision and error propagation), data management, inventory modelling and measurement and governance. An effective metal accounting practitioner must, as a minimum, have some understanding of all these topics and how they inter-relate, and how they relate to the requirements of the business.

The measurements, analysis and data that come from undertaking these metal accounting activities impact daily, weekly, monthly and yearly business decisions and impact credibility of company reporting to the market and customers. Activities include accurately measuring, monitoring, checking and improving performance, and identifying and quantifying bias and losses. These activities enable an organisation to meet its strategic and operational objectives.

In addition to equipping participants with a good understanding of the theory and practice of metal accounting, the course was designed to assist overcoming some of the common challenges industry practitioners face when trying to establish best practice at their sites. For example, it provides typical business cases to support procurement of improved measurement systems and 'quick wins', requiring minimal investment, to improve transparency of their existing metal accounting system.

The course has a basis in experiential learning (Kolb, 1984) or 'learning by doing' that is regarded as a valuable means of ensuring integration of knowledge demonstrated by learning outcomes. The learning objectives for the course are listed in Table 1 and explain what participants should be able to demonstrate on completion of the Professional Certificate in Metal Accounting.

TABLE 1
Learning outcomes for the Professional Certificate in Metal Accounting course.

Number	Course learning outcomes
1	Explain and critically analyse the principles of the AMIRA P754 Code of Practice for Metal Accounting
2	Identify current practices in metal accounting activities, and ways to improve practices in sampling, sample preparation, laboratory analysis, mass measurement, bias and error management, data management and reporting for metal accounting
3	Identify the requirements for sound data collection, management and storage
4	Recognise the potential risks of poor metal accounting practices
5	Relate the integrity of reported metal accounts with financial reporting and associated corporate governance principles
6	Explain how different disciplines interact constructively across the minerals industry to facilitate accurate and transparent metal accounting reporting

Course structure

The course structure is displayed in Table 2, containing six modules that examine a range of topics in metal accounting relevant in the minerals industry.

Subject matter experts, supported by senior education consultants, designed and developed the six modules over an 18-week content development phase, to test knowledge of the AMIRA code and its application. Modules were designed to promote interaction of students with their managers and reports in their work environments. The process entailed course design, production of modules, teaching and learning content, production of videos presenting subject material, videos including relevant industry expert experiences, tasks, learning activities, case studies, quiz activities, short answer questions, questions and answers (Q&A) sessions, writing the final assessment and rubric, online cohort facilitation and facilitation of virtual classroom meetings.

TABLE 2

Course structure for the Professional Certificate in Metal Accounting.

Module	Sub-modules
Week 1	
Module 1: Introduction to Metal Accounting and the AMIRA P754 Code of Practice	1.1. Metal accounting and the AMIRA Code 1.2. Benefits of metal accounting 1.3. The Competent Person
Week 2	
Module 2: Basic Statistical Concepts	2.1. Accuracy and precision 2.2. Quantifying error and uncertainty 2.3. Comparing quantities and variances, T-tests and propagation of error 2.4. Heterogeneity, fundamental sampling error, sampling nomogram and sampling variogram
Virtual Class #1 – Online meeting, facilitated by Subject Expert	
Week 3	
Module 3: Mass Measurement and Sampling	3.1. Mass flow measurements 3.2. Sampling 3.3. Measurement systems monitoring 3.4. Measuring mass
Final Assessment released (due date)	
Week 4	
Module 4: Sample Management, Sample Preparation and Laboratory analysis	4.1. Sample management and safety, health and environment 4.2. Sample management and preparation 4.3. Sample analysis and QA/QC
Virtual Class #2 – Online meeting, facilitated by Subject Expert	
Week 5	
Module 5: Data Analysis	5.1. Mass balancing requirements and methods 5.2. Handling inventory and data 5.3. Reconciliation
Week 6	
Module 6: Data Management and Reporting	6.1. Data storage and management principles 6.2. Reporting audience and objectives 6.3. Linking to financial reporting
Virtual Class #3 – Online meeting, facilitated by Subject Expert	
Weeks 7 and 8	
Final Assignment	Participants download, complete and submit the final assignment for marking
Successful completion of the course progressive assessments and final assessment results in award of a Professional Certificate in Metal Accounting	

The learning activities were aligned with what the learners should be able to demonstrate that they have achieved on completion of the course (Constructive Alignment – Biggs and Tang, 2011; Biggs, 2014).

For example, a learner should be able to demonstrate skills in application and evaluation of metal accounting best practices at their operation (eg Mine to Mill to process plant) and compare current practices at their worksite to the recommendations from the AMIRA code guidelines. In other words, the focus is on what students intend to do to apply knowledge after completing the module topics. This demonstrates 'work-integrated learning' in 'recognition of the workplace as a unique and valuable learning environment for students' (McLennan, 2008). In this way, students can apply learnings to their workplace, stimulating discussion and asking questions.

Activities and tasks were developed to give learners opportunities to share and reflect on metal accounting experiences at their workplace, giving and receiving constructive peer feedback on current practices.

The online platform facilitated delivery of automatically assessed and graded quizzes with instant feedback at the end of module to test learners' knowledge and application. Course facilitators provide additional feedback and encouragement along the way and facilitate online live virtual classroom activities.

A final assessment, aligned with the intended learning outcomes, is designed to consolidate the teaching and learning experience by providing participants opportunity to demonstrate application of knowledge learned during the course. Subject matter experts developed the formal final assessment that entails answering several discussion questions relating to a flow sheet of a plant depicting activities and associated risks in a base metal or precious metal scenario, with clear reference to the AMIRA P754 Code of Practice. Although the subject matter experts developed the rubric for final assessment, independent consultants undertake marking of students' final assessments uploaded onto the learning platform.

External auditing and feedback on the course learning outcomes, subject matter content, sequence of the modules and assessments, was provided by an industry-based expert to ensure the course was contemporary and addressed current issues in the minerals industry. The AusIMM Education Committee approved the final course content and structure.

Stage 2 – Delivery and student feedback on the Professional Certificate in Metal Accounting

The course is designed for a global cohort of participants, who require professional and flexible online delivery focused on 'time-poor' professionals working in the minerals industry, undertaking studies in addition to their family and work responsibilities.

Following completion of Stage 1 of the project, an 8-week cohort-based delivery of the course proceeded, using a blended mode including online portal technologies, learning activities such as discussions and tasks, live virtual classrooms with panel discussions and break-out group discussions with Q&A sessions (see example in Figure 1).

Learners commit to between four to five hrs per module through online platforms. Access is provided to a wide range of online materials, short video clips, reading assignments, case studies and metal accounting tools, interspersed between activities and resources. As well as participating in online learning, short response and multiple-choice Q&A are used to assess learner progress at the completion of each module topic. The mode of online learning provides opportunity for learners to work through content at their own pace but within a defined time, with ready access to information through digital technology rather than via traditional 'must attend' lectures.

Course participants are encouraged to undertake learning based activities in live virtual classrooms where subject matter experts facilitate online meetings to discuss with learners' examples of case studies and practical application of metal accounting activities. These live virtual classrooms were further developed in later delivery of the course, whereby 'break-out groups' are formed to encourage interaction of participants with subject matter experts and their peers to discuss examples of good application and poor application of metal accounting activities at their work environment. The virtual classrooms are structured to encourage learners to actively engage with ideas and issues, reflect on their learning, and to learn collaboratively. Three live virtual classroom sessions are delivered after completion of modules 2, 4 and 5 as a scaffold to reinforce further learnings.

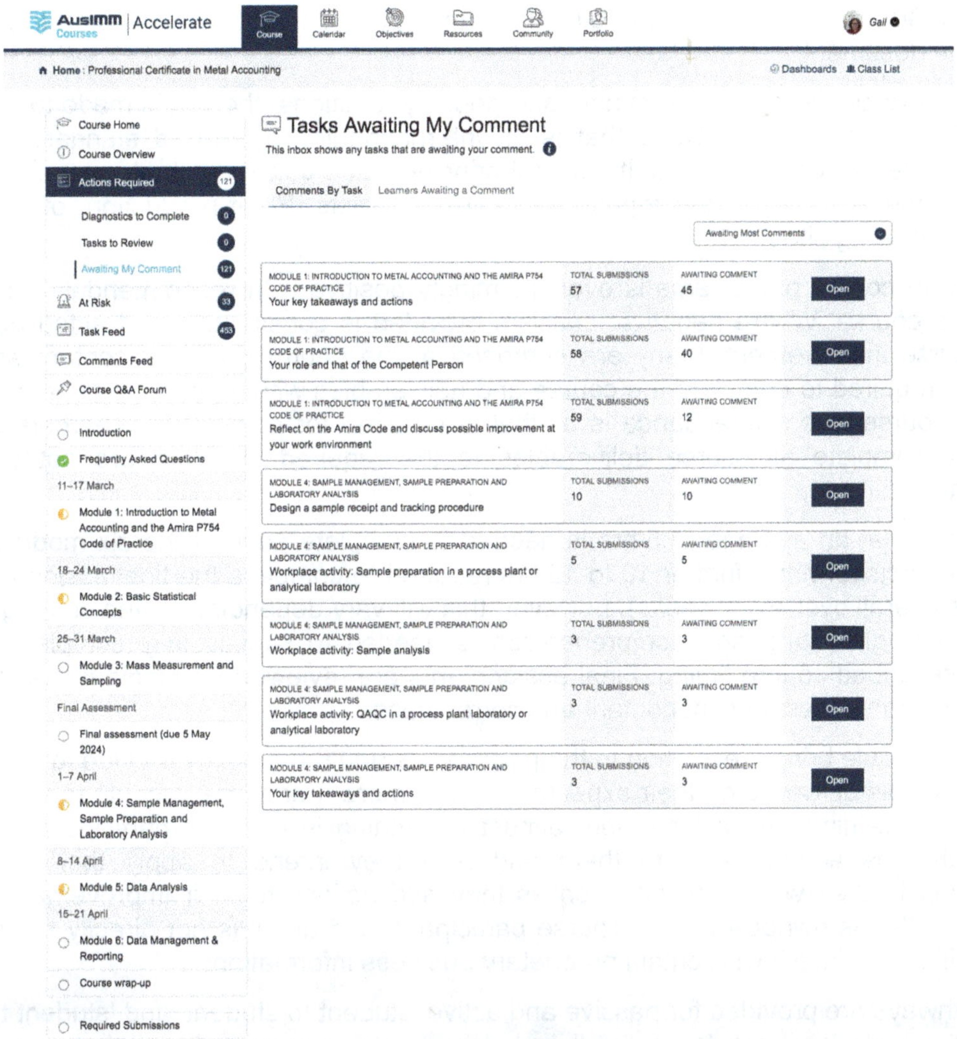

FIG 1 – Example of tasks and activities for the Professional Certificate in Metal Accounting.

To support 'work-integrated learning', course participants are strongly encouraged to identify immediate improvement opportunities for their own sites based on course content. In addition to reinforcing learning through tangible actions within the learner's own context, this allows course participants to potentially deliver immediate tangible value to their organisations, reinforcing the benefits of broader industry-wide competence in this area.

A post-implementation review is undertaken following each course delivery, followed by update and refinement of course material before the next delivery of the course. For example, 98 per cent of the participants, enrolled in the most recent delivery of the course, gave a positive recommendation for the course.

Learners' perspective on the course

While the majority of course participants come from metallurgical, analytical chemistry and metal accounting functions keen to boost their knowledge and capability, there is significant representation from the new project, mine and mine planning, geology and reserves reporting, geotechnical, marketing and finance functions. Participants come from mining locations across Europe, the Middle East, Africa, Asia, North and South America, Oceania and Australasia with organisational positions ranging from new graduates to group technical functions to executives and project consultants.

For many, this course is the first time they have been introduced to the voluntary AMIRA P754 Code of Practice and Guidelines, what it contains and means in terms of practical application and requirements and why the requirements are important. A full electronic copy of the Code of

Practice and Guidelines is provided to all participants although course focus is on the initial Code portion of the document.

For several, particularly those from mining and geology functions, the link is made to the Joint Ore Reserves Committee (JORC) Code that is an internationally recognised standard for reporting mineral resources and ore reserves. It sets out principles and guidelines that ensure transparency, consistency, and accuracy in the reporting of mineral assets. Plant production, of course, feeds into this for operating sites.

Feedback from course participants is overwhelmingly positive with recommendation for others to complete the course. Where negative, criticism provided is constructive and acted on as part of ongoing course improvement. Many are surprised at the depth of course content and the time commitment required to complete the course and deliverables within the defined period. It is not a 'tick a box' course where attendance is all that is required for completion. Learners who do not satisfactorily complete all course deliverables to the required standard are not awarded the Professional Certificate.

It is estimated that an average of 5 hrs is needed to complete each of the six modules including mandatory exercises with a further 10 to 12 hrs required to complete the final assignment (around 40 to 50 hrs overall over eight weeks), however, this will vary depending on the starting knowledge level. Sound English language comprehension is needed to successfully complete the course within the prescribed course timing. Final assignments are reviewed for plagiarism and the use of AI tools in addition to assignment content and correctness.

Each course module brings revelation to the participants and the majority provide feedback that the course overall has delivered on their expectations and more. For some, module topics refresh and reinforce prior learnings while for others almost everything is new. All participants identify how each module has added value to them and how they intend to apply their new skills or understanding to their workplace situation as they add actions to their improvement plans. The improvement plan is owned by each course participant, and detail is not directly shared with the class or facilitators since it will contain proprietary business information.

Multiple pathways are provided for passive and active 'student to student' and 'student to facilitator' interaction via topic feedback (open to all and closed), an open 'question and answer' forum and virtual classrooms with real-time participation. Students are slow or unwilling to participate in open question forums although this interaction increases as the course progresses. Many students realise that they are not alone with problems or issues that they face with metal accounting application in their workplaces. Other operational sites have similar problems. This finding helps builds solidarity between participants and increases sharing of encountered issues.

In addition to leading virtual classrooms in conjunction with a course manager, course facilitators directly and promptly respond to short answer exercises completed or feedback provided. Response may take the form of encouragement, acknowledgement, correction or clarification. Where possible facilitators strive to promote participants to think further about their issues, how they relate to business risk and the possible benefits of change.

Learning modules stress that key to sound metal accounting is the goodness of the underlying measurements, equipment and methodologies for making the measurements, assessing measurement quality, what can go wrong and why, how the data are handled and how substandard measurements impact business risk. Concepts identified as opening new understanding or appreciation are those of the Competent Person, heterogeneity and it's impacts, methods for monitoring and managing data quality, the importance of proper measurement system design, management and maintenance, how metal accounting and measurement impacts business risk and decision-making and Exception Reports documenting the risk where measurement best practice is not possible, amongst others. Reporting, data management and metal accounting software are touched on during the course, but emphasis is on the measurement layer.

In the final assessment, learners are asked to complete several discussion questions relating to a flow sheet of a plant depicting activities and associated risks in a base metal or precious metal scenario, with clear reference to the AMIRA P754 Code of Practice. Feedback from one external marker of the final assessments indicated that responses to questions 'offered up The Good, The

Bad and The Ugly' that ranged from distinction grades to unsatisfactory submissions following which participants could resubmit the final assessment for remarking.

It can be expected that time pressures play a part with some participants, as a well as comprehension where some learners appeared not to have 'read the question carefully', as given in the final assessment instructions, and English as a second language appeared problematic in some cases. Another external marker stated, *'Participants' understanding of the subject was relatively good. This tells me that the course design was appropriate for participants. The question sets were well thought out. However, for some participants, a bit more clarity on what was being asked was required. This may be due to some language barriers.'* The course coordinators were responsive to these concerns, and consequently the course material and final assessment was reviewed and updated to address the issues raised in the students' and external markers' feedback.

Limitations and challenges faced by the project team

While the AMIRA P754 Code of Practice was first released in October 2005, there was relatively little teaching material the team could draw upon. Eighteen years before development of this course, supporting articles and textbooks had been released – including the excellent publication by Morrison (2008) providing an introduction to metal balancing and reconciliation. Guidance for the main course content development essentially started with a brief description of the course structure provided by the AusIMM learning committee and related specialists. The actual course content was developed from scratch based on reference texts, personal experience and prior learning or training. Materials development took many more hours than allowed in the budget.

Due to the breadth of technical disciplines involved in metal accounting, determining how rigorously each topic should be covered balanced against the overall course length was an ever-present challenge. Module 2 (Statistics) has quickly become renowned for its depth and rigour! As a minimum, the team sought to provide enough context for each topic to ensure learners understood its importance and where to look for further information. Pending demand from industry, there may be subsequent courses developed to delve more deeply into important topics. For example, the correct design of mechanically correct samplers based on Sampling Theory, materials handling and the practicalities of engineering a sampler for reliability, is a significant topic in its own right.

Given the wide range of potential course attendees, content that focuses on 'work-integrated learning' is not always appropriate. For example, an exercise involving reviewing site measurement system calibration records cannot be undertaken by individuals not currently working at an operating site. The project team determined that the benefits to the participants who could undertake these exercises were more than sufficient to justify leaving them in even if not all participants could complete them. Following feedback from earlier cohorts, participants are now briefed on possible alternatives, in order to complete the learning exercise.

Insights from design and development of the course

Industry subject matter experts (SME's) with a wide range of skills and backgrounds, and who were available and willing to participate in the project, formed the project team in conjunction with a project leader and learning specialists.

The various topic modules were distributed to the SME's for development as appropriate to their skills and backgrounds. A senior project leader was appointed to ensure continuity and team stability with responsibility for overall project management including scheduling regular meeting with the team members and 'cracking the whip' to ensure the team stuck to the project and tight progressive deliverables schedule.

Engagement of a specialist 'Learning Management System' and 'Online learning' specialist to structure the course was key to success of the project. The specialist worked with the SME's to develop and populate the story board and coached the project team and facilitators in converting materials including quizzes, short assignments, video scripting and more traditional handout learning materials as suitable for online delivery. This included assisting SME's to script topics and prepare screen overlays and materials for video delivery both as structured presentations and via industry practitioner interviews. The AusIMM learning department worked with SME's to film video sessions and to prepare and finalise video graphics, overlays and transcripts for course delivery.

There is a huge amount of behind the scenes work and many team members are needed to produce and deliver a quality online course offering.

Team members engaged easily and enthusiastically as the project progressed and recognised and respected workload issues of the other team members. All project team engagement to keep work on track was conducted weekly via online meetings and at critical project milestones, emails and 'one-on-one' phone calls for communication of work in progress (WIP).

Establishing a document storage site for sharing and managing information reduced the risk of data loss and was vital for version control.

The Learning project leader and coordinators have continued to provide support of the team as the number of course deliveries has increased over time. Following each course delivery, debriefing occurs based on a cycle of reflection of what could be improved, review of quiz success and difficulties, areas participants struggled to successfully complete, review of feedback from course participants and direct feedback from SME's engaged to assess the final assignments. The result of this reflection is update of the course structure and modification of course content.

A key element to success of the course design and delivery is collaboration. This is expected to help maintain sustainability of the course delivery into the future.

Insights from delivery of the course

Use of the online teaching and learning platform for course delivery is very successful. The learning platform is easy and intuitive to use. There have been only limited issues identified from the online mode of delivery, based on feedback from participants and observations by the delivery team. One is suitable timing for online virtual classroom sessions that all course participants have opportunity to attend due to the global nature of enrolments. Another is the requirement for revision of some of the initial course material to improve clarity.

Very surprising has been the sample of participants, with very broad international participation and attracting international interest through professional bodies, such as AusIMM and other professional societies.

Methods of engaging with participants include SME facilitators responding to reflection or short answer assignments and follow-up questions or observations in the various modules as the course progresses and via live virtual classrooms sessions. There is also a central Q&A forum and emailing that is used by the AusIMM course moderators to welcome participants to the course, to alert participants as course modules and the final assignment open for completion, and reminders to enrol for virtual classroom sessions. Participants also use this to ask any questions they may have for response by SME's or other participants.

While we see that participants are genuinely interested in what happens at other sites and in other businesses, active questioning could be greater. To aid this, we actively encourage participants to show interest in sharing practice with others. Most of the sharing occurs at the start of the course where participants introduce themselves, what they do, why that are taking the course and what they seek to gain from their learning experience. Participants also share their feedback on what they have gained from each module or subtopic and how they intend to use this and why.

We encourage participants to clearly articulate an interest in gaining insight from others. There are some instances of this occurring in the general Q&A forum. We appreciate when participants in more experienced participants roles offer their views and feedback to others. Input from the full range of experience and roles across the student cohort is also of value. A real benefit to participants is provision of a ready-made network of contacts across the course cohort for future collaboration. After first delivery, the end of course survey 'recommend the course' rating was 98 per cent positive recommendation. Rating for the 'course met your expectations' had 84 per cent responses rated between very good and excellent.

The virtual classrooms have developed with each course offering. Initial classes involved all three SME's covering general topics to promote discussion. The latest course offering virtual classrooms utilise the SME's to cover particular exercises or concepts that previous participants have found more difficult to grasp, and to present a walk-through of the final assignment to assist understanding of requirements. Capturing and retaining spoken responses, feedback or verbal data from the virtual classrooms, is one area that can be built upon for future offerings.

CONCLUSIONS

- The objective of the Professional Certificate in Metal Accounting was to improve transparency, reliability and accuracy in metal accounting across the industry by introducing the voluntary AMIRA code principles and best practice guidelines.

- With the Professional Certificate in Metal Accounting, we have developed a new qualification that combines a strong understanding of metal accounting practices and its practical application in the minerals industry.

- A diverse team of subject matter experts, supported by senior education consultants, worked together to develop an authentic course.

- The course was designed for a global cohort of participants, who required a professional and flexible online delivery focused on 'time-poor' professionals working in the minerals industry, undertaking studies in addition to their family and work responsibilities.

- Challenges with respect to delivery, work-integrated learning activities and timeliness of feedback as to be expected in a new course, have been addressed with content refinement, updates and continuous improvement after each course delivery.

- A high degree of global industry engagement and approval by a diverse learners cohort is evident in the increased enrolments for each delivery of the Professional Certificate in Metal Accounting.

ACKNOWLEDGEMENTS

We would like to acknowledge the AusIMM for approval to publish this paper. In particular, Mel Cheong (Senior Learning Designer – AusIMM), Alison Bickford (Director and Principal Consultant – Connect Thinking® Pty Ltd), Kristy Burt (Senior Operations Manager, Courses – AusIMM), Katrina Wilkinson and Chelsea Coto (Online Learning Coordinators – AusIMM), Leon Lorenzen (Lorenzen Consultants) provided valuable insights during course design, development, and delivery, as well as the support of the entire AusIMM Online Learning team and the AusIMM course Marketing team.

We thank AMIRA International for permission to provide a full copy of the latest edition of the AMIRA P754 Code of Practice and guidelines to all course participants. AusIMM and SAIMM contributed by providing free access to several conference papers included as module reading resources. The University of Queensland is thanked for offering a discount for the Metal Accounting monograph for course participants who choose to purchase this volume.

Thanks also to the final assignment assessment team of Geoff Booth, Michael Braaksma, Cagri Emer and Sushant Dayal who also provide valuable feedback for course issues and possible improvements.

Finally, others including Newmont Corporation (John Cole), Peter Birnbaum (posthumous), Peter Munro, Rodolfo Espinosa-Gomez, Aidan Giblett, Heath Arvidson, Geoff Lyman (posthumous), Michael Dunglison, Carlota David-Howoses, Eliakim Tshiningayamwe, Johan Coetzee, Clint Armstrong, Ralph Holmes and Mark O'Dwyer variously provided valuable mentoring, discussion, opinion, or permission to utilise resource materials for input to the course.

REFERENCES

AMIRA, 2007. P754 2007 – Metal accounting, code of practice and guidelines: Release 3.

Biggs, J, 2014. Constructive alignment in university teaching, *HERDSA Review of Higher Education*, 1:5–22.

Biggs, J B and Tang, C S, 2011. *Teaching for quality learning at university: What the student does*, 4th edition, McGraw-Hill/Society for Research into Higher Education.

Gaylard, P G, Randolph, N, Wortley, C and Ralston, I, 2006. Design for Metal Accounting, *Journal of The Southern African Institute of Mining and Metallurgy*, 106:4.

Kolb, D, 1984. *Experiential learning: Experience as the source of learning and development* (Prentice-Hall Inc).

McLennan, B, 2008. Work-integrated learning (WIL) in Australian universities: The challenges of mainstreaming WIL, paper presented at the Career Development Learning — Maximising the Contribution of Work Integrated Learning to the Student Experience NAGCAS Symposium. Available from: <http://www.usq.edu.au/resources/nagcasaltc symposiumprereading120608.pdf> [Accessed: 30 July 2008].

Morrison, R D and Gaylard, P G, 2011. Applying the AMIRA P754 code of practice for metal accounting, in *Proceedings of MetPlant 2008,* pp 3–17 (The Australasian Institute of Mining and Metallurgy: Melbourne).

Morrison, R D, 2008. *An Introduction to metal balancing and reconciliation*, issue 4 of JKMRC monograph series in mining and mineral processing, 618 p (Julius Kruttschnitt Mineral Research Centre).

Oliver, B, 2010. *Teaching Fellowship: Benchmarking partnerships for graduate employability,* Australian Learning and Teaching Fellows. Available from: <https://altf.org/fellowships/teaching-fellowship-benchmarking-partnerships-for-graduate-employability>

Patrick, C, Peach, D, Pocknee, C, Webb, F, Fletcher, M and Pretto, G, 2008. The WIL (Work Integrated Learning) report: a national scoping study, final report, Queensland University of Technology. Available from: <https://eprints.qut.edu.au/216185/>

Peach, D and Matthews, J, 2011. Work integrated learning for life: Encouraging agentic engagement, in *Research and Development in Higher Education: Higher Education on the Edge* (eds: K L Krause, C Grimmer, M Buckridge and S Purbrick-Illek), papers from the 34th HERDSA Annual International Conference, pp 227–237 (Higher Education Research and Development Society of Australasia, Inc).

Radinsky, J, Bouillion, L, Hanson, K, Gomez, L, Vermeer, D and Fishman, B, 1998. A framework for authenticity: Mutual benefits partnerships, paper presented at the annual meeting of the American Educational Research Association.

Stephenson, J, 1998. The concept of capability and its importance in higher education, in *Capability and quality in higher education* (eds: J Stephenson and M Yorke), pp 1–13.

Yorke, M, 2004. Pedagogy for employability, The Higher Education Academy, Learning and Employability Series One.

Yorke, M, 2006. Employability in higher education: what it is – what it is not, Learning and Employability Series: Higher Education Academy.

Mill relining automation – applying lessons learned

S Gwynn-Jones[1], T A Ogden[2], J Bohorquez[3], D Sims[4], M Turner[5], C Kramer[6] and S Smith[7]

1. Global Applications Engineering Manager, Russell Mineral Equipment, UK.
   Email: stephen.gwynn-jones@rmeglobal.com
2. Senior Regional Applications Engineer, Russell Mineral Equipment, Toowoomba Qld 4350.
   Email: taylor.ogden@rmeglobal.com
3. MRD Engineering Team Leader, Russell Mineral Equipment, Toowoomba Qld 4350.
   Email: joel.bohorquez@rmeglobal.com
4. MRD Engineer, Russell Mineral Equipment, Toowoomba Qld 4350.
   Email: daniel.sims@rmeglobal.com
5. MRD Analyst, Russell Mineral Equipment, Toowoomba Qld 4350.
   Email: melinda.turner@rmeglobal.com
6. Product Manager THUNDERBOLT Products, Russell Mineral Equipment, Toowoomba Qld 4350. Email: christian.kramer@rmeglobal.com
7. Group Manager – RME Optimised Relining Solutions, Russell Mineral Equipment, Toowoomba Qld 4350. Email: steven.smith@rmeglobal.com

ABSTRACT

Replacing worn mill liners constitutes between 2–5 per cent of lost annual mill availability, making it one of the most expensive maintenance tasks associated with grinding mills. In critical commodities like copper, declining ore grades mean mill operators are increasingly requiring larger-diameter mill sizes (eg 36–40') to achieve sufficient throughput. The value for a 36–40' semi-autogenous grinding (SAG) mill can be between US$80k to 200k per hour based on London Metal Exchange (LME) prices of contained metal in ore.

Automating mill relining removes personnel from high-risk activities and allows faster machine movements to be applied, consistently and repeatedly, to reduce concentrator downtime. If automation can deliver a 1 per cent improvement (an additional 87 hrs of production) in annual mill availability for a mill-constrained copper plant, it could generate an additional US$9 M–18 M in production value per annum.

Leveraging its global original equipment manufacturer (OEM) knowledge and site experience, Russell Mineral Equipment (RME) has advanced the automation of mill relining. System development has taken place in close collaboration with early adopter customers, liner suppliers, and third-party reline crews. It has been a significant program, primarily because of the infrequency of relining events, typically occurring two to three times annually at a site, which limits opportunities for introducing improved processes and technologies. Additionally, in some cases, existing liner inventory must be depleted before automation-ready liners can be implemented.

Consequently, RME has taken a modular, systems engineering approach and focused on the roboticisation of field-proven relining technologies. This materially reduces implementation risks and costs for mill operators compared to unproven prototype equipment. Early adopter brownfield and greenfield sites have now completed over 30 relines using a range of RME's automated technologies.

This paper will provide an update on the progress of mill relining automation technologies and methodologies, and report on the quantitative safety and efficiency results achieved. Site data has been video-captured and analysed using time and motion studies combined with discrete event simulation technology. This paper will also share learnings from common implementation challenges that have led to promising developments in upcoming automated relining technologies, ensuring continued advancements in mill relining speed and safety.

INTRODUCTION

This paper builds on previous technical conference papers on relining automation from SAG 2019 (Smith *et al*, 2019), Mill Operators 2021 (Brander *et al*, 2021), SAG 2023 (Raharjo *et al*, 2023) and

SAG 2023 (Walton *et al,* 2023) by providing new data from the most recent site implementations and sharing additional insights.

Overview of the mill relining process

Mill relining is a critical maintenance process in mill operations in which worn liners inside grinding mills are exchanged for new liners. This is done both to protect the mill from damage and to ensure throughput targets are achieved.

An example of the typical mill relining process is described below and illustrated in Figure 1:

1. Pre-reline set-up:

   o Equipment is staged on the mill deck, including liners, bolts, nuts, torque guns, and other tooling.

   o Relining equipment undergoes pre-reline inspection and maintenance work by a qualified technician to ensure readiness prior to the shutdown.

2. Shutdown and grind-out:

   o Feed of ore into the mill is stopped, while the mill continues to spin. The existing ore inside the mill is 'ground out' to the appropriate level for reline activities.

   o The mill is stopped, isolated, and locked out.

   o Guarding is removed, and platforms and reline equipment is installed around the mill.

   o The feed chute is removed from the mill entry, and the feed conveyor discharge box is closed/sealed for the duration of the reline.

3. Mill relining machine insertion:

   o The mill relining machine (MRM) is driven from its storage location to the mill, where it is inserted into the mill and tied down in the operating position.

4. Nut-removal and liner knock-in:

   o Nuts are removed from accessible rows of liner bolts on the outside of the mill.

   o Hammers are used to knock in bolts and liners onto the charge inside the mill (Figure 1).

   o No personnel are allowed into the mill while liners and bolts are being knocked in.

5. Worn liner removal:

   o The reline crew enters the mill, and worn liners are removed from inside the mill using chains (Figure 1), or a grabbing claw, attached to the MRM.

   o Liners are transported out of the mill on the MRM.

   o A forklift collects worn liners from the rear of the MRM and takes them away.

6. New liner placement:

   o New liners are brought to the MRM by a forklift, which are then transported into the mill on the MRM.

   o The MRM picks up the new liners inside the mill and places them into position on the mill shell.

   o The reline crew installs nuts and bolts to fasten the new liners into place on the mill shell.

7. Inching:

   o All personnel exit the mill, the mill drive is de-isolated, and the mill is rotated to give access to new rows of bolts. The mill drive is then re-isolated and locked out.

8. Steps 4–7 are repeated until all worn liners have been removed and new liners have been installed.

9. The MRM is removed, the feed chute and guarding are re-installed, and the mill is re-started.

10. Post-reline activities:

   o After a period of operation, the mill is stopped again to torque the nuts to the manufacturer specification.

   o The relining equipment is cleaned and inspected prior to storage by a qualified technician, any service requirements are noted and planned to occur before the next relining event.

FIG 1 – Examples of typical activities during a reline: liner bolt knock-in (left) and worn liner removal (right).

Many of the factors that determine the safety and performance of mill relining result from design decisions made during the plant design, as explored by Van Rooyen *et al* (2023).

Examples of some of the residual risks in traditional relining are illustrated in Figures 2 and 3.

FIG 2 – A five kilogram liner bolt becomes a projectile inside the mill during the knock-in phase.

FIG 3 – Near miss incident: A worn liner weighing between three to four tonnes dropped during liner removal due to incorrect slinging practices. Anonymised figurines overlay the positions of real workers who were standing nearby on the mill charge during the incident.

TECHNOLOGY – ELIMINATING MILL RELINING HAZARDS THROUGH AUTOMATION

Despite the well-established mill relining processes and controls implemented in industry, residual risks still persist and can result in serious and tragic incidents. Safety incidents have been recorded during relining resulting in near misses, or worse, injuries and fatalities (Chinh and Chieu, 2024). These risks include confined spaces, working at heights, handling multi-ton suspended loads and working in close proximity to heavy mobile plant and equipment. Like so many other high-risk tasks in mining, a great deal of time and effort has been applied to controlling the hazards.

RME's approach to controlling the risks associated with mill relining is to eliminate or substitute the hazards as much as possible. The methodology applied has been to engineer ways to introduce automation technologies and methods using already proven and well-established relining systems. A prime example of this is the commonly-used RUSSELL 7-axis Mill Relining Machine (RUSSELL MRM), which can now be configured with a robotics-enabled software and hardware system called RME AutoMotion. Another example is known as THUNDERBOLT SKYWAY: Used outside the mill to automate the guidance and operation of liner bolt removal tools (THUNDERBOLT Recoilless Hammers) which knock-in worn liners and/or liner bolts. Examples of engineered systems that enable automated mill relining are illustrated in Figure 4. This incremental approach helps grinding mill operations de-risk the introduction of the new technologies, while enabling improvements in safety and performance.

FIG 4 – Examples of relining systems that enable automated mill relining and leverage familiar and well-used technologies. Left: THUNDERBOLT SKYWAY for automating the operation of liner bolt removal tools for knocking-in worn liners and bolts on a 38' semi-autogenous grinding (SAG) mill. Right: RUSSELL MRM with RME *AutoMotion* capable of automatically placing liners weighing over seven tonne for a 38' SAG mill.

RME's approach to reducing risk exposure in mill relining is to leverage automation to either eliminate or substitute hazardous manual tasks. These are the two most effective types of risk controls as illustrated by the hierarchy of controls in Figure 5.

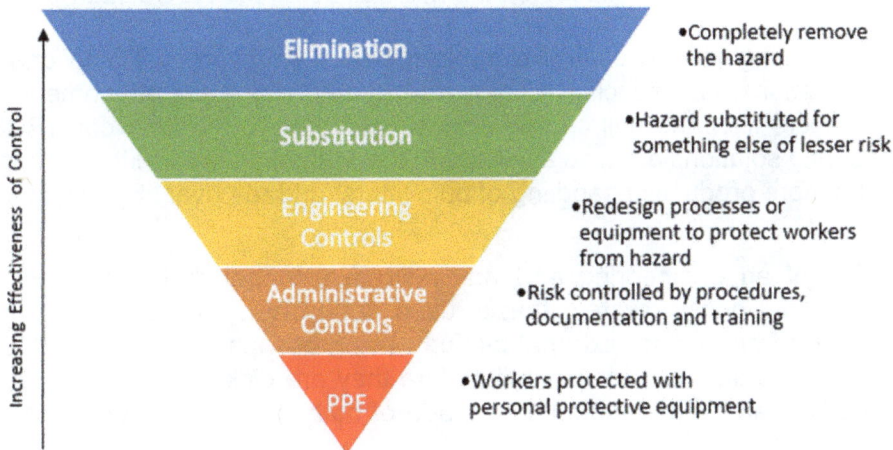

FIG 5 – Hierarchy of controls.

A key element in RME's methods for eliminating or substituting hazardous mill relining tasks has been achieved through what it refers to as 'INSIDEOUT' relining (RME INSIDEOUT Technology). INSIDEOUT is best described as an integrated suite of tools and methodologies (ie engineering and administrative controls) that are utilised at different stages of the relining process. These tools and methodologies are referred to as INSIDEOUT Placement and INSIDEOUT Removal. Their purpose is to enable mill operators to reduce risk inside the mill by performing relining tasks from outside of the mill, and in a way that also improves productivity. In this paper they will often be referred to as automation-enabling technologies to reduce the usage of product names.

An example of these engineering controls and methodologies are grapple attachments that eliminate the need for reline crews to work on the charge inside the mill during worn liner removal. One such engineering control (RUSSELL Forks Tool) consists of a custom-engineered tool that is held by the MRM's grapple, and works in combination with small liner modifications to support the MRM-liner interface. It is important to note that these liner modifications do not affect the grinding profile or wear life of the liner. However, in combination, the system enables the MRM to pull worn liners straight off the wall following bolt 'knock-in', thereby eliminating the need for hazardous manual liner slinging.

In addition to significant safety benefits, this engineered solution also provides performance benefits by eliminating the step whereby liners are knocked in onto the mill charge to await collection by the MRM. This tooling enables liners to be removed directly—and more efficiently—from the mill wall and transported to the liner cart. A visual comparison of liners being removed using chain slings and the Forks Tool at site 1 is shown in Figure 6.

FIG 6 – Comparison of worn liner removal using traditional methods (left), and RME INSIDEOUT Technology (right) which removes the need for people to enter the mill and sling heavy-weight liners during the removal phase. Left: Traditional worn liner removal using chain slings. Right: Worn liner removal directly off the mill wall using the RUSSELL Forks Tool.

Alternatively, another engineering control suitable for worn liner removal, and one that does not require liner modifications, is a specially-engineered hydraulic grapple attachment that can 'grab' knocked-in, worn liners from the mill charge and deliver them to the liner cart (RUSSELL Claw). Additional engineering solutions are also available in the form of hydraulic lifting grapple tools which allow for safer and more productive handling of pulp lifters, and removal of grate liners directly from the mill wall.

Another illustration of an engineering and administrative control that has been developed for utilisation is an external bolting system that is used during new liner placement. It enables the securing of new liners from a safe, external platform using encapsulated bolts (RME BOLTBOSS) that are inserted into the liner outside the mill before they are picked up and placed by the MRM. This solution does not rely on the liners being made of both steel and rubber, and works on most existing steel liner types.

To maximise the benefits of INSIDEOUT relining in terms of safety and productivity, mine sites should consider implementing these methodologies alongside an automated mill relining machine and modest liner design modifications for seamless relining processes. INSIDEOUT methodologies can still deliver significant risk reduction, however, even without a fully-automated mill relining solution.

RME's extensive OEM experience recognises that a one-size-fits-all approach might not be feasible or optimal for every mine site. Therefore, INSIDEOUT has been engineered for flexibility, allowing mine sites to deploy a solution appropriate for their specific needs.

This system-engineering approach allows individual tools and methodologies to be implemented incrementally during scheduled maintenance windows, delivering safety and productivity improvements without extending downtime or impacting mill availability. This approach also fosters smoother cultural adoption for relining crews, who can be trained concurrently as new tools and methods are introduced, facilitating a successful transition to safer and faster relining practices.

Finally, INSIDEOUT engineering and administrative controls offer a cost-effective and low technical risk solution for both new and operational RME Mill Relining Systems. RME's engineering studies have shown that these methodologies can eliminate the safety hazards associated with manual handling of heavy liners and working in confined spaces near moving machinery.

RESULTS – RECENT SITE PERFORMANCE USING AUTOMATED MILL RELINING SYSTEMS

Multiple mine sites globally have now used automated mill relining equipment and/or automation-enabling relining technologies in commercial settings to remove people from hazardous areas inside and outside the mill. This section summarises recent safety and productivity performance data from these real-world relining operations where the technologies have been utilised. By examining this data, RME aims to quantify the benefits of automated mill relining and provide implementation insights for mill operators and the comminution industry.

Method

RME utilises a rigorous and data-driven methodology for analysing and quantifying mill relining time and safety performance. To ensure a robust comparison and assessment of the impact of automated and automation-enabling equipment compared to traditional practises, this data-driven methodology was applied to analyse outcomes from the implementations. RME's methodology is:

- Video camera systems were installed inside and outside grinding mills at sites where traditional relining methods were utilised, and at sites where automated technologies were planned for future implementation. These cameras captured the performance of all activities and processes, exactly as they occurred during the full reline event. Video footage was captured during mill relines at sites as they were implementing automation-enabling technology.

- Video footage collected from all sites was reduced to obtain performance data:

  - 'Reducing' the data means inserting timestamps at the start and end of each activity based on fixed definitions. This is then converted into a data set of activity times. An example of this would be 'the time to secure a liner to the mill wall'.

  - Data was extracted on the time taken to perform individual activities. For example, this includes the time taken to attach slings and chains to a worn liner on the charge, and engaging INSIDEOUT grapple tools to a worn liner. It also includes the MRM transport times, such as how long it takes to deliver a worn liner to the awaiting liner cart, and how long it takes the MRM to pick up a new liner from the liner cart and transport it to its final location on the mill wall for installation, and others.

- The performance data captured was used to compare mill relines using traditional methods to those using automation-enabling technology in terms of:

  - Individual activity, phase and overall relining durations.

  - Person-hours of workers' exposure to hazards inside and outside the mill.

- In some cases, RME used the data as part of a study to help the mine site optimise their mill relining processes. Recommendations from these studies were provided to:

  - The relining crews from studied sites, recommending changes to their current processes and more efficient use of their existing mill relining tools.

  - Sites and liner suppliers, recommending modifications of liner design features to improve mill relining efficiency.

  - RME's engineering development team, recommending modifications to automation-enabling technology features in order to optimise relining methodologies in response to specific conditions observed during the studied mill relines.

Mill relines at sites using automated and semi-automated technologies

RME has presented a number of case studies at recent industry conferences, sharing performance data from customer sites that are implementing automated and/or automation-enabling mill relining technologies. This data, along with future predicted performance data, that has been quantified using discrete event simulation methodologies, for sites intending to implement automated technologies, are the basis for the performance data presented in this paper. These sites are summarised in Table 1.

TABLE 1

Reference list and sources of reline performance data from current and future sites.

Location and commodity	Mill quantities and diameters	Year technologies first implemented	Relevant relining technologies
Site 1 (South-east Asia) Copper/Gold	1x 38' 1x 34'	2021	RUSSELL 7 *AutoMotion* MRM THUNDERBOLT SKYWAY RME INSIDEOUT Technology
Site 2 (Central America) Copper	3x 40'	2019	THUNDERBOLT SKYWAY
Site 3 (Australasia) Copper/Gold	1x 38' 1x 34'	2021	RME INSIDEOUT Technology
Site 4 (Australasia) Gold	1x 24'	2019	RME INSIDEOUT Technology
Site 5 (Australasia) Gold	1x 36'	2023	RME INSIDEOUT Technology
Site 6 (South-east Asia) Copper/Gold	1x 38'	2025 (estimated)	RUSSELL 7 *AutoMotion* MRM THUNDERBOLT SKYWAY RME INSIDEOUT Technology
Site 7 (South America) Copper	3x 40'	2027 (estimated)	RUSSELL 7 *AutoMotion* MRM THUNDERBOLT SKYWAY RME INSIDEOUT Technology
Site 8 (South America) Copper	2x 40'	2026 (estimated)	THUNDERBOLT SKYPORT RME INSIDEOUT Technology

Figure 7 shows the number of liner movements performed by various automated and automation-enabling technologies, as analysed in this paper. These data points represent all liner movements captured by video footage and/or RME personnel in reline attendance at Sites 1, 2 and 3 over a three-year period from 2021–2023. It exclude relines outside of this time period, or those for which data was not able to be collected and analysed.

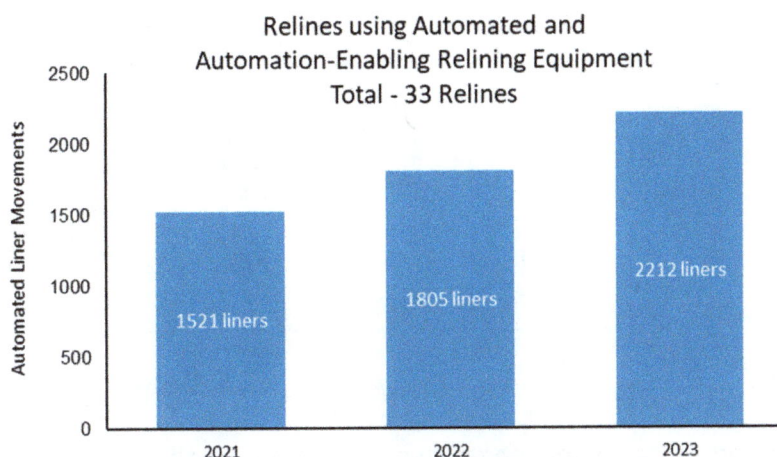

FIG 7 – Captured liner movements performed using automated and automation-enabling relining equipment from relines performed between 2021 and 2023. Data is from Sites 1, 2 and 3.

The liner movements that are included in Figure 7 are performed by one of the following four technologies:

- THUNDERBOLT SKYWAY
- RUSSELL 7 *AutoMotion* MRM
- INSIDEOUT Removal
- INSIDEOUT Placement
- Note that individual liners may be included more than once if moved by two or more different technologies as described above.

Figure 7 shows that 5538 liner movements were performed and analysed, across 33 relines in the three-year period between 2021 and 2023, representing a growth of ~20 per cent per annum.

Results from technology implementations to remove worker exposure to hazards — inside the mill

RME has captured mill relining productivity and safety performance data at several sites following implementation of RME INSIDEOUT Technology. This section summarises the performance data captured at Site 1.

To eliminate the need for personnel to enter the mill during liner exchange, Site 1 undertook a progressive implementation of technology upgrades and process modifications across multiple reline events. INSIDEOUT Placement methods were implemented first, so the site could perform liner installation without requiring reline crew on the charge. Subsequent relines incorporated INSIDEOUT Removal methods to enable the removal of worn liners, also without requiring reline crew on the charge.

INSIDEOUT Placement: The results from technology implementations to remove people from exposure to hazards inside the mill during the installation of liners is shown in Figure 8. This graph illustrates the time to install shell liners over three relines. It indicates that successful implementation and optimisation of this technology has resulted in liner placement times which are faster on average (represented by the 'x'), and of a smaller spread than times achieved using traditional liner installation methods.

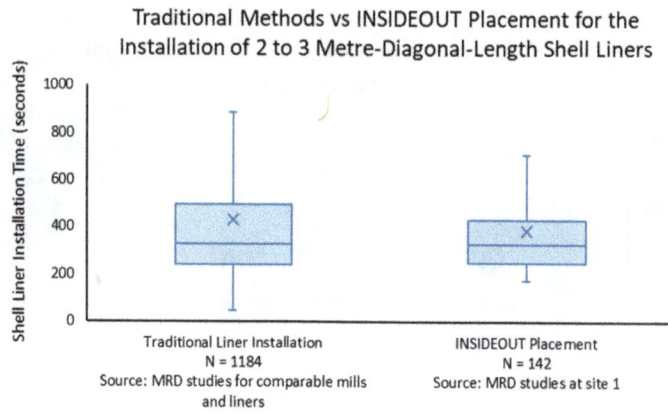

FIG 8 – Time performance comparison between traditional methods and RME INSIDEOUT Placement for installation of shell liners at Site 1. Outliers exceeding 1.5xIQR are omitted for clarity.

INSIDEOUT Removal (Figure 9) compares the average individual liner removal times for both traditional and INSIDEOUT Liner removal for three different liner types. Traditional liner removal included the use of chain slings to assist MRM collection of knocked-in worn liners and delivery to the liner cart. INSIDEOUT Removal used the RUSSELL Forks Tool to take knocked-in liners straight off the mill wall and deliver to the liner cart. The results show decreases in liner removal duration for two liner types upon initial implementation of the RUSSELL Forks Tool, and further decreases in all three liner types in subsequent uses. This decrease is because the technology is eliminating several manual steps associated with traditional removal practise, as well as improvements in crew familiarity and operating skills on the second usage. This demonstrates the potential for continued improvement over time from automated mill relining methods. Figures 10 and 11 show a visual explanation of the differences in process.

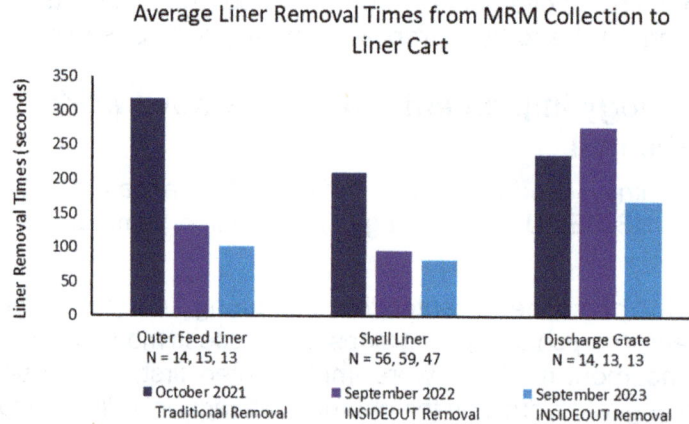

FIG 9 – Comparison of worn shell liner removal times using chain slings versus the RUSSELL Forks Tool at Site 1, over three similar reline operations.

FIG 10 – Shell liner removal from the mill charge to the liner cart using traditional chain slings, after the liner has been knocked from the wall onto the charge.

FIG 11 – Shell liner removal directly from the mill wall to the liner cart, using the Forks Tool grapple attachment.

Site 1's October 2021 reline used traditional chains to sling knocked-in liners from the charge and transport them to the liner cart. The average removal duration shown in Figure 9 includes the time taken to attach the liner to the MRM grapple using the chains, travel to the cart, position the liner on the cart and release the chains (see Figure 10). It also includes a small period of time during which the MRM was used to dislodge the liner from the wall, prior to its removal from the mill. It excludes idle time spent waiting for the liner cart to be in position.

Site 1's September 2022 and September 2023 relines used the Forks Tool to remove liners directly from the mill wall and transport them to the liner cart. The average duration shown in Figure 9 includes the period from when the liner is dislodged from the wall following the final bolt knock-in, to when the liner is released onto the liner cart (see Figure 11). It does not include Forks Tool attachment time, as this is usually completed during the knocking in of liner bolts. It does include some time spent waiting for the liner cart to be in position, as workers intentionally remained on the beam to conduct liner inspections prior to driving the liner cart into position.

The graph in Figure 12 illustrates the number of 'person-hours' during which Site 1's workers were inside the mill, and therefore exposed to all associated hazards, during their progression from traditional to automated-enabled methods of relining. 'Person-hours' were calculated by multiplying the average number of workers inside the mill on the charge by the length of time for each activity to be performed.

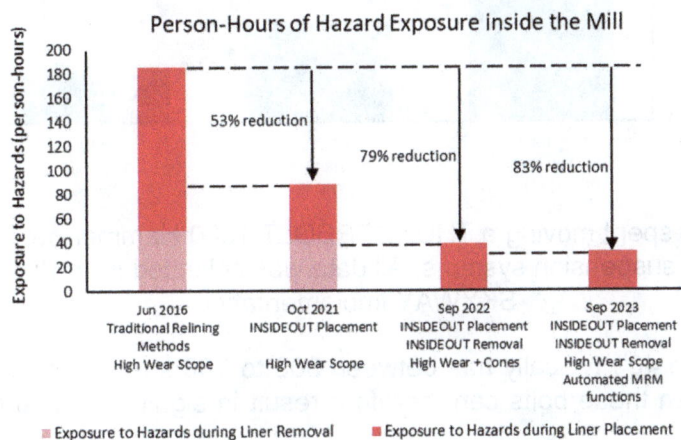

FIG 12 – Quantifying the safety benefits of automated relining: Reductions in exposure to hazards inside the mill during the liner exchange phases.

The first bar represents reline operations at Site 1 prior to implementation of automation-enabling technologies. These relines were performed using manual slinging techniques, as illustrated in Figure 10, and the manual placement of bolts into newly placed liners by reline crew working inside the mill. An average of 2.5 workers were observed inside the mill during the liner removal phase, and four people during the liner placement phase.

Site 1's October 2021 reline operation saw implementation of automation-enabling technologies (INSIDEOUT Placement) for the installation of new liners. This effectively removed all workers from the charge during this activity, significantly reducing the person-hours of hazard exposure during this phase. The remaining exposure times shown in Figure 12 for this reline, during liner placement, represent time spent manually installing rubber filler rings and repairing rubber on the mill wall.

Site 1's reline operations in 2022 and 2023 were performed using automation-enabling (INSIDEOUT) technologies for both liner placement and removal, enabling the full removal of workers from inside the mill (on the charge) during liner removal and completely eliminating hazard exposure during the liner exchange phases.

Results from technology implementations to remove worker exposure to hazards — outside the mill

RME has captured the average time duration for moving a liner bolt removal tool (THUNDERBOLT Recoilless Hammer) between consecutive shell bolts during the 'knock-in' phase of relining. This performance data, as illustrated in Figure 13, compares three different methods of hammer suspension at Site 2.

RME's most advanced suspension technology is SKYWAY. SKYWAY is a fixed plant gantry system around the mill's exterior that enables automated suspension and operation of the liner bolt removal tools, and supports safe operator platform modules for reline crews.

SKYWAY offers a number of performance and safety benefits. These include automated hammer movements, which allow for smooth and direct travel between bolts, and the elimination of small positioning adjustments and human inefficiencies seen during manual operation. Figure 13 shows the additional movement time reductions resulting from this system.

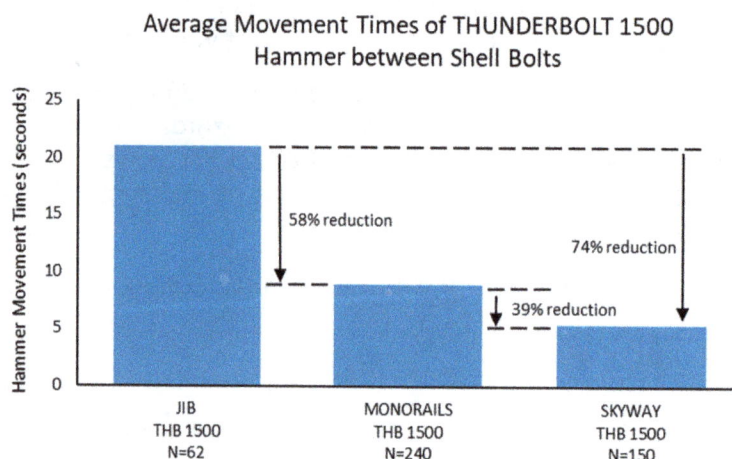

FIG 13 – Average time spent moving a THUNDERBOLT 1500 Hammer between bolts on the shell zone using different suspension systems. All data was collected from Site 2, before and after SKYWAY implementation.

A large SAG mill (36' to 40') typically has between 300 to 600 shell liner bolts. Reducing hammer movement time between these bolts can therefore result in significant total time savings over the duration of a reline.

Overall safety and performance impact of implementing automated relining technology

Site 1 best demonstrates the benefits of automated and automation-enabling mill relining technology implementation. This site currently utilises a full suite of such technologies and is therefore able to conduct all relining phases using these methodologies. As a result, Site 1 has significantly reduced the number of person-hours during which workers are exposed to hazards both inside and outside the mill, as shown in Figure 14.

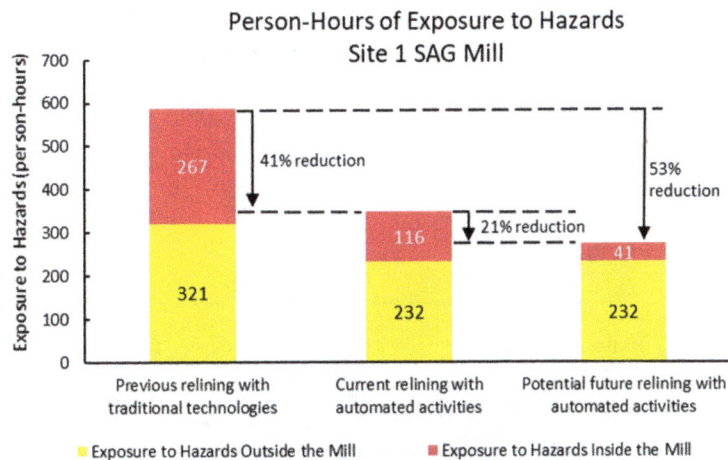

Person-Hours of Exposure to Hazards Site 1 SAG Mill

Exposure to Hazards Outside the Mill (yellow)
Exposure to Hazards Inside the Mill (red)

Previous relining with traditional technologies: 321 (outside), 267 (inside)
41% reduction

Current relining with automated activities: 232 (outside), 116 (inside)
21% reduction

Potential future relining with automated activities: 232 (outside), 41 (inside)
53% reduction

FIG 14 – Comparing the safety benefits at Site 1 during progressive implementation of automated technologies.

Figure 14 illustrates the reduction in hazard exposure outside the mill, enabled by SKYWAY which automates the knock-in of bolts and liners. The remaining exposure represents workers required for the removal and placement of nuts and washers. Figure 14 also shows the reduction in hazard exposure inside the mill, enabled by INSIDEOUT removal and placement. The remaining exposure represents workers currently required for other inside tasks such as the manual placement of rubber strips and wall rubber, and manual MRM operation. With continued engineering development and reline crew training and mentoring, further reductions in hazard exposure inside the mill are anticipated in future relines, as depicted in the far-right bar.

In addition to reductions in hazard exposure, implementation of automated technologies outside and inside the mill has provided other performance benefits for Site 1.

- Reline crew size reductions: The number of crew members required to complete mill relining was reduced from 22 workers per shift using traditional relining methods, to 14 workers per shift for relines using automated relining technologies.

- Reline duration reductions: Reline scopes during the 2015–2016 period involved different liner quantities to reline scopes performed during 2022–2023 (see Figure 15). This made overall relining durations difficult to compare directly. 'Minutes per liner replaced' provides a fairer comparison of relining speeds. Large-scope relines using traditional methods, typically resulted in overall relining durations exceeding 90 hrs, equating to average liner replacement speeds of more than 40 min per liner (see Figure 15). After implementation of automated and automation-enabling technologies, the reline time was reduced to 78 hrs, equating to an average liner replacement speed of 30 min per liner. This represents a reduction in average liner replacement time by more than 25 per cent.

Note that liner replacement speed is calculated using the reline duration, excluding non-reline time.

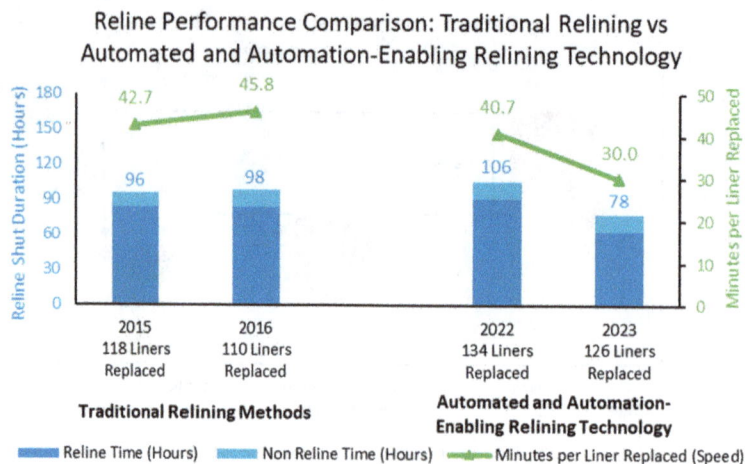

Reline Performance Comparison: Traditional Relining vs Automated and Automation-Enabling Relining Technology

FIG 15 – Performance benefits at Site 1 comparing traditional relining methods to automated and automation-enabling technology methods.

DISCUSSION – CHALLENGES AND LESSONS LEARNED IN AUTOMATING RELINING

The importance of cultural change programs for automated mill relining success

RME has identified that automating mill relining presents new challenges and learning curves for reline crews, especially during the early stages of adoption. With each progressive implementation of automation and INSIDEOUT methodologies, reline crews are not only operating new equipment, but also adopting new processes and work sequences. Providing sufficient mentoring and training to the reline crews involved in the automated relining operation has proven to be critical to the successful application and adoption of these technologies. By developing reline crew competencies in using the new processes and equipment, RME has not only ensured the newly-implemented automated technologies secured material productivity improvements through reduced relining time, but has also reduced the person-hours during which workers are exposed to risk.

RME has assessed the appropriate training of reline crews and site trainer requirements differently based on the level of automation being implemented. For Site 1, the complexity of adopting of a fully-automated mill relining system was managed by providing crew training at the RME headquarters in Toowoomba. A full size replica mill was constructed, and the site's automation system was installed, allowing for comprehensive 'real-world' training and practice to be conducted. This training minimised potential surprises when using the automated equipment in a real reline operation, and empowered the site trainers and reline crew for success. For Sites 2 to 5, RME provided on-site training and support to enable the successful implementation of automated relining equipment. For these sites, and other sites currently on the automation journey, RME also provided training at the RME Toowoomba headquarters, utilising some of the training facilities that had been constructed for the training of Site 1.

In addition to training at RME's headquarters, RME has supported the implementation of the fully-automated relining systems and process at Site 1 by having a specialist on-site during relining shutdowns. Similarly, RME has supported other sites' implementations of partially-automated relining through the presence of technical personnel on-site. The benefit of having RME training, engineers and technicians on-site during the implementation of such these new systems, is to:

- support the commissioning of the equipment

- support the change management

- provide expedited resolution to technical issues that may arise.

RME has attended relines at Site 1 throughout the three years during which it has been on the mill relining automation journey. This support has enabled Site 1 to implement the automated mill relining systems successfully through an incremental and staged approach over several relines. This has led to Site 1 successfully achieving the world's first reline in which liner installation was completed using automated MRM movements. This was achieved in 2023, with the support of RME personnel on-site.

The importance of mitigating human-automation risks with systems-engineering and implementation sequence for automated mill relining success

Operating automated equipment in close proximity to personnel is generally considered high-risk and unsafe due to the risk of uncontrolled machine movements. For this reason, when implementing fully-automated mill relining systems, it is critical to remove people from working on the charge prior to utilising the automated MRM. Removing people from machine proximity eliminates the inherent risks of human-robotics interaction. A roboticised MRM can move up to four times faster, with all seven axes of movement now synchronised in the automated MRM's programming.

To eliminate these risks and achieve fast and safe automated mill relining, RME needed to consider all the interdependent tasks across the entire mill relining process. This meant considering how changes to activities inside the mill would impact the associated activities outside, and *vice versa*.

This systems-engineering approach yielded to two key outcomes. First, it lead to the development and implementation of an 'integrated' advanced technology suite, including INSIDEOUT tooling as described earlier. These easy-to-adopt tools can be commissioned on-site, and reline crews can be trained for competent use, in preparation for the deployment and eventual use of an automated MRM. Second, the approach delivered key learnings into best practises for the staged 'implementation sequence' of automated tools and methodologies.

At Site 1, RME initially commenced the mill relining automation journey at the liner placement phase. This involved the implementation of the INSIDEOUT external bolting tool (RME BOLTBOSS) and the method for securing of new bolts and nuts from the mill's exterior, thereby removing people from inside the mill during new liner installation. RME had initially considered this task a suitable starting point as it required fewer liner modifications. However, RME soon learned that changing mill relining processes from traditional bolt insertion—which requires crew on the charge—to automation-compatible external bolting tools, necessitates new processes and tool competency. This meant that a cultural change program was required, and RME now understands that the liner placement phase is a difficult starting point for mine sites and reline crews from which to adopt automated relining.

This barrier to smooth adoption has been overcome with adjustments to the implementation sequence. RME now recommends mine sites starting the mill relining automation journey at the worn liner removal phase, specifically with the implementation of the grapple hydraulic grabbing tool attachment for the MRM grapple, (RUSSELL Claw, part of the INSIDEOUT removal family). This tool eliminates the need manual slinging of worn liners and removes the requirement for reline crews to work on the charge inside the mill during liner removal.

A case study documenting this liner removal methodology using the RUSSELL Claw, was detailed in a paper at the SAG 2023 conference (Walton *et al*, 2023). It outlines why and how mine sites are best to start the automation journey by addressing the liner removal process first. This is because the tooling and process changes proved to be easier to implement, with no need for liner design modifications. It also minimises crew training requirements and has limited impact on procedural changes. This case study also demonstrated that it resulted in faster uptake, user acceptance and adoption by crews and sites.

Figure 16 demonstrates the ease of implementing the RUSSELL Claw at Site 3 (in Table 1) and highlights the immediate benefits of faster relining speeds from the tool's first use. This enables reline crews and sites to achieve fast reductions in liner removal phase time, right from the first reline of their automation journey. These 'quick wins' foster a positive culture around the technology adoption journey for the site and crew, setting the stage for a successful automation journey.

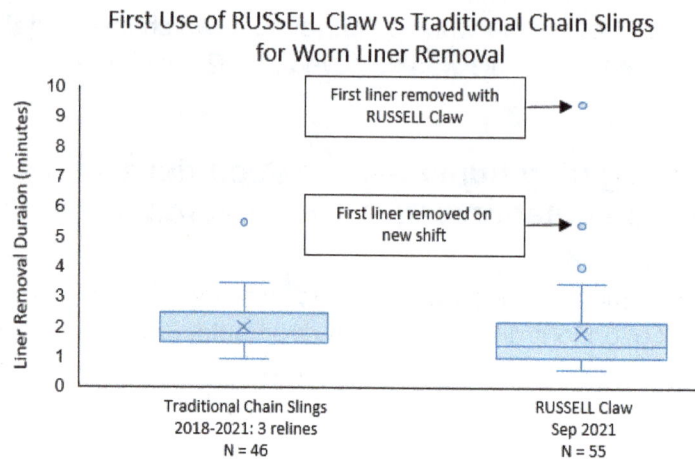

First Use of RUSSELL Claw vs Traditional Chain Slings for Worn Liner Removal

FIG 16 – Comparison of liner removal times using traditional chain slinging methods and the first usage of a hydraulic grabbing tool (RUSSELL Claw) connected to the MRM grapple. Data from both methods is comprised of feed lifters and discharge grate liners.

Additional liner removal tooling is required at a later stage in the automation journey to deploy and use the automated functions of the MRM. Therefore, familiarisation with the RUSSELL Claw's operation makes implementation of these additional tooling requirements smoother. The site and reline crew will be accustomed to removing worn liners without personnel on the charge. This was not the case for Site 1 which started the automation journey with liner placement.

RME's work with these different sites has also revealed other opportunities to eliminate in-mill reline tasks through engineering (tooling) or administrative (process) controls. In addition to the slinging and manual handling of worn liners, RME identified and developed solutions for the removal and placement of corner blocks and placement of rubber strips (liner widgets). These changes can also be implemented without making major modifications to liner design.

RME's experience with these implementations highlighted another critical safety principle when implementing automated equipment: establishing exclusion zones. These zones prevent personnel access to certain areas during automated machine operation. This is achieved through additional risk controls that includes use of a blue light warning system, and a light curtain at the entry of the mill which is interlocked with the machine functions. A paper detailing these further engineering and administrative controls for removing personnel from inside the mill during relining was presented at SAG 2023 (Raharjo *et al*, 2023).

Building on the successful implementation of automated and automation-enabled mill relining systems at multiple sites globally, RME is confident that a clear path towards fully automated liner exchange exists through the adoption of INSIDEOUT relining methodologies. Furthermore, this phased approach leverages proven technologies, minimising technical risks and costs associated with new technology adoption. Once these new processes are established at sites, they will pave the way for significant advancements in mill relining speed and safety.

Continuous improvement — practical tooling transport to enable efficient automated relining

RME's close collaboration with sites and reline crews during automation and automation-enabled technology trials and implementations, means it consistently gains valuable insights into improvement opportunities.

For example, at Site 1, the initial liner removal technology and methods implemented required four different tools for the removal of different liner designs, each brought into the mill on a separate tooling tray. Within the first few relines, RME identified an opportunity to improve the efficiency of this process. In response, RME engineered a universal 'one-tray' solution that reduced the time consumed by bringing the different tooling trays in and out of the mill via the liner cart during each removal phase,

This engineered solution is illustrated in Figure 17. The new solution allowed all four tools to be brought into the mill together at the start of the removal phase. The tray is then lifted onto the charge by the MRM, and remains there for the duration of the removal phase. The MRM operator then swaps between the tools simply, efficiently and with minimal delay. This improvement enabled further time savings on an already improved liner removal process.

FIG 17 – A universal 'one-tray' solution entering the mill on the MRM liner cart. This is used for efficient transport of multiple liner removal tools that are required for different types of liners.

Different challenges in automating relining on a greenfield site compared to a brownfield site

RME has been successfully implementing automated mill relining technologies and INSIDEOUT methodologies at both greenfield and brownfield mine sites since 2018. This multi-site experience has revealed different challenges associated with the implementation of automated relining technologies for each site type.

Brownfield sites often pose unique challenges for equipment, commissioning and process changes, due to the time-sensitive nature of processing plant shutdowns. Commissioning equipment within these strict shutdown timelines necessitates careful planning and execution. It also means the installation process may need to be broken down and separated into smaller manageable stages that can be integrated effectively within the maintenance window or normal site shutdown duration.

RME has also faced challenges with crane availability at brownfield sites, as the crane is a highly sought after asset during a shutdown. This means reline equipment commissioning must be executed in tightly-segmented time periods in order to work in with other crucial shutdown activities.

At greenfield sites, where automated relining technologies have already been installed, RME has been afforded extended periods of time to commission and test equipment, with full crane availability and priority during the construction phase of the site.

Through exposure to these multiple and diverse commissioning experiences at greenfield and brownfield sites, RME has developed a deep understanding of the potential challenges for future deployments. This insight has led to the development of robust implementation plans that proactively identify and mitigate risks, ensuring smooth execution and adherence to project schedules.

Importance of stakeholder collaboration for automated mill relining success

RME has learned from site experience that successful implementation of automated relining technologies relies on good collaboration with the liner supplier. They are an important stakeholder whose 'buy in' is important to the site's automation journey. Liner modifications are required for multiple stages of automation-enabling technologies to ensure compatibility between the liners and INSIDEOUT tooling.

A critical factor in Site 1's (Table 1) success was the strong working relationship established with the liner supplier throughout the technology implementation. This foundation streamlined the project and contributed to its successful outcomes. In the case of Site 1, the liner supplier has worked closely with both RME and the Site 1's team to make the required changes to liner designs and ensure compatibility with the processes and equipment required for automated mill relining.

RME understands from its experience at Site 1 the challenges a site could face if they were to change liner suppliers during the automation journey. It could result in setbacks in the progression of relining automation, the severity of which can vary depending on the project stage at which they occur. Early-stage issues might delay initial implementation phases, while challenges arising during later stages could disrupt established automation processes. This is because new liner suppliers need time to become familiar with the liner design requirements, and accurately design and manufacture the necessary changes.

To ensure seamless integration of mill liners with automated relining equipment, liner compatibility should ideally be a key consideration and specified during the tendering process. This can be supported by incorporating liner-MRM interface studies into this stage.

CONCLUSIONS

This paper has provided an update on RME's implementation of automated relining equipment across a number of mine sites globally. Quantitative data has been collected from these sites using video footage and analysed to demonstrate the following insights:

- The number of liners being replaced using automated relining equipment (or automation enabling equipment) is growing by a rate of approximately 20 per cent each year.

- INSIDEOUT Placement, an automation-enabling technology, improves safety by removing people from the mill during liner placement, and has been shown to achieve similar speeds to traditional relining methods.

- INSIDEOUT Removal, another automation-enabling technology, also improves safety by removing people from the mill during liner removal, and has been shown to speed up the liner removal process significantly.

- Site 1 demonstrated in its September 2023 reline event that it is possible to eliminate 80 per cent of the exposure to hazards inside the mill using INSIDEOUT tooling and automated mill relining technologies compared to traditional methods.

- Site 2 demonstrated that the knock-in speed can be increased by 74 per cent using THUNDERBOLT SKYWAY to automate the travel, positioning and operation of the liner bolt removal tools (THUNDERBOLT recoilless hammers).

- Site 1 demonstrated a significant improvement in overall relining speed (~20 per cent), enabling greater mill uptime, following the implementation of automation and automation-enabling relining equipment.

This paper has also documented the lessons learned during mill relining automation implementations, including experiences related to implementation sequencing, brownfield versus greenfield challenges, stakeholder collaboration and practical toolkit changes.

ACKNOWLEDGEMENTS

The authors would like to express our gratitude to RME for their support and permission to share the information in this paper. We are particularly thankful for the collaboration and dedication shown by RME's customers. Their commitment to driving the implementations of new technologies have significantly contributed to advancements in relining safety. This collective effort will help set new benchmarks for safety standards for mill operators and the mineral processing industry.

REFERENCES

Brander, D, Bohorquez, J, Gwynn-Jones, S and Turner, M, 2021. Automated mill relining – current progress and future, in *Proceedings of the Mill Operators Conference 2021*, pp 10–28.

Chinh, G and Chieu, P, 2024. Cement factory incident kills seven, injures three in northern province, *VnExpress International*. Available from: https://e.vnexpress.net/photo/news/the-cement-mill-involved-in-incident-that-killed-7–4737742.html [Accessed: 23 July 2024].

Raharjo, A, Wilmot, J, Manuel, A, Suhono, S, Smith, S, Herbertson, W and Bohorquez, J, 2023. PTFI and RME Collaboration: Technology Makes Mill Relines Safer and More Efficient, in Proceedings SAG Conference 2023.

Smith, S, Rubie, P, Hodges, J, Gwynn-Jones, S and Herbertson, W, 2019. Safer mill relining at PT Freeport Indonesia – The first automated mill relining system, in Proceedings SAG Conference 2019.

Van Rooyen, P, Bohorquez, J, Hodges, J and Gwynn-Jones, S, 2023. Designing grinding plants for mill relining to maximise mill availability – update on best practice, in Proceedings Comminution 2023.

Walton, J, Bohorquez, J, Gwynn-Jones, S and Salomon, J, 2023. Latest Developments in Mechanised Grinding Mill Relining: Site Trials and Simulation Results, in Proceedings SAG Conference 2023.

Optimising grinding circuit performance – a success story at Harmony Gold's Hidden Valley Mine with Rockwell Automation's Pavilion8™ MPC process optimisation technology

A Jain[1], T Tlhobo[2], L Silva[3] and F Anis[4]

1. Lead Advanced Process Control Engineer, Rockwell Automation, Perth WA 6103.
   Email: akhilesh.jain@rockwellautomation.com
2. Processing Principal, Harmony Gold, Milton Qld 4064.
   Email: teboho.tlhobo@harmonyseasia.com
3. Senior Process Control Engineer, Morobe Consolidated Goldfields Limited, Papua New Guinea.
   Email: leonardo.silva@harmonyseasia.com
4. Metallurgy Superintendent, Morobe Consolidated Goldfields Limited, Papua New Guinea.
   Email: fred.anis@harmonyseasia.com

ABSTRACT

Harmony Gold Mining Company Limited, a distinguished player in the gold mining and exploration sector, sought to enhance its operational efficiency and sustainability at the Hidden Valley Mine in Papua New Guinea. In collaboration with Rockwell Automation's industrial data science team, Harmony Gold implemented Factory Talk™ Pavilion8™ Model Predictive Control (MPC) in the semi-autogenous grinding (SAG) mill grinding process, showcasing a commitment to safe, profitable output through operational excellence.

The goal was clear: optimise the grinding circuit to achieve increased efficiency, throughput, and stability in SAG mill operations while ensuring consistent cyclone feed pressure. Facing challenges of ore variability and rising sustainability expectations, even a seemingly small increase in throughput can have significant positive effects on costs, efficiency, and productivity.

Engaging Rockwell Automation, Harmony Gold embarked on a comprehensive approach to optimise mill throughput and reduce process condition variability. The Rockwell Automation MPC team analysed 12 months' worth of historical data, identifying critical process parameters amenable to control and modification. Leveraging multivariable control, constraint handling, feedforward, decoupling, and dead-time compensation, Pavilion8 MPC Controller employed process models for optimised control actions.

The dynamic correlations within the grinding circuit were effectively managed by MPC by adjusting mill parameters such as mill speed and cyclone feed pump speed. MPC system achieved maximum achievable throughput while adhering to operational constraints such as power consumption, ore variability and availability. Pavilion8 also maintained stable cyclone feed pressure, reducing variability, and ensuring consistent cyclone operation.

Results exceeded expectations, with the MPC Grinding Circuit Application designed, implemented, and validated in less than six months. The project surpassed the initial throughput targets, achieving 20 per cent better than target improvement reduction in milling variability after a 30-day run, and more than double the target improvement over 221 days. Pavilion8 reduced the variability in feed rate by 35.1 per cent.

Additional achievements included a robust 90.8 per cent MPC uptime, indicating strong acceptance from operations, and consistent precision in maintaining mill weight closer to the target, a 41.2 per cent reduction in deviation from the target. The application sustained cyclone feed pressure within the desired range, displaying an 18.6 per cent reduction in variability compared to baseline performance. These milestones underscore the substantial impact and effectiveness of the Pavilion8 MPC Grinding Circuit Application at Hidden Valley Mine.

This success set the stage to rollout MPC technology into the rest of the processes, where there is opportunity to reduce variability and optimise productivity. The collaboration between Harmony and Rockwell Automation marks a continued commitment to operational excellence and sustainable mining practices.

INTRODUCTION

Hidden Valley Operation is an open pit mine located approximately 15 km from the township of Wau and approximately 90 km south from Lae, the capital of Morobe Province in Papua New Guinea. The mine was commissioned in 2009 to produce gold and silver from the Hidden Valley and Hamata pits (McLean and Watt, 2009; Burns and Peachey, 2014). Ore primarily from the Hidden Valley pit is crushed and transported via a 5 km pipe conveyor to the processing plant's coarse ore stockpile. Hamata ore is transported to the intermediate stockpile, this is then processed through two jaw crushers with the crushed product discharging to the coarse ore stockpile.

Ore milling for most processing plants is a significant cost, and often this represents a quarter of the processing operating expenditure. Hidden Valley processing plant is no different and its semi-autogenous grinding (SAG) milling process has gone through several optimisation programs to improve cost efficiency. Optimisation of the milling process in the later years, has been primarily focused on reducing milling costs by maximising throughput while maintaining the product size.

Since inception, the SAG mill control system has operated with several independent control variables (CVs) each focusing on achieving the target set values. This control strategy required the mill operator to watch over several variables to keep the mill operating at optimum. An experienced and well-trained operator may be able to maintain the required efficiencies. However, each operator would apply a different methodology to achieve the mill objectives of the shift which led to variability in the process and hence variability in results. Harmony approached Rockwell Automation for an advanced mill process control solution to try and reduce this variability, essentially having the best mill operator at all times.

Rockwell Automation, a global leader in industrial automation and digital transformation solutions, worked with the Hidden Valley Operators and Processing team to optimise the milling control using Pavilion8 Model Predictive Control (MPC) software. The FactoryTalk Analytics Pavilion8 MPC was developed in 1991 for the polymer industry and since has been implemented in many industries around the world to automate complex multiple variable processes. Pavilion8 is an intelligent layer built on top of existing control systems. Its purpose is to continuously drive plants toward multiple business objectives in real time.

The Pavilion8 MPC leverages real-time closed loop control to improve process performance by reducing variability in a process. Figure 1 offers a visual demonstration of how MPC generates benefits.

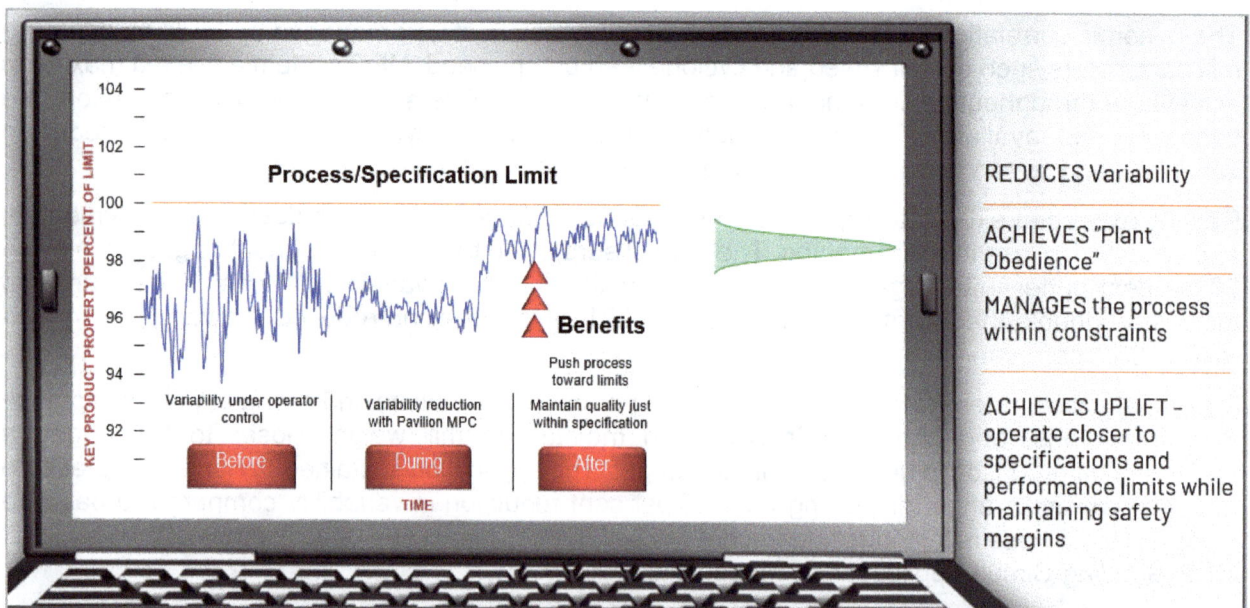

FIG 1 – How MPC generates benefits.

In mineral processing, reduced variability leads to:

- improved plant stability
- increased throughput
- enhanced and more consistent product quality
- reduced scrap, rework, and waste
- improved yield
- decreased specific energy consumption
- mitigated environmental risk
- lowered manufacturing costs
- enhanced profitability.

Giblett and Putland (2018), cited that the benefits were also associated with increased recovery or improved product grades associated with reduced variability in or better control of grind size. The Pavilion8 MPC application enhances control performance by automatically adjusting set points, typically handled by human operators. These adjustments are calculated using current conditions and a process model, considering the response of control elements and their flexibility. Operators retain control over enabling/disabling MPC and setting operational boundaries. However, the system autonomously optimises set point actions in real-time, considering various constraints and economic drivers, for effective disturbance rejection and optimal performance over time.

The objective at Hidden Valley Hamata Processing Plant was to find a technology that was proven and tested, that can scale across the processing plant, included linear and nonlinear processes, and is an open platform for operators, process engineers and metallurgists to maintain, monitor, and tune over time whilst also bringing process stability and automation.

HIDDEN VALLEY SAG MILL CIRCUIT DESCRIPTION

Ore from the Coarse Ore Stockpile is withdrawn using two variable speed apron feeders discharging onto a mill feed conveyor. The mill feed conveyor discharges ore into the SAG mill feed chute together with feed dilution water, gravity screen oversize, ore scats and primary cyclone underflow. Steel balls are intermittently added onto the feed conveyor belt at a set ratio to maintain a predetermined ball charge. The milled product slurry discharges to a 20 mm aperture vibrating screen, oversize material discharges onto a scats conveyor that goes through a magnetic separator. Magnetic material discharges to a scats bunker while non-magnetic materials are conveyed back to the mill via a series of conveyors. Screen undersize discharges to a discharge hopper together with dilution water, leach reactor solid tails and the gravity concentrator tailings, the contents of the mill discharge hopper are then pumped to a cluster of 12 cyclones (ten primary and two secondary gravity feed cyclones). The two secondary cyclones operate in a duty-standby configuration. The secondary cyclones underflow product is screened through a 2 mm aperture vibrating screen, with undersize reporting to the gravity concentrator and oversize product reporting to the SAG mill feed chute. The secondary cyclone overflow combines with the primary cyclone overflow, gravitating to the flotation circuit. Leach reactor solid tailings are pumped back to the discharge hopper while the pregnant liquor is pumped to the zinc precipitation circuit. Figure 2 shows a simplified SAG mill circuit flow diagram.

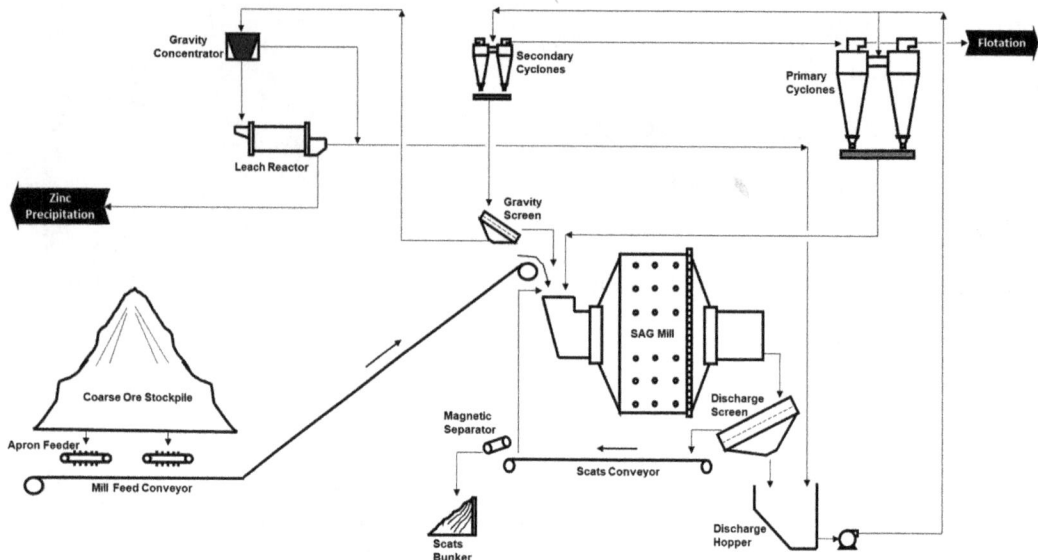

FIG 2 – Hidden Valley SAG mill ore flow diagram.

Pre-MPC SAG mill control system

The initial control system objective was mainly based on maximising mill feed rate while producing the target grind size for the preceding processes. Achieving both the throughput and product size would meet the required recovery efficiency and unit cost. There were several control parameters each with a set objective to maintain the overall milling efficiency.

The mill feed rate is controlled by regulating the apron feeder speed, this is based on either the mill load or mill power, whichever of the two is the lowest. A Hi-Hi mill power alarm will stop the apron feeders thereby forcing a mill grind out. This is similar for the mill load, when a Hi-Hi condition activates. The mill load is a user set value with a target range set within the distributed control system (DCS) for a Hi-Hi or Lo-Lo condition. Mill speed is a user input set point based on the relationship between power drawn and mill load relationship. Speed can be varied from 65 to 85 per cent of the critical speed.

Cyclone valves are regulated by the feed pressure. When the pressure rises to a specified level, a valve opens after a pre-determined time. Conversely, if the feed pressure falls below the set value, the valve closes after a set delay. The cyclone feed pressure changes based on the speed of the cyclone feed pump. The speed of the cyclone feed pump is regulated by the level in the SAG discharge hopper.

These controls are tuned through the use of a PID to minimise the variation, however, the fact that these are all independent creates variability. Often the operator would respond to each condition to stabilise the specific parameters, however, this would create another process deviation. The introduction of the leach reactor tails, would also present a unique challenge in stabilising the mill circuit.

There was a need to have a system that monitors all the variables and prioritises which controllers to modify and/or modifies all controllers simultaneously in real time. It was clear that a superior controller would yield the maximum feed rate and a reduction in variability.

PROJECT EXECUTION

Definition and planning

The first step in the Pavilion8 team's 'Value First' methodology is to complete a benefit study which confirmed reasonable handles for applying MPC technology. This approach involved meticulous analysis of 12 months of operational data with the objective and main circuit challenges in mind. Through comprehensive data analysis and evaluation, a clear valuable MPC project was defined. The results of the study culminated in a techno-commercial proposal that outlined a fixed price for project delivery, key performance indicators (KPIs) for success measurement, return on investment (ROI) calculations, and the basis of study calculations. Additionally, the proposal included a clear schedule for project delivery, defining the timeline for implementation and expected milestones. Moreover, the proposal outlined the scope of work, detailing the tasks and responsibilities undertaken, as well as the scope of work for Hidden Valley, delineating the collaboration and support required from Hidden Valley to ensure project success. Through this comprehensive proposal the teams had a clear roadmap for project execution, ensuring mutual understanding and alignment of expectations for successful project outcomes.

A project execution plan was set for 10 months, with the key milestones; Functional Specification Approval, Application Factory Complete, Deployment and Acceptance as outlined in Table 1.

TABLE 1

Hidden Valley grinding MPC project delivery schedule.

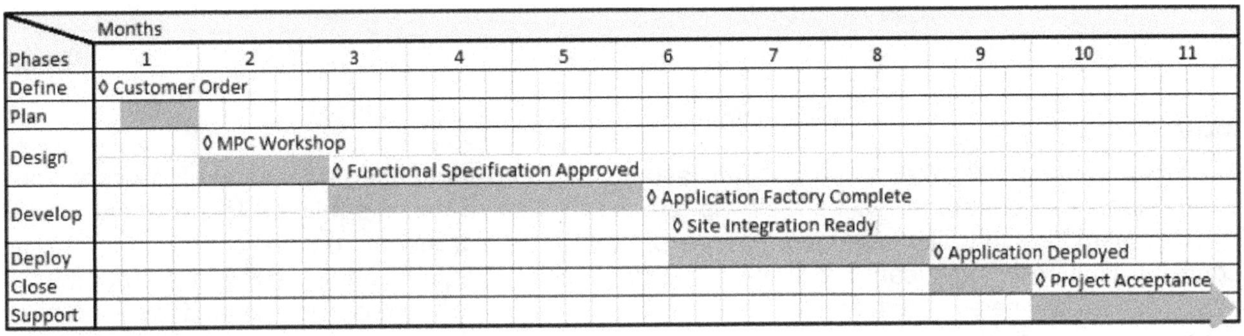

Phases	Months 1	2	3	4	5	6	7	8	9	10	11
Define	◊ Customer Order										
Plan											
Design		◊ MPC Workshop									
			◊ Functional Specification Approved								
Develop						◊ Application Factory Complete					
						◊ Site Integration Ready					
Deploy									◊ Application Deployed		
Close										◊ Project Acceptance	
Support											

The Benefit Study was completed, and clear benefits were confirmed. The KPIs for the project are outlined in Table 2.

TABLE 2

KPIs outlined in the project proposal.

KPIs	Units	Expected benefits	Comments
Throughput Increase	t/h	1–3%	Maximise milling rate
Reduction in SD of mill weight error (SP-PV)	t	20–50%	Stabilise mill operation and maximise runtime
Reduction in SD of cyclone feed pressure error (SP-PV)	kPa	30–50%	Optimise cyclone overflow particle size distribution

SD – standard deviation; SP – set point; PV – process value.

The three main objectives in this circuit were to maximise throughput subject to constraints, minimise SAG mill weight deviation from the target and to stabilise cyclone feed pressure. A simplified process flow sheet is illustrated in Figure 3.

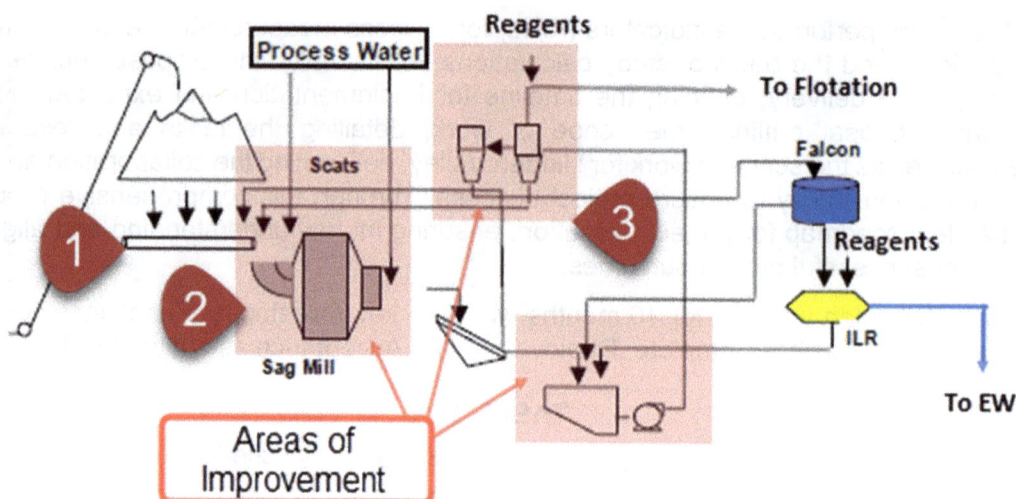

FIG 3 – Simplified grinding circuit at Hidden Valley (PNG).

Main challenges

In analysing the grinding circuit, two critical issues surfaced that caused a majority of the variability. First, suboptimal control of the Variable Speed Drive (VSD) within the mill led to inefficient weight management. Factors like unmeasured ore properties and transport delays complicated weight control, resulting in significant process variability. Second, frequent adjustments to the number of cyclones online were necessary to maintain stable discharge hopper levels and cyclone feed pressure. These fluctuations significantly impacted mill performance.

Design of MPC parameters

Based on the specifics of the Hidden Valley grinding circuit, process objectives and the inherent multivariable nature of the control problem necessitated a thoughtful approach. Analysis of the relationships between control, manipulated and disturbance variables (CV, MV and DV) was conducted. To mitigate interactions, the influence of an MV on multiple CVs were deliberately chosen, allowing compensation through other available MVs. Crucially, the primary control model aligned with the control objectives. In the context of MPC, stability is achieved by configuring control variables appropriately. For instance, consider the control matrix in Figure 4, mill weight is targeted within limits, mill power is controlled within a range, and scats are constrained to a high limit. These diverse control configurations yield positive degrees of freedom, enhancing MPC's effectiveness.

FIG 4 – Grinding circuit MPC matrix.

MPC site specific development

The development of MPC constituted a significant endeavour, involving meticulous analysis of historical and step test data. The step tests were conducted over several weeks to capture the true dynamics of process interactions comprehensively. Several other variables were systematically adjusted to evaluate their potential impact on the predefined control variables. These adjustments were executed to ensure that the resulting responses were of substantial magnitude.

Following these tests, dynamic models were identified leveraging the Pavilion8 Solution Builder. This software tool is indispensable for controller development, model identification, initial tuning, and tag mapping with DCS/PLC (distributed control system / programmable logic controller) systems within an offline environment. Subsequently, the identified models underwent rigorous testing under various scenarios to validate both the models and the MPC responses for grinding operations.

Overcoming the challenges

The Pavilion8 MPC suite offered a range of supporting tools, among which Pavilion8 Calcs played a crucial role. To address specific challenges, supporting calculations were developed. These calculations allowed the MPC system to regulate the mill motor speed at intervals of 10 mins in the absence of Variable Speed Drive (VSD). Consequently, the MPC adjusts the motor speed accordingly to attain the desired mill weight target, facilitating the maximisation of feed rate and subsequently enhancing production rates.

Conversely, cyclone calculations serve a proactive function by monitoring changes in the number of cyclones via the DCS. These calculations enable the MPC to dynamically adjust manipulated variables, ensuring enhanced stability within the cyclones.

MPC deployment

After a thorough review of the MPC responses, the MPC system was deemed ready for commissioning. Following the commissioning plan meticulously, each Manipulated Variable (MV) was commissioned individually, with careful attention to ensure production continuity.

To enhance operational flexibility, a DCS operator interface was developed, enabling operators to enable/disable the MPC system either partially or entirely. Additionally, operators could adjust variable limits directly from the DCS screens. Furthermore, the DCS interface was equipped to alert operators in the event of MPC shutdown due to external factors.

The deployment of the MPC solution was strategically executed to ensure bumpless control transitions between the DCS and MPC systems, guaranteeing uninterrupted operations.

OUTCOME

The overall objective was mainly to stabilise the mill circuit, with the added benefits of higher throughput while maintaining the overall mill product size. A reduction in the mill weight variability is indicated in Figure 5. Pre MPC the mill was operated in an overload load condition, this is also represented in the mill feed rate variance.

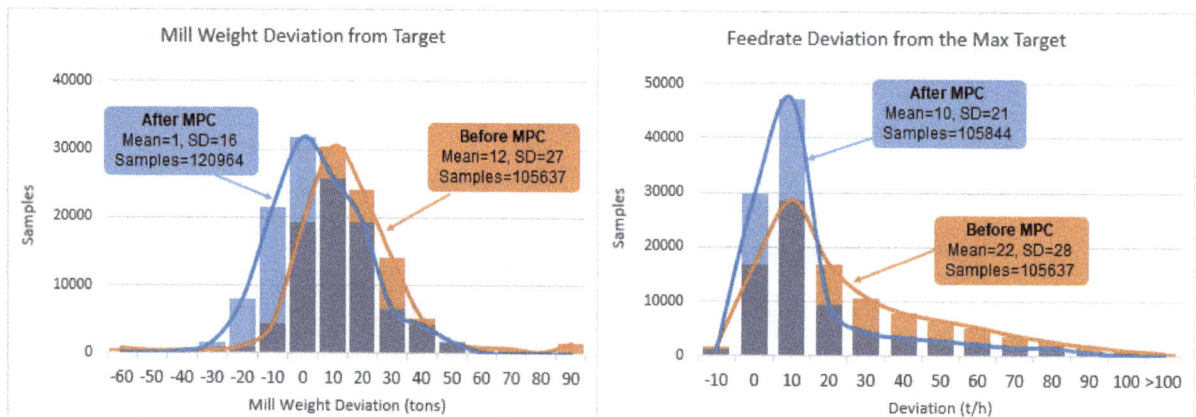

FIG 5 – Mill weight and feed rate deviation before and after MPC implementation.

Mill weight deviation graphs show that before MPC, to keep the mill weight on the target, feed rate was changed frequently which resulted in higher variability.

With MPC mill weight deviation from the target value decreased by 41.2 per cent which resulted in reducing mill feed rate deviation from the max target by 24.0 per cent. Reduced variability in the feed rate allowed the throughput to increase post MPC implementation within the first 30 days and a further increase over a period of 221 days.

Cyclone feed pressure deviation from set point reduced with the MPC implementation, having the number of open cyclones as a DV along with cyclone feed pump speed as an MV reduced the variability in cyclone feed pressure from its target. As a result, a 21.2 per cent reduction in variability was achieved as shown in Figure 6. In addition, the plot in Figure 7 shows the improvement in cyclone feed pressure control before and after MPC.

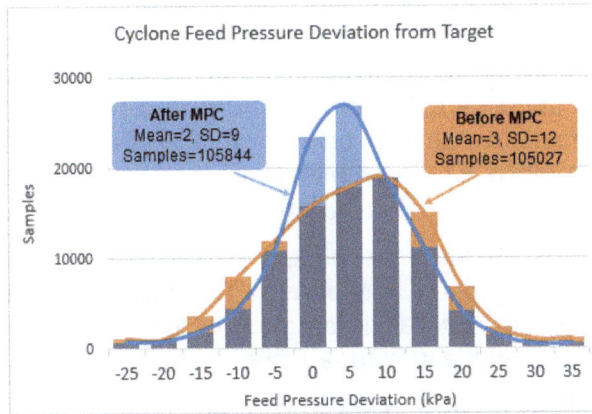

FIG 6 – Cyclone feed pressure deviation before and after MPC implementation.

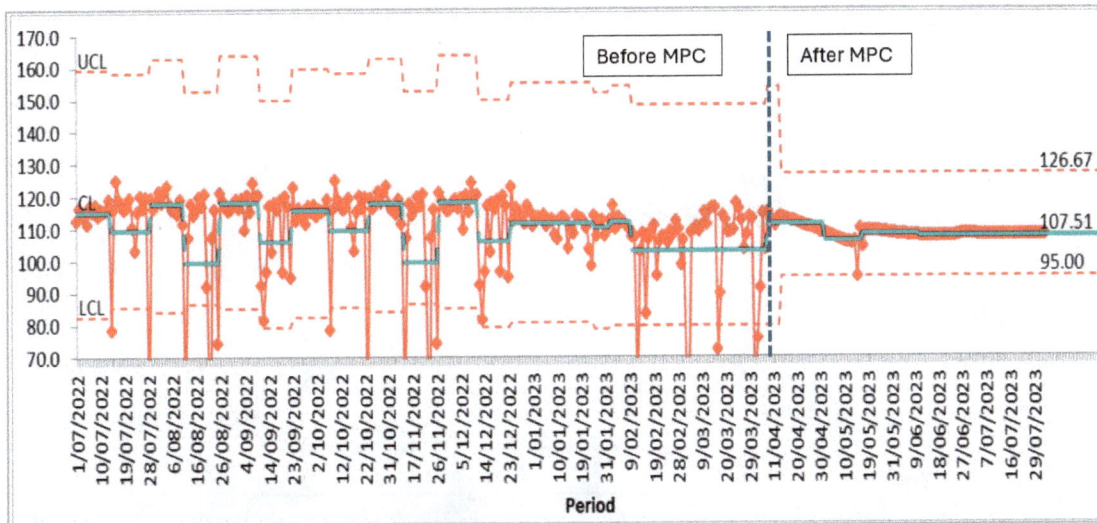

FIG 7 – Actual cyclone feed pressure before and after MPC implementation.

Operational acceptance is measured by tracking when MPC is switched on or off (whilst the mill is online). Typically, above 80 per cent MPC uptime is considered fair operational acceptance. 90.8 per cent MPC uptime was recorded as depicted in Figure 8, signifying strong operational acceptance.

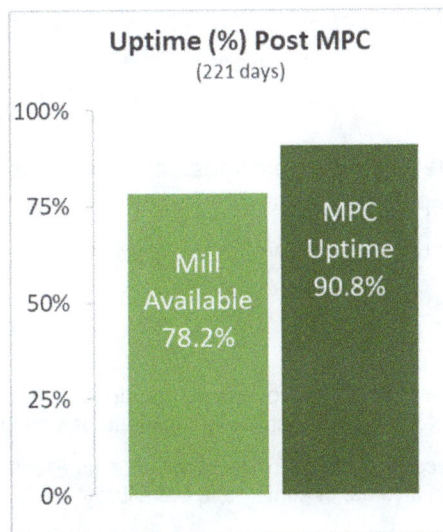

FIG 8 – Post MPC project mill availability versus relative MPC uptime.

CONCLUSION

Historically the adoption of MPC for milling circuits have proved to be superior to the traditional PID or fuzzy logic controller. The implementation of the model predictive controller at Hidden Valley has proven that this controller is able to stabilise t complex milling circuits with greater ongoing rewards. Through the implementation of the SAG mill MPC, mill feed rate and cyclone pressure variability were reduced resulting in an improvement in milling efficiency.

The project was able to be executed and completed faster than the proposed timeline due to team collaboration and ease of site technology acceptance. The positive outcomes and consistently high MPC uptime further solidified the cultural acceptance of MPC technology among Hidden Valley operators. This success has paved the way for advancing the technology to further processes.

Harmony Gold Management will be evaluating a roadmap for downstream processes, as depicted in Figure 9, leveraging the promising results and effectiveness of the Pavilion8 MPC Grinding Circuit Application at Hidden Valley Mine. These developments signal a bright future for enhanced operational efficiency and productivity on-site, underlining the substantial impact of MPC technology in driving operational excellence.

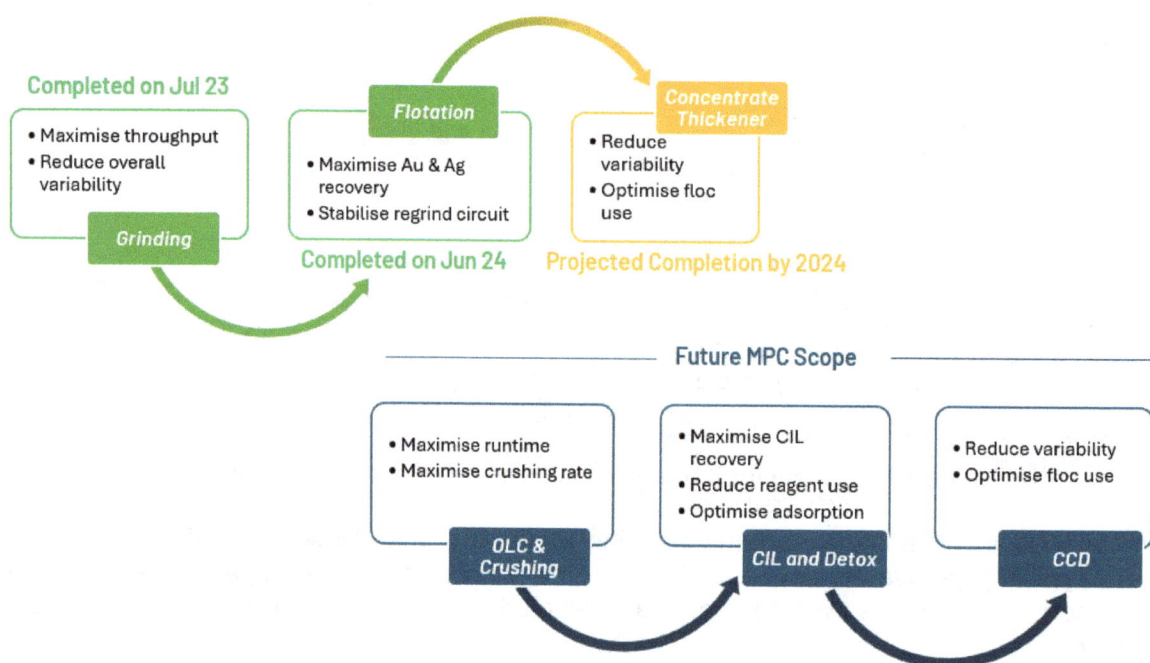

FIG 9 – Hidden Valley Mine MPC Projects: Completed: Green; FY2024 Completion: Yellow; and Future Scope: Blue.

ACKNOWLEDGEMENTS

Harmony Gold, the Hidden Valley Mine operational and processing teams, and Harmony Gold management; for their support to publish the recent successful collaboration.

Erin Daley, MPC Solutions Executive, Rockwell Automation; for her constant support to get this paper done.

REFERENCES

Burns, F and Peachey, G, 2014. Recent Process Improvements at Hidden Valley Gold Mine, in *Proceedings 12th Mill Operators' Conference 2014*, pp 43–54 (The Australasian Institute of Mining and Metallurgy: Melbourne).

Giblett, A and Putland, B, 2018. The basics of grinding circuit optimisation, in *Proceedings 14th Mill Operators' Conference 2018*, pp 43–52 (The Australasian Institute of Mining and Metallurgy: Melbourne).

McLean, E and Watt, J, 2009. Processing strategies for Hidden Valley Operations, in *Proceedings Tenth Mill Operators' Conference 2009*, pp 43–52 (The Australasian Institute of Mining and Metallurgy: Melbourne).

The effect of primary grind size on flotation recovery and the theoretical grade-recovery curve at Olympic Dam

Y Li[1], K Ehrig[2], V Liebezeit[3] and S Barsby[4]

1. Senior Geometallurgist, BHP Olympic Dam, Adelaide SA 5000. Email: yan.li@bhp.com
2. Superintendent Geometallurgy, BHP Olympic Dam, Adelaide SA 5000.
   Email: kathy.ehrig@bhp.com
3. Principal Geometallurgist, BHP Olympic Dam, Adelaide SA 5000.
   Email: vanessa.liebezeit@bhp.com
4. Senior Concentrator Development Metallurgist, BHP Olympic Dam, Roxby Downs SA 5725.
   Email: sarina.barsby1@bhp.com

ABSTRACT

Mineralogy underpins the metallurgical response in all mineral processing plants and Olympic Dam is no exception. The theoretical (mineralogically limited) grade-recovery curve, similar to the grade-recovery curve widely used by concentrator metallurgists, sets the separation limit that can be achieved under perfect plant conditions for any given feed mineralogy/liberation. Actual plant recovery will always sit below the theoretical grade-recovery curve for two reasons: 1) liberation measured from 2D sections of particles overestimates the true 3D liberation and 2) perfect separation doesn't occur in plants. Nevertheless, the distance between actual plant performance and the curve indicates how well the plant is operating, with a larger gap suggesting the potential for improvements to the plant conditions. The curve can also be used to rank the relative flotation performance of unblended future ores (ie geometallurgical samples) based simply on mineralogical/ liberation analysis on a size-by-size basis of samples ground to a specified grind size.

Grind size is selected to achieve satisfactory sulfide mineral liberation without excessive slimes generation and power wastage. Design primary grind size at Olympic Dam is 80 per cent passing 75 µm, with further regrind of rougher flotation concentrate to a P_{80} of 30 µm. To increase plant throughput, the primary grind size of flotation feed has been increased gradually through changes to cyclone configuration and mill operation. The consequent reduced sulfide mineral liberation in mill feed has a negative impact on flotation recovery but how much recovery is sacrificed by a coarser primary grind? The theoretical grade-recovery curve based on sulfide mineral liberation data can be used in conjunction with the plant data to quantify the effect of the coarser grind and determine the trade-off between recovery and throughput.

This paper describes the theoretical (or mineralogically limited) grade-recovery curve using Mineral Liberation Analyser results on monthly composite samples and compares the curve of before and after the coarse grind trial. It also relates the results generated from the geometallurgy laboratory program to the monthly composite program and highlights the impact of sulfide minerals grain size and liberation on achievable separation.

INTRODUCTION

The Olympic Dam iron-oxide copper-gold (IOCG)-uranium-silver deposit is one of the world's largest polymetallic deposits. The operation has been producing since 1988. The fully integrated processing plant produces final products (copper cathode, uranium ore concentrate, gold and silver bullion) at the Olympic Dam (OD) site. Main operating units include grinding, flotation, leaching, solvent extraction, smelting and refining.

Ore is extracted underground via sublevel open stoping. Stope size varies from ~100–500 kt with stope footprints of 25 to 30 m across and up to 200 m in height (Robertson et al, 2013). Ore is crushed before sending to the run-of-mine (ROM) stockpile. Mill feed is sourced from five to ten stopes daily or around 60 stopes annually with about 20 per cent of the material derived from lateral development (Ehrig et al, 2020).

Olympic Dam sulfide mineralogy

The primary sulfide minerals are pyrite, chalcopyrite, bornite and chalcocite, ranging in size from <20 μm up to several mm, with the average size of around 50–100 μm (Ehrig et al, 2019). Copper content varies greatly among different sulfide minerals as shown in Table 1.

TABLE 1
Summary for major Olympic Dam sulfide minerals.

Sulfide mineral	Formula	Cu (wt%)	Fe (wt%)	S (wt%)	Cu:S
pyrite (py)	FeS_2	0.00	46.55	53.45	0.00
chalcopyrite (cp)	$CuFeS_2$	34.63	30.43	34.94	0.99
bornite (bn)	Cu_5FeS_4	63.31	11.13	25.56	2.48
chalcocite (cc)	Cu_2S	79.85	0.00	20.15	3.96

There is distinct zonation of copper±iron sulfide minerals across the orebody. From the deposit margins progressing upwards and inwards:

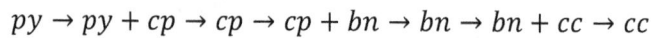

$$py \rightarrow py + cp \rightarrow cp \rightarrow cp + bn \rightarrow bn \rightarrow bn + cc \rightarrow cc$$

or simplified to:

$$py \rightarrow cp \rightarrow bn \rightarrow cc$$

Each sulfide can occur on its own or as binary pairs with the sulfide either immediately before or after in the above sequence. However, pyrite has not been observed in physical contact with bornite or chalcocite, and chalcopyrite has not been observed in contact with chalcocite.

Concentrator circuit summary

Milling occurs in two autogenous mills. The Fuller mill installed in 1995 is a 10.4 m diameter mill powered by a fixed speed dual pinion drive with a total power of 10.4 MW. It has a throughput of approximately 600 t/h of ore depending on ore type. The Svedala mill installed in 1999 is a 11.58 m diameter mill powered by a single 18 MW variable speed gearless mill drive. It has a throughput of approximately 1100 t/h of ore depending on ore type.

Figure 1 is a simplified flow sheet for the grinding circuit. Oversize material from each mill is directed to their own pebble crusher and undersize portion is fed into the circuit's classifying cyclone cluster. Cyclone underflow is redirected back to the mill for further grinding and cyclone overflow is sent to the common flotation feed tank as milling circuit product.

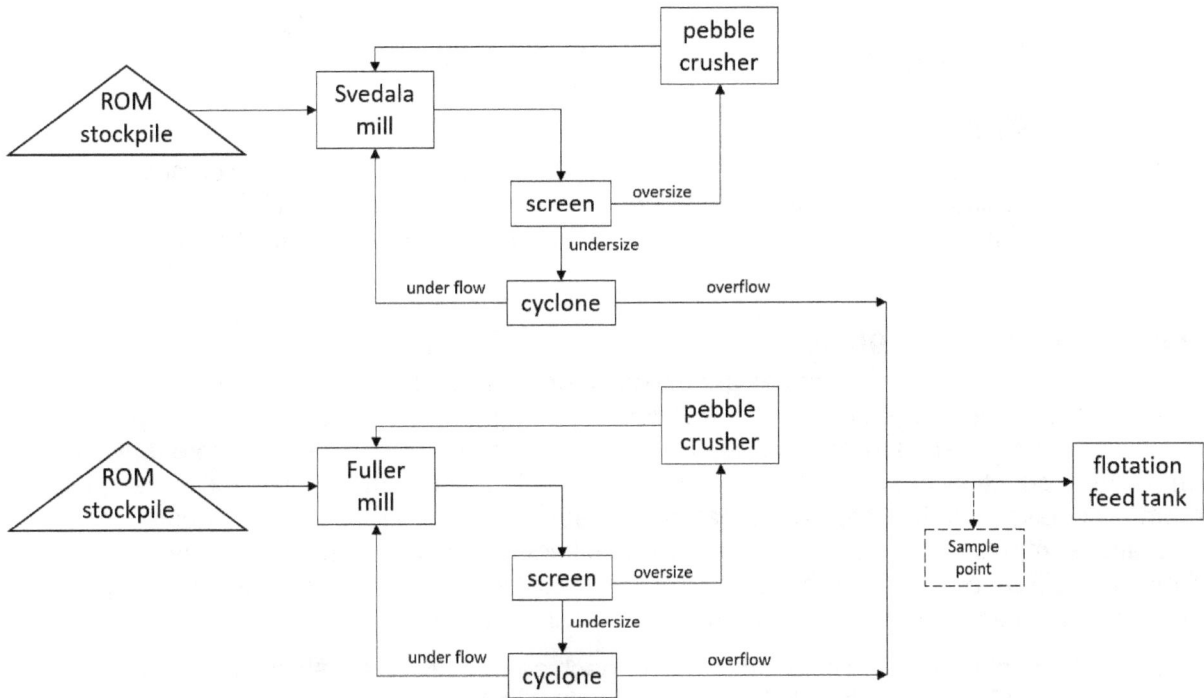

FIG 1 – Olympic Dam grinding circuit schematic.

Particle size of the grinding circuit or flotation feed is a critical parameter. Too coarse, sulfide minerals will not be sufficiently liberated and result in increased valuable mineral reporting to waste and sacrifice recovery. Too fine, time and power will be wasted on unnecessary grinding and excessive slimes will lead to low recovery. Based on the sulfide grain size at OD, the P_{80} was nominally set at 75 μm which maximises sulfide mineral liberation while minimising slimes generation and power wastage.

There are two lines of rougher-scavenger flotation. A portion of the rougher concentrate is sent directly to cleaner 2, while the remaining rougher and scavenger concentrate is treated in the regrind mill to about 80 per cent passing 30 μm. There are three stages of cleaning with recycle of cleaner tails to the preceding stage, and recycle of cleaner 1 tail to flotation feed (see Figure 2).

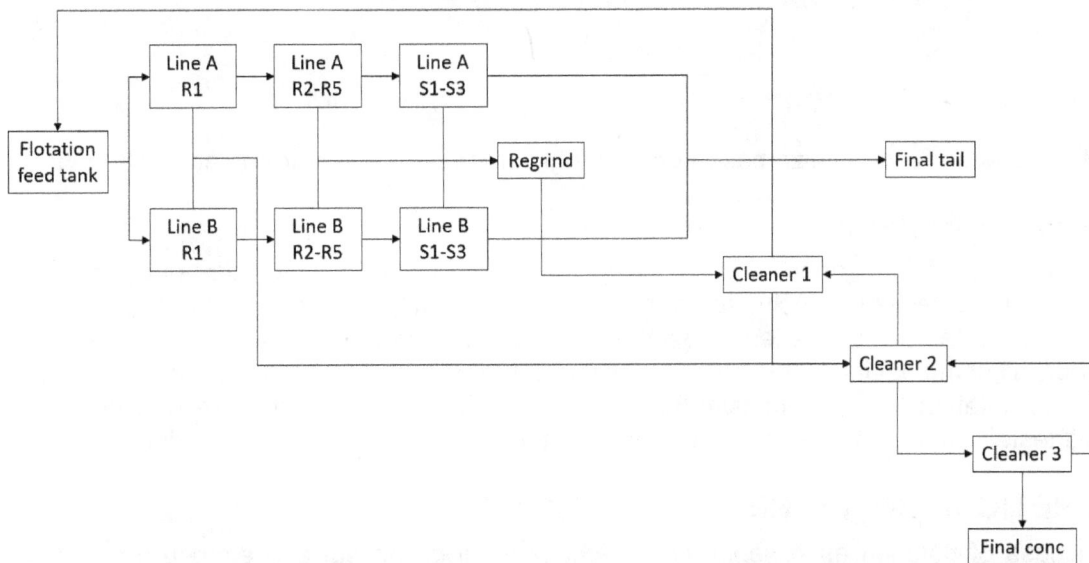

FIG 2 – Olympic Dam flotation circuit schematic.

In the middle of 2020, the concentrator team started to make adjustments to the Svedala cyclone set-up to increase the grind size with the aim of increasing mill throughput as part of a de-bottlenecking program. The main changes included vortex finder set-up and increasing the cyclone feed density target. The Svedala mill maximum throughput was first increased from 960 t/h to 1040 t/h in October 2020 then further increased to 1100 t/h in July 2021.

Geometallurgy program

There are two main test programs within the geometallurgy program: the monthly composite program and the future ore characterisation (FOC) program. The purpose of the monthly composite program is to monitor plant performance while the purpose of future ore characterisation program is to develop metallurgical models.

Monthly composite program

During normal operation, shift samples are collected on selected streams. The on-site lab processes the samples and puts a portion aside for the monthly composite. At the end of each month, the shift samples of each stream are composited according to the production rate to represent the entire month. As production critical streams, flotation feed, flotation concentrate and flotation tailings streams all have monthly composite samples created. There are a suite of tests on monthly composite samples (on a size-by-size basis) including detailed chemistry, quantitative X-ray diffraction (QXRD), and mineral liberation analyser (MLA) which provides modal mineralogy, mineral association and liberation. Refer to Gu (2003) for a full description of MLA.

The sizing data enables the calculation of P_{80} for each sample. Figure 3 shows the flotation feed P_{80} from January 2018 to November 2023. The P_{80} fluctuates from month to month but as the general trend it ranges between 60 and 70 μm for 2018–2020, increased to between 70 and 80 μm in 2021 then further increased to between 80 and 90 μm since 2022.

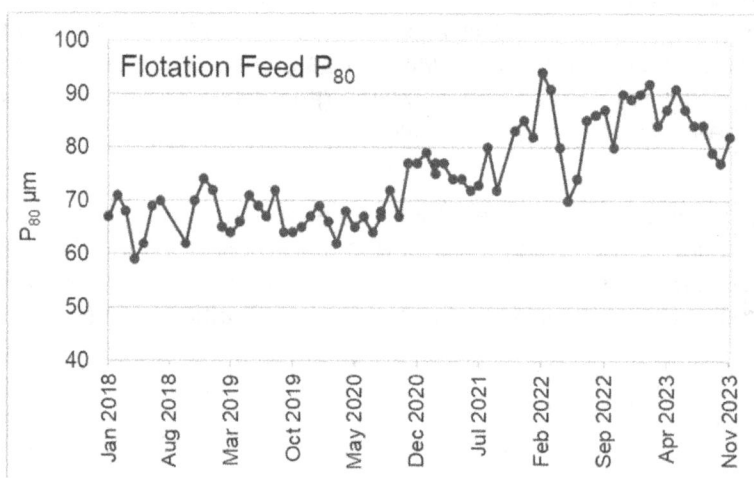

FIG 3 – Flotation feed P_{80} size based on monthly composite sample from Jan 2018 to Nov 2023.

Future ore characterisation program

The future ore characterisation program uses 20 m drill core intervals to carry out geometallurgical tests. This characterisation includes chemistry, MLA, QXRD, specific gravity, comminution tests, various flotation tests and tails leach tests. Since the aim of the FOC program is to test ore-related behaviour, all samples under this program are treated under the same conditions. For example, all feed to the flotation test is nominally 80 per cent passing 75 μm. There are now well over 1200 samples tested under the FOC program from all over the potential mining footprint.

THEORETICAL GRADE-RECOVERY CURVE

The success of flotation as a separation method requires the surface exposure of the valuable mineral.

A comparison of the plant performance with the best achievable performance can be made by considering the degree of liberation of the valuable minerals. As described by Johnson (2010), a theoretical (or mineralogically-limited) grade-recovery curve assumes that the only way gangue particles can dilute the concentrate grade is in composite particles with the valuable mineral – no fully liberated gangue reaches the concentrate. Similarly, only particles with some exposure (liberation) of valuable mineral can be recovered. No account is taken of kinetics. Therefore, the theoretical grade-recovery curve is not necessarily a realistic target but does provide a maximum limit. The relationship between sulfide-gangue mineral composite particle liberation and the grade-recovery curve is described by Cropp, Goodall and Bradshaw (2013) and reproduced in Figure 4. Actual plant performance always plots below the theoretical grade-recovery curve.

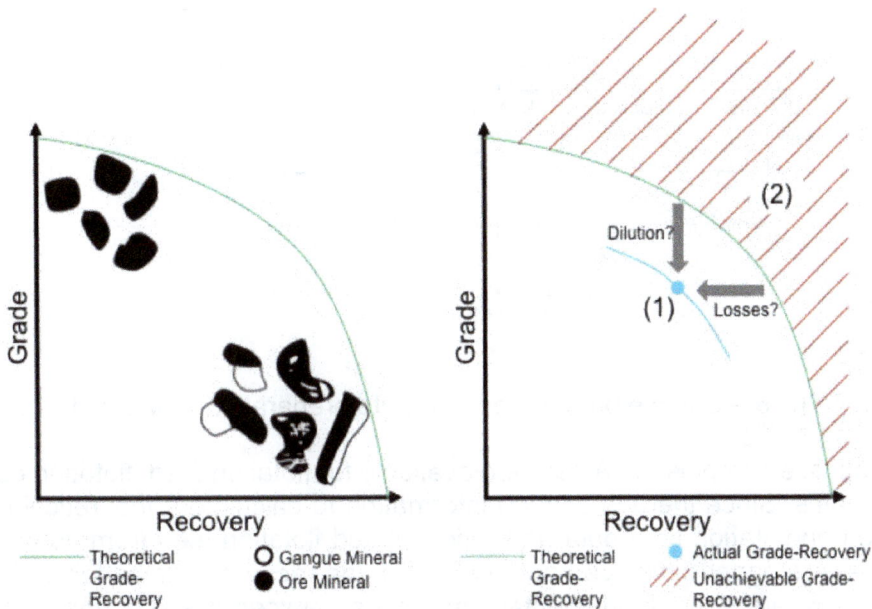

FIG 4 – (left): a schematic of how particle composition defines the curve; (right): how changes in operation condition will change actual grade and recovery (1) areas where changes in operating condition can improve performance (2) areas where improve in liberation is required to improve performance (copied from Cropp, Goodall and Bradshaw, 2013).

Johnson (2010) describes the method for converting mineralogical data, in our case generated by MLA, to a theoretical (mineralogically-limited) grade-recovery curve and this method has been applied in this paper.

Data preparation

Both monthly composite and FOC samples have very similar data preparation process (see Figure 5). The head sample is sized using screen sizing and cyclosizing to specified size fractions. Depending on the stream there might be no +106 µm or +75 µm fractions. Each size fraction has an assay and MLA fraction except for the minus C5 (-C5) fraction where particle size is too fine for MLA measurement and instead is characterised by assay and QXRD. The next size up (-C3+C5) is similar enough to -C5 fraction and it is acceptable to assume sulfide particles in -C5 size fraction have the same liberation and association characteristics as -C3+C5 fraction. MLA data of individual size fractions can then be 'rolled up' according to their weight percent to give a head mineralogy.

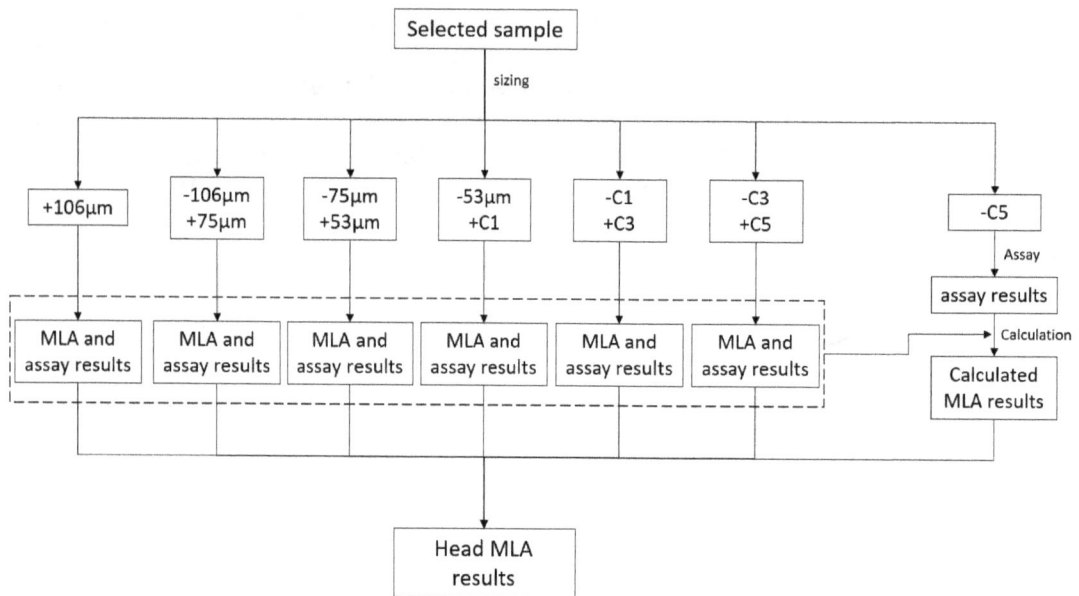

FIG 5 – Sample preparation and data preparation flow sheet.

For monthly composite samples, MLA data are available for flotation feed, flotation concentrate and flotation tails streams. Since there is enough information to calculate a theoretical grade-recovery curve both based on flotation feed data and reconstituted flotation feed from flotation concentrate and flotation tails, this paper developed both sets of grade recovery curves for monthly composite data. However, there are some fundamental differences between the sets: the regrind of rougher flotation concentrate stream means the reconstituted flotation feed takes account of the regrind while the flotation feed data doesn't.

FOC samples are different from monthly composite as there is only MLA data on flotation feed and flotation tails streams but no MLA data on flotation concentrate stream. Therefore, no reconstituted flotation feed can be calculated for FOC samples.

Monthly composite samples theoretical grade-recovery curve

Two monthly composite samples, one before coarse grind (Sample 1) and one after (Sample 2) are used in this paper to illustrate the impact of grind size on the theoretical grade-recovery curve. Detailed information of those two samples can be found in Table 2. Both samples have very similar abundance of sulfide minerals.

TABLE 2

Details of monthly composite samples used in this paper.

Sample	flotation feed P_{80} (μm)	Cu (wt%)	Cu:S	bornite* (wt%)	chalcocite* (wt%)	chalcopyrite* (wt%)	pyrite* (wt%)
Sample 1	64	2.4	1.5	1.6	0.5	2.5	0.3
Sample 2	92	2.1	1.4	1.5	0.6	2.5	0.4

* Measured by MLA.

Theoretical grade-recovery curves for combined sulfide minerals (ie py, cp, bn, cc) are plotted in Figure 6 for before and after coarse grind. Dashed lines represent reconstituted curve based on flotation concentrate and flotation tails while solid lines are based on actual flotation feed data. Blue lines are for sample before coarse grind and orange lines represent after coarse grind. For both flotation feed curve and reconstituted curve, increasing the grind size has shifted the theoretical grade-recovery curve downward. That means to obtain the target mineral grade in concentrate, flotation feed with coarser particle size needs to sacrifice recovery compared with feed of finer sizing.

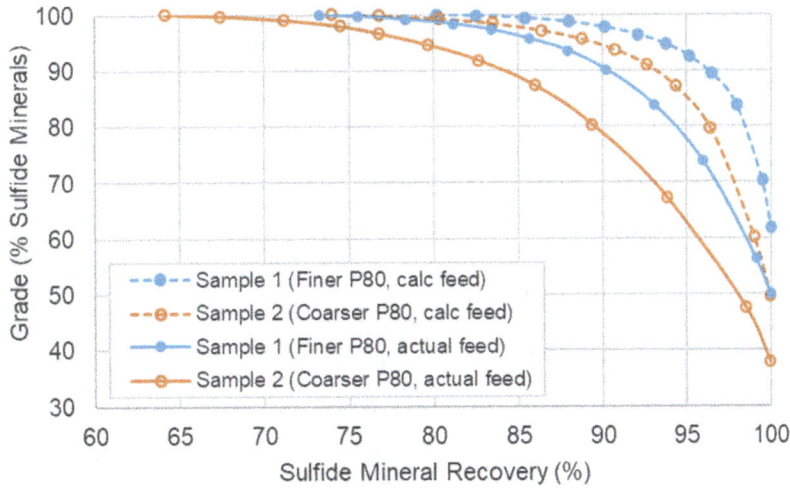

FIG 6 – Sample 1 and 2 theoretical grade versus recovery curves: solid line from actual flotation feed; dashed line represents reconstituted feed from flotation concentrate and tails.

Figure 7 plots reconstituted theoretical grade-recovery curve with achieved sulfide grade and recovery in the plant over the same time frame. Both plant samples sit below their corresponding theoretical grade-recovery curve and that reflects the difference between perfect separation and attainable separation. It also shows that the horizontal distance between the points is similar to the horizontal distance between the two curves. Hence the theoretical grade-recovery curve is a good predictor of the actual recovery loss. Increasing the grind size from Sample 1 to Sample 2 has increased the Svedala throughput from 800 t/h to 900 t/h while reduced recovery by 3 per cent.

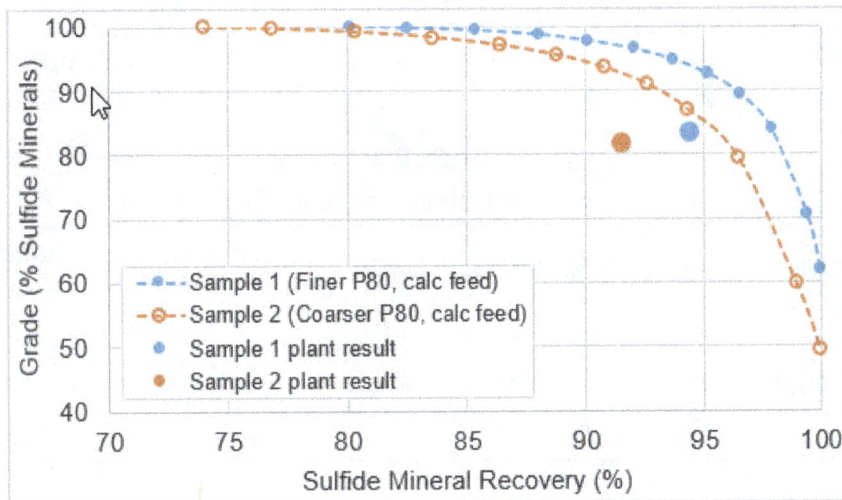

FIG 7 – Sample 1 and 2 theoretical grade versus recovery curves: reconstituted feed from flotation concentrate and tails, compares with plant results (single point).

Figure 8 replicates Figure 6 for Sample 1 by plotting reconstituted theoretical grade-recovery curve (dashed line) against flotation feed curve (solid line) for the same sample (Sample 1). Since regrind occurred for the rougher flotation concentrate, the lift in the grade-recovery curve from the solid line to the dashed line can be attributed to the regrind process. Grinding is an energy-intensive process and reducing the amount of material go through the mill is an effective way to reduce energy consumption and processing cost. Both Figures 6 and 8 provide mineralogical evidence that regrinding only the rougher concentrate, which is ~20–30 per cent of plant feed by weight, to a finer size can successfully improve liberation of sulfide minerals and achieve satisfactory recovery.

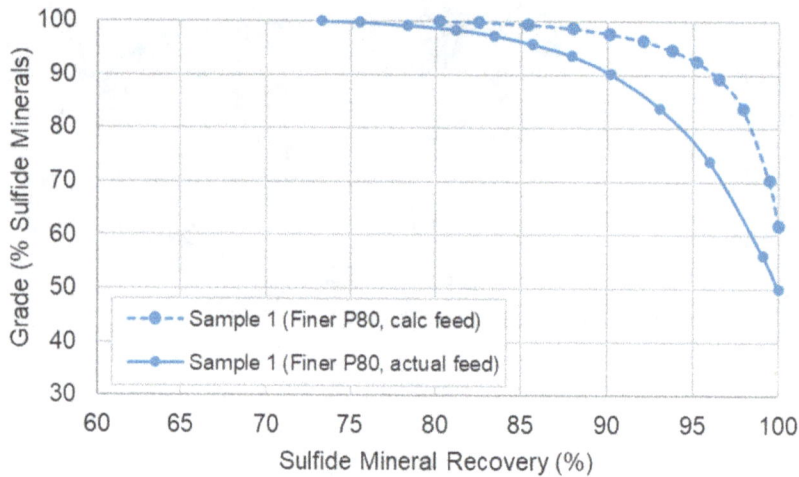

FIG 8 – Sample 1 theoretical grade versus recovery curves: solid line from actual flotation feed; dashed line represents reconstituted feed from flotation concentrate and tails.

Future ore characterisation samples theoretical grade-recovery curve

Plant feed at any given time is sourced from various locations across the deposit and will contain all four sulfide minerals. FOC samples on the other hand are 20 m intervals from resource dill core. The binary nature of sulfide minerals over short intervals results in the majority of FOC samples containing only one or two sulfide minerals. It is very rare for a FOC sample to have more than two sulfide minerals. Table 3 lists the composition of sulfide minerals for each FOC sample used in this paper along with its dominant sulfide mineral. All FOC samples have P_{80} of 75 μm, which is the nominal design grind size for Olympic Dam, and sits between Sample 1 and Sample 2. Figure 9 plots three FOC samples with Sample 1 and Sample 2 from the monthly composite program. FOC samples only have grade-recovery curves based on flotation feed, while Sample 1 and Sample 2 curves are based on flotation feed data as well to make the data sets consistent (no regrind).

TABLE 3

Sulfide mineral relative abundances for FOC samples.

Sample ID	py%	cp%	bn%	cc%	dominant sulfide mineral	sulfide P_{50} (μm)
FC0989	0%	1%	75%	24%	bornite	99
FC1076	0%	0%	13%	87%	chalcocite	107
FC1201	11%	89%	0%	0%	chalcopyrite	132

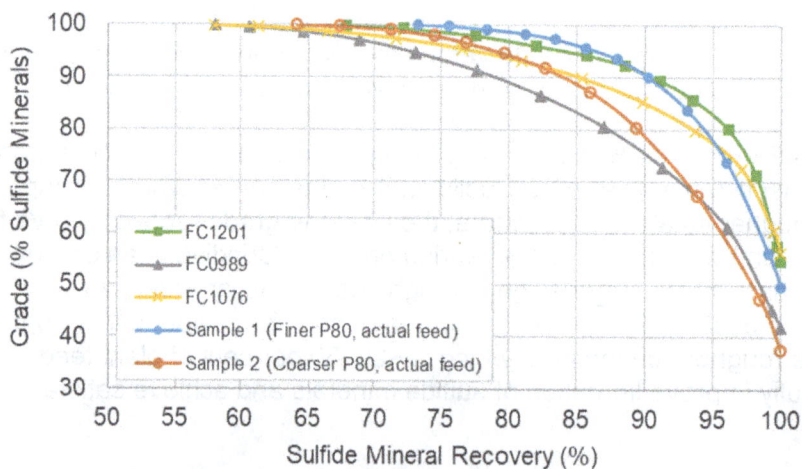

FIG 9 – Theoretical grade versus recovery curves for Samples 1 and 2 (monthly composite) and FOC samples (FC0989, FC1076 and FC1201). Solid lines represent actual flotation feed.

The FOC sample grade-recovery curve (Figure 9) is scattered around the monthly composite curves depending on dominant sulfide mineral type. Since monthly composite material is always a blend of a range of stopes, it can be considered as an 'average' while FOC samples sourced from 20 m drill core are more like an 'individual' sulfide mineral.

Figure 9 also shows that at the same grind size, the chalcopyrite rich sample (FC1201) can achieve better separation than either the chalcocite rich sample (FC1076) or bornite rich sample (FC0989). The sulfide mineral P_{50} for those three samples (see Table 3) indicates FC1201 has the largest sulfide grain size followed by FC1076 while FC0989 has the finest sulfide grain size. This means at the same grind size, FC1201 can achieve the best liberation for sulfide mineral and FC0989 has the worst liberation among all three samples. In general, chalcopyrite and pyrite have larger grain sizes compared with bornite and chalcocite (see Figure 10). Therefore, at the same grind size chalcopyrite and pyrite can achieve higher liberation than bornite and chalcocite. This also means that chalcopyrite and pyrite rich samples will be less sensitive to grind size while any change in grind size will have a much larger impact on bornite and chalcocite rich samples. There are other mineral-related factors that impact flotation recovery such as flotation rate and hydrophobicity, however the focus of this paper is on the mineral grain size and its effect on liberation. Olympic Dam flotation circuit is a bulk sulfide flotation process and doesn't try to separate one type of sulfide from the others and hence the grade-recovery curve with all sulfide minerals combined is more applicable.

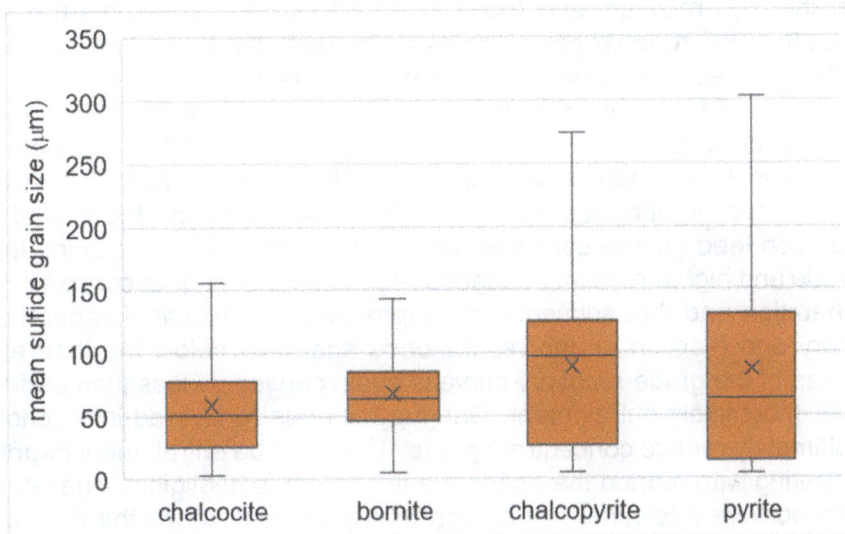

FIG 10 – Grain size distribution for sulfide minerals, all FOC data.

Laboratory cleaner flotation tests (FKC) are conducted on selected FOC samples. It mimics the plant process by regrinding rougher flotation concentrate before sending it to cleaner flotation stage. Figure 11 plots the theoretical grade-recovery curve from the flotation feed (calculated in the same manner as for Figure 9) with the 'actual' grade-recovery curve generated from the FKC test for FC1076 (where seven products are generated from three stages of cleaning).

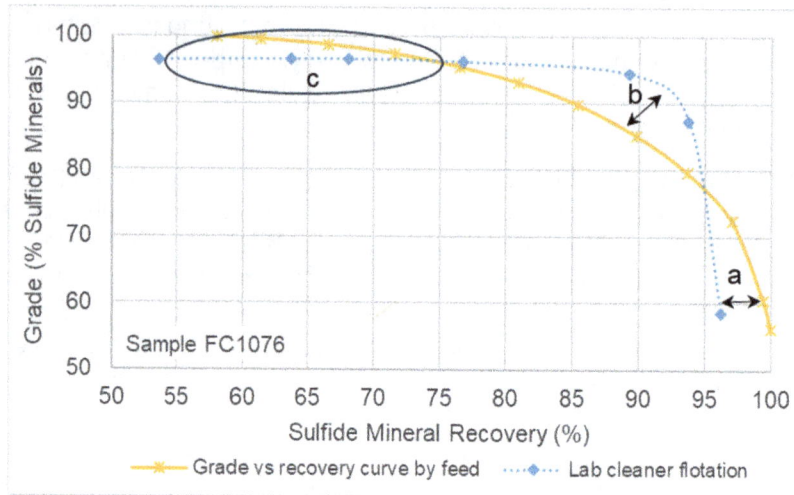

FIG 11 –Theoretical and actual grade versus recovery curves for FC1076. Solid line is the theoretical actual flotation feed while dotted line is the measured grade and recovery points from lab flotation test.

The data point at the high recovery end (point a) from FKC test represents the rougher flotation stage. The fact that the FKC rougher point locates underneath the theoretical grade-recovery curve represents less than perfect separation in the laboratory test (Figure 4: Johnson, 2010; Cropp, Goodall and Bradshaw, 2013). It could be reagent addition, circuit set-up or rougher flotation time during the test is too short to recover all sulfide minerals (noting a standard geometallurgy test is not optimised for each sample). The gap between the blue dotted line and the yellow solid line at section b can be attributed to the regrind step during the FKC test. Since the theoretical grade-recovery curve from the flotation feed (yellow solid line) doesn't take regrind into account, the FKC test can achieve higher grade and higher recovery because regrinding the rougher concentrate increases the sulfide mineral liberation and that subsequently improves the achievable separation. At the high-grade, low recovery end (section c), lab results once again fall below the theoretical curve. The difference at this part of the grade-recovery curve is again caused by less than perfect separation in the laboratory test, most likely entrainment. Gangue minerals recovered into concentrate through entrainment will ultimately reduce concentrate grade. There will be entrainment happening at section b as well but comparing with regrind the effect of entrainment is negligible. The lab test is very well controlled and can achieve a 'cleaner' separation than plant and hence the diluting effect through entrainment in the plant will likely be worse than the laboratory test.

Sulfide grade does not equal to Cu grade

Three dominant copper-bearing sulfide minerals exist in the deposit and each sulfide mineral has a different Cu content (Table 1). At any given sulfide grade, the actual Cu grade of a sample depends on the mix of the four sulfide minerals (py, cp, bn, cc). For example, at 100 per cent sulfides, Cu grade can range from 0 per cent for pure pyrite to 80 per cent for pure chalcocite (Table 1).

Figure 12 reproduces Figure 9 for the geometallurgy samples, but with copper grade on the y-axis instead of sulfide grade. This illustrates that knowledge of the sulfide minerals is required to understand the achievable copper grade in a mixed sulfide system (Table 2, Figure 12). At Olympic Dam, the copper to sulfur ratio (Cu:S) provides this link to the mix of sulfide minerals (Ehrig *et al*, 2019).

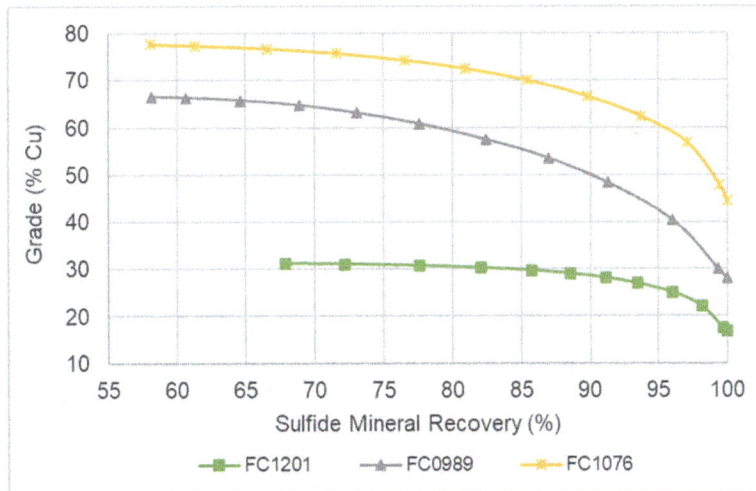

FIG 12 – Grade (%Cu)-recovery curve for chalcopyrite-pyrite (FC1201), bornite-chalcocite (FC989), chalcocite-bornite (FC1076) samples. Refer to Table 3 for the relative % of each sulfide mineral in the three FOC samples.

CONCLUSIONS

Theoretical (mineralogically limited) grade-recovery curves show the maximum, achievable recovery under perfect separation which is controlled by the liberation state and grain size of the mineral(s) of interest. The most effective method to increase liberation is to reduce grind size as evidenced by monthly composite data. Both plant operation and laboratory test work cannot achieve the separation defined by the curve owing to a range of practical factors such as reagent addition, flotation kinetics, entrainment. At different stages of flotation, different factors have larger impact on product grade and recovery.

Analysis of the theoretical grade-recovery curve at a 'fine' and 'coarse' plant grind shows the magnitude of the decrease in recovery at the same grade. This was reflected in actual plant operation. The information generated on throughput and recovery as part of the coarse grind trial will be used in future value chain assessments for Olympic Dam.

ACKNOWLEDGEMENTS

The authors thank BHP Olympic Dam for permission to publish this paper. We would like to acknowledge the geometallurgy team for their contributions to this paper: Benjamath Pewkliang, Edeltraud Macmillan, Michelle Smith, Sam Slabbert and Sasha Roslin. Special thanks to Benjamath Pewkliang for providing mineralogy data used in the paper.

REFERENCES

Cropp, A F, Goodall, W R and Bradshaw, D J, 2013. The influence of textural variation and gangue mineralogy on recovery of copper by flotation from porphyry ore – A review, in *Proceedings AusIMM GeoMet 2013*, pp 279–291 (The Australasian Institute of Mining and Metallurgy: Melbourne).

Ehrig, K, Liebezeit, V, Smith, M, Pewkliang, B, Li, Y and Macmillan, E, 2019. Predicting sulfide mineral abundances from drill core to smelter feed: Impact on the Olympic Dam value chain, in *Proceedings AusIMM International Mining Geology 2019*, pp 279–290 (Australasian Institute of Mining and Metallurgy: Melbourne).

Ehrig, K, Liebezeit, V, Smith, M, Pewkliang, B, Li, Y and Macmillan, E, 2020. Pre-concentration via waste rejection based on fragment size – Olympic Dam case study, in *AusIMM Preconcentration Digital Conference 2020*, pp 336–345 (Australasian Institute of Mining and Metallurgy: Melbourne).

Gu, Y, 2003. Automated scanning electron microscope based mineral liberation analysis, *Journal of Minerals Materials Characterization & Engineering*, 2(1):33–41.

Johnson, N W, 2010. Existing methods for process analysis, in *Flotation Plant Optimisation A Metallurgical Guide to Identifying and Solving Problems in Flotation Plants* (ed: C J Greet), Spectrum Series No 16, chapter 2, pp 35–64 (The Australasian Institute of Mining and Metallurgy: Melbourne).

Robertson, A, Grant, D, Liebezeit, V, Ehrig, K, Badenhorst, C and Durandt, G, 2013. Olympic Dam Mine, in *Australasian Mining and Metallurgical Operations* (ed: W J Rankin), Monograph 28 (3rd ed), pp 793–799 (The Australasian Institute of Mining and Metallurgy: Melbourne).

The journey of process improvement – Ok Tedi's vision

H Liang[1], L Brown[2] and B Wong[3]

1. MAusIMM, Senior Process Engineer, Mipac Pty Ltd, Brisbane Qld 4000.
   Email: hliang@mipac.com.au
2. Interface Manager – PAR Project, Ok Tedi Mining Ltd, Tabubil 332, Western Province, Papua New Guinea. Email: lyndah.brown@oktedi.com
3. MAusIMM, Operations Manager, JKTech Pty Ltd, Chelmer Qld 4068.
   Email: b.wong@jktech.com.au

ABSTRACT

Ok Tedi has been operational since 1984 and is the longest running open pit mine in PNG, producing copper, gold, and silver. In 2015, Ok Tedi conducted an internal gap analysis on the state of their existing data network and control systems to identify priority work fronts to de-risk the operation in order to achieve the business objective to safely deliver on life-of-mine production.

Various improvement projects were executed during 2015 to 2017 to mitigate the identified business risks such as upgrading of the network infrastructure and improving reliability of instrumentation and equipment on-site. However, declining ore grades and various production challenges leading into 2018 culminated in Ok Tedi experiencing low gold and copper recoveries due to unstable operation, resulting in operations reverting to manual control of the processing plant as they had lost trust in the process control system. The operating norm became one of reacting and firefighting, further compounding the situation was the lack of visibility and transparency of underperformance root cause, once again bringing to the limelight gaps identified in the 2015 review relating to insufficient digital maturity for business intelligence.

The Processing Improvement department at Ok Tedi was established in 2018 as part of the business improvement strategy. The group initiated various engagements with subject matter experts and concurrent improvement projects to address the poor performance and mitigate future business risks based on the previous gaps identified.

This paper discusses the process improvement journey Ok Tedi embarked upon in 2017 and some of the project initiatives executed by the technical services group and partnerships which ultimately resulted in recovery improvement uplifts for both copper (2.4 per cent) and gold (7.4 per cent).

INTRODUCTION

Ok Tedi Mining Limited (OTML) operates the Ok Tedi mine, situated 2000 m above sea level at Mount Fubilan in the remote Star Mountains of the Western Province of Papua New Guinea (PNG). The Ok Tedi mine is the longest running open pit copper, gold and silver mine in Papua New Guinea.

From first production in 1984 to the end of 2022, Ok Tedi has produced 5.17 Mt of copper, 15.9 Moz of gold and 36.4 Moz of silver (Ok Tedi Mining Limited, 2024).

The Ok Tedi project was to have an initial life-of-mine of 25 to 30 years. Around 2010, Ok Tedi's life-of-mine was extended to 2025 following extensive community consultation and revised mine plans. In 2021, the life-of-mine was further extended to 2032. As of 2023, the OTML board approved another extension with the current life-of-mine up to 2050.

CONCENTRATOR FLOW SHEET

The principal copper-gold ore deposit at Mount Fubilan consisted of a Leached Cap and supergene enrichment containing significant gold mineralisation, on top of the underlying porphyry copper-gold system (Jones and Maconochie, 1990).

Production of gold commenced in 1984 treating the original Leach Cap containing significant gold mineralisation. The original plant consisted of a single stage semi-autogenous grinding (SAG) mill followed by screening, with oversize treated using gravity recovery methods and the screen undersize undergoing cyanidation in a conventional carbon-in-pulp (CIP) leach plant (Newman, 1985).

The processing plant facilities have been progressively modified to suit the various available copper ore types as the mine developed, particularly around the milling, flotation, and concentrate handling areas. The first copper circuit (Train 1) was built in 1987, the same year gold bullion production ceased, with the second copper circuit (Train 2) completed in 1989 (England, Kilgour and Kanau, 1991).

A simplified schematic of the Ok Tedi copper processing flow sheet as at 2023 is illustrated in Figure 1, showing one of the two parallel and independent grinding and flotation circuits. As of the beginning of 2024, the concentrator flow sheet and process description will change going forward with the processing plant upgrades currently underway.

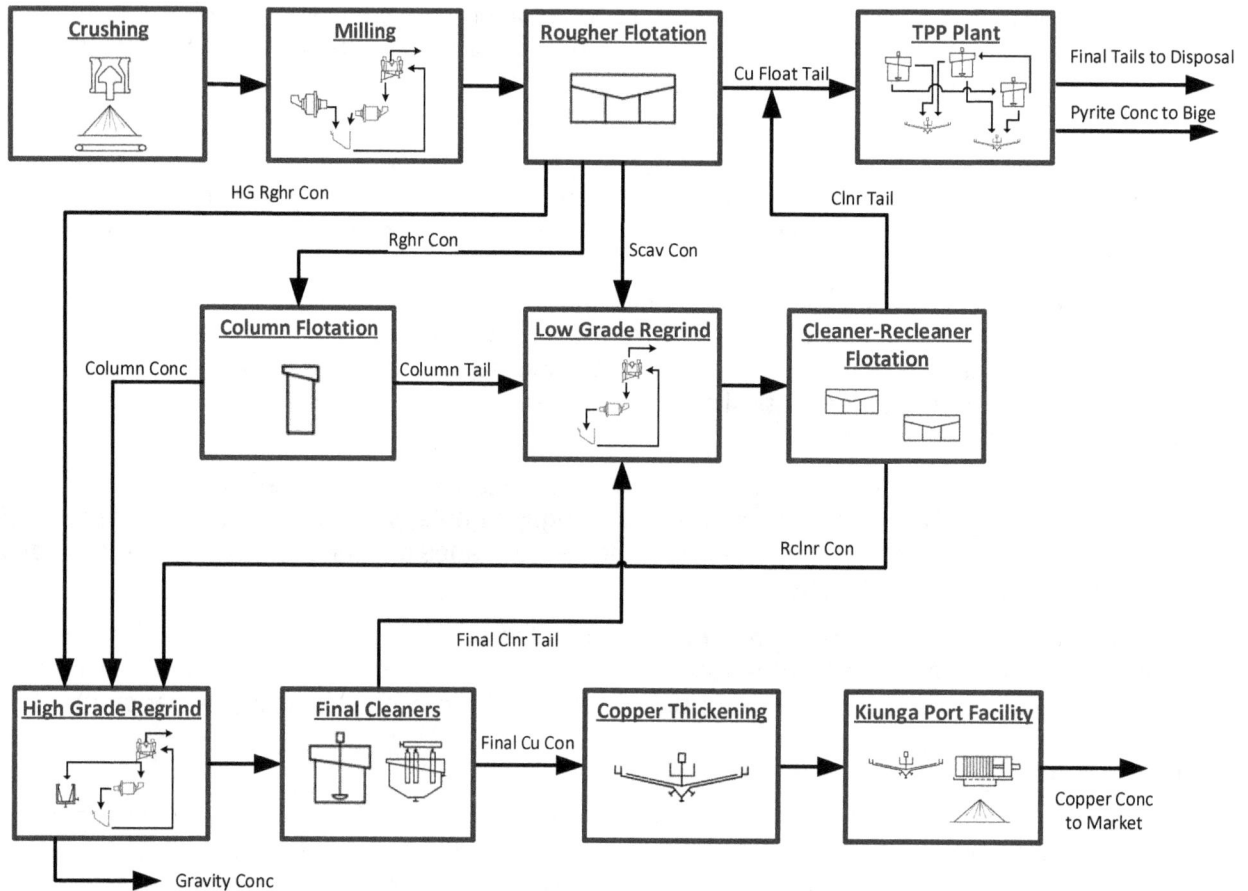

FIG 1 – Simplified block flow diagram of Ok Tedi Concentrator flow sheet as at the end of 2023.

Process description overview

Crushed ore is conveyed to two SAG feed stockpiles, with each stockpile feeding the corresponding processing trains. There are two independent grinding and flotation trains (differentiated as Unit1 and Unit2 for Train 1 and 2 respectively). The grinding circuit of each train utilises a SAG mill followed by two secondary ball mills to produce a flotation feed product size 80 per cent passing of 180 μm.

Each rougher-scavenger flotation train has 30 OK-38 (38 m³) flotation cells, arranged as two parallel banks of 15 cells (differentiated as damside and roadside). Concentrates from the rougher-scavenger cells report to the regrind and cleaner circuits via different routes: the primary roughers are diverted to the high-grade regrind circuit, the scavenger concentrate reports to the low-grade regrind circuit, and the secondary rougher concentrate reports to a column cell, with the column cell concentrate reporting to the high-grade regrind circuit and column tailings flowing to the low-grade regrind circuit.

The two regrind mills in each flotation train are run in closed circuit with a corresponding cyclone cluster. Regrind cyclone overflow from the low-grade circuit reports to a two-stage cleaning circuit. The first cleaner stage consists of 16 Denver DR500 flotation cells, with the cleaner concentrate further upgraded using eight Denver DR500 flotation cells as re-cleaners.

Re-cleaner concentrate reports to the high-grade regrind circuit. The cyclone overflow from the high-grade circuit reports to a single OK-100 (100 m^3) Tank Cell for the final stage of cleaning, producing the final copper concentrate. A portion of the high-grade regrind cyclone underflow is split to a single Knelson gravity concentrator, producing a separate gold/silver concentrate.

During periods of high fluorine content, a pair of Jameson cells in rougher and cleaner duty are used to treat the concentrate from the Tank Cell as a fluorine reverse flotation circuit. The Jameson rougher concentrate is pumped to the Jameson cleaner cell, with the Jameson cleaner concentrate reports to the copper tailings stream depending on the contained copper content. Jameson cleaner tailings is recycled back to the Jameson rougher, with the Jameson rougher tailings reporting to the copper concentrate thickener as final concentrate.

The thickened copper concentrate from the Ok Tedi mill is piped 156 km to the river port at Kiunga, where the concentrate is dewatered by thickening and filtration prior to storage in a purpose-built concentrate shed until it is loaded onto river vessels.

The combined copper scavenger and first cleaner tailings are pumped to the pyrite processing (TPP) plant, where the remaining sulfur in the form of pyrite is recovered via flotation. The pyrite concentrate is transported via a dedicated 128 km slurry pipeline and stored in specially engineered storage pits. TPP tailings is thickened prior to riverine disposal.

GOING FOR GOLD

In 2015, a Processing Improvement Program was initiated in the Processing Area. Plant performance had been declining in the lead up to 2015 in the key metrics of throughput, copper, and gold recovery (Figure 2). A combination of factors contributed to the decline such as lower head grades, increasing pyrite content in the plant feed, and plant downtime and maintenance issues (Figure 3). At the time there was a significant focus around increasing throughput and any recovery related projects were in the context of ensuring recovery was maintained while throughput was increased.

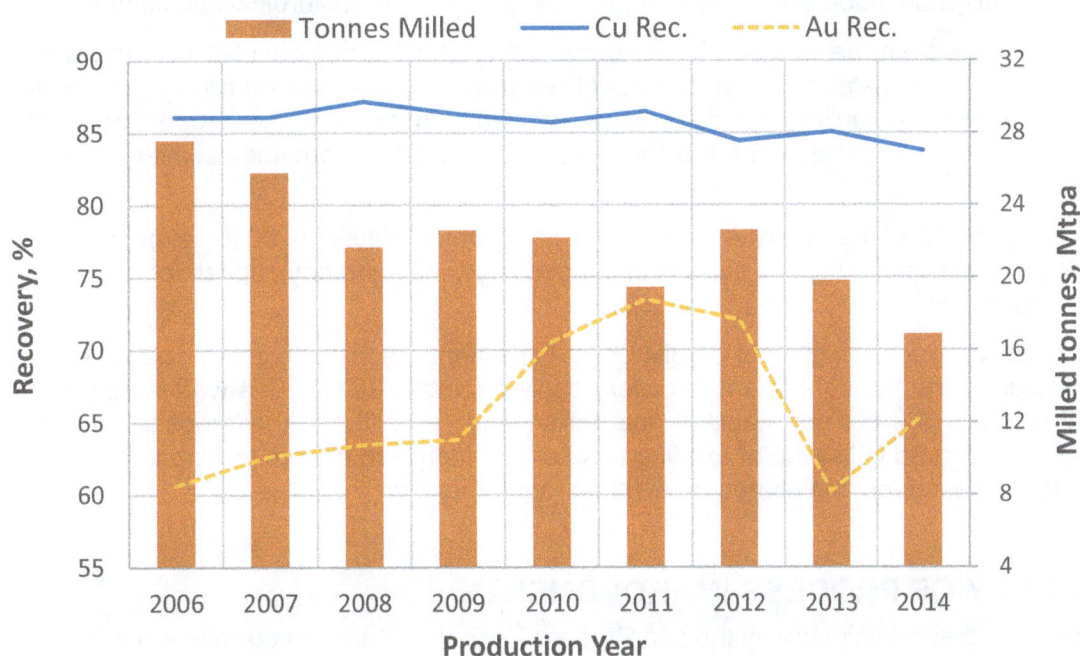

FIG 2 – Ok Tedi production snapshot between 2006 to 2014 (summary mill production report).

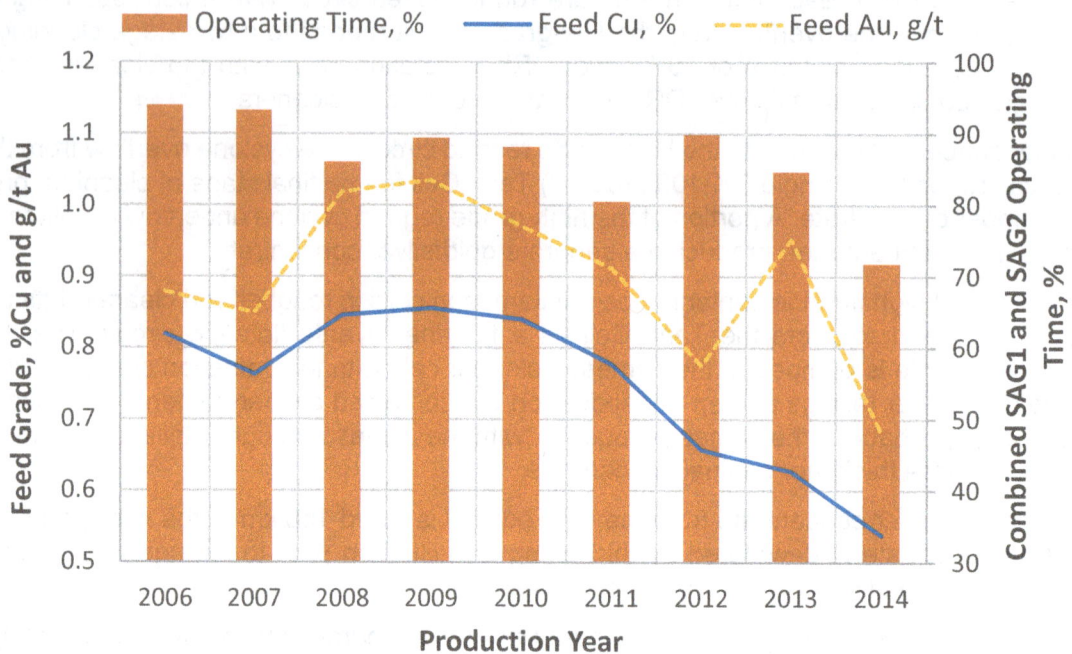

FIG 3 – SAG operating time and feed copper and gold grades between 2006 to 2014 (summary mill production report).

Compounding the metallurgical and fixed plant challenges, a gap analysis of the mill process control system and network infrastructure in 2015 highlighted the significant business risk posed to sustainably achieve the new life-of-mine production targets. Automated control of the Ok Tedi concentrator and TPP was via a hybrid of programmable logic controllers (PLC), Expert systems, and a distributed control system (DCS), of which it was found almost 40 per cent of the existing automation was bypassed or redundant. The ABB Bailey DCS contained legacy equipment and components that were no longer supported by the original equipment manufacturer (OEM).

In 2016, a strategic business plan was developed for Ok Tedi based around life-of-mine production using rigorous modelling and sensitivity analyses incorporating updated resource and reserve models, cut-off grade models, various mining plans, along with metallurgical and commercial models.

The key findings from the exercise in 2016 highlighted that under multiple mining scenarios, the unprecedented total material movement quantities required, meant maximising ore delivery to the processing plant was unlikely to be a lever available to achieve the production targets. This emphasised the importance and need for increased focus on improving recovery, particularly for gold.

As a result, the Gold Recovery Improvement Plan was initiated in 2016, exploring avenues to increase gold recovery within the context of the existing circuit and any capital upgrades required for plant modifications.

It was noted that achieving plant stability and consistently achieving plant operating targets will restore much of the throughput and recovery decline (Noble, 2017). This was underpinned by the need to address the current stability, operability, and reliability of the process and equipment, allowing for improved utilisation of existing assets. The identified drivers for improved gold recovery were largely based on maintenance of plant equipment or instrumentation and general plant infrastructure.

A JOURNEY OF PROCESS IMPROVEMENT

Significant progress was made by the site Process Technical Services team to spearhead the initial priority projects identified in the Gold Recovery Improvement Plan. Many of the projects were progressed or nearing completion around mid-2017 and were executed without additional external resources. Concurrently in mid-2017, an independent review of the site metallurgical performance was carried out by JKTech in addition to facilitating various plant surveys of the grinding and flotation area. The findings from JKTech identified additional high impact projects that would advance Ok Tedi's goal of improving overall plant performance.

To enable consolidation and prioritisation of the Process Technical Services Projects, Gold Recovery Improvement Plan Actions as reported by Noble (2017), and JKTech's recommendations into an effective implementation plan, the Process Improvements Team was established in 2018. This culminated in a total of 81 impact projects to address the issue of sub-optimal gold and subsequently copper recovery, with most of the projects focused on improving stability and reducing process variability.

There were considerable efforts by the site teams and other contracted subject matter experts involved between 2017 to 2020 towards the overarching goal of improving gold recovery. The main topics discussed in this paper centre around the process control and automation work fronts that drove the improvements to process stability.

THE ILLUSION OF (MANUAL) CONTROL

Mipac was engaged in April 2018 to conduct a process control review for Ok Tedi. The improvement opportunities identified in the review centred around physical automation projects to drive process stability, as well as improvements to process monitoring and key performance indicators (KPI) tracking by leveraging the plant historian data (AVEVA PI System). Several automation projects were identified, with the below selected improvements discussed in the following sections as they significantly contributed to reducing process variability for the operation:

1. Primary cyclone control.

2. Flotation level control.

Initially, there were various challenges with reliability of plant equipment and instrumentation which stalled the implementation of any revised control strategies. Various instruments required calibration or replacing before any automation works could commence. A small task force was established by the site Process Improvements team in Q3 of 2018 to commence with maintenance of the failing equipment, instrumentation, and processing plant infrastructure, enabling automation works to kick off by the end of January 2019.

Variability is my middle name

The Ok Tedi orebody consists of many different lithological types as reported in Smith, Horacek, and Sheppard (2007). The various geological units found within the deposit as of 2015 are categorised as follows for mine planning and production forecasting: Siltstone, Siltstone – Taranaki, Siltstone – Other, Limestone (Waste), Monzonite Porphyry, Monzodiorite, Endoskarn, Skarn, Pyrite Skarn, Oxide Skarn.

Kanau and Katom (1997) discussed the varying ore hardness work indices based on the different ore types. As a result of the inherent variability, each SAG mill can treat anywhere between 700 and 3000 t/h (McCaffery, Katom and Craven, 2002). As such, managing the hardness of the ore and the subsequent mill throughput rates that requires daily planning and management.

Stabilising the grinding circuit with a variable feed has been an iterative process from a process optimisation perspective, and significant works and developments have been evaluated in this space particularly around SAG mill control (Kanau and Katom, 1997; McCaffery, Katom and Craven, 2002; Savage, Rodriguez and Metzner, 2013), prior to the installation of variable speed drives in 2019.

Whilst SAG controls will not be discussed in this paper, the inherent variability of the feed to Ok Tedi's concentrator provides the context for the importance of mitigating flow disturbances downstream to the cyclones post milling, and subsequently to the copper flotation circuit.

Primary cyclone control

Mill discharge from the SAG and two ball mills on each train is combined into a common sump which supplies feed to the primary cyclone packs. The configuration of the two trains is similar, the main differences are the number of cyclone packs (Train 1: three packs, Train 2: two packs) and the cyclone models. Each cyclone feed sump is fitted with three cyclone feed pumps. The configuration differences are summarised in Table 1.

TABLE 1

Primary cyclone configuration Train 1 versus Train 2.

Operating line	Cyclone cluster	Feed pump
Train 1	CS01 (8 × Krebs gMAX26-H)	PP01
	CS03 (10 × Cavex 650CVX)	PP04
	CS04 (10 × Krebs gMAX26-H)	PP05
Train 2	CS01 (10 × Cavex 650CVX)	PP01
	CS02 (10 × Cavex 650CVX)	PP02 or PP03

Cyclone feed sump level control challenges

With the backdrop of the declining performance, operating with an ageing asset, and equipment reliability challenges, the cyclone feed sump controllers (and many controllers observed during the process control review) were operated in manual. This resulted in large variations in sump level and poor cyclone density control.

Limited functioning automation in this section of the plant introduced unnecessary variability downstream, aggravating the daily firefighting which had become the operating norm. The inherent variability of the ore feed resulted in highly variable flows transferring to the sump, then to the cyclone packs, and inevitably to the copper flotation circuit.

When sump levels got to below 10 per cent, operators would intervene and make manual changes to the cyclone feed pump outputs once the cyclone feed pressure became highly variable due to pump cavitation. Discussions with site personnel in 2018 noted frequent (almost monthly) replacement of pump impellers due to damage and loss of capacity, likely compounded by if not the result of frequent and prolonged cavitation.

Operational insight based on process data was lacking with PI tags created on an ad hoc basis or configured with sub-optimal PI tag settings to track key performance metrics. For example, at the time of review the cyclone feed flow controllers were not captured within the historian, limiting capacity to assess the control loop performance.

Primary cyclone pressure control challenges

Interlinked with the variable sump levels, from the process control review there was no automatic control utilised by operators for controlling the primary cyclones to pressure. There appears to have been attempts at incorporating controls previously, as Paki (2000) had reported the necessary instrumentation was installed to automate cyclone operation in October of 2000. The existing cyclone open-close sequence within the DCS was not based on cyclone operating hours to manage even wear but on varying delay times depending on the cyclone. As a result, the sequence if active, would preferentially open and operate two out of the eight available cyclones when the pressure spiked.

Manual operation of cyclones to maintain pressure added to the cognitive load on the operators as a significant portion of their shift was firefighting and manually controlling the circuit to the variations in feed entering the mill. Consequently, operators would fail to respond to most high and low operating pressure events and the associated classification impacts (variable flotation feed sizing) transferred downstream. With manual cyclone operation, uneven cyclone wear rates were evident, further compounding the variability in cyclone overflow sizing.

Operations had reported various downtime events historically that were tied to poor pressure control. Within a three-week period during Q1 of 2019, two events occurred that resulted in a cumulative downtime of 28 hrs on Train 1 (cyclone feed line bogging) and 17 hrs on Train 2 (cyclone overflow launder sanding).

Revised cyclone control strategy

There was existing logic in the DCS that utilised the level controller output to cascade a flow rate set point to the individual cyclone feed pump flow controllers. The required flow was determined via a function block that converted the controller output in per cent to the required flow rate. However, as each of the pumps had unique level controller output to flow rate relationship, this arrangement resulted in different flow rates for each cyclone pack at the same level controller output, and the level controller process gain would change depending on which pump was operating.

Resolving the sump level control was of critical importance and enables more consistent cyclone pressure control. A schematic of the revised control strategy is presented in Figure 4.

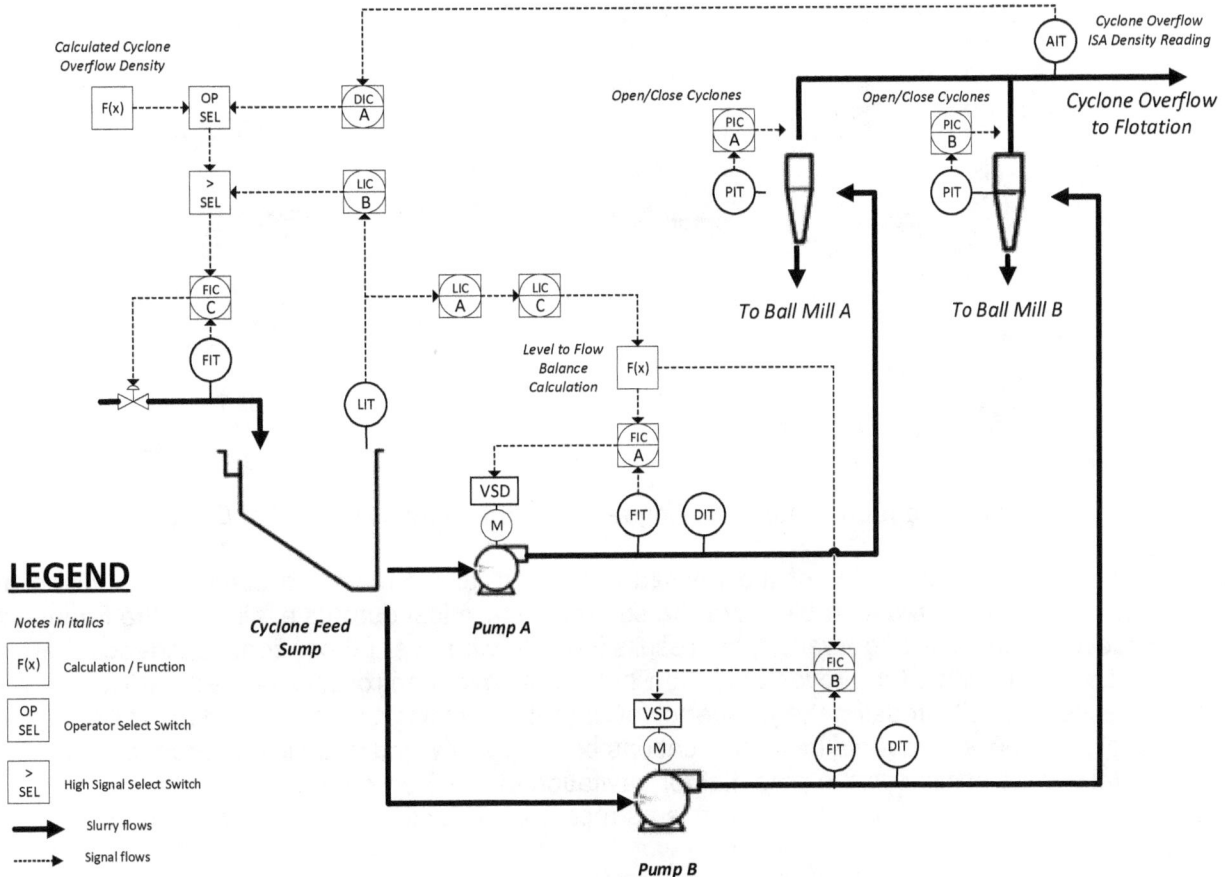

FIG 4 – Revised cyclone feed sump control strategy.

The primary control objective is to utilise the volume of the cyclone feed sumps as surge capacity to smooth out variations inflow downstream to flotation, without overflowing the cyclone feed hopper or allowing it to run empty. The primary level controller (LIC-A) is thus configured for averaging level control to maintain the sump level within an operating range (20 to 80 per cent) as opposed to tight level control (maintaining level to a set point target).

The LIC-A controller output then cascades to the balancing flow controller (LIC-C). The objective of the LIC-C controller is to balance flow distribution to the available cyclone feed pumps (FIC-A, FIC-B) and distribute the load between the cyclone packs.

The inclusion of a low-level controller (LIC-B) provides a temporary buffer from low sump level events by increasing the process water addition to the cyclone feed sumps if the level drops to a point where pump cavitation risk is likely.

Coupling this controller with the cyclone overflow density controller (DIC-A), both controller outputs (LIC-B, DIC-A) are passed through a selector switch which transfers the highest controller output to the process water flow controller (FIC-C) ie LIC-B should only drive the process water flow when a low sump level event is occurring, with the density control loop (DIC-A) as the main controller driving the process water flow.

Presented in Figure 5 is the distribution of the sump level bands for both trains comparing manual to automatic control (LIC-A controller status in AUTO = OFF or ON). The period of comparison is the start of Q1 of 2017 to the end of Q4 of 2021, with the proposed controls commissioned in Q4 of 2018. The data sets are extracted from the PI historian assessing total duration the measured sump level is LOW (<20 per cent), HIGH (>80 per cent), or within Target (20 to 80 per cent) range on a per production day basis. The data has been filtered for periods where the respective trains SAG mill is running with throughput greater than 500 t/h to broadly represent normal operating conditions.

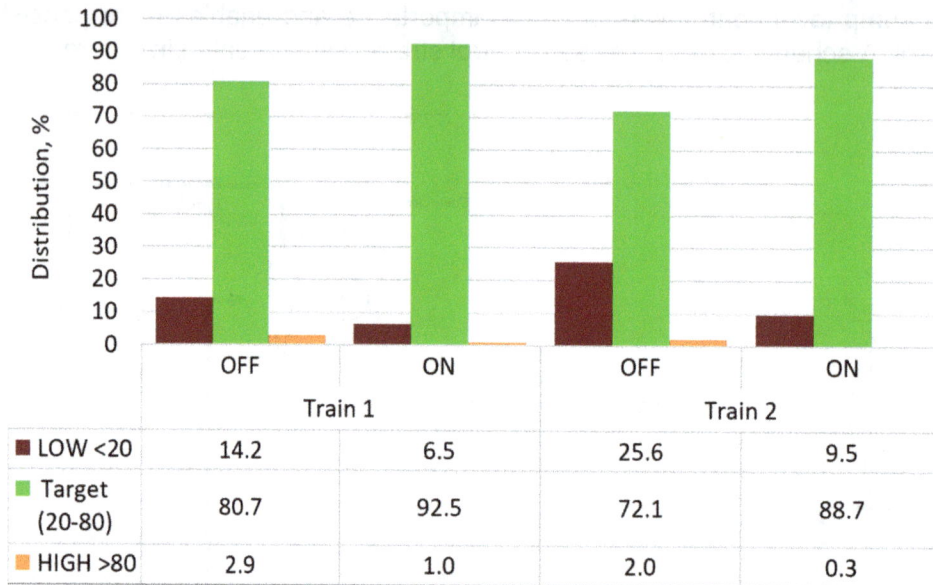

	OFF	ON	OFF	ON
	Train 1		Train 2	
■ LOW <20	14.2	6.5	25.6	9.5
■ Target (20-80)	80.7	92.5	72.1	88.7
■ HIGH >80	2.9	1.0	2.0	0.3

FIG 5 – Cyclone feed sump operating range distributions Q1–2017 to Q4–2021.

As noted, the primary objective of the revised controls is to maintain the sump levels within an operating range. Presented in Table 2 are the summary statistical outputs evaluating the frequency of sump levels within the target range, the results indicate with over 99 per cent confidence there is an 11.8 and 16.6 per cent improvement when the LIC-A level controller is utilised for Train 1 and Train 2 respectively. Increasing the frequency of operating withing the target band and minimising low and high sump level events have implications on operability and optimisation: reducing the low sump level events minimises the potential for cavitation of the feed pumps, and reducing the high-high sump level events means removing this as a constraint for limiting mill throughput to manage overflowing sumps.

TABLE 2

Summary statistics comparing cyclone feed sump level distributions within target range.

Parameter	SAG1		SAG2	
	OFF	ON	OFF	ON
Mean	80.7	92.5	72.1	88.7
Standard deviation	28.8	18.3	30.2	23.5
Data points	1597	220	1595	254
Degrees of freedom	1815		1847	
Standard error	27.7		29.4	
p-value (1-tail t-test)	4.48E-09		1.05E-16	
t-score	-5.9		-8.4	
Confidence interval	8.92E-09		2.09E-16	
Difference in means	11.8		16.6	

Flotation level control

Automated control of the flotation cell levels at Ok Tedi was achieved with the FloatStar system, and as a backup the ABB Bailey DCS system. Operator utilisation of FloatStar Level Stabiliser package was good, and performance was adequate in relation to controlling flotation levels. A potential risk raised in the review conducted by Mipac in 2018 centred around the robustness of the system, as it was operated from a standalone computer with no redundancy, and site had noted system failures had occurred previously.

Calling for backup

Come January 2019, the FloatStar server failed and was unable to be restarted. Level control reverted to the DCS based control using simple proportional-integral-differential (PID) blocks for feedback control, with some updates to the tuning parameters to minimise the observed level oscillations. Flotation cell level control was sub-optimal during the period with FloatStar offline, as the specific PID blocks configured in the DCS did not have the functionality of incorporating feedforward control, a critical component in pre-empting upstream disturbances that can significantly impact flotation cell level control.

While not configured, the existing ABB Bailey DCS software had additional function blocks which could be used to enable advanced features such as cascade control and feedforward. As this was implemented in the flotation level control at the TPP a few years prior with Mipac, Ok Tedi had reviewed the option of whether to recommission the FloatStar system or replace the existing PID blocks with the Advanced PID (APID) block within the Bailey DCS library.

Both options were expected to produce similar control performance, and ultimately the decision was made by site to build-up in-house capability and update the control blocks in the DCS, alleviating the need to upgrade the hardware, software, and associated licensing with recommissioning the previous FloatStar system.

Capability Development

The revised APID blocks were commissioned in late June 2019 by Mipac assisting the site process control team. To compare the performance of the different control modes on flotation cell level control a period of approximately one-month for each of the control modes was extracted from the PI historian as 5-minute averaged calculated data, filtered for periods where the SAG throughput is greater than 500 t/h to reflect normal operation.

The following periods were used for comparison on the performance of the FloatStar Level Stabiliser package, the original Bailey DCS PID blocks, and the updated APID control blocks with feedforward.

1. FloatStar Level Stabiliser – 1/9/2018 to 1/10/2018.

2. DCS PID Blocks – 1/5/2019 to 1/6/2019.

3. DCS APID Blocks with Feedforward (APID_FF) – 22/6/19 to 10/7/2019.

The level measurement (process variable, PV) against target (set point, SP) for the three data sets is presented in Figure 6 as a time-series chart. For brevity, only the first three rougher flotation cells with level control valves for Damside and Roadside on Train 2 are presented as a snapshot of the control improvements.

Additional PI trends and KPI tracking reports were co-developed with site to enable monitoring of flotation level performance. Tracking the error in level as a KPI also highlighted various flotation cells that required maintenance and investigation due to issues such as valve stiction, or improper zeroing of instrumentation. As an example, for the U2 RS Cell 6 trend shown in Figure 6, the trends enabled more streamlined identification of a pinch valve blockage/failure that needed to be addressed.

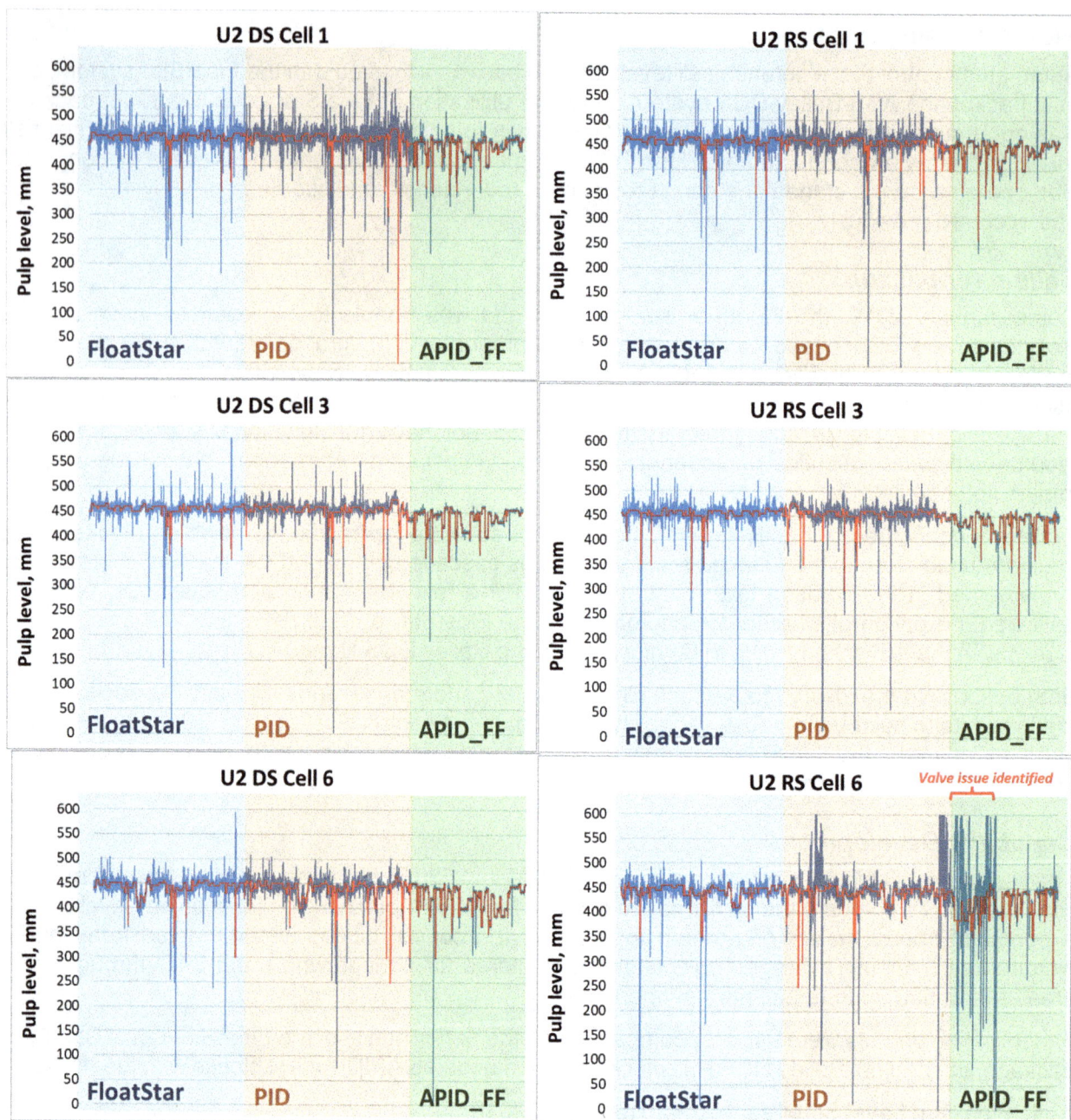

FIG 6 – Time series trends (3-month block) comparing measured level (process variable) to set point for the different control modes (FloatStar, PID, and APID_FF) on Train 2. DS = Damside, RS = Roadside.

Observing the trends of the measured level to the set point in Figure 6, it is evident the implemented APID blocks with feedforward can provide robust process control even with multiple set point changes. Overall, the new APID blocks within DCS showed superior performance to both PID and the previous FloatStar system (Figure 7). Excluding U2 RS Cell 6 due to the identified valve issue, the results demonstrate average reductions in the mean level error (PV minus SP) between 36 to 71 per cent (able to operate closer to target), and a corresponding reduction between 61 to 80 per cent in the standard deviation (tighter level control) using the APID blocks compared to the FloatStar system.

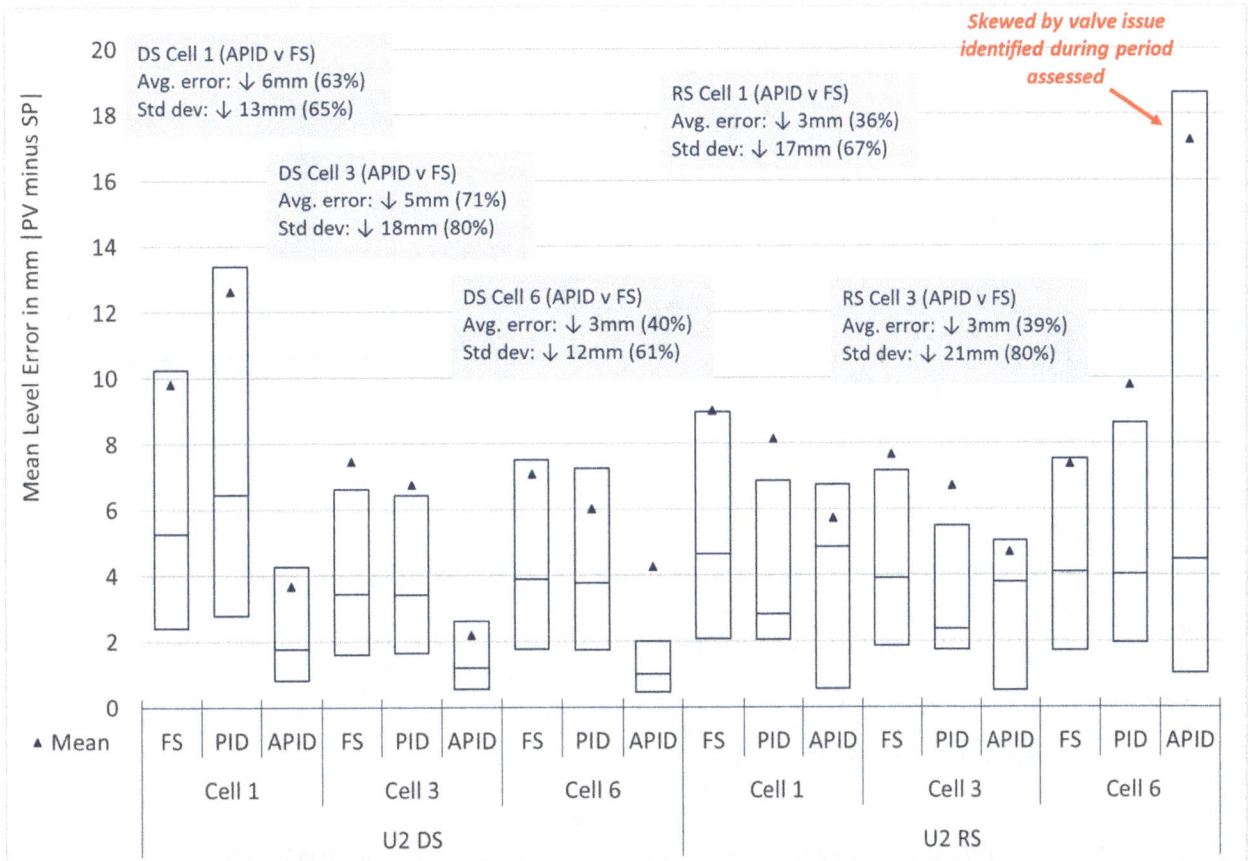

FIG 7 – Box-plot of error distribution (PV-SP) for the first three rougher cell level controls comparing FloatStar, PID, and APID_FF on Train 2. DS = Damside, RS = Roadside.

In principle, the control philosophy and strategy used for the APID blocks with feedforward and the FloatStar Level Stabiliser are similar, and the performance was expected to be comparable. The likely cause of the observed difference is lack of routine maintenance and retuning of the FloatStar system, which was believed to have last occurred in 2016 based on discussions with site personnel.

With the long operating history of Ok Tedi, process responses change over time and controller tuning needs to adapt with the change in process dynamics. Updating and implementing the APID solution reinforced one of the benefits to Ok Tedi of completing the control within the DCS system: developing in-house capability and enabling site personnel to maintain and improve the control.

RESILIENCE MAKES THE DREAM WORK

Addressing the gaps in the data systems to provide production insight and prioritising projects that addressed plant stability and process variability were key enablers for successful execution of the concurrent improvement projects undertaken by the Processing Improvements team and Process Technical Services teams. Implementation of the 81 impact projects from the Gold Recovery Improvements action plan (Noble, 2017) and JKTech's independent review were accelerated in 2018, with 82 per cent of the projects having been completed or in progress at the end of 2019 (Brown, 2020).

The successful completion and execution of the Gold Recovery Improvement plant initiatives has had a quantifiable effect on the overall plant performance at Ok Tedi. Results from the analysis using the daily production data extracted from MinVu are presented below. Illustrated by Figure 8 is the box plots showing the distribution of gold and copper recoveries by quarter between 2017 to end of 2021, with summary statistics comparing performance in 2017 to 2021 provided in Tables 3a and 3b for gold and copper respectively.

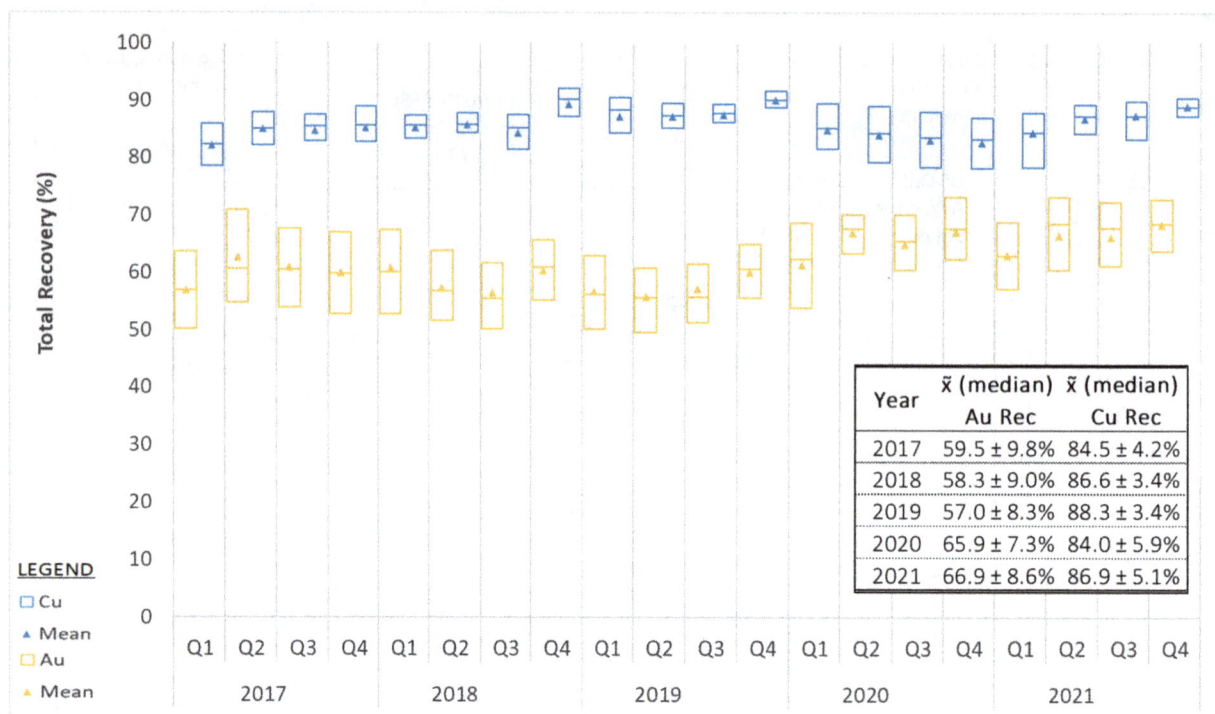

FIG 8 – Total gold and copper recovery between Q1–2017 to Q4–2021.

TABLE 3A

Summary statistics comparing gold performance metrics in 2017 to 2021.

Parameter	Au Rec (%)		Au in Feed (g/t)		Au in Final Float Con (g/t)	
	2017	2021	2017	2021	2017	2021
Q1 – median	56.9	62.8	0.64	0.64	13.7	23.6
Q2 – median	60.6	68.5	0.70	0.58	17.9	22.2
Q3 – median	60.5	67.8	0.62	0.43	16.8	22.7
Q4 – median	59.8	68.5	0.70	0.45	18.7	20.4
Mean of Q1-Q4	59.5	66.9	0.67	0.52	16.8	22.2
Standard deviation	9.8	8.6	0.17	0.19	3.40	4.87
Data points	348	334	360	342	313	337
Degrees of freedom	680		700		648	
Standard Error	9.24		0.18		4.23	
p-value (1-tail t-test)	1.9E-03		2.1E-02		2.8E-03	
t-value	-10.50		10.64		-16.33	
Confidence Interval	1.3E-03		2.8E-04		9.4E-04	
Difference in means	7.4		-0.14		5.4	

TABLE 3B

Summary statistics comparing copper performance metrics in 2017 to 2021.

Parameter	Cu Rec (%)		Cu in Feed (%)		Cu in Final Float Con (%)	
	2017	2021	2017	2021	2017	2021
Q1 – median	82.3	84.4	0.66	0.43	25.0	24.6
Q2 – median	84.9	87.2	0.65	0.40	24.4	25.2
Q3 – median	85.3	87.3	0.59	0.32	24.6	24.9
Q4 – median	85.6	88.8	0.57	0.42	24.0	24.8
Mean of Q1-Q4	84.5	86.9	0.61	0.39	24.5	24.9
Standard deviation	4.2	5.1	0.13	0.39	1.9	2.3
Data points	357	143	360	342	360	338
Degrees of freedom	498		700		696	
Standard Error	4.45		0.29		2.10	
p-value (1-tail t-test)	4.6E-02		2.8E-04		8.1E-02	
t-value	-5.44		10.12		-2.41	
Confidence Interval	2.0E-02		6.1E-06		1.3E-02	
Difference in means	2.4		-0.22		0.4	

Comparing the median recoveries for the 2017 calendar year to the 2021 calendar year gold recovery has improved by 7.4 per cent (with 99 per cent confidence) from 59.5 ± 9.8 to 66.9 ± 8.6 per cent, with copper recovery improving by 2.4 per cent (with 95 per cent confidence) from 84.5 ± 4.2 to 86.9 ± 5.1 per cent.

For both copper and gold in the rougher feed, the data indicates there has been a significant decrease in grades in 2021 compared to 2017 (continuing the declining ore feed trend). Gold content in rougher feed having decreased by 0.14 g/t (from 0.67 g/t to 0.52 g/t with 98 per cent confidence), and copper in feed decreasing by 0.22 per cent (from 0.61 to 0.39 per cent with 99 per cent confidence).

Furthermore, the gold and copper reporting to the final flotation concentrate has increased in 2021 compared to 2017. Gold content to final flotation concentrate increased by 5.4 g/t (from 16.8 to 22.2 g/t with 99 per cent confidence) and copper grade increasing by 0.4 per cent (from 24.5 to 24.9 per cent with 92 per cent confidence).

Based on the extracted production data, the observed improvements to the copper and gold recoveries appear independent of head grade as well as final concentrate grades. The data and outcomes presented have been validated independently by JKTech for the purposes of this paper.

CONCLUSIONS

Improved automation drives process stability, reduces operating risk and is key to enabling sustainable increases to production. As part of the evolving journey of process improvement undertaken by Ok Tedi in 2015, there has been a concerted effort to address the foundational gaps of the ageing asset to sustainably meet the life-of-mine production targets. Reverting to first principles metallurgy with a focus on improved plant and instrument reliability were significant enablers to meeting production targets in the context of increasing ore hardness, declining feed grades, and assets approaching end of life. The strategic drive and efforts towards the common goal by the site teams and contractors culminated in significant increases to gold and copper recoveries by 7.4 per cent and 2.4 per cent respectively.

ACKNOWLEDGEMENTS

The authors wish to thank the management of Ok Tedi Mining Limited for permission to present this paper, Mipac Pty Ltd and JKTech in their collaboration and assistance in the analysis of the presented outcomes in this paper, and to the Process Technical Services Metallurgy team and Process Improvement teams on the significant contribution and collaboration demonstrated to accomplish the projects that advanced the reported improvements to overall plant performance.

Further, the authors would like to express their appreciation and gratitude to the support from the external vendors and contractors throughout the process improvement journey (Alasdair Noble, JKTech, Mipac, Flottec, Magotteaux, Consep, SGS, MinVu).

Finally, special mention to Martin Randall and Anthony Porter for championing and implementing the various process control changes during the period recounted.

REFERENCES

Brown, L, 2020. Gold Recovery Improvement Plan 2018–2019 Review Report, Internal report.

England, J K, Kilgour, I and Kanau, J L, 1991. Processing Copper-Gold Ore at Ok Tedi, in *Proceedings of the PNG Geology, Exploration and Mining Conference,* pp 183–190 (The Australasian Institute of Mining and Metallurgy: Melbourne).

Jones, T R P and Maconochie, A P, 1990. Twenty five million tonnes of ore and ten metres of rain, in *Proceedings of the Mine Geologists' Conference*, pp 159–166 (The Australasian Institute of Mining and Metallurgy: Melbourne).

Kanau, J L and Katom, M, 1997. Operation and control of 32 feet diameter SAG mills at Ok Tedi, in *Proceedings of the Sixth Mill Operators Conference*, pp 37–54 (The Australasian Institute of Mining and Metallurgy: Melbourne).

McCaffery, K M, Katom, M and Craven, J W, 2002. Ongoing evolution of advanced SAG mill control at Ok Tedi, *Minerals and Metallurgical Processing*, May 2002, 19(2):72–80.

Newman, I C, 1985. The Ok Tedi Project, *Bull Proc Australas Inst Min Metall*, 290(5):67–72.

Noble, A, 2017. Gold Recovery Improvement Project Action Plan – May 2017, inter-office memo.

Ok Tedi Mining Limited, 2024. Home page. Available from: <http://www.oktedi.com/> [Accessed: 21 March 2024].

Paki, O K, 2000. Evaluation of CAVEX Hydrocyclone in the Ok Tedi Grinding Circuit, in *Proceedings of the Seventh Mill Operators Conference*, pp 33–46 (The Australasian Institute of Mining and Metallurgy: Melbourne).

Savage, C, Rodriguez, R and Metzner, G, 2013. Hybrid Process Control Strategy for the Ok Tedi Mining Ltd Semi-Autogenous Grinding Mill #2 – A Comparative Analysis, in *Proceedings of the Metallurgical Plant Design and Operating Strategies (MetPlant 2013)*, pp 202–214 (The Australasian Institute of Mining and Metallurgy: Melbourne).

Smith, M L, Horacek, D and Sheppard, I K, 2007. Optimisation of Ok Tedi's Medium-Term Production Schedules Using the Life of Business Optimisation System (LOBOS), in *Proceedings of the Large Open Pit Mining Conference*, pp 145–152 (The Australasian Institute of Mining and Metallurgy: Melbourne).

Applying geometallurgical principles for metallurgical sample selection

I T Lipton[1], P G Greenhill[2] and L Torres[3]

1. FAusIMM, Principal Geometallurgist, AMC Consultants, Brisbane Qld 4000.
   Email: ilipton@amcconsultants.com
2. FAusIMM(CP), Principal Consultant, AMC Consultants, Melbourne Vic 3000.
   Email: pgreenhill@amcconsultants.com
3. Senior Data Scientist, AMC Consultants, Melbourne Vic 3000.
   Email: ltorres@amcconsultants.com

ABSTRACT

Metallurgical sampling and the results of metallurgical test work form the quantitative basis for prediction of the processing behaviour of ore. These predictions are critical inputs for the design and capital and operating cost estimates for the ore processing plant, the economic evaluation of the project and the final investment decision.

Ore characteristics such as hardness, competency, mineral content, grain size and texture, control ore behaviour in a mineral processing circuit. Therefore, the principles for selection of metallurgical samples should be strongly guided by orebody geology. Furthermore, only a tiny fraction of the orebody will be tested before it is mined, so metallurgical test work is always data-poor. In contrast, the geological sample database is likely to consist of tens or hundreds of thousands of records.

A key principle of geometallurgy is to use the geological sample database to gain leverage from relatively few high-cost metallurgical tests. Applying geometallurgical principles to design of metallurgical sampling programmes aligns sparse test work data to abundant geological data and ensures that the data is suitable for the application of data science methods to derive robust predictions of ore processing behaviour.

The selection of composite, variability, and blended samples with respect to purpose, representativity, and predictive modelling is discussed. Application of machine learning to metallurgical sample selection using multivariate data is demonstrated using case studies. The vexed question of 'how many samples do we need?' is addressed and a data-driven solution is proposed.

INTRODUCTION

Metallurgical sampling and the results of metallurgical test work form the quantitative basis for prediction of the processing behaviour of ore. These predictions are critical inputs for the design and capital cost estimates for the ore processing plant, the estimates of mineral recoveries and operating costs, and the economic evaluation of the project.

Ore characteristics such as hardness, competency, mineral content, grain size and texture, control the behaviour of the ore in a mineral processing circuit. Therefore, the principles for selection of metallurgical samples should be strongly guided by the geology of the orebody. Furthermore, only a tiny fraction of the orebody will be tested before it is mined, so metallurgical test work is always data-poor. In contrast, the geological sample database is likely to consist of tens or hundreds of thousands of records which have the potential to improve understanding and prediction of metallurgical responses.

A key principle of geometallurgy is to use the geological sample database to gain leverage from the relatively small number of high-cost metallurgical tests. Applying geometallurgical principles to design of metallurgical sampling programmes aligns sparse test work data to abundant geological data and ensures that the data is suitable for the application of data science methods to derive robust predictions of ore processing behaviour at the local scale.

The authors propose that the use of geometallurgical principles and data science for metallurgical sampling and test data analysis provides a measurable and repeatable basis for sample-selection, driven by the geological characteristics of the mineralisation. It provides an alternative to the plethora of rules of thumb sampling procedures that have limited connection to the geological variability, nor the variability of the metallurgical responses.

What type of samples do we need?

Before any sampling or test work is commissioned the purpose of the test work must be clear so that the right sample type is selected. Drill core is a necessity for comminution testing and coarse particle beneficiation. For mineral recovery test work (leaching, flotation, fine particle separation) uncrushed drill core is preferred so that the sample can be crushed in stages to produce a particle size distribution that mimics the process plant design. Coarse rejects from core samples crushed in a well-controlled laboratory may also be suitable but Kormos *et al* (2013) note that assay rejects may contain excessive fines, so checks on particle size distribution should be carried out to confirm suitability. Reverse circulation drill cuttings may contain excessive fines which may bias the test results.

From a geometallurgical perspective, drill core is also preferred because much more geological data can be obtained by visual logging, hyperspectral scanning, geotechnical measurements, and non-destructive hardness testing than can be obtained from chip samples. Drill core data joined with metallurgical test results from the matching depth intervals forms the essential input to predictive Geometallurgical modelling.

There is an underlying and infrequently articulated idea regarding representativity. That is, that a representative sample can fully describe the behaviour of an ore deposit and provide sufficient information to confirm that the designed plant will operate in a uniform and predictable manner. The reality is that such a design will perform in a manner reflective of the variability of the ore feed over time and the extremes of the ore feed will perform very differently from the representative sample. The best scenario is a plant designed to perform within defined specifications across all ore types.

Composite samples (typically multiple ore intercepts from multiple drill holes) are used for developing the basic processing and test work flow sheet. Composites are made from blends of the ore types expected to be mined over the life-of-mine or other long production periods. Later, large composites (bulk samples) may be required to provide products for further testing such as concentrates or tailings. Composite samples may establish a feasible process flow sheet but they provide little information about the behaviour of individual ore types.

To design a process plant that can accommodate all the variation in ore characteristics, variability sampling is required. Variability samples should be selected to characterise the full range of geological ore types. The authors recommend selection of variability samples from continuous intercepts in single drill holes. This approach reduces the risk of accidentally testing a blend of ore types and also allows the metallurgical test work results to be considered within their geospatial context. However, it should be recognised that there is no intrinsic causal relationship between geospatial location and metallurgical behaviour.

The geometallurgical principle for variability sampling is to identify samples that cover the full range of the relevant multivariate characteristics of the orebody. 'Relevant' refers to those physical, mineralogical or chemical characteristics that have the potential to affect ore processing responses.

We define an ore type as mineralised material of economic interest with similar ore and gangue mineral assemblages, texture, and tenor (grade). In most orebodies, ore types have a spatial arrangement that is the product of the geological history of the mineral deposit. The term 'domain' is commonly applied to describe a volume of rock with similar geological characteristics. From a geometallurgical perspective, it is preferable that a domain consists of a single ore type so that, ultimately, estimates of ore processing response derived from testing samples of an individual ore type can be applied within the domain.

It is common practice to select metallurgical samples from three-dimensional, geographic locations with the aim of providing good spatial coverage of the deposit. This is understandable in the absence of a good knowledge of the spatial distribution of ore types and domains. However, three-dimensional coverage is not a logical requirement of metallurgical sampling. Ore processing response is intrinsically a function of geology not location. Therefore, from a geometallurgical perspective, it is more important to achieve satisfactory coverage of all individual ore types, than it is to select samples from all points of the compass. Once multivariate geological representativity is satisfied, aiming for broad spatial coverage provides a degree of insurance in case there are variations within ore types across the orebody that have not previously been recognised, such as competency increasing with depth or microscopic changes in grain size or texture.

Finally, grouping material into domains with similar characteristics can be of practical convenience for scheduling mine production and blending and may assist with metallurgical sample selection and geological interpretation.

Assessing representativity of multivariate characteristics

A modern drill hole database is likely to include diverse data types. Assay data commonly includes 30 or more elements, hyperspectral data may include quantitative or semi-quantitative analyses of minerals, and there may be bulk density, rebound hardness measurements and visually logged characteristics such as rock type or texture. With such a large number of features available, assessment of metallurgical sample representativity using traditional one, two or even three dimensional statistical or graphical tools is not efficient or reproducible.

Machine learning provides many tools for assessing sample representativity in multivariate terms. Unsupervised machine learning algorithms, where the algorithm is not directed by a *priori* assumptions about the outcomes, are useful for grouping multivariate geological data with similar characteristics.

We illustrate the principles of dimension reduction and group analysis using data from a porphyry copper project. The geological database consists of 28 734 samples of 2 m length, with multielement inductively coupled plasma – optical emission spectrophotometry (ICP-OES) assays, silicate alteration type groups derived from processing of Corescan™ hyperspectral data, sulfide mineral species estimated from elemental assays and partial-digest copper assays, and hardness measured with an Equotip rebound hardness tester. The objective of the analysis was to assess whether samples selected for flotation and comminution testing provided good multivariate coverage of the complete set of drill hole samples.

We applied the Uniform Manifold Approximation and Projection (UMAP) dimension reduction algorithm to reduce the high-dimensional data, to a small number of dimensions for viewing and grouping analysis. UMAP is a dimension reduction technique (McInnes, Healy and Melville, 2020) widely used in data science. UMAP was chosen for the study as it better preserves local and global structure than many other dimension reduction algorithms and the outcomes can be applied to multiple data sets, such as drill hole data and block model values. As the study was primarily concerned with flotation performance, we trained UMAP with the chalcopyrite, pyrite, bornite, and chalcocite (the estimates of chalcocite content include covellite) grades estimated from the assay data.

Useful groups may be visually identified by plotting the data according to the first two or three UMAP vectors but to make the process repeatable and independent of cognitive biases a clustering algorithm is usually applied to the transformed data to automate the identification and labelling of groups. The clustering algorithms evaluate the similarity of each data point to its neighbours. The number of groups is selected by the user, based on their usefulness: how well the groups are discriminated, their interpretability and meaningful relationships. We applied the K-Means clustering algorithm to the first two UMAP vectors and, based on experimentation, selected K = 7 as the most useful number of groups.

Figure 1 shows the data points projected in UMAP space, coloured according to the local density of the point cloud. The first two vectors generated by the UMAP dimension reduction were used as the x and y coordinates for visualising the data points. Data points that are similar in a multivariate sense appear close together in the UMAP space while those points that are dissimilar appear further apart. Even without looking at the raw data, the point cloud suggests three strongly discriminated groups. The K-means group boundaries were approximated using Dirichlet tessellation and the group centroids are shown as a black + symbol and labelled with the group number.

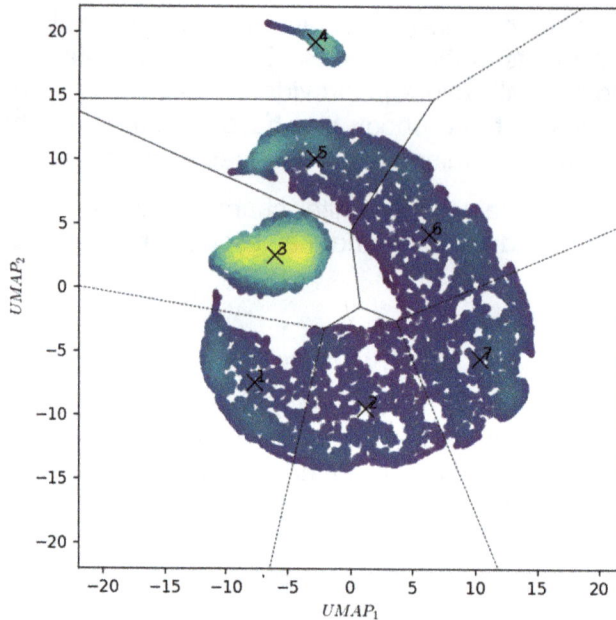

FIG 1 – Scatter plot showing the distribution of samples in UMAP space with groups for K = 7.

An important step in validating the selection of the variables used to train the UMAP transformation and the value selected for K, is to visually assess the distribution of all the variables across the K-means groups. The original multivariate data is retained in the UMAP output file, therefore the points in UMAP space can be interrogated in terms of any of the variables in the input data.

Figure 2 shows the data points coloured according to the four training features. Bornite-dominant and chalcocite-dominant samples are strongly discriminated by the UMAP transformation and K-means analysis. Pyrite and chalcopyrite content show a more complex, largely negative correlation with gradational changes. Further examination showed that the multivariate patterns observed in the UMAP space are consistent with the observations of alteration and mineralisation associations observed in the drill core. The seven groups provide a useful tool for ore type description, sample selection, and interpretation of sulfide domains in three dimensions.

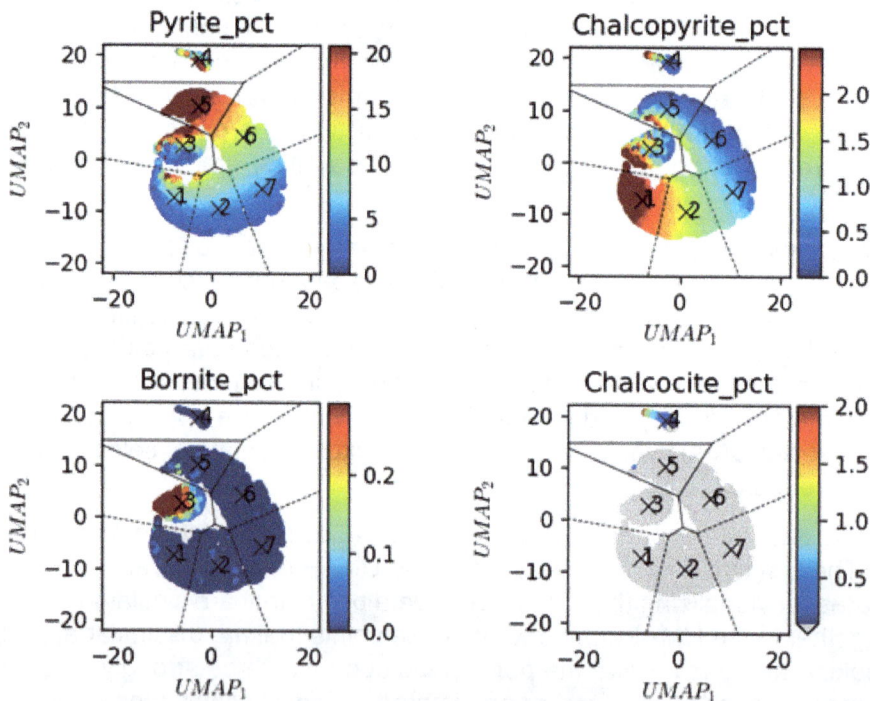

FIG 2 – Scatter plots showing distribution of bornite, chalcocite, chalcopyrite and pyrite in UMAP space.

A data-driven multivariate sample selection methodology

The UMAP dimension reduction and K-means grouping analysis provide a multivariate view of the data with which the question of multivariate representativity can now be considered in a convenient and reproducible manner. Figure 3 shows two possible selections of metallurgical samples plotted in UMAP space against the geological composite data. The selections provide reasonably good multivariate coverage of the seven identified mineralisation groups but there are notable gaps, illustrated by the nine larger green dots.

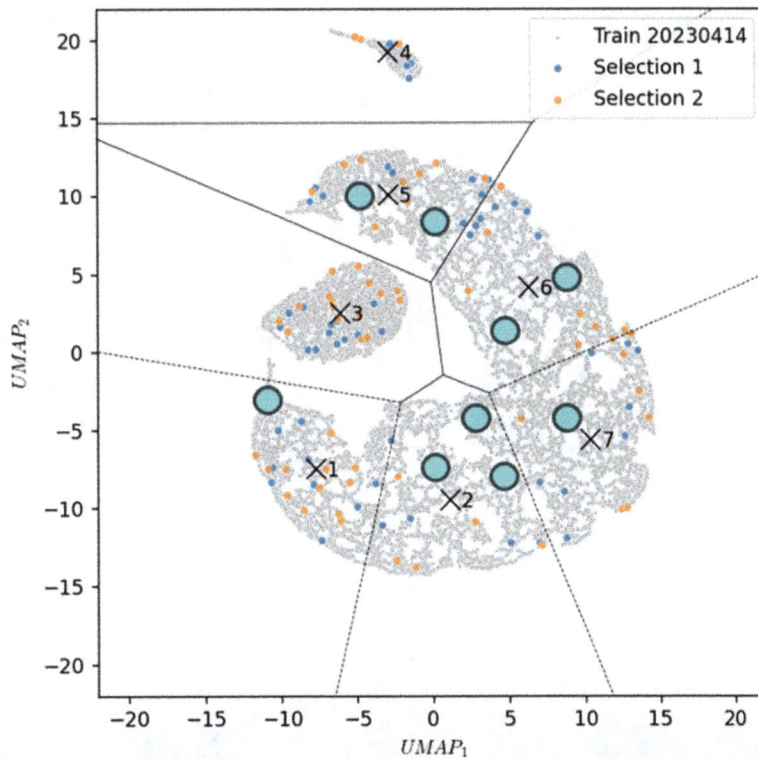

FIG 3 – Scatter plots showing metallurgical samples and gaps in UMAP space.

We evaluated the composition of the gaps by statistical analysis of the drill hole samples proximal to the gaps, identified using their UMAP coordinates. The boxplots in Figure 4 show that the compositional ranges of the gaps are sufficiently narrow, in most cases, to form specific, practical targets for further metallurgical sample selection.

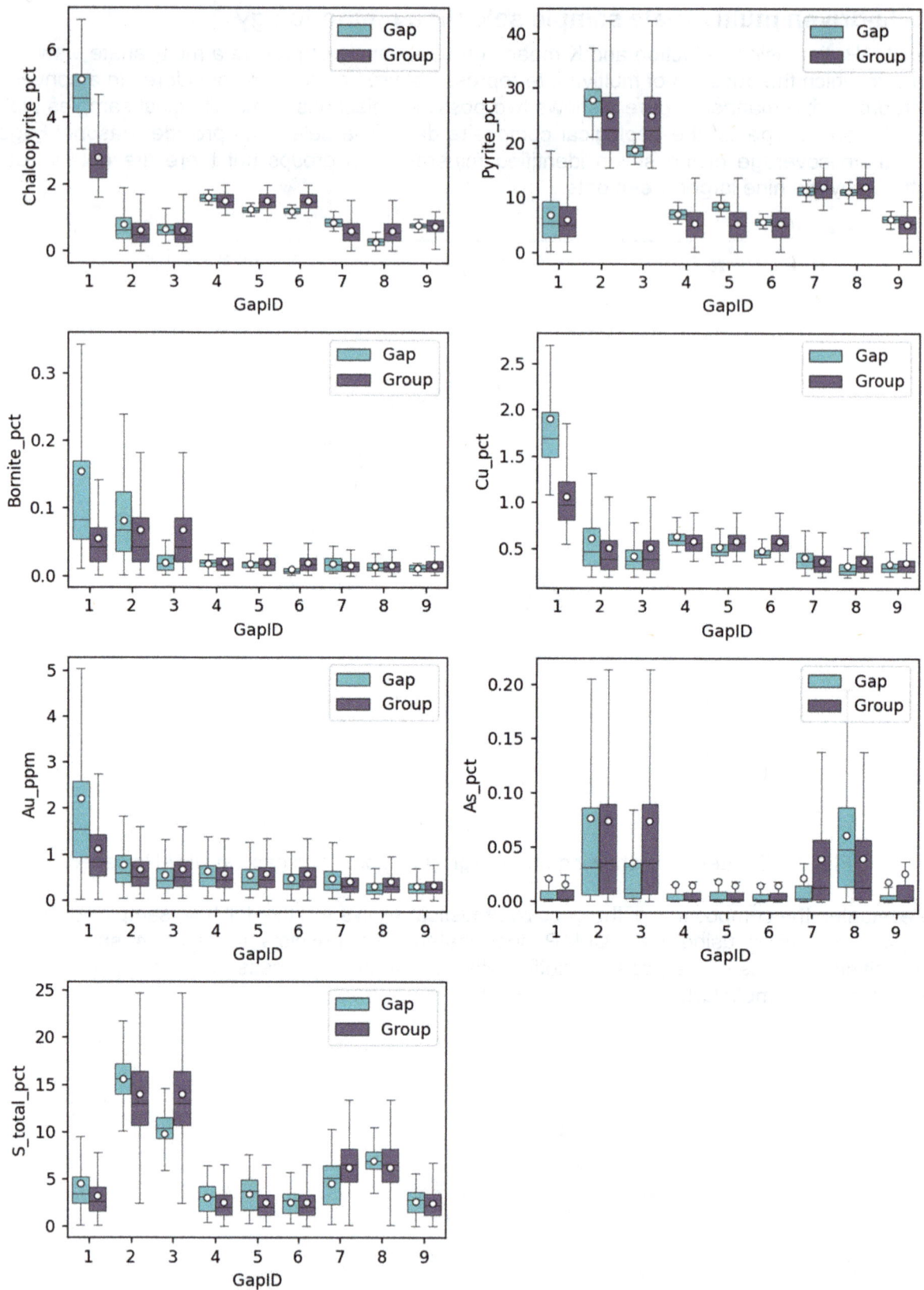

FIG 4 – Boxplots of Chalcopyrite, Pyrite, Bornite, Cu, Au, As and S_total, comparing compositions close to the gaps versus the whole K-means group.

How many samples do we need? – rules of thumb

When faced with deciding the minimum number of samples required for metallurgical test work, an important question is 'what are the results to be used for?' There are two primary applications in the geometallurgical space. First is for the purpose of process design and equipment selection. The second is for the development of a geometallurgical block model.

For process design and sizing of equipment, the primary objective is to design a series of unit processes to handle ore with average characteristics, at a specified throughput rate. It is also necessary to make sure the full range of ore characteristics is tested and is within the capability of the proposed design. The design outcome is a compromise.

By way of illustration, consider the process (simplified) to size a SAG mill. Samples are selected for testing on the assumption they represent the range of ore hardness (or more correctly the specific energies) to achieve the required size reduction and throughput. The samples are tested (drop weight tests, SAG mill comminution tests, etc) and the 85th percentile measurement is (usually) used as the basis of design. The outcome is a mill that performs well at the nominated specific energy and slightly less well (lower throughput or larger P_{80}) at higher energies that comprise 15 per cent or less of the ore, and slightly better at lower energies. In this sense, process design is generally targeting a point or one-dimensional answer and conventional statistics and rules of thumb typically guide the number of samples.

Rules of thumb are widely used and many variations have been proposed. Several of these rules consider the number of parameters used for regression modelling and/or are based on statistical concepts of variance in populations. Examples of published rules of thumb include:

- The 20:1 rule (Burmeister and Aitken, 2012) which states that the ratio of the sample size to the number of variables in a regression model should be at least 20 to 1. If a variable is categorical, each category should be counted as a variable. In other words, if a categorical variable has values of 'A' and 'B' it is counted as two variables, rather than one.

- An alternative method of sample number (N) calculation for multiple regression suggested by Green (1991) as: N > 50 + 8p where p is the number of variables.

- Ecological studies suggest N = 10 to 20 per predictor variable (Gotelli and Ellison, 2004).

- Jenkins and Quintana-Ascensio (2020) recommend N > 25 per regression but suggest N = 8 is adequate for a dependent variable with low variance.

- Sample numbers based on the variance of the input data set. However, input data with high variance could result in a very good predictive model if the model is controlled by a wide range of input values. Conversely, a lot of data may be required if a robust and accurate model is to be developed using only a narrow range (low variance) of input values.

A limitation of these rules is that they are largely based on the statistical properties of the input data and do not consider the properties of the output predictions. High variances of input variables will not necessarily lead to high errors in the output predictions. It is important to recognise that this is different to the requirement to adequately cover the ranges of the data to build the regression model.

For comminution, authors provide guidance based on study stage and test type (Meadows, Scinto and Starkey, 2012), number of ore types (Morrell, 2011; Giblett and Morrell, 2016), or per orebody (Doll and Barratt, 2011). The following table is reproduced from Meadows, Scinto and Starkey (2012) and suggests the recommended number of tests by each method.

TABLE 1

Number of grinding tests recommended for best practices.

Test	Number of tests recommended					Remarks
	1 Scope	1 PEA	3 PFS	4FS	5EPC	
Bond (BWi, RWi, CWi, Ai)	3	12	40	100	200	New drilling required to get samples
JKDWT	1	6	20	50	100	Limited by material Available
MacPherson AWI	1	2	6	15	30	Large composite samples req.
Protodyakonov	1	2	6	15	30	Large composite samples req.
SAGDesign	1	3	10	25	50	Composite or point samples
SPI	3	12	40	100	200	Composite or point samples
SMC	3	12	40	100	200	Point hardness samples only

Morrell (2011) and Giblett and Morrell (2016) state that at least ten samples per ore type are required to perform a meaningful statistical analysis at prefeasibility stage. Morrell (2011) further suggests that for the development of a geometallurgical model that is useful for the forecasting of daily grinding circuit throughput then the number of samples is 'at least an order of magnitude higher'.

For an entire geometallurgical mapping program, Williams and Richardson (2004) estimated that 100 to 300 metallurgical samples would be necessary, supported by at least 1000 mineralogy tests and 10 000 assays. In 2023 dollars, the difference between 100 and 300 flotation tests is of the order of $0.5 million, so many companies would be tempted to stop at 100 samples.

The development of a geometallurgical block model has a different objective; that of providing predictions of the variability of the ore that will be supplied to the process plant. The geometallurgical model represents localised ore and waste characteristics as a three-dimensional matrix of blocks and is based on the mineral resource block model. The modelled characteristics typically include metallurgical response variables such as mass pull, recovery, concentrate quality, work indices, throughput or specific energy. Machine learning and regression modelling can be used to predict the metallurgical response variables of typically sparse metallurgical test data and a much larger geological sample database. This approach derives maximum leverage from the high-cost metallurgical tests. The regression models can then be applied to the block model to provide block by block predictions of ore processing response based on the local geological data. These granular, quantitative inputs allow ore processing responses to be incorporated into mine planning, stockpile management and production scheduling for better control of feed to the process plant.

The importance of testing sufficient samples in this context has been captured by Garrido et al (2019) who states:

> In general, highly variable domains (high variance) are 'difficult' GMUs (geo metallurgical unit) to model or process and require more analysis than homogeneous domains. If a database has a poor sampling density, then the short-term models will be smoothed and will not represent the real variability of the mineral feed, losing the short-term predictability and, therefore, decreasing the metallurgist's ability to react preventively before the change of ore type occurs.

It is therefore the accuracy of the models built to predict short-term fluctuations in processing response that is important. This observation points to an alternative route to selection of the minimum number of samples required for metallurgical testing, based on the measured errors in the predictive models.

How many samples do we need? – let the data speak

The authors propose that in establishing the minimum number of samples required for evaluation of a domain or ore type, the principle should be to let the data speak. That is, once the available test work results produce robust predictive models that do not vary significantly as additional test work results are added, there is little benefit from testing more samples.

Error or uncertainty in a predictive model will, at some point, cease to be reduced by the addition of more samples and test work data. This is because the accuracy of the model will be limited by the accumulated errors associated with geological sample heterogeneity, subsampling, measurements of masses of samples and test work products, geochemical analysis, and errors introduced by operator behaviour or sub-optimal function of the equipment during the metallurgical test. In many deposits, important characteristics that impact processing response, such as grain size and texture, are not measured or captured in the geological database and their absence contributes to the total error of the prediction.

Therefore, the aim of the method proposed in this paper is to measure the errors in the predictions as the number of samples and test results increases and identify the point at which the addition of further data does not significantly improve the model. If the samples tested cover the full extent of multivariate ore characteristics and the predictive model is applied to the entire relevant drill hole database and/or the resource block model, it can be said that planning and evaluation will not be improved by further testing.

The challenge then is to identify useful measures of robustness, apply them to the predictive models and establish the criteria for the stage of study by which a model is accepted as being sufficiently robust.

Possible candidates include:

- Standard error of model must be similar to that of replicate testing. This is not a practical solution because there is rarely sufficient replicate testing data (if any at all).

- Adjusted R-squared metric. Whilst this is a commonly used measure of the performance of predictive models, it is not necessarily robust, and it does not provide a measure of error.

- Confidence limits based on the assumption that errors are normally distributed.

For this study, we chose the root mean squared error (RMSE) as our metric to compare predictive models. RMSE is a measure of the average error and by assuming that the errors are normally distributed, confidence intervals, with which the mining industry is familiar, may be estimated.

We tested the impact of increasing numbers of test work samples using a set of 41 Drop Weight index (DWi) test results from an iron oxide copper gold project. This is a data set that was considered possibly too small to demonstrate that the limit of useful metallurgical testing has been reached. For the test, we derived a much larger synthetic data set by using a simulation method known as the Probabilistic Graphical Model (PGM). The synthetic data set consists of 10 000 samples with the same multivariate statistical characteristics as the real set of 41 samples. For the purpose of this analysis, we nominated DWi as the target variable for prediction and nominated the following seven numerical features and one categorical feature to set-up a predictive model: gold, silver, sulfur, copper, iron, and magnetic iron grades, density, and lithology (categorical).

We considered a simple linear model (linear regression (LR)) and a Cubist model. In linear regression, the goal is to find the parameter values, or coefficients, that minimise the quadratic sum of the differences between the actual values and the estimated values. The Cubist algorithm is a machine learning model that incorporates decision trees and multivariate linear regression along with ensemble learning. Ensemble learning is a powerful method of training multiple models on different parts of the data set. The approach takes advantage of the wisdom of the crowd where the average of multiple models tends to be more accurate than any single model. The technique also helps to minimise the risk of over-fitting by training multiple models on different parts of the data. The Cubist model commonly performs well with non-linear and categorical geometallurgical data. Figure 5 is a comparison of actual DWi versus predicted DWi for the LR and the Cubist model.

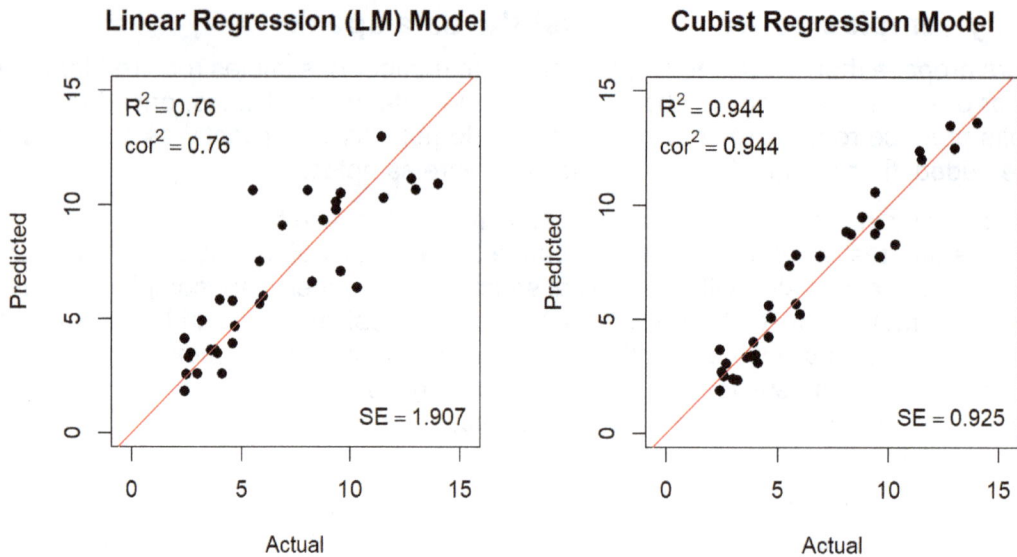

FIG 5 – Model predictive performance for DWi for the LR and Cubist models.

We generated LR and Cubist predictive models in an iterative procedure, increasing the number of samples and comparing the improvement in the models, using the RMSE regression metric. The samples were drawn randomly from the synthetic data set. Samples were increased in unit steps from 10 to 100, and steps of 10 from 100 to 500.

Figure 6 shows the RMSE for the LR and Cubist models with sample numbers increasing from 1 to 500. Both models improve rapidly before their performance stabilises. This transition is especially sharp for the LR model, where the RMSE stabilises at 2.3 and there is no significant improvement for more than 100 samples. The Cubist model continues to improve beyond 100 samples, asymptotically approaching an RMSE of approximately 1.4. This simple comparison shows that for this mineral deposit:

- Increasing the number of drop weight tests beyond 100 does not add value to the LR model. If LR is the selected regression modelling method, further drop weight testing is unnecessary.

- The Cubist model performs consistently better than the LR model. A set of 100 drop weight tests produces an RMSE of about 2.1. If this is an acceptable level of error, further testing is unnecessary. Improvement of the Cubist model is possible by carrying out more than 100 drop weight tests but the benefits, measured by RMSE, steadily diminish.

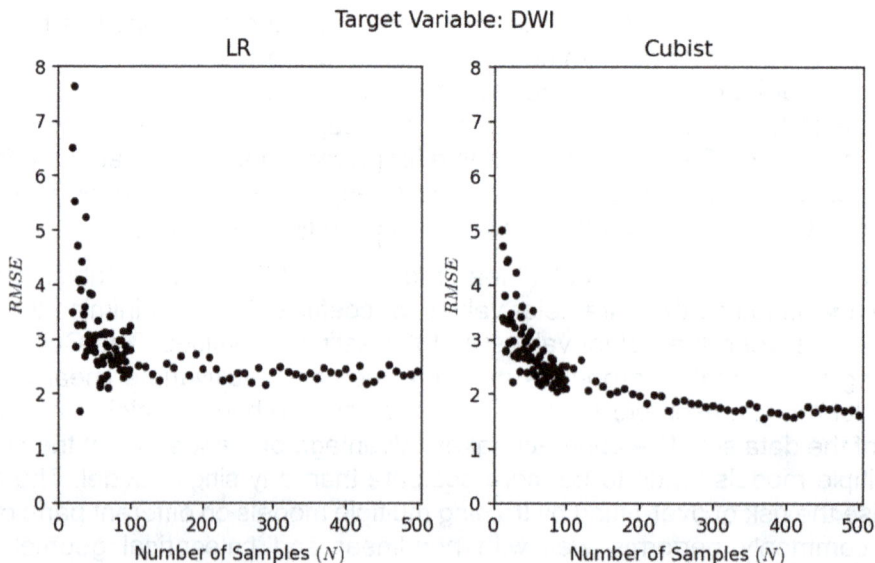

FIG 6 – RMSE values of DWi prediction against the number of samples. Linear regression (left panel) and cubist model (right panel).

The test demonstrates the principle that changes in RMSE as the number of samples is increased can be used to identify a threshold at which either the error of the predictive model is sufficiently low as to be fit for purpose, or further testing will not significantly improve the predictions. This threshold is data-driven and not reliant upon rules of thumb or *a priori* assumptions. The threshold will change depending on the processing response that is being predicted, the modelling algorithms that are being used and the number of variables/features that are used by the models.

The analysis measures the improvement in the predictive model that is expected from the addition of more samples to the original 41 test samples. It can also be used to select the number of additional samples required to reduce the uncertainty to a level commensurate with project requirements and risk appetite.

The ability to use dimension reduction techniques to assess sample representivity coupled with modelling to assess sample number needs forms a powerful diagnostic tool for the management of studies and assessment of new orebodies.

CONCLUSIONS

Well-designed metallurgical sampling programmes start with clarity about the objectives of the test work. Applying geometallurgical principles to the sample selection ensures that maximum leverage is obtained from the metallurgical test work. Data science and machine learning provide tools for the assessment of multivariate data that aid interpretation of relationships and avoid the limitations of univariate and bivariate statistical methods.

The trial outlined in the paper demonstrates that the impact of increasing sample numbers on the accuracy of linear regression and Cubist algorithms can be tested and the RMSE measured. For the target variable tested, drop weight index, both models clearly show a minimum number of samples beyond which the addition of further samples made little or no improvement to the error in the models. The RMSE declines more or less asymptotically to the total error associated with geological sample heterogeneity, experimental error and the error associated with ore characteristics not captured in the input data.

The use of geometallurgical principles and data science for metallurgical sampling and test data analysis provides a measurable and repeatable basis for sample-selection, driven by the geological characteristics of the mineralisation. It provides an alternative to the plethora of rules of thumb sampling procedures that have limited connection to geological variability nor the variability of the metallurgical responses.

REFERENCES

Burmeister, E and Aitken, L M, 2012. Sample size: How many is enough?, *Australian Critical Care*, 25(4):271–274.

Doll, A and Barratt, D, 2011. Grinding: Why so many tests?, in *Proceedings of the 43rd Annual Meeting of the Canadian Mineral Processors*, pp 537–556.

Garrido, M, Ortiz, J M, Sepulveda, E, Farfan, L and Townley, B, 2019. An overview of good practices in the use of geometallurgy to support mining reserves in copper sulfides deposits, in Procemin GEOMET 2019, Chapter 2: Geometallurgical Characterization and Modeling.

Giblett, A and Morrell, S, 2016. Process development testing for comminution circuit design, *Minerals and Metallurgical Processing*, 33(4):172–177.

Gotelli, N J and Ellison, A M, 2004. *Primer of ecological statistics* (Sunderland: Sinauer Associates).

Green, S B, 1991. How many subjects does it take to do a regression analysis?, *Multivariate Behavioral Research*, 26:499–510.

Jenkins, D G and Quintana-Ascencio, P F, 2020. A solution to minimum sample size for regressions, PLoS ONE 15(2):e0229345. https://doi.org/10.1371/journal.pone.0229345

Kormos, L, Sliwinski, J, Oliveira, J and Hill, G, 2013. Geometallurgical Characterisation And Representative Metallurgical Sampling At Xstrata Process Support, Annual Canadian Mineral Processors Operators Conference.

Meadows, D, Scinto, P A and Starkey, J H, 2012. Seeking Consensus – How Many Samples And What Test work Is Required For A Low Risk Sag Circuit Design, CIM 2012. Available from: https://onemine.org/documents/seeking-consensus-how-many-samples-and-what-test work-is-required-for-a-low-risk-sag-circuit-design

McInnes, L, Healy, J and Melville, J, 2020. UMAP: Uniform Manifold Approximation and Projection for Dimension Reduction. Available from: https://arxiv.org/abs/1802.03426

Morrell, S, 2011. Mapping Orebody Hardness Variability for AG/SAG/Crushing and HPGR Circuits, International Autogenous Grinding, Semiautogenous Grinding and High Pressure Grinding Roll Technology 2011.

Williams, S and Richardson, J, 2004. Geometallurgical Mapping: A new approach that reduces technical risk, in *Proceedings of the 36th Annual Meeting of the Canadian Mineral Processors*, pp 241–268.

The evolution of level control and its role in advanced control in the Northparkes flotation circuit

C P Lowes[1], M D Curtis[2] and M J Garside[3]

1. AAusIMM, Evaluation Metallurgist, Northparkes Operations, Parkes NSW 2870.
   Email: callan.lowes@evolutionmining.com
2. MAusIMM, Senior Evaluation Metallurgist, Northparkes Operations, Parkes NSW 2870.
   Email: matthew.curtis@evolutionmining.com
3. MAusIMM, Ore Processing Manager, Northparkes Operations, Parkes NSW 2870.
   Email: mitchell.garside@evolutionmining.com

ABSTRACT

The Northparkes processing plant is comprised of two parallel grinding modules each feeding a primary rougher flotation cell. The primary rougher underflows are then combined and processed through a single flotation bank consisting of four rougher and two scavenger 200 m³ tank cells. Since the commissioning of this circuit in 2018, achieving circuit stability had been challenging due to issues with the control valves and a lack of appropriate control philosophies.

The primary areas of concern were the actuators on the rougher cells and the pinch valves controlling the scavenger cells. For the rougher cells, it was found that torsion exerted on the internal darts by the slurry caused the feedback mechanisms to the positioners to loosen or even fail. Hence, the valves actuated at an uncontrolled rate. A feedforward control logic was implemented so the cells could react to upstream disturbances ahead of time. In late 2021, fit-for-purpose actuators were installed on these cells to improve the level control significantly, though the scavenger cells remained problematic.

A project was then initiated to replace the scavenger pinch valves with external dart valves. Computational fluid dynamics (CFD) analysis was conducted to determine the optimal piping layout and valve sizing for accurate level control. As a result, the level deviation from set point was reduced from around ± 50 mm to around ± 10 mm.

Success here has enabled implementation of the advanced real-time process optimiser ESTIMATA in the flotation circuit. ESTIMATA performs a real-time mass-balance around the circuit and calculates optimised set points based on adaptive fundamental models. These optimised set points are sent directly to the PID control loops. The results of extensive on/off trials have shown a 1.21 per cent recovery benefit when utilising ESTIMATA. This paper details the outcomes of these projects, including analysis of the process control improvements and plant survey data to quantify the metallurgical benefits achieved.

INTRODUCTION

Northparkes Operations (NPO) is an Australian copper and gold mine located in the Central West region of New South Wales, approximately 27 km north-west of Parkes. The operation is a joint venture between Evolution Mining (80 per cent, acquired in 2023 from CMOC Mining), Sumitomo Metal Mining Oceania (13.3 per cent) and SC Mineral Resources (6.7 per cent). The concentrator processes material from underground and open cut ore sources at a design rate of 7.6 Mtpa to produce a copper-gold concentrate for export.

The NPO ore processing plant consists of two parallel grinding modules which are equivalent in configuration but sized to process roughly 50 per cent higher throughput in Module 2. The circuits consist of a SAG mill followed by secondary and tertiary ball milling in closed-circuit configurations. The cyclone overflows at each module's respective tertiary milling stages are sent to the flotation circuit and processed as shown in Figure 1.

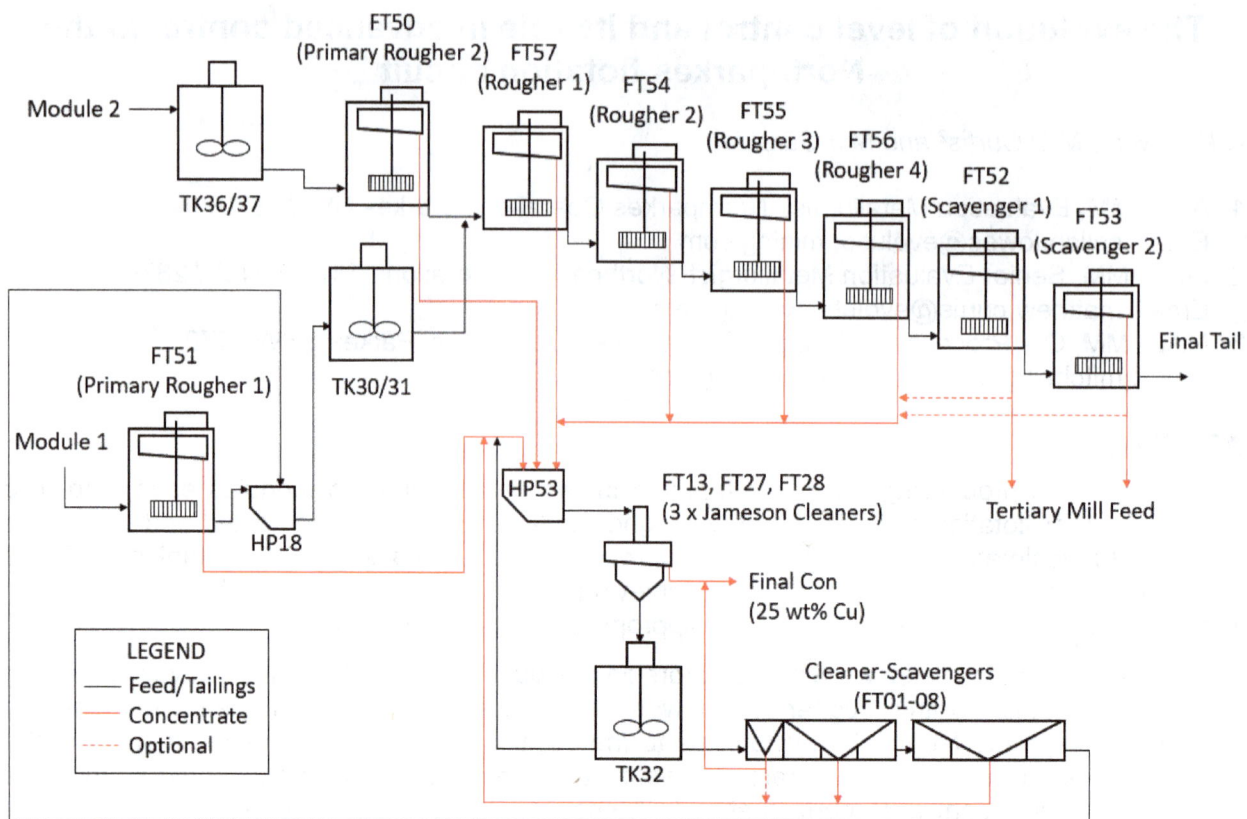

FIG 1 – Northparkes flotation circuit flow sheet.

Each module is first processed in a primary roughing stage, which is a 150 m³ tank cell for Module 1 and a 200 m³ tank cell for Module 2. The primary rougher underflows are then combined with the cleaner-scavenger tailings and directed to a single flotation bank consisting of four rougher and two scavenger 200 m³ tank cells. The combined rougher concentrates report to the cleaner circuit, while the scavenger concentrates are sent back to the tertiary mills for regrinding (with the option to direct to cleaner feed, which is mainly used during shutdowns). The cleaner circuit consists of three Jameson cells producing final concentrate and eight cleaner-scavenger 17 m³ mechanical cells. The first cell of the cleaner-scavengers (FT01) reports to final concentrate but has the option to be re-directed to the cleaner feed hopper (HP53). The final concentrate grade target is currently 25 wt per cent Cu.

The layout of the NPO processing plant prior to the commissioning of the above flow sheet has been described previously (Harbort *et al,* 2017). A major design change was made in 2018 where the primary rougher underflows were combined and processed in a new single rougher bank (FT54, FT55, FT56). The scavenger cells were also taken from scavenging the rougher tail from each module separately to being part of this single bank. In 2020, the capacity of the circuit was increased by installing FT57, and FT50 was moved and connected to the front of the bank as in Figure 1. From this point forward the cells will be referred to as Primary Rougher 1, Rougher 1, Rougher 2 etc as indicated in Figure 1.

While this circuit expansion allowed for increased capacity, circuit stability quickly became a major issue across the rougher and scavenger cells. A series of projects have since been implemented which have delivered significant improvements in level control and performance. This paper outlines the issues experienced and the outcomes of these projects, including how this has enabled the implementation of advanced process control strategies in the flotation circuit.

PROCESS CONTROL STRATEGY

The operating strategy in the flotation circuit revolves around the cleaner circuit, specifically optimising the mass pull rates into the cleaner feed hopper (HP53). This hopper receives feed from the rougher concentrates, cleaner-scavenger concentrates, and the cleaner tail recycle. A PID control loop regulates the cleaner tail recycle to maintain a hopper level so that the Jameson cell feed pressures can be kept constant without cavitating the pumps. If the cleaner tail recycle pump reaches maximum speed and the hopper level cannot be maintained, process water addition provides an extra degree of automated hopper level control. High rates of cleaner tail recycle or water addition to HP53 are an indicator of low rougher pull rates.

The flotation cell pull rates are controlled using a cascade froth velocity controller. The primary rougher, rougher and cleaner-scavenger cells are fitted with VisioFroth™ (Runge *et al*, 2007) cameras that measure the velocity of the overflowing froth. Based on the measured froth velocity, air rate set points are cascaded to the air flow controller to target a given velocity set point. The operators then set the velocity set points and froth depths based on the cleaner tail recycle rate and on-stream analyser (OSA) assay data. The froth depth and air flow rate in the Jameson cells can then be adjusted to target the required concentrate grade.

Figure 2 shows the frequency distribution of the cleaner tailings recycle for a range of cascade utilisations. The legend indicates the percentage of the rougher and scavenger cells which were in cascade and the number of data points that the trends are based on. In general, increased utilisation of cascade velocity control across the circuit prevents excessive cleaner tail recycle as there is a narrower distribution with higher frequency in the target range of 25–30 per cent.

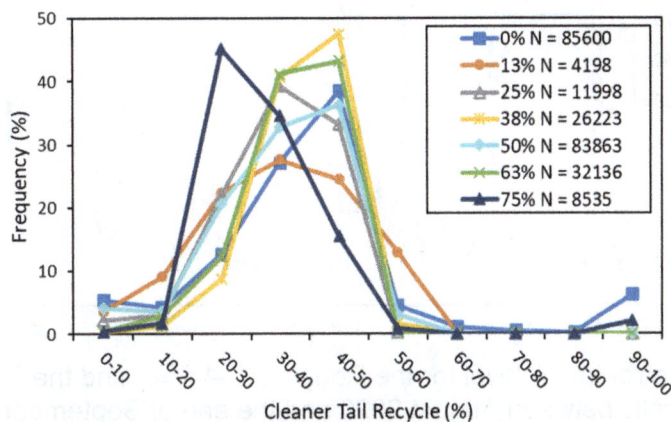

FIG 2 – Frequency distribution of the cleaner tailings recycle percent showing the effect of increasing cascade velocity control utilisation across the circuit. The legend indicates the percentage of the cells in cascade and the number of data points, N, from which the distribution was calculated.

Although the primary rougher, rougher and scavenger cells are all tank cells, one key difference is the final control elements for the level controllers. Roughers 1–4 use dual internal dart valves, while Primary Rougher 1 and Scavengers 1 and 2, until recently, used external pinch valves (Primary Rougher 1 single and Scavengers 1 and 2 dual valves). Primary Rougher 2 originally used an external pinch valve but this was replaced by a dual external dart valve when it was connected to the rougher bank during the expansion project. The cells normally operate with one valve in manual and the other in auto controlling to a level set point. The operators adjust the position of the manual valve to keep the auto valve within its operating range, though functionality does exist for dual auto operation.

STABILITY CHALLENGES FOLLOWING CIRCUIT EXPANSION

Issues with stability were noticed early on following the commissioning of the expanded flotation circuit. Figure 3 shows the daily average absolute error in cell level relative to set point, referred to as level error, for the rougher and scavenger cells over a two year period. Anecdotally, acceptable level control would be considered as less than ± 20 mm, while very good control would be around less than ± 10 mm. For Roughers 1–4, the cells experienced periods where the average error spikes to greater than 50 mm. For Primary Rougher 1 and Scavengers 1 and 2, the average error is very erratic and seldom do these cells operate with less than 30–40 mm level error.

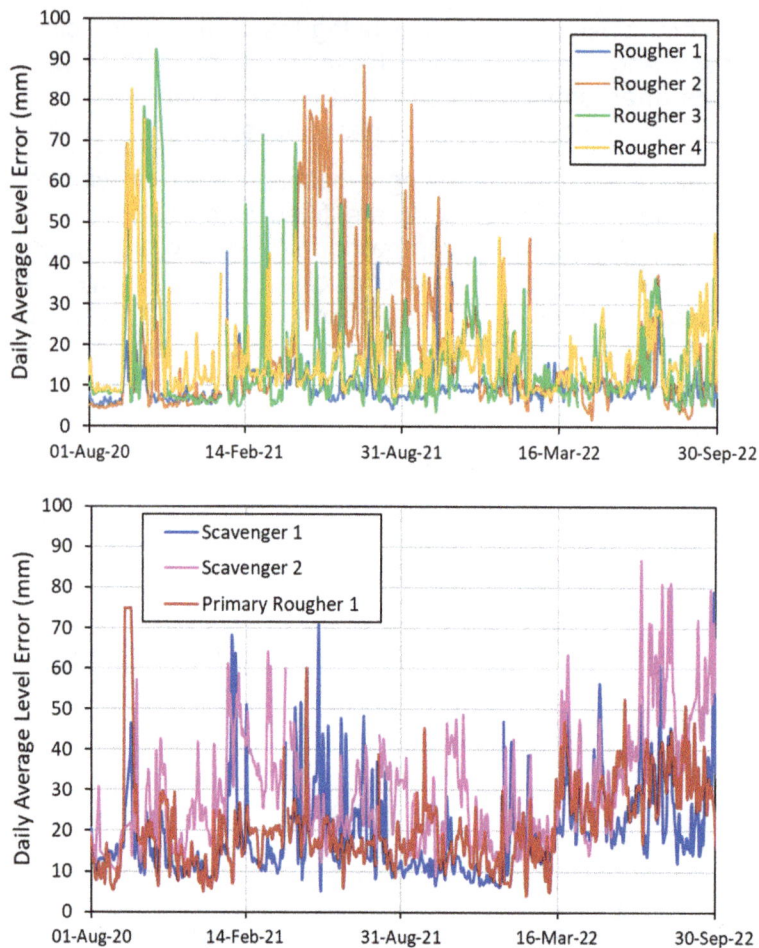

FIG 3 – Daily average error in cell level for the Rougher 1–4 (top) and the Primary Rougher 1 and Scavenger (bottom) cells between August 2020 and the end of September 2022. The data has been filtered to exclude plant shutdowns.

Circuit behaviour as shown in Figure 3 is problematic because the flotation cells effectively transition between two extremes: zero mass pull shortly followed by pulping the cell. The flow on effect is disturbances propagating through the flotation bank and flooding of the cleaner feed hopper. The net result is that the cells must be operated with velocity control offline, with the air addition set in auto control at a low set point and deep froth levels. These conditions are not favourable for recovery and velocity control cannot be utilised.

For Roughers 1–4, the issues were initially thought to be a result of the PID tuning parameters. Re-calibration of the positioners and feedforward control action provided some improvement, but ultimately the valves returned to actuating at an uncontrolled rate. Even in manual, the valves failed to hold their position. Inspection of the actuators and positioners consistently found looseness in the mechanical feedback mechanisms. In October 2020, one of the adaptor boxes which connected the dart stem to the piston rod on one of the Rougher 4 valves became disconnected. As a result, the extended piston rod impacted the adaptor box and the entire cylinder had to be replaced. In July 2021, the piston rod which was connected to the dart stem of one of Rougher 3's valves sheared off completely as shown in Figure 4a. These events suggested that the slurry flow past the dart plugs was creating torsion on the mechanical elements, causing them to loosen and even fail.

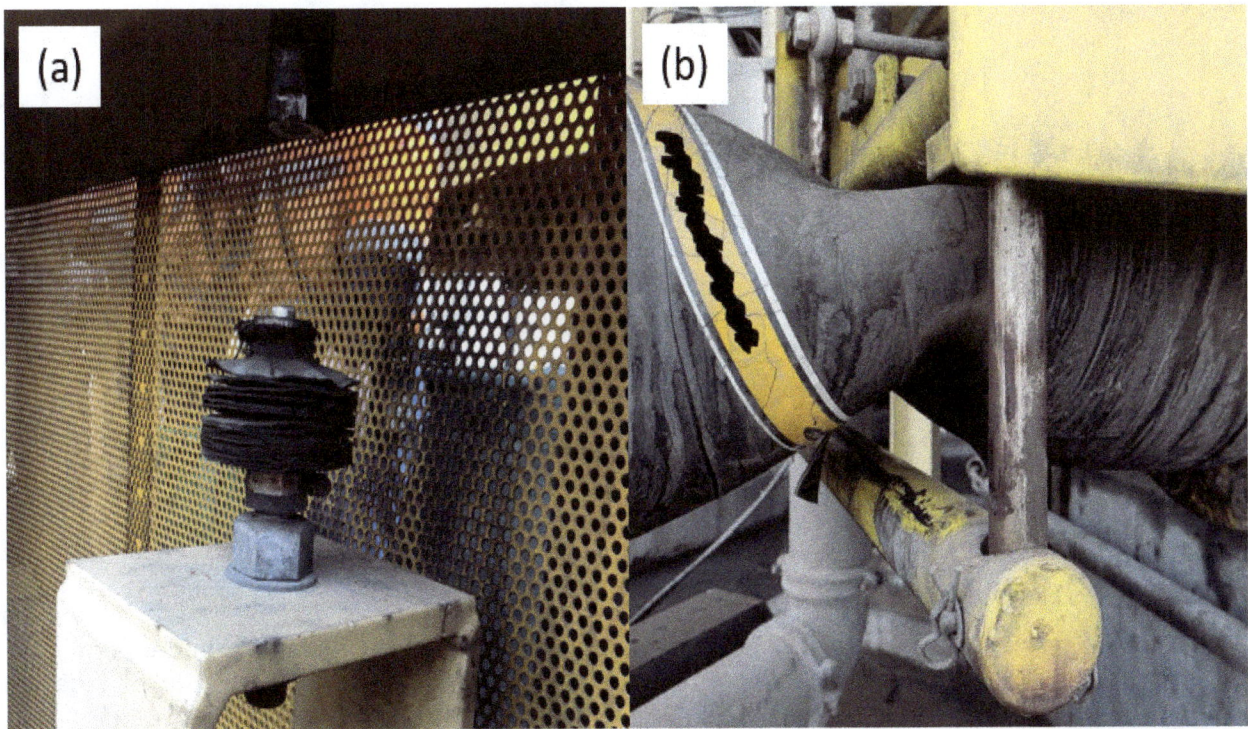

FIG 4 – (a) failure of the piston rod on one of valve actuators on Rougher 3; and (b) fatigue on the pinch sleeve of Primary Rougher 1 resulting in deadband in the controller action.

For Primary Rougher 1 and Scavengers 1 and 2, fatigue and memory in the pinch sleeves as shown in Figure 4b caused deadband in the controller action. This, combined with a ~30 sec full stroke (from 0 to 100 per cent) time on the actuator, made it difficult to achieve accurate level control. Scavengers 1 and 2 were particularly problematic given the disturbances they experience being at the end of a bank with two feed streams and negligible head difference between adjacent cells. Primary Rougher 1, however, only receives cyclone overflow and discharges into a hopper so the level control should be much simpler. Even for this application, the pinch valves were not a robust solution.

As part of the scavenger cell underflow valve upgrade described in following sections, a survey was conducted on Scavenger 2 to examine the metallurgical performance of such poor level control. Scavenger 2 was chosen as this was the worst performing cell at the time. Due to the oscillatory nature of the mass pull, a survey method was developed to constantly sample the lip, changing the sampler and recording the time as the sampler became full. Figure 5 shows the results of this survey, with a significant fluctuation in the solids rate, water rate and concentrate grade as a function of time. The solids rate also approaches and is maintained at zero for a total of 2 mins in the 10 min survey. This suggests that the cell operated with no mass pull for around 20 per cent of the time.

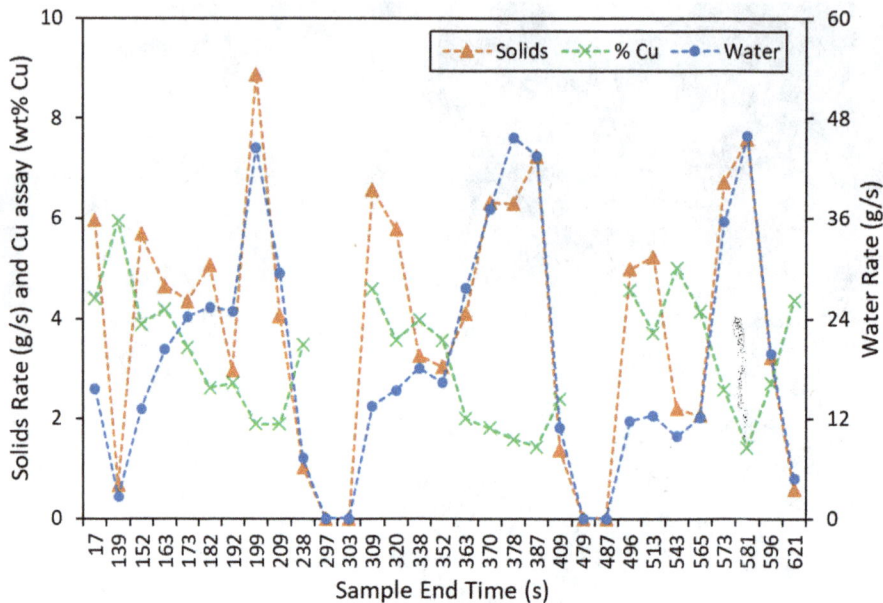

FIG 5 – Solids rate, water rate and Cu assay sampled from the lip of Scavenger 2 under poor level control conditions. Note that the data are plotted as a function of the samples taken with the cumulative time shown for each sample.

A second survey was then conducted sampling the lip of both scavenger cells. Here, the same approach as above was applied but the samples were collected as a single concentrate for each cell. The rougher tail (Rougher 4 tail) and final tail (Scavenger 2 tail) were sampled from the OSA, enabling the bank performance to be determined. The scavenger bank copper recovery was 4.6 per cent at 5.9 wt per cent Cu concentrate grade with a 0.08 wt per cent mass pull (mass-balance reconciled data). This concentrate grade is considered high for a scavenger application; material at this grade is better suited to cleaner feed but is the result of both cells operating for sustained periods with extremely low mass pull.

PROCESS IMPROVEMENTS

Rougher cell control philosophy and actuator upgrade

The control philosophy of the expanded flotation circuit initially included only PID functionality for level control. Following commissioning, it was noted that there was significant disturbance in the level, particularly towards the scavenger end of the circuit. Analysis of process data over a 4 hr period found that the variance in level relative to set point increased from 37 to 152 mm between Rougher 2 and Scavenger 2 (Rougher 1 was not commissioned at this stage). It became clear that since the flotation tanks operate independently from one another, the PID level control could not compensate for surges, or for changes in the air or level in other parts of the bank. Potential solutions which were investigated to limit downstream disturbances included adding process variable filtering and derivative action to the level controller. However, ultimately this led to a slow response and therefore wasn't considered a long-term solution.

Upon review, it was determined that the installation of a feedforward controller was required to allow downstream cells to react to the upstream disturbances ahead of time. Figure 6 shows a box plot of the level error for Rougher 3, Rougher 4, Scavenger 1 and Scavenger 2 comparing before and after the feedforward control implementation. The stability in Rougher 1 and 2 has improved at the expense of Scavenger 1 and 2. During the commissioning of the feedforward controller, it was found that many of the rougher underflow valves in manual were actuating at an uncontrolled rate attempting to find their position. This was the first exposure to issues with the positioning of the rougher cell control valves. These additional disturbances were passed on to the scavengers where the pinch valves struggled to cope. The recommendation from Manta controls following the commissioning of the feedforward logic was to investigate replacing the pinch valves with dart plugs.

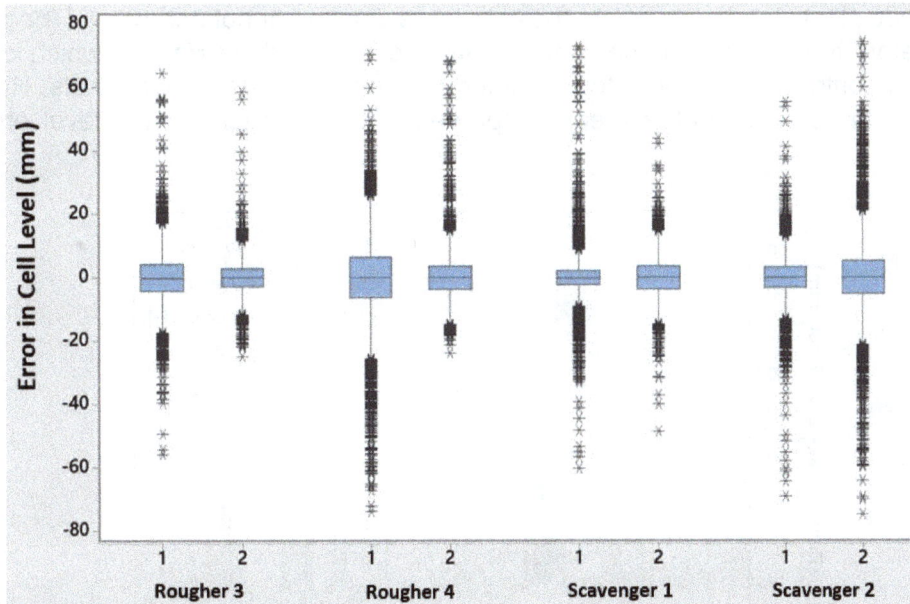

FIG 6 – Box plot of the error in cell level across the rougher and scavenger bank before (indicated as 1) and after (indicated as 2) implementation of feedforward control action. The stability in Rougher 1 and Rougher 2 has improved at the expense of Scavenger 1 and Scavenger 2.

The issues associated with the rougher cell actuators were primarily due to the mechanical linkages in the valve assembly and the mechanical feedback mechanism to the positioner. In April 2021, a trial was conducted on Rougher 4 using a new system in collaboration with SMC Australia. A floating joint was installed between the piston rod and adaptor box as shown in Figure 7a. The floating joint enabled some rotation of the dart stem without significantly impacting the cylinder internals. Further, the new system involved a positioner with magnetic feedback rather than mechanical (SMC Air Servo Cylinder IN-777) as shown in Figure 7b. The full stroke time was also reduced from around 30 to 4.5 secs. The trial resulted in a 56 per cent reduction in the level error which provided the justification for a capital upgrade of the entire bank.

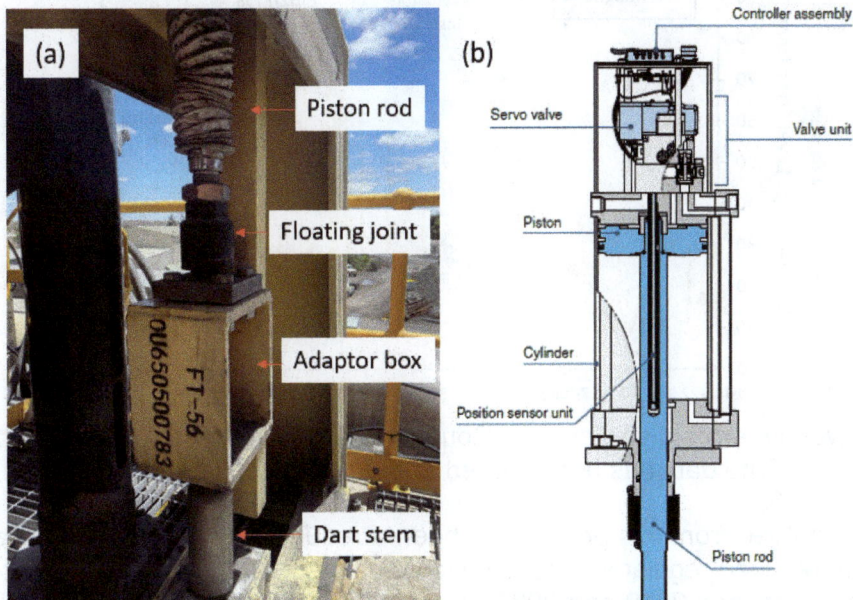

FIG 7 – (a) labelled picture of the dart stem adaptor box and floating joint arrangement for the rougher cell actuators and (b) schematic of the SMC Air Servo Cylinder IN-777 which was installed on the rougher cells (provided by SMC Australia).

The installation of the new actuator systems was completed in October 2022. Figure 8 shows the level error for Roughers 1–3 before and after the installation and commissioning. The data has been filtered to exclude plant shutdowns where the cells were emptied or not operational (hence very large level error). Before the installation, the control may be considered satisfactory which is the result of the feedforward controller and consistent maintenance on the positioner internals. However, after the installation, the level control is greatly improved, with less than 10 mm level error achieved consistently.

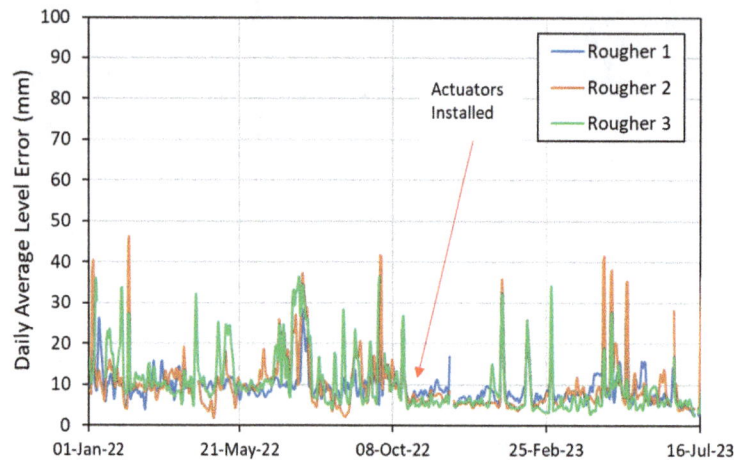

FIG 8 – Daily average error in cell level for Rougher 1, Rougher 2 and Rougher 3 following the upgrade of the actuators. The trends are shown between January 2022 and July 2023. The data has been filtered to exclude plant shutdowns.

For Rougher 4, where the level error data is shown in Figure 9, there is still some variation following the actuator install which was thought to be due to interaction with the poorly controlled Scavenger 1. However, following the Scavenger 1 level control valve upgrade, the performance was still not as good as the other roughers. Recent investigations found that the actuator speed setting was incorrect, and the floating joint not correctly aligned, on one of the valves. Correction of these issues seems to have provided some improvement.

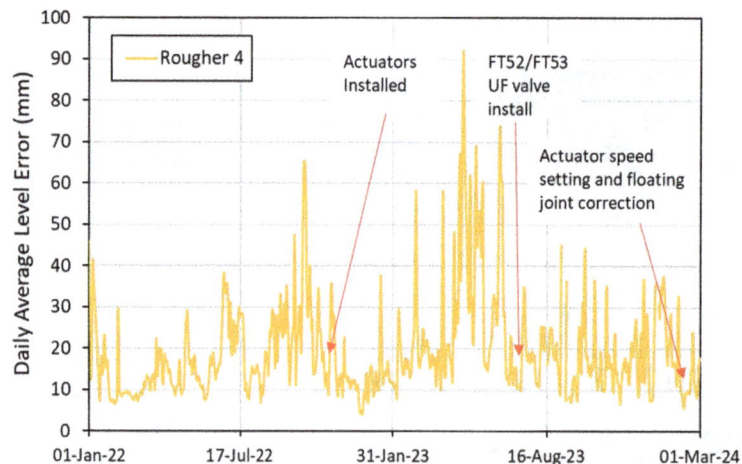

FIG 9 – Daily average error in cell level for Rougher 4 shown from January 2022 to March 2024. The data has been filtered to exclude plant shutdowns.

A further benefit realised from this project has been a reduction in the maintenance requirements. The initial system required regular inspection and adjustment to ensure satisfactory operation, with maintenance costs between 2019 and 2021 around $75 000 (including a full cylinder replacement on Rougher 4). From October 2022 onwards, maintenance costs have been $3000, mainly associated with air filter replacements and yearly positioner calibration. No functional failures have been observed.

Primary rougher and scavenger cell underflow valve replacement

For Primary Rougher 1 and Scavengers 1 and 2, issues had been recognised with the level control valves; however, there were several other factors identified which made accurate level control difficult. Scavengers 1 and 2 are positioned at the end of a bank of cells with a difference in head level between adjacent cells which causes them to interact. Furthermore, for Scavenger 2, the underflow is discharged into a hopper where the difference between the slurry level in the cell and slurry level in the hopper (referred to as the step height) was around 2.7 m. Compared to adjacent cells in the flotation bank, where the slurry level difference is less than 1 m, the pressure drop across the valve was significantly higher. Hence, small changes in the valve opening position had a greater impact on the level. This was also true for Primary Rougher 1, so, although there were minimal upstream disturbances, the pipe velocities were excessive which made accurate level control difficult.

A capital project was initiated to replace the valves on these cells. Options considered included internal dart valves, though the work required to retrofit this solution was found to be prohibitively expensive. A low-cost solution was achieved by installing a dual external dart valve (supplied by eDart) on Primary Rougher 2 when it was connected to the front of the rougher bank. As such, eDart were engaged during the design phase to look at solutions for the extra complexity described above.

Computational fluid dynamics (CFD) simulations were conducted by eDart to determine the optimal piping layout and valve sizing for accurate level control. The approach taken was to vary the valve size (and therefore pipe size) and step height to achieve the required flow at 50 per cent valve opening. This criterion was selected to provide some surge capacity and redundancy for future rate increases. The velocity profiles through the valve were then verified at the nominal, maximum (20 per cent increase on nominal) and minimum (Module 2 shutdown where rates are decreased by ~60 per cent) flows to ensure that there was no risk of excessive wear or sanding in the pipes.

For Scavenger 1, the recommendation was to change out the valves while keeping the pipe size the same (DN600). For Primary Rougher 1 and Scavenger 2, the pressure drop issue was addressed by implementing a back-pressure-pipe (BPP) design to decrease the step height to around 1 m. As a result, the pipe size was increased from DN400 to DN600, and a breather was installed on the valve discharge pipe to prevent suction on the dart plug. Figure 10 shows the BPP's and dart valves installed on Scavenger 2.

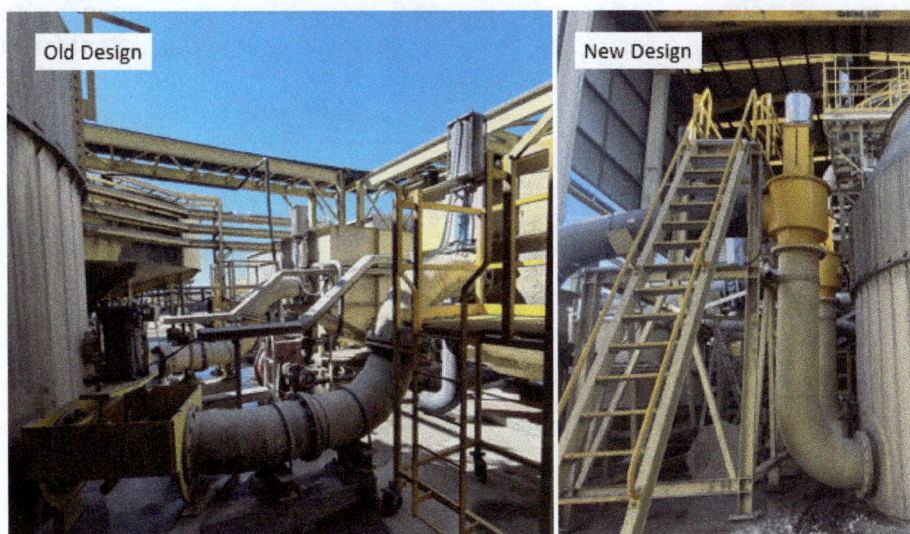

FIG 10 – Comparison between the old underflow discharge design (left) and the new design involving a BPP (right) on Scavenger 2. The pinch valves have also been replaced by dart valves and the pipe size has been increased from DN400 to DN600.

Figure 11 shows the level error for Primary Rougher 1 and Scavengers 1 and 2 from August 2022 through to March 2024. The construction and commissioning of the Primary Rougher 1 and Scavengers 1 and 2 level control valves were completed in separate shutdowns which are indicated on Figure 11. For Scavengers 1 and 2, the level error has been reduced from 30+ mm for both cells to around 10 mm consistently. The only major issue experienced was a failure of the air supply to one of the valves on Scavenger 1 which resulted in a period of poor level control (as indicated on Figure 11). Overall, this is a remarkable improvement in level control on cells where this is difficult to achieve due to their downstream position in the circuit. For Primary Rougher 1, the performance is even better as the level error has maintained around 5 mm following the install. This is significant as the primary roughers are a major driver of the overall plant performance given this is the only opportunity for floating fresh feed without the influence of recycle streams.

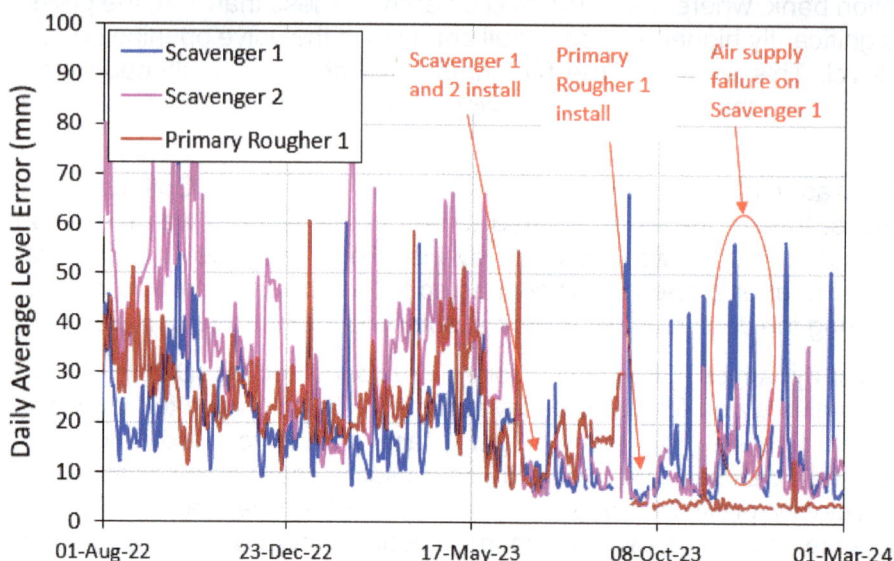

FIG 11 – Daily average error in cell level relative to the set point for Primary Rougher 1, Scavenger 1 and Scavenger 2. The trends are shown from August 2022 to March 2024 and key install dates for the underflow valve upgrade project implementation are indicated.

A follow up survey was conducted to quantify the effect of the improved level control on cell performance. The survey procedure outlined earlier was once again used to sample Scavenger 2 and then the overall scavenger bank under the same conditions. Figure 12 shows the results of the Scavenger 2 lip sampling survey. Compared to Figure 5, there is less variation in the solids rate with no periods of zero mass pull. As a result, the concentrate grade is very consistent for the duration of the survey.

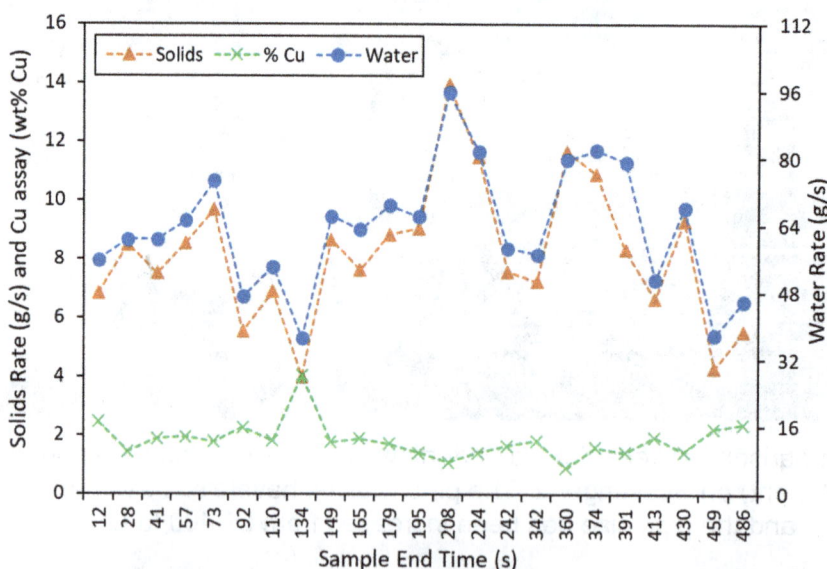

FIG 12 – Solids rate, water rate and Cu assay sampled from the lip of Scavenger 2 under improved level control conditions following the underflow valve upgrade. Note that the data are plotted as a function of the samples taken with the cumulative time shown for each.

Table 1 shows the results of the surveys before and after the level control valve replacement. In the Phase 1 results, the average of the complete Scavenger 2 lip sampling survey data further shows the reduced variation in concentrate grade, a result of the significant reduction in level error between the two surveys. In Phase 2, the concentrate grade from the scavenger bank is now more suitable for a scavenger application, and the mass pull has been increased by over an order of magnitude.

TABLE 1

Summary of the survey results comparing the before (Part 1) and after (Part 2) of the underflow valve upgrade project. Phase 1 is the average results of the Scavenger 2 lip sampling survey (Figures 5 and 12) while Phase 2 is the combined scavenger bank survey data.

	Part 1 – Original design	Part 2 – Dart valves
Phase 1 – Scavenger 2 lip sampling survey		
Solids rate (g/s)	3.9 ± 2.2	8.2 ± 1.8
Water rate (g/s)	18.9 ± 11.4	63.2 ± 11.4
Cu assay (wt% Cu)	3.25 ± 1.3	1.8 ± 0.4
Average level error (mm)	23	8
Phase 2 – Scavenger bank survey (performance on feed to bank)		
Mass pull (wt%)	0.08	1.1
Concentrate Grade (wt% Cu)	5.9	2.1
Cu Recovery (%)	4.6	18.9

To determine the copper recovery benefit resulting from the underflow valve upgrade, a multivariable regression was conducted using the daily production data to generate a recovery model. The model included parameters for feed grade, concentrate grade, solids throughput and a dummy variable to account for the on/off change. Table 2 shows the results of the regression analysis. The coefficients suggest that as feed grade increases so does recovery, while increases in concentrate grade and throughput reduce recovery. The copper recovery benefit of the level control valve upgrade is shown to be 1.10 per cent with greater than 95 per cent confidence.

TABLE 2

Results of a multivariable regression analysis conducted on the daily production data to identify the copper recovery benefit from the underflow valve upgrade project.

Regression Stats		
R^2	0.65	
Adjusted R^2	0.63	
Observations	103	
ANOVA F-value	2.6×10^{-21}	
Regression Model	**Coefficient**	**P-value**
Intercept	94.8	3.39×10^{-50}
Feed Grade (wt% Cu)	18.3	7.58×10^{-8}
Concentrate Grade (wt% Cu)	-0.33	8.19×10^{-4}
Solids Throughput (t/h)	-0.01	5.19×10^{-3}
Dart Valves (% On versus Off)	1.10	0.04

Model based optimisation process control

The cascade velocity controller was initially implemented via the Visiofroth™ expert controller. The measured average froth velocity and set point was calculated externally to the PLC and then sent in 30 sec intervals. Based on the error between the measured value and set point, the controller would change the manipulated variable (either air or level) by a set amount depending on the logic in place (eg if the error was greater than x cm/s then change air addition by y m^3/h). Thus, to get fine control, the underlying logic needed to be extremely complex. This made troubleshooting and tuning of the controller very difficult. Further, the 30 sec evaluation time often meant that circuit surges weren't responded to well which caused flooding of the cleaner hopper. As such, operator confidence in cascade velocity control was very low.

As a result, the cascade velocity controller was simplified by moving from the expert controller to a PLC based PID controller. The evaluation time was decreased from 30 secs to around 200 ms, resulting in a much more gradual response that was relative to the magnitude of the error. This made the controller more robust to process upsets. Further, the controller was also easier to tune as there was significantly less dead time.

The reduced complexity of the PID control led to greater acceptance and utilisation of the cascade velocity controller. However, the overall control strategy was still highly dependent on operator input for the selection of set points. Cascade velocity control requires operator judgement when selecting the velocity set points and flotation bank profiling. Many of the process stability challenges had largely been solved, enabling opportunities to increase the flotation process performance with improvements to the process control strategy of the flotation circuit.

NPO partnered with INNOVATION X to trial a real-time process optimiser ESTIMATA across the entire flotation circuit. ESTIMATA uses fundamental dynamic flotation models differing from other advanced controllers which provide set points based on variables such as concentrate mass flow (eg Seaman *et al,* 2021; Supomo *et al,* 2008; Runge *et al,* 2007) without consideration of the process fundamentals. Process measurements (OSA assays, flow metres, density gauges, weightometers etc) are used to calculate a circuit-wide mass-balance reconciliation, adapting the model parameters and determining optimised process set points.

During the NPO plant trial, ESTIMATA autonomously managed froth velocity set points for the rougher, scavenger, and cleaner-scavenger cells, as well as level set points for the cleaner cells. Subsequent improvements included integrating the Jameson cell air addition and feed pressure set points into ESTIMATA's control. The optimisation objective was to maximise recovery with the constraint of a minimum concentrate grade. Real-time process data was captured from the PLC into the Aveva Pi historian, where it was then consumed by the ESTIMATA optimiser (cloud based). Optimised set points were then returned to the on-site Aveva Pi instance and sent back to the PLC every 5 mins.

Figure 13 shows the concentrate grade (measured by OSA), total velocity, cleaner cell FT27 level, and cleaner tail recycle across four days spanning a single off/on transition selected from the trials conducted. The initiation of the ESTIMATA off/on-transition is marked by a broken line. Prior to activating ESTIMATA, the concentrate grade consistently exceeded the target, which at the time was 24.5 wt per cent Cu, with noticeable fluctuations. However, upon activating ESTIMATA, the mass pull in the rougher circuit (total velocity) increased without compromising concentrate grade. The optimiser reacted to feed changes over time, resulting in concentrate grade approaching the target with reduced variability.

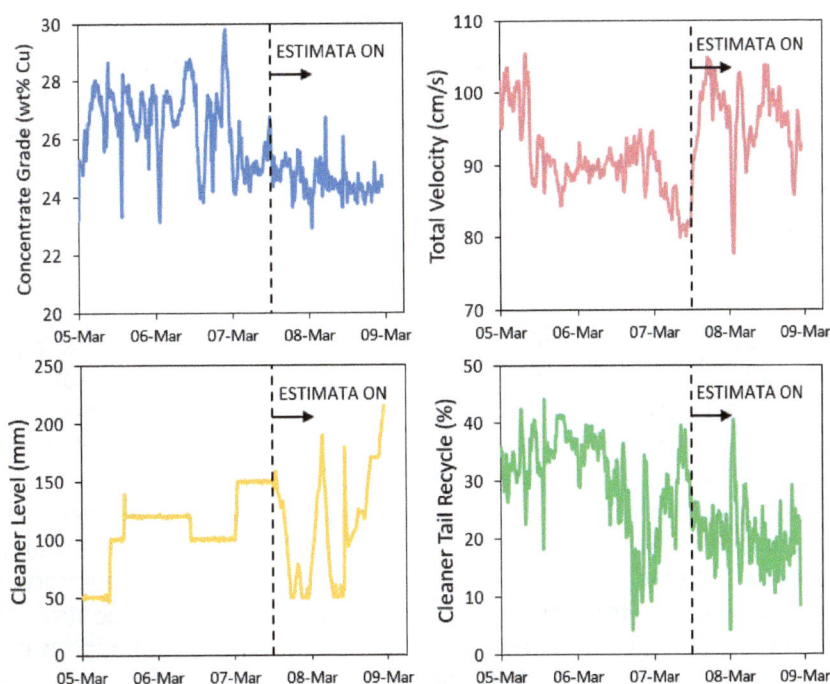

FIG 13 – Concentrate grade, total velocity, cleaner cell FT27 level and cleaner tail recycle over a four day period. The ESTIMATA flotation optimiser was turned on which resulted in concentrate grade closer to target with less variation.

The extensive trials, involving two test periods totalling 15 weeks of on/off transitions, were undertaken to evaluate the efficacy of ESTIMATA. The copper recovery benefit achieved was determined through a rigorous analysis combining simple on/off average differences, transition step response (paired data) and regression analysis (each using different data selections and variables). The analysis was conducted by INNOVATION X in collaboration with NPO and is summarised in Table 3. The individual trial analyses were then combined using a precision weighted average to provide the aggregate improvement estimate of 1.21 per cent copper recovery with a 1-tailed P-value of 0.00067 (99.92 per cent confidence that the increase is not due to random chance).

TABLE 3

Summary of the statistical analysis conducted to assess the copper recovery benefit achieved utilising ESTIMATA. Two trial periods were examined using multiple analysis techniques.

Analysis type	Trial period	Recovery increase (%)	Standard error	t-stat	P-value (1-tailed)	Observations	Variance
On/off	1	1.55	0.803	1.925	0.035	39	25.2
Regression	1	1.17	0.685	1.714	0.048	39	18.3
On/off	2	0.68	0.638	1.057	0.148	49	20.0
Transition	2	1.33	0.549	2.421	0.015	15	4.52
Regression	2	0.96	0.608	1.575	0.060	60	22.2

Initially, there were challenges regarding operator acceptance due to ESTIMATA's tendency to push the circuit beyond conventional operational norms. However, the trial demonstrated the feasibility and benefit of operating under these conditions, serving as a valuable learning experience for the plant operators. Subsequently, ESTIMATA has become the recommended operating strategy for the flotation plant. In fact, ESTIMATA utilisation and availability is now tracked as a production metric so that issues can be identified and corrected. This has placed more focus on OSA management in terms of calibration, sample availability and maintenance. Sample points which were previously redundant have also been reinstated to provide as much information as possible to the optimiser.

The site technical team interacts with the ESTIMATA system to improve the robustness of the optimiser. Experience has shown that it can be beneficial to impose limits on the range to which the optimiser can adjust certain variables. For example, operationally there can be situations where the launders back up and the cameras see a stationary froth at 0 cm/s. To prevent ramp up of the air rate to the absolute maximum, high limits are input for the air additions which are adjusted as needed. Since ESTIMATA operates via the PID control loops, these limits are visible to and easily adjusted by the plant metallurgists.

CONCLUSIONS

The expansion of the Northparkes flotation circuit introduced stability issues in the roughers and scavengers which were found to be due to the control valves and lack of appropriate control philosophies. The issues in the rougher cells were a result of using mechanical feedback in the positioners. Installation of a system that removed the mechanical linkages in the feedback and used a floating joint to absorb some of the torsion on the dart stem provided a robust solution. For the scavengers, the existing control valves were no longer fit-for-purpose when the cells were connected at the end of the larger bank. External dart valves were installed on these cells with a back-pressure-pipe design on the last cell to reduce the pressure drop across the valve. Feedforward control action was also found to be a powerful tool for disturbance rejection. The level control upgrade improved stability and performance which was verified with plant survey and production data. These projects were a key enabler for the implementation of model based optimisation control in the flotation circuit. The optimiser was found to push the circuit beyond what was considered normal operation while maintaining consistency in the concentrate grade to maximise recovery.

ACKNOWLEDGEMENTS

The authors would like to acknowledge the Northparkes metallurgy, fixed plant maintenance, sustaining capital and operations teams for their support during these projects. In particular, we would like to thank Esther Bruce and Matthew Vizard who have since moved on from the operation but made significant contributions. The support of SMC Australia, Manta Controls, MIPAC, Multotec, eDart and INNOVATION X during commissioning activities is also gratefully acknowledged. We would also like to thank Dr Peter Mills and Dr Gerard Duffy from INNOVATION X for their review and input on the ESTIMATA content in this paper.

REFERENCES

Harbort, G, Jones, K, Morgan, D and Sola, C, 2017. Integrating geometallurgy with copper concentrator design and operation, *We are Metallurgists, Not Magicians*, pp 37–53 (The Australasian Institute of Mining and Metallurgy: Melbourne).

Runge, K, McMaster, J, Wortley, M, La Rosa, D and Guyot, O, 2007. A Correlation Between VisioFroth™ Measurements and the Performance of a Flotation Cell, in *Proceedings of the Ninth Mill Operators' Conference*, pp 79–86 (The Australasian Institute of Mining and Metallurgy: Melbourne).

Seaman, D R, Li, K, Lamson, G, Seaman, B A and Adams, M H, 2021. Overcoming rougher residence time limitations in the rougher bank at Red Chris Mine, in *Proceedings of the 15th Australasian Mill Operators Conference*, pp 193–207 (The Australasian Institute of Mining and Metallurgy: Melbourne).

Supomo, A, Yap, E, Zheng, X, Banini, G, Mosher, J and Partanen, A, 2008. PT Freeport Indonesia's mass-pull control strategy for rougher flotation, *Minerals Engineering*, 21:808–816.

Optimisation from mine to mill to mine at Newmont Lihir gold mine

J Moilanen[1], A Remes[2], G Peachey[3] and J Kaartinen[4]

1. Director – Digital Solutions, Metso Finland Oy, Espoo 02230, Finland. Email: jari.moilanen@metso.com
2. Technology advisor – process modelling and simulation, Metso Finland Oy, Espoo 02230, Finland. Email: antti.remes@metso.com
3. Senior specialist – Process Engineering, Newmont Ltd, Lihir Island, New Ireland Province, Papua New Guinea. Email: gareth.peachey@newmont.com
4. Director – automation and concept development, Metso Finland Oy, Espoo 02230, Finland. Email: jani.kaartinen@metso.com

ABSTRACT

Understanding and optimising the material flows in the mine and processing plants are essential for operations to meet their productivity targets. Digital twin technologies create new insights and provides the operations with an opportunity to manage the production chain with variability in the ore feed, metallurgical processes and process equipment, and the business environment. Increased knowledge and situational awareness allow for better planning and control actions within the plant, resulting in better mineral recovery and process optimisation. Recipe matching to variable ore types allows for savings in energy, water, and chemicals per produced ton of product. With the ability to test run any process configuration and operating strategy before execution, the risk of environmental, financial or safety issues are greatly mitigated.

This paper describes a case study how a physics and AI based metallurgical digital twin was developed at Newmont Lihir gold plant in 2023. The case study discusses the key learnings and critical points in the ways of working in a digital twin project. The paper further discusses the site's technical architecture with existing systems for process control and IT infrastructure. Furthermore, the paper examines the future potential of closing the information loop from production back to the mine planning systems.

INTRODUCTION

Metallurgical digital twins are simulation models that encompass either the entire mine-to-metal value chain or specific process stages. The primary objective of these digital twins is to align process operating parameters—such as optimal feed capacity, target grades, and mineral recoveries—with the varying characteristics of different ore types and the availability of process equipment. By fine-tuning daily operating parameters, these digital twins contribute to achieving production goals while promoting sustainability through reduced energy consumption, efficient water usage, and minimised CO_2 emissions.

Typically, the metallurgical digital twin operates in tandem with Advanced Process Control (APC) systems or Process Optimisers specific to each process area as presented in Figure 1. The comprehensive plant-scale metallurgical simulation model anticipates the impact of ore blending and adjustments to APC target set points before they are implemented in actual plant operations. By interfacing with real time plant data, the digital twin reflects the current operational state and provides what-if predictions for plant performance (Remes et al, 2022). The concept of applying first-principle simulation models for online use (Pantelides and Renfro, 2013), and integrating them into a feedback manner to the real system fulfills the definition of digital twins as classified by Kritzinger et al (2018).

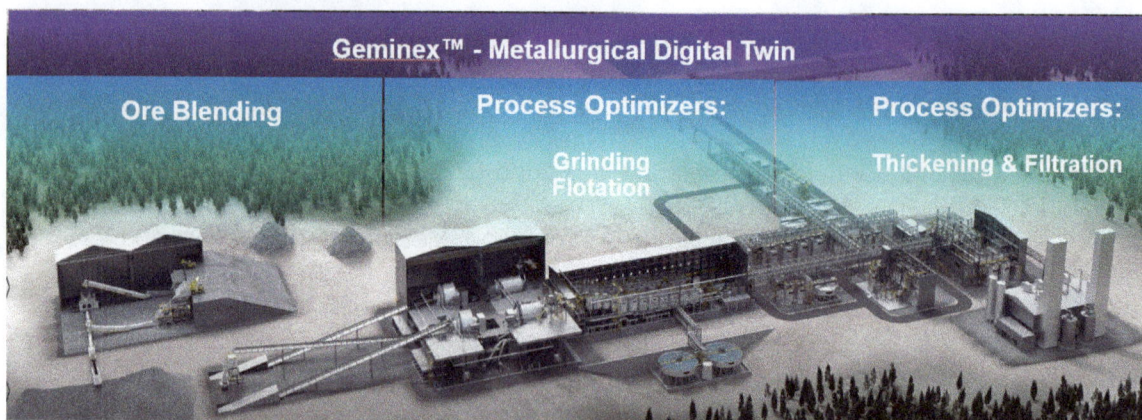

FIG 1 – Typical process optimisers, for which a metallurgical digital twin seeks the set point targets in a mineral concentrator plant.

Automated simulation model adaptation relies on machine learning techniques. This case example describes a metallurgical digital twin that was developed to Newmont Lihir gold plant in Papua New Guinea. This example provides an overview of how a science-based metallurgical digital twin is practically employed in an industrial context within a gold plant.

SIMULATION MODEL FOR METALLURGICAL DIGITAL TWIN

Ensuring the validity and prediction accuracy of a digital twin model involves several critical factors. Firstly, a substantial portion of the model's structure should rely on first-principle equations, with properties empirically calibrated (Lamberg, 2010). These models maintain accurate material balances and material inventories across various operating conditions. Calibration is typically performed for each distinct ore type, resulting in ore-specific metallurgical dynamic equipment modes. These equipment modes are connected according to process flow sheet diagrams, facilitating robust 'on the fly' simulated adjustments to circuit configurations and plant operating strategies.

Secondly, a digital twin simulation model leverages a live-data connection for real time feed information, intermediate process stream assays, whether online or from laboratory measurements, reagent addition rates, and all process control parameters. Additionally, the digital twin models incorporate measured slurry flows, densities and known water chemistry as observed in the plant. However, unknown disturbances will, unfortunately, always exist. This may lead into differences between the simulated process state and the real state over a long period of time.

Calibration of the simulation model to geometallurgical ore types

To address the challenge of time varying modelling accuracy due to changing ore types, the geometallurgical model benefits from defined feed ore types, and modelling their blending. Liipo *et al* (2019) presented a case study involving the characterising of complex copper-gold ores with high correlation to recovery results. In the context of ore blends during comminution there exists multicomponent population balance models. These models provide mineral specific Particle Size Distributions (PSDs) rather than relying solely on bulk mineralogy PSDs as demonstrated by Izart (2021).

Automatic adaptation of the simulation model

Simulation models of a digital twin need continuous monitoring to stay accurate. When discrepancies arise between plant data and simulated outcomes, an automatic adaptation algorithm will dynamically adjust the model parameters. For instance, variations in ore grindability or mineral flotation rate factors within the feed ore trigger real-time adjustments. The data flow schema for an adaptive digital twin is shown in Figure 2.

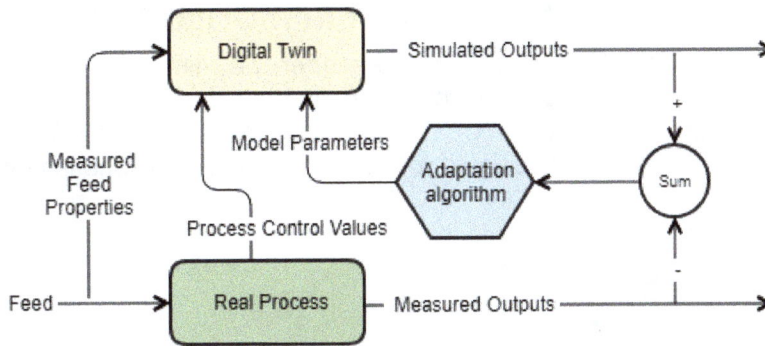

FIG 2 – Adaptation of a digital twin simulation to real process operation.

There are several adaptation algorithms that are available for industrial use as reported by Koistinen *et al* (2020), Friman and Airikka (2012) and Ohenoja *et al* (2023), among others. Common thread across these methods is that the same machine learning approaches can be employed both adapting simulation models during plant operation and calibrating models using offline process history data.

The principle of adapting first-principle semi-empirical models – as applied in case of the metallurgical digital twins – with machine learning algorithms, differs significantly from the training/adaptation process for data-driven black-box models. Typically, data-driven models are trained using extensive process data, without prior knowledge of their underlying cross-effects, physics, or chemistry. In contrast, metallurgical twins focus on specific Key Performance Indicators (KPIs) as their primary adaptation targets. These critical model adaptation parameters are carefully selected by simulation metallurgists. Additionally, the online plant data used for adaptation is purposefully pre-selected to suit the specific process, filtering out unnecessary variables. Table 1 presents typical adaptation target KPIs, required plant measurements and adjusted model adaptation parameters for concentrator plant digital twins.

TABLE 1

Typical selections for concentration plant adaptation targets, measurements, and parameters.

Target KPIs	Plant measurement for machine learning digital twins	Model adaptation parameters
Follow plant material and water balance	Solids and water flow rates, slurry densities, pumping and valve openings	Separation efficiencies, material inventory rates, equipment utilisation rates
Maintain grade and recovery targets	Online or laboratory chemical assays, reagent dosages, pH, air, level	Mineral composition, flotation kinetics per minerals, froth characteristics, entrainment, separation efficiencies
Prevent too high circulating loads and mass pulls	Froth cameras, sump levels, pumping rates	Froth recoveries, froth volume factors, air and water recoveries
Maintain target throughput and particle size	Mill power, mill charge, on-belt rock size, online or laboratory slurry size assays, cyclone pressure, cyclone tomography	Ore work index, grindability, breakage rates, ball loads, mill discharge rates, cyclone efficiencies
Prevent exceeding of dewatering capacity	Bed level, pressure, underflow (UF) density, flocculant dosage, rake torque, overflow (OF) turbidity, filter moisture	Settling rates, solids loadings, bed densities, filtration efficiencies

IMPLEMENTATION OF DIGITAL TWIN AT LIHIR GOLD MINE

Lihir gold mine in Papua New Guinea operates an open pit mine with refractory ore. The Lihir processing plant consists of material handling, comminution and flotation circuits, and pressure oxidation, counter current decantation (CCD) and carbon-in-leach (CIL) circuits with carbon stripping, electrowinning and smelting for gold recovery. A simplified process flow sheet is shown in Figure 3.

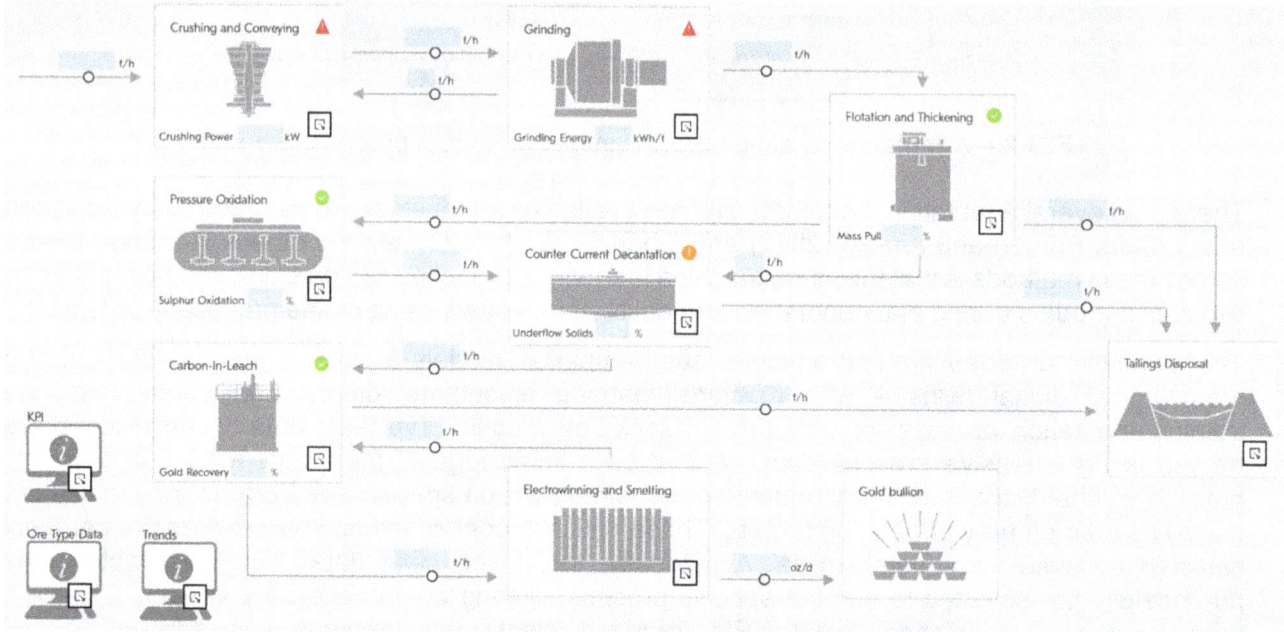

FIG 3 – Simplified flow sheet of Lihir processing plant.

Key aims from an operational and short-term planning perspective included:

- Eliminate the time lag between information received into the SAOC (Site Asset Operations Centre) for operational decisions/changes required:

  o Near real time virtual online analysis of feed stockpile characteristics and changes.

  o Help remove uncertainty around potentially erroneous assay results.

  o Faster identification and troubleshooting of operational issues.

  o Enable optimal operation to the conditions you have, not to the conditions that were planned.

- Detailed dynamic projections into the future of current feed and upcoming changes to assist with short-term operational planning, strategy, process routing and relevant set points.

Also, as part of a digitalisation project aimed at implementing Model Predictive Control (MPC) process optimisers for Advanced Process Control (APC), the project explored opportunities to unlock additional value using digital twin technologies, as illustrated in Figure 1. This looked at using digital twin outputs as VOAs (Virtual Online Analysers) to provide the MPC information about the process where it is impossible or impractical to install instrumentation.

Development of the digital twin in phases

The development of digital twin for the Lihir gold plant was carried out in multiple phases, and the project was managed with a stage gate model. Progress made in each phase was evaluated in stage gate meetings, and the once the results meet the predefined criteria the implementation moved on to the subsequent phase. This structured approach not only facilitated effective communication within the development team but also ensured a well-defined and successful implementation project. Figure 4 describes the main activities and required resources in the development project.

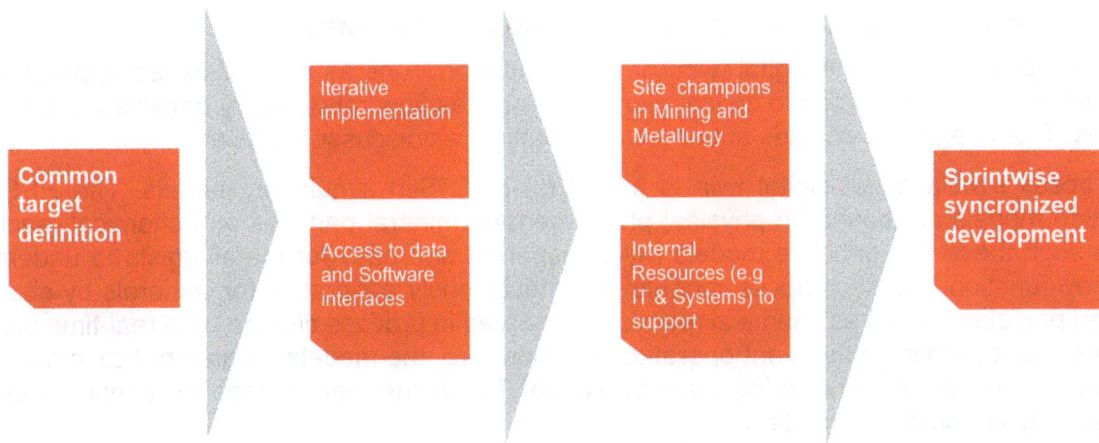

FIG 4 – Main activities and resources in the digital twin development project.

The journey toward creating a metallurgical digital twin began with a high-level vision: optimisation from the mine to mill and back to the mine. The goals included achieving higher throughput, enhancing gold recovery rates, improving operational efficiency, and tailoring ore feed blends based on material value. To realise this vision the focus was on establishing robust systems for monitoring and predicting the plant operations consisting of:

- mining data tracking to know plant ore feed in advance
- detailed plant model for production simulations with current asset set-up
- continuous monitoring of site performance, gold production rate and losses in tails
- feed ore blend for short- and long-term optimisation
- historical and economic performance linked into a dynamic block model for simulation of future performance
- use as operator training tool with existing process control system.

This would enable timely feed control and proactive plant actions to avoid disturbances. The key drivers and expected benefits of the digital twin project are summarised in Table 2.

TABLE 2

The key drivers of the digital twin project.

Operational efficiency	• Streamlining operations at the crusher, barge, diggers, and trucks.
	• Eliminating rehandle requirements and adopting targeted material haulage based on value.
Grade engineering	• Leveraging a dynamic block model to reduce grade control costs and enhance data accessibility.
	• Achieving grade uplift through targeted feed to the mill.
	• Enhancing grades by selectively blending materials based on their value.
Improved feed material rate and value	• Reducing throughput while maintaining the same metal output (resulting in reduced power costs and consumables).
	• Enhancing recovery rates by leveraging foreknowledge of material properties and optimising mill conditions.
	• Identifying material value changes through blending.
	• Ensuring that only high-grade positive-margin material is fed into the crusher.
Informed decision-making	• Gaining complete oversight of material flow and blends.
	• Adopting a proactive and planned approach to plant shutdowns and material adjustments.
	• Mining specific areas based on real-time material property changes tracked by thermal and material sensors.

Digital twin process models with geometallurgical data

The developed metallurgical digital twin for Lihir gold plant runs as a cloud-based application. Its simulation model is connected to critical plant data such as flow rates, slurry densities and reagent dosages. This integration enables automated adaptation, as discussed earlier.

The process simulation of digital twin is based on HSC Sim simulation models. The HSC Sim simulation models are based on physical phenomena of mineral particles and chemical reactions, and as such these first-principle models are transparent and easy for metallurgists to understand how they function. These process models are meticulously calibrated for minerals-by-size with liberated particles as well as chemical reactions observed in process studies. The real-time process simulation runs parallel to the plant operation. Furthermore, the models create predictive scenarios that can be initiated at point during plant operation. The model parameters are continuously and automatically adapted to plant data.

In a comprehensive evaluation, Izart (2021) assessed the accuracy of developed multicomponent population balance models for Lihir ore. The study focused on three distinct alteration types, resulting in a total of nine ore variants. To assess model performance, simulations were compared against a geometallurgical study conducted by Newmont. The relative differences between expected and simulated values varied based on the specific ore type:

- Cyclone Overflow P_{80}: from 0.5 per cent to 1.7 per cent.

- Flotation Mass Pull: from 1.7 per cent to 4.9 per cent.

- Flotation Gold (Au) Recovery: from 1.0 per cent to 3.1 per cent.

- Flotation Sulfur (S) Recovery: from 0.6 per cent to 2.4 per cent.

These findings provide valuable insights into the performance of the population balance models, shedding light on their applicability and potential impact on operational efficiency. The HSC Sim models are based on minerals; to define them Lihir deposit has two main types of mineralisation:

- A refractory potassium feldspar-sulfide mineralisation with sub-micron gold included within the sulfides (refractory). In this type of mineralisation, the average sulfide grade is 6 per cent.

- A low-sulfidation quartz-chlorite-bladed anhydrite association with occasional free gold.

The mine block model, derived from these mineralisation's, distinguishes nine structural domains determined by fault boundaries. It also identifies five alteration zones: Argillic Clay, Advanced Argillic, Leached Soak Domain, Boiling Zone, and Anhydrite Sealed.

The developed HSC Sim model based digital twin for Lihir gold mine can simulate the process with all these nine structural ore domains and their blends. This process simulation can be done both dynamically for short-term simulations and in static mode for long-term production planning. By incorporating these complex geological factors, the digital twin provides a comprehensive and accurate tool for optimising the plant operations.

User interface of digital twin

The user interface of the digital twin was specifically designed to be intuitive for operators and metallurgists. It features flow sheet views for each process area, providing detailed information for process monitoring with soft sensors. Additionally, it includes scenario analysis views, where users can create their own simulation scenarios and compare the results.

These features allow for a comprehensive understanding of the processes and facilitate effective decision-making. Figure 5 provides examples of these process flow sheet views, illustrating the interface's user-friendly design and its capacity for detailed process monitoring and scenario analysis. In these, users can navigate in drill in to several levels of process areas and sub-processes, starting from the whole plant view (Figure 3). This design ensures that users can effectively interact with the digital twin, enhancing operational efficiency and productivity.

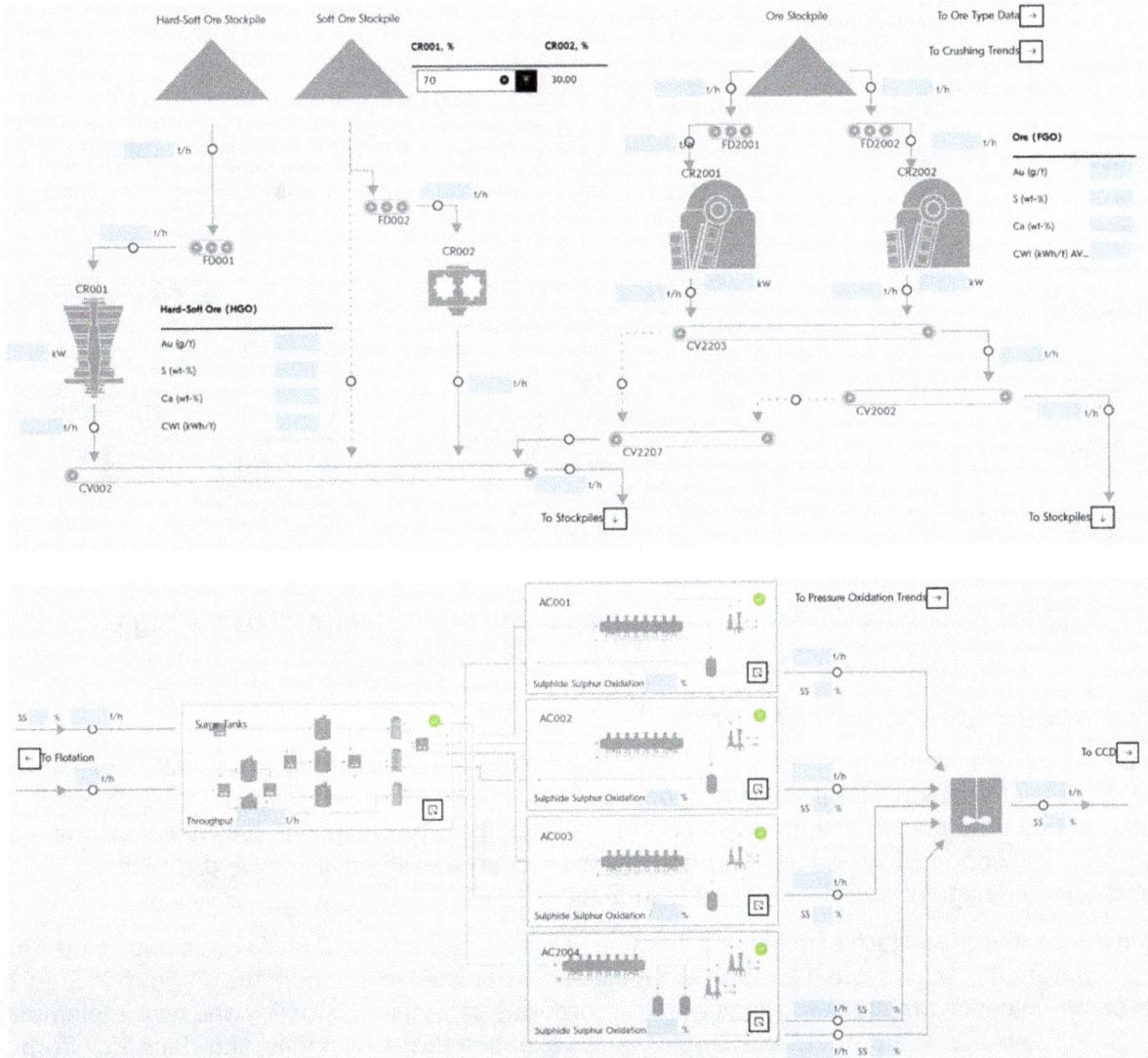

FIG 5 – Flow sheet display examples of process areas. Pictures from crushing and conveying and pressure oxidation (POX).

Simulation scenarios for informed decisions

The digital twin for Lihir gold mine allows the plant metallurgists and operators create their own what-if scenario simulations that cover the entire process. Alternatively, the users can simulate only selected process areas with alternative operating parameters. The results of these simulations are displayed using user-friendly comparison tools for operational decisions. Figure 6 provides an example of simulated forecasts for oxidation degree for autoclaves AC1 to AC4. These simulations consider varying oxygen feed rates, slurry feed rates, and options for bypassing the preceding flotation line. This level of detail and flexibility in the digital twin simulations ensures a comprehensive understanding of potential outcomes, thereby facilitating informed decision-making in the metallurgical process.

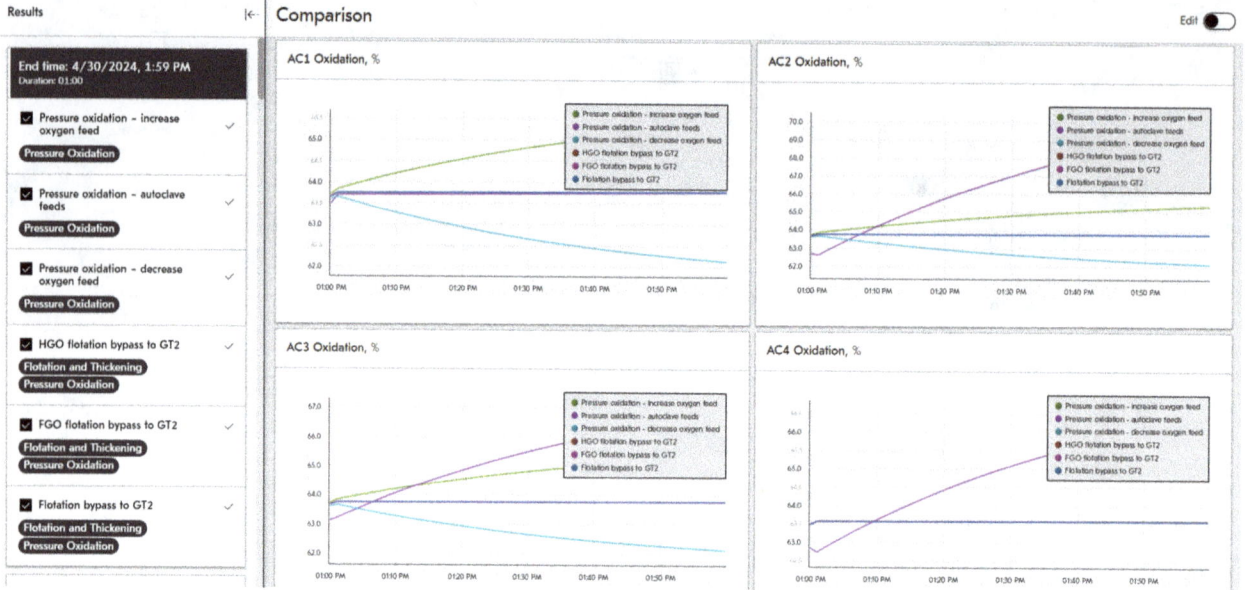

FIG 6 – What-if scenario simulations of varying process parameters to oxidation degree in AC1-AC4.

Cybersecure data connectivity

Digital twins are capable of handling and generating potentially sensitive data, which may encompass information about current operations and plans. This information is confidential and should not be disclosed to unauthorised parties. The optimal approach is to fully integrate the digital twin and its connectivity within the end user's corporate IT policies and practices, where cybersecurity design principles are already in place.

For instance, the digital twin of the Lihir gold mine operates on the Microsoft Azure cloud. It interfaces with data from the edge system through a data lake, as depicted in a schematic diagram in Figure 7. This communication architecture aligns with the corporate cybersecurity policy and was implemented smoothly. However, the process of mapping and validating the data points and data flow from the data lake to the digital twin required significant effort.

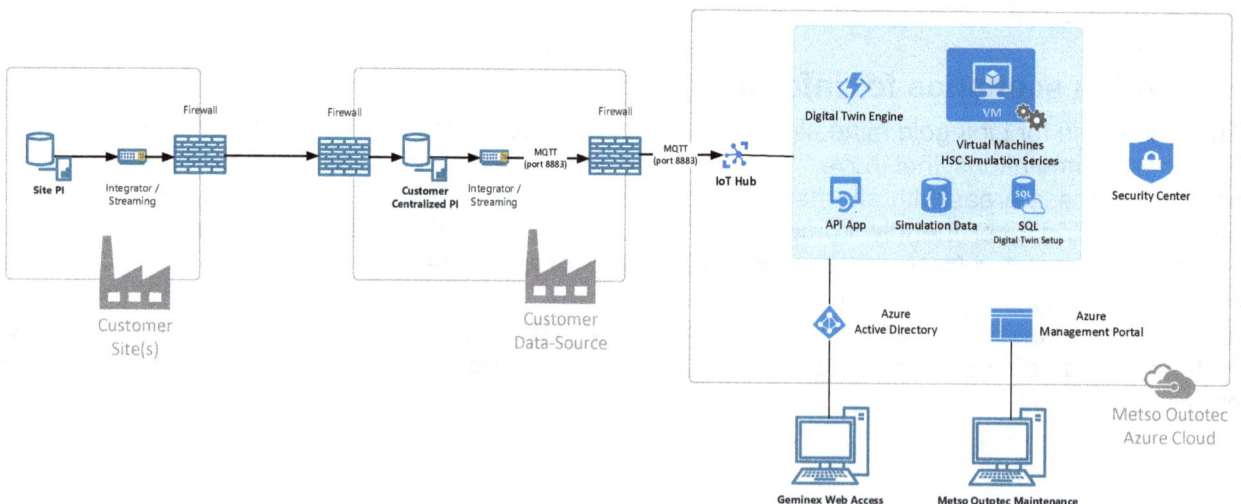

FIG 7 – Connection from edge to cloud in alignment with Newmont corporate cyber security policy.

Access to the digital twin is controlled through Single Sign-On (SSO) authentication. This ensures that only authenticated users can access the digital twin, providing effective management of access to the data it provides. This approach ensures the security and integrity of the data generated by the digital twin.

Current status and the next steps for the digital twin

The technical deployment of the digital twin at the Lihir gold mine has been completed. However, operational commissioning is pending due to recent personnel changes among key individuals at the mine. These key users play a crucial role in ensuring successful adoption of the digital twin and achieving the expected benefits. Consequently, the targets and goals set at the project's outset are still in progress and have not yet been verified.

In short-term the next steps of deployment include tasks to:

- Complete auto integration of crusher feed characteristics to enable use and testing of the stockpile model.

- Ensure all equipment running tags exist in distributed control system (DCS) and are working correctly to ensure the twin accurately reflects the current state of the plant at all times.

- Review the way data is supplied from Newmont to Geminex and to see whether it can be more robust and less time delay.

- Develop and test auto optimising algorithm (whether by automatically running multiple scenarios to determine max gold production over a time period or other method) and recommend plant settings.

Longer term plan includes expanding the system for automated value chain optimisation which leverages implemented communication interface with mine planning and ore tracking systems. This expansion will enable feed material optimisation considering the full production value chain and its constraints. This holistic approach ensures that all aspects of the production process are considered, leading to more efficient and effective operations.

Moreover, the communication of processing performance for known parcels of ore will facilitate the information flow back to the resource models. This feedback loop is crucial for continuous improvement, as it allows for adjustments to be made based on actual performance data. Figure 8 presents a mine model with ore information. This map can include KPIs metrics such as recovery rates and energy efficiency for each type of ore for developing operational efficiency and productivity.

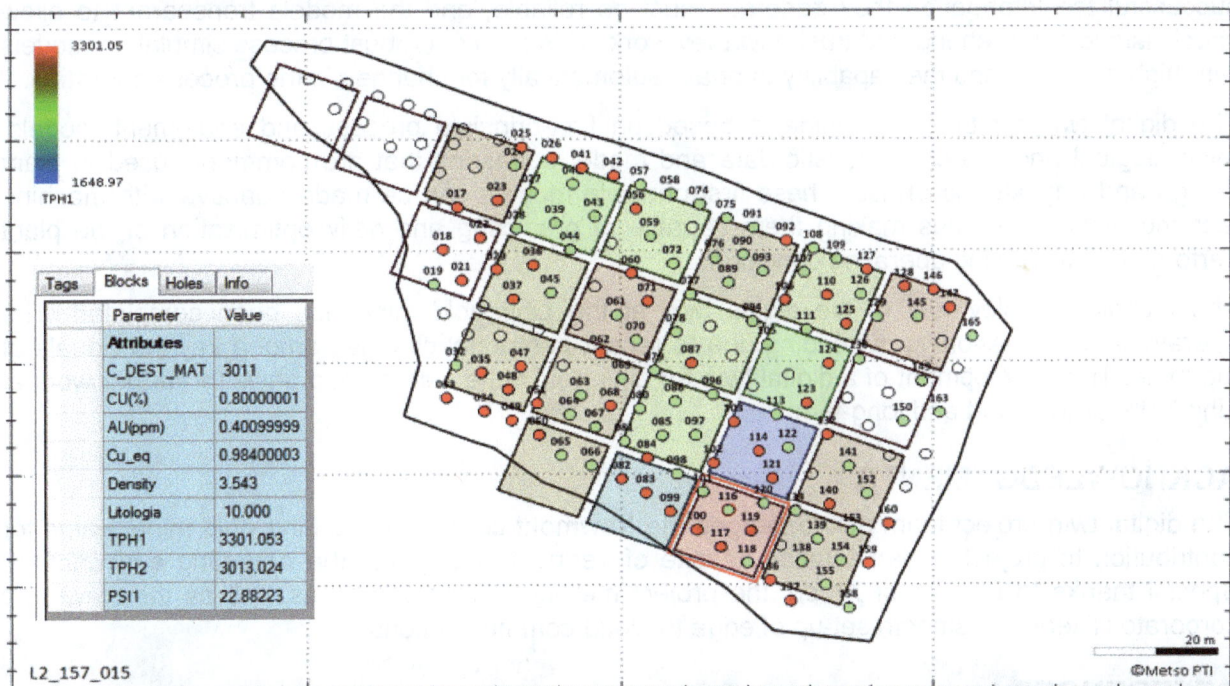

FIG 8 – Data feedback map from mine to mill to mine with block ore data and their processing characteristics, showing metallurgical digital twin model KPIs.

Furthermore, the developed digital twin is not only a tool for operational efficiency but also a powerful resource for sustainability. It is designed to interface seamlessly with process sustainability data, providing real-time information and simulation scenarios. This capability is particularly useful for life cycle assessment (LCA), a method used to evaluate the environmental impacts associated with all the stages of a product's life. The digital twin can simulate various scenarios, allowing for a comprehensive LCA that considers different aspects of the mining process (Talikka, Remes and Horn, 2019).

Key learnings of the development project

Development of the digital twin for Lihir gold mine started as a research project to study how the simulation technologies and process models can be employed to create value in mining operations. A shared vision of the opportunities created commitment and drove the project forward, creating a sound basis for the research project to proceed in subsequent development phases. The global COVID pandemic, high turnover of key people, and shifting priorities slowed down the project. A strong commitment by Lihir, however, backed by a compelling vision carried the project to a fully functional digital twin at Lihir.

One learning from the development project is that it may be beneficial to start the project with areas that make a clear difference, rather than trying to embrace all at once. This will ensure that the real problems are addressed first to engage and excite the users.

An identified risk of using a highly functional digital tool is that operators may quickly lose their skills to manage the process in abnormal fault conditions and getting back to manual control by DCS only may be difficult. The 'challenge when technology is not working' is a real risk that can happen at any time. This could be the case as an example due to issues with instrumentation, communication failures, or external factors such as power outages. Mitigation of this risk could be a continuous training program with simulations for moments when the digital tools are not working. The digital twin can create a simulation environment to practice handling abnormal conditions, thereby maintaining operator skills and readiness for any situation. This approach ensures that operators remain proficient and confident in managing the process, even in the absence of digital tools.

CONCLUSIONS

Metallurgical digital twins are powerful tools for plant monitoring and predictive simulations. To be successful the information they generate must be reliable, and the models transparent to allow specialists to comprehend and trust how they work. This requires robust process simulation models with high accuracy and the capability to adapt automatically for changes in the process operation.

The digital twin for Lihir gold mine is based on first-principle process and equipment models, mineralogical and ore characteristic data and model equations that are commonly used in plant design and engineering phase. These first-principle models can be made adaptive with machine learning techniques, thus making them capable of predicting and daily optimisation of the plant performance in varying operating conditions.

The technical deployment of the digital twin at the Lihir gold mine has been completed. The operational commissioning is pending due to recent personnel changes among key individuals at the mine. The development of a digital twin for Lihir gold mine sets a good basis for further work at Lihir both for the short and long-term.

ACKNOWLEDGEMENTS

The digital twin project team acknowledges the Newmont corporate and Lihir gold mine teams for contribution to project implementation despite of frequent changes in the operating environment. Special thanks to Dr. Samar Amari, the project manager at Newmont, as well as the Newmont corporate IT team for smooth set-up of edge to cloud communications.

REFERENCES

Friman, M and Airikka, P, 2012. Tracking Simulation Based on PI Controllers and Autotuning, in *Proc of the 2nd IFAC Conference on Advances in PID Control*, pp 548–553. https://doi.org/10.3182/20120328-3-IT-3014.00093

Izart, C, 2021. Integrated simulation for improved mineral recovery process, PhD Thesis, Trinity College Dublin, School of Natural Sciences, Discipline of Geology, 151 p.

Koistinen, A, Ohenoja, M, Tomperi, J and Ruusunen, M, 2020. Adaptation framework for an industrial digital twin, in 61st SIMS Conference in Simulation and Modelling, virtual.

Kritzinger, W, Karner, M, Traar, G, Henjes, J and Sihn, W, 2018. Digital Twin in manufacturing: A categorical literature review and classification, in *Proceedings of the 16th IFAC Symposium on Information Control Problems in Manufacturing, INCOM 2018*, pp 1016–1022. https://doi.org/10.1016/j.ifacol.2018.08.474

Lamberg, P, 2010. Structure of a Property Based Simulator for Minerals and Metallurgical Industry, in Proceedings of the 51st Conference on Simulation and Modelling (SIMS), 5 p.

Liipo, J, Hicks, M, Takalo, V-P, Remes, A, Talikka, M, Khizanishvili, S and Natsvlishvili, M, 2019. Geometallurgical characterization of South Georgian complex copper-gold ores, *SAIMM Journal*, 119:333–338.

Ohenoja, M, Koistinen, A, Hultgren, M, Remes, A, Kortelainen, J, Kaartinen, J, Peltoniemi, M and Ruusunen, M, 2023. Continuous adaptation of a digital twin model for a pilot flotation plant, *Minerals Engineering*, 198:108081.

Pantelides, C C and Renfro, J G, 2013. The online use of first-principles models in process operations: Review, current status and future needs, *Comput Chem Eng CPC VIII*, 51:136–148. https://doi.org/10.1016/j.compchemeng.2012.07.008

Remes, A, Hultgren, M, Kortelainen, J and Moilanen, J, 2022. Plant operational strategy with Metallurgical Digital Twin, in *Proceedings of International Mineral Processing Congress Asia Pacific 2022*, pp 1313–1317 (The Australasian Institute of Mining and Metallurgy: Melbourne).

Talikka, M, Remes, A and Horn, S, 2019. Technical, economic and environmental assessment of gold mine operations using advanced simulations, Conference in Minerals Engineering, p 9.

Strategic ore planning – using metallurgical performance data to drive mine planning for operational optimisation: case study 1 (oxides versus sulfides)

B Newcombe[1] and G Newcombe[2]

1. FAusIMM, Principal Geometallurgist, OptiFroth Solutions Pty Ltd, Orange NSW 2800. Email: biancanewcombe@hotmal.com
2. Director, OptiFroth Solutions Pt Ltd, Orange NSW 2800. Email: geoffreynewcombe@gmail.com

ABSTRACT

The objective of strategic ore planning is to ensure that mining operations can be planned with a view to ensuring optimal financial performance of the overall site. This requires an in-depth understanding of the processing implications of the varying ore sources that are to be treated. This integrated approach requires inputs from all site departments and a coordinated effort in interpretation of the data generated.

At many operations, mill feed material comes from a number of different mining zones and can be blended or batch treated through ore processing operations. Different ore types may have unique processing requirements that need individual optimisation both physically and chemically. When blending is necessary the optimal mix of ores must be found to maximise overall processing performance, or an optimised understanding of the processing needs of fixed blends should be determined.

This paper looks at how this optimisation can be practically achieved through integrated use of geological, mineralogical and metallurgical properties in conjunction with mine planning for life-of-mine (LOM) forecasting and strategic flow sheet optimisation. A case study is presented in which soft oxide ores are required to be blended with hard sulfide ores through a single processing route.

Analysis methods used will be discussed and the strategies developed for each site to enable processing of problematic ores will be presented. The implications to mining operations and LOM forecasting will also be discussed.

INTRODUCTION

A well establish gold mining operation undertakes regular and ongoing assessments of future ores from both current and proposed mining operations. Information gathered from these ore assessments (geological and metallurgical) is used for long-term operational planning and forecasting of performance. The ore processing facility receives ore feed material from both an underground mine and several open pits (each of which is at varying stages of its operational life) and is required to treat a mixture of hard rock (sulfide ores) and soft rock (oxide ores) simultaneously. The open pits produce ore that is considered to be a soft oxide (containing clay) at the commencement of mining, which transitions to hard rock with depth, containing both 'free' and sulfide associated Au. The underground mine produces hard rock of various host lithological styles, also with both 'free' and sulfide associated Au. The ore processing team have reported that difficulties are encountered in achieving throughput and Au recovery targets when the amount of soft feed material exceeds, anecdotally, 30 per cent by weight. This negative impact was believed to correlate with the amount of clay contained in the ore.

As the operation has been in production for a number of decades there are well established laboratory standard tests and a large suite of operating plant data from which to baseline assessments of 'new' ores. Extensive work has previously been undertaken with this site to determine the lab-to-plant offset, to allow factors to be applied to laboratory results for predicting meaningful plant recoveries.

An expansion of mining activities has necessitated the assessment of ores (both soft and hard rock types) that are to be produced from three new open pits. The commencement of mining from each pit is staggered, which will limit the amount of soft rock being sent to the processing facility at any one time, as the material from each pit transitions from soft to hard rock with depth. There is likely to be fluctuations in the balance between hard and soft material over time, in accordance with the mine schedule and an understanding of the implications of these fluctuations on processing throughput and Au recovery was required.

A geometallurgical drilling program was undertaken to obtain samples that are both spatially relevant and represent all lithology types within each planned mine shell. To date 80 individual samples have undergone standard testing as part of the geomet program. These samples also offer a range of head grades on which to base a grade-recovery assessment. Each discrete geometallurgical sample is tested through a standardised protocol of test work and in addition to this, various ratios of ores identified as potentially problematic were analysed with a view to providing mine-planning a guideline for problematic ore processing and assist in recovery and throughput loss mitigation, whilst ensuring the mining strategy could be optimised to reduce the need for large stockpiles and the associated costs.

The ore processing facility utilises a typical flow sheet for Au extraction for mixed ore types. Comminution is performed by a semi-autogenous grinding (SAG) – Ball – Cyclone circuit which incorporates flash flotation on cyclone underflow. Conventional flotation methods are employed to recover sulfide associated and free Au. Cyanide leaching is performed on both the flotation tails and flotation concentrate streams to extract gold for the direct production of bullion on-site.

A metallurgical test work program and plant data analysis have been carried out to assess the impact of blending various ore types and these analyses will be discussed and presented separately in subsequent sections. Metallurgical testing has been undertaken as part of a larger ore variability program which incorporates discrete tests for each major process within the processing facility (comminution; flotation; float tail leach and flotation concentrate leach). The ore variability program (geomet) utilised discrete samples from each mine, ensuring representivity of all known lithology types and varying head grades. Individual ore samples are selected in conjunction with the site geology team to ensure good spatial coverage is achieved across potential mining zones, ensuring all potential host rock types are tested, a range of feed grades are tested, and various mineralogical types are covered.

The standard testing protocol for assessing variability between ores used in this program incorporated the following:

- head sample geochemistry
- feed mineral liberation analyser (MLA) on a select group of samples
- Au deportment work on a select group of samples
- kinetic flotation tests for 30-minute duration
- grind sensitivity tests for flotation recovery at 3 P_{80}'s
- concentrate leach kinetic tests at high and low NaCN additions
- concentrate leach fine-grind sensitivity tests at 3 P_{80}'s
- flotation tails leaching at 3 P_{80}'s
- gravity recoverable gold assessment
- comminution evaluation involving SMC, Bond Work index (BWi) and Ai
- cyanide detox testing
- water quality impact testing.

The results of the standard protocol are then used to baseline the expected ore performance when blending of different ore types is undertaken by calculating the expected Au metal units which should be proportionately recovered from each individual ore. Where there is a large difference in the calculated versus actual recovery, an interaction is deemed to have occurred. Results of the standard testing protocol are further used for mill simulation work to determine the likely throughput impacts that the mine plan will have as various ore types are produced. This work can then be used as part of the scheduling optimisation for mine planning purposes.

PROBLEM STATEMENT AND PROJECT OBJECTIVES

Problem statement

Processing of ores from multiple sources is observed to impact both plant Au recovery and throughput rates. This is believed to be associated with the amount and type of clay minerals within the soft, oxide ore types. The mine is embarking on a long-term mine plan and needs to optimise open pit sequencing to maximise operational performance.

Defined objectives

Determine the likely cause of observed processing variability when treating oxide ores through a targeted laboratory test work program utilising ores from the geomet variability test program.

Determine hydrometallurgical implications caused by 'clay' containing ore zones (this includes chemical and physical (rheological) processing issues).

Use predictive mill simulations based on the mine plan to determine future throughput rates and consequent recoveries likely to achieved, using the geomet database as the primary input.

PLANT DATA ANALYSIS SUMMARY

Plant data for a four year period was analysed to assess Au recovery performance of the concentrator under various operating conditions, with the purpose of isolating and assessing a period where oxide material was being feed into the processing facility at varying rates. This would potentially allow an understanding of the impact of this material type on the overall operation of the plant. All ore for this analysis is sourced from Open Pit #1 in the existing operation. The findings generated by this analysis were then used to compare the findings of the laboratory test work program and also to evaluate the likely performance of future ore sources, using mill simulation work in conjunction with the current mine plan (ore schedule).

Data was provided by site from the process historian as either 24 or 12 hourly averages. All data has undergone a high-level filtering process to remove both mill and crusher shutdown periods. Days where mill run hours are less than 20 hrs were also removed to mitigate any data skewing by circuit ramp-up and ramp down periods. Assessment of metallurgically reconciled head grades and recoveries was also undertaken as part of the data filtering process to ensure days where metallurgical accounting anomalies are removed (shifts where assays were missing for example). Of the 1430 days, a run of 399 days was available for assessment after data filtering was complete. This period is reflective of both variable oxide addition rates and a fixed operating mode of the circuit.

Key findings of this assessment were:

- The proportion of oxide in the feed was determined *not* to be an independent variable against both head grade and throughput.

- An increasing amount of oxide in feed simultaneously increased circuit throughput and decreased flotation feed grade.

Regression analysis has shown that for the data provided both tonnes per hour (t/h) and Au head grade influence Au recovery. Table 1 provides the summary statistics for this analysis. Figure 1 presents the plant data with the model using a fixed average throughput rate. For the purposes of the analysis presented here, the plant Au recovery model for the period under consideration is given by Equation 1. Equation 1 is considered to be a reasonable representation of the plant data provided and will be utilised for forecasting purposes within the mine plan in the subsequent discussion (NB: Equation 1 is not used by site for forecasting purposes; the complexity of the models in use by site would justify discussion in a separate paper. Equation 1 has been generated for ease of discussion in this paper).

$$\text{Au Rec \%} = 95.60 - 0.01 * \text{t/h} + 12.99 * Ln\,(\text{Au g/t}) \tag{1}$$

TABLE 1

Summary regression statistics.

	Value		Coefficient	Std error	P-Value
Multiple R	0.717	**Intercept**	95.598	1.759	1.1E-185
R²	0.513	**TpH**	-0.010	0.002	2.77E-10
Adjusted R²	0.511	**Ln (Au g/t)**	12.990	0.692	1.04E-56
Std error	1.866				

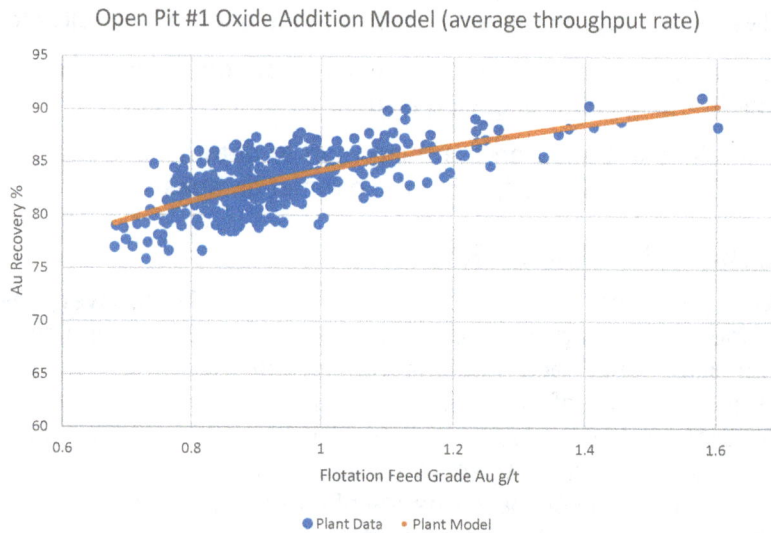

FIG 1 – Plant model with oxide material from Open Pit #1.

LABORATORY PROGRAM

The site geomet program was established during 2019 and is an ongoing program of work that uses a standardised testing method on all ore samples tested. To date 80 individual samples have undergone this standard testing. In order to identify potentially problematic mixtures of ores, an additional laboratory program was run in parallel to the standard geomet test suite. This program used varying mixtures of hard sulfide ores with softer oxide ores to determine whether interactions were occurring between the ore types, and also the maximum level of mixing that could be tolerated before negative interactions became severely detrimental to Au recovery.

Standard geomet test program

The schematic in Figure 2 shows the metallurgical test work process that samples are required to go through to allow meaningful performance models to be developed. This process has been developed and implemented by the author elsewhere for ore evaluation purposes and has provided other operations with world-leading geo-metallurgical models (Newcombe, 2022).

Samples undergo hardness testing (SMC and Bond ball mill work index (BBWi)) and standardised flotation and cyanide leaching on both flotation tails and reground flotation concentrates. Gravity recoverable Au (GRG) assessment is performed on head samples and is used as part of the total ore assessment, rather than the flow sheet assessment. This approach to GRG has been used by the author on a number of projects for many years and has been found to provide more stable results, as opposed to attempting to incorporate the gravity extraction within the laboratory workflow (which would involve slurry rehandling and sampling between stages). As both flotation concentrates and tails undergo cyanide leaching, the final recovery output incorporates all cyanidable Au, and given the duration of leaching the GRG component should be sufficiently extracted.

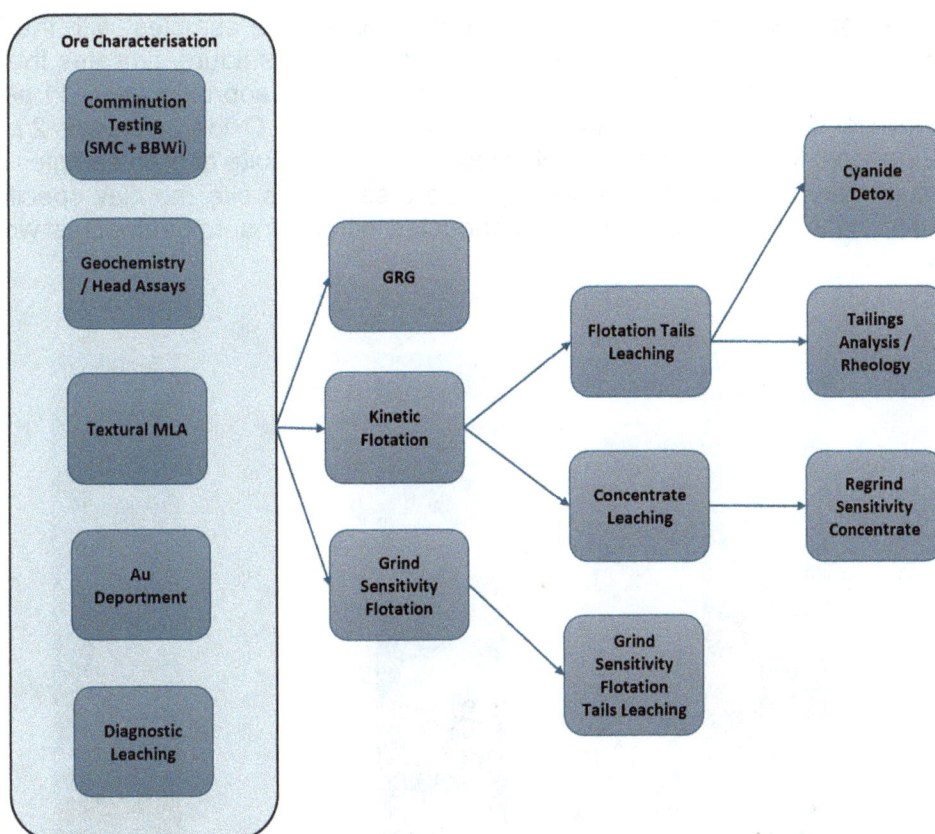

FIG 2 – Laboratory flow chart for geometallurgical modelling.

The detailed results of the testing of individual ore samples utilised for this program will not be presented here due to the volume and complexity of data generated. Discussion will focus on the findings of the blending of specific ores and the likely implications thereof.

Ore interaction test work (blends)

Ores for strategic blending were selected to simulate a likely mill feed blend in future years. Varying levels of hard versus soft rock have been tested to determine the optimal amounts of each ore type and establish if a theoretical maximum amount of a given ore type exists, where further addition beyond this point causes severely problematic processing conditions. To assess the impact of increasing soft ore proportion with future ore types, blends have undergone standardised testing with either 10, 20 or 30 per cent soft oxide ore in the feed mix, with the balance being made up by hard sulfide ore from a well characterised region of the existing mining operation. Current mine schedules have an average of 20 per cent soft oxide ore in the mill feed blend, and this will be considered the base case for comparison. Sulfide ore material is indicated as 'Hard Rock' in tables and represents an unblended mill feed. All ore samples have been sourced from the geomet variability program and we therefore also have known end points (100 per cent) for each ore type. Summary head assay data for the three oxides to be tested and their source pit is presented in Table 2, along with information for the major sulfide ore constituent used in blending.

TABLE 2

Ore sample key and assay summary.

Sample ID	Source	Au g/t	Ag g/t	Cu (ppm)	S %
Oxide 1	Open Pit 3	1.05	0.84	171	0.02
Oxide 2	Open Pit 4	0.90	0.42	496	0.04
Oxide 3	Open Pit 2	0.78	0.39	334	0.10
Hard Rock	Open Pit 1	0.77	1.27	204	0.77

High level mineralogy has also been performed on these oxide ore samples, with the Quantitative X-ray diffraction (QXRD) data for each presented in Figure 3. This figure indicates that only Oxide sample 1 contains significant amount of 'clay minerals', at approximately 11 per cent, and 10 per cent Kaolinite. Oxide 2 contains 14 per cent kaolinite, while Oxide 3 contains 2 per cent 'clay mineral' and 2 per cent kaolinite. Subsequent investigations on a suite of grab samples indicate that smectite is the primary clay mineral present in oxide ores on this site (no clay specific XRD was performed as part of the general investigations conducted in the standard geomet test work scheme).

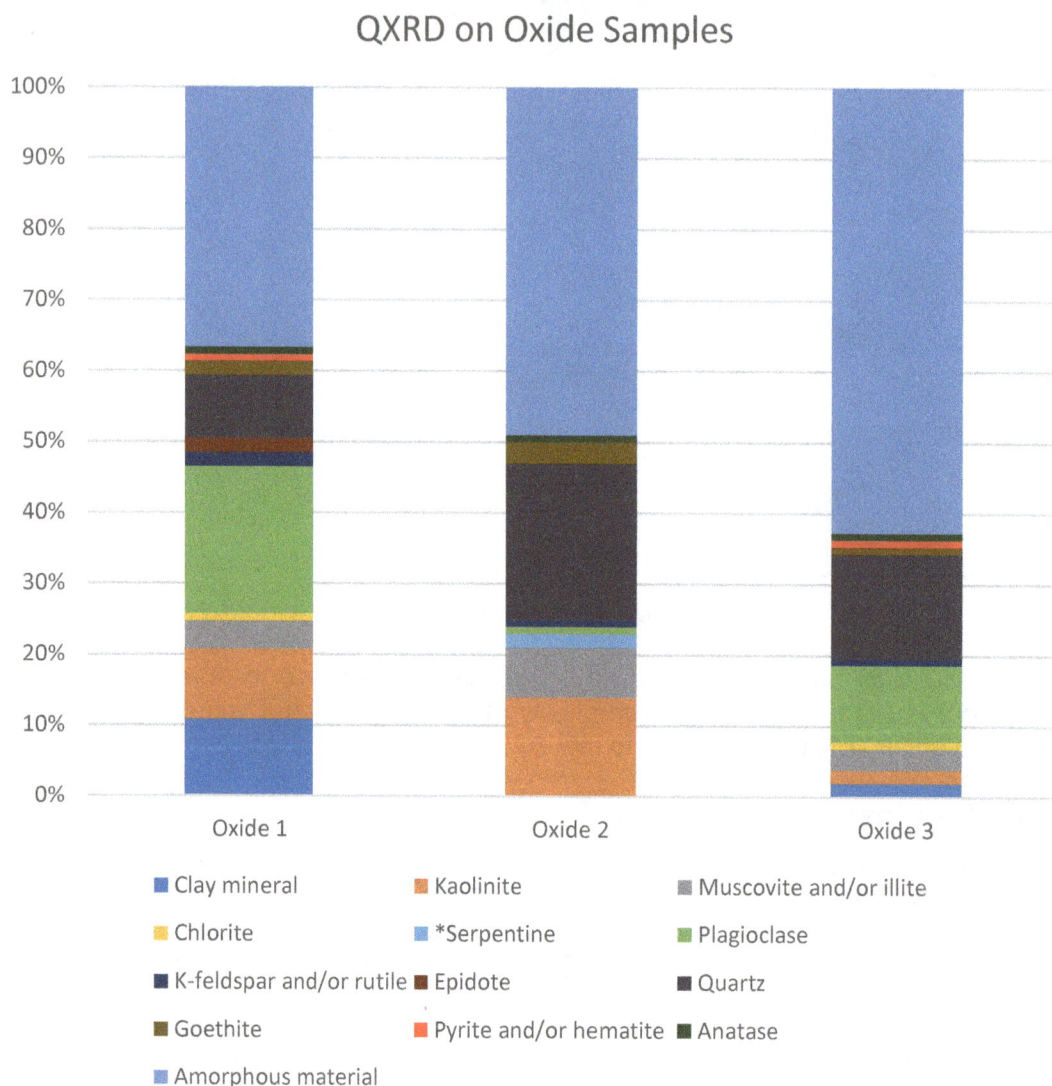

FIG 3 – QXRD for the oxide samples.

The blending program for each soft (oxide) ore is represented by Table 3, each sample has its own test sequences (a total of 27 tests being completed). The 'Base Case' condition represents the scheduled average oxide addition of 20 per cent soft ore, based on the mine plan current at the time of testing. The mill feed P_{80} typically averages approximately 200 µm, hence 200 µm is the base case P_{80} for comparison. As the addition of soft ore causes changes to the product P_{80}, tests have also been conducted at 150 and 250 µm to assess the impact of varying feed P_{80} on overall Au recovery. Testing of each blend within the matrix is done as per the geomet standard sequence, as previously shown in Figure 2, to allow a comparison against the zero and 100 per cent oxide ore in feed conditions. Base case tests are done in triplicate, in line with all standard geomet testing (triplicates), with the other tests in the matrix being done in duplicate due to cost limitations.

TABLE 3

Blending program for each oxide ore tested.

Oxide proportion	10%	20%	30%
P_{80} = 150 μm	☑	☑	☑
P_{80} = 200 μm	☑	Base case	☑
P_{80} = 250 μm	☑	☑	☑

Discussion

Each individual ore sample has been put through the standardised test work series, which allows each individual ores Au performance to be assessed prior to any blending taking place. This testing incorporates gravity Au recovery, flotation performance and leaching performance of both the flotation concentrate and tails streams (separately). Whilst a myriad of other tests are performed, these four tests are representative of the major processing units for Au recovery on-site. Summary data for each ores response to these four standard tests is presented in Table 4. The performance of the hard rock sample is typical for ores of this type. Oxide 3 showed the worst performance, Oxide 1 was the best performing oxide overall, while Oxide 2 had better flotation performance, but poorer tails leaching performance when compared to the other oxides, hinting at a difference in Au deportment. Oxide 1 is noted to have a higher Au and Ag head grade comparatively, which may have assisted in its better overall performance.

TABLE 4

Individual ore performance comparison au extraction/recoveries.

	Gravity rec %	Flotation rec %	Con leach %	Float tail leach %	Combined rec estimate %
Oxide 1	20.2	57.3	99.1	67.8	83.1
Oxide 2	15.1	71.3	99.1	34.8	78.3
Oxide 3	12.4	44.8	95.4	59.7	73.9
Hard Rock	28.6	91.4	80.3	55.4	81.7

For each mix of ores tested, a theoretical (predicted) Au recovery is calculated based on the individual performances of the input ores. This gives a mathematical estimation of what the recovery performance should be if no negative (or positive) interactions are occurring when the ores are mixed together. These estimations are based on input metal units (not mass proportion alone) to accommodate variances in head grade. Where the predicted and observed recoveries are within ±2 per cent of each other, ores are assumed to have no interactions (within experimental error). Where the deviation is greater than ±2 per cent a positive or negative interaction may be detected. If a trend of either negative or positive recovery impact is shown for a certain ore blend, then it is considered highly likely that this trend will present in the ore processing facility. Figure 4 provides a summary of the recovery impact observed for each feed P_{80} tested, for each of the three different oxide ores tested. This data indicates that not all oxides are impacting performance to the same extent, and there is potentially preferential grinding occurring in the laboratory mill, causing higher losses at finer P_{80}'s under the standard test conditions used. An added complexity to this analysis is that the varying proportions of the input ores are changing the head grade into each test. Variations in head grade are presented in Table 5.

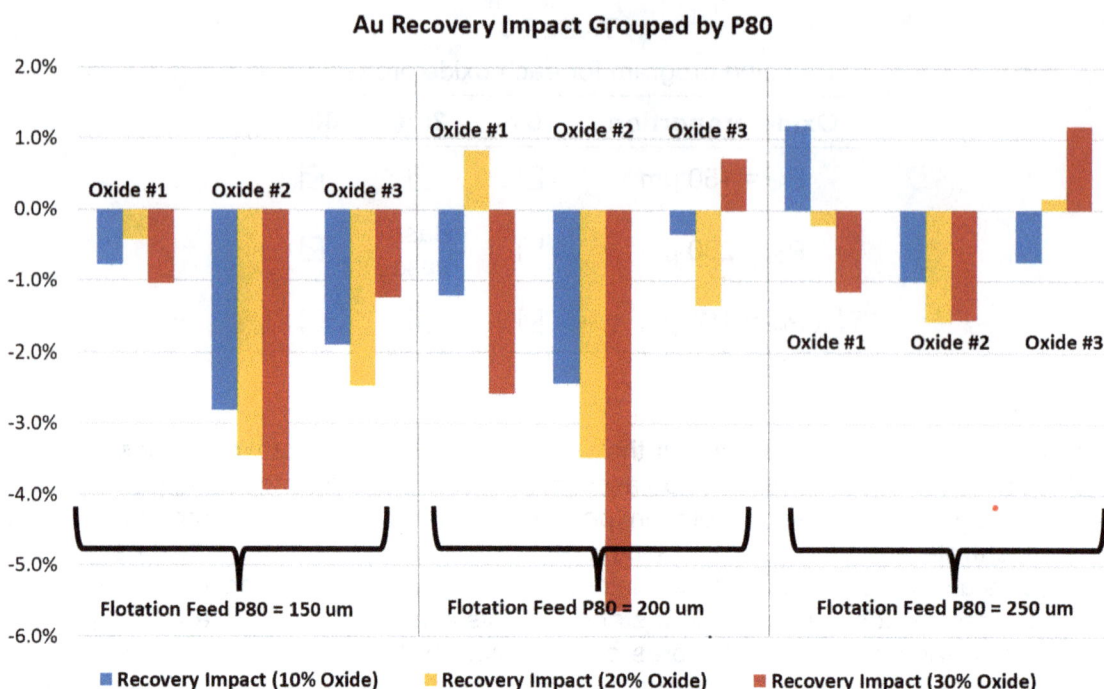

FIG 4 – Summary Au recovery impact.

TABLE 5

Implication of changing head grade with blending.

Blend	% Oxide	Assay Au g/t
Oxide 1	10	1.21
Oxide 1	20	1.22
Oxide 1	30	1.08
Oxide 2	10	0.93
Oxide 2	20	1.01
Oxide 2	30	1.16
Oxide 3	10	1.08
Oxide 3	20	0.93
Oxide 3	30	1.04

The head grade of Au is known to influence the recovery achieved by a sample; this has been well established from comprehensive laboratory programs as well as analysis of plant operating data. The majority of tests indicate a lower than predicted recovery of Au when the ores are blended, however Oxide 2 shows a sustained negative impact under all conditions tested. This impact is highlighted in Figure 5, which shows the generalised impact of increasing P_{80} grind size and proportion oxide in feed on the Au recovery response. Further investigations were subsequently conducted to determine if any ore specific characteristics could be used to identify why Oxide 2 is notably different and by extension, which ore zones within the mine-space may present higher processing risks.

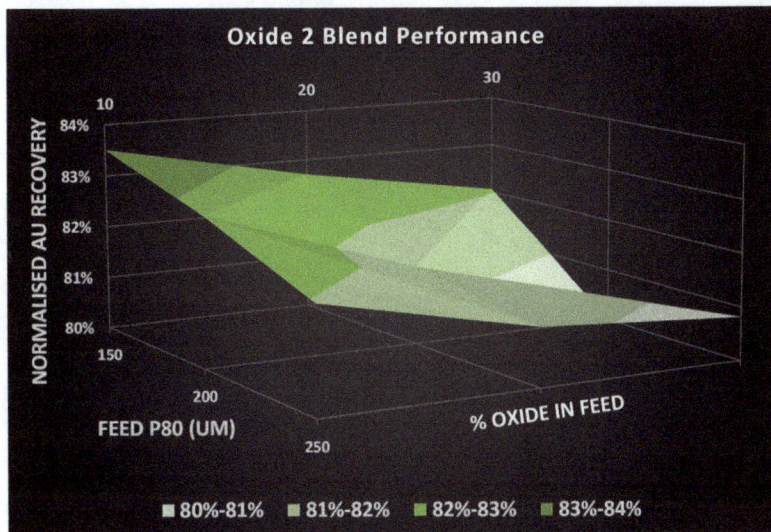

FIG 5 – Summary Au performance with varying feed P_{80} and oxide proportion.

Ore investigations

From evaluation of data from all test work sources, two factors were identified as having the largest potential impact on ore processing performance: rheology and Au mineralogy.

Rheology

The potential treatment of new soft oxide material has generated concerns over slurry rheology and whether pumping and dewatering systems currently installed would be sufficient to handle the planed load. A further concern was whether this new material would impact mixing in the Au extraction circuits and cause a decrease in recovery.

Individual ore testing highlighted the significantly higher slurry viscosities of the oxide ores, and dewatering tests at the laboratory scale were noted to be significantly impaired for these particular ore types. As these oxides will not be treated through the processing facility in isolation the blends (10, 20 and 30 per cent) for the 200 μm feed P_{80} underwent viscosity and settling tests (using site water and the incumbent flocculent). Figures 6 and 7 present the data generated through blending ores and indicate that with increasing proportions of oxide material, both slurry viscosity and settling rates are impacted. It was determined that at very high proportions of certain oxides, mixing may become impeded, and an alternative flocculant may be required in the thickeners on-site to achieved suitable overflow clarities, particularly for Oxide 2. Testing of both dispersants and alternative flocculants was recommended to site for consideration in a future test work program to quantify and potential improvements.

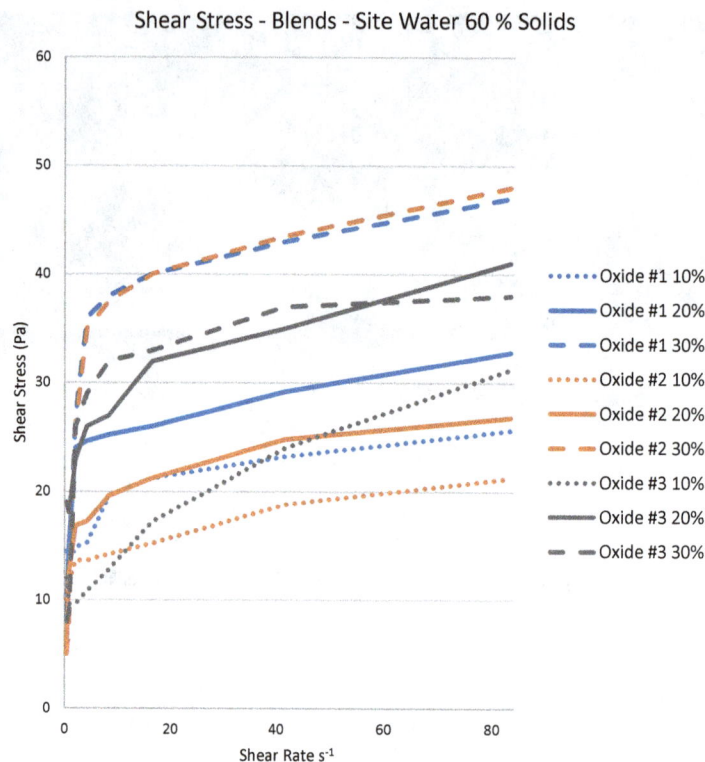

Shear Stress - Blends - Site Water 60 % Solids

FIG 6 – Viscosity impact of oxide blending.

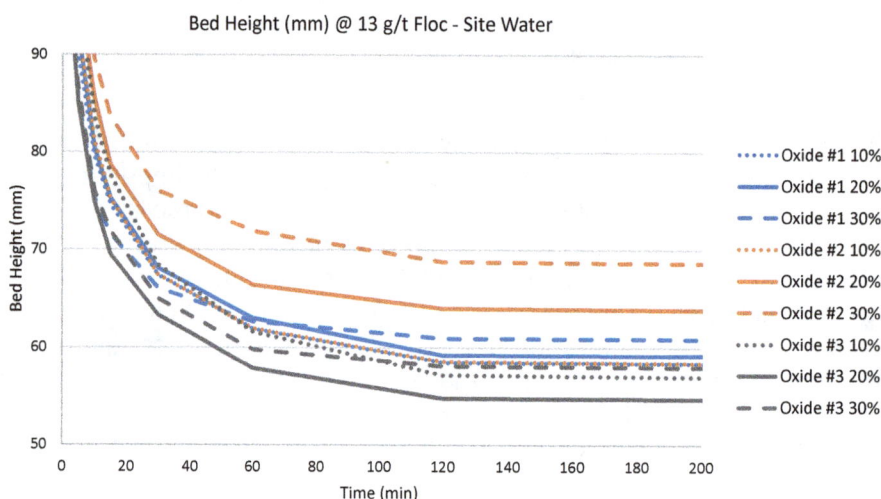

Bed Height (mm) @ 13 g/t Floc - Site Water

FIG 7 – Settling tests of oxide blends.

Au mineralogy

Au deportment data and ore feed characteristics have been investigated to understand how future oxides might vary from those currently being treated. Attempts were made to develop recovery algorithms for the oxide ores as a group, however no head-grade based model could be developed. This is contrary to the hard rock types treated on-site, which have well established and defined behaviours and robust predicative Au recovery models. Consideration of clay speciation, mineralogy and rheological responses could not provide insight into the inability to define Au performance characteristics. Spatial grouping of samples provided the means to allow further interpretation to be achieved. Whilst there are many nuances to the Au mineralogy of the ores from each of the mines, broadly speaking, two Au deportment categories were identified for the soft ores:

1. **Categorisation based on Au:Cu** (gold to copper ratio) in feed.

2. **Encapsulated and Refractory Au** is present in larger amounts in spatially discrete regions.

Oxide ores were able to be classified in terms of the Au:Cu ratio in feed, as shown in Figure 8. Two groups were identified, a low Au:Cu group in which recovery was observed to vary independently of the ratio, and a 'high' Au:Cu group, in which the recovery increased with increasing Au:Cu in feed. It was noted that all oxide samples taken from open pit 4 were all in the low Au:Cu group (Sample Oxide 2 was from this area), whilst the oxides from the other mining areas contained ores in both groups (high and low Au:Cu).

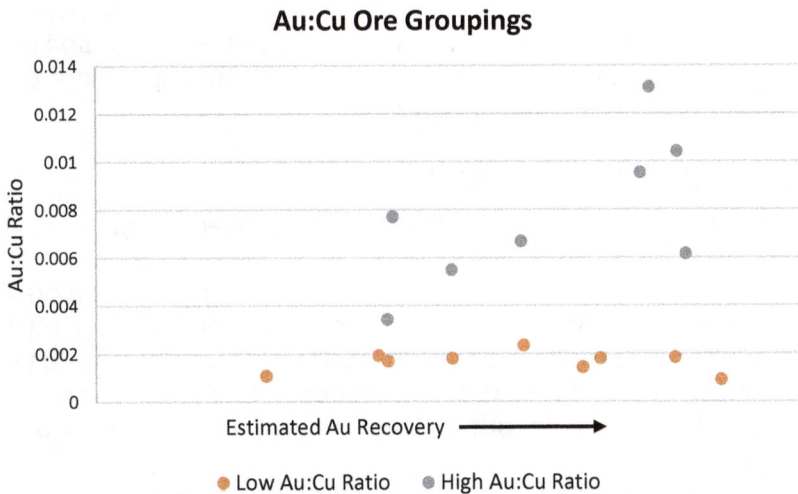

FIG 8 – Grouping Au with respect to Cu in feed.

The amount of Au present as refractory and/or encapsulated at the target feed P_{80} of 200 µm also has a direct influence on the overall recovery. In this case, refractory Au is <7 µm in size, primarily in pyrite although rarely Au can also present in Cu sulfides and arsenopyrite. Encapsulated Au is Au that is enclosed by host rock minerals (silicates).

Figure 9 shows a steady decline in recoveries with increasing amounts of refractory and encapsulated Au. Analysis of this data by spatial location indicated that the oxides with the highest amounts of refractory and encapsulated Au were predominantly from Open Pit 4 (four of the five samples over 15 per cent in Figure 9 were from this area, indicated in pink). This region is also where sample Oxide 2 is sourced from in the blending program.

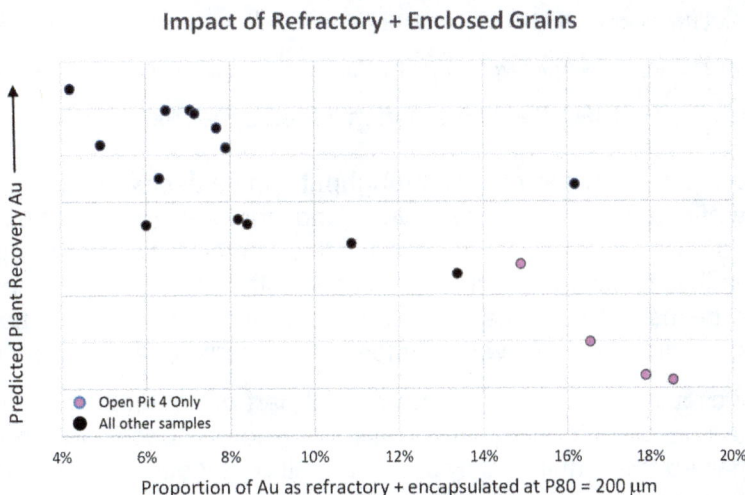

FIG 9 – Impact of refractory + encapsulated Au.

Dedicated Au deportment work has since shown that oxide samples from Open Pit 4 contain some of the highest levels of encapsulated or refractory Au (15 to 26 per cent). Oxides from open pits 2 and 3 also contain refractory and encapsulated Au, however there is a continuum of Au performance results observed from these pits, which are in close spatial proximity to each other at the northern end of the current open pit 1. Open pit 4 is to the south of the current pit, and for recovery forecasting purposes will be treated as a separate entity to the other pits, in alignment with discussions held with site geology, indicating that material changes are occurring within the ore system in the southerly direction.

The strong correlation between the amount of encapsulated and refractory Au and the overall recovery likely to be achieved has resulted in open pit 4 incurring an average 3 per cent downgrade in predicted Au recovery. No recovery downgrades due to Au mineralogy need to be applied to other mining areas. Without the ability to spatially isolate the Au characteristics and consequent performance of open pit 4, the site was anticipating a downgrade across all soft oxide ore types. The work here has shown that this was not necessary and limited the impact to a certain region which can be spatially isolated within mine planning systems.

Utilisation of data within the mine plan

The proposed mining schedule for a 12-year period is summarised in Figure 10. This schedule has an initial addition rate of 4 per cent oxide from current mining operations, the total amount of oxide from all sources was capped at 25 per cent in this scenario. The problematic ore identified from the geomet and ore interaction test work is the oxide material coming from Pit #4, which will be presenting to the mill continually over the period under consideration. In order to understand whether planned mining rates need to be changed or pit sequencing altered to mitigate potential issues arising from the addition of this particular ore, further investigation was required.

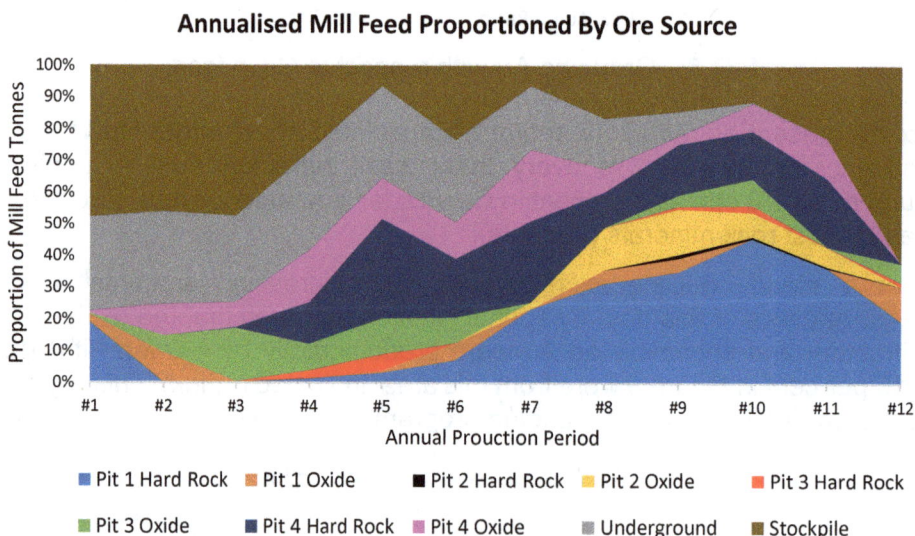

FIG 10 – Mine plan mill feed schedule.

The impact of the varying ore sources on mill throughput can be assessed using mill simulations and the geomet database of results for each ore source (and lithology type within each ore source). The output of this simulation work can identify variation in flotation circuit P_{80} if throughput is maintained, or if throughput is unable to be maintained (power constrained), what the impacts will be. This information can then be used for forecasting purposes, with the P_{80} prediction being utilised for recovery estimations, as circuit Au recovery is impacted by variations in flotation feed P_{80}.

The utilisation of mill circuit simulations forms a critical part of developing strategic mine plans for optimisation of processing circuits, however for this stage to be successful, the site needs to have developed and maintained the required databases of mill parameters and hardness characteristics of the ores being processed over time. The site under consideration here has sufficient data inventory to allow this to occur, and it is recommended that this be maintained through an ongoing program of routine belt cut sampling and analysis.

Figure 11 indicates that when throughput is maintained and the mine schedule of Figure 10 is utilised, the average P_{80} will vary between a low of 165 μm and a high of 201 μm. This information can now be used for forecasting LOM recoveries, by providing the P_{80} input into the recovery algorithms where required. Figure 12 uses Equation 1, throughput rate, feed grade and recovery adjustment factors such as the -3 per cent applied to open pit 4 to provide a basic example of the output of forecast ounces and recovery. Figure 12 indicates when higher recovery can be expected, versus the gradual decline toward mine closure. Mine planning has used the data generated by metallurgy to determine the overall impacts of changing pit sequencing, grade-shell boundaries and the treatment of low-grade stockpiled material to optimise the overall output for the life of an operation. When information from all available sources is tied together, a powerful tool is generated for companies to more accurately predict and optimise cash flow variations.

Annual Mill Simulation - Summary Data

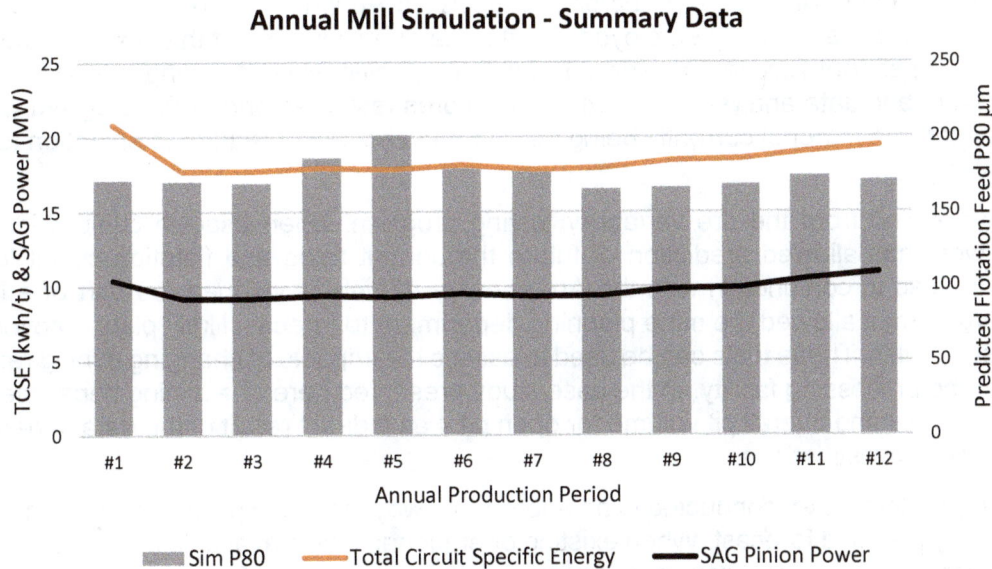

FIG 11 – Mill simulation based on mine plan.

Au Production Forecast

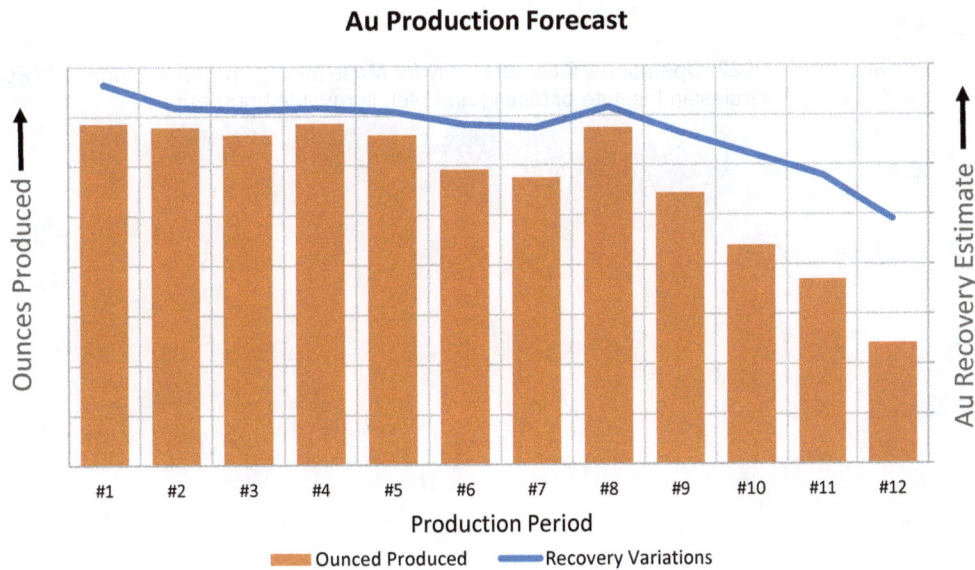

FIG 12 – Production forecasting.

CONCLUSIONS

A key finding of the work performed as part of this study is that not all clays are equally problematic, and any recovery downgrade factors should not be universally applied, as the processing issues encountered here could be limited to specific mining areas. These insights demonstrate the value a systematic and wholistic geometallurgy program can add with key mine schedule drivers included in whole of mine optimisation solutions.

Testing of blends with increasing proportions of soft, oxide material showed a generalised trend of increasing slurry viscosity. Material from open pit 4 (Oxide 2) was noted to have the most significant negative impact on slurry viscosity, and settling rates under the conditions tested. Although sample Oxide 2 has less 'clay mineral' than Oxide 1, this sample presented with the worst rheology impacts, which may have contributed to the lower recoveries observed as a function of mixing issues. Further mineralogical investigation of this sample is recommended, including a clay specific XRD study to identify the minerals that may be causing this rheological effect.

The ore samples tested from open pit 4 were also observed to have lower Au recoveries than ore from other mining areas. This difference is a function of both the Au mineralogy and properties of the host rock material which may be causing issues with slurry handing. Au from open pit 4 is noted to have larger amounts of refractory and/or encapsulated Au which will be difficult to recovery via the conventional means currently employed on-site, the overall impact of this was determined to be on average a 3 per cent recovery downgrade for that specific area of mining operations (pending further test work and data analysis). The other future ores tested as part of the program were found to be consistent with material currently being treated and should not require intervention to maintain recovery rates.

The data generated from the ore variability testing program, when used in conjunction with mill simulation work has allowed prediction of future throughput rates and flotation circuit feed P_{80}'s. When this is used in conjunction with the Au recovery models developed (as part of a larger ore variability program) it allowed the mine planning department to forecast likely plant recoveries using their mine schedules. These tools can be used to assess the impacts of changing mining rates and/or locations on the processing facility. In the case study presented here, the mining department altered both their schedule and overall pit volume for open pit 4 as a direct result of the data generated from the metallurgical investigations.

Ore variability testing, when conducted in an appropriate way, can be a powerful tool for an operation to appropriately plan and forecast. When existing operational data is available and of suitable quality, the integration of data sets can provide an immeasurable benefit to an operations overall performance.

REFERENCES

Newcombe, B and Newcombe G, 2022. Operational Geometallurgy for Mass Mining, in *Proceedings of IMPC Asia Pacific 2022*, pp 422–436 (The Australasian Institute of Mining and Metallurgy: Melbourne).

Metallurgical test work – the beginning of the end or the end of the beginning?

A J H Newell[1], P D Munro[2] and K Fiedler[3]

1. FAusIMM(CP), Manager, Metallurgy and Process Engineering, RPMGlobal, Brisbane Qld 4000.
   Email: anewell@rpmglobal.com
2. Principal Consultant, Mineralis Consultants Pty Ltd, Taringa Qld 4068.
   Email: pmunro@mineralis.com.au
3. Director, Mineral Processing Consultants Pty Ltd, Flaxton Qld 4560.
   Email: kelvinfiedler@mineralprocessing.com.au

ABSTRACT

This is the last paper in a series of three papers reviewing metallurgical test work practice. The first paper focused on the 'before phase' while the second paper dealt with the 'during phase'. This paper provides observations and guidelines about 'closing out' a test work program and how the unexpected issues encountered during the program can be satisfactorily dealt with.

The primary deliverable is the test work report prepared by the test work facility. Ensuring that the report has met the requested requirements is essential. At the onset of discussions relating to the selection of a test facility, a clear understanding should be established as to the content provided by the laboratory in relation to reporting test work, and importantly whether the report author will include any interpretation and commentary of the results presented. Report feedback from the test facility can be extremely useful in placing the results in the context of other test work programs of similar nature that have been conducted by the facility. An important evaluation of the test work program is whether the test work and the associated report is 'fit for purpose' and has successfully navigated the clash of the drivers (budget, schedule, quality and outcomes).

The basis for the interpretation of results and how they may be used is presented, including design criteria, scale-up, flow sheet development, materials handling and equipment selection and sizing.

Managing potential post-test work 'blues' are discussed including the likely increasing requirements for the resolution of tailings and environmental issues, satisfying specialist consultants, equipment vendor testing requirements, marketing samples as well as the testing of production composites to satisfy project financiers and potential purchasers of the final product.

INTRODUCTION

Good quality test work and rigorous objective interpretation of the data obtained underpins the success of the mining industry.

This was succinctly captured in the quotation 'Make your mistakes on the small scale (ie test work) and your profits on the large scale' coined by Professor Arthur Taggart over a century ago.

When poor test work program outcomes occur, often due to a number of causes, projects typically fail causing substantial capital losses as well as damaged corporate and personal reputations. Poor test work outcomes result in inappropriate flow sheets, unsuitable process engineering designs and unsuitable equipment selection and sizing.

Surprisingly, poor test work outcomes remain a problem for many projects. This paper, in conjunction two previous papers in the series (Newell, Munro and Fiedler, 2018, 2021) attempt to provide practical guidance to what should be a relatively straightforward and scientifically driven exercise.

CONFIRMATION OF PURPOSE

It is important to confirm that the aims of the test work have been achieved and, whether the test work program was undertaken to support a study or solve a technical problem in an existing plant, there are common features of both objectives:

- Understanding the metallurgical behaviour and characteristics of the ore types and proposed mining blends to be treated.

- Identifying the required operating requirements to achieve the proposed metallurgical outcomes.
- Demonstrating a robust and pragmatic flow sheet.
- Generating design engineering data (design criteria, mass, water and energy balances) to support the selection and sizing of equipment.

It is crucial to ensure that satisfactory records are developed and stored for both current studies and any future undertakings. This includes not only the test work report(s) from the various test work facilities, but also internal memos and discussions (eg the establishment of a 'Decisions Register') as critical decisions as to the direction of the test work are sometimes made during the test work program.

All too often, test work reports, data sheets, consultants' comments etc, are stored in a variety of places. Consolidation and safe storage of documentation is essential.

To this end, it is recommended that an audit of the test work program and its' findings are commenced prior to the completion of the test work program.

Such an exercise will greatly assist in wrapping up the test work program, ensuring that all the goals have been either achieved or identified risks satisfactorily mitigated, and in managing the post-test work 'blues' and any associated politics.

The outcome of the audit may require that more test work is needed. However, it is better to recognise this and have it performed before the test work program is closed out, and most desirably, before all the available sample is consumed.

Sample availability, particularly if additional drilling is required, can be an outcome determining step.

AUDIT

In addition to the routine evaluation of results as they are distributed by the test work facility, the inclusion of an audit needs to be wide-ranging, thorough and interrogative to ensure that the objectives are clear and the correct tests and analytical methods are being adhered to. The audit process provides a timely assessment of 'Are we on track?'.

An audit should consider the following topics:

- Design of test work program:
  - Have sufficient characterisation tests been conducted on the ore types?
    - Liberation requirements.
    - Comminution properties:
      - Were a sufficient number of samples used to develop a satisfactory understanding of how ore hardness varies spatially across the deposit(s)?
      - Has the correlation of metallurgical test work been undertaken against geological data such as weathering profiles and *in situ* density or geotechnical data such a Rock Quality Designation (RQD) and Point Load Index (PLI)?
    - Mineralogy:
      - Feed size fraction by size fraction assays.
      - Was mineralogy conducted on the concentrate?
      - Was mineralogy conducted on the tailings – nature of metal losses and input into tailings impoundment design?
    - Other appropriate characteristics, such as oxygen demand, rheology, flocculant screening etc.
  - Has a sufficient range of samples been tested for gravity recovery, principally for gold ores?
    - Have feed grade-recovery and mineralogical-recovery relationships been established?
  - Was site water used, or its' potential effect recognised, and was recycled laboratory flotation water used, especially in the final stages of testing eg Locked Cycle Tests (LCTs)?

- o Based on LCTs, has feed-grade recovery relationships been established for each ore type?
  - Is the final concentrate grade-recovery relationship known for key ore types?
- o For flotation test work, were sufficient LCTs performed?
- o Were the tailings dewatering, geotechnical and geochemical properties for each ore types established?
- Ore types and proposed mining blends:
  - o Have all the ore types and key proposed mining blends been tested?
    - Ore types – check with the geologists: further drilling may have revealed new ore types that require testing:
      - Conversely, test work may reveal a new ore type (eg high talc) that needs to be included in the resource model, which flows into the mine schedule and the mining blends for the plant.
    - Key proposed mining blends – check with the mining engineers: the mine schedule may have changed (it is usually the last item to be decided on in any study – is it ever frozen?) and the nature of the proposed mining blends may be different.
    - In both cases, more test work may be required. It is always good practice to aim for more sample quantity than what the test work program requires at first sight.
  - o Were sufficient samples available for the test work?
    - Hopefully this was noted early in the test work program and addressed, although the requirement for repeat and additional testing may seriously deplete sample reserves and require additional drilling.
- Test work results:
  - o Was the test work rigorously conducted?
    - Can the optimised operating conditions be reasonably translated to an operating plant?
    - Were the test work conditions, key slurry parameters (pH, Oxidation-Reduction Potential (ORP), Dissolved Oxygen (DO), etc) and observations satisfactorily captured?
      - Were any photographs taken?
      - These photographs should include unusual test results, such as froth conditions, settling or filtration behaviour etc.
  - o Do the test work results make sense?
    - Were the key test work results reproducible? How many repeat test were conducted?
    - What was the experimental error?
    - Were the test results analysed in the context of the experimental error established by the laboratory?
    - Could so called 'spurious' or 'artefact' results be real?
  - o Did assaying satisfactorily meet the test work requirements?
    - Were the requested elements and species assayed? This can take considerable time and involve several analytical laboratories and methods for full base metal concentrate speciation.
    - Was the assaying accuracy satisfactory? If an analysis was below the detection limit, was it re-assayed using a different method with a lower detection limit?
- Test work reports:
  - o Were the interim test work reports satisfactory and correctly capturing:
    - Sample details and preparation:
      - Were photographs provided of the as-received samples, closing out the chain of custody?
      - Was what was despatched reconciled with what was received at the test work facility?
      - Were the individual sample lots weighed upon receipt?

- Test work procedures and equipment details and equipment operating conditions eg flotation percent solids, agitator speed etc.
- Results and supporting calculations – sometimes a 'dumb' spreadsheet (no calculations in cells are included) is received, requiring a laborious re-work of the input data to allow checks to be performed.
- Photographs of products.
- Observations – these can be invaluable eg drill core received at a laboratory for flotation test work covered in hydraulic fluid.
- Interpretations.
- Conclusions and recommendations:
  - Are there any difference in opinions as to the conclusions drawn?
  - Are they consistent with the data?
    - There has been a disturbing tendency over a number of years by budget constrained third parties to regurgitate text from test facilities without a rigorous review of the results, conclusions or recommendations.
- Good practice is for the client to include a map identifying where the samples came from and asking the laboratory to include it in an Appendix marked 'Provided by Client'.
  - Is an overall test work report required that captures all of the test work program?
  - Has an electronic copy of the report(s) been supplied for digital storage and clearly states the author, date of issue and the revision number of the document?
- Were test work products and intermediate products used for environmental characterisation?
  - Check that selected samples have been sent to the designated testing facility, and that they have actually been received. The transfer of samples from the primary test work facility to other specialist test facilities (eg for tailings geochemical and geotechnical assessment) is typically required in the preparation of a Feasibility Study.
  - Check with the environmental team that the results are satisfactory and fit for purpose.
- Interaction and relationship with test work facility:
  - Document the experience and the people involved as an internal memo.
  - Has the schedule been adhered to?
  - Have the invoiced costs been supplied in a format that is easily understood and aligns with the test work program, with all 'extras' being able to be clearly identified?
- Ensure that all test products, including tailings, and any remaining feed samples are retained for a number of months following issuing of the final report. Retrieving test products from a laboratory for additional investigation (eg mineralogy) or assaying post completion of the test program (eg for potential penalty elements in a base-metal concentrate, or for an alternate assaying method to be used) can sometimes be required.
- What feed samples and test work products including tailings, and in what quantities, remain at this test work facility and others?
- What will happen to the feed samples and test work products, including tailings?
  - Return to site.
  - Storage at test work facility for future testing, possibly in cold storage and under nitrogen.
  - Disposal (which generally will incur a fee).
- Site water:
  - At this and other test work facilities, what site water remains and in what quantities?
    - Return to site.
    - Can it be discharged to waste?

THE BUDGET

Management of the budget is naturally an incremental exercise and checked after each stage of the test work program with, hopefully, no 'surprises'. The interim test work facility invoices need to be carefully checked against the original quotation, recognising and substantiating any variations arising from requested additional test work or changes in the scope of work.

Any necessary or requested changes in the scope of work should be formally supported by a quotation. Hence it is extremely important that any 'extras' are noted in minutes of regular meetings with the test facility.

There should be sufficient contingency to cover any of these unforeseen excursions during the test work program, noting that assaying costs associated with additional test work can be significant, and often can cost more than the actual work to perform the test. However, major changes in the scope of work may require an increase in the budget, which will require internal support and approval. These changes may affect project timing, especially if unresolved issues delay the engineering design.

Typically, these major scope changes are relatively rare and arise from external events such as the discovery of more ore types, the higher than expected occurrence of a concentrate penalty element or changes to the mine schedule, where the mining yearly composites change.

Check the 'hidden' costs, such as sample/test work product storage and disposal and/or transport costs to and from site.

POST-TEST WORK 'BLUES'

While a number of factors can influence the mood post-test work completion, it is important to ensure that there are no regrets, such as some aspect or sample insufficiently tested to allow a reasonable decision to be made.

By reviewing the circumstances and execution of the test work program (the 'Audit'), at least reasonable grounds can be established to support the test work program outcomes and any apparent shortcomings.

Successfully designing and managing a test work program is no mean feat and deserves adulation and support for a job well done.

INTERPRETATION AND SCALE-UP

The interpretation of results, and thus confidence in the interpretation, is only as good as the quality of the test work design, procedures, implementation, outcomes and assaying. The subsequent interpretation of results should be relatively straightforward and often self-evident with well-designed and executed test work programs, delivering good quality results.

The scale-up of test work data undertakes two forms, namely for equipment sizing (derived from the generation of a design criteria and mass and water balance) and for metallurgical performance prediction, the latter used in the process design criteria and the financial model. Note that the financial model is typically the key outcome of studies and the basis for project investment, and typically comes under intense scrutiny by potential financiers through their appointed consultants.

COMMINUTION

Interpretation of comminution data is relatively straightforward at a high level if enough samples with sufficient spatial diversity have been tested allowing an assessment of the impact of lithology, alteration and weathering. The mantra is 'rock type controls the throughput'. The key message is always test more rather than less samples since there is no room for 'surprises' when the plant becomes operational, particularly later in the project's life when it is realised that another crusher or ball mill is required to be installed into a plant footprint that didn't allow for such a development.

It is wise to engage consultants specialising in the selection and sizing of comminution equipment such as mills, and particularly rod and attrition mills for the processing of graphitic ores.

Equipment vendors are also knowledgeable, and in some cases, such as High Pressure Grinding Rolls (HPGRs), would undertake pilot plant test work, size and specify equipment and provide process and wear guarantees.

Note that Morrell (2019) provides an excellent guide to the design of a comminution test work program.

The key data used in the selection and sizing of comminution equipment are as follows:

- Crushing:
  - Unconfined Compressive Strength (UCS, MPa) the value of which determines whether a jaw crusher would be suitable and if not, a gyratory/cone crusher would be required; in the event that a jaw crusher would be suitable, whether a single or double toggle would be required. Note that the rarity of the latter offered by equipment vendors these days means that ores with high UCS values may require larger units.
  - Bond Abrasion Index (Ai, g) is useful for calculating wear rates, mill media consumption and identifying materials of construction (eg manganese steel) and can rule out roll toothed crushers.
  - Taking this data into account, the size of crushers, particularly primary crushers, are based on the feed size – so, for most small to medium sized projects, make sure that the mining engineers are aware of this and keep the top size below 800 mm (or else have a grizzly and a robust rock breaker stationed at the Run of Mine (ROM) bin!).
  - In the light of the previous comments, it is difficult to see the value of conducting Bond Crushing Work Index (CWi) test work; Morrell's (2019) review on the wide range of comminution values that are determined shows that the CWi test results can be unreliable.
- Mills:
  - As noted, this has become a specialist area and both consultants and vendors typically undertake the interpretation of test work data.
  - Milling comminution parameters include the Bond series (Rod [BRMWi] and Ball [BBMWi]), MacPerson Autogenous Work Index, Starkey semi-autogenous grinding (SAG) Design test as well as the SMC Test® series (drop-weight index (DWi), A*b, ta etc):
    - The determination of the BRMWi is often omitted from comminution test work. This is a pity since the ratio of the BRMWi to BBMWi can be useful guide to the potential build-up of pebbles in SAG milling.
  - SMC Test® data is particularly useful since it can be used in the design of several comminution devices.

While the design is typically focused on the 80th percentile values, a thorough knowledge of the mine schedule and the corresponding comminution characteristics are required to ensure that extended periods of excessively hard or soft ore will not be processed that would impact the production rate or result in potential equipment damage eg soft ore resulting in rod or SAG mill liner damage.

PRE-CONCENTRATION

The potential for pre-concentration is dependent upon the mineralogy and how it varies with ore types. For pre-concentration to be effective, there is a requirement for a useful degree of liberation of mineral values and/or gangue to occur at coarse particle sizes.

This raises the issue of having statistically meaningful sub-samples which requires a large amount of sample mass.

Pre-concentration test work such as classification with screens or hydrocyclones, gravity concentration by jigs and Dense Medium Separation (DMS), magnetic separation etc readily describes the behaviour of individual particles. However, it is difficult to confidently predict the metallurgical outcome of applying the processing technique on the production scale when only core samples are available.

Assuming that suitable samples have been used, the test work results for these pre-concentration methods are relatively straightforward to analyse. The typical approach is based on a trade-off where the economic mineral recovery is compared to the mass recovery. Depending upon the economic mineral and its' value, 90–95 per cent recovery with at least 30 per cent mass rejection may be considered satisfactory, particularly if 'troublesome' or very hard, non-economic mineral bearing species preferentially report to the reject stream.

GRAVITY CONCENTRATION

Gravity test work data, principally for gold, can be misinterpreted for several reasons:

- The standard test work is based on Gravity Recoverable Gold (GRG) and represents the absolute maximum amount of recoverable gold from the sample tested. The laboratory scale equipment operates with a 'g' force greater than used in plant installations The GRG results simply does not translate into reality with available equipment and the actual processing conditions such as circulating loads.

- The gold may be GRG, however it may not make a satisfactory concentrate for sale or further treatment (eg gold in pyrite).

- The samples tested in nearly all cases are unlikely to be representative of ores that will be mined over the life-of-mine (LOM), particularly where the gravity gold proportion of an ore exceeds 20 per cent of the gold present.

- The ratio of the concentrate mass to the feed mass in test results can be higher than in an operating plant, which can lead to an over statement of recovery in plant operation.

- A particular issue, as with ore sorting, is determining the quantity of fines that will bypass the pre-concentration process. Koppalkar *et al* (2011) found that around 44 per cent of the GRG gold can exit in the hydrocyclone overflow and bypass gravity concentration.

Gravity gold, typically greater than 63 microns in size, can vary significantly in both size and location and has been termed 'spotty' gold by geologists. Generally, there is a relationship between the gold feed grade and the amount of gravity gold present, which is worth determining if the gravity gold exceeds 20 per cent of the gold present, or the hydrocyclone underflow grade is more than a factor of 10 above the grinding circuit feed grade.

Audits of gold gravity operations show that the test work gravity data (GRG) is not achieved in practice and is between one third and two-thirds of the GRG recovery (Koppalkar *et al* (2011). Hence, in assessing GRG results, a discount of 50 per cent is typically used.

A particular metallurgical 'howler' is the use of mercury amalgamation test work results to predict gravity gold recovery. This conflates 'free gold' occurrence with gravity recovery; it is unlikely that a 10 micron gold particle beaten flat in a grinding circuit can compete with a 150 micron pyrite particle.

ORE SORTING

Ore sorting, which is enjoying a surge of interest, requires pilot plant test work. With the development of new technologies, instrument applications and sophisticated controls, it has become a specialised area controlled by the vendors, who undertake pilot plant test work, size and specify equipment and provide process guarantees.

It should be noted that ore sorting, like many mineral processing techniques, is limited to specific size ranges. This needs to be kept in mind when assessing the overall benefit. For example, it is not uncommon for up to 40 per cent of the feed to be less than the minimum size for an ore sorter to operate satisfactorily.

The paper by Wraith, Resta and Welmans (2019) on the applicability of ore sorting in an operating plant is essential reading.

FLOTATION

The most important piece of flotation test work data is the Locked Cycle Test (LCT) result and interpreting this data requires detailed attention. A LCT is a bench scale attempt at imitating a continuous flow sheet and competent execution of a LCT may become problematic with multiple stages of flotation and regrinding and the associated recycle streams (eg many graphite flow sheets). It can also be impacted by minerals that quickly oxidise (eg some forms of pyrrhotite and marcasite). Nonetheless, short of conducting a pilot plant, a LCT is a very useful tool in demonstrating the flow sheet and likely recirculating loads and is definitive in predicting the production plant overall recoveries (after applying an appropriate discount) and concentrate grades.

The interpreted LCT data informs process design criteria, mass and water balances, as well as product recovery and grade and thus requires a reliable interpretation methodology.

LCT test work should only be conducted once the process operating conditions for satisfactory metallurgical performance have been determined and optimised by batch flotation test work eg primary grind size, reagent regimes, pH, % solids, rougher and cleaner residence times, concentrate regrind size, number of cleaning stages etc.

LCT results represent the best possible metallurgical outcome for the sample tested because it has been conducted under ideal conditions:

- No short circuiting of material.
- Dedicated and timely operator scraping of froth from the cell.
- More 'energy/work' (eg kWh/m^3) is applied than can be achieved on the plant scale.
- Feed materials is more uniform.
- Tighter feed size range (eg less fines from the use of laboratory rod mills and more efficient size classification (screens versus hydrocyclones).
- Satisfactory reagent conditioning practised (often not employed in modern plant designs).
- No wear on cell impellor and stator.

A LCT result is only valid if 'stability' has been achieved, in general, if the number of cycles required is equal to the number of separation stages plus one. Ounpuu (2001) provides a useful discussion on the interpretation and application of LCT results.

With 'stable' LCT data, the feed grade-recovery relationship can be prepared. It is important to understand how the feed grade affects the metallurgy and how sensitive recovery is to both the feed grade and the concentrate grade. This is why it is good practice to ensure that some concentrate grades are generated higher than what would be planned in an operating plant. It is much more credible to a project financier's reviewer to have interpolated concentrate grade-recovery data than to have extrapolated data.

This relationship is also important in the estimation of the ore reserve where the cut-off grade needs to be determined as well as for revenue calculations in the financial model. It strongly recommended testing samples with grades around the cut-off grade since they may have better recoveries than estimated by the feed grade-recovery relationship.

The methodology and generation of the feed grade-recovery algorithm (at a given concentrate grade) is generally the most scrutinised aspect of a feasibility study by metallurgical reviewers from potential project financing organisations. The algorithm(s) is retrofitted into the mining block model in the preparation of a feasibility study to determine what is feed ore, what is waste and what might be possibly stockpiled as 'low-grade' or 'marginal' ore, and what recovery and grade of concentrate will be produced for sale (the revenue stream).

In the construction of a feed grade-recovery relationship, it is important that recoveries are based on the same concentrate grade, since higher recoveries are achieved with lower concentrates, which is a strategy adopted in many copper operations particularly if gold is an important component (or can, depending on metal prices, become the majority component) of the revenue stream.

A concentrate grade-recovery relationship needs to be interpreted from the data and the recoveries adjusted accordingly. This can be an exhaustive and rigorous task and sometimes curve fitting may require different algorithms at a critical feed grade transition point.

In terms of flotation scale-up, two key parameters need to be considered.

The first scale-up factor concerns the metallurgical recovery, where a deduction of 2 per cent is generally applied based on LCT results and incorporated into the Financial Model. In the case of a pilot plant, the deduction is typically 1 per cent. The justification for this discount arises from the factors previously discussed as well as the likelihood that the sample tested would not be encountered during the life of the operating plant.

The second scale-up factor concerns the flotation residence time, which arises from the less intense energy input (kW/m^3) of industrial flotation equipment. The following scale-up factors are typical of those used in pyrite/gold, base metal and graphite flotation. Note that the laboratory flotation test work is typically conducted in conventional mechanically agitated cells.

When conventional mechanically agitated cells would be selected for the full-scale plant, the scale-up factor has been established based on historical experience and is typically 2.0 to 2.5 times the laboratory residence time for 'simple separations', such as a porphyry copper rougher flotation duty.

The scale-up typically needs to be increased with finer particle size distributions and with more severe the depressant regimes (eg massive pyrite ores or talcose ores), sometimes by between five to ten times in cleaning duties.

With the introduction of tank cells, a scale-up factor between 2.5 and 3.0 is typically used. Lower intensity means longer flotation residence times which often translates into larger or more flotation cells. Again, vendors should be approached if this style of equipment is being considered.

A similar approach is required with other proprietary technologies, such as Glencore's Jameson Cell, the Eriez's HydroFloat® and StackCell®, Jord's NovaCell, WoodGrove's Staged Flotation Reactor (SFR) and Direct Flotation Reactor (DFR) Cells and the Metso Concorde Cell, who may undertake pilot plant test work to provide a basis for the appropriate sizing of their equipment.

With regard to flotation columns, there is no need to apply to apply a scale-up factor to the laboratory residence time. The shearing of bubble-mineral aggregates is virtually non-existent and the need for minerals to re-attach to bubbles is not required.

While the key design consideration for columns is the carrying capacity (Ca, $t/h.m^2$) (Newell, Cantrell and Dunlop, 1992; Newell, 1989), it is worth obtaining vendor input if columns are planned to be used.

A principal benefit of flotation columns is their froth washing capability which has been adopted in other flotation machines (eg Jameson Cell). It is noted that the potential improvement that froth washing may offer can be tested in a conventional batch flotation test operated at very low percent solids (eg 5 per cent), where entrainment can be minimised compared to the standard test.

In addition, conducting a size fraction by size fraction analysis of the standard test concentrate can provide data on both the amount of entrainment and the species reporting to say the minus 5 micron size fraction.

The concentrate lip carrying capacity (t/h.m) of the proposed flotation unit also needs to be confirmed for the likely range of concentrate production rates for that flotation stage, together with consideration of the size distribution of the concentrate.

LEACHING

Hydrometallurgical data, especially leaching, does not generally require scale-up. The chemical reactions should occur to the same extent over a similar time in a vessel of any size., as long the processing conditions such as particle size range, reagent concentrations, pressure, temperature, residence time, viscosity, redox potential, pH, PO_2 etc are achieved in practice. In some cases the additional hydrostatic head in an operating plant can provide improved dissolution compared to laboratory sized equipment.

A feed grade-recovery relationship should be evident from the test work results and generally, a 'fixed' tailings grade identified.

While no scale-up factor is applied, it is common for a deduction of 1 per cent to be applied to gold recoveries in a Carbon-In-Leach (CIL) and Carbon-In-Pulp (CIP) gold processing plant employing cyanide leaching. This reflects the likely losses due to loaded carbon fines (attrition), carbon activity and ash losses during regeneration as well as general solution losses.

DEWATERING

Test work results for thickening and filtration are relatively straightforward to interpret.

Care must be taken to provide 'fresh' samples as there have been several instances where high clay dried solids have been re-pulped and used for thickening test work, which significantly understated the required settling rate and thus the required thickener area.

In the application of dewatering results to the sizing of equipment, two factors come into play reflecting the need for a conservative design in the concentrate and tailings dewatering areas and to minimise the potential for bottlenecks to production.

These are design allowances, typically 10 to 20 per cent and the use of the maximum likely throughput these pieces of equipment are likely to experience over the life of the operation.

Concentrate thickening and filtering results are considered generic for each concentrate type. For example, a chalcopyrite copper concentrate of a given particle size distribution and a given grade would exhibit very similar thickening and filtering characteristics anywhere in the world.

To that end, concentrate dewatering data could be assembled by concentrate type with a range of both particle size distributions and grades for use in selecting and sizing equipment. While this test work could be omitted from at least Pre-Feasibility Studies, equipment vendors would request test work to size equipment for a Feasibility Study following which, equipment tendering would progress.

The same logic and action could be applied to Transportable Moisture Limit (TML) data, which typically requires 8 kg of concentrate for testing.

As previously noted, it is strongly recommended the dewatering properties of the tailing of the ore types are characterised and the equipment sizing is based on the worst case. This recommendation is made since mine schedules may change or blending practices are not achieved or followed in practice and the tailings of some ore types may have poorer settling or pumping properties, resulting in a bottleneck and the need to lower milling throughput.

The dewatering properties of tailings are expected to have a major influence on plant design as the industry moves to filtering and dry stacking to minimise wet storage of tailings. The presence of clay in tailings can have a significant impact on the filtration requirements, where the unit filtration area can exceed that for a much finer concentrate.

Given the likely high importance of tailings dewatering unit operations, it is unfortunate that the mining industry has not developed a standard set of tests to determine the outcome drivers as has been achieved with comminution.

CONCLUSIONS

To satisfactorily close out a test work program, as much care and thought that went into establishing the test work program and managing the test work facility, needs to be applied.

Progressive reporting of the test work and results should be undertaken during the program from the test facility to the responsible personnel for the test work program to ensure that there are no 'Oh My Goodness' ('OMG') moments at the completion of the test work program.

Conducting an audit of the status of the test work program will address all of the key test work target criteria and confirm that the test work program is progressing satisfactorily to achieve the stated outcomes or confirm the validity of a change of direction of the program.

Progressively assessing results, successfully addressing the audit, satisfactorily reconciling the expenditure and communicating the test work outcomes with the management should ease any concerns that may arise upon completion of the test work program.

REFERENCES

Koppalkar, S, Bouajila, A, Gagnon, C and Noel, G, 2011. Understanding the discrepancy between prediction and plant GRG recovery for improving gold gravity performance, *Minerals Engineering*, 24(6):559–564.

Morrell, S, 2019. How to Formulate an Effective Ore Comminution Characterisation Program, in *Proceedings of the MetPlant Conference*, pp 51–75 (The Australasian Institute of Mining and Metallurgy: Melbourne).

Newell, A J H, 1989. Column Flotation – an Engineer's Perspective, Tunra/Amdel Column Flotation Workshop, Adelaide, Australia.

Newell, A J H, Cantrell, R A and Dunlop, G A, 1992. Some Considerations In Column Flotation Pilot Plant Testwork, in *Proceedings of the Extractive Metallurgy of Gold and Base Metals*, pp 253–257 (The Australasian Institute of Mining and Metallurgy: Melbourne).

Newell, A J H, Munro, P D and Fiedler, K, 2021. Metallurgical Testing – The Pain Continues, in *Proceedings of the 15th Mill Operators' Conference*, pp 397–443 (The Australasian Institute of Mining and Metallurgy: Melbourne).

Newell, A, Munro, P D and Fiedler, K, 2018. Metallurgical test work – between a rock and a hard place, in *Proceedings of the 14th Mill Operators' Conference*, pp 15–33 (The Australasian Institute of Mining and Metallurgy: Melbourne).

Ounpuu, M, 2001. Was that Locked Cycle Test Any Good, in *Proceedings of the 33rd Annual Meeting of the Canadian Mineral Processors*, pp 389–404 (Canadian Institute of Mining, Metallurgy and Petroleum: Ottawa).

Wraith, B, Resta, J and Welmans, J, 2019. Recent Improvements in ore sorting at the Renison Tin Concentrator – target 1 Mt/a, in *Proceedings of the 15th Mill Operators' Conference*, pp 339–351 (The Australasian Institute of Mining and Metallurgy: Melbourne).

Practical considerations in the design of flotation process twins for set point decision automation at MMG Dugald River Mine

W Smith[1], A Malcolm[2], I Goode[3], G Bai[4] and C Martin[5]

1. MAusIMM, Global Practice Lead Digital Twins, Hatch, Brisbane Qld 4064.
   Email: warwick.smith@hatch.com
2. Senior Metallurgist, MMG, Dugald River Qld 4824. Email: alec.malcolm@mmg.com
3. Manager Strategy, Innovation and Development, MMG, Dugald River Qld 4824.
   Email: iain.goode@mmg.com
4. Senior Process Data Scientist, Hatch, Brisbane Qld 4064. Email: george.bai@hatch.com
5. Superintendent Metallurgy, MMG, Dugald River Qld 4824. Email: catherine.martin@mmg.com

ABSTRACT

The combination of the increasing complexity of minerals processing operations, ongoing skills shortages and growing volumes of under-utilised plant data is resulting in increased demand for automation solutions aimed at improving decision-induced process variability. Since commissioning in 2017, Dugald River Mine (DRM) has implemented several improvements to their Zn flotation circuit resulting in a high baseline recovery performance. As with all flotation plants, some process variability remains, which is partly attributed to day-to-day set point decision-making. In this work, the authors present a process twin based approach to flotation set point decision automation and outline the design considerations applied to the DRM Zn rougher-scavenger. Process twins for flotation operations model process performance in near real time as a function of both feed properties and process control variables, and subsequently prescribe the optimal combination of process set points. Approaches to set point determination may vary from static ranges per variable, through to model predictive control (MPC). Traditional MPC installations are often limited by unrepresentative ore types during step-testing and the lack of a holistic multivariate process model which captures the combined interactions between multiple process variables. The authors followed a best-practice empirical modelling framework to examine a variety of machine learning techniques to model rougher-scavenger outputs as a function of both circuit inputs and control variables. As a 'sense' check, the practical framework included multiple reviews against standard metallurgical theory and understanding, to ensure robustness and improve confidence in the determined value improvement. Following preliminary training, a pair of linked deep neural networks were able to model a large proportion of concentrate and tail grade test-set variability for the rougher (R^2_{con} = 0.71, R^2_{tail} = 0.61) and scavenger (R^2_{con} = 0.77, R^2_{tail} = 0.72). The models were embedded in a twin simulation environment which enabled evaluation of alternative set point combinations for various feed classes. The results demonstrate potential recovery improvement via more consistent set point determination, and the capability of such systems to be integrated within the control layer for advanced open-loop or closed-loop control.

INTRODUCTION

DRM zinc flotation circuit

The Dugald River concentrator was commissioned in September 2017 and achieved its first 10 000 t shipment of Zn concentrate by December 2017. The process flow sheet, shown in Figure 1, contains an initial carbon preflotation circuit (to remove naturally hydrophobic carbonaceous material) followed by standard sequential Pb-Zn flotation. Due to simplistic nature of the flow sheet minimal performance ramp-up issues were expected.

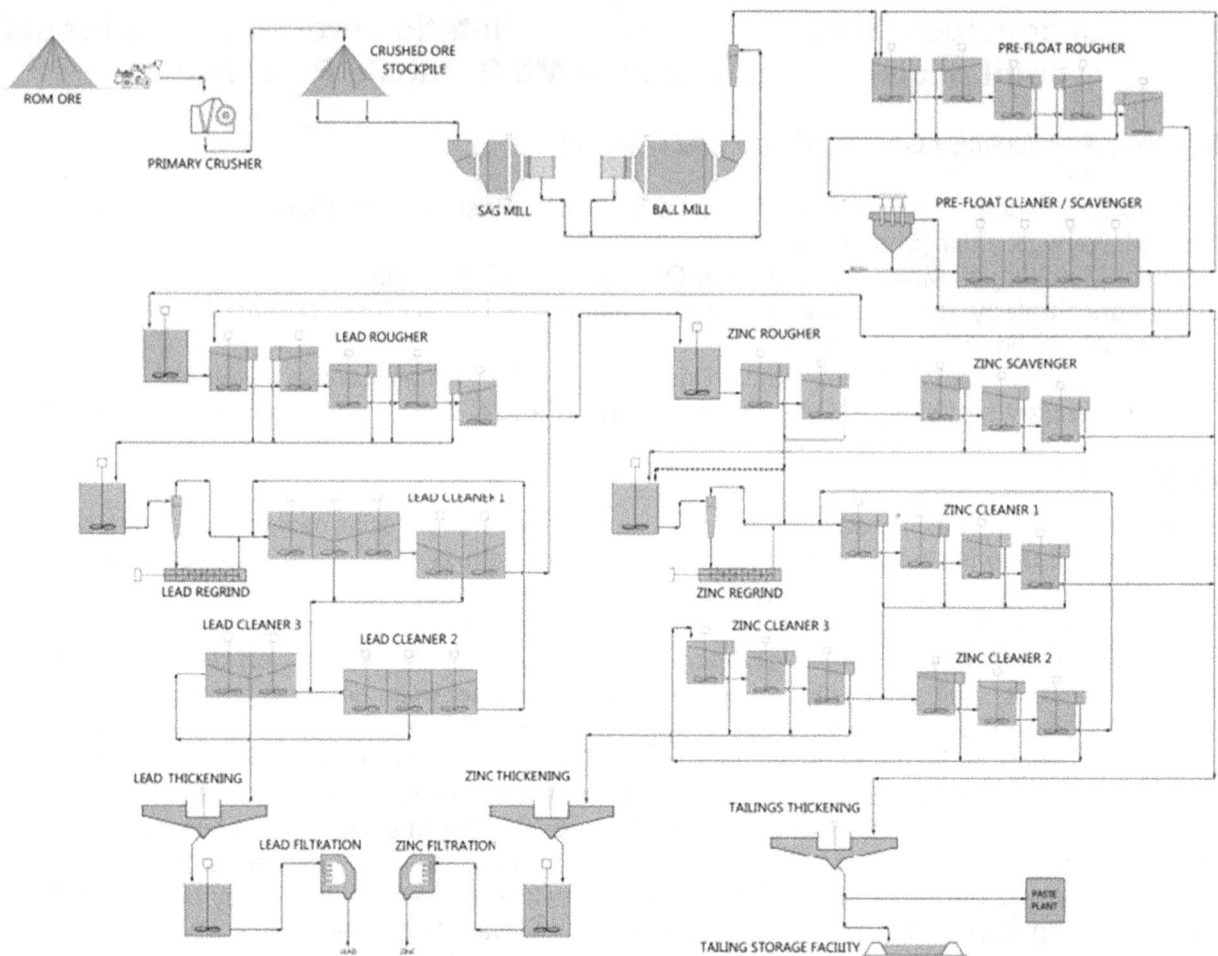

FIG 1 – The Dugald River Mine overall plant flow sheet.

Flotation performance efforts were restricted in 2018 as focus shifted to ramping up throughput above design capacities to counter the underperforming Zn feed grade. With the throughput meeting expectations, attention from 2019 then turned to Zn recovery and concentrate grade through two separate avenues; reduction of Zn lost to the carbon preflotation concentrate and overall performance of the Zn flotation circuit. These two areas were targeted as they represented the largest Zn losses to tailings in 2018 at 6.3 per cent and 9.3 per cent respectively (with respect to flotation feed).

Through 2019 and 2020 several improvements to the Zn flotation circuit performance were targeted:

- Circuit operation strategy and set points – understanding and developing guidelines for plant operators around standard flotation circuit control variables such as reagent addition, air flow rates and recirculating load management.

- Installation of additional collector addition points as it was identified staged dosage improved circuit stability and performance.

- Introduction of significant circuit automation through regulatory process control, inclusive of cascade loops to control reagent addition based on circuit feed rates and grades as well as ratio-based set points to proportion reagents and air to individual cells.

- Commencement of regrinding of the Zn scavenger concentrate to improve liberation prior to the cleaning circuit.

The listed improvements did not result in any discernible Zn flotation circuit recovery improvements, however there was a clear increase in final Zn concentrate grade of between 2.1 per cent and 3.8 per cent sphalerite in concentrate at 95 per cent confidence, when reviewing monthly mineralogical analysis. From a value perspective this was significant as 1 per cent sphalerite in Zn concentrate has an equivalent net smelter return value of approximately 0.5 per cent rougher scavenger Zn recovery.

Further minor improvement initiatives targeting the Zn flotation circuit operation were undertaken in 2021 to refine and optimise within the limitations of the current instrumentation and process control system, however the operational strategy has largely remained unchanged since then.

By 2021 efforts to reduce Zn losses to preflotation resulted in a decreased loss from 6.3 per cent to 3.5 per cent with respect to flotation feed. However, losses from the Zn circuit (scavenger and cleaner tails) remained relatively dominant at 7.1 per cent. With an ongoing focus on process automation, the Zn flotation circuit, and specifically the rougher-scavenger circuit, was identified as a logical candidate to benefit from a control-integrated digital twin to support operators with set point decision automation.

After a review of process digital twin providers throughout 2021, work began on how a process twin would be developed, validated, implemented, and ultimately maintained. This paper introduces the concept of a flotation process twin for set point automation purposes and summarises a subset of the operational and technical considerations made in the preliminary design of the Zn rougher-scavenger twin. Although not exhaustive, the practical considerations discussed by the authors include:

- general operational preparedness
- circuit selection to de-risk design
- pre-modelling efforts to de-risk design
- empirical modelling practices
- methods for evaluating of the business value prior to progressing to implementation.

Flotation process twins for set point automation

The digital twin has gathered multiple definitions in the past two decades (Jones *et al*, 2020) due to the breadth of its potential industrial applications. The term 'digital twin' derives from industrial product life cycle management in the early 2000s and was initially defined as the virtual representation of actual physical products, which contains information about the products at the time of production (Grieves, 2014). The key characteristics of industrial digital twins include:

1. Near real-time connectivity from the physical world to the virtual entity.

2. A sufficient level of fidelity in the virtual entity to enable a reasonable change in the physical world to be realised. That is, the digital twin must contain the required functionality to enact a change on a production process in a meaningful time frame.

3. Bi-directional, near real-time connectivity from the virtual entity back to the physical world, to the specific systems or processes capable of executing the change.

'Closing the loop' between the physical and virtual worlds via twinning cannot only drive consistency in time-sensitive production decisions, but also highlight errors in the current understanding and measurement of the specific physical process being replicated, and identify opportunities and risks in the way a production process is operated (Grieves and Vickers, 2017).

Within the flotation process, the majority of low-level, high-frequency production decisions being made by operators and metallurgists are concerned with the correct configuration of the process to manage recovery and final concentrate grade, as a function of the ore type presenting to the plant. Upstream geometallurgical properties (assays, lithologies etc) and comminution circuit performance (throughput, particle size distributions) are typically known drivers of flotation performance. However, these properties are also dynamic in nature and thus may require varying treatment strategies to maximise overall productivity of the circuit. Conversely, variability or inconsistency in operational decision-making can drive material cumulative losses of otherwise recoverable valuable mineral.

Flotation process twins therefore play the role of monitoring the current recovery performance drivers upstream from the flotation circuit, continuously predicting the flotation response as a function of both the operator's current control variables and the upstream drivers and returning the optimal set point suite (according to a defined process objective) to the operator or the control system in a timely manner for execution within the process.

Furthermore, flotation process twins can play an additional organisational role beyond near-real-time production decision-making. By encoding and visualising the key metallurgical control relationships within the process, and demonstrating how these relationships are impacted by geometallurgical parameters, the twin serves as a mechanism to both capture validated empirical knowledge of the process to assist with technical upskilling and potential knowledge loss, and identify where empirical knowledge deviates from theory.

Despite existing at different levels of the process control hierarchy, there is often a perceived overlap between digital twin technology and traditional model predictive control (MPC) technologies which similarly aim to provide supervisory set point automation on continuous processes. Some of the common shortfalls in MPC systems which can be supported by overarching digital twin technology include:

- System training or 'step-testing' on limited ore types which are not representative of all metallurgical domains or the relevant history of the process.

- The usage of linear, bivariate matrices of control and response variables which don't reflect the multivariate non-linear relationships often observed in flotation and similar processes.

- The focus on optimising local components of a circuit, while ignoring a potential master objective function for the overall flow sheet.

- The lack of transparency on the current level of system inaccuracy, and a lack of transparency on the confidence in predictions and recommendations.

- The inability to leverage the system for technical and operational knowledge capture.

PRACTICAL CONSIDERATIONS IN TWIN DEVELOPMENT

Prerequisites to improve chances of success in the future twin deployment

The final performance improvement derived from the digital twin is dependent on the operator's utilisation of set point recommendations returned from the twin. Having a high-fidelity model and optimiser for the physical system is meaningless if the operational groundwork has not been laid to enable an effective partnership between the operator and the model-based system. If the utilisation of recommendations is poor post deployment, early predictors of this outcome could include:

- The lack of a disciplined 'scientific' approach to the overall control of the process manifest as high process variability (ie structured observation of cause and effect during process control is not particularly valued).

- The lack of any precedence in adhering to simple models which advise or directly drive critical control variables (eg reagent specific dosage controllers).

- Inconsistency in treatment approaches for broad feed types, or a general inability to discern feed change impacts due to noise within the process.

- An incorrect understanding of how the twin outputs will interface with the control layer or the operator's everyday decision-making time frames.

Addressing the above issues are not explicit quantifiable prerequisites to starting digital twin development, however achieving a degree of operational and technical standardisation can help drive a culture of reducing process variability and greatly improve the likelihood of digital twin adoption and therefore return on investment.

Following earlier flow sheet and process control improvements, the DRM Zn rougher-scavenger operates with a relatively stable recovery distribution (μ = 96.4 per cent, σ = 1.1 per cent, with regard to Zn rougher feed). The baseline stability has been derived from the consistent usage of reagent control models (principally for collector and activator) which vary as a function of feed grades, mill throughput rates and upstream Pb circuit performance. The existing control philosophy requires relatively fixed management of froth depth to minimise unnecessary variability. Lime addition to the rougher feed is regulated by a standard pH controller. Adherence to these example practices provides a reasonable level of baseline stability, while generally still providing sufficient set point variance to enable process modelling and optimisation.

In addition, the usage of MPC and mill filling level models to manage SAG mill process control assisted in building a precedence for other model-based control applications across the flow sheet.

Considerations for target circuit selection

Developing an online process twin capable of acting on the physical plant for an entire flotation circuit or concentrator is large and complicated undertaking. However, by following a hierarchy of control from plant wide optimisation down to individual controllers, the flow sheet may be broken down into logical components which represent individual process models to be connected into a broader twin system over the longer term. The process for prioritising and selecting initial flotation process nodes for twinning may vary on a case-by-case basis. For Dugald River, the key considerations at this stage of design included:

- The estimated value to the business of tightening or automated performance the target circuit.

- The relative size of the model control volume as judged by the number of inputs, outputs and sub-processes within the boundary.

- The ability to measure, calculate and validate recovery and concentrate grades via on-stream analysis and laboratory composites at the model boundary.

- The ability to establish a meaningful objective function for the target circuit, and whether that objective could potentially negatively impact other downstream processes.

- The current levels of process variability, and the degree to which operational decision-making complexity is likely driving process variability.

- Whether metallurgical responses between control variables and dependent variables are observable within the plant data. This is often not defined (or perhaps at best anecdotally understood) at the time of initial circuit selection and is discussed further.

Basic assessment of the above criteria resulted in the DRM team selecting the Zn rougher-scavenger as the preliminary focal point for flotation circuit twinning.

Model framework and its impact on operational decision-making

The problem statement for the DRM Zn rougher-scavenger twin design can be summarised as follows; for a given set of unique upstream inputs, what is the optimal combination of process control variables which optimises the objective function of the Zn rougher-scavenger. In other words, as the plant feed and milling circuit varies, how do the operator's control variables around the Zn rougher-scavenger need to vary to maintain near-optimal performance. This problem statement implies the candidate features which need to be explored and (if justified) fitted to the model, the potential lagging required from upstream signals, and also the responsiveness required by the system.

Multiple model frameworks were considered and iteratively discarded during early design works (eg single). One guiding principle was the desire to keep model scopes small and approximately reflective of the actual process flow sheet. At the cost of a greater number of models to be managed in the future production system, keeping model scopes small prevents the potential for introducing spurious noise into the regression process which could negatively impact the shape of the individual control curves encapsulated by the model. Keeping model scopes small and focused to the problem statement further assists real-time error management against available on-stream data, chaining models together to achieve broader process objectives, and optimisation computation speed.

Figure 2 outlines the individual model boundaries for the rougher and scavenger and the overall rougher-scavenger utilised during the twin design. The variables shown do not reflect the breadth of process variables explored during the overall design, but importantly highlights the difference between major recovery drivers upstream of the relevant process node, and the operators control variables to be optimised per input condition. The treatment and preparation of model inputs, including the management of data measured at different time frames within the process is covered under Data Quality and Quantity.

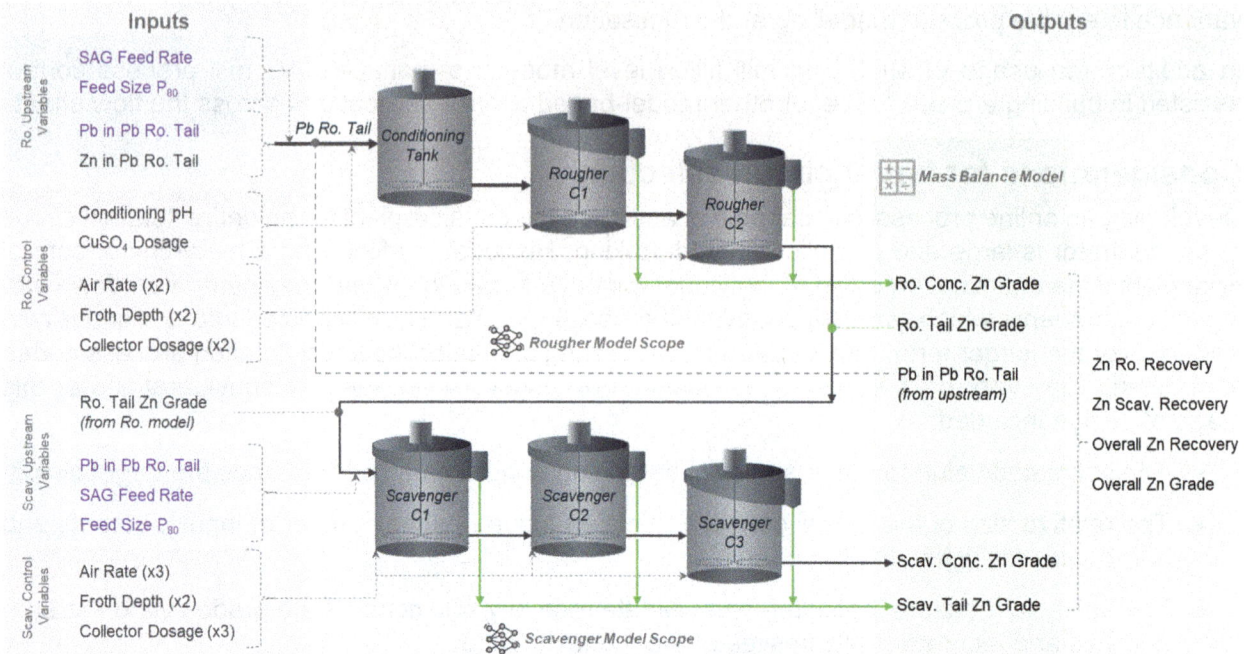

FIG 2 – Model framework for the DRM Zn rougher-scavenger.

For simplicity, the initial objective function set for the DRM rougher-scavenger was Zn recovery maximisation while maintaining concentrate grade at the incumbent point. At a higher level of the Dugald River plant hierarchy (which would require additional connected process models), this objective function can be superseded by a broader production objective, or even an economic objective such as net smelter return which would leverage the grade-recovery relationship.

Determining the modelling paradigm

The type of problem statement outlined in the previous section lends itself well to machine learning techniques due to:

- The need to keep the models 'lightweight' to support the computation associated with near-real-time optimisation (for set point determination).

- The non-linear and multidimensional nature of the flotation process.

- The need to capture potentially obscure recovery drivers within the underlying models (eg Pb circuit collector usage and dynamic downstream impacts of residual collector).

- The need to achieve a level of accuracy and inferential stability that will stand up to dynamic real time plant conditions.

However, in the same way that 'first-principles' models must be calibrated and tested against raw real-world data to achieve an acceptable level of accuracy, machine-learning models can and must leverage first-principles thinking to ensure the model design captures the desired ore response patterns. Utilising metallurgists with site specific and general operational experience in C-Pb-Zn sulfide flotation circuits is crucial to achieving reasonable model development outcomes, and allows the usage of data-driven approaches that are not limited by any pre-conceived notions of process variable relationships.

It is important to emphasise for the DRM application (and for process twins in general) that simply simulating the process in near-real-time is insufficient. A layer of decision logic (such as an optimiser) is required to continuously act on the process model to prescribe an action back to the physical world. Due to the size of the multidimensional space implied by the DRM flotation twin model structure, and the need to potentially iterate through many thousands of initial conditions to converge upon a robust inference, a data-driven or machine learning approach must be leveraged. Recent works involving the usage of physical process simulations for SAG mill throughput predictions have similarly observed the computational benefits of re-fitting the process model to a lighter machine learning construct without material alteration to the simulation accuracy (Bensoussan *et al*, 2022; Amini *et al*, 2023).

Data quality and quantity

Training large-scale data-driven models, like neural networks, is a complex process that often requires large quantities of data. This is particularly true for models designed to handle specific tasks where the training process must begin with an unweighted network (Soekhoe *et al*, 2016). For the Dugald River data set, approximately 18 000 observations remained post-cleansing, filtering and aggregation.

An essential aspect of data preparation for models with short interval focus, like those used in the flotation process, involves considering the mean residence time and the typical delay between input and output measurements. This consideration is critical because aggregating data beyond the circuit residence time can mitigate timing discrepancies. However, the aggregation interval introduces a compromise by diminishing the natural variability present in the input data and reducing the number of observations available for model training and testing. The authors made use of first principles estimates of mean residence times at various stages in the flow sheet to provide a starting point for the aggregation interval, while structured modelling experiments were utilised to directly measure the sensitivity of the model accuracy to varying residence time assumptions. The final aggregation interval utilised for the individual rougher and scavenger models was 30 mins.

Figure 2 not only implies the potential for measurement timing discrepancies across the rougher-scavenger, but also the presence of signals well upstream of the flotation circuit, such as SAG mill feed rate and grinding circuit P_{80} (via on-stream particle size analysis). Such measurements were lagged by a fixed number of 30 min intervals, determined by earlier mean residence time calculations across grinding, carbon preflotation, and Pb flotation circuits. Further work needs to be performed at the next level of production modelling to validate the correct techniques for lagging different types of upstream signals, however the simple fixed interval approach provided reasonable raw correlations.

The volume of data necessary for training a model not only depends on the model size and its complexity but also significantly on the quality and accuracy of the raw data being used. Often, small but high-quality data sets will outperform large noisy data sets. Although flotation digital twins can be developed without the usage of on-stream analysis, the availability of well-maintained on-stream analysis data can markedly improve the twin development process, but more importantly, provide the means of continuously validating the error between physical and virtual processes during production. The DRM on-stream data underwent conventional preprocessing and filtering techniques to eliminate missing and implausible data. The processed data was further refined and corrected by comparing it with laboratory composites and shift reconciled production data. Minor discrepancies were observed in parts of the data for scavenger concentrate which require correction, however Figure 3 demonstrates typical good agreement between the on-stream analyser (Thermofisher Multi-Stream Analyser, MSA) and reconciled shift data required for the calculated rougher recovery dependent variable.

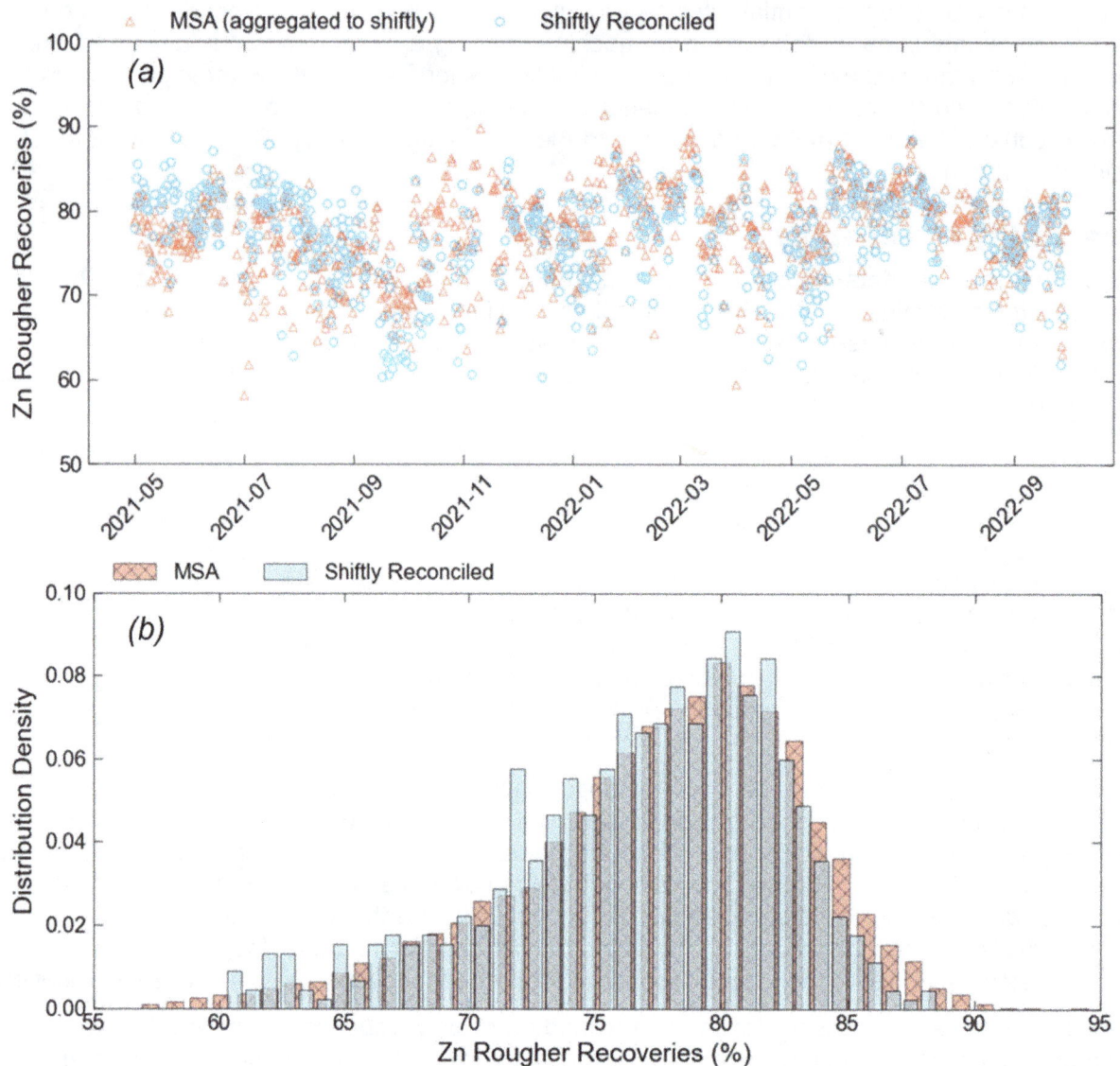

FIG 3 – A comparison of Zn rougher recoveries (with regard to Zn rougher feed) calculated by MSA data and shift reconciled data: (a) Timeseries; (b) Distribution.

Understanding key metallurgical and process control relationships

The ability to capture metallurgical relationships between the process inputs (ie upstream variables and control variables) and the dependent outputs (eg concentrate grades, tails grades, recoveries) is a key criterion for a flotation process model that is used for inferring control set points. If the typical metallurgical relationships are not somewhat observable within the raw data, then there is potentially little value in progressing to an expensive process modelling exercise. Despite the obviousness of this statement, many modelling teams skip over this important metallurgical practice. In addition, if it can be determined that an important control relationship (eg metal recovery versus collector dosage rate) dynamically 'shifts' in a logical way as a result of an upstream dimension (feed ore assays, grind size), then a fundamental basis exists for optimising the process as a function of the changing feed.

The Dugald River data set exhibits typical weak to moderate relationships between the upstream variables and flotation performance indicators. Figure 4 provides examples of these relationships which highlight the typical impacts of Pb and Zn in feed on rougher recovery and overall rougher-scavenger recovery.

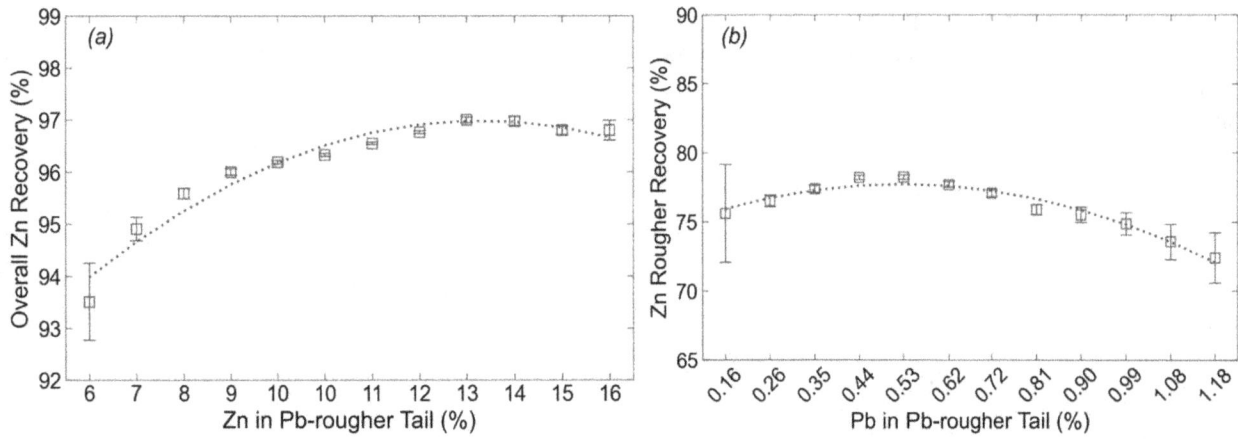

FIG 4 – (a) Overall Zn recovery versus Zn in Pb-rougher tail; (b) Zn rougher recovery versus Pb in Pb-rougher tail.

During exploratory analysis, the authors determined that important control relationships within the raw plant data do change dynamically as upstream conditions vary. For instance, in terms of scavenger recovery, the optimal collector dosage to scavenger cell 2 shifts with changing feed size, as shown in Figure 5. Partitioning the bulk dosage curve into simple P_{80} quartiles demonstrates, in accordance with theory, that as P_{80} decreases and available mineral surface area increases, the optimal collector dosage point with respect to scavenger recovery increases. In reality, there are many other dimensions to these control curves which will need to be captured by the eventual model, but this basic example highlights the potential recovery loss of operating at a sub-optimal point on the dosage curve for seemingly small changes in primary grind size.

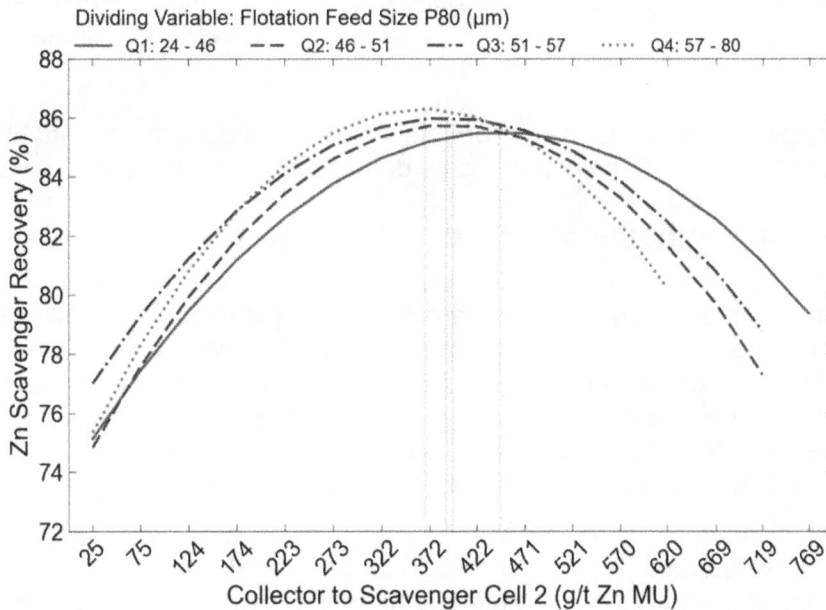

FIG 5 – Overall Zn recovery versus collector to scavenger cell 2, with data divided into four groups (Q1: first quartile, Q2: second quartile, Q3: third quartile and Q4: fourth quartile) by feed particle size, P_{80}. Trend lines represent the 2nd order fitting line of each subset data.

Another interesting point of validation from the raw metallurgical relationship review is that a good inference for a given control variable on the rougher does not necessarily result in a good outcome for the scavenger, and therefore the overall rougher-scavenger node. Taking the collector to rougher cell 2 as an example, this staged collector addition in reality serves the scavenger (likely due to conditioning time or carry through from the rougher). Figure 6 confirms that optimal inferences should be made where possible with respect to the outer process objective, not the individual nodal recoveries. Multiple observations were made on other control variables supporting the concept that operational strategies to minimise the intermediate Zn rougher tail does not achieve good overall grade-recovery results and does not make best usage of available flotation capacity. This argument could potentially be scaled out to the wider rougher-cleaner node, the broader concentrator, and so on.

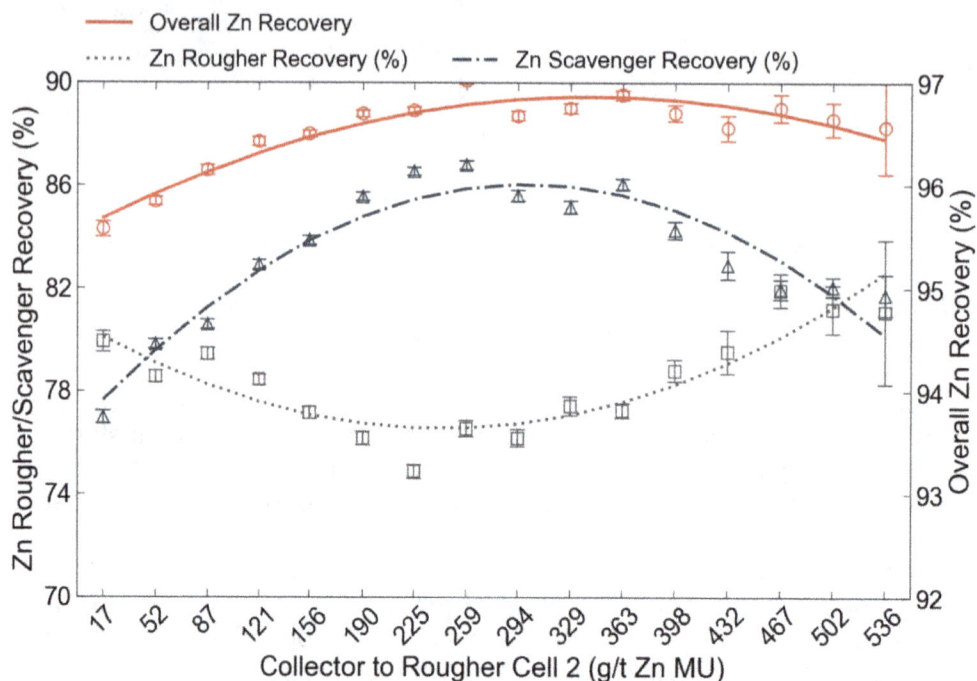

FIG 6 – Rougher, scavenger and combined rougher-scavenger Zn recoveries versus collector to rougher cell 2. Markers equate to the mean value at each axis level; trend lines represent the 2nd order fitting line.

Following data preparation and exploration of independent and dependent variables, the workflow focused on how to best evaluate the models to be developed for the rougher-scavenger twin.

Considerations on predictive model development and evaluation

The development of data-driven models, particularly for applications like the DRM flotation process model, involves an iterative, experimental process of adjusting the model design, and measuring the effectiveness of the change against consistent metrics and benchmarks. Moreover, the ultimate value of a model in control applications lies not necessarily in its predictive power but more so in the ability to make robust and sensible inferences from the model. This implies that, from a model evaluation perspective, techniques are required to simultaneously track traditional modelling accuracy and the metallurgical sensibility of inferences as the modelling iterations are carried out.

In this work, mean absolute error (MAE), root mean square error (RMSE), coefficient of determination (R^2) were used for tracking regression model accuracy, with the former two maintaining the units of the predicted variable. Training and testing error determination was carried out by standard k-fold cross-validation techniques with sensitivity conducted on the size of k.

In addition to selecting error metrics, it is also useful to set guiding targets for these metrics before commencing modelling work. Setting a guiding target helps avoid the potential situation of overinvesting time in a model which is unlikely to meet the required standard. For MAE and RMSE targets, initial error targets were set by following a crude heuristic of half the standard deviation of the respective dependent variable.

To gauge the relative value of increased complexity of machine learning algorithms, very low-complexity benchmarking models, such as 'select-the-mean' and linear regressions were employed during rougher and scavenger model development. If highly complex algorithm types exhibit similar accuracy to a simple regression or fixed constant, then there is no value in the increased complexity. This step was particularly relevant in the DRM context, as the modelling effort focused heavily on deep neural network usage over traditional machine learning algorithms. Table 1 shows the metrics of tested algorithms in this work, and the result suggested that neural networks were able to outperform other algorithms and maintain balanced training and testing errors.

TABLE 1

Metrics of subset of tested algorithms, in predicting rougher-scavenger Zn recovery (with regard to Zn rougher feed). various classes of algorithms have been included here for reference.

Algorithm	On data set	MAE	RMSE	R^2
Neural Net	Train	0.42	0.60	0.72
Neural Net	Test	0.44	0.61	0.72
XGB	Train	0.57	0.77	0.53
XGB	Test	0.59	0.82	0.50
Linear Reg	Train	0.60	0.85	0.44
Linear Reg	Test	0.61	0.85	0.45
SVM	Train	0.61	0.90	0.36
SVM	Test	0.61	0.91	0.38
Select-the-mean	Train	0.72	0.95	-
Select-the-mean	Test	0.74	0.98	-

However, as noted above, accuracy metrics alone should be treated with caution for models intended to be used for control purposes. For instance, a model capturing 80 per cent of the performance variance ($R^2 = 0.8$) in a process might be more useful and valuable than one capturing 90 per cent ($R^2 = 0.9$). This counterintuitive idea stems from the fact that high prediction accuracy alone doesn't guarantee the model has learned anything useful or causal. In the context of process control, the interpretation of metallurgical inference potentially holds greater significance than accuracy metrics. A comprehensive review of model captured metallurgical relationships between input variables and dependent variables were conducted and discussed in the following section.

Assessing what the twin's models have 'learnt'

As stated, a critical characteristic of a digital twin is its ability to enact a change back onto the physical world. That implies that additional layers of decision logic are required to derive sensible set point decisions from the twin's process models. This further implies that the models capture or 'learn' the relevant patterns between control variables and the dependent variable(s) to be optimised. To complete the model validation process, each individual control variable must be re-examined against both the predicted and dependent variables, to assess how well the 'shape' of the model complies with the plant data observed during earlier exploration.

By overlaying the raw recoveries and predicted recoveries against each control variable, multiple benefits can be realised, including:

- Providing a means for metallurgical 'sense-checks' (ie does the shape of the model match with operational and technical experience).

- Improved identification of specific regions of poor accuracy in the input space, where either a lack of training data or poor-quality data have influenced the learning outcomes.

- Early identification of the approximate position of constraints for the upcoming set point optimisation design.

For the Zn rougher-scavenger data, a thorough metallurgical relationship review and model comparison indicated that collectors generally demonstrated an excellent match between the model prediction and historical data. Predicted pH trends and air rate trends also matched raw data well, particularly within the normal operating ranges for these variables. Froth depth comparisons were tight on available data but lack of variation in the input data likely limited the models ability to learn a reasonable pattern for this variable.

Figure 7a, the overall Zn recovery versus collector to rougher cell 1 provides an example of very close agreement between the process model and the raw data, across a relatively large operating window. All collector dosages highlighted similar levels of matching between 0 g/t Zn MU to 500 g/t Zn MU, while a slightly larger difference merged at higher dosage points the where the data volume for learning was lower (above 500 g/t Zn). Figure 7b provides a further example of Zn recovery versus air rate for scavenger cell 2 and again suggests that there is reasonably good agreement between the process model and the raw data within the main operating window. In both examples, the approximate position of the optimal control variable range has been encoded by the model.

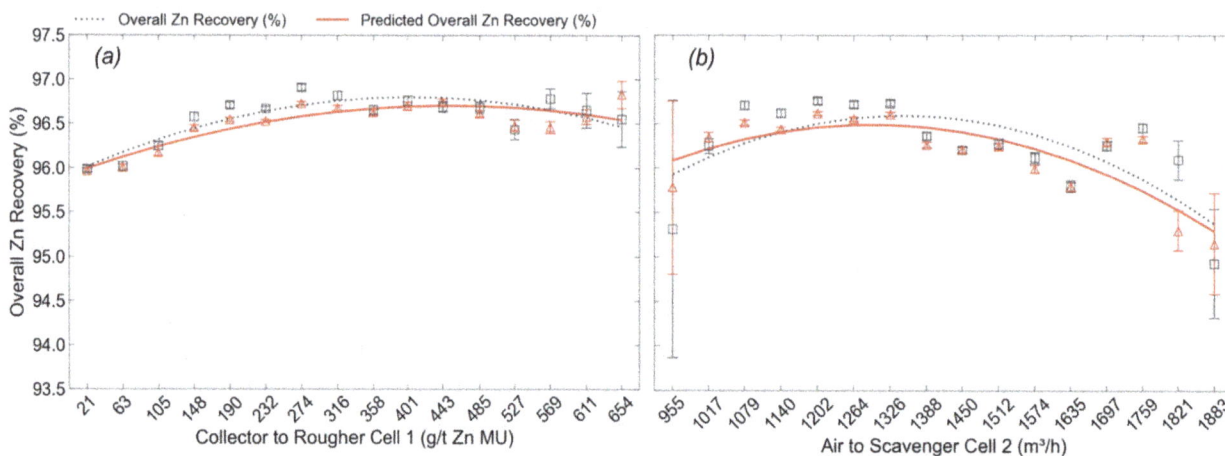

FIG 7 – A comparison between the original rougher-scavenger recovery (with regard to Zn rougher feed) and model predicted rougher-scavenger recovery versus (a) collector to rougher cell 1; (b) air to scavenger cell 2. markers in the plots indicate the mean value at each axis level, and trend lines represent the 2nd order fitting line of the data.

The early model development iterations for both the Zn rougher and Zn scavenger demonstrated the ability of the neural networks to capture important process control information in areas with sufficient data. Further insight was provided by this practice on future data collection requirements, and potential locations for control variable constraints in later optimisation. Due to the speed and non-invasiveness of modelling activities such as these, the authors believe that such practices can have a substantial impact on general control of operations through to the advanced process control design and accuracy maintenance.

Interfacing with the twin's models and inferring improved control set points

With the preliminary process model validation complete, the focus of the rougher-scavenger twin design returned to what the potential value of value of this system could be to the business in terms of recovery and concentrate grade impacts.

Various methods exist for completing this task ranging from a simple manual exercise of making step-wise set point selections for different feed types, through to the more complex approach of designing of an optimiser to solve for the optimal combination of set points at each time step. The advantage of the former method is the ability for a metallurgist to deeply inspect process scenarios and observe patterns in set point inferences across discrete feed types, at the cost of potentially missing a more 'global' solution. The advantage of the optimisation approach is the ability to infer set points for any potential combination of upstream variables without needing to discretise the data into arbitrary groups, at the cost of removing the 'expert' human interpretation of the best set points. Due to timeliness requirements for set point determination, a deployed twin ultimately requires this approach.

The Zn rougher-scavenger design followed the manual method initially, and required classifying the Pb Rougher Tail into four discrete classes representing high and low combinations for Pb and Zn grades in Pb rougher tail. The Pb and Zn assays were split at their median, however other approaches exist to segmenting these feed types, including making use of existing, potentially independent geometallurgical classes, and subsequently creating treatment strategies for each class. As this is a data-driven exercise, some care must be taken with this process: a large number of feed classes can potentially create operationally irrelevant groups, and reduce the data point count and confidence around model inferences.

A custom interface to the neural network models was created allowing a metallurgist to iteratively infer semi-optimal control set points for each feed. The iterative set point determination was conducted stepwise. That is, after making a selection for the pH set point for example, the pair of linked neural networks were re-evaluated, and the user was presented with an updated view of the process control relationships. Selections continued until the maximum recovery converged to a stable point with sensible set point selections (without negatively impacting concentrate grade). Multiple iterations of manual fine tuning by the metallurgist were typically required before a final set point suite for a given feed class was determined.

Once this process was completed for all feed classes (high Pb and low Zn, high Pb and high Zn etc) the system had produced a complete set of twin-derived set points for simulating against historical operator set point decision-making, and measuring the potential performance benefit.

RESULTS AND DISCUSSION

Estimated recovery and concentrate grade improvements

By combining the inferred new set points with its corresponding upstream feed condition into the process model, the potential recovery and concentrate impacts can be estimated, and statistically compared to the actual performance achieved at the time on the operator selected set points. Figure 8 shows the differential between the semi-optimised and original recovery data across the chosen feed classes outlined in methodology.

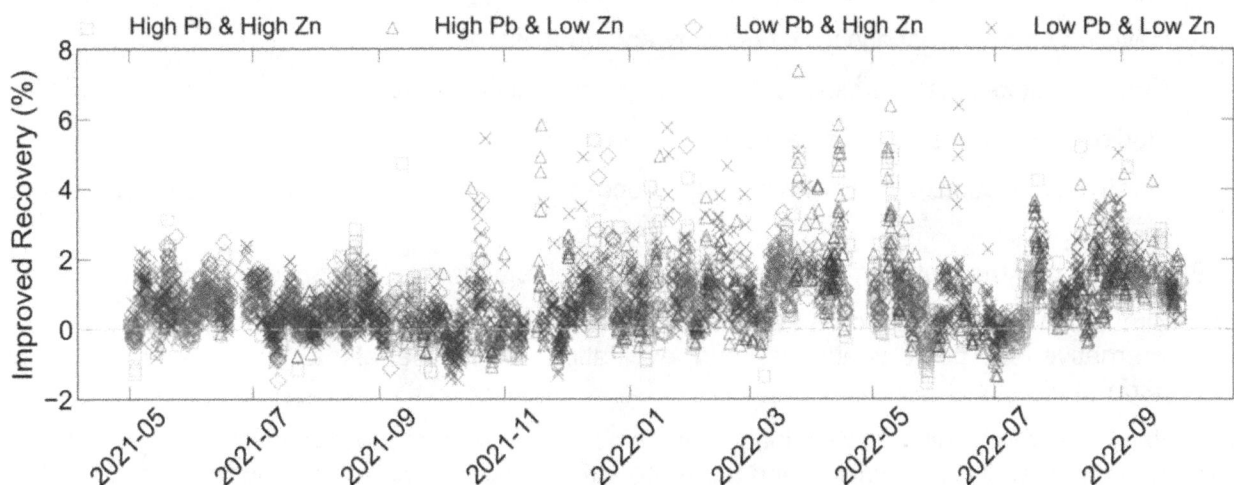

FIG 8 – Difference between actual and optimised recovery across the four discrete feed classes.

It is worth observing that in some cases, the applied model set points derived from the process twin's initial design models result in a negative benefit. This is due to a combination of factors including:

- The initial accuracy of the developed models.

- The metallurgical relevance of the four feed classes selected for optimisation.

- The quality of the manually inferred set points from the twin, and how well these relate to the general feed classes. This issue can be overcome with the usage of a well-designed optimisation layer.

- The reality that in some specific cases, the twin's recommendations may not exceed the incumbent plant performance. This highlights the importance of ongoing transparency of model accuracy when deployed in real-time.

However, Figure 8 demonstrates the value of decision automation – ie through the consistent application of just four set point strategies, an algorithmic approach to set point determination can outperform operator decision-making over the long-term. The final statistical difference across all four feed classes (when compared to actual data) at 95 per cent confidence was:

- +0.80 per cent absolute increase in overall rougher-scavenger Zn recovery with respect to Pb rougher tail (ie Zn rougher feed);

- a decrease in Zn scavenger concentrate grade of around -0.83 per cent, along with a +1.45 per cent, increase in rougher concentrate grade, resulting in a +0.54 per cent increase in combined concentrate grade. The strategies derived from the twin's models which resulted in the changing concentrate grade balance between rougher and scavenger are briefly discussed in the following section.

The magnitude of the recovery result, on the stated raw performance distribution (μ = 96.4 per cent, σ = 1.1 per cent, with regard to Zn rougher feed) was considered a reasonable and positive result from the design stage.

Reflections on the core twin design considerations

The preliminary twin design methodology followed in this paper covered a variety of structured metallurgical and analytical activities, including:

1. Assessment of the readiness of the Dugald River team to operate successfully and collaboratively with a process twin.

2. Selection of a target process node for twinning.

3. Iterative design of the process modelling framework.

4. Consideration of the modelling technologies required.

5. Review of and corrections to data quality and quantity.

6. Definition of key metallurgical and process control relationships.

7. Model performance metric selection and model baselining.

8. Iterative validation of the twin's early process models via traditional model performance metrics.

9. Validation of the metallurgical and process control relationships encapsulated within the twin.

10. Simulation of the twin's core systems to improve recovery across a range of feed types via alternative set point strategies, and evaluation of the benefit to the business of that performance uplift.

It is notable that the training and evaluation of the neural network process models represents only a fraction of the overall methodology, and there are a multitude of simple but important pre-modelling stages which, despite often being skipped by modelling teams, can help guide the success of the design process.

Following a structured methodology similar to the above places the metallurgist at the centre of the twin design process, and allows for a variety insights to be gathered throughout the design, irrespective of the design outcomes. For example:

- Irrespective of the feed type, the linked neural network models consistently guided the metallurgist towards control settings which maximised overall rougher-scavenger performance by raising the rougher tail grade and the rougher concentrate grade. The benefits here were consistently two-fold. Firstly, higher scavenger feed grades tended to make better usage of available scavenger flotation capacity. Secondly, the elevated rougher concentrate grade appeared to provide a 'grade buffer' allowing more aggressive control set points to be pursued on the scavenger without negatively impacting the combined concentrate grade. This observation was commensurate with operational experience.

- Despite different feed types exhibiting different specific values for 'down-the-bank' set points (eg staged collector, air rates), the profile or gradient determined by the neural networks was consistent, and provided deeper insight into the effectiveness of different types of control variables down the rougher-scavenger.

Regarding improvements to the methodology, experiments subsequent to this work have demonstrated that greater process modelling gains can be achieved through deeper interrogation and cleansing of raw process data in comparison to tuning model parameters themselves, which tended to provide diminishing returns. The methodology above could be improved by considering the weighting of each activity on a case-by-case basis and ensuring sufficient time is allocated to ongoing data refinement.

Future direction and anticipated challenges

From an operational perspective, the subsequent steps following this work include:

1. The development of a business case utilising the outputs from the study to assess the viability of progressing to implementation.
2. Completion of IT/OT architectural designs, risk assessments and approvals to support the future implementation.
3. Completion of detailed process modelling aimed at achieving production-grade robustness and stability within both the machine learning models and overarching set point solver.
4. Completion of the wraparound application development required to facilitate the digital twin data flows and set point orchestration back to end-use systems in a timely manner.
5. Completion of any operational change management, training and adoption support strategies required to ensure operator upskilling and acceptable utilisation of the system.
6. Detailed testing, deployment, commissioning and acceptance stages.
7. Development of the required twin maintenance strategies and model change approval processes required to sustain the value delivered from the system, including establishment of any technical support requirements.
8. Ongoing performance review and evaluation against the business case assumptions.

The above steps represent a six month program of work, excluding ongoing monitoring and maintenance. There are several challenges anticipated with the above activities which will generate new learnings in the twin development and deployment journey.

Firstly, the process of taking a set of desktop-evaluated models and transitioning into live production models will undoubtedly uncover further areas for process model improvement. A greater level of model validation and testing (beyond the standard approaches presented in this paper) will likely be required before a stable set of models is achieved for the process twin.

Secondly, per the digital twin definition followed in this work, the output set point recommendations from the rougher scavenger twin will need to be integrated back the specific systems or processes capable of executing the change. By securely deploying the twin's components to a point accessible by control layer, the set points will be made available for live advisory guidance to operators. To maintain a human-in-the-loop style of control, the preferred approach will be to give operators the choice to accept or reject set point recommendations, rather than allowing direct control. This initial approach represents just one component of the works required to steadily build trust between operator and twin over time.

Thirdly, the twin's process models will require regular ongoing maintenance to deal with both minor changes (eg instrumentation drift) and major changes (eg changing ore sources) within DRM over time. Standardising and simplifying the approach to retraining the models on new data, performing complete rebuilds, testing and approving candidate models and pushing approved models into production will become increasingly important over time.

Fourthly, given the Zn rougher-scavenger represents just one major process node within a larger plant flow sheet, the broader future opportunity with this technology is to scale the approach to other circuits requiring advanced decision automation, with the goal of connecting the system to achieve wider-area and potentially whole-of-plant process objectives. The greater the scope covered by the process digital twin, the greater the required level of system maintenance and longer-term management, implying that continued upskilling will be needed by the site team to realise the overall vision. The Dugald River processing team represent a steadily growing trend towards adding advanced control and modelling skills to traditional plant metallurgist capabilities.

CONCLUSION

In this work, the authors presented a definition for a flotation digital twin and outlined a series of practical considerations in the design of a rougher-scavenger flotation twin for the Dugald River Mine Zn circuit. Adhering to a structured methodology during the design process both de-risked the overall design and improved the team's technical understanding of the preliminary design aspects. The key aspects of the methodology presented in this paper focused less on the 'IT-related' activities (eg machine learning development and architecture design), and more on the operational, metallurgical and business perspectives of successful twin design. Important early considerations during the design process included establishing a rationale for the target circuit for twinning, specifying a process model framework which will support timely operational decision-making, selection of the types of modelling technology to be utilised, and identifying the presence of important metallurgical and process control patterns within the data as a stage-gate prior to any modelling or optimisation. Further critical considerations during early process modelling works included establishing consistent metrics and model benchmarks and validating the metallurgical sensibility of the process control patterns captured by the model against plant data. The final evaluation technique was conducted through manual (rather than solver-based) inspection of the digital twin's linked neural networks. The evaluation identified a material benefit to both rougher-scavenger recovery and combined concentrate grade. The twin design assessment recommended in proceeding to near-real-time connectivity and validation with the objective of enabling operators with advanced accept/reject advisory control for all rougher-scavenger flotation set points.

ACKNOWLEDGEMENTS

The authors would like to acknowledge both MMG and Hatch for the opportunity to conduct this work document the approach within this paper, and in particular, the support of the Dugald River metallurgy and process control team for their invaluable input and expertise.

REFERENCES

Amini, E, Koh, E, Beccera, M, Jara, C and Beaton, N, 2023. Development of Machine Learning Models using IES-ModelNet application for fast simulation of operation response to blocks of the resource block model, *26th World Mining Congress (WMC2023)* (CSIRO).

Bensoussan, A, Li, Y, Nguyen, D, Tran, M, Yam, S and Zhou, X, 2022. Chapter 16 – Machine learning and control theory, *Handbook of Numerical Analysis* (eds: E Trélat and E Zuazua), pp 531–558 (Elsevier).

Grieves, M and Vickers, J, 2017. Digital Twin: Mitigating Unpredictable, Undesirable Emergent Behavior in Complex System, *Transdisciplinary Perspectives on Complex Systems* (ed: F Kahlen), pp 85–113 (Springer).

Grieves, M, 2014. Digital Twin: Manufacturing Excellence through Virtual Factory Replication, White paper, Issue 12014, pp 1–7.

Jones, D, Jones, D, Snider, C N, Yon, J and Hicks, B, 2020. Characterising the Digital Twin: A systematic literature review, *CIRP Journal of Manufacturing Science and Technology*, 29A:36–52.

Soekhoe, D, Van Der Putten, P and Plaat, A, 2016. *On the impact of data set size in transfer learning using deep neural networks* (Springer International Publishing: Stockholm).

Operation updates

A review of operating data over 50 years of secondary crushing prior to AG/SAG mills

R Chandramohan[1] and G Lane[2]

1. FAusIMM, Global Technical Director – Operations and Process Optimisation, Ausenco, Vancouver BC, Canada. Email: rajiv.chandramhan@ausenco.com
2. FAusIMM, Principal Consultant, Ausenco, Brisbane Qld 4000. Email: greg.lane@ausenco.com

ABSTRACT

Since their first application in the late 1950s, autogenous grinding (AG) and semi-autogenous grinding (SAG) mills have provided the modern mining industry with the workhorses used for most high-throughput comminution applications. Secondary crushing is a means to debottleneck the AG/SAG mills when processing coarse-competent feed. In the early years of AG/SAG milling, secondary crushers were used in the flow sheet to manage feed size for smaller diameter AG/SAG mills and allow high capacity. As bigger shell diameters and larger motors were adopted, secondary crushing options were rarely considered in greenfield flow sheet design but instead regarded as an option to deconstrain an undersized AG/SAG mill when treating high competency or coarse feed.

Feeding AG/SAG mills with secondary crusher products changes the breakage behaviour and classification of the ore in the mill when compared with mills fed with primary crusher products. This requires changes to the ball and total mill filling levels, process control, shell liners and discharge system designs.

This paper reviews the last 50 years of published data on comminution circuits that use secondary crushing before AG/SAG mills. The key objectives are to assess:

- options for optimising the impact of critical size fractions in the mill feed

- the opportunities to reduce the grinding media consumption, therefore minimising the overall Greenhouse Gas Emissions (GHGe) footprint for high-throughput applications

- the impact of ore competence on optimum mill load (rock and steel) composition for maximising throughput and project value.

INTRODUCTION

Autogenous Grinding (AG) and Semi-Autogenous Grinding (SAG) mills are the modern-day workhorses in most hard rock mining. Since the early 1960s, their use has steadily grown due to their operational versatility and adaptability to handle a wide range of ore sources in many different circuit configurations. Comminution flow sheets comprising AG/SAG mills can treat various feed size distributions and ore competencies fed from primary gyratory, secondary or tertiary cone crushers, high-pressure grinding rolls (HPGRs), pebble crushers, scalped screens or run-of-mine feed (such as those implemented in South Africa in the gold and platinum industry) (Chandramohan et al, 2023; Chandramohan, Lane and Morley, 2021; Powell, Morrell and Latchireddi, 2001).

Two key ore factors influence the throughput in AG/SAG mills: the feed size distribution and the competency (Morrell and Valery, 2001; Lane, 2007). These factors influence the breakage rates for a particular operating condition, in turn affecting the resultant AG/SAG mill-specific energy. It is well documented that reducing the feed size distribution to a SAG mill reduces the mill's specific energy, thus increasing throughput. However, reducing the feed size to an AG mill can result in loss of 'grinding media' and a reduction in throughput. Most secondary crushing installations installed on comminution circuits since the 1970s are retrofits; due to not achieving the design throughput or as a critical component of cost-effective expansion.

Staged crushing before SAG milling can significantly increase circuit throughput and relieve SAG mill throughput constraints due to competent or coarse feed. For expansion projects requiring a significant throughput increase (eg doubling in throughput capacity), an analysis is necessary to determine the overall project capital expenditure required for the additional crushing facilities versus installing an entire new AG/SAG mill line, factoring in the life of the mine ore source, location, footprint and other specific project factors.

Giblett and Putland (2018) and Chandramohan, Lane and Morley (2021) summarised optimisation strategies for throughput-constrained SAG mill circuits. They highlighted key opportunities related to feed size distribution changes (data consolidated in Table 1). The biggest throughput gains are from pre-crushing (secondary or tertiary feed to the SAG mill).

TABLE 1

Opportunities for size reduction of the feed to the AG/SAG mill.

Focus	Performance benefit	Description	References
Mine to mill	12–18%	The fragmentation of the hard ore is increased using high-intensity blasting (HIB), through a combination of increased powder factor and reduced blasthole spacing. The fragmentation from HIB is fines-rich, with product size distribution curves similar to those of the ores with lower rock quality designation (RQD). HIB is a common practice with ores that have extreme competency (Axb < 35), aiming to reduce the top-size distribution in the primary crusher product.	Confluencia, Porgera, KCGM, Batu Hijau
Pebble crushing	5–15%	Pebble crushing is a cost-effective solution to debottleneck the AG/SAG mill. Pebble crushing reduces the critically sized pebbles that the mill does not process. This pebble-size feed distribution typically ranges between P_{80} 30 and 60 mm. The pebble crusher's closed-side setting is best set to crush below the mill's discharge screen's mean aperture. The crushed product from the pebble crusher is either recirculated back to the SAG mill or pushed forward to the ball mills, depending on the constraint in the grinding circuit.	Candelaria, Confluencia (and most semi-autogenous ball mill crushing circuits (SABC))
Pre-crushing	10–100%	Pre-crushing incurs significant capital cost to debottleneck the AG/SAG mill. Fundamentally, pre-crushing reduces the feed size of the SAG mill feed partially or fully to achieve significant throughput gains. Most pre-crushing circuits prior to AG/SAG involve secondary crushing. However, there are some circumstances for extreme ore competencies; three crushing stages are required for size reduction. Installing high-pressure grinding rolls (HPGR) as the third crushing stage has significant benefits. The product from HPGR can be sent to the ball mill circuits, bypassing the SAG mill entirely.	Two-stage – Porgera, Granny Smith, Kidston, Mt Rawdon, Geita, Arsarco Ray, Phoenix, Copper Mountain, Three stage – St Ives (early plant), Cadia, Peñasquito

During the staged development of Cadia from a SABC circuit to an open-circuit secondary crush, SAG, pebble crusher and ball mill (2C SABC) circuit, higher throughputs were achieved in secondary crush, SAG and ball mill (2C SAB) mode than in SABC mode using the same equipment – the pebble crushers were redeployed to a secondary crushing duty (Engelhardt *et al*, 2011). Engelhardt *et al* (2011) noted that the SAG mill discharge-grate apertures need modification (reduced) to manage the pebble generation and maintain mill load for the secondary crushed feed.

A consequence of the size reduction of the feed to the SAG mill is increased product grind size distribution – where a coarser circuit grind size can result in lower metal recoveries if additional ball mill capacity is not added. In mine-constrained scenarios, where a SAG mill is constrained by ore supply, a potential optimisation step may be to maximise the energy efficiency in the mills by coarsening the feed size distribution to the mill and reducing the steel media charge level in the mill load to increase fines generation (Siddal and Putland, 2007; Lane, 2007; Chandramohan, Lane and Morley, 2021). A coarser feed with reduced steel media consumption can reduce operating costs if the ore is of medium competence (if competence is too low, coarse feed breaks rapidly leaving no autogenous media in the mill; if the competence it too high, coarse feed breaks too slowly resulting in lower grinding efficiency). However, the cash flow in these mine-constrained operations is severely impacted by a lack of metal production, which is a function of throughput.

Siddal and Putland (2007) highlighted that '*Treating 100% of secondary crushed is difficult in anything other than a purpose-built and controlled circuit. Therefore, using partial secondary crushed feed best satisfies the operating requirements of stability and balance.*' Overall, the partial secondary crushed feed can provide the following benefits:

- increased grinding efficiency.
- use of coarse rock as media, reducing the requirement for steel media.
- fully utilise the available installed power by balancing the energy between the grinding stages.

Siddal and Putland (2007) concluded that it is important to tailor the correct coarse and fine feed blend through partial secondary crushing based on the power split between the SAG and ball mills for retrofit and expansion projects.

Powell *et al* (2015) proposed that the primary crusher product's mid-size fraction (approximately -90 +20 mm) can be screened and processed separately from the rest of the mill feed to maximise throughput, increase the metal grade through ore sorting, and reduce operating costs by reducing the grinding media volume in the SAG mill. The premise of their conceptual work is similar to the work undertaken at Trollius (Sylvestre, Abols and Barratt, 2001) and Fimiston (Nelson, Valery and Morrell, 1996), where the mid-size fraction was that grinding media consumption could be significantly reduced by feeding the SAG mill with a blend of coarse and fine secondary crushed feed, emphasising that fully-secondary crushed SAG feed is less effective.

Hanhiniemi's (2023) PhD thesis followed Powell *et al*'s (2015) concept with detailed techno-economic modelling of two SAG-based comminution circuits (Barrick's Cortez and Newmont's Cadia) with feed modified by high-intensity blasting ore blending strategies and partial pre-crushing. The SAG mill was operated with varying ball charge levels and grate apertures to determine the impact on throughput and circuit grind size. The trade-off factored circuit modification costs, wear rates and improved recoveries at reduced grind sizes. The simulation outcomes for Cadia showed that partial mid-size crushing of the SAG mill feed and operating with a lower ball charge and higher mill filling provided the highest net present value (NPV) – which predicts a lower throughput than the current 2C-HPGR-SABC operation. Hanhiniemi's (2023) outcomes assume that the efficiencies apparent with low to moderate competence ores reported by Powell are also applied to competent ores.

A key aspect of treating fine versus bimodal feed in a SAG mill is the ability to draw the power for grinding to achieve the target throughput. The mill load is 'more reactive', and it can be more challenging to establish and maintain a target mill filling if the coarse material in the mill feed has been removed. The breakage residence time of primary crushed feed differs from secondary or tertiary crushed feed in a mill, a function of ball load, mill speed and slurry discharge (Chandramohan,

Lane and Morley, 2021). There are operational changes that can increase the grinding residence time of fine feed at a fixed ball load in an SAG mill, such as:

- operating with a higher slurry discharge density.
- reducing grate aperture and open area or removing pebble ports.

However, an analysis needs to assess coarser versus fine feed to the SAG mill, factoring in mill load and media consumption. Most secondary crushing circuits installed to date are to debottleneck the SAG mill for competent and coarse feed. With that objective in mind, the question is:

> 'Can throughput be maintained if the rock load increases (and ball load decreases)? And what is the efficiency improvement?'

Based on a review of technical papers relating to secondary crushing before AG/SAG mills, this paper aims to address the following focusing questions:

- How is the impact of mill feed size distribution on critical size fractions optimised?
- What are the opportunities to reduce the grinding media consumption, thereby minimising the GHG footprint for high-throughput applications?
- How does ore competence impact the optimum mill load (rock and steel) composition to maximise throughput and project value?

Evolution of secondary crushing flow sheet development

Most secondary crushing operations installed since the introduction of AG/SAG mill circuits are retrofits, primarily necessitated by the debottlenecking of SAG mill capacity due to design-related capacity issues or expansion considerations in later project development.

When AG/SAG milling was first introduced in the minerals processing flow sheet in the late 1950s, the selection and sizing of mills were typically based on pilot plant trial data (Chandramohan *et al*, 2023). SAG mill pilot plants require large samples of the target ore. Ore characterisation test work and the mill sizing calculations were supported by pilot plant-based data. Design parameters mimicked the pilot plant trials configuration, such as ball charge and total charge, closing screen apertures and operating mill speeds. Given the limitations associated with obtaining large and representative samples, the early installed AG/SAG mills were susceptible to poor performance due to higher ore competence. Some of these projects were debottlenecked using secondary crushing retrofits (for example, Fimiston, Porgera, Granny Smith, Kidston, Trollius).

Advancements in ore characterisation techniques and increased data resolution using grindability and single particle comminution tests reduced project risks associated with SAG-based circuit design (Chandramohan *et al*, 2023). However, several SAG-based projects still succumbed to poor performance, and as a result, some financiers became less confident in AG/SAG-based circuits (Staples *et al*, 2015). These typical issues related to poor performance are one or a combination of:

- unrepresentative sampling
- poor testing protocols
- inappropriate interpretation of data and definition of design criteria
- inaccurate mill sizing methods and/or
- poor project management decisions.

Staples *et al* (2015) concluded that scepticism about AG/SAG mill performance is largely due to poor mill sizing techniques and appropriate benchmarking when designing circuits to treat competent ores. Bailey *et al* (2009) and Lane (2007) commented that in some cases, poorly complied JKSimMet models, interpretation of the data for competent ores, and general lack of experience in understanding the previous project failures are some of the causes of poor AG/SAG mill sizing. Staples *et al* (2015) recommended:

> If Axb values are below, say, 40 or DWI values are above 6.5, due consideration should be given to the inherent competency of the rock and benchmark data relating

ore competency to SAG mill-specific throughput. The scale-up of pilot plant mill-specific energy to large-scale operation requires careful consideration of the pilot mill discharge function and associated breakage rates external to the mill in SABC mode. As a minimum, careful benchmarking of design outcomes against available data in the public domain is strongly recommended.

A controlled feed size distribution is critical to achieving the design throughput and product circuit grind size distribution in AG and SAG mills (Chandramohan *et al*, 2023). Pebble or secondary crushing provide two methods for managing the impact of critical size in AG and SAG mills.

The secondary crushing circuits prior to SAG milling have been operated in three modes, depending on competency and feed size distribution of the feed to the crusher:

- **Open circuit without pre-screening** – Most smaller circuits treating competent ore that is fines deficient have used direct-fed secondary crushers where primary crushed ore directly feeds the open circuit secondary crusher. This is the lowest capital cost option and is suited to smaller applications treating competent ore.

- **Open circuit with pre-screening** – Most secondary crushers are operated in an open circuit with a pre-screen. Pre-screening the fines is important to minimise the 'ring-bounce' of the crusher anvil due to reduced void spaces in the ore in the crusher chamber. High fines in the crusher feed result in premature failure of the bearings/bushes and support bolts caused by additional stresses.

- **Closed circuit** – Closed circuit operation is required when maximum throughput-increases are necessary. The screen is used to remove all coarse material, maximising SAG mill throughput. If the screen is fine enough, the operation of the SAG mill can become similar to that of a ball mill.

Secondary crushing before AG milling (or low ball load (<8 per cent balls) SAG milling) can balance the mill load using optimal proportions of fine and coarse rock in the mill charge. Bueno, Foggiatto and Lane (2019) conducted extensive pilot plant trials on competent feed (Axb <30 and Bond Work index (BWi) ~ 20 kWh/t) in various circuit configurations to determine the optimum feed size distribution for grinding efficiency. One of the trials included pre-crushing the -75 + 25 mm fraction (critical size fractions) in the feed to determine the mill performance. Bueno, Foggiatto and Lane (2019) concluded that pre-crushing the 'critical size fraction' improved the milling efficiency by reducing the production of ultra-fines and produced a product size distribution's exponent closer to 0.5 (typical of stage crush, rod and ball mill circuits). Bueno, Foggiatto and Lane (2019) emphasised that maintaining optimal mill load is crucial in AG milling with pre-crushed feed; therefore, adequate mill and feed blend control was critical. Projects such as Trollius have utilised this approach to treat competent ores with AG and lower ball load SAG milling (Sylvestre, Abols and Barratt, 2001).

Most secondary crushing facilities are installed before the coarse ore stockpile (COS), feeding the AG/SAG mill (Figure 1). Depending on the duty, these facilities process the entire primary crusher product, or partial feed is scalped using an overflowing bin or diverter chute. Tertiary crushing to SAG mill flow sheets is rare and aims to debottleneck the AG/SAG mill for extremely competent ores. The third stage in tertiary crushing can be cone crushers or high-pressure grinding rolls (HPGR).

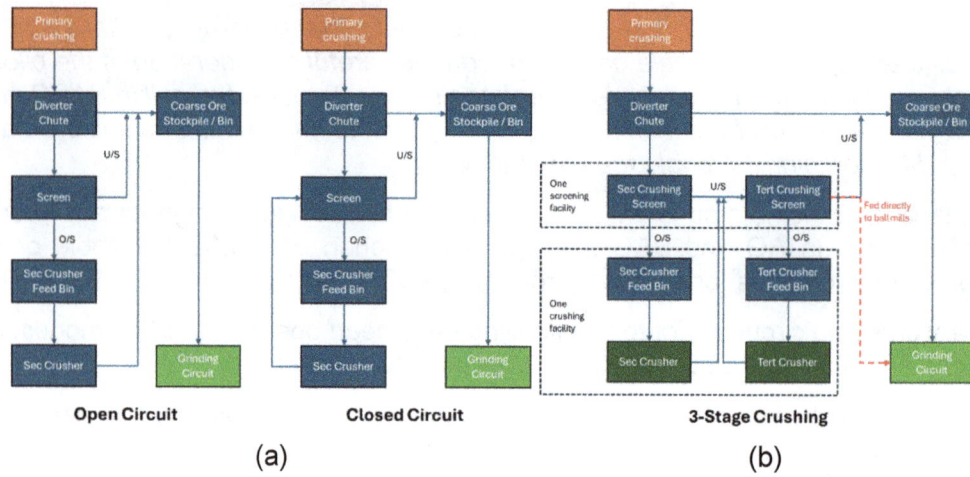

FIG 1 – Schematics of full and partial secondary crushing flow sheets in open and closed circuit configuration: (a) full or partial secondary crushing; (b) tertiary crushing.

Siddal and Putland (2007) provided an order-of-magnitude capital cost estimate for varying secondary crushing circuit configurations (Table 2). They stated that open-circuit intermediate or partial crushing installed off the mill feed conveyor (ie after COS) is not suitable for top-end competent ores (Axb<30) as uncrushed coarse feed will restrict SAG mill throughput when the crushing circuit goes offline. Therefore, crushing circuits should be installed before the COS for these top-end ores. However, for low competence ores (circa Axb>50), there are benefits of installing partially crushing circuits off the mill feed conveyor to dynamically balance the SAG mill load, throughput, and fines generation.

TABLE 2

Secondary crushing selection criteria adapted after Siddal and Putland (2007).

Secondary crushing configuration	Order of magnitude CAPEX*	Advantage	Consequence
Open circuit, no screening	Low CAPEX	-	Requires balancing the primary crusher's and secondary crusher's CSS's to maintain throughput at reduced size distributions – which results in more intermediate feed to the AG/SAG mill
Open circuit, scalping screening	Medium CAPEX	-	The secondary crusher may be restricted by the primary crusher product's top-size a
Open circuit, intermediate crushing	Medium-High CAPEX	Allows for finer secondary crusher CSS setting	Not suitable for extremely top-end competent ores. The amount of secondary crushed material is influenced by the primary crusher product.
Closed circuit product screen	High CAPEX	No intermediate-size generation	-

* Capital Expenditure.

Table 3 summarises operations with secondary and tertiary crushing facilities operating under partial or full SAG feed crushing and under open and closed operations. The data were sourced from published conference papers and journals. Most flow sheets have an allowance for feed bypass for unscheduled maintenance. In this scenario, the grinding circuit may receive the coarser primary crushed feed. In some operations, contract crushing is used to reduce coarse feed to the SAG mill. Coarse rock is usually scalped and processed from the sides of the stockpile – the product from the contract crushers is returned to COS or fed directly to the SAG mill. Generally,

- There is an increase in circuit throughput, and the circuit grind size coarsening.

- Secondary crushing is most effective with competent ore.

- The average ball charge used in the SAG mill ranges between 12–15 per cent v/v.

- Both the SAG mill and ball mill components have equal power. However, for tertiary crushed feed, larger ball mills (installed power) are required due to the coarsening of the transfer size sent to the mills (>60 per cent power split), and the SAG mills have much higher ball charge levels (>15 per cent v/v).

- Full-feed secondary crushed operations present the SAG mill with a mean F_{80} of 45 mm.

- Partially secondary crushed SAG mill feed ranges between F_{80} of 60 and 90 mm, depending on the proportion of the crush.

TABLE 3

Reviewed secondary crushing projects (SAG-based circuits).

Project Name	Full or Partial	Location of Secondary Crushing	Open or Closed Circuit	Throughput, kt/d	Circuit	Axb and BWI	Total installed grinding power, MW	SAG and Ball power split, %	Circuit Feed and Product, mm, um	SAG mill ball load, %v/v	Reference
Porgera	partial	Before COS	open	15	2C-SAB	n/a	18	50:50	90, 106		Thong, Pass and Lam, 2006
Granny Smith	full	Before COS	closed	8	2C-SAB	n/a	7.9	49:51	30, 150		Thong, Pass and Lam, 2006
Kidston	full	Before COS	open	19	2C-SAB	n/a	8	50:50	24, n/a		MacNevin, 1997; Nedham and Folland, 1994
Mt Rawdon	both	Before COS	open	9	2C-SAB	n/a	8	50:50			Putland, Siddal and Gunstone, 2004
Geita	partial	Before COS	open	16	2C-SAB	n/a	18	50:50			Mwehonge, 2006
St Ives	both	Before COS	closed	9	2C-SAB	45, 15.1	7.3	35:65	21, 161	15	Atasoy, Valery and Sklalski, 2001
Trollius	partial	Before COS	open	15	2C-SAB	37, 10	9.6	54:46	152, 95	10	Sylvestre, Abols and Barratt, 2001
Ascarco Ray	partial	Before COS	open	11	2C-SAB	23, 13.4	20	50:50	89, 111		McGhee et al, 2001
Meadowbank		Before COS	open	8	2C-SABC	n/a	8.1	46:54	25,106	15	Muteb and Fortin, 2015
Goldex		Before COS	open		2C-SAB	33, n/a	6.6	50:50	35, 140		Allaire et al, 2011
Pheonix	partial	Before COS	open		2C-SABC		27.7	49:51		12	Castillo and Bissue, 2011; Giblett, 2014
Cadia	full	After COS	closed	80	2C-HPGR-SABC	36, 22	52	38:62	12, 200	18	Waters et al, 2018; Lane, Seppelt and Wang, 2018
Edna May	full	Before COS	open	5	2C-SABC	48, n/a	5.7	35:65	35, n/a	14	Dance et al, 2014
KCGM (Fimiston)	both	Before COS	open		2C-SAB	32, n/a	19.7	60:40	39 and 117, 230	12 - 15	Nelson, Valery and Morrell, 1996
Peñasquito	partial	After COS	open	90	2C-HPGR-SABC	25,18	86.6	44:54	75, 110	17	Linde et al, 2019
Tarkwa	partial	Before COS	open		3C-SABC	55, n/a	28	50:50		18	Mainza et al, 2015
Jundee	full	Before COS	open	12	2C-SABC	29, 24	5.6	67:33	33, 106		Matson, Tyrell and Jackson, 2018
Kansanshi (sul)	full	Before COS	open	38	2C-SAB	51, n/a	19.5	72:28	70, 118	20	Powell et al, 2015
Kansanshi (mx)	full	Before COS	open	29	2C-SAB	48, n/a	6.2	58:42	100, 232	18	Powell et al, 2015
Malartic	full	Before COS	open	55	2C-SABC	38, 16.1	55.4	35:65	55, 65		Staples et al, 2015
Copper Mountain	full	Before COS	open	35	2C-SABC	n/a, 22.3	38.4	33:67	n/a, 150	15	Staples et al, 2015
Confluencia	partial	After COS	closed	95	3C-SABC-B	35, 16	56	42:58	53, 240	17	Beccera, M and Amelunzen, 2012
Mt Milligan	full	Before COS	open	60	2C-SABC	21, 20	48	46:54	60, 220		Staples et al, 2015

A tale of two circuits

Newmont's Cadia and Peñasquito operations are examples of comminution circuits that expanded their throughput capacity due to increased ore competence. Significant changes were made to the SAG mill operating philosophy (ball charge, rock load, discharge system) to manage the changing feed size distributions.

Cadia

The feed to the Cadia concentrator changed from an open pit to a more competent and harder underground block cave ore. This required debottlenecking of the SAG mill. Initial plant trials indicated that throughput could be nearly halved without modification to the feed size distribution (Engelhardt *et al*, 2011). Three-stage crushing, including secondary crushing in a closed circuit and an HPGR operating with edge recycle (2C-HPGR-SABC), was installed (Figure 2).

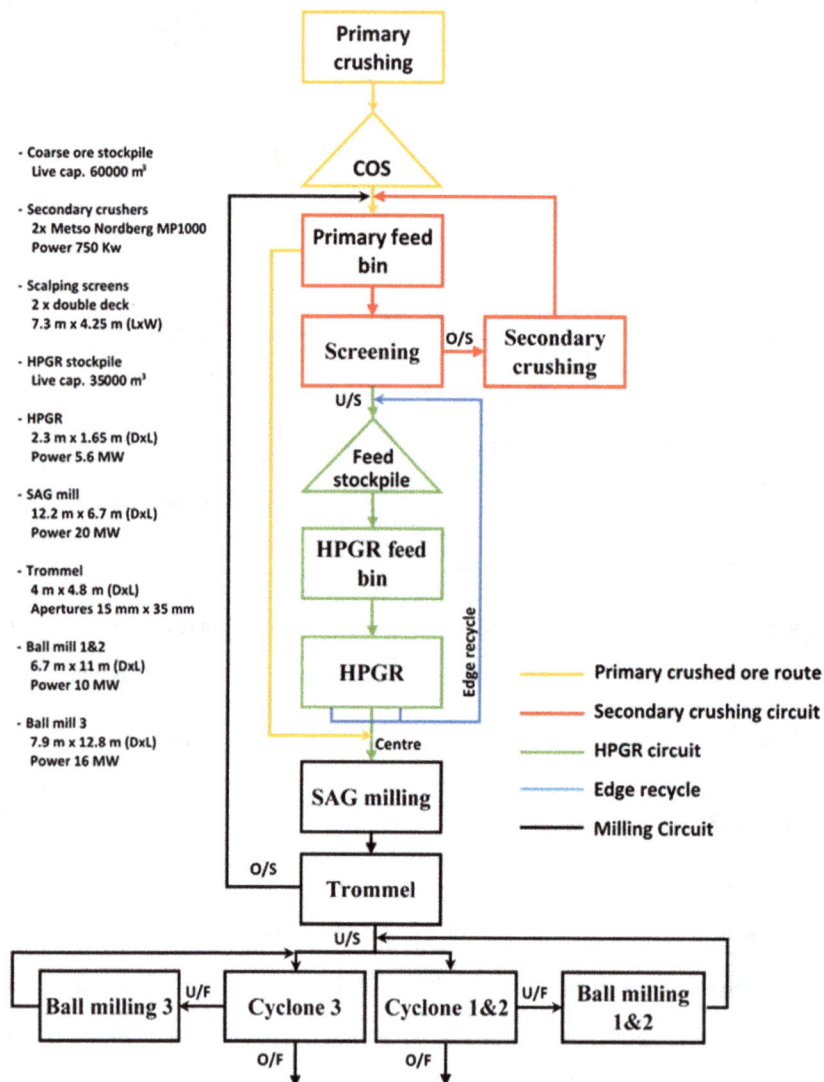

FIG 2 – Cadia Valley operations design (after Lane, Seppelt and Wang, 2018).

Table 4 and Figure 3 highlight the incremental performance change of the Cadia circuit before and after expansion. Feed size distribution is the key factor driving the increase in the SAG mill throughput. The product from the HPGR, with an F_{80} of 18 mm, changes the behaviour of the breakage characteristics in the SAG mill, requiring the mill's ball charge level to be increased to draw the power.

TABLE 4

Cadia's performance over the years (after Engelhardt, Lane and Powell, 2014; Lane, Seppelt and Wang, 2018; Geoghegan and Haines, 2023).

	Design (pre-expansion) 1998	Survey 2008 (Cadia Open Pit	Survey 2008 (Cadia East UG ore – blasted)	Design 2011 (expansion)	Survey 2016	Survey 2019
Circuit Configuration	SABC	SABC	SABC	2C-HPGR-SABC	2C-HPGR-SABC	2C-HPGR-SABC
BBWI, kWh/t	17.5	17.5	21.4	21.5	20.7	19.7
Axb	35	33	27	27	33	29
SAG mill fresh feed, t/h	2100	2308	1482	2455	2817	3239
SAG mill F_{80}, mm	120	110	80		16.5	17.6
SAG mill discharge trommel undersize, um					1840	1640
SAG mill power draw, MW	19.5	18.7	15.7		16.0	19.4
SAG mill ball charge level, %v/v	14	14	14		14	17
SAG mill specific energy, kWh/t	9.2	8.1	10.6	6.9	6.1	6.0
Ball mill specific energy, kWh/t	12.9	8.5	13.2	13.0	12.4	9.8
Circuit grind size P_{80} (cyclone O/F), um	150	175	140	150	124	200

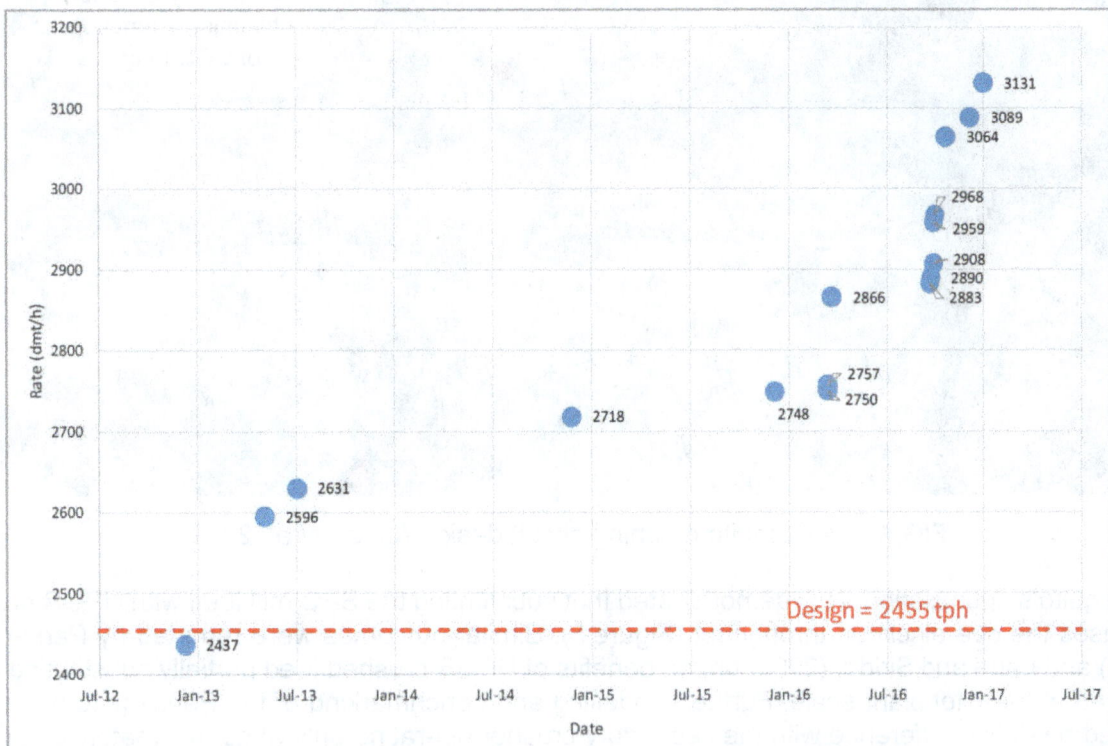

FIG 3 – History of Cadia of performance treating >50 per cent Cadia East Ore (after Waters *et al*, 2018).

Waters *et al* (2018) commented that due to the 'unknown behaviour of the hybrid feed' in the 40 ft diameter SAG mill, a technical decision was made based on two operating philosophies of the ball charge levels and rock filling:

- Running the SAG mill with lower ball charge (<13 per cent v/v) levels and high rock filling (>7 per cent v/v) using the middle feeder from the crushed ore bin, which delivered coarse, secondary crushed rock.

- Running the SAG mill with higher ball charge (>15 per cent v/v) and lower rock filling (<7 per cent v/v) to maximise breakage of the competent and hard Cadia underground ore.

Waters *et al* (2018) summarised that the final decision to operate the Cadia SAG mill with a higher ball charge was based on:

- Higher throughput rates were achieved operating with the higher ball load and lower rock filling.

- A higher ball load stabilised the mill for feed variations relating to throughput, size and hardness.

Peñasquito

Peñasquito is an example of secondary crushing and HPGR units retrofitted to the SABC flow sheet due to the inability to consistently reach design throughput targets due to ore competence. The retrofitted feeder to the secondary crusher is adjacent to the two parallel reclaim feeders feeding the primary crushed product to the two SAG mills (Figure 4). The secondary crusher product screen oversize is sent to the pebble crushers, and the screen undersize is combined with the pebble crusher product and sent to the HPGR unit, which has the option of supplementing the feed to the SAG mills or feed directly to the ball mills.

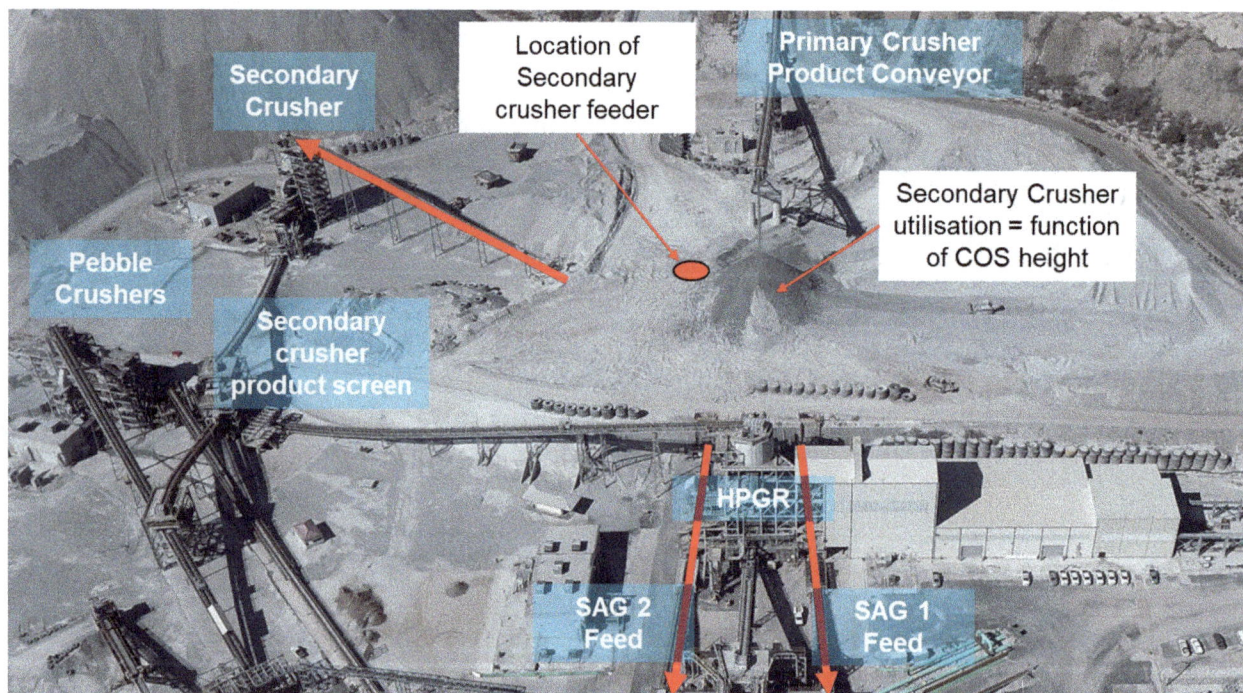

FIG 4 – Peñasquito crushing circuit design (Linde *et al*, 2019).

Peñasquito's optimisation work demonstrated that substituting the SAG mill feed with HPGR product increases the overall circuit throughput (Figure 5). Similar outcomes were reported by Parker *et al* (2001) and Lane and Siddal (2002) on the benefits of HPGR crushed feed partially substituting SAG mill feed at the pilot plant scale. Further modelling and benchmarking of the Peñasquito plant data showed a relative difference with the secondary crusher operating only at approximately 8 per cent and with HPGR operating in tertiary crushing mode, adding 7 per cent in throughput (relative throughput difference shown in Table 5). Unlike the Cadia SAG mill, minimal changes were applied to the Peñasquito SAG mill internals, primarily due to the changing feed composition, which depends on the selected operating strategy. Therefore, the SAG mill grates and liners were chosen for the competent coarse feed – directly from the primary crushers.

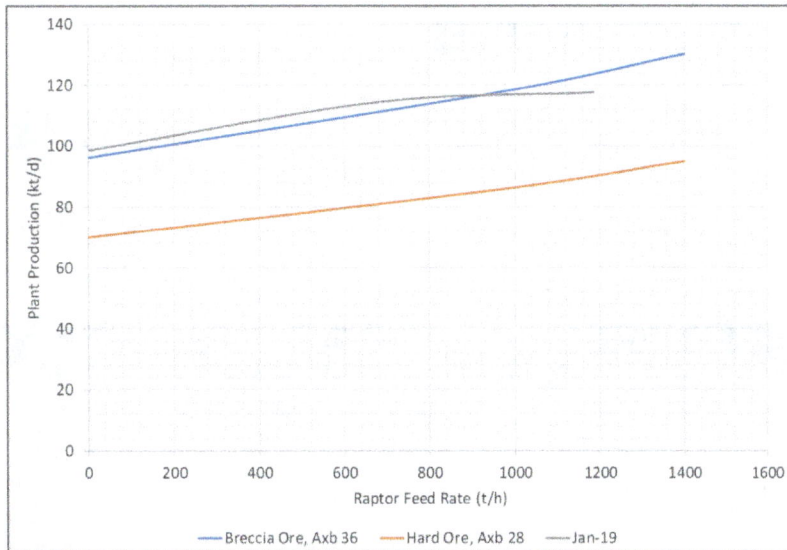

FIG 5 – Peñasquito's secondary crusher feed impact on overall plant throughput (after Linde *et al*, 2019).

TABLE 5

Relative throughput difference for varying operating strategy, Linde *et al* (2019).

	HPGR OFF	HPGR ON
Secondary Crusher ON	-8%	baseline
Secondary Crusher OFF	-15%	-16%

As part of the optimisation work undertaken at Peñasquito, the operating philosophy of the SAG mill was changed to maximise throughput for the competent feed (Linde *et al*, 2019). During a crash-stop survey, it was observed that the SAG mill was operated with high rock filling, with the total charge estimated at 39 per cent v/v (Figure 6). The high operating mill filling levels were primarily driven by the steeper shell liner design selected for the duty. The mill's operating philosophy operations was to protect the shell liners for the coarser feed, which, as a result, congested the mill severely and limited throughput.

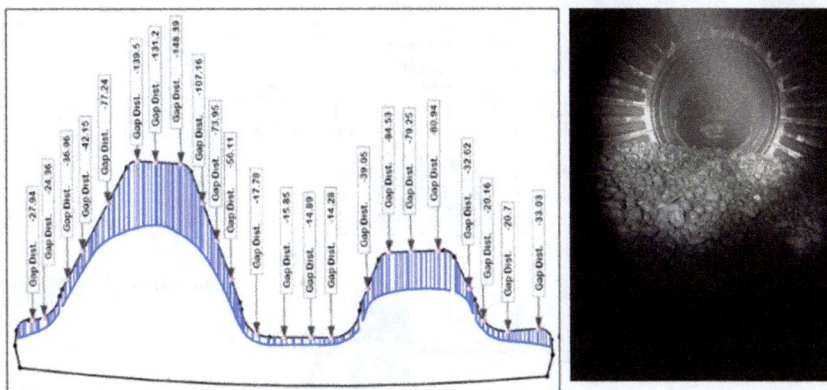

FIG 6 – Peñasquito's SAG mill shell liner design and survey filling levels (after Linde *et al*, 2019).

SAG mill load trials were conducted to determine the optimum ball load and mill filling levels (Figure 7). The ball charge levels were increased from 14 per cent to 18 per cent v/v with reduced rock load. The mill speed was reduced to prevent direct shell impacts, resulting in improved energy efficiency at the higher feed rate (reduced SAG mill-specific energy at reduced power draw).

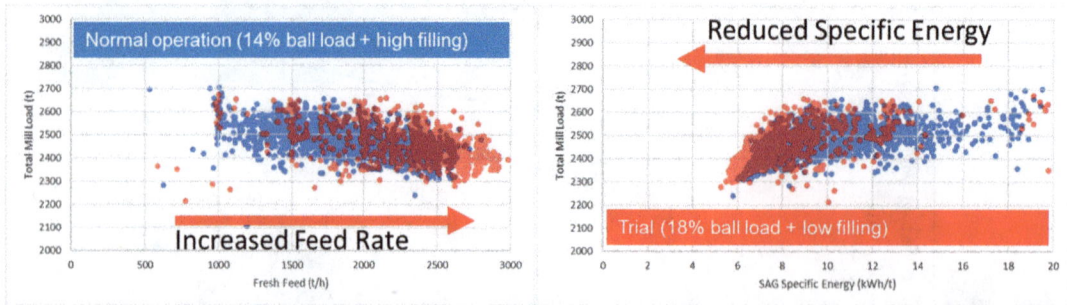

FIG 7 – Peñasquito SAG mill load trials (after Linde *et al,* 2019 – presentation).

IMPACT OF MILL LOAD COMPOSITION ON EFFICIENCY AND THROUGHPUT

Powell *et al*'s (2015) paper, and later Hanhiniemi's (2023) PhD thesis' premise, is that operating the SAG mill with a lower ball charge and higher rock load when processing secondary crushed feed can achieve significant cost savings when compared with Cadia's mode of operation with high ball charge and low rock load. They propose that the SAG mill can be optimally operated with pre-screened feed, where only the mid-size fraction is pre-crushed and processed through the SAG mill with a lower ball charge (Figure 8). The proposed flow sheet is more complicated, requiring multiple conveyors, screens, and bins than a typical secondary crushing circuit. They suggest that the overall net savings through reduced operating costs at the lower ball charge and associated increased throughput and metal production would increase NPV compared with a 2C-SABC operating with a higher ball charge and lower rock charge.

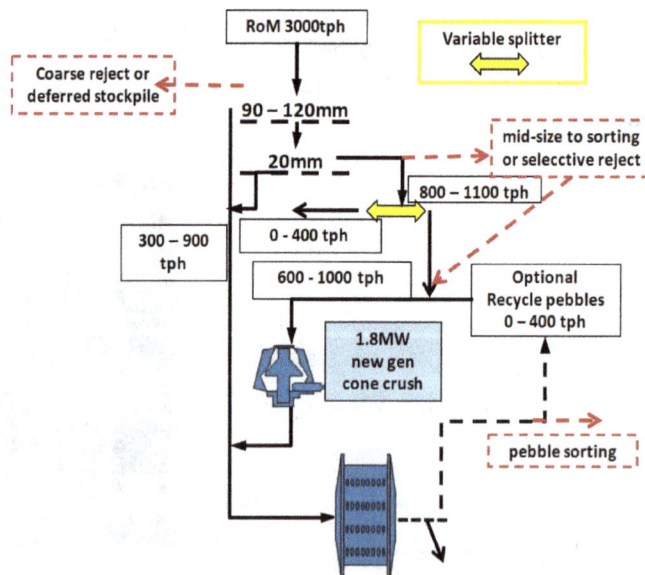

FIG 8 – Mid-size pre-crushing circuit, (after Powell *et al,* 2015).

Powell *et al*'s (2015) circuit configuration is a variation of the Trollius flow sheet (Figure 9) The Trollius AG mill was partially secondary crushed, where the -6 inches + 1 inches (-152 + 25 mm) primary crusher product was scalped and crushed below 1.5 inches (38 mm). The AG mill operated with an 8 per cent v/v ball charge and high rock filling (total filling ranging between 22–26 per cent v/v) and closed with a pebble crusher operating with a closed side setting (CSS) < 0.5 inches

(~12.5 mm). Sylvestre, Abols and Barratt (2001) noted that although the coarse ore competency was considered hard (Axb <40), the measured BWI was atypical for this type of ore, ranging in the soft end of the JK Database <10 kWh/t.

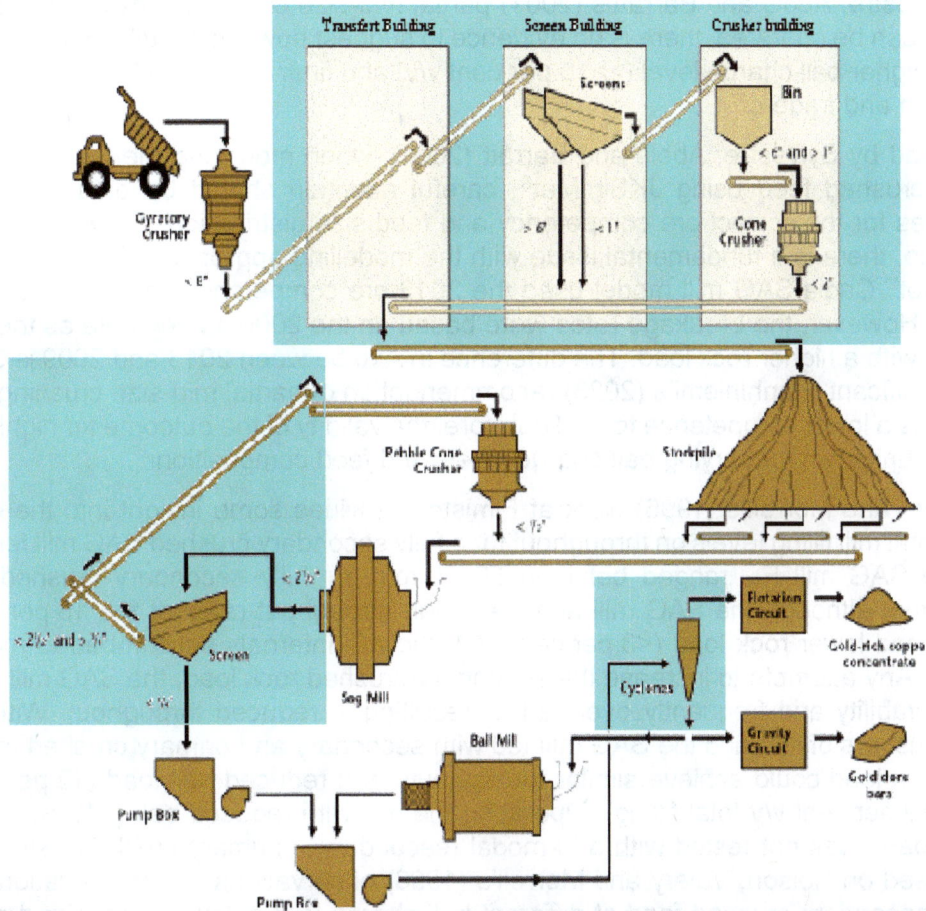

FIG 9 – Trollius mill flow sheet (after Sylvestre, Abols and Barratt, 2001).

To demonstrate the benefits of partial pre-crushing feed in the Trollius SAG mill, a JKSimMet® model was developed based on a series of surveys taken in 1998, before the installation of the pre-crushing circuit. Sylvestre, Abols and Barratt (2001) highlighted that the model correctly predicted the beneficial effect of pre-crushing but did not completely account for the high pebble generation rates experienced after start-up. This meant that the operator needed to throttle back the mill feed rate due to the overloading of the pebble recycle conveyor. Sylvestre, Abols and Barratt (2001) postulated that the limitations of the JKSimMet® model, especially modelling the effects of pre-crushed feed in the SAG mill, were due to:

- The SAG mill model developed using the survey data before pre-crushing was too pessimistic, depressing the breakage rates of the critical size (12.5 to 50 mm).

- The model assumed a finer feed size distribution in the mill feed than at start-up.

The following additional changes to the circuit were required once the partial pre-crushing circuit was in operation:

- A larger pebble crusher and conveyor upgrades to manage the high pebble generation rate. Pebble generation doubled in capacity from 90 t/h to 175 t/h.

- A larger cyclone feed pump to manage the high ball mill circulating loads. The transfer size to the ball mill increased from 1300 µm to 2400 µm.

- Changes to the pre-crushing screen apertures due to the changing ore hardness. Throughput dropped by 4 per cent, resulting in an increase in the bottom deck aperture to 2 inches (50 mm).

Sylvestre, Abols and Barratt (2001) concluded that partial secondary crushing with scalped feed at Trollius increased throughput by 35 per cent and reduced operating costs by 19 per cent compared to the previous SABC flow sheet.

Although Sylvestre, Abols and Barratt's (2001) partial mid-size crushing provides insights into how such a circuit can be operated, there is no evidence to suggest how the circuit would have performed with a much higher ball charge level (>>10 per cent v/v) and finer feed or with primary crushed feed, for comparison and trade-off.

As experienced by Sylvestre, Abols and Barratt (2001), when modelling the SAG mill load for the partially pre-crushed feed using JKSimMet®, careful attention should be paid to calibrating the breakage rates for the correct ore competency and feed size distribution. For Hanhiniemi's (2023) work on Cadia, there is a fundamental issue with the modelling approach used for the simulations. The JKSimMet® Cadia SAG mill model used the 2013 ore competency value for the performance assessment. However, the breakage rates were based on the 2009 survey data as the Cadia SAG mill operated with a higher rock load. The difference in Axb between 2013 and 2009 is 35 versus 56 – which is significant. Hanhiniemi's (2023) recommendation of partial mid-size crushing for Cadia is biased towards a lower competence feed. Therefore, the validity of the outcome for highly competent Cadia feed is unknown for varying ball charge levels and feed composition.

Nelson, Valery and Morrell's (1996) work at Fimiston provides some insight into the effect of ball charge and total mill filling levels on throughput for a fully secondary crushed SAG mill feed (Table 6). The average SAG mill F_{80} ranged between 50–80 mm for fully secondary crushed feed. They commented that although the SAG mill achieved high throughput rates at the 15 per cent v/v ball charge level and lower rock load (<3 per cent v/v), the mill internals succumbed to metal peening and damage. Any attempts to increase the secondary-crushed rock load, the SAG mill experienced excessive instability and frequently overloaded, resulting in reduced throughput. With one of the secondary crushers offline and the SAG mill fed with secondary and primary crushed feed, a higher and stable rock load could achieve similar throughputs at a reduced ball load (12 per cent v/v ball charge and 19 per cent v/v total filling). Operating the mill with reduced ball (<12 per cent v/v) and higher rock loads was not tested with a bi-modal (secondary + primary crushed) SAG mill feed at Fimiston. Based on Nelson, Valery and Morrell's (1996) observations on the operation of the SAG mill with full secondary crushed feed at different ball charge levels, lower throughput rates (<1070 t/h) are expected for a bi-modal feed with decreasing ball charge and higher rock load levels for the same power draw.

TABLE 6

Fimiston SAG mill performance during commissioning (Nelson, Valery and Morrell, 1996).

Feed rate (t/h)	Secondary crushed feed to SAG	SAG mill F_{80} (mm)	Ball load (%v/v)	Total load (%v/v)	SAG mill power draw (kW)	SAG Ecs (kWh/t)	%<106 µm (%)	T80 (mm)	Ball mill Ecs (kWh/t)	Circuit P_{80} and ball mill circ load (µm, %)	Total grinding Ecs (kWh/t)
350	Full	39	5	20–22	7075	-	-	-	-	-	-
475	Full	39	11	20–22	7860	-	-	-	-	-	-
620	Full	39	12.5	20–22	8700	-	-	-	-	-	-
1070	Full	39	15.5	16–17	8395	7.85	35	1.9	7.24	230, 300	15.1
1086	Partial	117	12	19	9408	8.66	36	1.5	6.53	158, 306	15.2

As shown in Table 6, the SAG mill product size is coarser for the fully secondary crushed feed than for the partial secondary crushed feed. Likewise, the circuit grind P_{80}s increase for the fully secondary crushed case. Interestingly, the ball mill circulating loads were similar in both operating conditions. Nelson, Valery and Morrell (1996) did not comment on why both conditions achieved similar circulating loads at different circuit P_{80}'s. One would have expected a higher ball mill circulating load with a coarser transfer size unless there were issues with pumping constraints. Suppose the operation of the classification and ball mill circuits could be optimised (densities, feed pressure and changes to cyclone internals) to achieve similar circuit P_{80}s. In that case, the question is: *Can the SAG mill fed with partially crushed feed achieve similar power draw and throughput as the case for a fully crushed SAG mill feed, and will there be a significant difference in the transfer size?*

Nelson, Valery and Morrell (1996) indicate that a SAG mill operating with a lower ball charge and coarser rock feed improves cash flows if an equivalent or higher throughput is achieved. This requires the total mill filling to be increased to achieve similar mill power, and the coarse rock in the mill forming the rock charge is efficiently broken and does not form 'critical size' material that limits throughput.

CONCLUSION AND SUMMARY

Secondary crushing can be used to debottleneck a SAG mill-based circuit. It is well understood that secondary (and tertiary) crushing in hard rock applications debottlenecks the SAG mill by increasing throughput for competent ore.

Siddal and Putland (2007) and Lane (2007) commented on the importance of balancing coarse and fine feed to maintain throughput and achieve the final circuit grind in assessing operations with secondary crushing retrofits. Ultimately, an analysis is required to determine the optimum proportion of coarse and fine SAG mill feed based on full or partial secondary crushing. It is well understood that changing the ball to rock in the mill charge impacts the SAG mill throughput and the product size distribution.

Powell *et al* (2015) and Hanhiniemi (2023) suggest through their modelling work that optimum SAG mill throughput and product grind size can be achieved by balancing the proportion of secondary crusher product in the Cadia SAG mill feed and with appropriate ball charge levels, therefore minimising the operating cost and reducing the GHG emissions impact of SAG milling. This approach has been demonstrated by the partial mid-size crushing installed at Trollius (Sylvestre, Abols and Barratt, 2001) and Fimiston (Nelson, Valery and Morrell, 1996).

Although Sylvestre, Abols and Barratt (2001) demonstrated the benefits of partial mid-size crushing at Trollius with increased throughput rates and reduced operating costs, there is no comparison of operation with a higher ball charge and reduced rock load. Sylvestre, Abols and Barratt (2001) summarised: *'Partial pre-crushing has benefits, but it is not suitable for all ores and should be considered for hard ore applications.'*

Nelson, Valery and Morrell's (1996) work at Fimiston shed light on the behaviour of the SAG mills fed with fully and partially secondary crushed feed at different ball charge and mill filling levels. Their work showed that:

- SAG mills fed with fully secondary crushed feed and operating with high ball charge levels and lower rock loads achieved higher throughput than mills operating with lower ball charge and higher rock loads at similar power draws, albeit that high rock loads and full SAG mill power draw could not be achieved with low ball loads.

- Partially substituting the secondary crushed feed with coarser rock (primary crusher product) and operating the SAG mill with a lower ball charge achieved similar throughput rates as the case for high ball charge plus fully secondary crushed feed. The SAG mill operated at a higher power draw due to the improved stability imparted by the coarse rock. For the same SAG mill power draw, the throughput would be lower with the higher rock charge.

- Partially substituting the secondary crushed feed with coarser rock (primary crusher product) and operating the SAG mill with a lower ball charge was a more energy-efficient outcome.

Cadia's operating experience has been that:

- Higher throughput rates were achieved operating with the higher ball load and lower rock filling. However, the lower ball charge and higher rock filling level were never tested.

- A higher ball load and operation as a 'primary open circuit ball mill' stabilised the mill for feed variations relating to throughput, size and hardness.

Peñasquito's SAG mills process partially secondary, and tertiary (HPGR) crushed feed. The SAG mill load trials conducted at site reaffirmed Cadia's operating strategy: high ball charge and low rock load result in higher throughput. The constraint during the SAG mill load trial was the shell liner design, which restricted maximising the operating mill speed.

The key question for this paper was:

'Can throughput be maintained when treating competent ore if the rock load increases (and ball load decreases and is there efficiency improvement?'

The operational evidence suggests that generally:

- SAG mills operating with higher ball charge and lower rock load achieve higher throughput.

- Energy efficiency can be improved by operating with lower ball and higher rock loads if the rock load does not create a 'critical size' issue in the AG/SAG mill.

Further evaluation of the above is warranted for both debottlenecking strategies and greenfield plant design. Care should be exercised when using simulation packages for modelling circuit outcomes that are not based on the same ore or use radically different feed size distributions. In short, the 'Jury is still. out'.

ACKNOWLEDGEMENTS

The authors thank Ausenco for the support and permission to publish this paper.

REFERENCES

Allaire, A, Runnels, D, Sylvestre, Y, Fournier, J and Robichaud, F, 2011. Increased SAG grinding capacity at Goldex secondary crushing of SAG mill feed, in Proceedings of an International Conference on Autogenous and semi-Autogenous Grinding Technology, SAG 2011, Vancouver.

Atasoy, Y, Valery, W and Sklalski, A, 2001. Primary versus secondary crushing at St Ives (WMC) SAG mill circuit, in Proceedings of an International Conference on Autogenous and semi-Autogenous Grinding Technology, SAG 2001, Vancouver.

Bailey, C, Lane, G, Morrell, S and Staples, S, 2009. What can go wrong with comminution circuit design?, in *Proceedings of the Tenth Mill Operators Conference*, pp 143–149 (The Australasian Institute of Mining and Metallurgy: Melbourne).

Beccera, M and Amelunzen, P, 2012. A Comparative analysis of grinding circuit design methodologies, in Proceedings of Procemin 2012, Santiago.

Bueno, M, Foggiatto, B and Lane, G, 2019. Single Stage Autogenous Grinding Revisited, in Proceedings of an International Conference on Autogenous and semi-Autogenous Grinding Technology, SAG 2019, Vancouver.

Castillo, G and Bissue, C, 2011. Evaluation of secondary crushing prior to SAG milling at Newmont's Phoenix operation, in Proceedings of an International Conference on Autogenous and semi-Autogenous Grinding Technology, SAG 2011, Vancouver.

Chandramohan, R, Lane, G S and Morley, C, 2021. Operational consideration for hybrid SAG-HPGR comminution circuits, presented at the Minerals Engineering Institute's Comminution Conference 2021 (online conference).

Chandramohan, R, Foggiatto, B, Lane, G, Meinke, C, Ballantyne, G, Reeves, S and Staples, P, 2023. A Review of SAG Milling – History of Mill Selection and Testwork Analysis, in Proceedings of an International Conference on Autogenous and semi-Autogenous Grinding Technology, SAG 2023, Vancouver.

Dance, A, Atheis, D, Williams, S and Taplin, D, 2014. Grinding circuit improvements at Evolution Mining's Edna May operation, in Proceedings of the 12th Mill Operators Conference (The Australasian Institute of Mining and Metallurgy: Melbourne).

Engelhardt, D, Lane, G and Powell, M, 2014. Cadia Expansion – The impact of installing high pressure grinding rolls to a semi-autogenous grinding mill, in Proceedings of the 12th Mill Operators Conference (The Australasian Institute of Mining and Metallurgy: Melbourne).

Engelhardt, D, Robertson, J, Lane, G, Powell, M S, Griffin, P and Cadia, 2011. Expansion – From Open Pit to Block Cave and Beyond, in Proceedings International Autogenous and Semiautogenous Grinding Technology.

Geoghegan, C and Haines, C, 2023. Operational debottlenecking of the Cadia 40 foot SAG mill through constraints mapping analysis, in Proceedings of an International Conference on Autogenous and semi-Autogenous Grinding Technology, SAG 2023, Vancouver.

Giblett, A, 2014. Operational experience and performance assessment of secondary crushing prior to SAG milling at Newmont Phoenix operation, SME Annual Meeting , Denver.

Giblett, A and Putland, B, 2018. The basics of grinding circuit optimisation, in Proceedings of the 14th Mill Operators Conference (The Australasian Institute of Mining and Metallurgy: Melbourne).

Hanhiniemi, J, 2023. Techno-economic multicomponent analysis of comminution using minerals processing simulators, PhD Thesis, University of Queensland.

Lane, G and Siddal, B, 2002. SAG milling in Australia – Focus on the future, Metallurgical Plant Design and Operating Strategies (The Australasian Institute of Mining and Metallurgy: Melbourne).

Lane, G, 2007. Some observations regarding SAG Milling, in Proceedings of the Ninth Mill Operators Conference (The Australasian Institute of Mining and Metallurgy: Melbourne).

Lane, G, Seppelt, J and Wang, E, 2018. Quantifying the Energy Efficiency Transformation at Cadia Due to HPGR Crushing, in Proceedings of the 14th Mill Operators Conference (The Australasian Institute of Mining and Metallurgy: Melbourne).

Linde, P, Erwin, K, Chandramohan, R, Tweed, D, Lane, G, Staples, P, Hille, S, Foggiatto, B, Awmack, J and Patterson, B, 2019. Optimisation opportunities at Newmont Goldcorp's Peñasquito Operation, in Proceedings of an International Conference on Autogenous and semi-Autogenous Grinding Technology, SAG 2019, Vancouver.

Parker, B, Rowe, P, Lane, G, Morrell, S, 2001. The decision to opt for high pressure grinding rolls for the Boddington expansion, in Proceedings of an International Conference on Autogenous and semi-Autogenous Grinding Technology, SAG 2001, Vancouver.

Powell, M S, Morrell, S and Latchireddi, S, 2001. Developments in the understanding of South African-style SAG mills, *Minerals Engineering*, 14(14):1143–1153.

Powell, M, Mainza, A, Hilden, M and Yahyaei, M, 2015. Full pre-crush to SAG mills – the case for changing this practice, in Proceedings of an International Conference on Autogenous and semi-Autogenous Grinding Technology, SAG 2015, Vancouver.

Putland, B, Siddal, B and Gunstone, A, 2004. Taking control of the mill feed: case study – partial secondary crushing Mt Rawdon, in Proceedings of MetPlant 2004 Conference (The Australasian Institute of Mining and Metallurgy: Melbourne).

MacNevin, W, 1997. Kidston Gold Mines case study: Evolution of the comminution circuit, in Proceedings of the Conference on Crushing and Grinding in the Mining Industry (IIR Conferences Sydney).

Mainza, A, Bepswa, P, Nutor, G, Arthur, S, Obri-Yeboah, J, Atiawu, H and Lombard, M, 2015. Improved SAG Mill Circuit Performance due to Partial Crushing of the Feed at Tarkwa Gold Mine, in Proceedings of an International Conference on Autogenous and semi-Autogenous Grinding Technology, SAG 2015, Vancouver.

Matson, K, Tyrell, S and Jackson, B, 2018. The Application of Secondary Crushing at the Jundee Mine, in Proceedings of the 14th Mill Operators Conference (The Australasian Institute of Mining and Metallurgy: Melbourne).

McGhee, S, Mosher, J, Richardson, M, David, D M and Morrison, R D, 2001. SAG feed pre-crushing at ASARCO's Ray concentrator: Development, implementation and evaluation, in Proceedings of an International Conference on Autogenous and semi-Autogenous Grinding Technology, SAG 2001, Vancouver.

Morrell, S and Valery, W, 2001. Influence of feed size on AG/SAG mill performance, in Proceedings of an International Conference on Autogenous and semi-Autogenous Grinding Technology, SAG 2001, Vancouver.

Muteb, P and Fortin, M, 2015. Meadowbank SAG Mill Throughout Ramp-Up, in Proceedings of an International Conference on Autogenous and semi-Autogenous Grinding Technology, SAG 2015, Vancouver.

Mwehonge, G, 2006. Crushing practice impact on SAG milling: addition of secondary crushing circuit at Geita Gold Mine, in Proceedings of an International Conference on Autogenous and semi-Autogenous Grinding Technology, SAG 2006, Vancouver.

Nedham, T and Folland, G, 1994. Grinding Circuit Expansion at Kidston Gold Mine, SME Annual Meeting, Albuquerque.

Nelson, M, Valery, W and Morrell, S, 1996. Performance characteristics and optimisation of the Fimiston (KCGM) SAG mill circuit, in Proceedings of an International Conference on Autogenous and semi-Autogenous Grinding Technology, SAG 1996, Vancouver.

Siddal, P and Putland, B, 2007. Process design and implementation for secondary crushing to increase milling capacity, in Proceedings of SME Annual Meeting, Denver.

Staples, P, Lane, G S, Braun, R, Foggiatto, B and Bueno, M P, 2015. Are SAG mills losing market confidence?, in Proceedings of an International Conference on Autogenous and semi-Autogenous Grinding Technology, SAG 2015, Vancouver.

Sylvestre, Y, Abols, J and Barratt, D, 2001. The Benefits of Pre-Crushing at the Inmet Troilus Mine, in Proceedings of an International Conference on Autogenous and semi-Autogenous Grinding Technology, SAG 2001, Vancouver.

Thong, S, Pass, D and Lam, M, 2006. Secondary crushed feed before SAG milling – An operators perspective of operating practices at Porgera and Granny Smith gold mines, in Proceedings of an International Conference on Autogenous and semi-Autogenous Grinding Technology, SAG 2006, Vancouver.

Waters, T, Rice, A, Seppelt, J, Bubnich, J and Akerstrom, B, 2018. The evolution of the Cadia 40 foot SAG mill to treat the Cadia East orebody: a case study of incremental change leading to operational stability, in Proceedings of the 14th Mill Operators Conference (The Australasian Institute of Mining and Metallurgy: Melbourne).

Pioneering coarse particle flotation – transformative insights from five years of operation

C Haines[1], L Vollert[2], W Downie[3], I Heath[4] and R Ghattas[5]

1. MAusIMM(CP), Superintendent – Metallurgy, Newmont Cadia Operation, NSW.
   Email: campbell.haines@newmont.com)
2. Principal Metallurgist, Technical Services Processing, Newmont Corporation, Brisbane Qld 4000.
3. Specialist Metallurgical Projects and Planning, Newmont Cadia Operation, NSW.
4. Plant Metallurgist, Newmont Cadia Operation, NSW.
5. Manager – Ore Processing, Newmont Cadia Operation, NSW.

ABSTRACT

The mining industry has seen a shift towards sacrificing metal recovery for power intensity by simply coarsening primary grind. There is an economic limit to this practice which can be overcome by the adoption of new technologies. For Cadia, this meant a shift away from conventional, power intensive fine grinding flow sheets towards the adoption of coarse particle flotation technology. This new style of flow sheet has potential to deliver significant reductions in concentrator footprint, power and water demand and enable the future use of environmentally preferential tailings storage options like dry stacking or co-mingled deposition. Newmont Cadia took a significant step forward in proving the technical viability of coarser processing for base metals when a full-scale trial of the Eriez HydroFloat™ units were commissioned in August 2018 in a tailings scavenger duty. The success of the trial installation provided Newmont with an alternative technology case when pursuing mill expansions. Subsequent Newmont studies concluded that the expansion of coarse particle flotation (CPF) capacity delivered an improved business case over additional fine grinding and a second CPF project progressed to execution. Post completion of the expansion project in 2022, 75 per cent of concentrator feed at Cadia is treated through coarse flotation systems. This has enabled Newmont to leverage this technology to exploit the material properties of the separated streams. The separated, fines deficient, tailings sands have value as embankment construction material and are a key basis of the tailings future at Cadia. This paper will examine the current reduction in power intensity being delivered by coarse flotation at Cadia and how the site tailings opportunity to further leverage the technology is being developed.

INTRODUCTION

The globe is marching towards a low-carbon economy that will require significant investment in renewable power generation, energy efficiency and electric vehicles. The rapid scaling of these technologies will be extremely material-intensive and demand for critical metals like copper is forecast to grow at a rate significantly higher than any demand increases seen over the past 30 years (Pickens, Joannides and Laul, 2022). However, supplying the resources for the carbon-neutral economy cannot be achieved by simply applying the same mining means in the same way. For minerals processors, there requires a shift away from conventional, power intensive fine grinding flow sheets. Some of the key emerging trends in minerals processing are:

- dry comminution technology.
- coarse particle separation (bulk sorting, particle sorting, coarse flotation).
- alternative tailings management practices (filtered/dry stacked sand-slimes split tailings, tailings valorisation).
- closing the loop on water and reagents.

These trends and technologies can drive efficiency improvements in minerals processing plants, helping to meet the growing demand in a sustainable and responsible manner. The Technology Readiness Level (TRL) is used to assess the maturity level of a particular technology. The technology levels range from TRL 1 (lowest level where research may have begun to be translated into applied research and development) to TRL 9 (highest level where the technology has been proven under operating conditions). In the case of coarse particle separation, industry is at TRL 9 (ICMM, 2022). The mineral processors who can then leverage new technologies to exploit the full value stream opportunity from mine to tailings will be competitively placed to sustainably operate into the future.

Sustainability in the context of ore processing refers to the measures taken by mining companies to ensure that minerals are extracted and refined in a way that minimises damage to the environment and the local communities in which they operate. This includes reducing greenhouse gas emissions, reducing waste and tailings, preserving water resources, and ensuring the health and safety of employees and wider communities (Vollert, Haines and Downie, 2023). Project evaluation processes need to be disciplined when prioritising these longer term objectives; which can be diluted during an early stage net present value project optimisation (Lane *et al*, 2019). One of the largest life cycle costs to an operation will be the tailings and water facility; through construction, operation and into closure. Consideration of the life of facility size, stability, water recovery and operability should be incorporated into the mineral processing flow sheet equipment selection trade-offs. As large tailings or water savings may be overlooked in pursuit of conventional flow sheet ideas, ideas that can also curtail the economic mine life early. The mining industry fails on average two tailings' dams per annum globally; high profile failures in Brazil in 2015 and 2019 resulted in significant fatalities and involved major mining companies (Williams, 2020). The loss of trust in industries' ability to safely manage tailings is placing significant pressures on the sustainability of operations, both future and existing. The Chilean government banned upstream tailings dam raises in 1970 following a significant tailings failure at El Cobre (Barrera and Caldwell, 2015); government regulators and company boards are increasingly drawing a line in the sand that they will not permit or approve upstream designs. Cadia operation suffered a tailings dam failure in 2018, which catalysed a prohibition on future upstream raises within the company. Insurers and investors are also paying closer attention, all of which impacts on the sustainability of mining operations or projects (Williams, 2020).

Copper and gold mining account for over 50 per cent of the tailings generated globally per annum. The opportunity exists now within these industry sectors to pioneer new, more sustainable, tailings solutions as the commodity demand increases. South American operations are already innovating tailings dam design and demonstrating the economic value of water efficiency to project development (Cacciuttolo and Valenzuela, 2022). From demonstration scale innovative solutions such as hydraulic dewatered stacking at El Soldado (Newman *et al*, 2023) to implementation of a hybrid water recovery tailings solution at the 90 ktpd Caserones operation. The potential operations' sustainability improvements through these innovations in tailings solutions shouldn't be overlooked. Equally, with increasing research into the possible valorisation opportunities, transformative ideas are coming to the tailings space.

The Australian market for Sand and Aggregates is A\$700M/a and over A\$100M/a in national exports (Segura-Salazar and Franks, 2023a). Internationally, demand for aggregates has tripled over the last two decades as emerging economies develop, and mature economies undertake, expansive infrastructure renewal projects. After water, these materials represent the second largest resource extracted and traded on a volume basis (Segura-Salazar and Franks, 2023a). 'Ore-Sand' has been coined as a new product opportunity with green wings, with Vale S.A. finding a market for its 'Ore Sand' produced from the Brucutu iron ore mine in Minas Gerais, Brazil. Researcher partners, the University of Queensland and University of Geneva, demonstrated the greenhouse gas (GHG) emissions from equivalent 'Ore Sand' generated by Vale S.A. was up to ten times lower than traditional sources (Figure 1) (Golev *et al*, 2022). Extensive 'Ore-Sand' testing completed on Quebec sand (Benarchid *et al*, 2019), Vale S.A. sand (Golev *et al*, 2022) and Cadia Sand (Segura-Salazar and Franks, 2023b) showed in all cases that the 'Ore-Sand' product was non-toxic, non-acid forming and meets the AS 2578.1: 2014 standards for construction sand typically used in concrete.

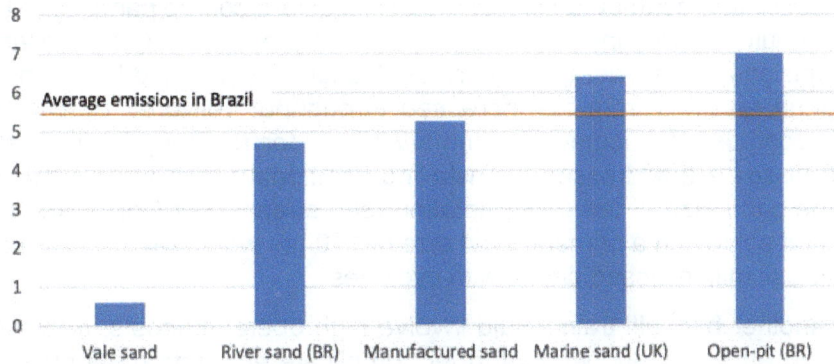

FIG 1 – GHG Emissions from Sand Extraction (grams of CO_2-equivalent per kg sand) (Golev *et al,* 2022).

Separation of coarse particles using coarse particle flotation is no longer to be considered novel. Proven at scale in isolation, the opportunity presented now is to leverage the material properties of a coarsened grind across the value chain.

COARSE PARTICLE RECOVERY – TECHNOLOGY TRADE-OFF – BALL MILL VERSUS HYDROFLOAT™

In 2015, an application to the New South Wales Department of Planning and Environment to modify the approval for the Cadia East Project was submitted. Aiming to increase the permitted upper limit for the processing plant's throughput. The subsequent study phases, publicly known as the Cadia Expansion Project (CXP), evaluated several expansion scenarios concerning the mine's capacity, processing facilities, and major infrastructure. Ultimately, the project scope for Concentrator 1 was divided into two stages:

Stage 1 aimed to ensure continuous production at Cadia while increasing processing throughput capacity. This stage involved the development of Panel Cave 2–3 (PC2–3), an upgrade to the mines material handling system, and debottlenecking of the Concentrator 1 comminution circuits.

Stage 2 focused on further boosting processing capacity with enhancing gold and copper recovery compared to Stage 1. This stage included the extension of coarse particle flotation.

During Stage 1 of the CXP, it was determined that the grinding circuit throughput for Concentrator 1 was constrained by the available grinding power and the size of the grinding circuit's product. The study concluded with some capital modifications plant throughput could be achieved at the expense of flotation feed size. However, with the current conventional tank cell flotation circuit and a coarser primary grind, gold and copper recovery would decrease, as illustrated in Figure 2 (Vollert, Haines and Downie, 2023).

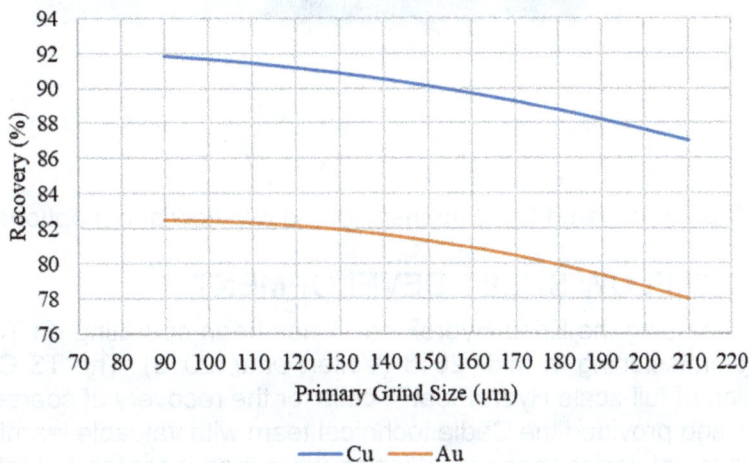

FIG 2 – Grind Size versus Recovery for Cadia East Ore (without coarse particle flotation (CPF)) (Vollert, Haines and Downie, 2023).

The two particle classes that contribute most to these losses are coarse particles (+150 μm) with low surface exposure of sulfides and fine particles (-7 μm) that are liberated but challenging to recover due to surface occlusions and slower flotation rates (Runge, Tabosa and Holtham, 2014). With the coarser primary grind resulting from the increased throughput delivered by CXP Stage 1, these coarse, low-grade composites became the primary target for recovery improvement in Stage 2. Traditionally, such losses are addressed by installing additional secondary or tertiary comminution power to reduce the grind size of the entire flotation feed stream. For Concentrator 1, the concept study determined that achieving a primary grind size of 110 μm would require a fourth grinding train, including a 9.5 MW ball mill in closed circuit with cyclones.

However, adding another ball mill train would involve high operating costs, primarily driven by the high cost of power in Australia, and would further complicate efforts to meet sustainability targets. By the time of the CXP pre-feasibility study in 2018, an alternative to the additional ball mill train had been successfully demonstrated through the metalliferous mining industry's first full-scale trial of coarse particle recovery using coarse particle flotation (CPF) technology at Cadia. Concentrator 1 operates three parallel ball milling-rougher flotation circuits, known as Trains 1, 2, and 3. In 2018, the newly commissioned Train 3 (T3) CPF circuit began recovering coarse, value-bearing composites from conventional flotation tailings without the need for additional upfront power to reduce particle size for better mineral liberation (Vollert et al, 2019).

A factor of merit analysis concluded that extending CPF technology to Trains 1 and 2 (T1/T2) was the preferred option for the feasibility study. Although there was perceived to be increased technical risk due to the relative immaturity of CPF technology for sulfides, the analysis showed a more than 45 per cent reduction in power consumption would be achieved equating to ~50 per cent reduction in CO_2 emissions per kilogram of copper produced when all scope 2 and 3 emissions are considered (Figure 3). Combined with an overall recovery benefit across various throughput scenarios, CPF presented a stronger economic case. Other advantages of CPF over fine grinding, such as its impact on tailings storage and water recovery, were qualitatively considered, although no financial value was attributed to them. In retrospect, these considerations were critical for Cadia's long-term sustainability as it enabled tailings optionality that would have been otherwise closed had a conventional ball milling design been pursued.

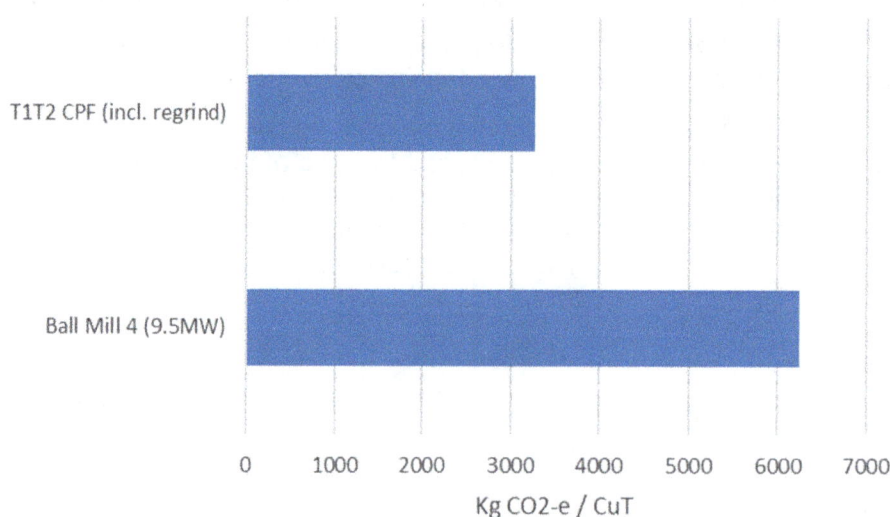

FIG 3 – Estimated GHG intensity of the two flow sheet options.

TRAIN 1 AND 2 CPF FLOW SHEET DEVELOPMENT

Coarse particle flotation using the Eriez HydroFloat™ has been operating on T3 of Concentrator 1 at Cadia since its commissioning in mid- 2018 (Vollert et al, 2019). The T3 CPF project was an industry first application of full-scale HydroFloat™ cells for the recovery of coarse composite copper and gold from tailings and provided the Cadia technical team with valuable learnings around design, operation, and metallurgical performance all of which were then incorporated into the CXP Stage 2 feasibility study for extension to Train 1 and 2 (Jaques et al, 2021). Two significant design philosophy changes applied to the Train 1 and 2 design were the removal of Crossflow™ teeter bed separators ahead of the HydroFloat™ units and incorporation of dedicated CPF concentrate regrind and cleaning to the flow sheet.

The CrossFlow™ demonstrated at full scale to produce a very efficient cut with minimal fines bypass, however the benefit was outweighed by operational stability issues and reduced circuit flexibility. A dual stage cyclone classification circuit was pursued for Train 1 and 2. A loss in classification efficiency and increased fines presentation to the HydroFloat™ was a technical risk with a dual stage cyclone circuit (Figure 4). The technical risk being primarily excess fines leading to increased viscosity in the freeboard zone resulting in unselective recovery of coarse fractions to concentrate due to their inability to settle. The HydroFloat™ unit process is a teeter bed separator with air injection to prompt bubble particle attachment. Any fine (-75 µm) particles present will be fluidised into the overflow, in addition to creating settling and/or selective separation issues due to changes in cell viscosity. This was derisked by completing necessary field trials and laboratory testing on Cadia East Ore during the project development phases (Vollert, Haines and Downie, 2023). The standard Newmont design for future CPF installations will use dual stage cyclones to prepare HydroFloat™ Feed instead of CrossFlow™ Classifiers.

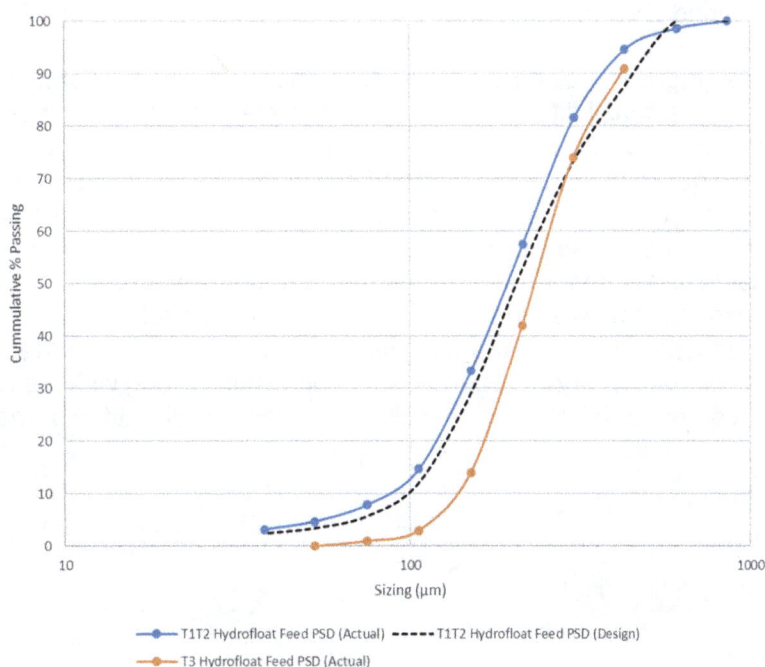

FIG 4 – HydroFloat™ Feed PSD's.

The decision to upgrade the HydroFloat™ concentrate via a dedicated regrind and cleaner flotation cell is aimed at maximising CPF circuit recovery whilst also reducing the mass flow back into the existing Train 1 and 2 regrind and cleaner circuit. Leveraging the learnings from the T3 demonstration CPF circuit where concentrate regrind treatment is in combination with conventional rougher-scavenger concentrate. This creates a suboptimal bi-modal size distribution feeding the conventional cleaning regrind circuit which reduces the overall liberation achieved through the regrind circuit (Figure 5). The Train 1 and 2 CPF circuit Jameson cell targets 80 per cent recovery of the copper and gold from HydroFloat™ concentrate into 2 per cent of the mass, producing approximately 2 t/h of final CPF concentrate at concentrate grade of ~18–20 per cent copper.

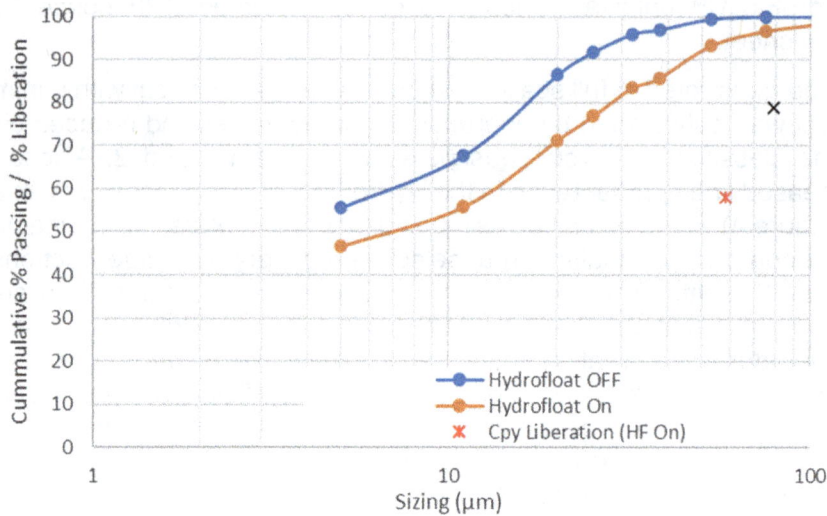

FIG 5 – T3 Regrind Cyclone Overflow with HydroFloat™ Concentrate On versus Off.

Lab flotation test work (Figure 6) using the dilution method to simulate Jameson cell performance was conducted during the feasibility study on bulk samples of coarse concentrate from the T3 HydroFloat™, with varying regrind product size distributions. The goal was to determine the regrind product size and hence quantum of dedicated grinding power needed to achieve sufficient liberation for upgrading the coarse concentrate. As anticipated, the flotation recovery of copper from the HydroFloat™ concentrate increased with a decrease in grind size, which improved mineral liberation. To produce a final concentrate copper grade directly from the CPF circuit, a regrind product size of 24 µm would be necessary. However, the target recovery can be attained with a grind size of 38 µm. With the significant reduction in mass, the intermediate-grade stream can then be returned to the existing T1/T2 cleaner circuit for further regrinding and upgrading along with the rougher concentrate stream. However, design optionality was also included to easily redirect the Jameson concentrate to final concentrate.

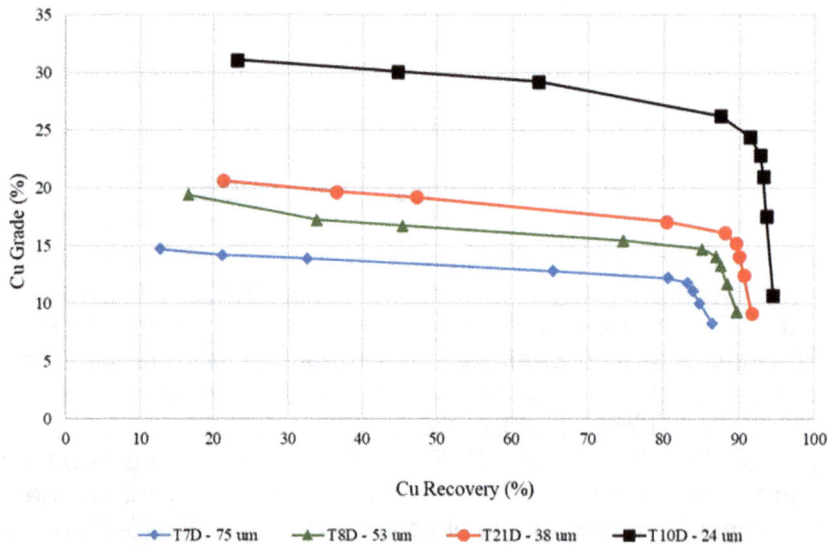

FIG 6 – Cadia East Ore; HydroFloat™ Concentrate Regrind Copper Grade-Recovery Curve (Vollert, Haines and Downie, 2023).

A Vertimill® was selected for the regrind duty mill due to the product target size, reduced footprint and commonality with existing mills on-site. Equipment sizing for a Vertimill® in this duty is challenging. The fines deficient feed and significant reduction ratio introduce uncertainty in the results of the different laboratory methods. Mill sizing was assessed using a number of different methods during the Feasibility stage, determining to exclude the Metso Jar mill test results and take the average of the specific energy from the Nippon-Eirich Tower mill test, Levin Test and Modified Levin Test corrected for Vertimill® (Vollert, Haines and Downie, 2023). The mill selected and installed was a VTM 4500 (3350 kW), an increase in the mill sizing due to uncertainty in the mill power demand and the known under sizing issues inherent with vertical screw type mills (Figure 7). The actual operational specific energy is higher than predicted, with the operational work index 10 per cent above design. However, due to the conservative equipment selection no detrimental impact to regrind product size has been observed.

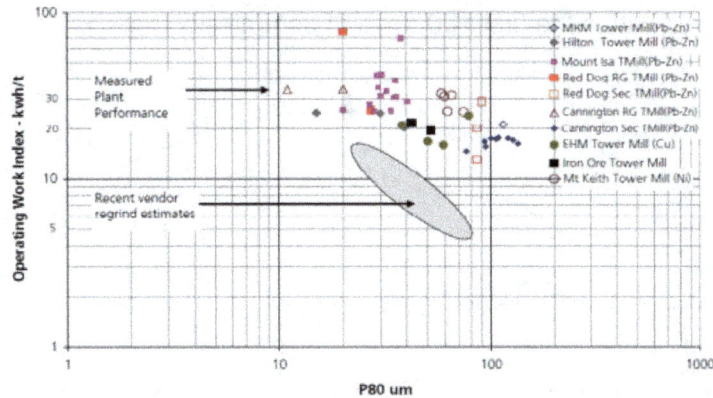

FIG 7 – Tower Mill Work Index Summary (Pease, 2010).

The Train 1 and 2 CPF circuit (Figure 8) consists of two stages of cyclones in series to remove -75 µm particles before presentation to four 4.26 m (14 ft) HydroFloat™ cells to recover the coarse value mineral particles into a concentrate. This concentrate stream is then dewatered before it is ground in a dedicated closed circuit regrind Vertimill® and then further upgraded in a Jameson cell before returning it to either the existing T1/T2 cleaner flotation regrind circuit or final concentrate. The tailings stream from the Jameson cell is returned to the primary cyclone feed hopper. The coarse tailings from the HydroFloat™ cells combine with the fine material from the primary, secondary and dewatering cyclone overflow streams to make up the final plant tailings which is thickened and pumped to the tailings storage facility.

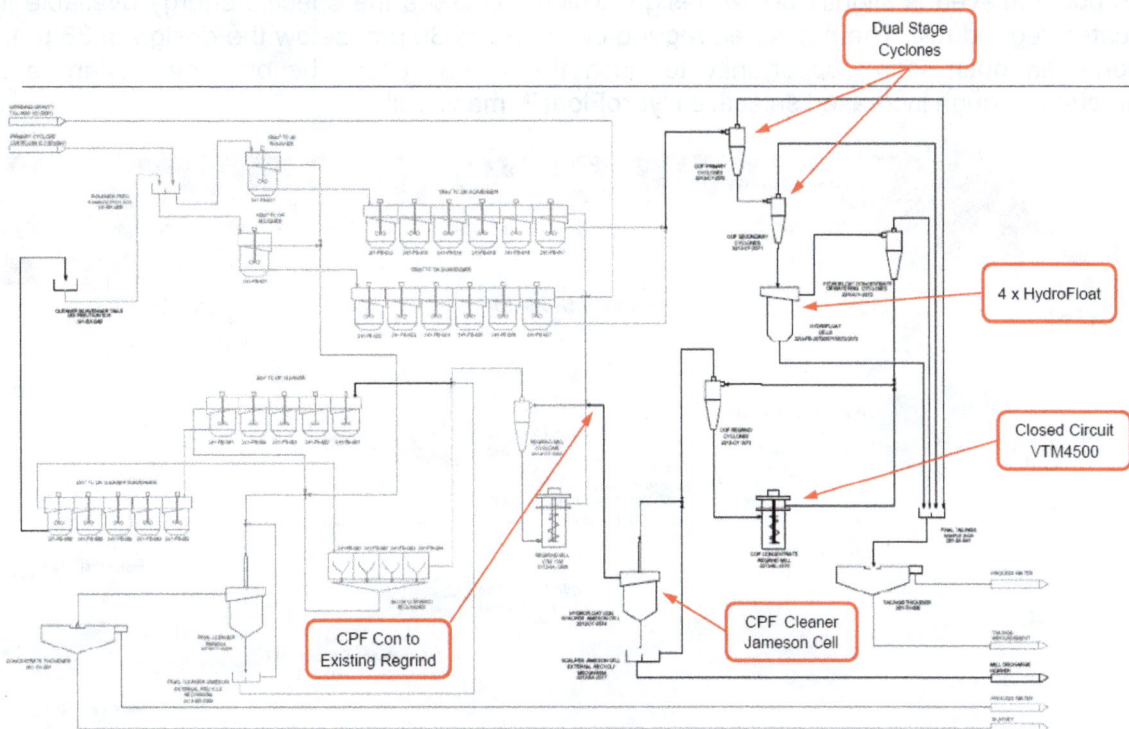

FIG 8 – Train 1 and 2 CPF Circuit Flow sheet.

TRAIN 1 AND 2 METALLURGICAL BASELINE PERFORMANCE

Application of the lessons learnt from the T3 demonstration CPF circuit (Vollert, Haines and Downie, 2023; Jaques *et al*, 2021) into the design of the T1/2 circuit has delivered an improved circuit which exceeds design and was ramped up to design recovery within three months (Figure 9). The new T1/2 CPF plant was handed over from the project team to operations for slurry commissioning in November 2022. The mantra for slurry commissioning was 'go-slow to go-fast' and an extended 4-week period was taken during commissioning to robustly test the full system again on water. The process control sequencing and instrument tuning was given significant attention prior to any slurry introduction. The design CPF circuit target of 30 per cent gold and copper recovery and 93 per cent availability was achieved within the first 100 days of commissioning.

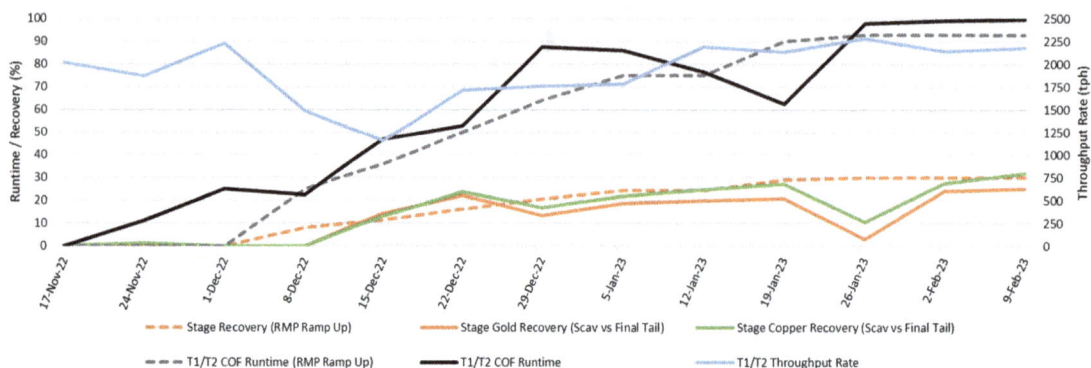

FIG 9 – Train 1 and 2 CPF ramp up curve, 7-day aggregated data.

A detailed metallurgical survey was completed on the 8th February 2023 to baseline the circuits performance prior to commissioning completion. Selective recovery of the target metal species through the flow sheet is illustrated in the Sankey charts (Figures 10 and 11). The circuit is achieving 0.1 per cent mass recovery at >30 per cent copper recovery to Jameson cell concentrate. Above the design target of 30 per cent copper recovery. The concentrate grade delivered from the single Jameson cell is 20–24 per cent copper at an enrichment ratio of 25; above the design concentrate grade target of 18–20 per cent. A high enrichment ratio and above design concentrate grade is enabled by above design mineral liberation achieved by the dedicated regrind mill. The HydroFloat™ mass pull achieved is slightly below design, which increases the specific energy available in the dedicated regrind mill. The measured regrind circuit P_{80} is 30 µm, below the design of 38 µm. This presents an optimisation opportunity to push the overall circuit beyond the design recovery parameters through increased selective HydroFloat™ mass pull.

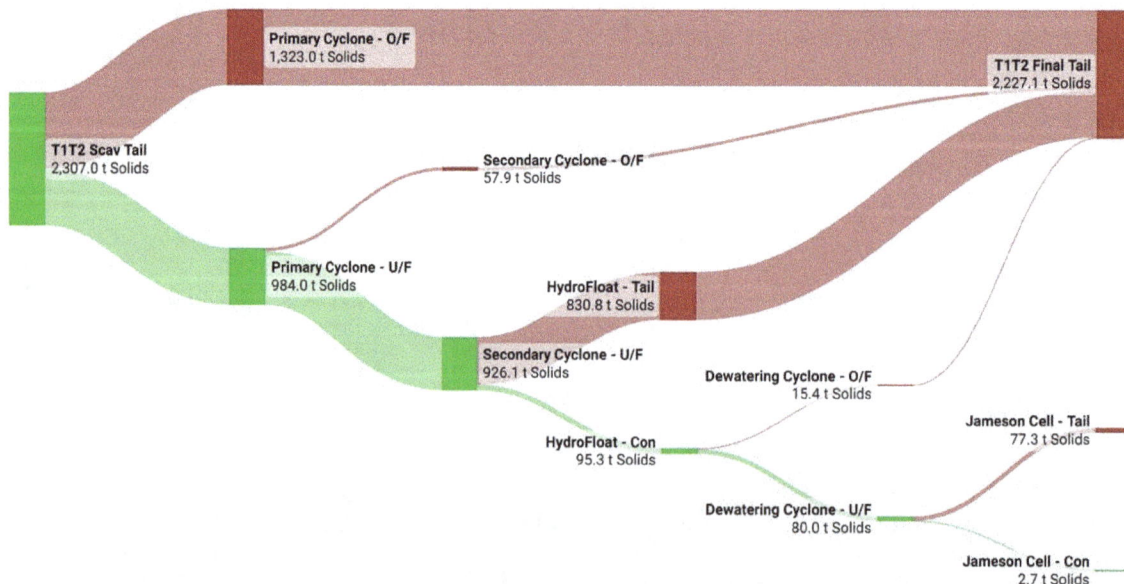

FIG 10 – CPF mass recovery by unit operation.

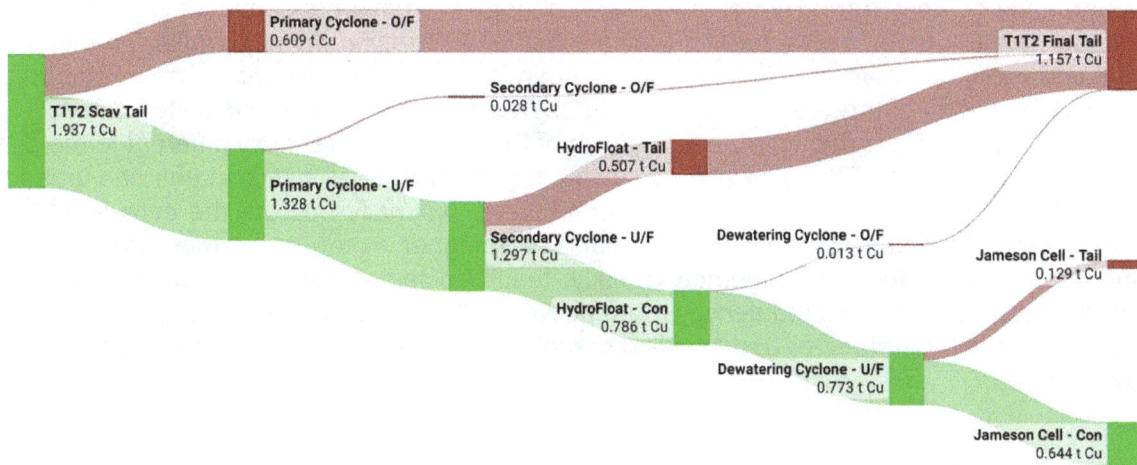

FIG 11 – CPF Copper Metal (t) recovered by unit operation.

The T1/T2 CPF design recovery was determined from the T3 metallurgical size by recovery. The T1/T2 CPF circuit has delivered above design recovery on a size by recovery basis for coarse particles (+150 μm) and below design for fine particles (-106 μm) (Figure 12). The target recovery size classes for the CPF circuit are +150 μm. Particles in the size classes below 150 μm should be recovered within the conventional circuit, with any losses in these size fractions addressed within the conventional circuit itself.

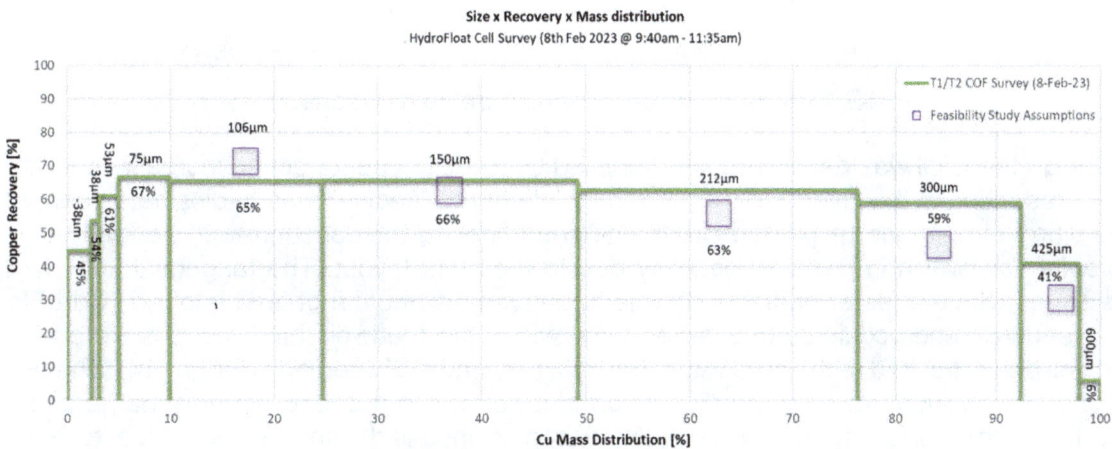

FIG 12 – Train1 and 2 CPF size by recovery.

TRAIN 1 AND 2 CIRCUIT OPTIMISATION

The successful deployment of the T1/T2 CPF circuit has shifted the liberation-recovery relationship, flattening the trade-off between recovery and grind size for copper and gold. The flattening of the recovery-grind size trade-off opens the possibility to challenge the economic breakeven point for throughput in T1/T2, beyond the design expectations of the circuit. However, several areas existed for operational optimisation that would lead to improved metallurgical outcomes. A dedicated optimisation process was completed over 12 months, targeting the key areas identified in collaboration with Eriez, Group metallurgy and site.

The optimisation program focused particularly on the following key areas:

- Process Control and Instrumentation improvements targeting stability.
- HydroFloat™ Parameters, Feed Properties and Wear.
- Circuit Water Balance Optimisation.

Not all aspects of this optimisation work will be discussed in this paper. However, the methods taken to address the changing operational behaviour due to the increased fines fraction will be discussed.

Increasing fines content to the HydroFloat™ feed changes the hydrodynamic conditions within the HydroFloat™ cell. A teeter bed separator typically has an increasing density profile with depth from the concentrate lip. However, the -75 µm fines entrained will accumulate within the freeboard zone due to the fluidisation of the teeter bed. This is observable in the Train1 and 2 HydroFloat™ units where the density profile is flat after the first 30 cm of the cell depth, distinctly different to the Train 3 HydroFloat™ (Figure 13). This increase in density higher in the cell necessitates changes in the operational strategy and instrumentation requirements to optimally operate the cells in T1/2. The notable changes included repositioning the ball floats higher in the cells, modifying secondary cyclone internals to reduce the underflow density, feed and teeter water addition to a determined maximum and controlling cell feed density more actively. The results of these changes improved the cell surface dynamics and teeter bed stability, contributing towards the incremental circuit recovery improvements.

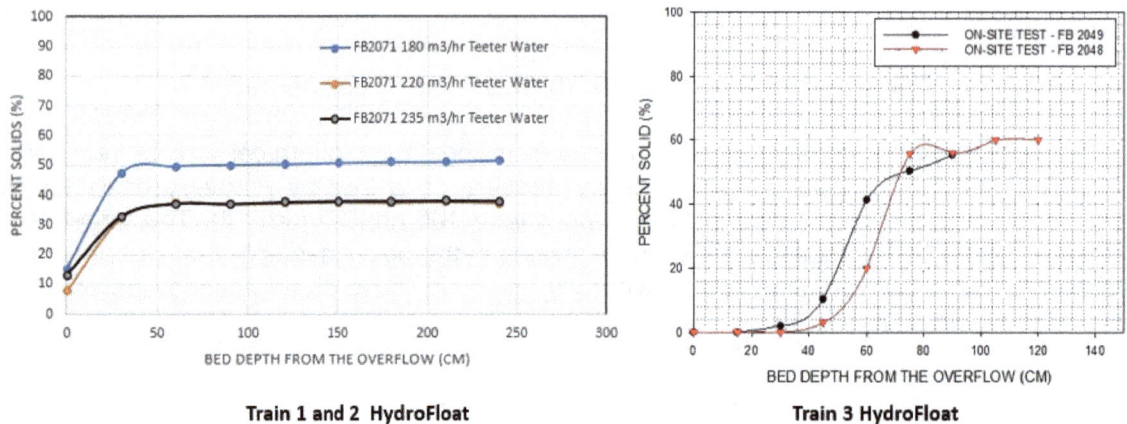

Train 1 and 2 HydroFloat   **Train 3 HydroFloat**

FIG 13 – HydroFloat™ density profile comparison.

A new design feedwell was developed and supplied by Eriez to counteract an increase in feed fines (-75 µm) presenting to the HydroFloat™. The principle of the feedwell was to dissipate energy from the feed slurry stream and help short circuit entrained fines to the concentrate. The feedwell is fed via a gravity flow which impacts tangentially due to the plant layout. The tangential flow dynamic caused issues with the teeter bed stability due to uneven distribution of feed into the HydroFloat™ from the feedwell, and accelerated localised wear within the feedwell. Modifications were made to the feedwell and retrofitted to the cells within the first 12 months of operation (Figure 14). The notable design changes included moving the dispersion plate closer to the feed pipe discharge point and ceramic lining the surfaces. An orifice plate was also installed into the feed pipe to assist in centralising the discharging flow onto the dispersion plate. The 'splash-skirt' was also removed as it was observed to be trapping entrained air from the slurry entering the feedwell and creating localised boiling, a phenomena counterproductive to recovery of coarse particles. The air was speculated to be entrained due to lack of energy dissipation occurring in the former design. The phenomena has not been witnessed since the skirt was removed.

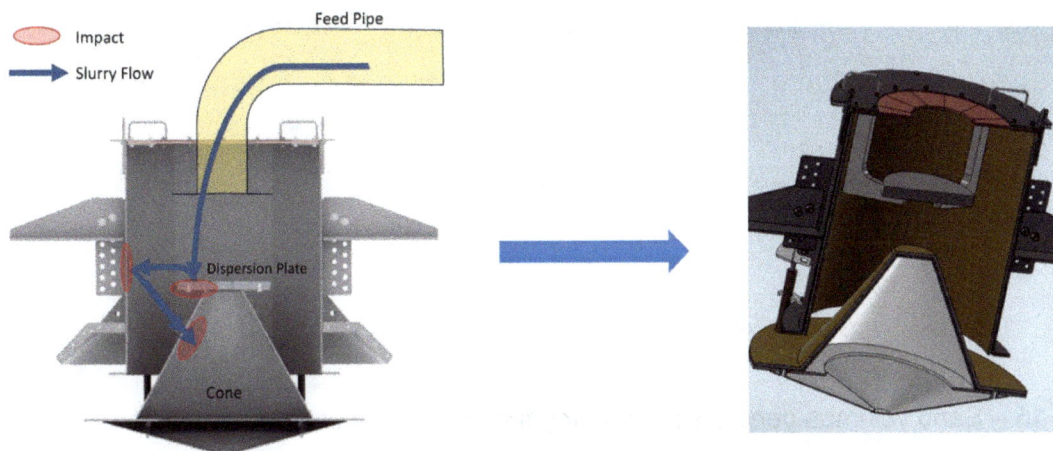

FIG 14 – 14 ft HydroFloat™ feedwell supplied by Eriez.

The expansive program of optimisation work built on the foundation of a robust plant design, which enabled design recovery within 100 days of project water commissioning completion. After 12 months of optimisation monthly HydroFloat™ recovery is 15 per cent (rel) above the design point for copper. To date the circuit copper recovery has outperformed gold recovery by 2–3 per cent (abs), both remain well above design. The driver of the slight recovery bias for copper over gold has not yet been defined, with diagnostic metallurgical work ongoing.

FUTURE TAILINGS AND WATER OPPORTUNITIES

Improving mine tailings management is an industry imperative to a sustainable future. Newmont is committed to the Global Industry Standard on Tailings Management with the goal of zero harm to people and the environment. Conventional wet tailings dams are increasingly difficult to permit, build and operate. Embankment design is increasingly moving towards downstream or centreline methods for wet dams; requiring large volumes of competent, non-acid generating rock fill. The block caving mining technique used on the Cadia East Orebody generates minimal levels of development waste rock. Sub-economic underground drawpoints are turned off and the waste left *in situ*. This creates very low cost mining but presents long-term challenges to tailings dam construction. The existing tailings storage facilities at Cadia were constructed using waste fill from the Cadia Hill pit which ceased operations in 2014. The surface tailings storage facilities do not have sufficient capacity to meet the operations life-of-mine tailings demand meaning Cadia will require additional storage in the future.

The salient long-term decision-making during the technical trade-off for CPF versus Ball Milling has enabled alternative options for tailings deposition and dam construction. Cadia is now applying to extend its operating license from 2031 to ~2050, inclusive of an extension to existing southern tailings storage facility (Minesoils Pty Ltd, 2024). The life of facility tailings storage extension will be constructed using a sand embankment, a method common in other areas of the world, particularly Chile. The quantum of sand possible to produce is proportional to the flotation feed size. In the case of nominal operations at a P_{80} of ~200 µm, 33 per cent of the tailings can be classified as 'Sand' suitable for embankment construction. This quantum is enough to meet the embankment design requirements of 27 per cent, inclusive of design factors. However, in the alternative ball mill case where a target P_{80} of 110 µm was proposed, 22 per cent of the tailings may be suitable as sand (Figure 15). This is not sufficient volume for embankment construction and alternative material would need to be sourced and placed at additional cost.

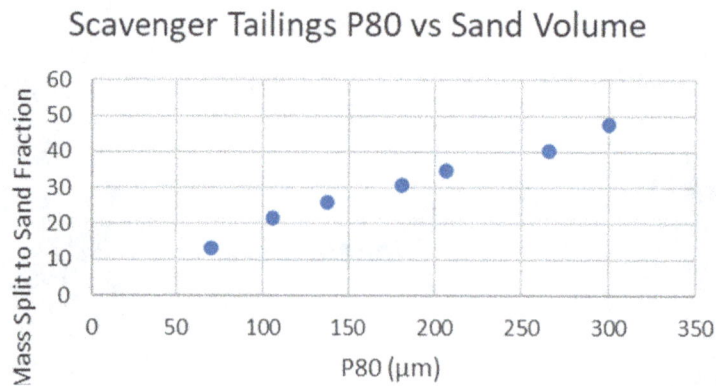

FIG 15 – Sand volumes generated assuming the efficiency curve of the Cadia CPF circuit.

Tailings dam design has historically been one of four typical types; conventional wet tailings (un-thickened), thickened tailings, paste tailings or filtered tailings. The drivers of design selection are a combination of regulatory, economic and environmental. A key environmental and operational consideration is the site's water efficiency. Tailings dam design and management will dictate water recovery and subsequent demand of the site (Table 1). The water demand is measured as the metres cubed per tonne (m^3/t) of tailings deposited, or more recently shifting to metres cubed per tonne of valuable metal recovered. The latter metric better showcasing the reducing water efficiency observed as operations scale throughput higher to combat declining feed grades (Lane *et al*, 2019). The majority of an operation's tailings return water is sourced from the supernatant, with minor seepage collection. Factors impacting water loss are principally entrainment to tailings, evaporation and uncontrolled seepage loss (Williams, 2020). Whilst it is acknowledged in the literature that conventional or thickened dams water recovery can match paste thickened, this is often restricted to the early years of operation when the dam rate of rise is high and the tailings surface stays wetted. After 10 years of operation when tailings embankment rate of rise slows, water recovery will drop, straining an operations water cycle during what are typically lower grade production years (Lyell, Copeland and Blight, 2008). South American operators Lundin at Caserones and Capstone at Mantos Blancos have been pioneering a new method of dam design which reduces the footprint and increases the water recovery relative to a traditional wet dam. Referred to as hybrid technology, the free-drainage properties of sand is exploited to uplift the total water recovery by separating the sand from the slimes and subsequently thickening the slimes (Barrera and Caldwell, 2015). Similarly Anglo American at El Soldado is pioneering hydraulicly dewatered tailings (HDS) at a demonstration scale as an engineered co-disposal technology that increases tailings consolidation and reduces water entrainment (Newman *et al*, 2023).

Table 1

Tailings design water demand best practice (Barrera and Caldwell, 2015).

Tailings dam design option	Water demand m^3/t
Conventional (Un-thickened)	0.36–0.7
Thickened	0.3–0.6
Paste Thickened	Not reported
Filtered Thickened	0.2–0.4
Hybrid – Sand Slimes Split Tailings Technology (SSSTT)	*0.3–0.4*

Cadia has been practicing in-pit tailings deposition since the 2018 northern tailings facility wall failure. Cadia's water demand during this period ranged between 0.56–0.75 m^3/t, averaging 0.61 m^3/t (Newcrest Mining Limited, 2024). Noting that external water was harvested following the 2019 drought, biasing the data high in the proceeding period. A demand of 0.47 m^3/t during the drought years of 2018–2020 is considered the site best current practice water usage. The proposed life of facility tailings dam will utilise hybrid sand-slimes split tailings technology. If best practice water demand can be achieved, the water savings by moving to a technology of this nature is between 1800–4300 ML//a of water measured against the baseline 0.47 m³/t.

A project shortcoming of the coarse particle recovery circuits at Cadia was failing to realise the value of keeping the CPF separated sand apart from the fines all the way through to tailings storage. The CPF design produces a tailings stream with a lower fines (-75 μm) content than can be achieved in a conventional tailings cyclone station, due to multiple pre-classification stages followed by hydraulic classification in the HydroFloat™ cell (Figure 8). A typical specification for sand product requires it to have less than 20 per cent passing 75 μm in the sands (measured by screen analysis), Figure 4 is an example of the HydroFloat™ feed distribution from Cadia CPF, demonstrating less than 10 per cent of the mass in that stream is passing 75 μm. A view during CPF circuit design should be taken to maintain a separate valuable sand stream or design in this philosophy for future cases. In the case of Cadia, the sand stream from the HydroFloat™ will deliver a positive revenue impact to the operation through capital efficiency found in tailings construction. Operations utilising the block cave mining method in particular may be able to improve their operations economic and environmental sustainability by changing tailings embankment construction methodology by leveraging CPF.

Due to the design shortcomings in the CPF facilities at Cadia, separation of sand requires a brownfield retrofit. A pre-feasibility study focusing on brownfield modification to the existing CPF circuits was completed in 2024 and is progressing into Feasibility in 2024/2025. This project was initiated to find capital efficiency in the base case life of facility tailings project, which completed Pre-Feasibility in 2021, and recommended building a new dedicated sand separation plant and sand embankment dam. However, following the successful commissioning in 2022/2023 of the T1/2 CPF, renewed focus was applied to the sand separation achieved by the existing HydroFloat™ cells and an options study was initiated in 2023. Several options were considered, utilising conventional technologies for material separation and transportation, namely pipes, pumps and cyclones. The design incorporates optionality to switch back to the nominal thickened tailings flow sheet on the run. The selected process block flow diagram is detailed in Figure 16.

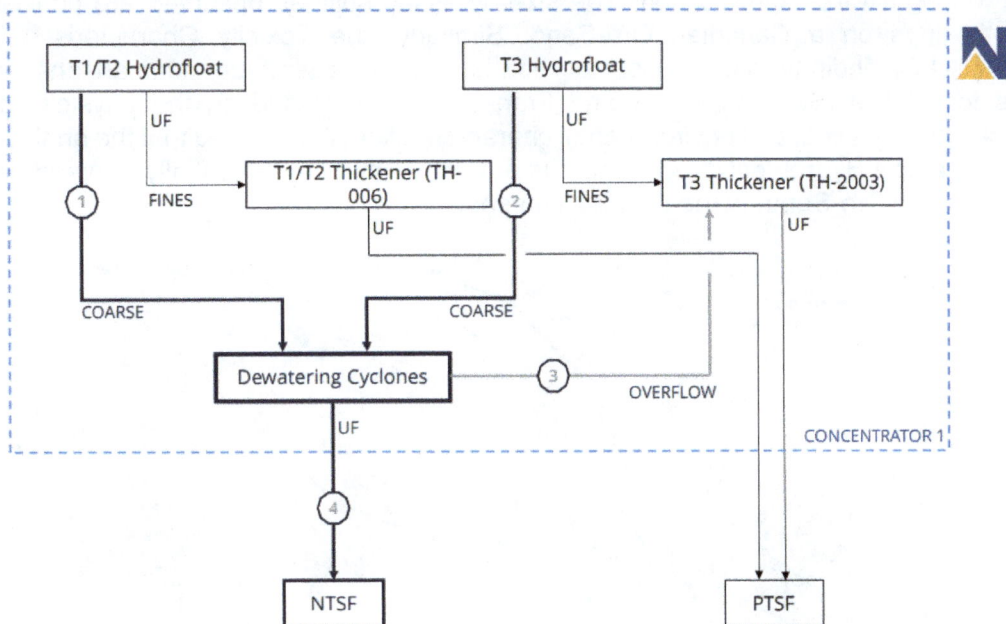

FIG 16 – Concentrator 1 coarse tailings separation and slimes thickening flow sheet utilising existing process equipment.

1. New pipeline for T1/T2 to a new dewatering cyclone located within the footprint of the existing concentrator.

2. New pipeline for dewatering cyclone overflow to the existing T3 tailings thickener.

3. New cyclone overflow pipeline of fines to T3 tailings thickener.

4. New coarse sand hopper, pump station and pipeline to the tailings storage facility.

Following project completion Concentrator 1 will transition from producing two thickened tailings streams at a nominal P_{80} of 200 μm to three tailings streams. The existing two tailings thickeners remain and transition to a 'slimes' thickening duty at a ~40 per cent lower flux rate (m²/t/h) and >90 per cent wt -75 μm feed when operating as per Figure 16. The remaining 'coarse tailings' stream generated is ~15 per cent wt -75 μm, well within the geotechnical placement parameter of 20 per cent wt -75 μm which is pumped directly to the tailings facility at ~60 per cent wt solids. The process fundamentals for the design proposed are robust. Streams 1 and 2 present for dewatering at densities <30 per cent wt solids and <20 per cent wt -75 μm, enabling efficient classification and minimal fines entrainment to underflow. This contrasts with conventional cyclone sand facilities operating at feed densities >40 wt per cent and higher fines contents. The design currently can produce sufficient sand to meet the embankment requirements for the life of facility tailings dam embankment. Creating opportunities for tailings placement such as terraforming sand stockpiles, increasing the tailings embankment factor of safety or finding a market for the excess 'Ore-Sand'.

TAILINGS VALOURISATION

'Ore-Sand' is a relatively new term to describe the value that can be derived from mineral processing tailings or mine waste streams. Vale S.A. has recently produced Ore-Sand at the Brucutu iron ore mine in Minas Gerais, Brazil (Golev *et al*, 2022). In 2022, samples from Cadia's T3 CPF facility were collected and tested at the University of Queensland to characterise physical, chemical, geotechnical, mineralogical, and environmental properties (Segura-Salazar and Franks, 2023b).

Cadia CPF product (Figure 17) is close to complying with the grading requirements for applications such as asphalt paving (grading 3) and is within the grading envelopes for natural sand in concrete as recommended by the Australian standard (AS 2758.1: 2014). Modal mineralogy determined sulfide species present as pyrite and chalcopyrite at 0.03 wt per cent and 0.15 wt per cent respectively. Sulfides are deleterious to concrete manufacturing and European standards set limits of Sulfur (exclusive of pyrrhotite) at <1.0 wt per cent S_{tot} (Benarchid *et al*, 2019); Cadia's sand product is well within this specification. The size-by-assay sulfides observed are consistent with Benarchid findings on a Canadian Ore-Sand. Similarly, the Toxicity Characteristic Leaching Procedure (TCLP) findings are of inert or non-hazardous based on comparisons with local regulations for metals (Segura-Salazar and Franks, 2023b). Based on the physical, chemical, mineralogical, geotechnical and environmental characterisation, it is inferred that the analysed sandy sample from Cadia's HydroFloat™ equipment is a material that can potentially be used as a novel alternative sand (ie Ore-Sand) in the construction sector.

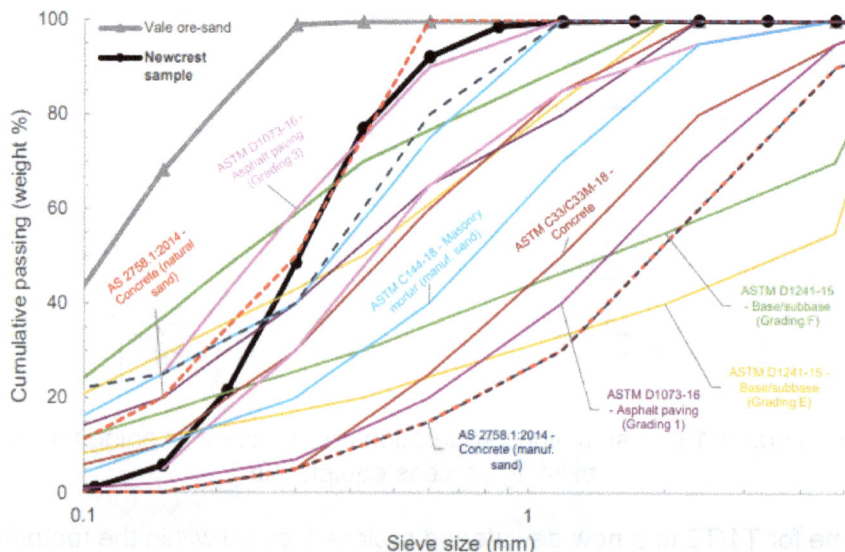

FIG 17 – PSD of Cadia CPF sample (Black line) versus specifications of construction sands (Segura-Salazar and Franks, 2023b).

There are, however, some barriers to be overcome for the uptake of Ore-Sand in practice, such as customer support for the product, transportation economics and government support. Selection of the right transportation mode and identifying the right markets will be the defining variable in making Ore-Sand a sustainable and economic alternative for any potential producer. The analyses of Ore-Sand's greenhouse gas (GHG) footprint conducted on Vale S.A. product found it to be ten times lower than conventional sand quarrying methods (Golev *et al*, 2022). The same research also demonstrated that a delivery distance greater than 35 km using small trucks will erode the GHG emission abatement achieved. Selection of the right transport mode is critical to a sustainable product. Ore-Sand will not be viable for all mining operations due to their distance to market and logistical complexities which create inefficiencies from the GHG perspective. These are variables which compound to make Ore-Sand a less competitive and less sustainable alternative. Cadia is 22 km by road to the nearest train terminal, and from there ~300 km to the nearest export port in Sydney, additionally there is demand for sand from central west regional infrastructure projects which may present opportunities. Cadia is currently permitting an extension to the operating license from 2031 to ~2050. Whilst the customer market and logistical pathway is not fully defined now, the foundation will exist to test 'Ore-Sand' as an additional revenue stream within the lifetime of Cadia.

CONCLUSIONS

Improving the operations' sustainability within the context of ore processing requires operators to invest in research and development and be open to new ideas. Early adoption and demonstration scale research and development at Cadia with regards to coarse particle flotation enabled a more sustainable process selection to be made during the plant expansion. The selection of coarse particle flotation not only delivered a 48 per cent reduction in greenhouse gas emission potential versus the fine grind alternative but also unlocked a more sustainable tailings future. Coarse particle flotation circuits in a scavenger duty can utilise dual stage cyclones as HydroFloat™ feed preparation and should consider dedicated HydroFloat™ concentrate regrind and cleaning. The coarse particle flotation circuit is also an effective sand separation process, the value or potential value of the sand stream should be considered during project development phases as a future opportunity.

The technology readiness of coarse particle flotation using the HydroFloat™ is proven in base metals and should no longer be considered a novel technology. The opportunity now is innovating towards more sustainable processing and tailings dam designs from the opportunity presented.

ACKNOWLEDGEMENTS

The authors would like to acknowledge the following contributions:

- The Cadia Operations, metallurgy, maintenance, and engineering teams, for their hard work, persistence and achievements.

- David and Brigitte Seaman, for helping see the bigger picture possible from coarse particle flotation.

- Eriez Flotation Division, for their ongoing collaborative support with Cadia to deliver exceptional coarse particle flotation results.

- Kym Runge and Lizette Verster with their team from the Julius Kruttschnitt Mineral Research Centre (JKMRC).

- Newmont Mining Corporation for permission to publish this paper.

REFERENCES

Barrera, S and Caldwell, J, 2015. Reassessment of best available tailings management practices, in Proceedings Tailings and Mine Waste 2015 (Tailings and Mine Waste: Vancouver).

Benarchid, Y, Taha, Y, Argane, R, Tagnit-Hamou, A and Benzaazoua, M, 2019. Concrete containing low-sulphide wase rock as fine and coarse aggregates: Preliminary assessments of materials, *Journal of Cleaner Production*, 221:419–429.

Cacciuttolo, C and Valenzuela, F, 2022. Efficient Use of Water in Tailings Mangement: New Technologies and Environmental Strategies for the Future of Mining, *Water*, 14: 1741.

Golev, A, Gallagher, L, Vander Velpen, A, Lynggaard, J, Friot, D, Stringer, M and Franks, D M, 2022. Ore-Sand: A potential new solution to the mine tailings and global sand sustainability crises, Final Report, version 1.4, The University of Queensland and University of Geneva.

ICMM, 2022. Tailings Reduction Roadmap.

Jaques, E, Vollert, L, Akerstrom, B and Seaman, B, 2021. Commissioning of the Coarse Ore Flotation Circuit at Cadia Valley Operations – Challenges and Successes, in *Proceedings of 15th Australasian Mill Operators' Conference*, pp 124–139 (The Australasian Institute of Mining and Metallurgy: Melbourne).

Lane, G, Hille, S, Pease, J and Pyle, M, 2019. Where are the opportunities in comminution for improved energy and water efficiency, in Proceedings SAG Conference 2019 (CIM: Vancouver).

Lyell, K M, Copeland, A M and Blight, G E, 2008. Alternatives to Paste Disposal with Lower Water Consumption, in Proceedings Paste 2008 (Paste 2008: Kasane, Botswana).

Minesoils Pty Ltd, 2024. NSW Independent Planning Commission: Cadia Continued Operations Project, Gateway Application Report. Available from: https://www.ipcn.nsw.gov.au/cases/ 2024/08/cadia-continued-operations-project [Accessed: August 2024].

Newcrest Mining Limited, 2024. Sustainability Reports, Newcrest website. Available from: https://www.newcrest.com/investor-centre/results-reports [Accessed: August 2024].

Newman, P, Brunton, M, Lopez, A, Burgos, J and Purrington, J, 2023. Success with Hydraulic Dewatered Stacking at the El Soldado Demonstration Facility, in Proceedings of Tailings and Mine Waste 2023 (Tailings and Mine Waste: Vancouver).

Pease, J, 2010. Elephant in the Mill, Proceedings of XXV IMPC (IMPC: Brisbane).

Pickens, N, Joannides, E and Laul, B, 2022. Red Metal, Green Demand, Wood Mackenzie Horizons.

Runge, K, Tabosa, E and Holtham, P, 2014. Integrated Optimisation of Grinding and Flotation Circuits, in Proceedings 12th Mill Operators Conference (The Australasian Institute of Mining and Metallurgy: Melbourne).

Segura-Salazar, J and Franks, D M, 2023a. Ore-Sand Co-production from Newcrest's Cadia East HydroFloat™ Reject: An Exploratory Study, Brisbane: University of Queensland.

Segura-Salazar, J and Franks, D, 2023b. Aggregates market in Australia, Brisbane: University of Queensland.

Vollert, L, Akerstrom, B, Seaman, B and Kohmuench, J, 2019. Newcrest's Industry First Application of the Eriez HydroFloat™ Technology for Copper Recovery from Tailings at Cadia Valley Operations, Proceedings of 10th International Copper Conference, COM MetSoc: Vancouver.

Vollert, L, Haines, C and Downie, W, 2023. Coarse Separation – An Enabler for Improving the Sustainability of Ore Processing, in *Proceedings 26th World Mining Conference*, pp 2891– 2903 (WMC 2023: Brisbane).

Williams, J D, 2020. The Role of Technology and Innovation in Improving Tailings Management, *In Towards Zero Harm: A compendium of papers prepared for the global tailings review* (ed: B Oberle, D Brereton and A Mihaylova), pp 64–84, Global Tailings Review: St Gallen, Switzerland.

Gruyere Gold Mine Western Australia – Part 2 – ramp up to operations, a journey of comminution, optimisation and challenges

G Heard[1], J Stevens[2], M Becker[3] and R Radford[4]

1. Senior Metallurgist, Gold Fields, Perth, WA 6000. Email: goldie.heard@goldfields.com
2. Senior Metallurgist, Gold Fields, Perth, WA 6000. Email: john.stevens@goldfields.com
3. Lead Metallurgist MAusIMM (CP), Orway Mineral Consultants, Perth WA 6167. Email: michael.becker@orway.com.au
4. Principal Specialist Metallurgy, Gold Fields, Perth WA 6000. Email: reg.radford@goldfields.com

ABSTRACT

The Gruyere Gold Project was discovered by Gold Road Resources Limited in October 2013 on the Yamarna Belt, 200 km east of Laverton in Western Australia. As a low-grade mineral resource, the feasibility study for Gruyere, completed in October 2016, culminated in the selection of a primary crush semi-autogenous grinding (SAG)-ball circuit with recycle pebble crushing (SABC). An engineering, procurement, and construction (EPC) contract was executed in June 2017, with construction commencing in early 2018 and commissioning of the SABC circuit commencing in early 2019.

Since 2019, the Gruyere Joint Venture (Gruyere JV), a collaboration between Gold Road Resources Ltd and Gold Fields Ltd, has been dedicated to the ramp-up and operations phases of the project. Operational ergonomics, advanced process control, throughput optimisation, mechanical availability, and metallurgical recovery consistency have all seen substantial improvements. In 2023, Gruyere JV successfully processed 9.4 Mt/a of ore, produced 320 000 gold ounces, and extended the mine's life through the open pit expansion project.

This paper investigates the continuous test work, subsequent modelling, and improvement projects carried out during the ramp-up and early operational phases, including:

- Implementing mechanical enhancement projects and modifications to the original design to alleviate wear and availability concerns.

- Conducting ongoing comminution studies and benchmarking surveys of the grinding circuit.

- Fine-tuning SAG and ball mill liner designs, optimising grinding media, and refining post-reline ramp-up strategies.

- Enhancing the pebble crushing circuit's capacity through a series of circuit and equipment upgrades.

INTRODUCTION

Gold Road Resources Limited discovered the Gruyere orebody in October 2013, which is situated on the Yamarna Belt approximately 200 km east of Laverton in Western Australia. The Gruyere feasibility study was completed in October 2016. The study envisaged a large-scale open pit mine feeding a 7.5 Mtpa processing plant producing an average of 270 000 ounces (oz) of gold a year over an initial 13-year mine life. In November 2016, Gold Road entered a 50:50 Joint Venture with Gold Fields Ltd (Gold Fields) to form the Gruyere Joint Venture (Gruyere JV), managed by Gruyere Management Pty Ltd (GRM)—a wholly owned subsidiary of Gold Fields.

The construction of the Gruyere processing plant commenced in the first quarter of 2018, with commissioning taking place in 2019 and the first gold poured during the second quarter of 2019. Since 2019, Gruyere has been dedicated to the ramp-up and operational phases of the project. Production outputs from the processing plant are now exceeding the design production, with 9.4 Mt processed through the plant to achieve 320 000 oz in 2023. The accomplishments at the site can be attributed to enhancements in operation ergonomics, advanced process control, throughput optimisation, mechanical availability, and metallurgical recovery.

In this paper, the authors analyse the site test work, modelling and improvement projects undertaken during the ramp-up and early operational phases, shedding light on the methodologies and outcomes of these initiatives.

Improvement initiatives in the reliability of the Gruyere processing plant are not the focus of this paper and are not discussed by the authors. It is worth noting that the Gruyere maintenance teams have focused on equipment reliability and shutdown scheduling improvement projects during the ramp-up phase of the Gruyere processing plant. Due to the inherently abrasive nature of the ore processed through the plant, the maintenance team has been working through the challenges that this has posed, contributing to circuit availability and increased throughput.

PROCESS FLOW DESCRIPTION

The Gruyere circuit is a conventional primary crush – semi-autogenous ball mill crushing circuit (SABC) with a pre-leach and tailings thickener. The gold recovery includes a standard carbon-in-leach (CIL) with a split AARL elution circuit. A stand-alone gravity circuit is fed from the mill discharge hopper with gravity concentrate treated via an Inline Leach Reactor (ILR).

The Gruyere comminution flow sheet includes primary crushing followed by a coarse ore stockpile (COS). The primary crusher is an FLSmidth TSU 1400 × 2100 gyratory crusher fitted with a 600 kW motor. Crushed ore is reclaimed from the COS via three apron feeders under the stockpile, discharging ore onto the semi-autogenous grinding (SAG) mill feed conveyor. The SAG mill is a single Outotec 10.97 m (36') diameter × 5.79 m (19') effective grinding length (EGL) mill equipped with a 15 MW dual pinion drive (Lovatt *et al*, 2023). The SAG mill discharges onto a single vibrating screen fitted with 14 mm aperture screen panels. The screen oversize is conveyed to the pebble crusher storage bin, where pebbles are reclaimed via two belt feeders, each reporting to a Metso HP4 cone crusher fitted with 315 kW drives. The pebble crushers are designed to treat 200 t/h each at a closed side setting (CSS) of 13 mm. The pebble crushing circuit was upgraded in 2023 by installing an additional Metso MP800 cone crusher with a 600 kW drive. The existing HP4 crushers are now utilised when the MP800 is offline for scheduled liner rebuilds. Crushed pebbles are returned to the SAG mill feed conveyor.

The SAG mill discharge screen undersize reports to the common mill discharge hopper, along with the ball mill discharge. These are pumped to the cyclone cluster for classification, which is fitted with 12 × 650 mm hydrocyclones. The cyclone overflow reports to the trash screens followed by the leach feed thickener, while the cyclone underflow reports to the FLSmidth 7.93 m (26') diameter × 10.82 m (35.4') EGL overflow discharge ball mill for further size reduction. The ball mill is also equipped with 15 MW dual pinion drive, common with the SAG mill. The flow sheet includes the ability to return part of the cyclone underflow to the SAG mill feed chute if required. The grind size P_{80} target is 125 µm.

A portion of the combined mill discharge is pumped from the mill discharge hopper to the gravity circuit. A baffle in the mill discharge hopper is designed to separate the SAG mill and ball mill discharge, with the gravity feed drawn from the ball mill side to assist in the pre-concentration of the gravity feed. The gravity circuit comprises two vibrating screens and four 48 inch Knelson concentrators. Tailings from the gravity concentrators are combined with the gravity screen oversize and returned to the mill discharge hopper on the SAG mill discharge side of the baffle. The gravity concentrate reports to a Gekko ILR, with pregnant solution reporting to the gold room, and washed leach residue returned to the mill discharge hopper. The Gruyere circuit process flow is shown in Figure 1.

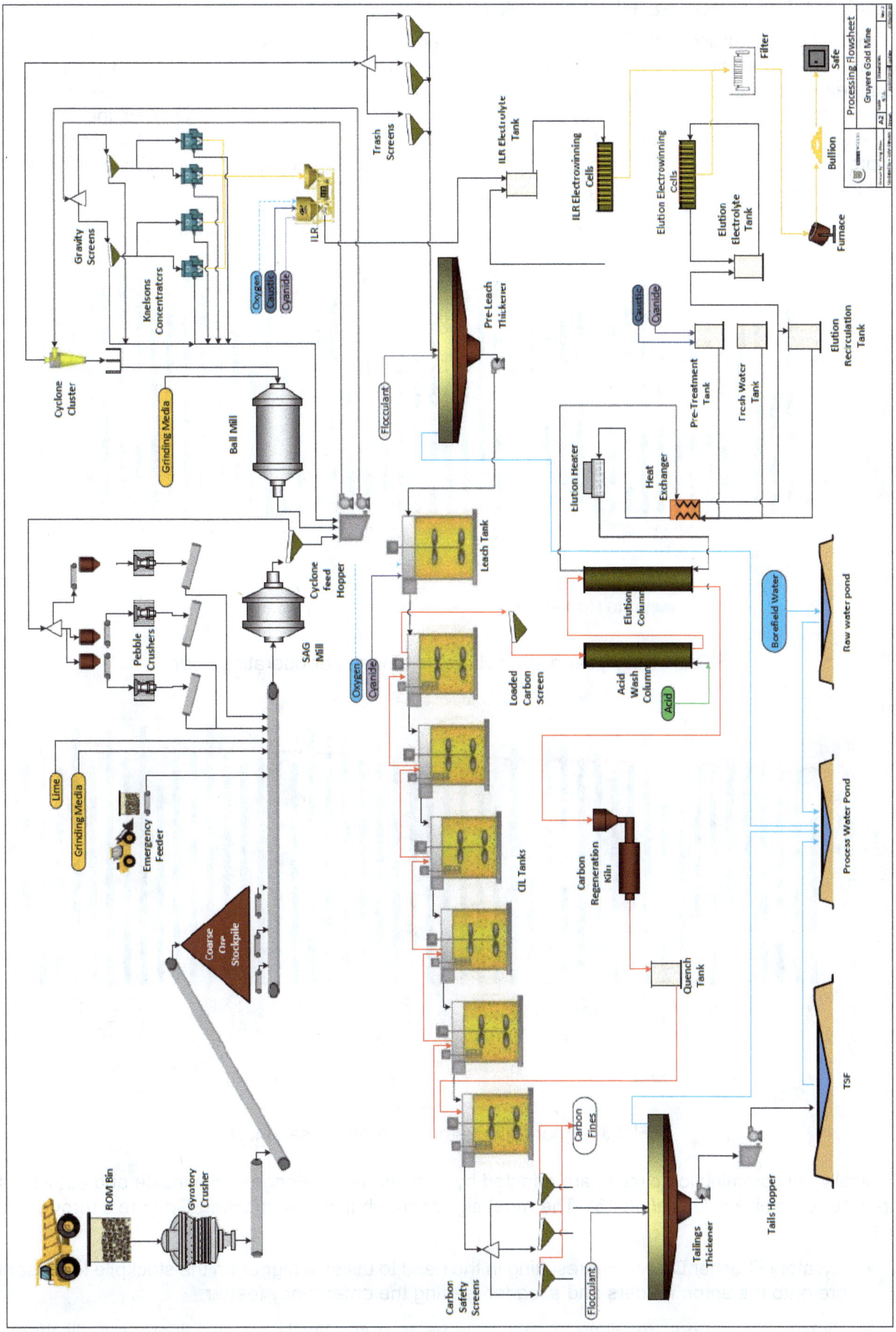

FIG 1 – Gruyere process flow diagram.

CIRCUIT HISTORICAL PERFORMANCE AND CHALLENGES

Circuit debottlenecking, reliability improvements and circuit upgrades have enabled Gruyere to steadily increase circuit throughput rates and SAG mill utilisation. Figures 2 and 3 present the four year historical trends in the SAG mill throughput and utilisation at the Gruyere processing plant. These figures illustrate the monthly average tonnes per operating hour (TPOH) for the SAG mill (Figure 2) and the utilisation of the SAG mill (Figure 3). The trends reflect the progressive improvements achieved through collaborative efforts by the Gruyere team, which aimed to increase mill throughput despite processing more competent fresh ore.

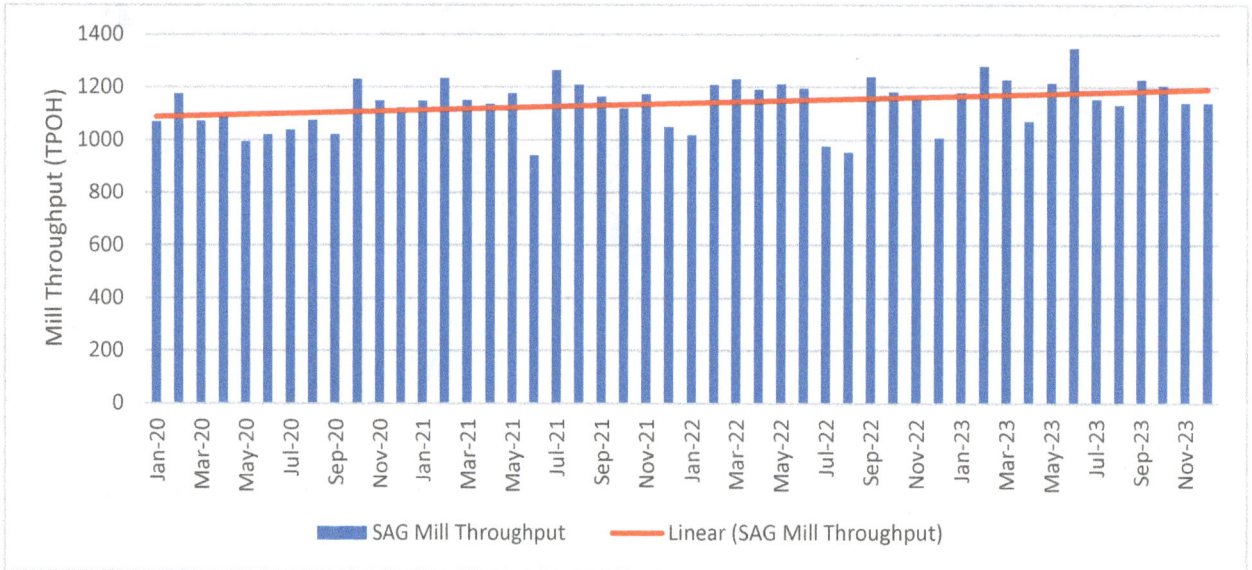

FIG 2 – Four-year historical trends tonnes per operating hour.

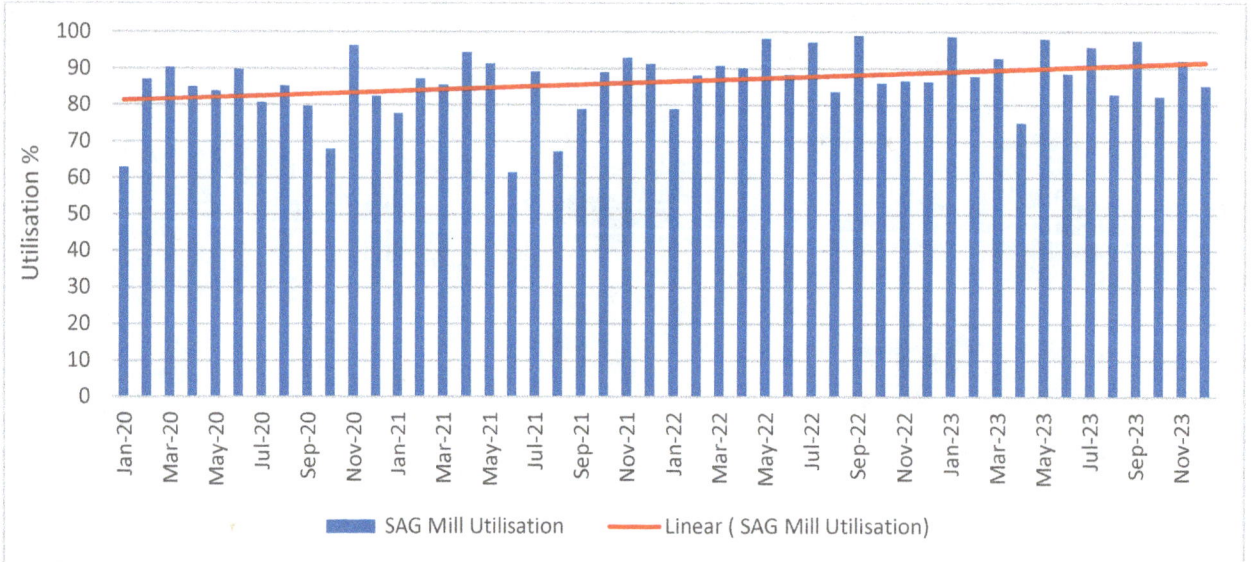

FIG 3 – Four-year historical trend utilisation.

Typically, all comminution circuits are affected by operational challenges both inside and outside of operator control (Kock *et al*, 2015). The most significant challenges affecting Gruyere's throughput are:

- Gyratory Crusher Downtime, resulting in the need to utilise a digger on the stockpile to present ore onto the apron feeders and a loader feeding the emergency feeder.

- Highly abrasive ore, resulting in excessive wear in crusher liners, mill liners, mill discharge hopper and chutes.

- Post-reline ramp-up, affecting SAG mill power draw and breakage.

- Ball mill downtime, resulting in reduced throughput operating in a single-stage SAG circuit configuration.
- Downstream thickener and pump considerations, restricting the overall circuit feed rate.

Circuit benchmarking

Since the inception of the Gruyere circuit commissioning, circuit benchmarking via regular full-circuit surveys has assisted in monitoring key performance parameters and guided the effectiveness of implemented optimisation projects (Radford *et al*, 2021). Table 1 summarises the key survey parameters during these evaluations in line with published methods (Siddall, Henderson and Putland, 1996; Scinto, Festa and Putland, 2015; Putland, 2006).

TABLE 1

Gruyere survey data summary.

Parameter	Unit	Dec 2019	Mar 2020	Nov 2020	Oct 2021	Jun 2022	Jun 2023
Feed blend							
Fresh	%	-	-	100	100	100	100
Transitional	%	57	40	-	-	-	-
Oxide	%	43	60	-	-	-	-
BWi	kWh/t	11.8	13.2	16.2	16.7	17.1	18.7
Axb	-	56.7	44.1	32.5	37.5	34.8	37.1
Ore specific gravity (SG)	t/m^3	2.59	2.61	2.68	2.69	2.69	2.70
Throughput, dry	tph	984	1272	1082	1191	1138	1286
Pebble extraction, dry	tph	13	149	158	188	173	298
Feed F_{80}	mm	121	117	76	101	83	98
Grind P_{80}	μm	113	136	108	139	120	153
SAG specific energy	kWh/t	11.9	10.0	12.3	10.5	11.3	11.0
Pebble crusher (PC) specific energy	kWh/t	0.0	0.1	0.2	0.2	0.1	0.4
Ball mill (BM) specific energy	kWh/t	13.7	10.6	12.9	9.7	11.6	10.4
Total specific energy	kWh/t	25.7	20.8	25.4	20.4	23.1	21.8
SAG mill pinion power	kW	11 739	12 776	13 278	12 534	12 910	14 162
Ball mill pinion power	kW	13 513	13 538	13 718	11 487	13 221	13 348
Pinion power efficiency (theoretical / actual)	%	56.8	76.3	80.6	90.9	87.7	85.5
Power utilisation	%	84.2	87.7	90.0	80.1	87.1	91.7

In Table 1 the theoretical Bond circuit energy consumption is compared to the actual absorbed circuit energy to gauge the overall circuit efficiency. 100 per cent efficiency signifies that the actual circuit power draw is just as efficient as the theoretical calculation, with no added circuit inefficiencies. Since the 2019 baseline survey, there has been a consistent improvement, apart from a slight reduction in the June 2022 and June 2023 surveys, likely associated with variability in cyclone overflow samples. The power utilisation compares the duty power to the installed mill motor power.

Notably, the ore hardness has increased as the mining pit has been developed, aligning with a higher proportion of fresh feed. Recent circuit debottlenecking studies (Orway Mineral Consultants (OMC) Report 8,177, May 2020) indicate that the current circuit configuration has a maximum theoretical throughput of 1357 t/h for a 90 per cent fresh feed blend. Downstream volumetric flow constraints, such as trash screen flux rates, leach feed, and tailings disposal pumping capacity restrict higher throughput rates.

SAG feed conveyor upgrades

The SAG mill feed conveyor CV03 design parameters were 1530 t/h total conveyor capacity, which included 1237 t/h of primary ore and 293 t/h of crushed pebbles. Debottlenecking exercises have enabled plant new feed rates up to 1475 t/h of oxide ore. Throughput at these levels represents the maximum capacity of the conveyor, limited by motor size and current draw. Several instances of conveyor high-amp trips have been observed.

The current CV03 head chute design experiences high maintenance and operational challenges. It exhibits significant wear, exacerbated by uneven water distribution, which leads to preferential wear on one side of the SAG feed chute.

The SAG feed conveyor was upgraded in 2022 with a new drive, gearbox, and variable speed drive (VSD) to the following design parameters: 2172 t/h total capacity, a 42 per cent increase on the initial design, 1782 t/h primary ore, and 400 t/h pebbles. The speed was increased from 2.0 m/s to 3.2 m/s which improved start conditions by lowering the total weight on the belt and preventing lifting during loaded start-ups.

The head chute underwent an upgrade incorporating a new rock box to accommodate the increased conveyor speed and enhanced wear liners. The water addition point was also relocated to the rear of the feed box, where water flows over a weir to prevent preferential flow-through the SAG feed chute.

SAG MILL LINER DEVELOPMENT

Premature SAG liner wear prompted the need to focus on SAG mill liner design. The SAG liner design was an iterative process, with the design changes aiming to achieve a six month reline frequency equivalent to a 5 Mt liner set.

The first SAG mill reline was in January 2020, which was earlier than originally planned due to the accelerated wear of the feed end head liner and feed end shell liners. The worn liners were replaced with the commissioning set of liners, which were a like-for-like set of the original liners. This premature wear led to design changes in the feed end head and feed end shell liners.

The first design changes were installed in the second SAG mill reline in June 2020; the key design changes included:

- Feed end middle and outer liners combined as one; this was a reline time-saving strategy as it removed 16 liner pieces from the design.

- The addition of white iron inserts into the feed end lifter section to provide improved wear resistance, rather than increase the lifter height and, therefore, the amount of extra weight needed.

- Feed end and middle shell liner lifter height and plate thickness increased in the high wear zone.

- Rubber filler rings were added for handling safety and to reduce the weight of steel items. Due to high wear, these rubber filler rings were later changed to include a steel insert.

To assist with reducing the reline times, the shell liners were changed to a single row to allow more lifter rows to be installed per inch (turn) of the mill; reline crews found that they could only install one row of the double chord shell liners as the reach was too high, whereas they could install three rows of the single chord liners. This reduced the number of times the mill had to be inched in a reline. The change to the single row also reduced the weight of the liner pieces, allowing for the shell design to change from three sections (feed, middle, and discharge) across the length of the mill to two sections (feed and discharge) across the shell. These modified shell liners were installed in the October 2020 reline.

Following accelerated wear in the feed end and the discharge end of the shell, the liner lifter heights and plate thickness were iteratively increased after each reline. Efforts were made to reduce the overall steel weight in the mill by installing a rubber modular discharge design in August 2021. This reduced the weight by around 18 t. The increases in shell liner thickness added an additional 114 t from the original set until the August 2022 design. This resulted in more weight of steel being installed in the SAG mill, leading to tighter liner spacing to height ratios of only 1.1. The typical design range for the spacing-to-height ratio is 1.7–2.0 (Toor and Brennan, 2022). These design changes were made with the target of achieving six monthly relines, a target which had not been achieved. Tight liner spacing ratio can also lead to packing between the liners, which was evident in a recent SAG mill crash stop during the June 2023 comminution survey (Figure 4).

FIG 4 – Photo of packing in the SAG mill, June 2023 crash stop.

Considering the SAG liner designs increased weight in the mill but had not achieved the targeted six month reline interval, the operational approach was revised to adopt a shorter four month lifespan or a throughput equivalent to 3.5 Mt. This adjustment in strategy aimed to synchronise with the shutdown schedule of other parts of the processing plant as the site maintenance teams were focused on improving wear life and shutdown schedule for the circuit. Consequently, the revised strategy enabled the liner design to prioritise throughput efficiency, ensuring that the liners are replaced with sufficient lifters still intact to maintain charge throw at the end of their life. Figure 5 is a photo of the SAG mill discharge liner showing no lifter left at the end of the liner life. Gruyere engaged Hatch to review the current design and recommend a revised design based on this new operational strategy, targeting a shorter liner life while facilitating a more rapid throughput ramp-up post-reline to offset the additional downtime.

FIG 5 – Worn SAG discharge end liner.

Conservative design changes were proposed for the first iteration of the new design, with an increase in feed end plate thickness and a reduction in feed end lifter height proposed. The discharge end plate thickness was also proposed to be reduced. This led to spacing-to-height ratios increasing from 1.1 to 1.33–1.38, which was still considered conservative for wear but still relatively tight spacing for design ranges.

Hatch completed simulations to assess the performance of both the existing and proposed shell liner designs over the full life cycle by simulating scenarios corresponding to new, 50 per cent worn, and 100 per cent worn liner profiles. Figure 6 shows that the proposed design had an initial uplift in throughput and faster ramp-up. The simulations indicated a 4 per cent increase in annual throughput if the simulated higher mill speeds and power draws were realised.

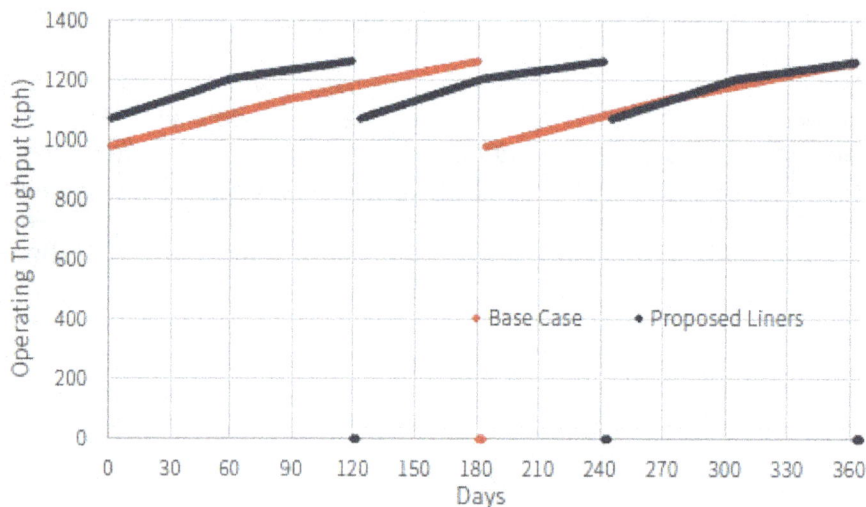

FIG 6 – Annualised throughput curves for the existing and proposed liners (Toor and Brennan, 2022).

The new design targeting a four monthly reline strategy was installed in April 2023.

SAG mill post reline ramp-up

The SAG mill post-reline ramp-up strategy has been fine-tuned since 2020, influencing variables such as SAG mill power draw, speed, and weight, resulting in better utilisation.

Initially, a strategy was pursued to extend the time between reline shutdowns by increasing the liner thickness (mass). As a result of premature wear, the feed and discharge end reline became out of sync. A notable learning point from this was that the increased liner mass offset the SAG operating charge at the same bearing pressure. The effect can be seen in throughput during 2021 and 2022, resulting in challenges in optimising the SAG weight and speed set points to match the SAG mill breakage requirement, which could have benefited from a higher SAG rock load. This was addressed towards the end of 2022 when the post-reline SAG weight set point was pushed close to the maximum bearing pressure limits.

The clear progression of the ramp-up strategy can be seen in Table 2. Once optimised, the Gruyere SAG mill achieves a high post-reline power draw (>13.8 MW) by managing the SAG speed, ball charge, and weight set points to optimise the impact breakage for the given liner wear state assisted by using MillROC (Oblokulov *et al*, 2021). Figure 7 gives an example of the Gruyere SAG mill running at both 'New' liner conditions (30° face angle, 14 per cent ball charge (BC), 30 per cent total load (TL) and 77 per cent critical speed (Nc) for an approximate total charge mass of 630 t) and 'Worn' Liners (45° face angle assumed, 14 per cent BC, 22 per cent TL and 80 per cent Nc with for a total charge mass of 520 t) used to maximise the impact breakage (Putland and Sciberras, 2019).

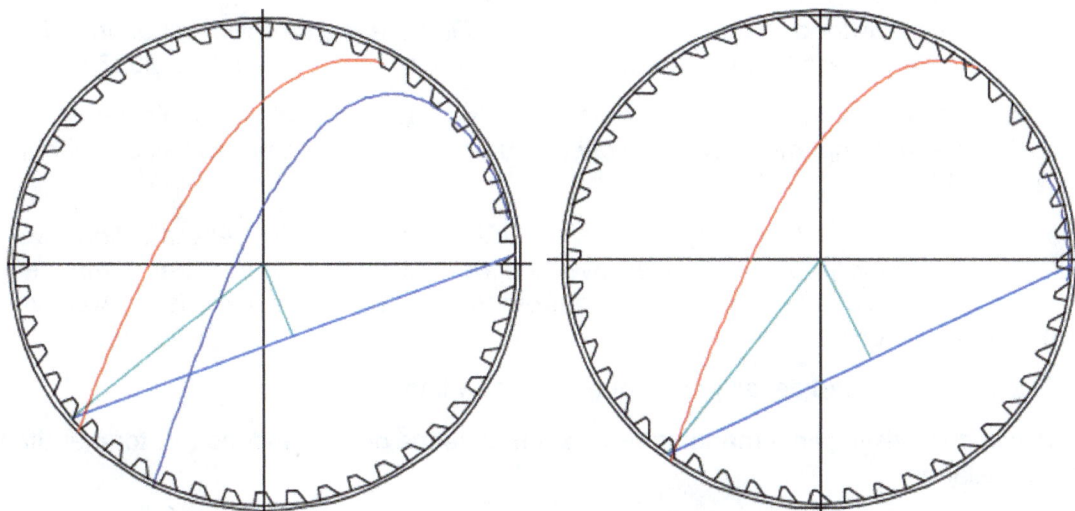

FIG 7 – Indicative trajectory (fresh liners left, worn liners right).

The net effect is a better utilisation of the available SAG mill power and breakage. Table 2 summarises the key reline data, which shows a measured improvement in the time required to achieve the target power draw optimised equipment utilisation.

TABLE 2

Reline ramp up summary.

Liner install date	Reline scope	Ramp up time, days	Avg SAG power, kW	Avg SAG weight, t	Avg SAG speed, %Nc
27 Feb 2021	Full	52	12 775	597	76.8
01 Jun 2021	Partial	55	12 453	608	74.9
14 Aug 2021	Full	68	12 989	603	72.6
26 Feb 2022	Partial	50	12 333	612	70.9
23 Apr 2022	Partial	64	13 129	585	74.8
05 Aug 2022	Full	52	13 314	652	76.8
10 Dec 2022	Full	60	13 254	657	76.7
15 Apr 2023	Full	**25**	13 637	627	77.1
5 Aug 2023	Full	**17**	13 903	615	77.9
2 Dec 2023	Full	**24**	13 274	603	72.8

Future SAG liner development

Gruyere's next steps in liner development are through a collaboration with Comminution and Transportation Tech Inc (CTTI) to move to a new high-performance liner system. Through discrete element method (DEM) modelling this project aims to fast-track a redesign of the SAG liners to achieve the following:

- An increase in comminution circuit performance (higher throughput, finer grind, and/or use less power) due to more effective transfer of energy into the charge and eliminating grate peening and pegging.

- Increase the overall liner system life and continue to target reline intervals closer to the original six-month target whilst maintaining the quick ramp-up in production.

- Reduce grinding ball consumption due to reduced ball-on-liner impacts throughout the liner life.

The project has been divided into four broad phases. Gruyere is currently embarking on the first phase and the information-gathering stage. The project's completion date is late 2025:

- Analysis of the existing mill process data to set the DEM simulation conditions, such as mill filling, ball-to-ore filling, mill rotational speed and power draw, and particle size distributions for ore and balls.

- Analysis of the current liner system using DEM simulations to generate liner wear rate calibrations and morphology changes over the wear life cycle for the liner components that match the observed wear performance from scans and observations of removed liner components.

- DEM simulation-based development of a CTTI new liner system.

- Work with the existing liner manufacturer to complete the design and models for manufacturing and installation.

SAG mill grate development

In efforts to increase SAG mill throughput, maximising the pebble extraction rate has been a key focus in debottlenecking the circuit. The original SAG mill grate configuration consisted of 32 segments with an 18 mm design aperture to manage the SAG charge level with the start-up oxide and transitional feed blend. Once fresh feed was introduced, initiatives to optimise the pebble extraction rate included increasing the 18 mm aperture grates to 45 mm and then further increasing them to 70 mm. This increase in aperture size, however, resulted in an interaction issue between the grate and the turbo pulp lifter resulting in pegging of the grates and restricting the pebbles and worn media to be able to exit the SAG mill. Figure 8 shows the design interaction, as well as the extent of grate pegging, experienced.

SAG Discharge Grate

Turbo pulp lifter

Interaction between grate and turbo
pulp lifter

70mm media and rocks getting stuck (pegging) in
the tight gap between the grate port and the turbo
pulp lifter behind, restricting the pebbles
discharged from the mill

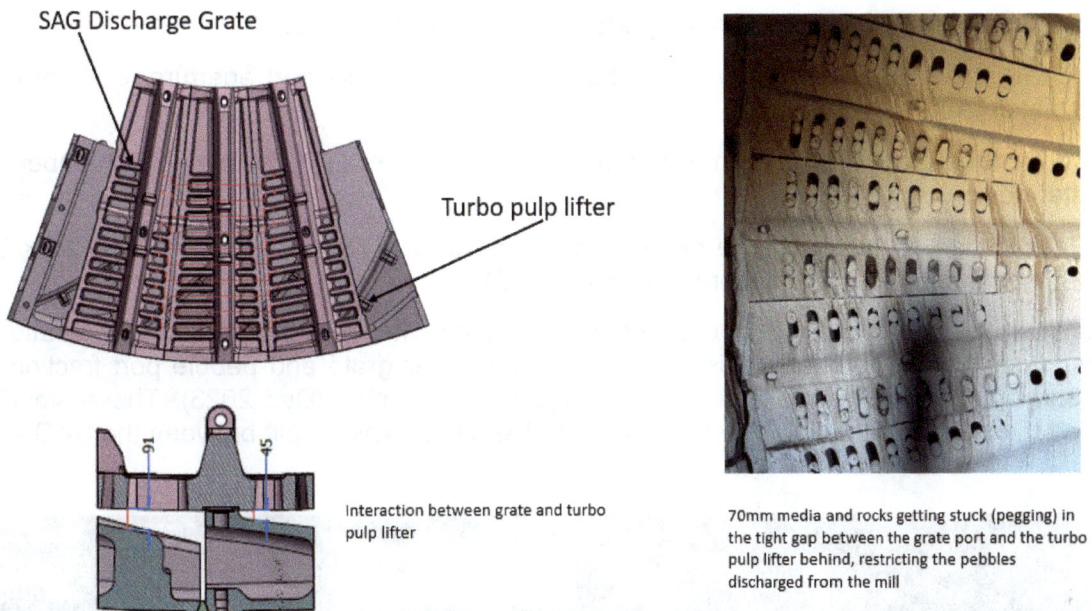

FIG 8 – Image showing turbo pulp lifter and grate interaction and pegging of the grates.

The progression of the grate designs is shown in Figure 9, an offset design proposed by the liner supplier was installed in February 2022. No improvement was observed in this design compared to the previous grates regarding the percentage of pebbles rate to fresh mill feed rate. The offset design achieved an average of 16.5 per cent of pebble rate to fresh mill feed rate. As part of the engagement with Hatch on the shell liner designs, Hatch also reviewed the SAG grate design and proposed a design that included angled ports in the area not obstructed by the turbo pulp lifter (TPL). This design was installed in August 2022 and improved the average pebble rate to fresh feed to 23.5 per cent.

The next iteration of the grate and discharge design was to transition to a matched curved grate arrangement instead of the TPLs. The curved arrangement was installed during the April 2023 SAG reline. This change increased the open area from the current Hatch grate from 9.5 per cent to 10 per cent but the effective open area increase was higher due to removing the TPL interaction. The average pebble rate to fresh feed achieved since the change to the curved design is 28.2 per cent.

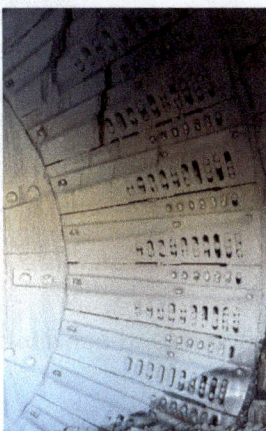

Offset Radial Design   *Hatch Radial Design*   *Curved Design*

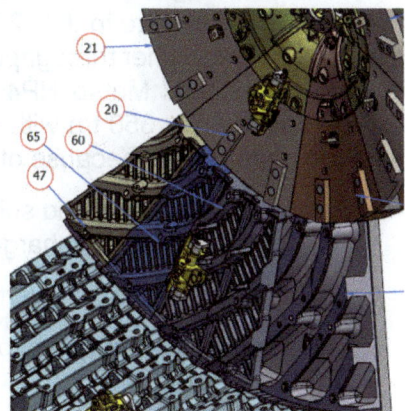

FIG 9 – Image of the progression of the grate designs.

A summary of the optimisation initiatives to increase the pebble extraction rate are:

- Increasing the pebble port fraction to 25 per cent or 8 × 45 mm segments (Aug 2019).

- Increasing the pebble port aperture up to 70 mm (Aug 2020).

- Increase the relief angle on grate apertures from 6° to 8° (Feb 2022).

- Update the grate design aperture to 60 mm and the pebble port aperture to 80 mm (Hatch radial design, Aug 2022).

- Convert the grate design from twin chamber curved lifters (TPL) to a single chamber curved lifter design (Apr 2023).

- Optimisation of the discharge grate combination to 8 × 75 mm segments and 24 × 55 mm segments and stockpile feeder ratios (Dec 2023).

Figure 10 shows the change in circuit pebble extraction rate over this period of operation. The greatest sustainable improvement is attributed to the larger grate and pebble port fractions (Aug 2022) and optimisation of the grates and stockpile feeder ratios (Dec 2023). These significantly improved the pebble extraction rate and balanced the circuit power split between the SAG and ball mills.

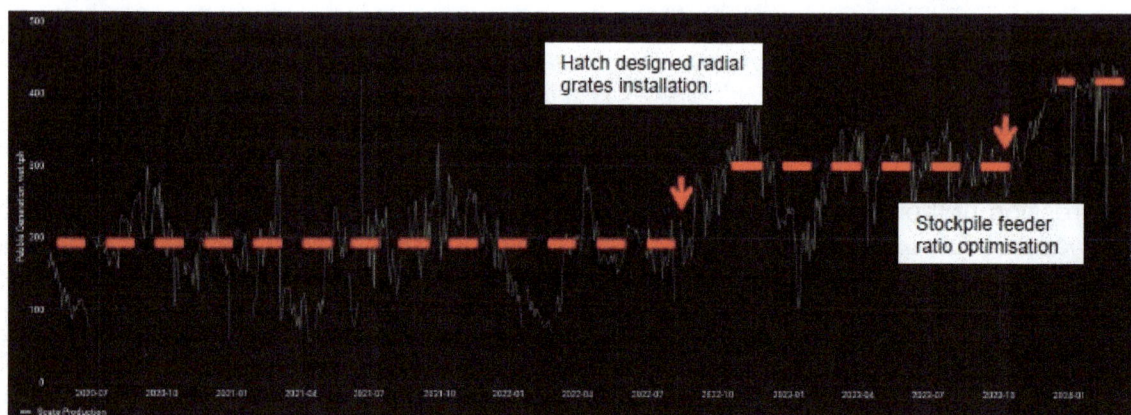

FIG 10 – Pebble extraction rate improvement.

Pebble crusher circuit upgrade

From the commencement of production, the Gruyere pebble crushing circuit (Figure 11) has experienced issues, including poor crusher availability and elevated chute wear. On oxide ore, the availability of the pebble crusher circuit had a limited impact on overall plant throughput. The change to fresh ore and higher throughput rates has seen the pebble crusher circuit issues become a source of SAG mill downtime and periods of low production. With the increase of the SAG mill grates to 70 mm to treat fresh ore in July 2020, the pebble crusher circuit throughput increased to a range of 250–350 t/h. This higher throughput brought to light the limited design and functionality of the crusher circuit. The installed Metso HP4 crushers could typically treat 200 t/h but cannot achieve duty-standby operation at 350 t/h with the shared feed bin configuration. Significant SAG mill downtime has been experienced because of numerous operational issues and design issues; these include:

- Regular overloading and spillage of the pebble crusher circuit feed belts CV04 and CV05 and the pebble crusher discharge belt CV06 and CV07.

- Jamming of the 3-way diverter gate in the feed bin.

- Metal detect events and bypass report direct to sump – major rehandle and refeed activity for process operators.

- No bin overflow resulting in variable feed rates to the SAG mill.

- Unable to isolate individual crushers due to the shared feed bin.

- The method for moving belt feeders is slow and cumbersome, increasing the rebuild time.

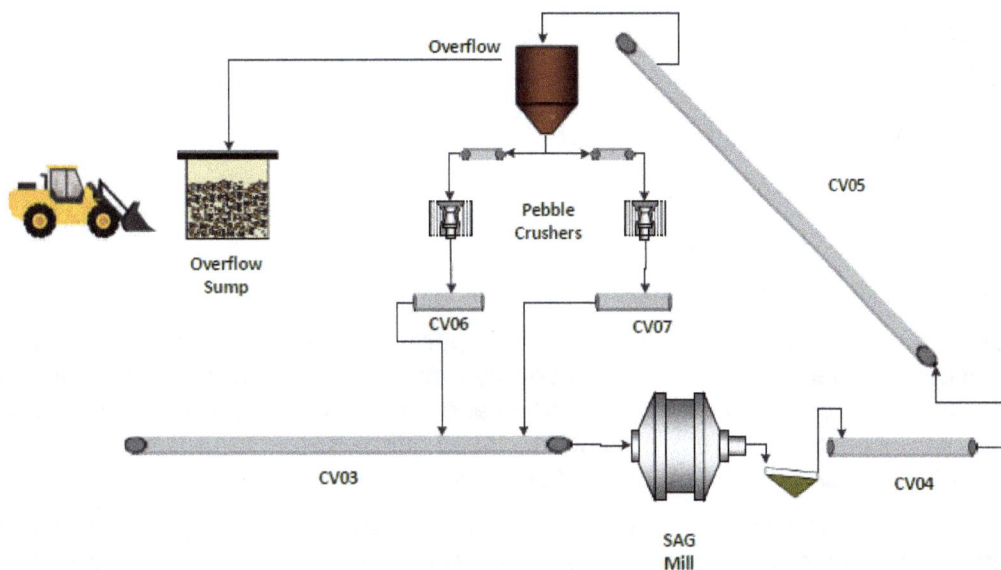

FIG 11 – The original pebble crusher flow diagram.

The abrasive nature of the fresh ore resulted in a short liner life for the pebble crushers. At the time a liner change was required every seven days and took 12 hrs to complete.

A pebble crusher upgrade project was initiated in 2020 with the aim of upgrading the circuit for reliable operation at a design throughput of 384 t/h. This represented a 21 per cent increase over the original design. The project scope was based on the preferred option study which included a bypass conveyor and upgraded bin with overflow to the SAG mill feed conveyor (Figure 12). The project also addressed accessibility and maintainability to reduce the time to compete a rebuild and improve the ergonomics of the maintenance task.

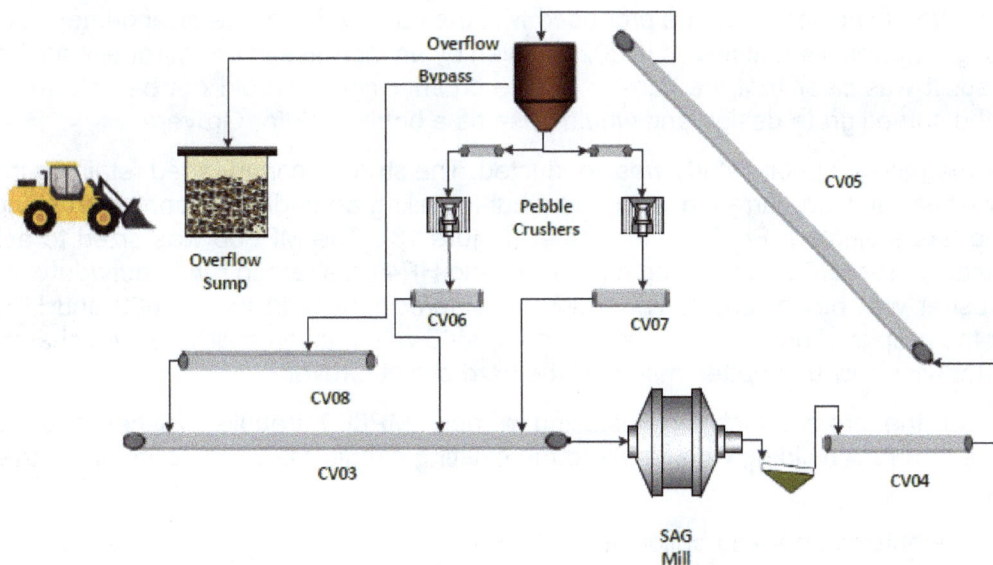

FIG 12 – The pebble crusher flow diagram after the circuit upgrade project.

The three key objectives of the pebble crusher circuit upgrade project were:

1. Consistent product return to the SAG feed conveyor CV03 to minimise the variability from the circuit on SAG mill stability and throughput.

2. Improve the availability and utilisation of the pebble crusher circuit.

3. Improve the maintainability of the pebble crusher circuit.

The scope of the project included:

- New CV08 bypass conveyor

- New diverter gate and bypass chute

- Upgraded bin with intrinsic overflow

- CV04 and CV05 drive, gearbox and VSD upgrades

- CV06 and CV07 drive upgrades

- Hydraulic isolation of belt feeders

- Maintenance access frame for crusher rebuilds.

By 2022 pebble crusher liners were only lasting 140 operating hours or around six operational days. Only one crusher was being run at a time, driven by maintenance costs and resourcing due to the frequent liner changes. During processing surveys and trials in 2022, the opportunity was taken to trial running both pebble crushers ensuring no bypassed pebbles were returned to the SAG mill. The trials indicated that the SAG mill throughput could be increased by a range of 112 t/h up to 130 t/h.

Hatch was engaged to model the throughput rate when running one pebble crusher verses running with all the pebbles being crushed (running the two pebble crushers). The modelling was completed at different hardness (A*b) and Bond ball mill work index. The modelling showed that irrespective of the ore hardness, the impact of running both pebble crushers and not bypassing pebbles to the SAG mill resulted in a 111–118 t/h increase in mill throughput and aligned with what has been observed in the plant trials.

From September 2022 the pebble crusher operational strategy was changed to run both pebble crushers. The maintenance teams increased liner stock and resource capacity to allow for this operational strategy until the MP800 pebble crusher was installed in December 2023.

Pebble Crusher MP800 installation

Replacement of the TPLs with a curved grate design effectively increased the open area of the grates, flooding the existing HP4 pebble crushing circuit with feed. Modelling from Hatch suggested that up to 400 t/h of pebbles would be produced with the curved discharge arrangement. A decrease in HP4 throughput was also observed in 2022, driven by an increase in ore hardness as the Gruyere pit deepened. It was clear that the current pebble crusher circuit would not be able to handle the change to the curved grate design and would pose as a bottleneck for Gruyere.

An order-of-magnitude options study was conducted. The study recommended retaining the existing two HP4 crushers and the current arrangement but installing an additional pebble crushing building to accommodate a single MP800 cone crusher (Figure 13). The MP800 was sized to achieve the 400 t/h as a duty crusher, thus allowing for the existing HP4 crushers to run in duty/duty mode while the new crusher was being rebuilt. The study considered that both the MP800 and HP900 were suitable for the proposed duty, but ultimately, the MP800 was chosen as it was a mechanically more robust and heavier-duty unit better suited for the hard ore at Gruyere.

The scope of the project included installing a new MP800 pebble crusher and associated infrastructure in a new building with tie-ins to the existing circuit. Major components of the upgrade included:

- MP800 pebble crusher and building

- Belt feeder and bin

- Hydraulic isolation gates for MP800 and existing HP4 crushers

- 50 t gantry crane

- New transformers and switch room

- New CV09 feed and CV10 discharge conveyor

- Extension of the existing CV08 bypass conveyor to accommodate CV10

- Upgrade existing CV04 and CV05 conveyors to 480 t/h (max) with troughing idlers.

The MP800 was tied into the existing pebble crushing circuit in December 2023, with commissioning occurring in January 2024.

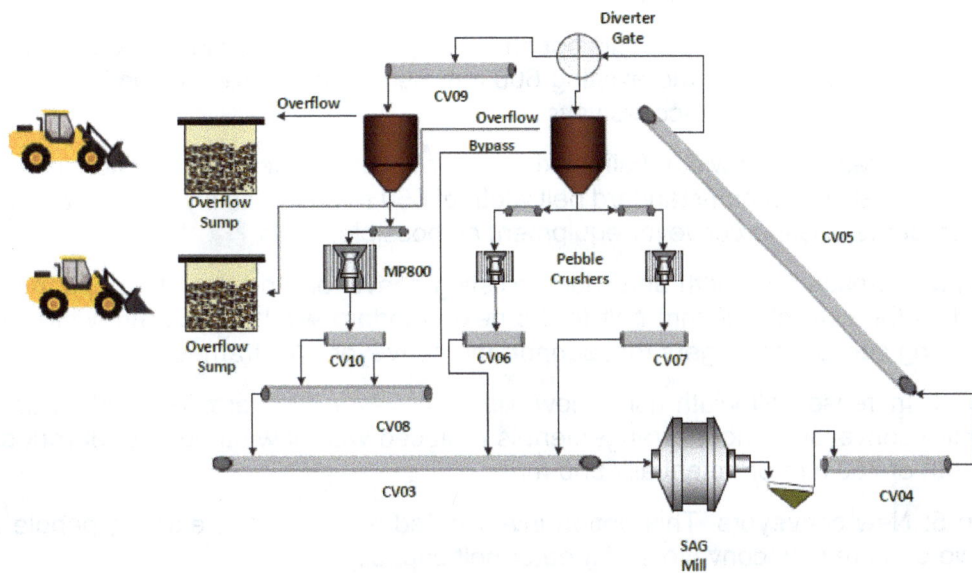

FIG 13 – The pebble crusher flow diagram after the installation of the MP800.

Future upgrades for the pebble crushing circuit

As part of the MP800 pebble crusher installation project, the existing pebble crusher conveyors, CV04 and CV05 were upgraded from their previous nominal capacity 320 t/h (384 t/h max) to 400 t/h (480 t/h max). However, an increase in operational throughput requirements beyond 500 t/h during SAG mill surge conditions is resulting in CV04 and CV05 becoming overloaded. This overloading is causing:

- Downtime events stemming from significant spillage, resulting in adverse impacts on production.

- Increased belt damage due to excessive spillage, necessitating frequent maintenance.

- An increase in the depth of material on CV05 is contributing to metal detection issues, leading to bypass issues at the pebble crushers further exacerbating the issue.

- Frequent damage from metal in the pebble crusher feed, due to the belt magnet being unable to reliably remove metal.

- Excessive spillage requiring permanent clean-up personnel, having safety and cost implications.

The new MP800 is capable of up to 600 t/h, however, to ensure surge capacity without excess spillage, the pebble crushing circuit is limited to approximately 450 t/h by the capacity of CV04 and CV05. This is posing a bottleneck risk for the comminution circuit. The Gruyere team has since initiated a project to upgrade the volumetric capacity of the pebble crusher conveyors.

In addition to increasing the conveyor volumetric capacity, the upgrade plans to address the following maintenance and performance items:

- Improve the CV04 and CV05 tramp magnet removal efficiency.

- Provide adequate belt volumetric capacity to avoid the spillage of pebbles from an overly full belt.

- Replace the existing CV04 head chute with an improved design, as the current design contributes to the poor tramp metal placement, reducing the tramp metal removal efficiency. In addition, high-liner wear requires continual maintenance.

The rapidity by which the proposed upgrade can be executed during phases for the design, procurement, and safe installation during shutdowns with minimised downtime is also a priority for the project. Five options were identified for preliminary evaluation:

- Option 1: Increased belt speed. This option investigated the potential to increase the conveyor belt speeds while retaining the existing 600 mm belt width, thereby minimising the impact on the conveyor structure and accessways.

- Option 2: Increased belt width, using existing conveyor stringers. This option investigated the potential of installing a non-standard belt width of 650 mm or possibly wider, whilst maintaining as much of the existing conveyor equipment as possible.

- Option 3: Increase belt width and widen existing conveyor stringers. This option investigated upgrading the current 600 mm belt to the next standard width of 750 mm whilst maintaining the existing conveyor stringers that support the conveyor idler frames.

- Option 4: Increase belt width using new wider conveyor stringers. This option assumes that the entire conveyor stringer arrangement is replaced with new stringer steelwork suited for a belt width of 750 mm or potentially 900 mm.

- Option 5: New conveyors. This option investigated replacing the existing pebble conveyors with two or more new conveyors of greater belt capacity.

These five options were defined to a level sufficient to enable a multi-criteria analysis (MCA) workshop. From this workshop, Options 1 and 5 were eliminated from further evaluation, and Options 2, 3, and 4 were selected for further investigation.

The options study concluded that Option 3 should proceed. In addition to achieving the highest overall MCA score, the option held notable advantages with respect to constructability and meeting key maintenance activities.

Option 3 includes installing a new 750 mm conveyor belt on both conveyors within the existing structure where practical. The two existing conveyors will also require wider pulleys, head and tail frames, new chutes, and ancillary equipment.

The CV-005 drive would be replaced entirely, and the CV-004 tramp magnet is also proposed to be upgraded with a larger, more powerful unit to improve tramp removal efficiency. Due to CV-004's short length, a standalone solution of completely replacing the remaining stringer was also agreed upon.

The installation of the proposed CV04 and CV05 upgrades is planned for the first half of 2025.

Ball mill liners

The first ball mill reline occurred in January 2021, when the commissioning set of liners was installed. The first set of liners lasted 11.7 Mt.

The following liner design changes were implemented:

- The head liners and shell liners were changed from Chrome-Moly to white iron liners; this was to increase the liner life by an estimated 30 per cent.

- The feed end and discharge end inner and outer head liners were combined into one single liner. This was proposed as wear was occurring across both liners near the joint line of the two pieces. This change also removed the radial joint which is currently showing heavy washing.

- The filler rings were incorporated into the feed end and discharge end shell liners to reduce the number of pieces and make them safer for relining. This also eliminated the gaps that were seen in the mill.

- The bolt hole was moved to the plate section to provide a solid lifter and reduce preferential wear at bolt holes.

- The bolts were changed to spherical-seated bolts to reduce the point stress at the bolt hole. Due to white iron hardness, the bolts' torque setting was reduced to further reduce the risk of cracking.

The white iron liner set was installed in January 2022. Three months later, an inspection of the mill internals found three cracked head liners, two on the feed end and one on the discharge end. The cracks in the three liners were all in the same position, the first inner bolt hole, as seen in Figure 14.

FIG 14 – Feed head liner crack position and orientation.

In August 2022, an inspection of the ball mill revealed several shell liners with sections of the high lifter missing due to significant impacts. Most of the damaged liners were in rows four and five of the shell; it appeared a large portion of a shell lifter had detached and collided with neighbouring liners. It was suspected that initial damage was caused by a chute liner from the feed chute. With a total of 16 liner pieces damaged in the mill, it was deemed necessary to perform a full reline in October 2022, earlier than initially planned.

The installation of a new set of white iron liners was carried out, with an anticipated wear life of approximately 12 months. Subsequently, for the next set to be installed in August 2023, the lifter height and plate thickness were increased to achieve an extended wear life.

In December 2022, two months following the installation of the second set of white iron liners into the ball mill, an internal inspection again revealed 29 cracked liners. A notable difference in installation between the first and second sets of white iron liners was an increase in bolt torque, raised to 2900 Nm from the previous 2100 Nm. This adjustment was made with the aim of reducing the incidence of head liner bolt breakage. Over the course of the preceding eight month campaign, numerous bolts had become loose and eventually failed due to fatigue.

The third set of white iron liners was installed in August 2023; with the lessons learned from set number two, this set was installed using a lower torque setting of 2350 Nm. To date this set has not seen any cracked liners from the inspections. It is worth mentioning that the torque applied to the first two chrome moly commissioning sets was notably higher, at 4850 Nm, and resulted in some head bolts becoming loose on these sets. To address the issue of the loose and broken bolts, the feed end and discharge end heads will be changed back to a two-piece as it is suspected that there is not enough flexibility in the white iron.

Ball mill media size optimisation

Working with the ball mill media supplier Magotteaux, Gruyere undertook a ball mill media size optimisation study. At the time, the ball mill was being topped up with 100 per cent 60 mm media, with an F_{80} 2.1 mm being achieved from a May 2021 cyclone survey. The objective of the study was to evaluate whether 80 mm grinding media was required to effectively grind the coarsest particles in the ball mill feed and what the optimum percentage of 80 mm media top-up was.

The top-up sizes tested were:

- 100 per cent 60 mm (reference)

- 75 per cent/25 per cent 60 mm/80 mm

- 50 per cent/50 per cent 60 mm/80 mm

- 100 per cent 80 mm.

The optimisation study's results identified that the introduction of 80 mm balls into the top-up charge unquestionably improved the speed of reduction of course particles. Increasing the portion from 25 per cent to 100 per cent continued to improve the grinding rate of the coarse fractions greater than 300 µm. Figure 15 shows an inflection point of 300 µm. Under this size fraction, the grinding rate deteriorates when 80 mm balls are used.

This need not be a concern if the particles are already at or below the grinding circuit target product size; however, it is a concern for particles in the intermediate size fractions, in Gruyere's case, +150–300 µm. The grinding rate of the intermediate particles (+150–300 µm) with an increasing portion of 80 mm balls did not appear to deteriorate further. The grinding rate of the fine particles -150 µm was highest with 100 per cent 60 mm top-up. Expending energy to selectively grind this fraction is undesirable as they are already at, or below the target grind size. In this regard, the use of 80 mm balls is not detrimental.

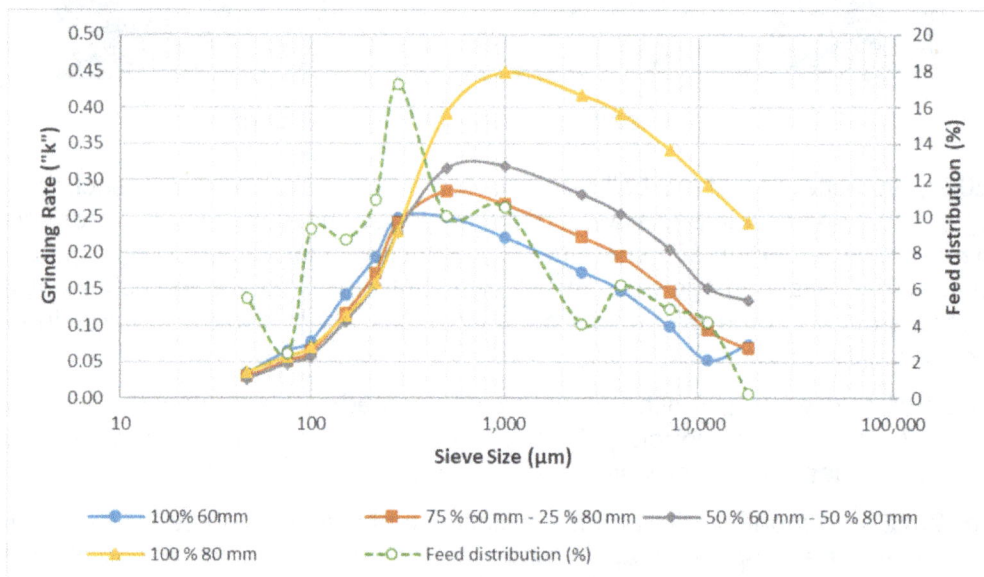

FIG 15 – Grinding rates (k) after 6.85 kWh/t of grinding energy (Myllynen and Bruwier, 2021).

The efficient grinding of intermediate and fine particles is less dependent on the physical media top up size and more affected by the surface area of the grinding charge. Simply put the more grinding surface area, the better the grinding efficiency of the intermediate and finer fractions. Increasing from 60 mm media to 80 mm media reduces the grinding surface area. Changing to an 80 mm top-up size might be highly effective for grinding the coarse particles, but it comes at the expense of grinding surface areas and, thus, the grinding rate of the critical intermediate-size fractions. Considering all this, Gruyere decided to retain a 50 per cent portion of the 60 mm media to maximise the grinding rate of the intermediate particles and went to a 50/50 60 mm/80 mm top-up profile for the ball mill.

Ball mill high chrome media change

In March 2021 Gruyere commenced adding high chrome media Duromax®T MGT media into the ball mill to leverage the benefits offered by high chrome grinding media. In March and April 2022, a marked ball test (MBT) was performed in the Gruyere ball mill. During the MBT three high chrome alloys were tested against forged steel media. Figure 16 demonstrates that the high chrome media performed significantly better in the Gruyere ball mill compared to forged steel, which had a wear rate of 48 per cent higher than Duromax®T MGT. Both Duromax®U MGT and Duromax®K MGT, which have high chrome content, performed better for wear resistance than the current Duromax®T MGT; however, following financial evaluation, it was decided to continue to use the current alloy.

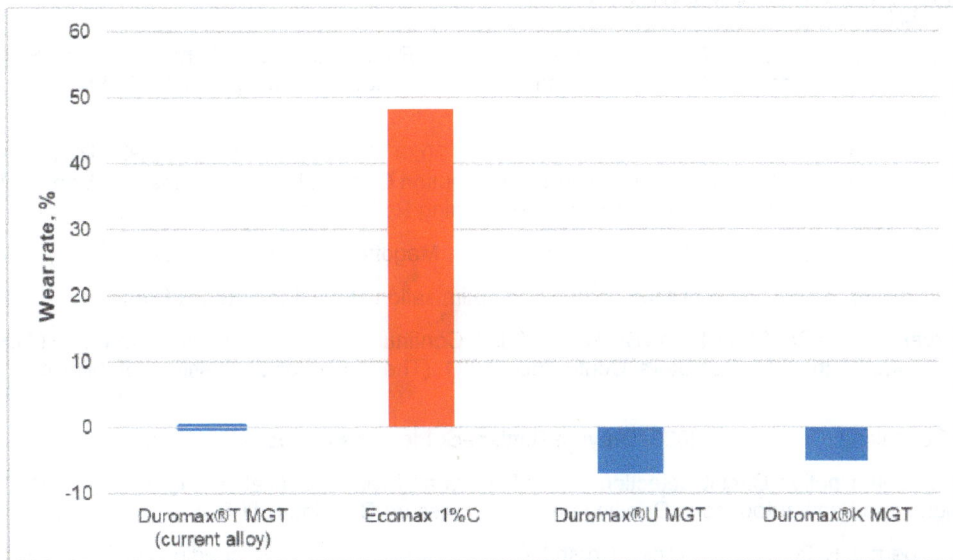

FIG 16 – Difference in wear rate relative to Duromax®T MGT (Myllynen, 2022).

As part of the change to high chrome media, pre- and post-slurry pulp surveys were completed. This pulp survey showed that following the purge to Duromax®T MGT high chrome media there was a shift to higher pulp potentials and higher dissolved oxygen concentrations. These were attributed to the use of high chrome media, which, through a decrease in grinding media consumption, an increase in corrosion resistance, and inherently containing less iron than forged steel media, resulted in an average of 25 per cent reduction in ethylenediaminetetraacetic acid (EDTA) extractable iron contamination in the pulp. This decrease in EDTA extractable iron did not deliver a measurable reduction in cyanide consumption.

CONCLUSIONS

The Gruyere processing plant initially designed to process 7.5 Mt/a of fresh ore, achieved 9.4 Mt/a in 2023 through progressive improvements and debottlenecking initiates. Regular comminution benchmarking surveys have played a crucial role in monitoring key performance parameters, guiding the effectiveness of implemented comminution optimisation projects. Regular health checks conducted through surveys serve as invaluable tools for metallurgists to benchmark performance, identify areas for improvement, and showcase the improvements achieved from optimisation activities.

Key debottlenecking activities around the pebble crushing circuit and SAG internal discharge configuration have resulted in incremental improvements in the pebble discharge rate. These debottlenecking activities had cascading effects, prompting the capital expenditure for the installation of a new MP800 pebble crusher and future works planned to address the additional pebble production on the pebble handling conveyors.

Focusing on wear and increasing the time between relines resulted in liners that were outside the typical design range. Taking a step back and changing the strategy to target more frequent relines and focusing on throughput optimisation assisted Gruyere in achieving increased throughput by managing the SAG speed, ball charge, and weight set points to optimise the impact breakage for the given liner wear state.

ACKNOWLEDGEMENTS

The authors wish to acknowledge the support and contribution of the Gruyere operations and maintenance teams and the numerous professionals who continuously contribute to and support operations.

REFERENCES

Kock, F, Siddal, L, Lovatt, I, Giddy, M and DiTrento, M, 2015. Rapid Ramp-Up of the Tropicana HPGR Circuit, in Proceedings of Advances in Autogenous and Semi-autogenous Grinding and High Pressure Grinding Roll Technology, SAG 2015.

Lovatt, I, Becker, M, Putland, B, Radford, R and Robinson, J, 2023. Trade-Off Realities in HPGR vs, SAG Milling – A Practical Comparison of Tropicana and Gruyere Comminution Circuits, in Proceedings of Advances in Autogenous and Semi-autogenous Grinding and High Pressure Grinding Roll Technology, SAG 2023.

Myllynen, M, 2022. Post-purge Pulp Chemistry Characterisation, Magotteaux Report GJV221012.

Myllynen, M and Bruwier, A, 2021. Ball mill top up ball size optimisation study, Magotteaux Presentation GJV211208.

Oblokulov, O, Green, S, Van Der Spuy, D and Becker, M, 2021. Continuous improvement at Edikan (with MillROC support), in Proceedings of the Mill Operators Conference 2021 (The Australasian Institute of Mining and Metallurgy: Melbourne).

Orway Mineral Consulting (OMC), May 2020. Gruyere Bottleneck Identification Update, Report No. 8117.

Putland, B, 2006. Comminution Circuit Selection – Key Drivers and Circuit Limitations, in Proceedings of Advances in Autogenous and Semi-autogenous Grinding and High Pressure Grinding Roll Technology, SAG 2006.

Putland, B and Sciberras, R, 2019. Hard Rock – Crush It Or Let It Break Itself?, in Proceedings of Advances in Autogenous and Semi-autogenous Grinding and High Pressure Grinding Roll Technology, SAG 2019.

Radford, R, Lovatt, I, Putland, B and Becker, M, 2021. Gruyere Gold Project Western Australia, part 1 – from design to commissioning, in Proceedings of the 15th Australasian Mill Operators Conference 2021, pp 302–322 (The Australasian Institute of Mining and Metallurgy, Melbourne).

Scinto, P, Festa, A and Putland, B, 2015. OMC Power-Based Comminution Calculations for Design, Modelling and Circuit Optimization, in Proceedings of 47th Annual Canadian Mineral Processors Operators Conference, pp 271–285.

Siddall, B, Henderson, G and Putland, B, 1996. Factors Influencing Sizing of SAG Mills from Drill Core Samples, in Proceedings of Advances in Autogenous and Semi-autogenous Grinding and High Pressure Grinding Roll Technology, SAG 1996.

Toor, P and Brennan, L, 2022. SAG Mill Liner Optimisation Study, HATCH Report H365117-00000-200-230-0003.

Mastering clay and oxide ores at St Ives Gold Mine – easing circuit constraints and bottlenecks with time-honoured comminution, leach CIP and tailings circuit wizardry!

E Mort[1], M Simpson[2], B van Saarloos[3] and R Radford[4]

1. AAusIMM, Processing Superintendent, Gold Fields Australia, Perth WA 6005.
   Email: emily.mort@goldfields.com
2. Project Metallurgist, Gold Fields Australia, Perth WA 6005.
   Email: mikayla.simpson@goldfields.com
3. Project Metallurgist, Orway Mineral Consultants, Perth WA 6004.
   Email: blaze.vansaarloos@orway.com.au
4. Principal Specialist Metallurgy, Gold Fields, Perth WA 6000. Email: reg.radford@goldfields.com

ABSTRACT

The St Ives Gold Mine processing plant is designed to process 4.2 million tonnes per annum (Mt/a) of predominantly fresh ores at a nominal gold head grade of 2 to 5 g/t. Over time, stockpiles containing approximately 2.5 Mt of clay, oxide, and transitional ores have accumulated, with the prioritisation of treating higher-grade fresh open pit or underground ores. Historically, the plant feed blend contained approximately 10 per cent oxide ores. The clay/oxide ores at St Ives are relatively 'sticky' and have high viscosity when finely ground in slurry form.

The recent reduction in the availability of open pit mined ores has resulted in the need to process larger volumes of clay/oxide stockpiles (including recently mined shallow open pits containing oxide/clay ore, such as Delta Island), requiring blending up to 30 per cent. In the fiscal year 2023, St Ives processed more than 1 Mt of oxide/clay material, with an additional 2.5 Mt planned for processing in 2024 and 2025.

The treatment of oxide/clay ores caused significant challenges with the processing plant, including material handling issues on the run-of-mine (ROM) pads, haulage, and crushing; and performance issues with the single-stage SAG mill, trash screens, slurry pumps, and tailings thickener.

In response, the St Ives team has implemented a series of operational adjustments, circuit and equipment upgrades and modifications, test work programs, and milling circuit surveys and studies to ensure that the plant maintains its production targets while processing the higher proportion of oxide/clay ores in the feed blend.

This paper discusses St Ives' operational challenges in processing the oxide/clay stockpiled ore and provides an overview of the successful strategies implemented and the lessons learned.

INTRODUCTION

St Ives Gold Mine (SIGM) is located approximately 80 km south-south-east of the regional city of Kalgoorlie-Boulder and approximately 630 km east of the capital city of Perth in Western Australia (refer to Figure 1). The site extends over 60 km of the prospective Norseman-Wiluna Archaean Greenstone belt. St Ives operates two underground mines, Invincible and Hamlet, and a series of small open pits, including Neptune and Pistol Club. Over the years, approximately 2.5 Mt of lower-grade ore stockpiles from various open pit mines have accumulated.

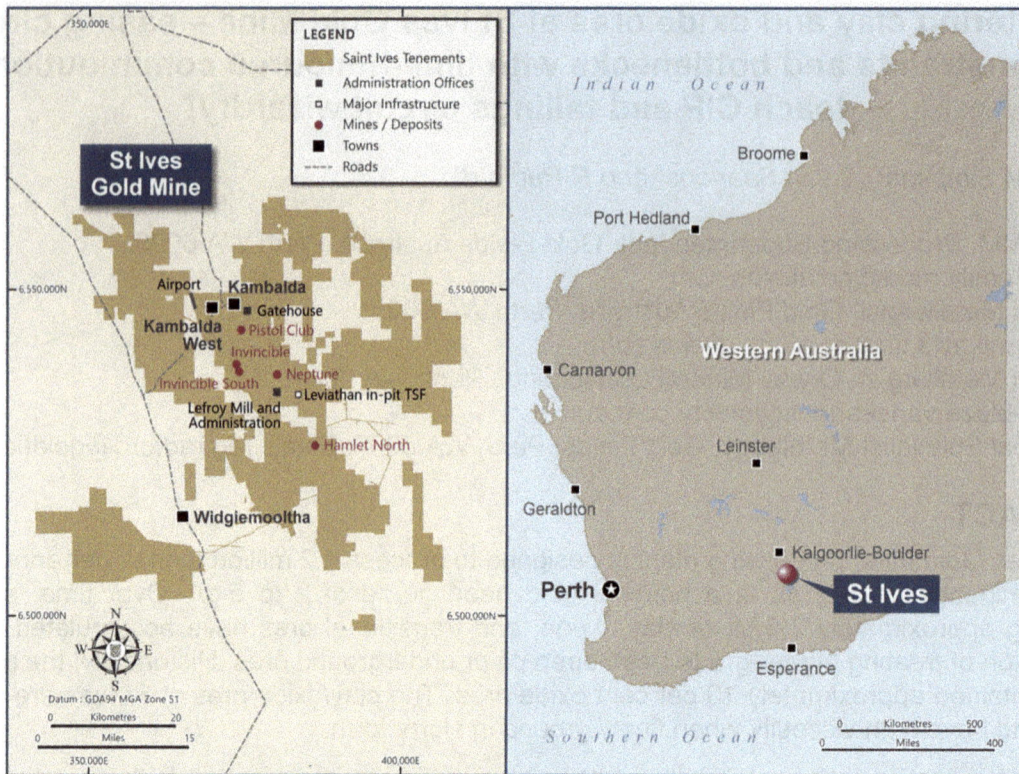

FIG 1 – St Ives' location.

Commissioned in 2005, the 4.2 Mt/a plant consists of single-stage crushing, single-stage semi-autogenous (SAG) milling, gravity concentration, leaching, carousel carbon-in-pulp (CIP) adsorption, Anglo-American Research Laboratories (AARL) elution, and tailings thickening. Figure 2 outlines the St Ives processing plant process flow sheet.

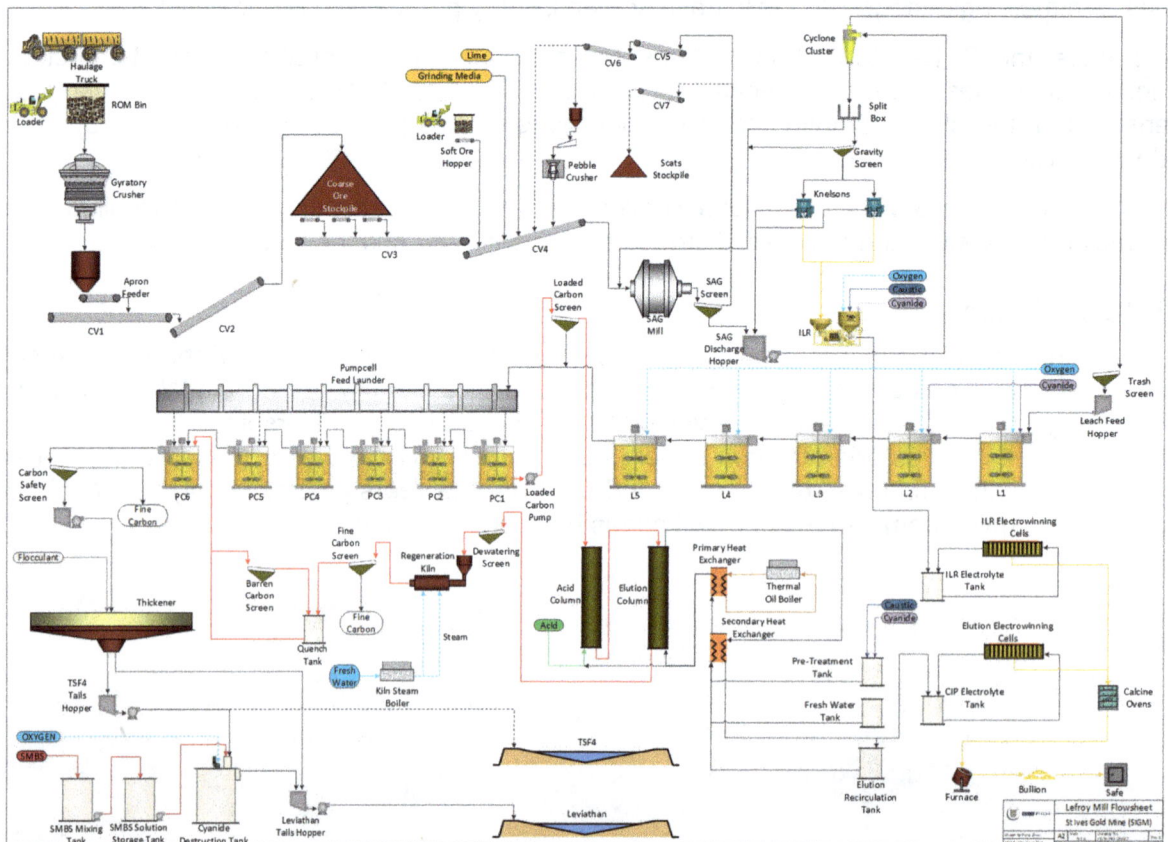

FIG 2 – St Ives process flow sheet.

St Ives targets an annualised gold production rate of approximately 360 000 ounces, requiring a plant throughput of approximately 4.0 Mt/a. Despite the plant's ability to exceed this capacity (historically processing 4.8 Mt in 2020), it has maintained a steady throughput rate with the required scheduled maintenance shutdowns. It has consistently achieved its gold production targets.

Due to the recent ramping-down of open pit mining, processing more of the accumulated low-grade oxide/clay stockpiles has become necessary. The 'sticky' and high viscosity characteristics of these stockpiled ores have created operating challenges for the processing plant associated with stockpile reclamation, transportation, crushing, milling, slurry screening, pumping, and thickening.

St Ives has overcome these challenges to optimise the circuit and increase the blend content of oxide/clay ores being processed. This has included replacing the linear trash screens with horizontal vibrating screens, implementing static screening of ROM ore ahead of milling, process and operational adjustments to comminution and thickener circuit parameters, instrumentation installation, additional rheology test work programs, process control optimisation, and ongoing milling circuit surveys.

This paper discusses the improvement work carried out by the St Ives team and the learnings experienced.

COMMINUTION

Blend requirements

Early in 2022, the plant production forecast was reviewed to determine the proportion of oxide/clay stockpiles (locally termed 'MOPS', an abbreviation of Mine Ore Pads) that must be processed through the plant to achieve the 2022 target gold production. It became evident at that time that higher proportions of oxide/clay ores would need to be processed than originally expected. This required a re-focus by St Ives to improve the capability of the plant to achieve this.

Static screening

Based on historical experiences of crusher/chute blockages while processing MOPS ores through the crushing circuit, static screening was investigated to provide an acceptable product top-size, allowing the material to be fed via the emergency feeder directly into the SAG mill without causing blockages by oversized material.

While the static screening investigation was underway, the proportion of oxides in the crusher feed blend was also increased. To determine the possible blend proportion, known, more favourable oxide MOPS was used. It was quickly determined that the crusher and associated chutes could not handle the higher proportion of oxide/clay material without blockages.

The static screening process commenced in June 2022. The MOP material was hauled to the soft ore pad and fed via a front-end loader (FEL) onto the static screen. The oversized material was stockpiled, and the undersized material was fed into the SAG mill via the emergency feeder. This volume of screened ore fed was quantified utilising the FEL onboard weighing system (Loadrite) to confirm the blend proportion.

Soft and difficult ground conditions associated with the MOPS areas and the transport route presented challenges, and ultimately, it was established that 40 tonne single-axle articulated dump trucks (Moxy trucks) were required.

FIG 3 – Static screening on soft ore pad.

Due to limited space on the soft ore pad, the oversized screened material (30 per cent) was periodically hauled back to the 'Redback MOP'. Relocating the static screen to the MOP area eliminated the additional cost of double-handling the oversize back to the MOPS. Figure 4 shows an aerial of the Redback MOP relative to the soft ore pad.

FIG 4 – Aerial photo.

Although feeding the screened oxide material through the emergency feeder was a significant improvement compared to feeding through the crushing circuit, it also had its own challenges. Blockages of the grizzly bars with the clay material were frequent, and blockages of the bin itself also required periodic manual air spearing to clear. Several mill downtime events were caused by oversized rocks getting through the grizzly bars.

FIG 5 – Emergency feeder blockage.

Alternative clay size reduction options were investigated due to the high costs associated with the static screening activity and the inefficiency of the process due to the build-up of stockpiles of oversized material remaining.

Total Rockbreaking Solutions Allu Transformer bucket ('Allu Bucket') was identified in mid-2023 as a potential performance improvement and cost reduction opportunity. The Allu bucket was hired for a trial period, and once commissioned in November 2023, it demonstrated a capacity to process more than 4000 t per day. It generated a product suitable for directly feeding into the mill emergency feeder, with a remnant product ideal for feeding the main plant crusher. Plans are in place to purchase a permanent Allu Bucket due to its effectiveness and the cost savings demonstrated during the trial.

FIG 6 – Allu transformer bucket in operation.

FIG 7 – Allu transformer bucket.

SAG milling

The St Ives comminution circuit consists of a closed-circuit SAG mill measuring 11 m in diameter and 6 m in length, equipped with a variable-speed 13 000 kW motor. The design power draw is 11 380 kW at a mill speed of 78 per cent of critical speed, ball charge of 8.5 per cent, and total charge of 24 per cent. The design capacity of the SAG mill motor will enable operation at a total charge level of 30 per cent.

Crushed ore is fed directly into the mill feed chute and discharges over a vibrating screen. The screened oversize material (pebbles) is recycled through a crusher and returned to mill feed. The SAG mill operates with a fixed mill feed rate, with the SAG mill speed automatically adjusted to achieve a fixed mill load set point. A MANTA SAG Cube is installed and controls main parameters such as mill throughput, mill speed, mill load, cyclone pressure etc, to maintain steady operations.

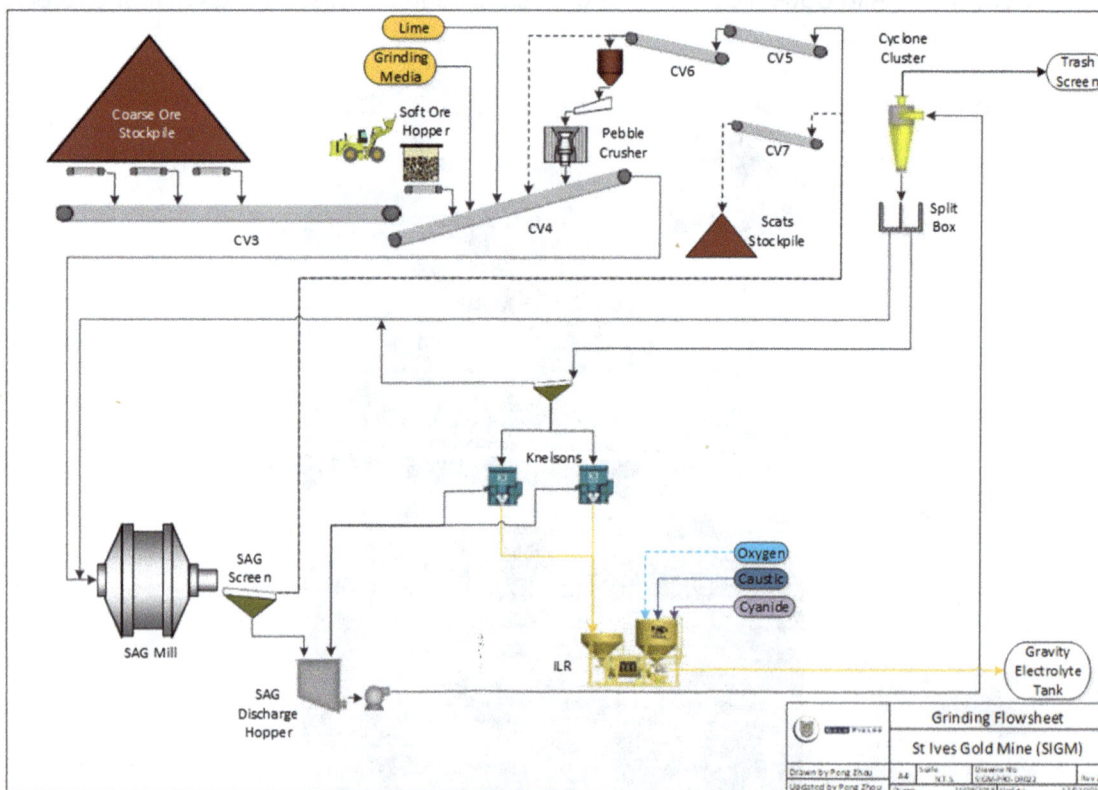

FIG 8 – St Ives comminution circuit.

During the August 2022 scheduled mill shutdown, the SAG discharge grates were changed to reduce the open area from 7.3 per cent to 6.2 per cent, based on recommendations from the 2021 grinding survey study report (Putland, Zawadski and Sciberras, 2021). This improved the retainment of the smaller grinding media, helping achieve a finer grind, with this being a single-stage SAG mill.

In September 2022, a new oxide ore source, Delta Island 3, was fed to the mill at approximately a 20 per cent blend. The Delta Island ore source was relatively recently mined and stockpiled. While the viscous nature of the material was known through metallurgical test work, the oxide component had not previously been fed to the mill.

Following the installation of the finer mill discharge grates, it was evident that the new grates were causing flow restrictions and slurry pooling within the mill, indicated by the frequent mill weight loading-up and power-drops on the control system. Further to this, two mill overload incidents occurred during September and October.

It was determined that, in addition to the feed blend composition, several different factors and decisions led to these mill overloading events:

- The operation of the mill feed water in auto, as opposed to cascade, allowing the milling density to fluctuate up and down without correct control.

- The calculated milling density was based on an outdated circuit configuration, which was used when the mill feed water was switched into cascade mode (adjusted to maintain a constant density).

- Poor loop tuning on the SAG feed water controller, preventing the effective cascade control of the milling density.

- The reluctance of mill operators to adjust the milling set point outside their 'standard' range.

- The changeout of grates to reduce the open area.

- The mill control loop, which uses speed to control load, was poorly tuned and did not react quickly or aggressively enough as the mill load increased.

The new Delta Island oxide material proved extremely viscous at the 70–73 per cent milling density, which was atypical for oxide blends at the time. It required a much lower milling density to flow freely through the mill and discharge screen. The operators were unfamiliar with operating at the low densities required and were reluctant to adjust the feed water.

The solution to this issue involved a multi-pronged approach to address the individual problems identified. Firstly, the grate configuration change was partially rolled back, replacing four 20 mm panels with 40 mm. This increased the open area back to 6.7 per cent.

The milling density calculations were also updated to reflect the current circuit configuration. The operators adjusted the feed water set point while still operating in auto to target a specific density loosely.

Perhaps most importantly, the operators gained confidence operating in the new conditions. Although the operations team had been trained in processing fresh feed, they were still trying to achieve a difficult grind size based on the throughput and the mill's capacity.

Education and training on the management of densities, mill speeds, cyclone operation etc, were important and gave the operations team the confidence to allow the controller to run the higher water additions necessary to achieve the lower densities. In addition, regular monitoring of the discharge screen was implemented to help operators make density set point decisions. For example, if the slurry on the mill discharge screen appeared too viscous, the water addition was increased until the parcel of problematic ore passed.

Upgrade work was undertaken on the SAG cube controller, which included new overload protection and an alternative classification circuit control philosophy (control of cyclone pressure versus cyclone feed density control).

The milling density cascade control loop was re-tuned. The mill speed control loop was also re-tuned; and then re-tuned again, twice over the next 12 months. Before the mill speed control loop re-tuning, operators began manually controlling the mill speed by increasing the minimum speed set point when the mill load was observed to be climbing rapidly. This effectively minimised the extent of overloading by increasing the speed within 5 mins instead of the 10–15 mins observed when the controller was allowed to operate without interference.

The cyclone feed water control valve was identified as oversized, so a replacement was sourced. The valve size was reduced from 8 inches to 6 inches, enabling more effective process control performance.

Implementing these strategies allowed the mill at St Ives to process the high-viscosity oxide ore with minimal further downtime attributable to the material performance in the comminution circuit. The load stability improved marginally after the first tuning, with the reaction time of the controller improving; however, the bulk of the improvement was achieved during the second tuning.

During early 2023, the slurry was still occasionally leaking out the SAG mill feed chute, indicating a limited understanding of the charge levels within the mill. The mill was partially ground out during the grate replacements, and mill power models showed that the likely ball charge in the mill was lower than the operation was targeting at that time (actual 5–6 per cent, target 8–10 per cent). As a result, the targeted weight set point yielded a much higher volumetric fill, and the ball addition was increased substantially. The mill was later crash-stopped and ground out during a survey in June 2023, and it revealed that despite the increased ball addition, the total load was still much higher than expected (around 34 per cent, very close to the feed chute level).

As the liners wore, a fixed weight set point replaced the mass lost by the liners with slurry, effectively increasing the volumetric filling. An estimate of the liner mass loss over the course of the liner life, developed by Orway IQ, helped to provide a modified weight set point that would correct for the change in liner weight due to wear, maintain a fixed volume within the mill and prevent unexpectedly high total loads. The constant volumetric fill helped to minimise the slurry pool occurrences, which appeared to be exacerbated by the high level of slurry sitting above most of the grate apertures.

Sampling and test work

Grinding survey

In December 2021, the St Ives processing team subscribed to Orway IQ's Mill Remote Optimisation Consulting and Coaching (MillROC) to assist the site team via continuous online consulting. The subscription provides an online platform of real-time operating data available to the site team and MillROC consultants to enable timely circuit performance improvements.

A mill survey was conducted to gather a baseline of current operating parameters and to model the circuit to determine an operating strategy and circuit improvements for processing this oxide material. Key recommendations from the Orway Mineral Consultants (OMC) grinding survey report that were implemented were adjustments to grinding media sizes, ball charge, and cyclone vortex finder diameters.

The recommendations were implemented over a seven-month period from June 2023 to January 2024, and the site was able to achieve the desired P_{80} of 125 µm.

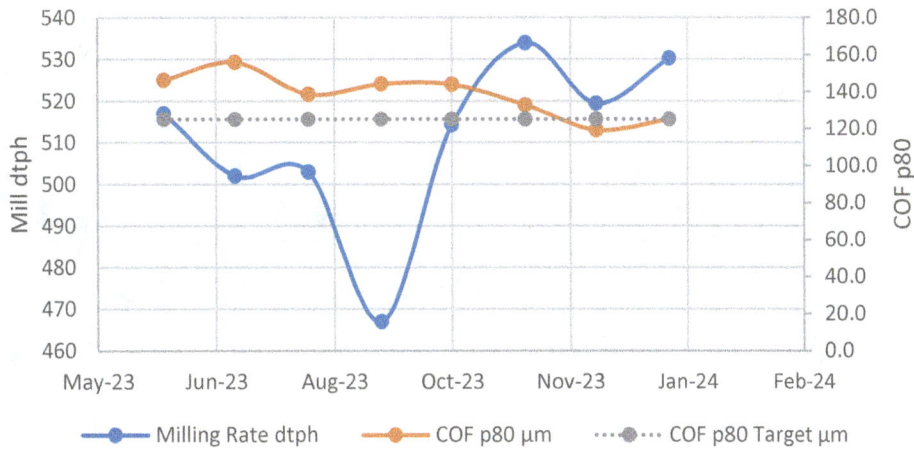

Plant performance following survey recommendations

FIG 9 – Plant performance following survey recommendations.

- June 2023 commenced 50 per cent/50 per cent 105 mm/125 mm grinding media addition.
- July 2023 ball charge target increased to 15 per cent.
- November 2023 190 mm vortex finders installed.
- January 2024 170 mm vortex finders installed.

Ore rheology test work

Neptune and Delta Island MOPS sampling campaigns were undertaken, and 26 stockpile samples were sent to ALS Metallurgy for additional rheology test work. The test work aimed to establish baseline data for ore viscosity and differentiate between the oxide ores. The data was used to determine which stockpiles would be troublesome when fed to the mill and allow sufficient blending to mitigate downstream issues.

FIG 10 – Stockpile sampling.

Of the 26 stockpiles sampled, ten exceeded the typical 100 cps viscosity limit for centrifugal pumping at 40 per cent solids (see Figure 11). These stockpiles must be blended in small quantities to facilitate plant tailing pumping, particularly when the site targets a thickener density set point greater than 50 per cent. All the Delta Island stockpiles were evidently problematic, and some Neptune stockpiles exceeded the viscosity limit by more than 50 per cent.

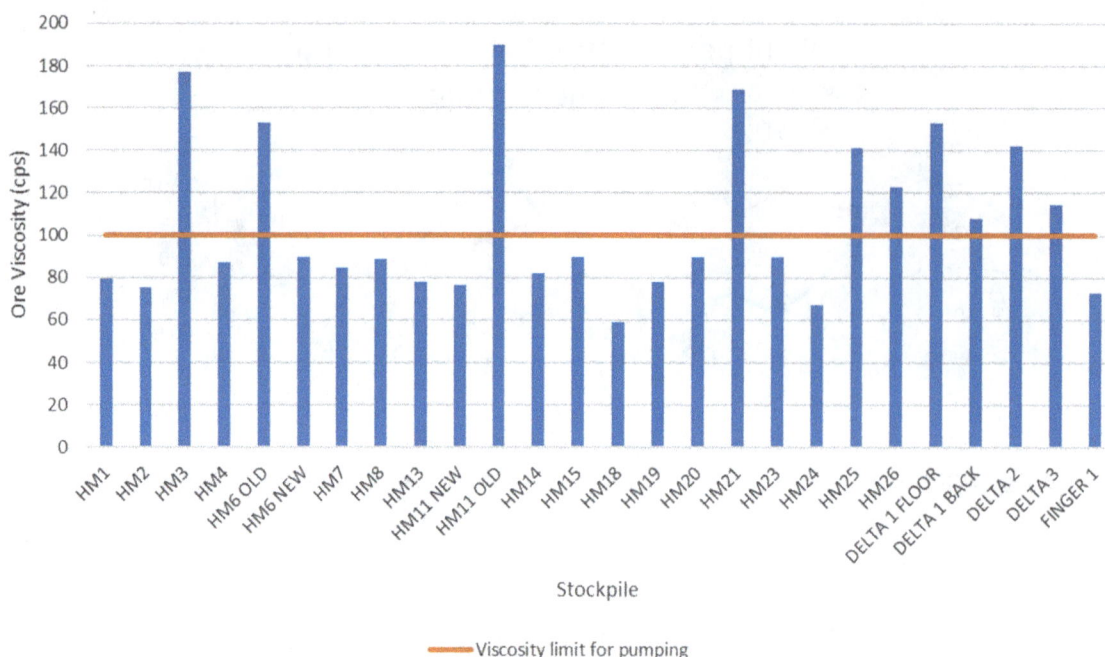

FIG 11 – Ore viscosity at 40 per cent solids.

Recommendations from a review of test work results included:

- Conserve lower viscosity stockpiles to allow sufficient volume for effective blending.

- Problematic ore sources should be blended in smaller proportions.

- Communicate with operations before feeding the problematic ore.

- Set points should be adjusted at the mill and tailings thickener to pre-empt viscosity issues.

The continuation of sampling campaigns of oxide stockpiles was based on these results.

LEACHING/ADSORPTION

The St Ives leach circuit comprises five 3400 m^3 leach tanks with a total leaching residence time of 17.5 hrs. The cyclone overflow slurry from the primary cyclone cluster flows to the trash screen feed splitter box and onto single-deck vibrating screens. Cyanide is dosed into leach tanks 1 and 2, and oxygen is introduced to each leach tank via a down-shaft addition system.

The adsorption circuit comprises six 200 m^3 carbon-in-slurry (CIP) Kemix Pumpcell® adsorption tanks, providing a residence time of approximately 13 mins per tank. The cells are configured to operate in a carousel arrangement and to achieve counter-current carbon movement relative to the slurry.

As the linear trash screens had previously been replaced with horizontal vibrating screens, viscosity and flow-through the screens had minimal effect on production. Extreme events were managed by running both screens in duty mode. The main effect on production in the leach/adsorption circuit was the increase in cyanide consumption due to the elevated levels of cyanide-soluble copper in the oxide feed.

Before introducing oxide ore, cyanide consumption in the leach circuit was typically 0.28 kg/t; however, in August 2023, it reached 0.49 kg/t. Lime and cyanide consumption varied considerably depending on the ore source, impacting the plant's operating cost. Aside from cost inflation, the major concern with the elevated cyanide consumption was the stringent tailings weak acid dissociable (WAD) cyanide limits to align with Gold Fields being voluntary signatories to the International Cyanide Management Code (ICMC).

Although cyanide set points within the leaching circuit were minimised, the copper in the feed was still causing high WAD CN concentrations. St Ives has a CN destruct plant installed; however, it cannot treat the full flow of the tailings stream. Further testing and sampling were needed to determine which MOPS were high in copper and nickel content. This information was then used as part of the blending strategy.

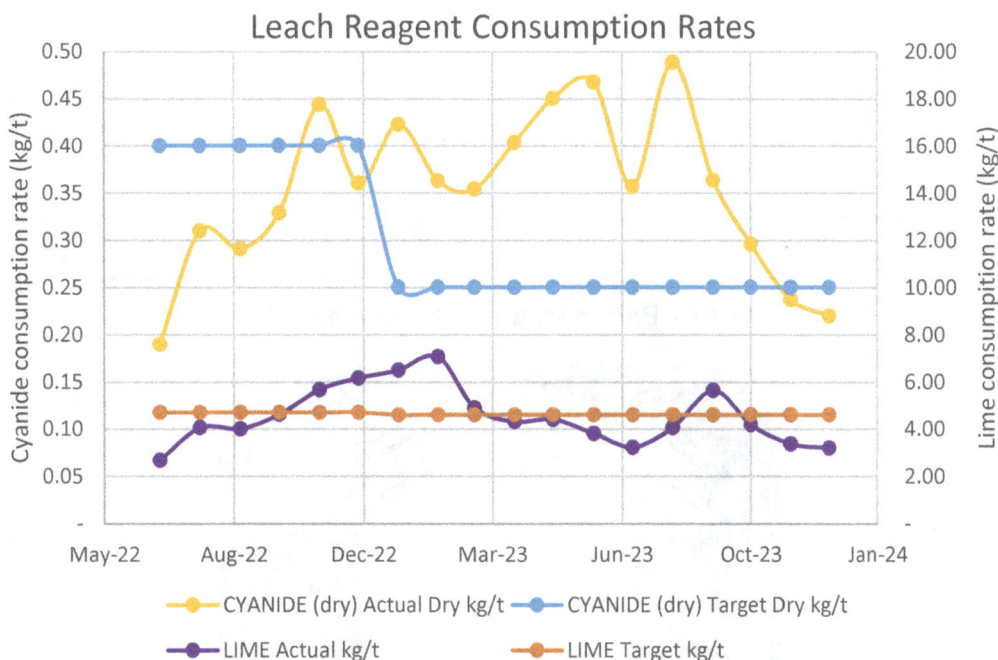

FIG 12 – Leach reagent consumption rates.

In addition to the variability of cyanide consumption with the ore sources, the site also identified an improvement opportunity with the return of the barren eluate solution (containing high remnant cyanide concentrations with high alkalinity) from the CIP and ILR electrolyte storage tanks directly to the leach tanks. The circuit's design required the CIP and ILR tanks to be emptied before commencing the next elution batch, which then resulted in the high cyanide concentrated solution (1000–40 000 ppm NaCN) being transferred at high flow rates over a short 1 to 2 hr period, to the leach tanks. This eluate transfer process created regular instability in the leaching circuit's cyanide dosing rates (Figure 13).

To improve and stabilise the cyanide control in the leach tanks, the site installed a barren recycle tank which receives all elution and gravity electrowinning barren eluate solutions. The barren eluate tank has the storage capacity to fill and empty while the solution transfer pumping rate is controlled at a near-constant level over a 24 hr period (Figure 14). This allows the barren eluate to be transferred to the leach circuit at a steady and consistent rate, which helps to stabilise cyanide concentration and pH control.

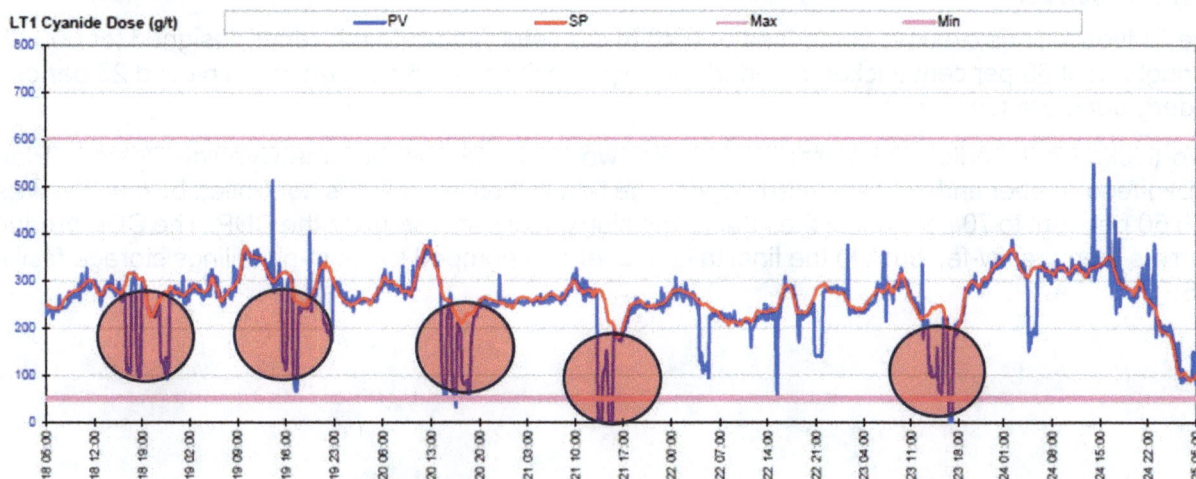

FIG 13 – LT1 cyanide dosing rate fluctuations due to barren eluate return.

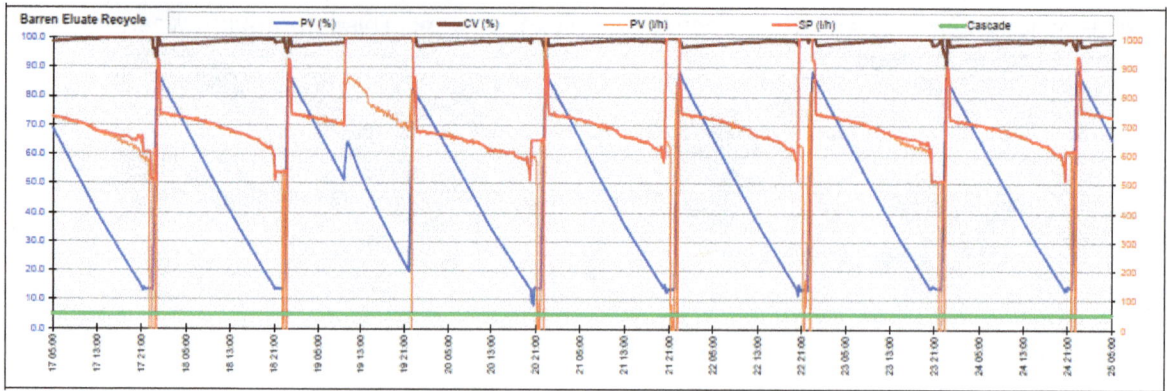

FIG 14 – Barren eluate tank recycle control.

FIG 15 – Elution and gravity barren eluate recycle tank.

This scenario exemplifies how attempting to resolve a leach circuit instability issue, such as cyanide control, solely through process control can result in an overly complex solution that remains suboptimal. Instead, it is advisable to reassess the circuit design, implement suitable engineering solutions, and subsequently simplify the process control logic to achieve a permanent and more efficient outcome.

THICKENING

The St Ives tailings circuit is comprised of a 40 m diameter high-rate thickener designed for 600 t/hr throughput at 55 per cent thickener underflow pulp density on a 75 per cent open pit and 25 per cent underground ore feed blend.

Two thickener underflow lines gravity feed into two separate hoppers: the Cyanide Destruct Plant (CDP) feed hopper and the final tails hopper. The flow to both hoppers is controlled by pinch valves, with 50 per cent to 70 per cent of the total tailing slurry flow transferring to the CDP. The CDP product slurry is then gravity-fed back to the final tails hopper and pumped to the in-pit tailings storage facility (TSF).

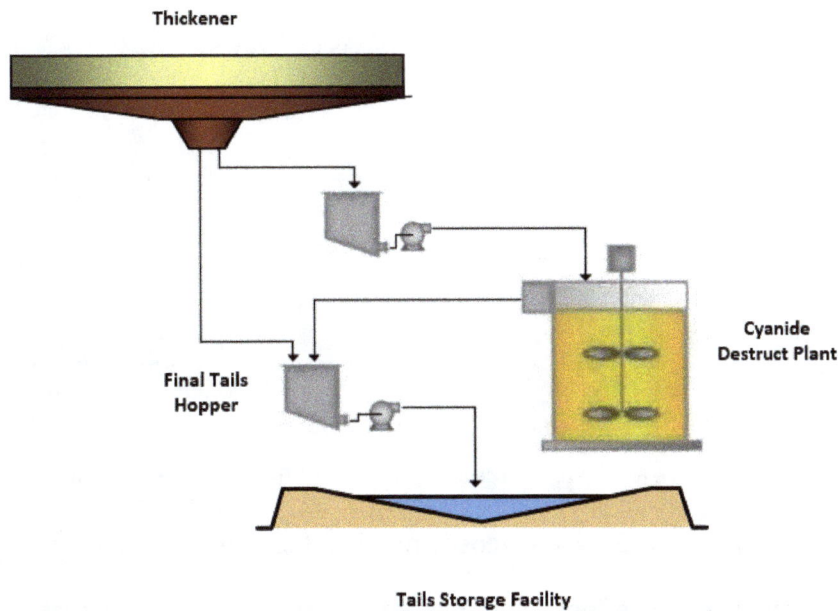

FIG 16 – Tails thickening, detoxification and storage facility.

A MANTA thickener control system is installed at St Ives, which utilises outputs from the MANTA Sub instrumentation, including the heavy mud level, interface level, settling band, and clarity. Rake torque, bed pressure and underflow density are used to manipulate the thickener underflow set point, while flocculant flow rate addition is controlled using the overflow clarity and thickener settling band. Pinch valves are employed on the tailings underflow pipe to control the final tailings hopper level. The slurry is gravity-fed to the final tailings hopper and pumped to the in-pit tailings storage facility.

Underflow restriction

Slurry viscosity through the thickener underflow became an operational issue after the milling circuit bottlenecks were resolved in October 2022. The pinch valve on the thickener underflow is controlled to maintain a set point level in the final tailing's hopper; however, there were several instances whereby the valve would open 100 per cent but could still not provide enough flow to manage the thickener bed inventory. The thickener underflow slurry was so viscous that the slurry could not flow-through the thickener underflow lines and gravity feed to the final tailing's hopper. Dilution water was placed on the thickener underflow line to combat this.

While adding water to the thickener underflow line would allow the slurry to flow, the density metre was located after the dilution water addition point, meaning that the recorded density did not reflect the actual thickener underflow density. This caused problems with the control system, which erroneously responded to reducing density by closing the discharge valve.

The second underflow line was utilised to eliminate the need to add dilution water. This line feeds the CDP tank via a hopper and pump, with the CDP discharge slurry gravity-fed to the final tailing's hopper. A trial was undertaken to demonstrate that it was possible to stop reagent dosing into the CDP tank when cyanide destruction was not required and continue to use the CDP to facilitate thickener discharge flow to the tailing's hopper.

The application of a viscosity modifier was also trialled on-site. It was used when flow-through the underflow lines or final tailing's pumps was impacted. A pumping system was installed to dose the viscosity modifier into the thickener underflow lines, where dilution water was usually added. Tests indicated that the viscosity modifier increased flow rates by 38 per cent and decreased Marsh viscosity by 28 per cent. The viscosity modifier also assisted with pumping tailings to the TSF.

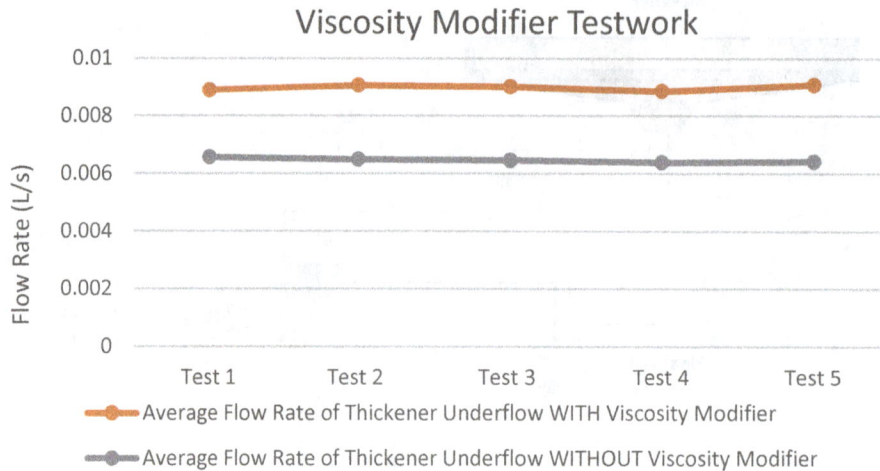

FIG 17 – Viscosity modifier test work.

Whilst the viscosity modifier was effective on most oxide ore types, several sections of Delta Island oxide ores did not respond to the modifier. This was demonstrated during a short period when a high proportion of Delta Island oxide was fed to the plant. Flow rates from the thickener reduced considerably, but all tested modifier dosing rates did not improve thickener underflow flow rates.

Tails pumping restriction

Once the flow rate bottleneck was removed from the thickener underflow lines, it became apparent that the tailings pumps could not pump the slurry with the higher viscosity at the required throughput rates to the TSF.

Key operating set points on the thickener controller were revised to prevent the slurry from being over-thickened. The tailings density and bed level set points were lowered to reduce the solids inventory within the thickener.

Once the MANTA Thickener Cube controller was optimised for the oxide ore type and the site gained confidence in blending the high-viscosity ores, it was identified that the existing tailing's pump impellor could be upgraded to provide enough spare capacity for the operating periods of high-oxide ores.

Flocculant optimisation

With the increased oxide component in the feed, periods of poor settling in the thickener were common. As described previously, the flocculant was controlled by the settling band measurement from the MANTA Sub installed on the thickener. A calorimeter is also installed, which provides a settling time measurement. However, the measurement is not connected to any flocculant control systems.

During periods of poor settling, sliming in the thickener overflow was common and large periods of mill downtime were experienced to allow the thickener inventory to be pumped out. The flocculant supplier was engaged to perform flocculant tests to determine if the supplied flocculant was still the most suitable given the changes in ore feed. A second flocculant was determined to provide improved settling rates and overflow clarity, so a plant trial was conducted.

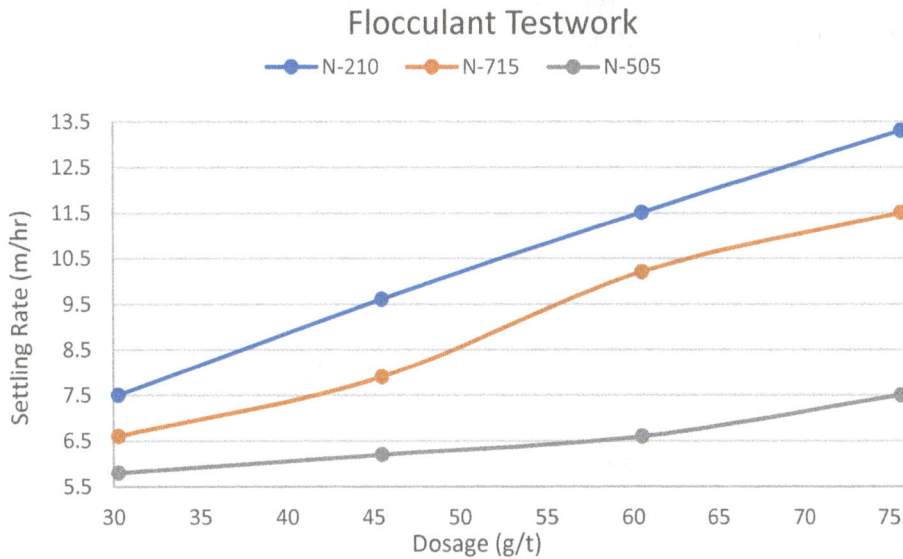

FIG 18 – Flocculant test work.

Trial flocculant bags were loaded into the dry flocculant storage silo and mixed in the on-site flocculant mixing plant. It was immediately apparent that the trial flocculant was not mixing in the plant's raw water, creating 'fish eyes' or undissolved clumps of flocculant. It was determined that the flocculant settling rate tests were originally performed by the flocculant supplier using process water (pH 8) to dissolve the product, while the flocculant mixing plant on-site uses acidic raw water with a pH of 3. As a result, the plant trial was suspended, requiring the flocculant mixing system to be cleaned out and the system returned to the original product.

FIG 19 – Flocculant trial flocculent 'fish eyes'.

One of the main factors behind the poor settling rate was the limitation to the amount of flocculant site was able to dose to the thickener. With one flocculant pump running, the maximum addition was 8 m³/hr, and the flow rate was subsequently increased to 10 m³/hr by utilising the second flocculant pump. When pumping flocculant full time over 8 m³/hr, the flocculant mixing plant was unable to keep up with the demand. The flocculant strength was increased from 0.025 per cent to 0.040 per cent to reduce the batch mixing frequency required. Ultimately flocculant consumption increased from 16 g/t in 2021 to 45 g/t during some periods in 2022.

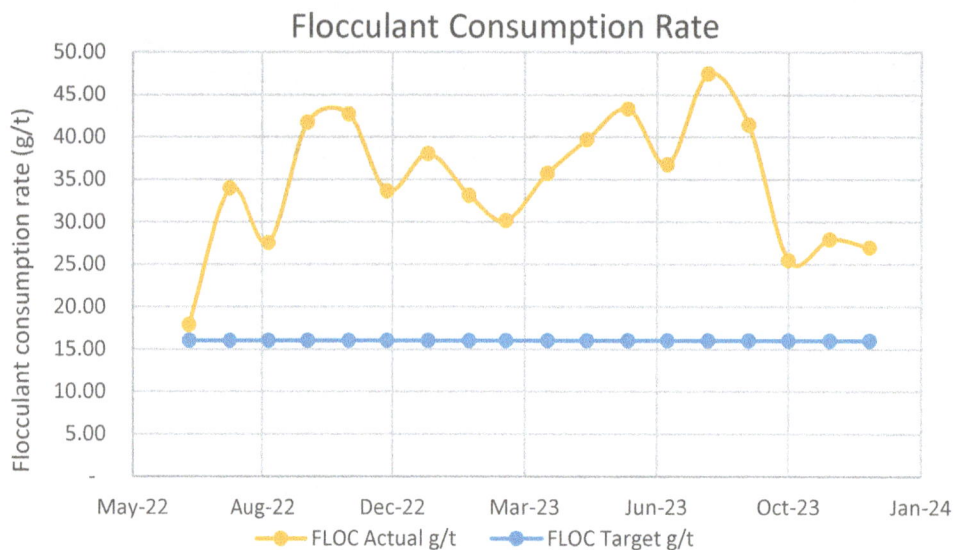

FIG 20 – Flocculant consumption rates.

CONCLUSIONS

Due to the declining availability of fresh (hard rock) open pit sourced ores at St. Ives, it became required to process higher quantities of oxide/clay ores sources from long-term stockpiles and recently mined shallow oxide/clay open pits (including Delta Island) to maintain historical and future target gold production requirements. These oxide/clay ores at St. Ives typically exhibit high viscosity characteristics and high variability in viscosity and cyanide-soluble copper concentrations.

This change in the mill feed blend required the operating team at St. Ives to respond quickly to achieve the required gold production rates.

Successful changes implemented to enable the plant to increase capacity to handle these more problematic ores include:

- Crush and screen oxide/clay ores through the Total Rockbreaking Solutions Allu Transformer bucket and feed directly to the SAG mill via an emergency feeder, thereby bypassing the main plant primary crusher.

- Sampling and undertaking viscosity test work and copper and nickel analyses on the mined ore stockpiles to quantify the different ore sources' viscosities and potential WAD CN generation capacities to inform a dedicated mill feed blending program.

- Re-tune and optimise the MANTA Cube SAG mill expert control system and the mill feed water feed controller.

- Establish a means to improve SAG mill steel and total load estimation to account for changes in mill weight caused by steel liner wear between liner changes.

- Optimise the SAG mill discharge grate configuration to account for the higher viscosity of the mill slurry contents.

- Implementation of visual monitoring of the SAG discharge screen to inform the selection of the SAG mill density control set point.

- Optimisation and repair of control valves and other plant instrumentation that became essential for the ongoing operation of the plant.

- Effectively using the capability and know-how of external comminution experts, such as that provided via the MillROC program implemented at St Ives and undertaking regular grinding circuit surveys.

- Upgrading fine aperture linear trash screens to vibrating screens to improve capacity for handling higher viscosity slurries.

- Improving the stability and the control performance of the leach circuit's free and WAD cyanide concentrations by installing a surge tank to hold the returning eluate cyanide solution between the batch processes of carbon stripping/ILR electrowinning and the continuous cyanide leaching process.

- Increasing the capacity of the existing flocculant reagents mixing and dosing facilities.

- Improving the capacity of the thickener discharge lines to allow increased flow of higher viscosity slurries.

- Optimising existing centrifugal pumps to maximise capacity for pumping higher viscosity slurries.

- Use of viscosity modifiers to reduce viscosity to improve slurry transport properties.

Further to the improvements made, key learnings included:

- The importance of exploring technologies beyond the norm and reacting swiftly to challenges, enabled by undertaking the regular MillROC discussion sessions involving a multi-faceted team of experts, which proved invaluable in maintaining consistent focus and providing external guidance.

- Conducting mill surveys, promptly addressing issues, and collecting baseline data were crucial in adapting to a changing operational environment.

- Undertaking product plant trials necessitates a thorough understanding of test work parameters to avoid detrimental impacts to plant productivity from the trial.

As a result of these strategies and adaptations, the processing plant can now regularly treat 30 per cent oxide/clay ore content in the feed blend at high throughput rates, achieving plant targets for grind size and recovery and ultimately meeting ounce production targets.

ACKNOWLEDGEMENTS

I would like to acknowledge my co-authors Mikayla Simpson and Blaze van Saarloos for their assistance in writing this paper and Reg Radford for his guidance and support throughout the operating challenges at St Ives and his contribution to this paper. I also acknowledge Gold Fields Australia's approval to publish this paper.

REFERENCES

Putland, B, Zawadski, A and Sciberras, R, 2021. St Ives Gold Mine Site Visit Review and Survey, Orway Mineral Consultants report 8175-01 Rev 0.

Navigating the operational challenges at Campo Morado – a success story

R Whittering[1], A Bill[2], C Meinke[3], J Mendoza[4], A Villalvazo[5] and R Chandramohan[6]

1. FAusIMM, Principal Process Engineer, Ausenco, Vancouver BC, Canada. Email: richard.whittering@ausenco.com
2. Lead Process Consultant, Ausenco, Vancouver BC, Canada. Email: adrian.bill@ausenco.com
3. Senior Process Consultant, Ausenco, Vancouver BC, Canada. Email: connor.meinke@ausenco.com
4. Senior Process Engineer, Ausenco, Vancouver BC, Canada. Email: jorge.mendoza@ausenco.com
5. Senior Metallurgist, Campo Morado, Mexico. Email: avillalvazo@campomorado.com
6. Global Technical Director – Operations and Process Optimisation, Ausenco, Vancouver BC, Canada. Email: rajiv.chandramhan@ausenco.com

ABSTRACT

The Campo Morado Mine in Guerrero State, Mexico, was initially commissioned in 2009 and operated by Farallon Resources. Over the years, the mine has exchanged owners to address the poor metallurgical performance and plant-related design issues. In 2017, Campo Morado was acquired and restarted by Altaley Mining Corporation (now Luca Mining Corporation) after the mine was put under care and maintenance by Nyrstar due to the downturn in the mining industry in 2015.

Campo Morado is an underground multi-metal mine with an original processing capacity of 1500 tons of ore per day. Mineralisation at Campo Morado occurs in complex volcanogenic-style massive sulfide (VMS) deposits. The Campo Morado deposit consists of several massive sulfide-horizons, hosting polymetallic (base metal and precious metal) mineralisation within a complex, layered felsic to intermediate volcanic sequence. The metals of interest include zinc, copper, silver, gold, and lead.

The operational challenges at Campo Morado are multiple: poor mine fleet availability and reliability resulting in inconsistent ore production, poor ore blending, maintaining a consistent metal grade feeding the process plant, operational complexities due to staged polymetallic processing at the ultrafine grind, and persistent maintenance issues culminating in poor overall performance.

Since 2022, Ausenco has been involved in various optimisation activities supporting the Campo Morado team. A systematic approach to project identification, evaluation, and implementation has achieved continuous performance improvements, leading to a progressive improvement in flotation performance.

The paper discusses the process of continual metallurgical improvements at the Campo Morado concentrator, such as developing a Mine to Mill program to maintain consistent metal grades feeding the plant, a geometallurgy program to characterise the deposit, the modernisation of the flotation circuit, development of a more robust flotation circuit flow sheet, and operational training to navigate the processing complexities at Campo Morado.

INTRODUCTION

The Campo Morado Project hosts several polymetallic massive sulfide deposits containing zinc, copper, silver, gold and lead mineralisation. The Project area is in the north-eastern part of Guerrero State, Mexico, 160 km south-west of Mexico City. The Property is in the municipality of Arcelia, 20 km south-east of the city centre.

In 2017, Campo Morado was acquired by Altaley Mining Corporation (now Luca Mining Corporation) with a restart of the operation to full-scale mining and mill processing end of 2017. Since the restart, commercial production has increased to 2060 tons per day.

Campo Morado timeline:

- 1810 to 1821:
  - Mexican War of Independence from Spain.

- General Vicente Guerrero paid his soldiers with copper coins from crudely smelted and refined copper ore derived from near-surface oxidised mineral deposits at Campo Morado.

- 1885 to 1940:

  - Reforma Mining and Milling Co. commenced exploration and mining/smelting of high-grade silver veins from the Reforma deposit.

  - The 1903 through 1910 production totalled 3387 kg gold, 125 230 kg silver and 4157 t lead; a minor part of which came from the Naranjo oxide deposit.

  - Mining halted around 1912 at the time of the Mexican revolution.

  - From 1914 to 1916, coins were minted at Campo Morado under the revolutionary General Jesus H Salgado.

  - Intermittent mining operations occurred from 1920 through 1927.

  - From 1937 through 1940, sporadic mining and development of oxide and minor sulfide ore took place.

- 1973 to 1977:

  - Union Oil subsidiary Minerals Exploration Company and its successor, Moly Corp, rehabilitated underground workings and drilling of core from the historic Reforma and Naranjo Oxide deposits.

- 1995 to 2010:

  - Farallon Resources Limited optioned the Campo and began exploration around the old Reforma Mine deposit area in November 1995.

  - The first drill hole was completed on the Reforma Deposit in June 1996 and within 24 months, several previously unknown massive sulfide bodies were either discovered or delineated by drilling between 1996 to 1998.

  - Then in late 1998, exploration work was put on hold for six years due to unfavourable metal prices.

  - Exploration resumed in August of 2004 and the extensive diamond-drilling program continued. In June 2005, a surface drill hole intersected a high-grade zone of the G9 deposit, which led to a refocus of interest on this area and the delineation of the G9 deposit. This program led to the discovery of several similar deposits nearby.

  - A metallurgical test program was initiated in 2005. Various companies were commissioned to carry out metallurgical and processing test programs on samples of G9 mineralisation from the zones to be mined. Tests were also carried out on the resultant flotation products. By the end of that year, a pilot plant test program had been completed. Test work continued in 2006 and 2007 and expanded in scope.

  - In April 2007, the company received its primary mine permit required for full mine and mill construction and operation of its G9 project at Campo Morado. Acquisition of equipment to build the G9 mill started in 2007 and by the latter half of the year, construction of the mill was underway. Commissioning of the mill took place in September 2008 with a nominal throughput of 1630 tons per day.

  - By July 2008, the first drill hole collared from the underground was completed. Stoping started in the North Zone of the G9 deposit in September 2008.

  - The first zinc concentrates were shipped in bulk by truck on 12 October 2008, to the port of Manzanillo in Colima State.

  - On April 1, 2009 Farallon announced commercial production at the G9 Mine.

- 2011 to 2017:

- o Nyrstar purchased Farallon Resources in 2011, the G9 mine was subsequently called the Campo Morado Mine.

- o Nyrstar commenced several laboratories and site-based programs to improve the operations in late 2013 to address the poor metallurgical performance of the deposits and plant-related design issues.

- o Nyrstar operated Campo Morado until December 2014. From January 2015, the mine was placed on care and maintenance due to deteriorations in metal prices and security concerns in the State of Guerrero.

- o The closure at Campo Morado and care and maintenance status remained in effect until the property was sold to Altaly Mining Corporation (formerly Telson Mining Corporation).

GEOLOGY AND MINERALOGY – EFFECTS ON PROCESSING

Mineralisation at Campo Morado occurs in a series of volcanogenic-style massive sulfide (VMS) deposits (Lorinczi and Miranda, 1978). The massive sulfide horizons host polymetallic (base metal and precious metal) mineralisation within a complex, layered sequence of felsic to intermediate volcanics.

The metals of interest include zinc, copper, silver, gold, and lead. Five Campo Morado mineral deposits have been extensively drilled: G9, El Largo, Reforma, Naranjo and El Rey. The concentrator is currently supplied ore from the G9 deposit, which consists of several mining zones including A9, A8, Muñeco, and Southwest. There are several other, less well-defined mineral occurrences.

Mineralogy

The Campo Morado Cu-Pb-Zn deposits are highly variable and complex fine-grained ores which are commonly known as refractory massive sulfides. Refractory massive sulfide ores are used as a term in a wide variety of contexts and there is no universally accepted definition (Bulatovic, 2007a, 2007b). The term refractory massive sulfide ores are used to distinguish these ores from other massive sulfides, where the treatment process is extremely complex (New Brunswick and Kuroco mines). In most cases, these ores are usually compared to Kuroco (Japan) and New Brunswick (Canada) ore types.

Grinding to achieve sufficient surface area liberation can be aided with automated mineralogy. By measuring the liberation and surface area exposure of minerals and gangue, an initial target grind for flotation can be established, avoiding trial and error metallurgical testing.

For grinding and flotation, mineralogy and textural association are the major influence on the grades and recoveries of the flotation concentrates.

The combination of mineral composition, liberation, and textures allows for the development of predictive models which can be compared to plant data or metallurgical test results to identify opportunities for improvement.

Table 1 displays of mineralogical data from other operating polymetallic deposits. A few other operating mines have mineral characteristics like that of Campo Morado. Some of these mines can be found in the Iberian pyrite belt located in Spain and Portugal. Two other well-known operations are Century and McArthur River. While these two mines are described as sedimentary hosted lead–zinc deposits, the mineralogy, liberation, and intensity of sphalerite interlocking are very similar to the El Largo zone.

TABLE 1

Mineral liberation of comparable operations.

Deposit	Grind	Two-dimensional liberation – %				
	µm K80	Cs	Ga	Sp	Py	Gn
Rey de Plata	85	50	50	78	70	93
Trout Lake	75	70	-	70	75	90
Louvicourt	65	75	-	75	80	85
Iberian 1	65	40	30	45	45	60
Iberian 2	60	40	35	40	75	45
Sullivan	60	-	65	63	88	92
Scuddles	50	65	60	75	80	90
Tizapa	50	70	55	75	80	70
Jerome	50	60	-	65	75	90
Cayeli	40	60	-	65	70	80
Mount Isa	40	-	72	74	85	86
Caribou	35	55	55	60	70	75
Century	35	-	55	40	41	74
McArthur River	35	-	30	45	65	60
Benambra	35	60	-	60	75	90
El Largo (average)[A]	**30**	**39**	**19**	**37**	**63**	**70**
G9 (average)[A]	**40**	**51**	**26**	**57**	**69**	**85**
SW (average)[B]	**35**	**64**	**46**	**59**	**73**	**89**
A9 (average)[B]	**35**	**67**	**51**	**72**	**77**	**91**

Note: Cs – Copper; Ga – Galena; Sp – Sphalerite; Py – Pyrite; Gn – Gangue. A – Data from ALS Kamloops (2005, 2006). B – Data from ALS Kamloops (2024).

Both the Century and McArthur River concentrators utilise a primary grind size of about 35 µm K80 and zinc regrind discharge sizes of 7–9 µm K80. The current Campo Morado configuration can achieve similar primary grind sizes as demonstrated by the recent operating data. However, the regrind size achieved by Century or McArthur River is not routinely achieved at Campo Morado.

Mineralogy is critical to developing knowledge of the processing characteristics of the Campo Morado deposits and is an essential part of the geometallurgical program. For example, galena has an average grain size of 6 microns and is typically locked within pyrite, as shown in Figure 1.

The problem statement is that the bulk rougher circuit uses a pyrite depression system, which also depresses the galena because it is associated with pyrite. Once the galena is recovered to the bulk concentrate, the cleaning circuit achieves a high-stage recovery because the regrind circuit liberates the galena from pyrite; however, without adopting a different bulk rougher reagent scheme, the overall galena recovery will always be reduced significantly.

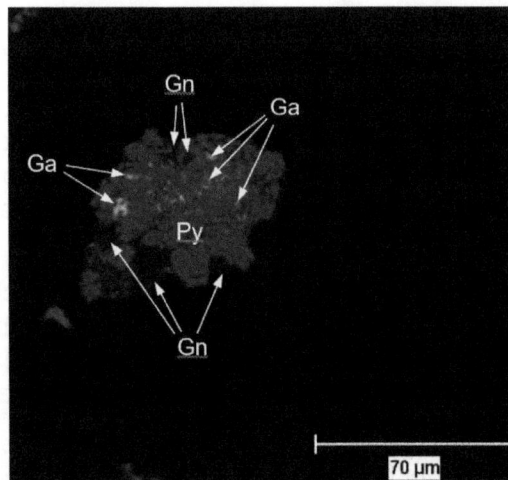

FIG 1 – Southwest bulk rougher tailings sample (ALS Kamloops, 2024).

Ultimately, the efficiency of the separation will be dictated by the physical constraints of mineral liberation. Once addressed, the chemical aspects of separating the minerals by flotation can be optimised.

Gold and silver

Despite the relatively high gold grades in many of the Campo Morado mineralised zones, recovery of gold to payable concentrates has been relatively poor. Historical studies have examined the deportment of gold, both with direct mineralogical assessments and chemical deportment models.

The recoverable gold and silver were shown to behave like galena and chalcopyrite in the flotation process, with preferential recovery to the copper and lead concentrate (ALS Kamloops, 2024). The deportment of gold and silver to the copper concentrate are summarised in Table 2.

TABLE 2

Summary of gold and silver deportment to the copper concentrate.

Sample	Assay, % or g/t			Recovery, %		
	Cu	Ag	Au	Cu	Ag	Au
Muñeco	17.9	1776	46.4	82.8	52.5	35.5
A8	11.6	4760	85.8	80.8	66.8	65.1
A9	26.9	684	7.11	93.8	51.8	43.5
Southwest	23.1	470	12.3	85.5	38.8	46.7

Mineralogy has identified that gold and silver recovery to the copper and lead concentrates are in the form of an electrum and an electrum with either mercury or lead. The remainder of the gold was interlocked with pyrite or complex multiphase particles. Gold in this form is at odds with flotation recovery to a copper, lead, or zinc concentrate.

The key finding from the gold deportment review is that the current body of Campo Morado's metallurgy and mineralogy would suggest that increasing both copper and lead recovery will increase both silver and gold recoveries.

RESULTS OF METALLURGICAL TEST WORK

Geometallurgy is an important addition to any mining operation, large or small, as it aims to improve the Net Present Value (NPV) of an orebody (Dominy, O'Connor and Xie, 2016). It is known to increase site stakeholder communications and collaborations (Jones and Morgan, 2016; Suazo *et al*, 2011). Creating the right environment for knowledge sharing while concentrating on improved data acquisition and interrogation will result in confident data integration into the Mine-Value Chain. All these aspects create better business optimisation, better staff utilisation, and better targeted key performance indicators.

Historically, the site team has not undertaken a geometallurgy approach to developing the deposits within the Campo Morado deposits. Instead, the site team has adopted a 'Reactive' approach, whereby batch laboratory flotation test work has been conducted on plant samples during periods of poor metallurgical performance.

Variability is central to the geometallurgy effort. Appropriate variability characterisation through good sampling and adequate drilling will help lower the risk of mischaracterisation and enable more effective project planning. The historical Campo Morado metallurgical work has been less than representative, and no full variability program has ever been undertaken.

Only recently have a few grab samples been collected from the mine workings and submitted for batch laboratory flotation test work, but these results have not been incorporated into the production planning.

Routine geometallurgical testing of the current mining ore zones known as Southwest, A9 and Muñeco is underway. Figure 2 presents the Leapfrog model with the main deposits currently included in the geometallurgy program.

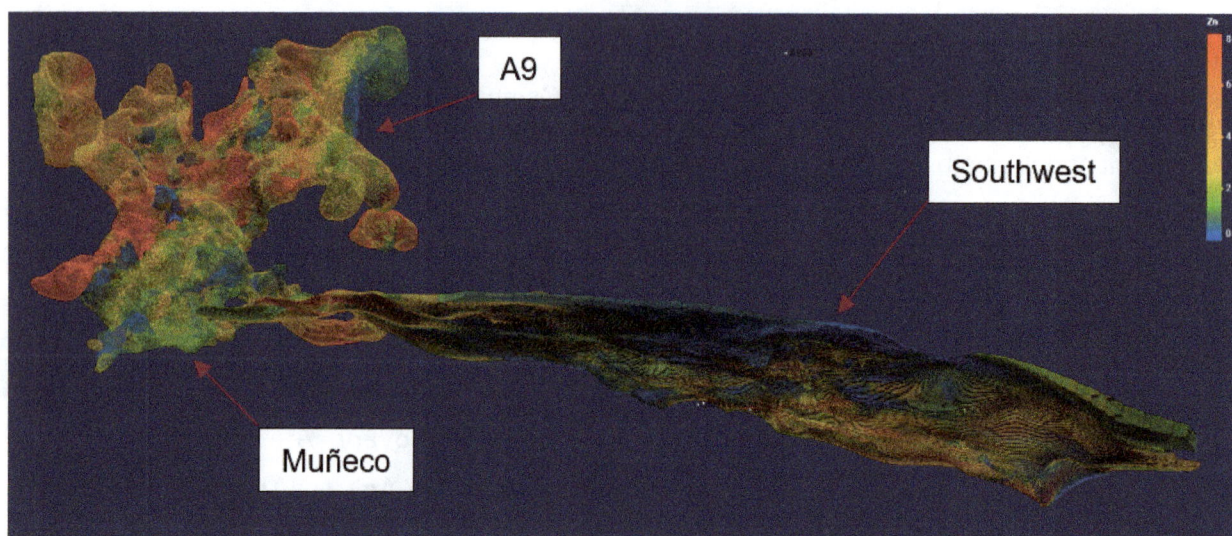

FIG 2 – Campo Morado Geological model of Southwest, A9, and Muñeco ore zones (Leapfrog model, Luca Mining Ltd, Campo Morado deposit model, 2024).

The objectives of the geometallurgy program include:

- The ore characterisation of the resources with 'infill sampling' and, when combined with the spatial modelling, the resulting block model and mine schedule will provide the critical ore feed schedule to the processing plant.

- Developing the spatial ore characteristics of the resource will result in a more reliable model of the plant and production capacity.

- Develop operational geometallurgy, support life-of-mine (LOM) planning, and provide information to develop future studies (eg deleterious elements, gold, expansion etc).

This includes coordinating routine meetings with the geologists and mining engineers to determine the weekly blending strategies to maximise metallurgical performance.

Ore types that do not respond to the typical Campo Morado flow sheet conditions are reviewed with the geologists/mining engineers and then separated for blending or processing with further flow sheet optimisation.

The geometallurgy program required a complete rethinking of the laboratory test work procedures to better align with the actual plant operation. Additional laboratory equipment has been purchased to improve the operational efficiency of the site metallurgical laboratory, including purchasing new primary and regrind mills, pH probes, screens, micro syringes, and refurbishing of the flotation cells.

The site Metallurgical Research Team (MRT) have undertaken training at ALS Kamloops to expand their understanding of the laboratory equipment and procedures needed to test the various ore types mined at Campo Morado.

The Campo Morado laboratory now has similar capabilities to those of a commercial metallurgy laboratory, like ALS Kamloops. This allows faster turnaround times and significantly low costs to implement the geometallurgy program.

Breakage domains

The Bond Ball Work Index (BBWi) machine located in the site metallurgical laboratory was used to test 42 samples from the Southwest, A9, and Muñeco ore zones. The BBWi valves' range was between 9.6 and 14.4 kWh/t.

A multivariate and statistical analysis was completed to develop ore breakage domains. Figure 3 presents the predicted versus actual BBWi valves for the ore breakage domains.

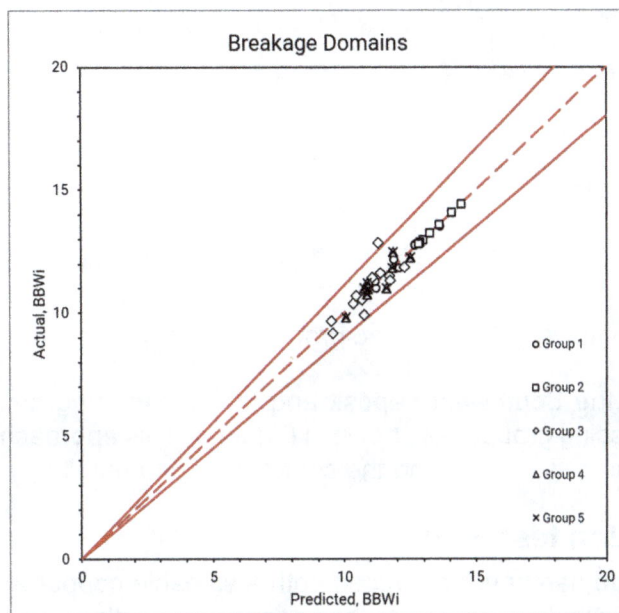

FIG 3 – Campo Morado ore breakage domain models.

The ore breakage domains were included in the block model to improve the predictability of the plant throughput and flotation feed particle size, which impacts the base metal recoveries.

Metallurgical domains

The Southwest deposit contributes most of the ore processed through the concentrator during 2024 and has been the focus of the geometallurgy test work. The Southwest deposit contains a very wide range of copper, lead, and zinc grades, as shown in Figure 4.

The geometallurgy test work has identified three metallurgical domains, including:

- Group 1 samples performed well using the current bulk flow sheet, with more than 70 per cent copper recovery and appreciable lead recovery greater than 65 per cent.

- Group 2 contains low lead grades and responds very well to the current bulk concentrate flow sheet, with copper recoveries typically more than 85 per cent achievable.

- Group 3 contains an appreciable amount of total organic carbon and does not respond well to the current bulk concentrate flow sheet without improved preflotation circuit performance.

Additionally, several samples had poor cleaner circuit performance because of ultra-fine-grained carbon liberated during the regrind process.

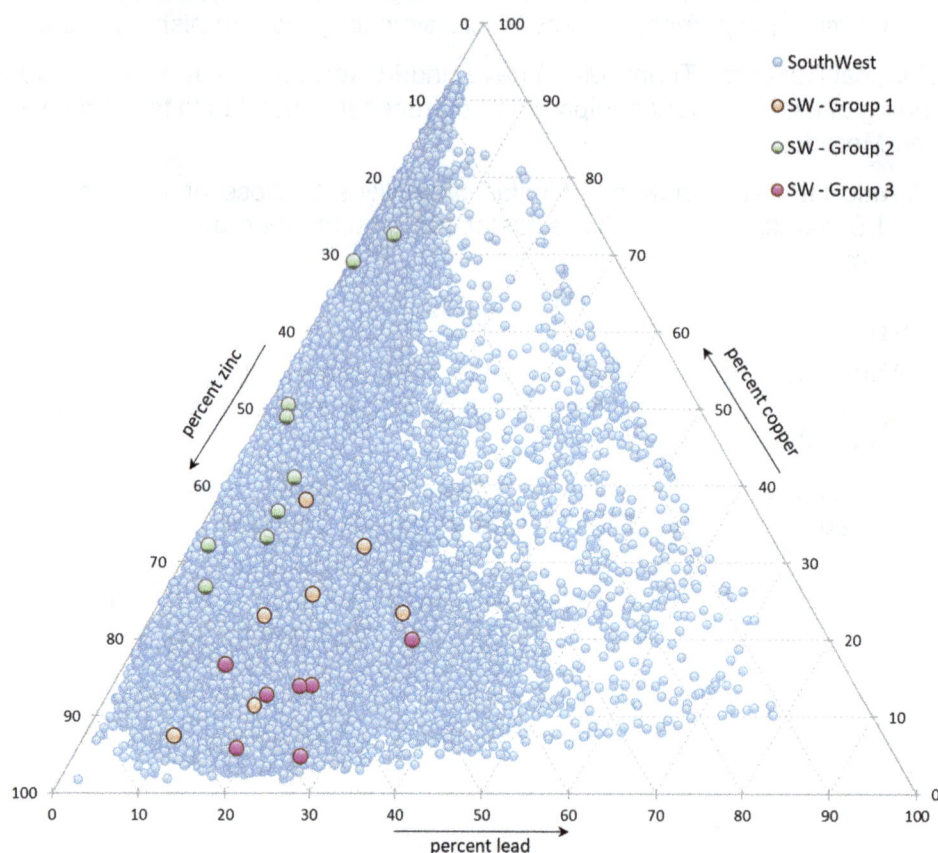

FIG 4 – Southwest deposit feed grade and ore type ternary diagram.

The elemental data from the Southwest deposit and the ternary diagram is used to highlight the different metallurgy processing groups, as shown in Figure 4. This approach has been used to assist with the Mine to Mill blending strategies and the selection of geometallurgy samples.

Copper-lead separation test work

Historically, Campo Morado had never produced both a saleable copper and lead concentrate and, therefore, had produced a single bulk copper-lead concentrate with an appreciable quantity of gold and silver credits.

The original Campo Morado flow sheet consisted of bulk copper–lead flotation method, followed by zinc flotation from the bulk tailing. The copper–lead separation involving lead depression and copper flotation was performed on the upgraded bulk concentrate. This method is the most used in the treatment of copper–lead–zinc ores (Bulatovic, 2007a, 2007b).

The original flow sheet development test work was conducted at ALS Kamloops in 2009 and incorporated the SO_2–starch method for the copper–lead (Lentz and McCutcheon, 2006), albeit very limited test work was conducted. This method is used in the treatment of disseminated massive sulfide copper–lead–zinc ores (New Brunswick, Canada; San Nicolas, Mexico). In addition, this procedure also included preconditioning the pulp with starch or dextrin at elevated temperature (65–85°C), cooling the pulp in the presence of SO_2 to a pH of 5.0–5.5, followed by copper flotation (Kuroko, Japan).

A new flow sheet development test work program was undertaken at ALS Kamloops in 2024 to investigate the conventional bulk copper-lead separation flotation methods and more recent developments in flotation reagents. Copper-lead separation tests were completed at a regrind 80 per cent passing of 6 microns and using dextrin, sodium carboxymethyl cellulose (CMC), ammonium persulfate (APS), sodium metabisulfite (SMBS) and high temperature conditioning (Kuroko, Japan).

Copper-lead separation test work was completed at ALS Kamloops (ALS Kamloops, 2024) using parent samples collected from the Southwest, A8, A9 and Muñeco deposits. A blend composite representing the Q1 2024 mine production plan was constructed using the parent samples. The blend composite was used for the flow sheet development and copper-lead separation test work.

The flotation results were evaluated by the mass balance of components and separation products, as well as by plotting recovery-recovery, as shown in Figure 5. The obtained results allow for a comparison of the upgrading selectivity of copper and lead.

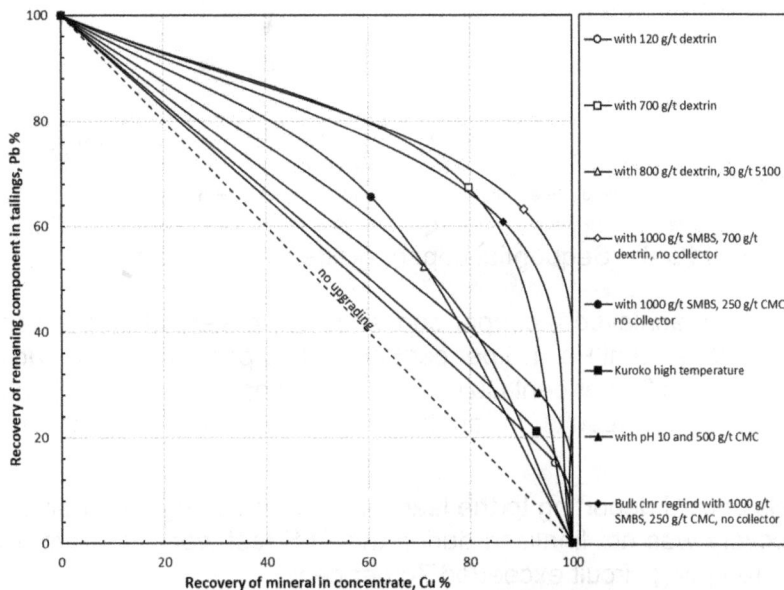

FIG 5 – Blend composite copper-lead separation test work results.

In these experiments, the best results of upgrading of copper were obtained in the presence of dextrin and SMBS in the absence of collector. The copper concentrate recovery was 67 per cent with concentrate grades of 20 per cent Cu, 5.7 per cent Pb, 1.3 per cent Zn and 29 per cent Fe and lead concentrate recovery was 28 per cent with concentrate grades of 1.5 per cent Cu, 6.9 per cent Pb, 2.7 per cent Zn and 33 per cent Fe.

Although the outcomes from the conventional copper-lead separation tests were not promising, similar performance was achieved at other sites (Liang *et al*, 2021) when using starch. It wasn't until a modified starch was used that significant improvements in the copper-lead separation were achieved.

Subsequently, the bisulfite sequential copper–lead flotation method, in which the copper and lead are sequentially floated from the bulk rougher concentrate to produce separate copper and lead concentrates, achieved superior flotation performance compared to conventional copper-lead separation methods.

The laboratory tests consisted of bench-scale open circuit flotation tests, which are presented in Figure 6. Using this method, the principal lead depressant during copper flotation is SMBS. The copper is then floated using Thionocarbamate (A5100). In the lead circuit, lime or soda ash and cyanide are used as the principal modifiers with a xanthate collector.

Figure 6 presents the results of the copper grade-recovery curves for the blend composite and mine zone samples.

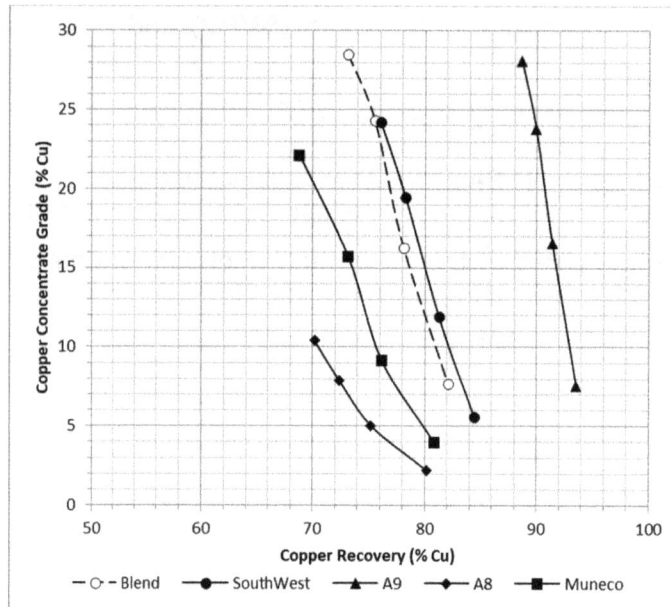

FIG 6 – Sequential copper grade-recovery curves.

The sequential flow sheet achieved a copper concentrate recovery of 73 per cent Cu at a grade of 28.5 per cent Cu and 1.8 per cent Pb, 1.3 per cent Zn and 28 per cent Fe. Reducing the concentrate grade from 28 per cent Cu to 20 per cent Cu (like the bulk concentrate quality), increased the copper recovery to 77 per cent and results in another 5 per cent gold and silver recovered to the concentrate.

Due to the low levels of lead deporting to the lead circuit when using the open circuit flotation tests, the overall lead recovery was not finalised during the ALS test work program. However, the stage recovery of the lead roughing circuit exceeded 70 per cent.

The overall lead recovery is dependent upon the bulk rougher circuit recovery. Increasing the iron recovered to the bulk concentrate directly increases the overall lead recovery as more lead reports to the ultrafine regrind and sequential separation circuit.

The ALS test work has identified a copper-lead separation flow sheet that was previously not achievable at Campo Morado. The performance is correlated to the mineral grain size and iron content.

The copper-lead separation flow sheet is currently being tested in the Campo Morado laboratory.

In addition, a new flow sheet development program and reagent optimisation test work demonstrates further upside in the metallurgical performance of the Campo Morado ore zones.

PROCESS DEVELOPMENT

Several improvements projects have been investigated in the past to improve the metallurgy and increase revenue income. These projects make up the Campo Morado Improvement Project (CMIP) and are discussed in the sections below.

Met research team

The Metallurgical Research Team (MRT) is tasked with performing routine field measurements (eg flotation and regrind circuit particle size and flotation circuit surveys) and conducting bench-scale metallurgical flotation tests. Their function is paramount to understanding the processing response of future deposits and investigating causes of poor plant performance.

The key activities of the MRT include:

- The bench-scale flotation test work conditions were not aligned with the current process flow sheet. The MRT is in the process of aligning the procedures.

- The geometallurgy program is led by the MRT, as most of the geometallurgical activities align with the scope of work to be either undertaken or coordinated by this team.

- Carbonates and other deleterious elements affect the metallurgical performance and concentrate quality. The MRT has a program to confirm the assays to be included in the routine metallurgical reporting to assist with identifying operational constraints.

- The MRT has managed the copper-lead separation test work underway at ALS Kamloops. ALS Kamloops was chosen previously due to its historical work on the Campo Morado ores, which gave it the most understanding of how the ore responds to flotation.

- Opportunities to improve gold recovery have been included in the geometallurgy program and hence coordinated by the MRT.

The MRT has gained valuable experience through engaging experienced metallurgical laboratories like ALS Kamloops. Key personnel from the MRT have been present during the various test work programs at ALS Kamloops to gain insight into best practices when processing complex VMS deposits.

Mine to mill planning

The mine has shifted from operating 'hand-to-mouth' operation, where feed variations to the process plant significantly affected metallurgical performance. As the mine is combining ore from different areas with different Cu-Pb-Zn, Cu-Zn, Cu, and Zn ore types, these behave very differently in the process plant and can ultimately result in suboptimal performance when mixing different ore types.

Figure 7 presents the new Mine to Mill strategy at Campo Morado. This strategy uses both software and operational activities to improve ore scheduling and predictability of the ore characteristics.

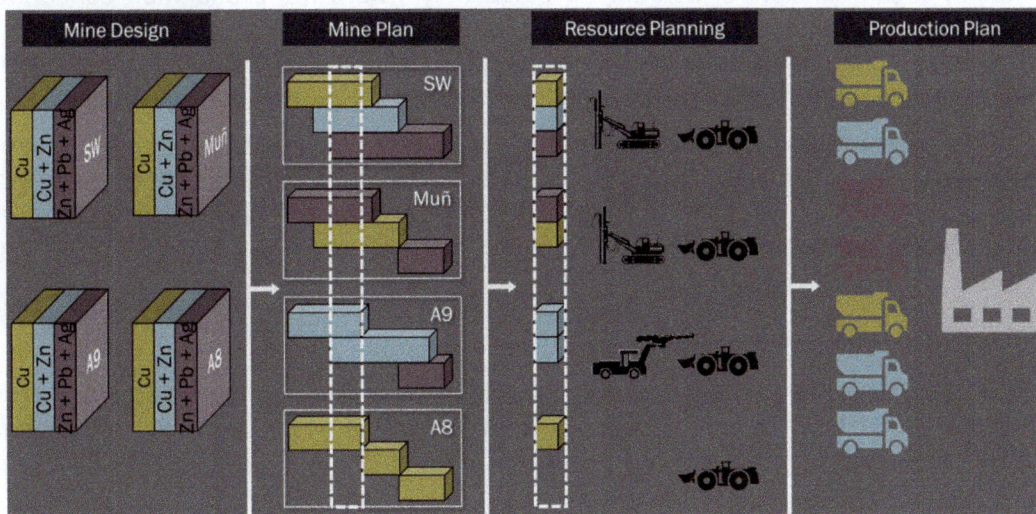

FIG 7 – Campo Morado Mine to Mill operating strategy.

Leapfrog modelling software is being used for the geometallurgical model. The current model considers:

- Adding an attribute to all stopes and development that is a representation of the critical variable causing the issues for the mill (ie liberation, carbonates etc).

- Set quantity limits (upper and lower) for that variable to target.

- Schedule using that quantity limit (needs to be granular – from daily to weekly).

In doing those steps, the scheduler will target the variable to stay within the range set and allocate production to areas and maintain it. The variable will then be in the acceptable range provided the data in the block model/mine design is accurate.

In operations, to support the definition of local metallurgy, the following methodology is under development:

- First, understand what the problem mineral/component/variable is (the one defined above).

- Then inspect how that data is represented in the block model, design, and schedule (how granular and old is the data?).

- Local probe drilling and sampling could be used to make the data much more recent and localised. This data would be much more accurate than the block model. However, this depends on being able to assay the probe drilling results quickly (and could get expensive).

- Check the accuracy of the probe drilling and assaying against the mill feed results (to ensure the data being collected is accurate).

- Incorporate this drill/assay data into the mine planning software (this is described in the software idea above).

The Mine to Mill implementation plan is aligned with numerous activities to develop the necessary inputs to the block model and mine scheduling software and will be implemented by Q3 2024.

Automated reagents dosing systems

The existing reagent systems are in poor condition, and very few pumps are in operation. The site has been replacing the existing pumps slowly due to the excessive cost of the reagent pumps.

Insufficient reagent addition and control is a major contributor to poor plant performance. To mitigate this issue, a new collector, frother and bulk reagent dosing systems will be installed. Dosing will be automatically adjusted based on metallurgical domains from the geometallurgical model.

Preflotation circuit

The preflotation circuit consists of a roughing stage only where the preflotation concentrate was discarded directly to tailings, however, this stream contained significant amounts of copper (>5 per cent), lead (>8 per cent), and zinc (>6 per cent). Typically, carbonaceous preflotation circuits incorporate a cleaner stage (Century, Penasquito, Red Dog, Mt Isa etc).

The MRT has conducted laboratory cleaner tests which demonstrate significant improvements in base metal rejection from the preflotation concentrate. The project will likely assess the application of the Jameson cell technology for the cleaning circuit duty as demonstrated improved flotation circuit performance at other mine sites (Smith *et al*, 2008, Rantucci, Akroyd and Grattan, 2011, Runge *et al*, 2021).

Bulk conditioning tank

The laboratory test work has demonstrated that sufficient conditioning time is required to maximise bulk concentrate production. The CMIP will retrofit an existing 20 m^3 flotation cell at the head of the bulk rougher flotation banks to be used as a conditioner tank.

The conditioner tank will also be modified to enable control of the flow rate between the two bulk rougher flotation banks. Because one bank has more capacity than the other, this will lead to shorter residence time and lower recovery.

The installation of the conditioner tank will also utilise a gravity feed design which will eliminate the need to pump slurry, thus reducing operating and maintenance costs.

Bulk concentrate surge tank

The operation of the bulk regrind mills and subsequent copper-lead separation circuits depends on steady-state operation, which is affected by the plant throughput and feed grades.

The existing 60 m^3 lime distribution tank will be converted into a surge tank to provide buffering between the bulk rougher and sequential copper-lead separation circuits. The surge tank ensures that the target regrind particle size of 80 per cent, passing 6 microns, is achieved to affect the necessary separation efficiency of copper and lead into separate concentrates.

Two-stage bulk regrind

The Stirred Media Detritor (SMD) mills are well proven in the fine grinding application of polymetallic concentrates, like Campo Morado. There are four SMD355 (ie 355 hp), two in each of the bulk and zinc regrind circuits.

The CMIP reinstalled the original SMD mills to be operated in series. Operating the SMD units in series allows the use of two different media sizes, which will maximise the grinding efficiency of the regrind circuit and achieve the necessary sequential flotation feed size of 80 per cent passing 6 microns. Matching the grinding media size to the grinding application can substantially reduce the specific energy required.

Sequential copper and lead flotation cleaning circuits

New sequential copper and lead flotation cleaning circuits will be installed using the existing tank cells from the partially constructed copper concentrate circuit, which is located beside the polymetallic concentrator.

Further details of the new two-stage cleaner copper flotation cells are described in the flow sheet and equipment section of this paper. Similarly, a new three-stage lead cleaning circuit will be installed beside the new copper cleaner flotation cells.

Flotation cell refurbishment

The process plant was constructed in 2009, and the flotation cells and pump hoppers are in poor condition (ie corroded steel and worn liners). The deterioration of the flotation cells is affecting the slurry suspension and causing short-circuiting of slurry, which is impacting metallurgical performance.

The equipment which will be refurbished initially includes:

- Copper and Lead roughing flotation banks.
- Zinc rougher line 1 flotation banks.
- SMD discharge launders.
- Copper circuit concentrate hoppers.
- Lead circuit concentrate hoppers.
- Tailings and concentrate thickeners.

Additional modifications will be undertaken during the refurbishing of the flotation cells and include installing new flotation cell mechanisms and air control as described further.

New flotation cell NextSTEP™ mechanisms

Laboratory and plant surveys identified a root cause for low recoveries: the poor recovery of fine particles in the flotation banks. Over the past 20 years, improvements in the flotation cell mechanism design and new technologies have been developed to process complex VMS ore deposits (Muller and Law, 2018).

The poor condition of the flotation cells and mechanisms as shown in Figure 8.

FIG 8 – Existing flotation cell mechanisms.

The CMIP includes retrofitting next-generation flotation cell mechanisms in the sequential copper/lead roughing and zinc rougher flotation banks. The NextSTEP™ mechanism delivers superior flotation cell performance compared with the current Door Oliver free-flow mechanism.

New flotation cell air control

Air control is an important component of the flotation cell operation (Cooper *et al*, 2004). Poor air control of the flotation cell aeration rate affects the recovery, grade, and throughput.

Plant operating data highlights the root cause of poor recovery and concentrate grade, which is inadequate flotation cell air control. To overcome this issue, all the flotation cells will be retrofitted with an air flowmeter and actuated control valves to automate the flotation cell's air control. A new control philosophy will be incorporated within the existing Supervisory Control and Data Acquisition (SCADA) system to maximise both the recovery and grade of the different circuits, especially when the plant receives variable feed grades.

Zinc circuit modifications

As the zinc cleaner circuit was designed for an 18 per cent zinc feed grade and the life-of-mine feed grade is less than 3 per cent, there was sufficient time and the number of cleaning stages to achieve a cleaner circuit stage recovery >95 per cent. However, the zinc cleaning circuit will be reconfigured to operate with four cleaning stages to mitigate periods of high iron feed sources and achieve improved zinc concentrate grades.

In addition, the zinc regrind circuit hydrocyclones will be repaired and put back into service to improve the operation of the zinc regrind circuit and achieve a regrind circuit 80 per cent passing lower than 9 microns.

FLOW SHEET AND EQUIPMENT

The Campo Morado process flow sheet is shown in Figure 9.

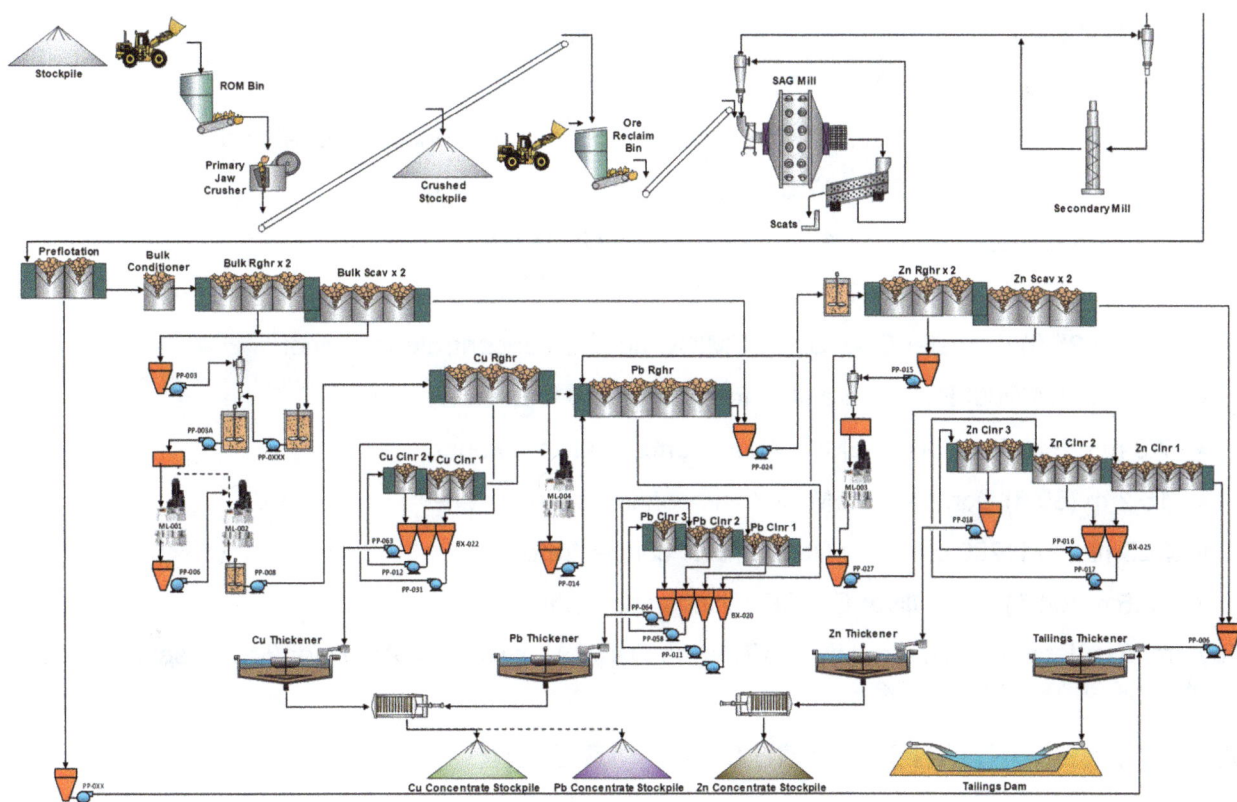

FIG 9 – Campo Morado flow sheet.

MAJOR EQUIPMENT ITEMS

Major equipment items in the concentrator are:

- 30' × 42' Lippmann-Milwaukee jaw crusher with 149 kW (200 hp) motor with a closed side setting of 150 mm for a P_{80} of <140 mm.

- Crushed ore stockpile with 6000 t and reclaim hopper of 140 t live capacity.

- 6.7 m (9 ft) inside diameter × 2.7 m (22 ft) EGL (effective grinding length) Farnell-Thompson semi-autogenous grinding (SAG) mill with 1500 kW motor (six per cent ball load). The mill is fixed speed and operates at 73 per cent of critical speed. The FLSmidth LoadIQ smart sensor technology is installed on the SAG mill.

- 1.83 m (6 ft) wide × 3.66 m (10 ft) long Schenck Model H1–1830 SAG mill discharge screen fitted with 31.8 mm square apertures.

- Metso Vertimill VTM-1000-WB secondary mill with 746 kW (1000 hp) motor.

- 2 × Metso RCS-20 20 m³ tank cells for preflotation rougher flotation.

- 9 × Dorr Oliver DO300-UT 8.5 m³ and 4 × Metso RCS-10 10 m³ cells for bulk rougher flotation.

- 1 × 60 m³ bulk rougher concentrate surge tank.

- 2 × Metso SMD 355 with 265 kW motors for regrinding bulk rougher concentrate.

- 3 × Dorr Oliver DO300-UT 8.5 m³ cells for copper rougher flotation.

- 2 × Metso RCS-10 10 m³ cells for copper cleaner 1 flotation.

- 1 × Metso RCS-10 10 m³ cells for copper cleaner 2 flotation.

- 1 × Metso SMD 355 with 265 kW motor for lead rougher regrind.

- 1 × Metso RCS-20 20 m³ cells for lead cleaner 1 flotation.

- 2 × Metso RCS-5 5 m³ cells for lead cleaner 2 flotation.

- 1 × Metso RCS-5 5 m³ cells for lead cleaner 3 flotation.

- 12 × Dorr Oliver DO600-UT 17.5 m³ cells for zinc rougher flotation.

- 1 × Metso SMD 355 with 265 kW motor for zinc rougher concentrate regrind.

- 4 × Dorr Oliver DO600-UT 17.5 m³ cells for zinc cleaner 1 flotation.

- 3 × Dorr Oliver DO600-UT 17.5 m³ cells for zinc cleaner 2 flotation.

- 3 × Dorr Oliver DO600-UT 17.5 m³ cells for zinc cleaner 3 flotation.

- 7.6 m (25 ft) diameter Dorr-Oliver EMICO copper concentrate thickener.

- FLSmidth M900FB pressure filter for copper and lead concentrate.

- 7.6 m (25 ft) Dorr-Oliver EMICO lead concentrate thickener.

- 15.2 m (50 ft) Dorr-Oliver EMICO zinc concentrate thickener.

- FLSmidth M1200 pressure filter for zinc concentrate.

- 19.5 m (64 ft) Dorr-Oliver EMICO tailings thickener.

Key process data are monitored by two Thermo Fisher Multi-Stream Analyser (MSA), seven streams in total, for online measurement.

PLANT PERFORMANCE SINCE COMMISSIONING

Bulk concentrate

Bulk concentrate day and night shift recovery data between June 2023 and April 2024 is presented in Figure 10.

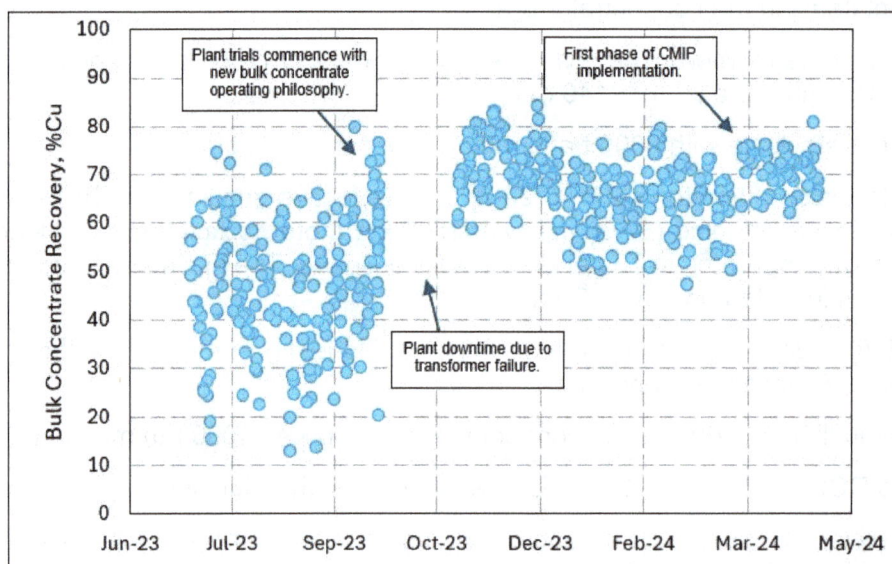

FIG 10 – Campo Morado bulk concentrate recovery production summary.

Bulk concentrate production prior to October 2023 was highly variable, ranging between 12 and 75 per cent copper recovery. Although copper and lead concentrates were produced, both concentrates were combined and sold in single bulk.

A plant trial was conducted in late September 2023 using the laboratory reagent conditions and removing the lead bulk rougher cells from the circuit. The average plant feed grades during the plant trial were 0.8 per cent Cu and 0.65 per cent Pb. A bulk concentrate was produced with >70 per cent copper recovery and >56 per cent lead recovery. The concentrate produced was >19 per cent Cu and <9 per cent Pb.

The trial was cut short due to issues with low ore supply (poor ore blending), low process water availability, and transformer failure on-site.

The actions undertaken to achieve the improved bulk circuit performance include:

- Providing consistent feed grade (ore blending) and throughput.
- Blending the ores to provide a feed ratio of Cu 1: Pb < 1
- Reconfigure the circuit as a conventional bulk roughing and cleaning circuit to produce a copper-lead concentrate (ie bulk concentrate).
- Increased the reagent additions based on metallurgical test work (historically operated at low level to mitigate over-dosing).
- Maintain regrind product size of 80 per cent passing 8 microns.

The first phase of the CMIP implementation included repurposing existing 20 m³ flotation cells for the bulk cleaner duty. The larger flotation cells and simplification of the cleaner circuit resulted and improved bulk cleaner circuit stability resulting in improved copper recovery.

The plant data highlights the impact of plant stoppages attributed to no ore feed. The shutdown and restarting of the plant significantly reduce concentrate production due to the instability of the operation.

The recovery data also highlighted the treatment of another ore type during November which achieves superior bulk concentrate recovery and the need to separate the different ore types to maximise bulk concentrate production.

Zinc concentrate

Zinc concentrate day and night shift recovery data between June 2023 and April 2024 is presented in Figure 11.

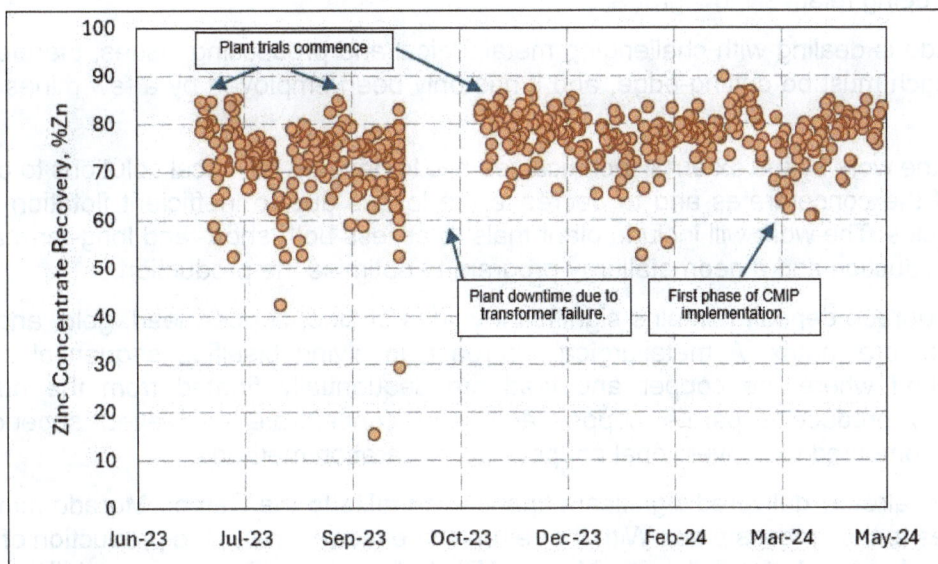

FIG 11 – Campo Morado zinc concentrate recovery production summary.

Laboratory tests were conducted using zinc circuit feed samples to investigate the potential improvements with zinc circuit stage recovery. The comparative plant-laboratory results showed that >95 per cent zinc stage recovery was achievable in the laboratory (with a shorter laboratory residence compared with the plant using a scaleup factor of 2.5) whereas, during the plant operation, the zinc stage recovery was measured at 86 per cent zinc rougher stage recovery.

Following the laboratory test work, the following changes were implemented, including increased copper sulfate and xanthate dosages, lower rougher pulp levels, and higher aeration rates. These changes resulted in the overall zinc recovery increasing from 75 per cent to 86 per cent and the final concentrate grade increasing from 46 per cent to 56 per cent.

During the trial, the zinc regrind product size of 12 microns was achieved, providing sufficient liberation to achieve high zinc concentrate grades. As the zinc cleaner circuit was designed for an 18 per cent zinc feed grade, there was sufficient time and a sufficient number of cleaning stages to achieve a cleaner circuit stage recovery of >95 per cent.

These results were achieved during an 8-hour trial. The trial was stopped due to a significant reduction in the zinc feed grade as the circuit was overdosed. It was also observed that poor mixing was occurring with more than half the DO600 flotation cells. The cells were subsequently drained and inspected. The poor mixing resulted from the excessive wear on the flotation cell mechanisms. A decrease in flotation rate is typically expected with poor flotation cell mixing; however, as the zinc rougher circuit has more than 50 mins of slurry residence time, the impact of poor flotation cell mixing has not been observed to date.

Zinc concentrate production is clearly affected by the upstream bulk circuit performance. Lower copper/lead content in the bulk concentrate results in more zinc metal being recovered. Increasing the zinc deportment to the bulk concentrate will reduce the overall zinc circuit recovery.

The first phase of the bulk concentrate impacted the zinc circuit recovery because of the failure of the concentrate pump in the zinc circuit and the re-tasking of an existing pump, which directed zinc directly into the modified bulk cleaner circuit. Once the pump was redirected, the zinc deportment to the bulk concentrate reduced back to normal levels. In addition, the plant was also direct fed with an ore type, which requires different operating conditions within the zinc circuit to restore zinc circuit performance.

CONCLUSIONS

The refractory VMS deposit, in Ausenco's experience, has always involved an energy and reagent-intensive flow sheet due to the complexity of the mineralisation, with the resulting product concentrates being relatively low-grade.

Campo Morado is dealing with challenging metallurgical and processing issues; therefore, the site team's approach must be cutting-edge, and it has only been employed by a few mines globally so far.

The focus of the work by the external technical team is to develop technical solutions to optimise the production of the concentrates and to decrease the losses due to inefficient flotation of fine and ultrafine particles. The work will include plant trials to assess both short- and long-term alternatives to improve production and a geometallurgy program to optimise the production.

The Campo Morado deposit contains significant values of zinc, copper, lead, gold, and silver in a massive pyrite ore matrix. A metallurgical approach involving bisulfide sequential copper–lead flotation method where the copper and lead are sequentially floated from the bulk rougher concentrate to produce separate copper and lead concentrates achieved superior flotation performance compared to conventional copper-lead separation methods.

The CMIP has already delivered significant financial benefits to the Campo Morado mine with only minor changes to the process plant. With the refurbishment equipment, the production of a separate copper and lead concentrate, and a new Mine to Mill strategy supported by geometallurgy, the CMIP will deliver a robust business model for the Campo Morado mine.

In addition, several other improvement projects will further increase the profitability of the Campo Morado mine.

ACKNOWLEDGEMENTS

The authors thank Luca Mining Corporation for permission to publish this paper.

REFERENCES

ALS Kamloops, 2005. A preliminary assessment Campo Morado Project – El Largo Deposit: Evaluation of distributed regrinding on metallurgical Performance – KM1651, report.

ALS Kamloops, 2006. A preliminary assessment of response: The G9 Deposit – Report 2 – KM1772, report.

ALS Kamloops, 2024. Metallurgical Testing for Campo Morado Mine – KM7078, report.

Bulatovic, S M, 2007a. Flotation of Copper-Lead-Zinc Ores, in *Handbook of Flotation Reagents, Chemistry, Theory and Practice Flotation of Sulfide Ores*, pp 367–400 (Elsevier: Amsterdam).

Bulatovic, S M, 2007b. Interaction of Inorganic Regulating Reagents, in *Handbook of Flotation Reagents, Chemistry, Theory and Practice Flotation of Sulfide Ores*, pp 153–184 (Elsevier: Amsterdam).

Cooper, M, Scott, D, Dahlke, R and Gomez, C, 2004. Impact of air distribution on banks in a zinc cleaning circuit, *CIM Bulletin*, pp 1–6.

Dominy, S, O'Connor, L and Xie, Y, 2016. Sampling and testwork protocol development for geometallurgical characterisation of a sheeted vein gold deposit, in Proceedings Third Geometallurgy Conference (The Australasian Institute of Mining and Metallurgy: Melbourne).

Jones, K and Morgan, D, 2016. On-site process mineralogy – linking departments and disciplines with the common goal of increasing plant performance, in Proceedings of the 13th Mill Operators' Conference (The Australasian Institute of Mining and Metallurgy: Melbourne).

Liang, Z, Li, G, Wei, Z, Wu, W, Huang, X, Wang, J, Cui, L, Ni, X and Zhong, S, 2021. Replacing Dichromate with Polysaccharide Depressant in Cu-Pb Separation: Lab Bench Tests and Plant Trials in Zijin Mining, *Mining, Metallurgy & Exploration*, 38(5).

Lentz, D and McCutcheon, S, 2006. The Brunswick No. 6 Massive sulfide deposit, Bathurst Mining Camp, Northern New Brunswick, Canada: A synopsis of the geology and hydrothermal alteration system, *Exploration and Mining Geology*, 15(3–4), July-October 2006.

Lorinczi, G and Miranda, V, 1978. Geology of the massive sulfide deposits of Campo Morado, Guerrero, Mexico, *Economic Geology Journal*, 73(2):180–191.

Muller, M and Law, H, 2018. High shear rotor trails in the Mount Isa Mines Copper Concentrator, in Proceedings of the 14th Mill Operators Conference (The Australasian Institute of Mining and Metallurgy: Melbourne).

Rantucci, D, Akroyd, T and Grattan, L, 2011. Carbon prefloat improvements at Century Mine, in Proceedings of the Metallurgical Plant Design and Operating Strategies Conference, MetPlant 2011 (The Australasian Institute of Mining and Metallurgy: Melbourne).

Runge, K, Brito e Abreu, S, Evan, C and O'Donnell, R, 2021. Analysis of the Mt Isa Copper Concentrate Preflotation Circuit using advanced diagnostic techniques, in Proceedings of the 15th Mill Operators' Conference (The Australasian Institute of Mining and Metallurgy: Melbourne).

Smith, T, Lin, D, Lacouture, B and Anderson, G, 2008. Removal of organic carbon with a Jameson Cell at Red Dog Mine, in Proceedings of the 40th Annual Meeting of the Canadian Mineral Processors.

Suazo, C J, Hofmann, A, Aguilar, M, Tay, Y and Bastidas, G, 2011. Geo-metallurgical modelling of the Collahuasi grinding circuit for mine planning, in Proceedings of the Procemin Conference.

Optimisation and improvement

Managing froth tenacity at BHP Carrapateena

F Burns[1], J Reinhold[2], J Seppelt[3], S Assmann[4] and G Tsatouhas[5]

1. Superintendent – Metallurgy, BHP, Carrapateena, Adelaide SA 5950.
   Email: fraser.burns@bhp.com
2. Senior Metallurgist – Project, BHP, Carrapateena, Adelaide SA 5950.
   Email: jacqueline.reinhold@bhp.com
3. Processing Manager, BHP, Carrapateena, Adelaide SA 5950. Email: joe.seppelt@bhp.com
4. Technical Manager – Mining, InterChem, Abbotsford Vic 3067.
   Email: sassmann@interchem.com.au
5. Business Manager – Mining, Interchem, Abbotsford Vic 3067.
   Email: gtsatouhas@interchem.com.au

ABSTRACT

Carrapateena is an iron ore copper–gold (IOCG) deposit hosted in a brecciated granite complex located in the Gawler Craton, South Australia. The ore is processed by a conventional sulfide flotation concentrator, producing a copper gold silver (Cu-Au-Ag) concentrate with chalcopyrite and bornite as the main copper bearing minerals. The flotation circuit includes a fine regrind to below 20 µm by 2 × 1.6 MW Metso HIG mills and operates with hypersaline water drawn from local wellfields.

Since commissioning in 2020, Carrapateena has experienced operational issues from tenacious froth stability. Identified contributing factors include ultrafine particle size, flotation water quality and sulfide mineralogy mix of the ore. Froth stability has led to several challenges across the circuit, including froth transport in launders, hoppers; froth pumping performance; flotation cell control; concentrate thickener performance. Ultimately the need to manage froth in the flotation and thickening circuits often would constrain milling throughput.

Carrapateena has embarked on a multi-year journey to address these challenges involving the development of froth mitigation operating strategies; modifying and expanding plant equipment to better handle froth volumes, the design and installation of froth busting equipment; and partnering with the site principal reagent supplier to explore alternative and additional reagents to provide the options required for the operating teams to regain froth control.

This paper details the successes (and failures) of this froth management journey and outlines plans for the future.

INTRODUCTION

Carrapateena is a copper–gold deposit, located 460 km north of Adelaide, South Australia. Sulfide mineralisation hosted within the hematite breccias of Carrapateena is a combination of dominantly fine-grained disseminated to medium-grained chalcopyrite with a discrete high-grade zone of bornite; rare chalcocite, digenite, and covellite in hematite breccias. The initial Sub-Level Cave (SLC) footprint focuses initial mining activities to an initial bornite-chalcopyrite dominant core of the resource. Further expansion to full Block Cave (BC) operation is presently under development and will expand mining activities into an increasing portion of chalcopyrite and chalcopyrite-pyrite domains.

A sulfide flotation concentrator produces a Cu-Au-Ag concentrate for sale to copper smelters. Construction of the concentrator was completed in late 2019, slurry commissioning began in Q4 2019 with the first concentrate produced and filtered in December that same year. The concentrator, as commissioned in 2019, consisted of a rougher flotation bank, a High Intensity Grinding (HIG) mill for rougher concentrate regrind to a P_{80} of 20 µm, a Jameson cell as a cleaner scalper, and a three-stage conventional cleaning circuit. Jameson cell and conventional third cleaner concentrates were combined to form the final Cu-Au-Ag concentrate. Throughout 2021 and 2022, the concentrator flotation and regrind circuits were expanded and reconfigured to provide greater cleaning capacity and flexibility to uplift concentrate quality through improved rejection of ultra-fine grained (~8 µm average grain size) deleterious non-sulfide gangue (NSG) minerals. While the detail of the

Carrapateena flow sheet is presented by others (Reinhold *et al*, 2023), of note is the expanded regrind circuits can produce a P_{80} below 15 μm, which was assessed to be a likely future target size to deliver the desired concentrate quality. A simplified process flow diagram is presented in Figure 1.

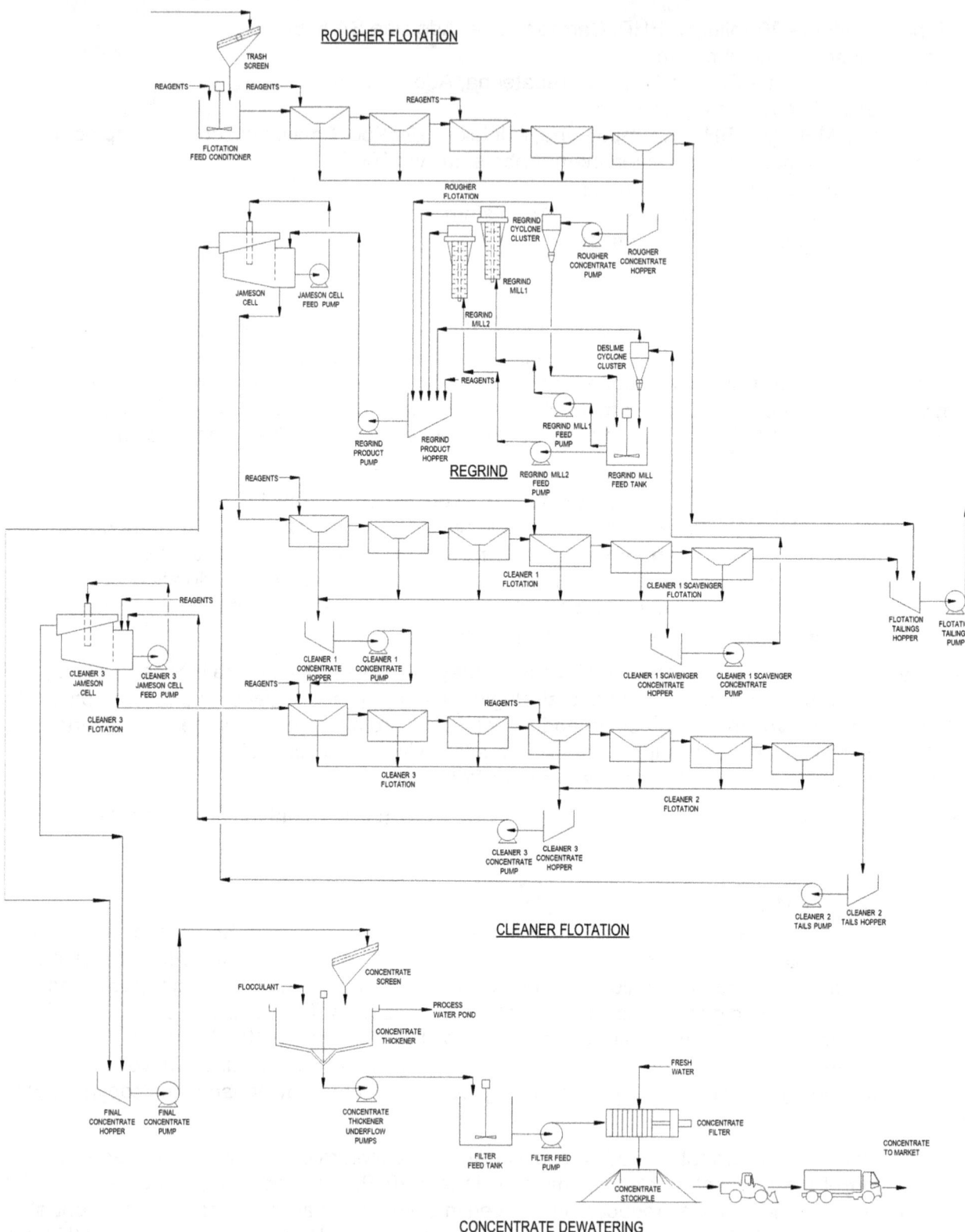

FIG 1 – Flotation process flow sheet.

As mining progresses to deeper sections of the orebody, particularly the BC footprint, the mill feed will see an increase in the proportion of chalcopyrite and pyrite. However, chalcopyrite and pyrite dominate ores are not restricted to later stages of the mine life and have been present in the mill feed since commissioning. Mining front progression under the SLC method is primary determined by cave flow requirements, when combined with a lack of surface stockpile facilities for blending, has led to a mill feed that can see material short-term variation of feed grades and sulfide mineralogy mix, from effectively bornite only to chalcopyrite only feed, and typical some blend in between.

Water for the flotation concentrator is drawn from two local well fields. The water from these sources are hypersaline and when combined with a negative site water balance, due to high evaporation rates, has led to process water with consistent total dissolve salt (TDS) quality of >80 000 ppm.

Since commissioning in 2019, Carrapateena has experienced operational issues from tenacious froth stability. Froth stability has led to several challenges across the circuit, including froth transport in launders, hoppers; froth pumping performance; flotation cell control; concentrate thickener performance. Ultimately the need to manage froth in the flotation and thickening circuits often would constrain milling throughput.

Identified contributing factors to froth tenacity include:

- ultrafine regrind particle size
- process water quality (salinity)
- specific sulfide mineralogy (particularly chalcopyrite dominance) of the mill feed.

This paper details the successes (and failures) of Carrapateena's froth management journey and outlines future plans to address these continuing issues.

OBSERVATIONS OF FROTH TENACITY AND ITS IMPACTS

Carrapateena ramped up to, and beyond, design throughput rates with a deliberate intent to test known, and to draw out unknown, plant bottlenecks ahead of the full ramp up of the mine supply. Accordingly, the impacts of froth tenacity were realised very early in Carrapateena's commissioning, this section details these early observations of how froth tenacity presented and the impacts these had on performance.

Early laboratory test work

Early indications of high froth stability at Carrapateena were first observed during pre-commissioning test work conducted in the site metallurgy laboratory. These tests were the first able to be conducted with real samples of site sourced raw water.

These tests included laboratory tests comparing site bore water to fresh water. In these tests, ore samples were ground in a 5 L rod mill to a P_{80} of 75 µm, and floated in a 5 L Alsto lab flotation cell using the Carrapateena's standard 10 min kinetic rougher flotation test procedure with Sodium Ethyl Xanthate (SEX) and Isopropyl Ethyl Thionocarbate (IPETC) as collectors, and frother dosed as required.

The rougher-grade recovery response of these tests, presented in Figure 2, demonstrated higher mass pulls and lower grades for the bore water tests over fresh water tests. Test notes indicated significantly more stable froth for bore water, with limited frother required compared to the fresh water tests.

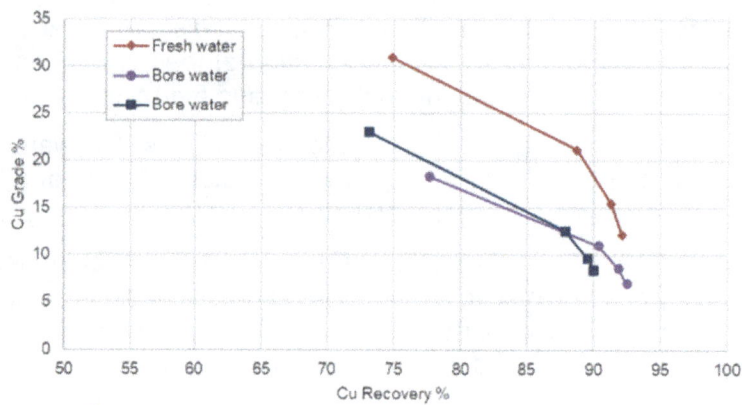

FIG 2 – Rougher flotation lab test – bore water versus fresh water.

Subsequently, a separate laboratory testing programme explored the impact of regrind size on cleaner flotation. In these tests, samples of rougher concentrate were reground to P_{80} targets ranging from 30 µm to 8 µm, by a lab stirred ceramic bead regrind mill, and then floated in a 5 L Alsto lab flotation cell using the Glencore Technology provided Jameson Cell simulation test procedure involving three stages of dilute (<10 per cent solids) mechanical cleaning.

Figure 3 compares the froth appearance during these tests at P_{80} of 25 µm and P_{80} of 8 µm. Test notes indicated that scraping of froth was unnecessary at the finest regrind sizes, and concentrate was resistant to breakdown in concentrate trays. It was clear from this work, that regrind size contributed significantly to froth stability.

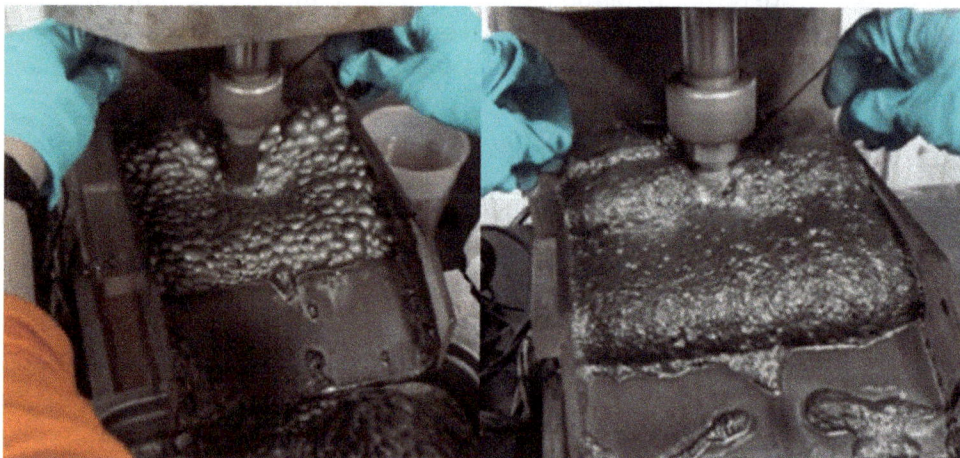

FIG 3 – Cleaner flotation lab test – effect of regrind size on froth stability: (left) P_{80} of 25 µm; (right) P_{80} of 8 µm.

Frother demand

An early indication during commissioning, was the very low required dose rate to achieve froth formation and stability within the flotation circuit in comparison to feasibility test work programmes. Secondly, there was little resolution in frother dose rate response, a significant impact on the cells was observed for minor dose rate changes. Jameson cell performance in test work relied on the optimisation of frother addition, but in the case of the Carrapateena Jameson Scalper, little to no frother was added in the field. This was a result of downstream froth stability and, whilst not optimised, reasonable scalper performance.

Water recovery

Mass balances of Jameson Scalper cell surveys also showed a large variance in water recovery in comparison to the design. The Carrapateena design mass balance anticipated a concentrate density at 30–35 per cent solids off the Jameson Scalper cell, however in reality the actual density was typically 10–15 per cent. This above design water recovery put notable unplanned pressure on downstream equipment, in particular the final concentrate pump and concentrate thickener, elevating the required duty points often resulting in spillage.

Regrind size

The relationship between froth stability to regrind size seen in early laboratory test work, was replicated at plant scale. During the regrind mill maintenance days, the change in flotation response was rapid and stark. Frother dosage requirements increased, water recovery reduced (although remained above design levels), and froth induced spillages reduced (although not completely).

The opposite responses were also noted when ultrafine regrind trials were held, where regrind product size set points were reduced from the nominal P_{80} of 20 μm down to P_{80} less than 15 μm. These trials were held with the intent of understanding concentrate grade and quality relationships with regrind size, and to identify regrind power capacity limits.

Chalcopyrite and mineralogy

The extent of froth tenacity was noticed to vary across time. As operating history accumulated, it became clear that the extremes of froth tenacity were almost exclusively correlated with chalcopyrite dominant mil feed. While not all chalcopyrite feeds would result in frothing episodes, but nearly all frothing episodes occurred during with a chalcopyrite dominant mill feed.

In efforts to identify mineralogical causes of the froth incidents, numerous samples were collected for various analysis. Of particularly interest here, was whether there were 'troublesome' trace minerals more prevalent in frothy versus non-frothy samples.

A series of samples were collected in 2022 of both final concentrate froth and slurry during three periods of high froth tenacity to compare the mineral abundance and sizing of particles present in the 'top of froth' phase versus the 'whole of slurry' phase. Samples were air dried, sized and dispatched to QUT for quantitative X-ray diffraction (QXRD) and clay analysis.

Comparing the P_{80} across the three sets of samples, on average the froth samples finer compared to their respective slurry sample, with a 5 per cent decrease in mass observed in the +20 μm size fraction and a 6 per cent increase of mass in the -C5 fraction. This further supported the lab test work that a finer particles size promote froth stability. Review of the copper to sulfur ratios, confirmed by the mineralogy, showed a larger portion of chalcopyrite present across all size fractions in the froth phase versus the slurry phase. Only trace amounts of clay were present in any of the samples however large portions of unidentifiable, 'amorphous' material were present, the proportions varied across all samples and size fractions between 10 and 39 per cent.

Shift samples, one set from a particularly high froth stability day in the rougher bank, were submitted to Bureau Veritas for QXRD and amorphous content determination. Field observations showed the concentrate to be particularly brown or red in colour suggesting an increase in hematite content and poor copper grade. Upon analysis, the concentrate grade was no poorer in quality during the high froth stability shift, and like previous results, a higher proportion of chalcopyrite was present in the feed and concentrate compared to the lower frothy shift. Further detailed analysis in comparison to the QUT results was able to identify the presence and quantity of other minerals such as halite, chlorite, Mica, Dolomite and K-Feldspar. The largest proportion of which was Mica at 10 per cent in the flotation feed in both the frothy and non-frothy days. All samples still contained between 12 and 20 per cent of further amorphous material and the higher froth stability shift did not yield a higher amorphous content.

Ultimately, the various QXRD studies yielded inconclusive leads on identifying a specific potential troublesome trace mineral(s) in the concentrate.

Froth transport

Managing the tenacious froth when being transported from one unit operation to another became a significant challenge; and notably worsened as the concentrate progress through the flow sheet to final concentrate. The Carrapateena circuit was design and constructed with numerous standard froth mitigation design. Deep concentrate launders, tangential hopper entry nozzles, large continuous bore pump suctions, selection of froth centrifugal pumps, typically from the Weir Warman AHF series.

Despite these standard and sensible measures, froth spillage was omnipresent. Concentrate flowing in launders would not fully breakdown when flowing and accumulate till the point of spilling at any obstruction, such as downcomers from additional cells, or launders transitions into the hopper entry nozzles.

Concentrate hoppers would often overflow with froth, level indicators would register a high level, only for the pumps to speed up in response, but show little impact on pumped flow rates. Review of flow and pump power draw data indicated the concentrate flow rate would often be erratic and below set point with the pump speed increasing or already at maximum, yet still not achieving target flow. In these instances, hoppers were found to be filled entirely with froth, rather than slurry, affecting pump efficiency. This was most routinely observed in the Cleaner 2 concentrate and Final concentrate hoppers.

During severe racing episodes, with a duration range of a singular hour, to numerous shifts, milling rates would be lowered: firstly, during the episode, to reduce flows to within the transport capacity of the circuit; and secondly, after the episode had passed, to provide capacity for the circuit to accept sump returns from the subsequent housekeeping efforts.

Pump wear was found to be higher than anticipated for the concentrate pumps. On several occasions, the final and rougher concentrate pumps failed well before their minimum maintenance frequency of 13 weeks. Failures were due typically to cavitation pitting and erosion, and eventual failure of pump impellers. Higher cost hyper-chrome alloys were quickly pursued to improve wear life. An additional challenge to managing reliability of the pumps, was that the frequency and severity of frothing episodes could vary from maintenance campaign to campaign and would often lead to contradicting wear measurements during inspections and ultimately difficult to predict expected wear life for the wetted pump components.

Concentrate thickening

The impacts of froth tenacity were most severely felt at the concentrate dewatering circuit, centred around a Outotec 15 m High-Rate Thickener fitted with 2 m diameter vane feed well and 1.2 m feed deaeration tank.

When a frothing episode was underway, a layer of froth, which would often be described more suitably as fine bubble foam, would form on the entire surface of the thickener, such that no clear water would be visible. The auto dilution port, which should have enabled clear surface water to enter the feed well to support flocculant dilution, often had froth passing the opposite direction out of the feed well and was a major source of froth to thickener surface.

The thickener was fitted with a froth baffle designed to prevent froth from leaving the surface of the thickener into the overflow launder, but this was largely ineffective. Froth built to such a height that it would either overflow the top of the baffle, or to such a depth be drawn from the bottom of froth layer, and pass with the clear water to the overflow launder. At its worst, the thickener could be seen having froth overflow the launder onto the ground below around the entire circumference.

Once froth had passed the baffle to the overflow, these solids accumulated, in either the:

- overflow launder, constraining launder capacity
- lowest horizontal section of the buried gravity pipeline that returned supernatant to process water dam, restricting flow capacity of this pipeline
- process water dam; restricting live capacity of the dam.

Ultimately these combined effects restricted water removal from the thickener, raising the liquor level in the thickener, which when combined with lower than anticipated thickener feed lower %solids, had two significant impacts:

- Increasing hydraulic back pressure on the gravity feed thickener feed system, resulting in spillage from the thickener feed hopper

- Spillage of clear liquor out of the thickener overflow pipe inlet box to the bund was commonplace Figure 4.

FIG 4 – Supernatant spillage from the thickener overflow pipe inlet box.

Flocculant consumption required to achieve adequate settling rates proved in practice to be well above those anticipated by design, or any test work since on deaerated slurry. For instance, dynamic thickener test work completed by both the equipment and flocculant supplier laboratories on completely deaerated, 'stale' concentrate samples typically only required dose rates of 20 g/t to achieve 65 per cent solids underflow density. Whereas, flocculant tests on fresh aerated concentrate would require three times as much, refer Figure 5, to achieve good clarity and the greatest compaction. The in-practice implication of these high dosing rates, which could be five to six times laboratory, is that flocculant pumping capacity was rapidly exceeded. Several flocculant pumping upgrades were completed to satisfy the demand.

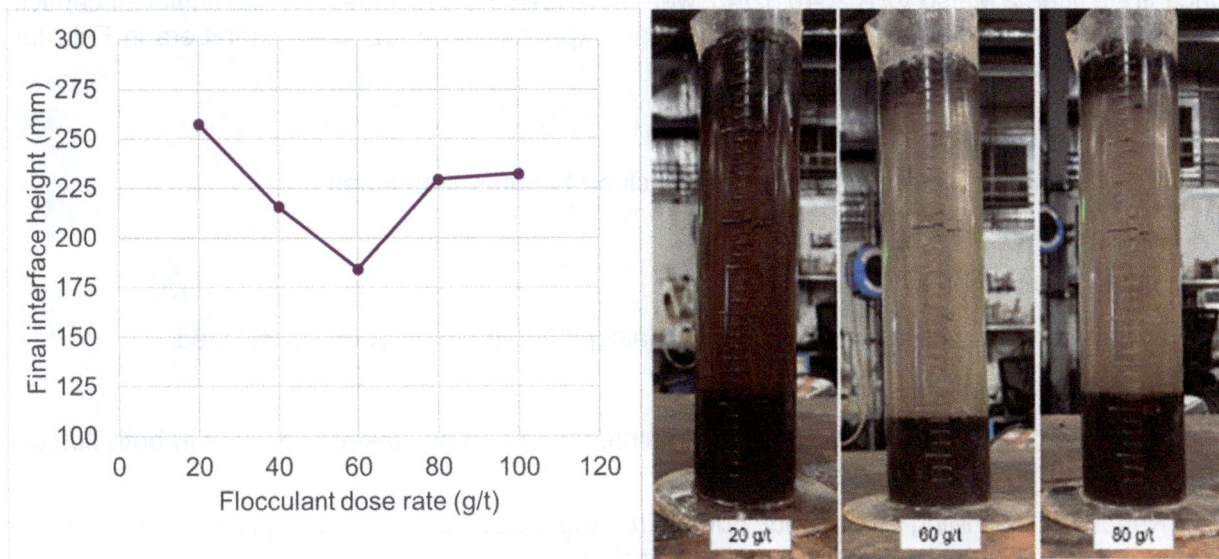

FIG 5 – Thickener feed flocculant dosing rate testing: (left) Final compaction versus dose; (right) final clarity versus dose.

Thickener samples collected during frothing episodes, are shown in Figure 6. Thickener overflow samples displayed clearly flocculated particles still in suspension with a 'fluffy' appearance. These particles could remain suspended in, or float to, the surface even after hours in a still container. Signifying that much of the thickener clarity issues were related to entrainment of buoyant flocs that contained entrapped air in the rising water. This air entrapment was suspected to be likely a driver for the excessive flocculant rates required to achieve acceptable clarity.

FIG 6 – Aeration of concentrate thickener: (left) underflow sample; (right) 'Fluffy' solids in overflow samples.

Eventually, operations technicians operated the thickener with a significantly lower than normal bed mass to compensate for poor settling and regularly varying feed stream into the thickener. The lower bed mass gave the thickener more freeboard and time to accept a disturbance due to a sudden froth tenacity or feed rate change, before the onset of a poor clarity overflow. However, this, combined with significant air entrapment in the underflow (Figure 6), led to a thickener underflow typically at 50 per cent solids compared to the designed value of 65 per cent solids, needed for efficient operation of the concentrate filter downstream. At this stage, thickener operation had become more about managing froth formation and reducing spillage than it had producing a suitable slurry underflow density for filtration. A result of the lower thickener densities is that filling durations of the concentrate filter, a Metso VPA 2040 32/40, were extended; however the filter had sufficient capacity to accept longer cycle times, particularly after an expansion from 32 to 40 chambers in February 2021.

Summary of factors affecting stability

Identified contributing factors to froth tenacity include the aforementioned:

- ultrafine regrind particle size
- process water quality (salinity)
- specific sulfide mineralogy (particularly chalcopyrite dominance) of the mill feed.

Impacts of this forth tenacity included:

- Regular spillage, leading to excessive maintenance and housekeeping effort in both flotation and concentrate dewatering circuits.
- Reduction of milling throughput during frothing episodes to cease spillage; and/or return acceptable concentrate thickener overflow clarity.
- Loss of concentrate solids to process water dam sludge; requiring routine dam dredging.
- Increased pump wear, and with less predictable wear life, requiring higher cost wear components.

FROTHER OPTIMISATION

Carrapateena commissioned using InterChem INTERFLOAT F308 as its frother. This reagent was selected as it was equivalent to the frother used throughout Carrapateena's laboratory development studies, where it provided adequate performance. Further it was equivalent to the frother in use as the nearby BHP Prominent Hill concentrator treating a similar IOCG ore.

A commissioning progressed, it became evident that excessive frothing plagued the cleaner circuit. Further within the rougher circuit, the sparing use of frother was required to generate required froth stability. Often the entire flotation circuit was operated for extended periods, without any frother addition. For instance, the minimum frother addition was often considered to generate froth characteristics that were too stable. Consistent feedback from operations technicians was that F308 was not providing the sufficient resolution of dosage response to effectively control stability.

Accordingly in early 2020, existing frother supplier, InterChem, was engaged to identify an alternative frother that would return a degree controllability to frother dosing.

Frother trials

The frother assessments were completed only at plant trial level, without prior lab screening, as most froth stability issues only emerged at scale. Trials involved operating the plant with complete replacement of the incumbent frother. Trial assessment was largely qualitative in nature, via visual assessment of froth build-up, feedback from the field technicians and monitoring frequency of total operation of the flotation circuit without frother.

The first trial commenced in April 2020, where the first weaker frother F236N replaced F308 over a trial period of five days. Observations during the trial were:

- Cleaner 2 and Cleaner 3 concentrate launder no longer fully engulfed in froth.

- Concentrate hoppers frothing and spillage reduced, but not removed:

  o Cleaner 2 concentrate hopper no longer frothing overflowing.

  o Final concentrate hopper still required sprays to break the froth.

- Froth stability in Rougher and Cleaner 1 banks was maintained, but with higher dose rates.

- No change to Concentrate Thickener frothing issues, with continuous froth sprays still required.

Review of dosage rates during this trial, revealed an increase in dose rate up to 50 per cent (from an admittedly low base) at the rougher bank was required to maintain froth stability. Further, dosing to the Cleaner 1 Jameson cell, was required where previously it was largely not utilised.

Whilst these improvements were noted; none of the major operational issued were entirely resolved, as such another trial was coordinated with a further weakened frother blend. In the meantime, F236N was incorporated into the site's reagent suite, replacing F308.

The second trial was held in Oct 2020. The trial frother was a blend of the in-use INTERFLOAT F236N, with increasing dilution by INTERFLOAT F100. This blending increased the alcohol content, paired with a reduction of the glycol content. A similar result was achieved as the previous trial. Improvements were noted, but not to the extent that was desired, and a further trial was coordinated with InterChem.

This third trial was commenced in Jan 2021 and concluded in Feb 2021, using INTERFLOAT F228. This frother further weakened the blend, with primary change being the remaining glycol components moving fully to lower molecular weight glycols. This reagent was equivalent to the frother in use at the nearby BHP Olympic Dam concentrator. The trial again showed improvements in minimising excess froth, with higher dose rates (Figure 7) indicating more dose response resolution available to the operations technicians.

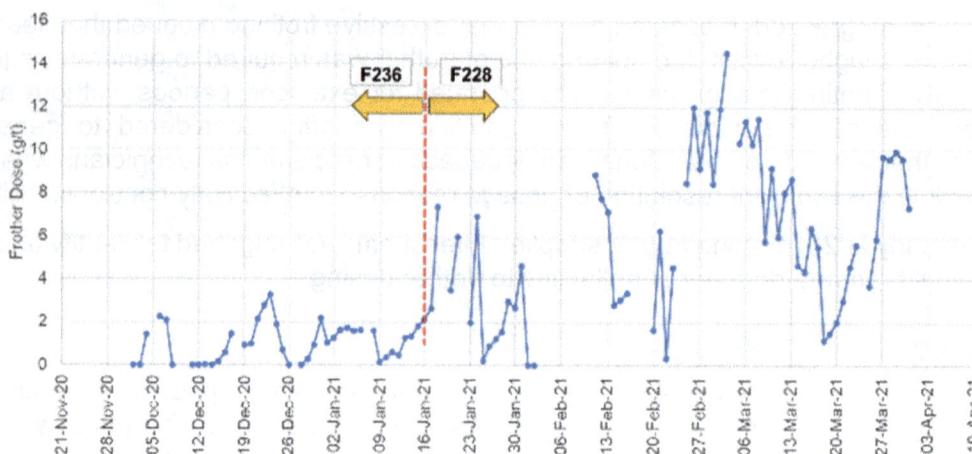

FIG 7 – F228 frother trial – total dose rates (g/t).

Frother optimisation outcome

Ultimately the decision was made in Feb 2021 to transition permanently to the F228 product. This final blend:

- Represented a frother blend, as near as possible to a pure alcohol based frother; without transitioning to a dangerous goods classified product.

- Markedly improved the frother dose rate response characteristics for the flotation technicians, allowing frother to be used more consistently across the circuit.

- Reduced frequency and severity of frothing events, particularly of hopper and launder transport issues.

However, this did not represent a full resolution of froth tenacity issues at Carrapateena. Major frothing episodes continued despite the ongoing use of a weaker frother and separate actions were required to further address these events.

ADDRESSING FROTH HANDLING AND TRANSPORT

With froth tenacity clearly becoming a persistent, long-term challenge for Carrapateena, without a clear line of sight on a method of prevention; attention and effort were placed on adopting mitigations, especially relating to spillage and housekeeping of the plant.

Hopper and launders

Concentrate pump hoppers and gravity flow of concentrate in launders and pipes were the obvious manifestation of stable froths as spillage. Frothy, aerated flows in launders often exceeded the lauder capacity, with froth accumulating at any 'pinch points' or restrictions in the launders. Aerated flows in gravity pipes, often became air-locked due to the progressive accumulation of gas pockets at, particularly at pipe reductions or on long near-horizontal runs. Unimpeded accumulation of froth on top of the slurry surface in hoppers inevitably ended in spillage due to either simple froth build-up to the overflow invert, or a loss of pumping efficiency due air presence at the pump.

Modifications to mitigate these issues involved the installation or froth spray bars on hoppers and launders; and breathers on pipework where air locks were present.

Innumerable spray nozzle types were trialled for various effectiveness, from repurposed fire hose nozzles through to duck feet nozzles. Ultimately the use of circular hollow cone, eg pig tail nozzle, gained favour and widespread adoption in this duty. Similar spray bar designs varied widely, but again iterated to a preferred design that facilitated maintenance of spray heads – typically of HDPE pipe construction to reduce weight, and hose connected for ease of movement. Further designs that placed multiple sprays at differing heights above the slurry level were implemented, so the sprays were effective across wide range of hopper levels, which vary over time.

Additional breather retrofits were targeted on any long gravity flows pipelines, or any lines where a pipe reduction existed, including obvious candidate such as final concentrate pump suction pipe, concentrate thickener feed pipe, and also unexpected froth capture streams pipe including primary cyclone overflow transfer pipe, and concentrate thickener overflow pipe.

Spray bar and breather installations while reasonably effective, however led their own issues. Sprays generated significant hypersaline mist with concentrate that tended to quickly cover all nearby equipment in salty concentrate scale. Breathers, while being initially sized correctly to not hydraulically overflow, would often themselves be subject for froth build-up and eventual froth spillage. Hoppers lids were designed and fitted to various concentrate hoppers to contain overspray mist (Figure 8) and breathers extended to discharge to appropriate nearby locations/process vessels (Figure 9).

FIG 8 – Hopper modifications: (left) spray bar; and (right) retro fitted lid.

FIG 9 – Breather installation: (left) long shallow con thickener overflow pipe; (right) before reducer on con thickener feed pipe.

Instrumentation and process control to assist pumping

To tackle the loss of final concentrate pumping efficiency, during frothing events significant manual water addition into the hopper was used by operations technicians as a tried-and-true measure. The water addition often allowed the pump to regain some efficiency and improve pumping flow rate, occasionally sufficient to regain control of hopper level. A downside of this technique lay with additional flow reporting to an already stressed, at-capacity concentrate thickener, exacerbating any issues presenting there. The amount of water added in this manual technique was decided by the area technician's best judgement, and required routine monitoring to ensure water wasn't still being added after a frothing episode had past.

In order to help balance delivery of improved pumping performance; and the addition of minimal water to not unnecessarily overload downstream capacity, a water-assisted pumping control philosophy was devised to manage the water addition (Figure 10).

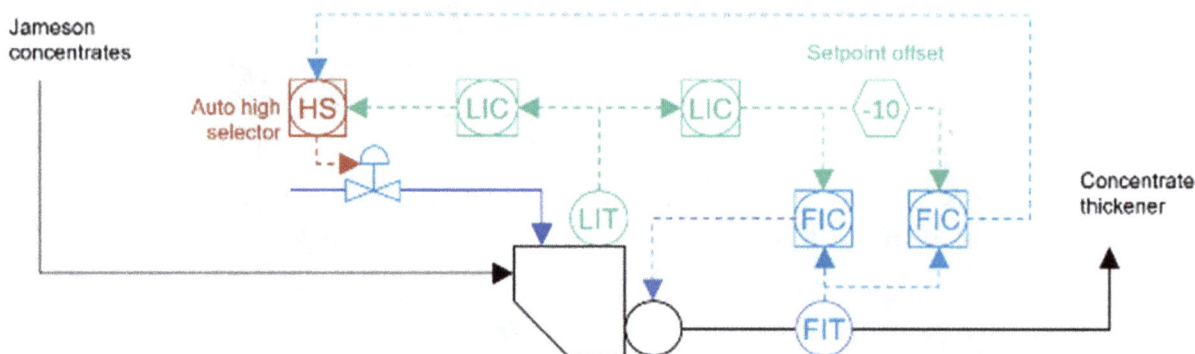

FIG 10 – Concentrate hopper control philosophy – with water assisted pumping controller.

First, in June 2021, a flow metre was installed on the final concentrate piping. A PID controller was built in to control the flow by manipulating pump speed. This primary flow set point was cascaded from the primary hopper level controller. A second flow controller was also configured, that manipulated water addition valve to achieve a secondary low flow set point that was offset below the primary set point by a fixed value, typically 10 m³/h. When pumping efficiency was lost, invariably the achieved concentrate flow would deviate below the primary set point, once the offset secondary set point was reached, water would be automatically added to the hopper.

The benefit of the secondary controller was only the minimum required amount of water to assist pump was used. Further, once pumping efficiency improved the flow would increase above the secondary set point, causing the water addition to reduce until fully ceased when no longer required. This addition of water by the secondary controller often provided early indication of a froth build-up in the hoppers and was a preventative measure rather than a reactive one following froth spillage from overflowing hoppers.

Concentrate pumping capacity upgrades

A further avenue pursued to reduce hopper spillage was to upgrade pumping capacity of the Cleaner 2 and Final concentrate pumps. Intent was to provide greater raw capacity to overcome, at least to some extent, degraded pumping efficiency. Both pumps underwent (multiple) motor and pulley upgrades to increase the upper capacity of the units (Table 1).

TABLE 1

Concentrate pump upgrade details.

	Cleaner 2 concentrate pump		Final concentrate pump	
	Original	Current	Original	Current
Motor power rating (kW)	7.5	15	18.5	30
Max impeller speed (rev/min)	1245	1380	920	1035
Maximum flow (m³/h)	50	90	85	220

Continuous Air Removal System (CARS) pumping

In addition to upgrading raw pumping power, upgrades to improve the efficiency of froth were explored in partnership with the manufacturer Weir Minerals. These centred on the modification of the final concentrate and cleaner 2 concentrate pump design.

The Continuous Air Removal System (CARS) upgrade to the standard Weir Warman AHF froth pump removes gas from the froth in a staged process, see Figure 11. Firstly, the flow inducer blades of the impeller help draw the froth slurry into the pump while, at the same time, inducing slurry pre-rotation in the suction piping. Pre-rotation generates an initial air-slurry separation, where air accumulates at the centre of suction pipe. Secondly, the centralised air passes through the vent holes in the back shroud of the impeller, a feature unique to the QUAR impeller used in CARS pumps. Finally, a flow inducer, which is an expeller located in gas collection chamber behind the main impeller, moves the collected air through a vent pipe, typically directed back into the pump hopper.

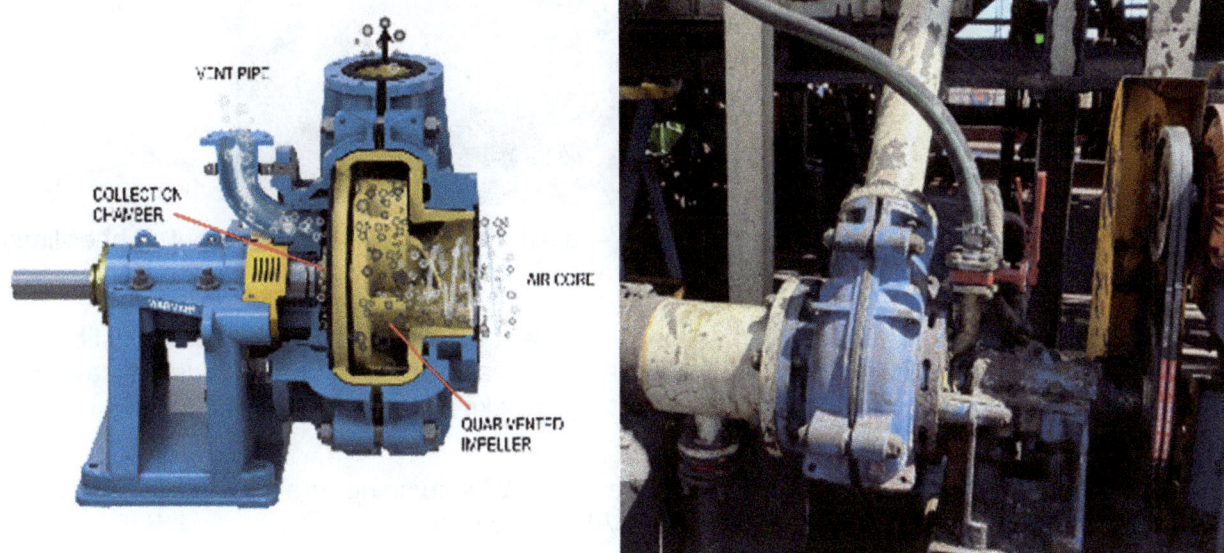

FIG 11 – CARS pump: (left) diagram (image provided by Weir Minerals Australia Ltd); (right) initial installation.

A trial conversion was sourced for the Warman 4AHF final concentrate pump. To accelerate the trial start date, components were purchased in then off-the-shelf materials. In particular, the QUAR impeller was supplied in a standard wear alloy, whereas a hyperchrome alloy was the incumbent material to combat high rates of air induced cavitation erosion, and the collection chamber was sourced as a metal lined part. The potential for an early failure – defined as failure prior to the standard 13 week maintenance shutdown cadence – was an acknowledged and accepted risk of the trial, in order to bring forward an understanding of the pumping performance improvements.

In September 2021, the conversion kit was put in to service, and clearly provided a pumping efficiency boost. Table 2 presents a comparison of two 6 hr snapshots of pump performance during frothing episodes, the presented comparison was considered as characteristic of the performance during the trial. The CARS pump typically was able to achieve greater flows, at lower pump speed and motor power draw, while requiring less or no assistance water to dilute the froth.

TABLE 2
CARS pump performance during trial – average of 6 hr period.

		AHF pump	CARS pump
Concentrate flow	m³/h	121	128
Pump speed	% speed	92	77
Dilution water valve position	% open	34	0

However after two weeks of service, the pump developed a leak through the collection chamber. Upon inspection, Figure 12 (left), the unlined collection chamber had suffered severe erosion, leading to the premature equipment failure. Interestingly, the standard alloy QUAR impeller showed no signs of wear, and presented as new. Given the pumps qualitative improvement in performance, Weir were engaged to improve the wear resistance for the collection chamber, eventuated in a rubber-lined collection chamber for additional trials. The rubber-lined chamber performed exceptionally well and has since regularly completed 26 weeks of service, Figure 12 (right).

FIG 12 – CARS collection chamber: (left) unlined – failed after 14 days service; (right) rubber-lined – planned replacement after 6 months service.

During the subsequent trial, the QUAR impeller also present with improved wear characteristics. The continuous removal of entrained air from the impeller appeared to have greatly reduced cavitation erosion, and the life of the impeller was extended to 26 weeks even with the standard alloy material in use.

After adoption and continued good performance of the CARS pumping arrangement, a conversion of Cleaner 2 concentrate pump, a Weir Warman 3AHF, was instigated.

DEFOAMER ADOPTION

Chemical means to reduce froth stability became a focus of attention from mid-2021. Initially this effort was focused on addressing possible mineral instigators of froth stability through use of dispersants, particularly the presence of ultra-fine non-sulfide gangue minerals, for instance clay-like specular hematite which was known to be present in patches throughout the ore was targeted in this phase. This pivoted to then modifying the surface tension of the water-air interfaces through use of defoamers, which was felt to be more effective against a broader range of possible causes for froth stability.

Dispersant trial

Sodium silicate was selected for a dispersant plant trial after initial sighter laboratory tests indicated some potential improvement in NSG rejection from the concentrate. The dispersant trial intent was to mitigate froth stability that was potentially being instigated by the presence of ultrafine (particles less than 5 μm) NSG minerals.

A trial was held in July and August 2021, and involved dosing of 1.4 g/L sodium silicate into the regrind product hopper which in turn fed the Cleaner Scalper Jameson feed stream. This location was selected due excessive froth stability primarily (but not exclusively) being observed in the Jameson concentrate.

To conserve the limited amount of on hand dispersant stocks, no dosing occurred during normal running, and was only commenced upon the onset of the frothing 'episode'. Initial dose rate was set at ~500 g/t, this was rapidly increased to 750 g/t during the first frothing episode, after no response was observed at the initial dosing rate.

Across the month-long trial, several frothing episodes were encountered. Unfortunately, no consistent and repeatable improvement in froth transport characteristics was observed when the dispersant was dosed. Similar laboratory flotation tests (using the standard Jameson Cell simulation procedure) on Jameson feed before and after dosing commenced presented no repeatable improvement concentrate quality. Consequently, dispersant trials were abandoned after the full consumption of all stock on hand for the trial.

Defoamer trials

After completion of the dispersant trial, focus pivoted towards chemically destabilising froth regardless of cause mechanism utilising defoamers. Defoamers have an affinity for the air-liquid interface and spread rapidly on the foam surface. This destabilises the foam lamellas, which causes rupture of the air bubbles and breakdown of surface foam (or froth in flotation applications). Defoamers generally work by either disrupting the foam structure in the stabilising film via increased drainage and creating disorder at the interface (like selected InterChem INTERFOAM® defoamers) or by using hydrophobic particles to bridge the liquid channels (lamella) between the bubbles (eg wax dispersion and oily droplet defoamers) rupturing the film causing foam collapse.

Several defoaming agents, both in market and alternative, were explored in both lab and plant settings. Defoamers sourced for assessment included Solvay Oreprep D275, Interchem INTERFOAM A, and fusel oil (sourced as by-product from local wineries).

An initial screening test program dosed defoamer to the start of a standard concentrate settling tests. This quickly identified that fusel oil in fact exacerbated frothing, with almost all flocculated concentrate solids forming a floating foam layer rather settling in standard cylinder tests, refer to Figure 13. This product was not taken further. The Oreprep and INTERFOAM products were taken forward to plant trial, with both reagent families showing good performance in lab testing.

FIG 13 – Final concentrate settling tests with various defoamer products added.

First plant trials commenced in November 2021. Initial dosing location was to the concentrate thickener feed box. Dose rates were set by the area technician, based on observations of froth build-up on the thickener surface or of froth overflowing the thickener feedbox. Trial assessment was largely qualitative in nature, via visual assessment of froth build-up and feedback from the field technicians.

The initial trial was conducted with the INTERFOAM A. This product rapidly reduced froth buildup on the concentrate thickener, and reduced frequency of overflows of the thickener feed box. However, the positive results did require high dose rates (+500 g/t concentrate) which quickly consumed the trial quantities of the product. Subsequently, a second trial was conducted dosing Oreprep D275. This product performed similarly well, but also at high dose rates. Upon considering the similar performance, D275's dangerous goods classification, and D275's more limited supply availability (at that time), a decision was made to progress further trials only with INTERFOAM product line.

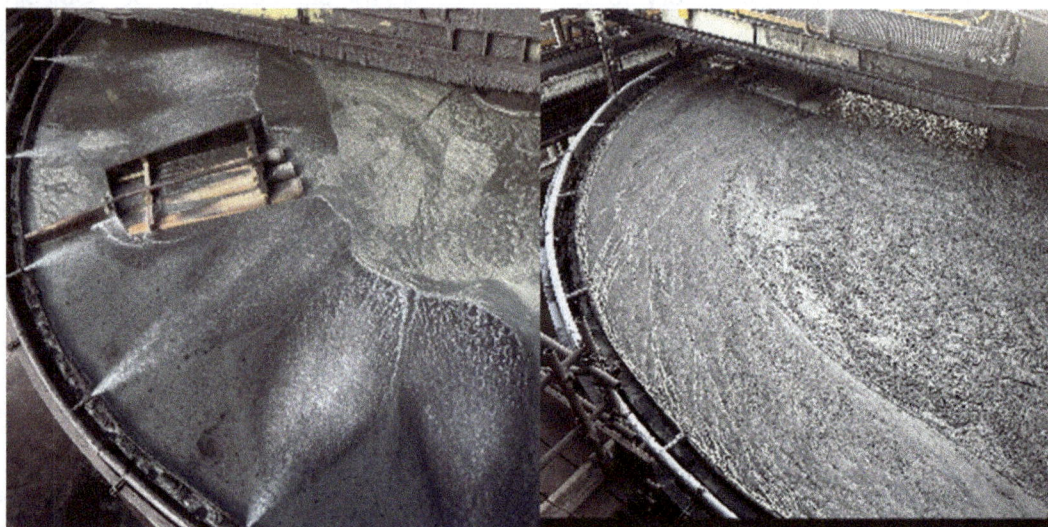

FIG 14 – Final concentrate thickener surface during frothing episode: (left) With defoamer; (right) without defoamer.

In early 2022, to reduce required dose rates, an alternative defoamer blend, INTERFOAM B, was sourced for further trials.

Dispersibility is an important consideration, the defoamer needed to migrate to the surface of the bubble to work, but if too incompatible with the contact fluid it will become less useful. Alternatively, if highly compatible it would be highly dispersed in the fluid and not control foam as effectively and may contribute to foaming. InterChem's INTERFOAM A is ester based while INTERFOAM B is a glycol type, both offering effective foam control additives at different performance degrees but differ in terms of level of hydrophobicity. A is more hydrophobic than B and won't disperse in water. It's lower solubility in water should, in most cases, create a better foam control product at different dosage requirements. In contrast, B will disperse in water and is also soluble in some oil-based materials.

The later trials of INTERFOAM B reaffirmed the effectiveness of defoamer of reducing frequency and severity of frothing episodes, with additional indications that defoamer affected thickener performance by improving clarity and increasing thickener underflow densities.

During this trial the dosing location was relocated upstream to the final concentrate hopper, with the intent to alleviate frothing at the hopper, improve pumping performance and allow additional conditioning and mixing time for defoamer with the thickener feed slurry. Addition to the hopper was a successful change, Figure 15 presents the onset of defoamer dosing having an immediate impact on hopper level and pump performance. However, relocation of the doing point to a location further away from the reagent storage and dosing pumps, uncovered challenges related to the high viscosity of defoamer. During the trial, pump and tubing failures were not uncommon due to high line pressures required.

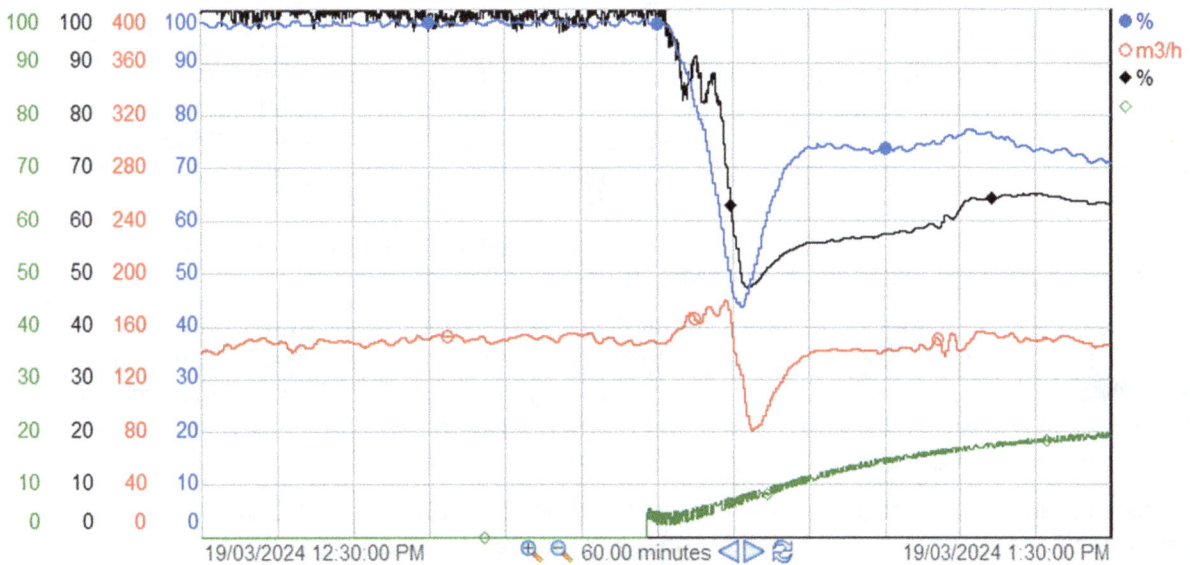

FIG 15 – Final concentrate hopper: Level (% – Blue), Pump speed (% – Black) and Flow (m³/h – Red) with Defoamer addition (L/hr – Green).

During trial planning, an identified concern was whether defoamer would build in the process water, ultimately affecting flotation circuit performance. Discussion with suppliers indicated the defoamer active components had relatively short life after dosing and would be expected to 'burn off' within typical process water dam residence time. Further during the pretrial laboratory screening phase, minor defoamer additions – representative of recycled content in process water – to standard rougher flotation test indicated no observable impact on recoveries. Ultimately, throughout the trial period no impacts on flotation recovery were noted.

Defoamer implementation

From March 2022, the decision was made to permanently incorporate defoamer as part of the Carrapateena reagent suite. Initially, dosing was directly from 1000 L IBC totes. Through 2023 engineering and installation was completed of a permanent storage tank and IBC decant stand; associated control system connected dosing pumps and flowmeters; and emulsifying static mixers.

FIG 16 – 3D model of defoamer storage and pumping area.

To counter the impact of high defoamer viscosity noted during the trial phase, this permanent dosing arrangement included static mixing and dilution system to emulsify the defoamer with process water. Pump selection was changed from typical diaphragm dosing pump that was used for the trial, to higher pressure rated peristaltic dosing pumps. This proved to greatly relieve stress on the dosing pumps and piping system and is believed to improve dispersion of the defoamer within the concentrate slurry. A further change with the implementation of the permanent system, was the dosing location was changed from the final concentrate hopper, to the two Jameson cell launders that the feed in to this hopper. Again this change in aid of improved mixing and reducing froth entrapment during the passage of slurry through pipe work to the hopper.

More recently, defoamer dosing was extended to the Cleaner 2 concentrate launder in February 2024, due to routine overflowing. Dosing to this location, which is within the flotation circuit, is used sparingly at very low dose rates, and only when truly required. This is to reduce impacts on the immediately downstream Cleaner 3 Jameson cell. Moderate dose rates have been noted to have a visible impact on the appearance of the froth in that cell.

The permanent installation is integrated with the plant control system (PCS) and is configured to dose defoamer as a g/t ratio with the final concentrate solid t/h. Operating technicians regularly adjust this set dosing rate based on observation of the concentrate thickener for froth loading. Dosing rates greatly reduced during non-frothy periods due to manage reagent costs. Figure 17 presents how this dosing rate is dependent largely on the prevailing sulfide mineralogy. Lower defoamer rates are typically required for treatment of bornite dominate ores, while dose rates often increase in excess of 1000 g/t during extended froth episodes, which most often coincide with chalcopyrite dominate ores.

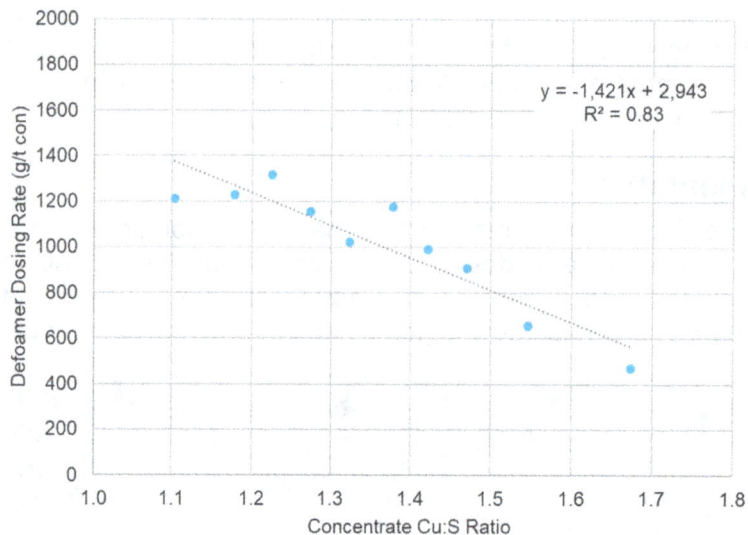

FIG 17 – Average shiftly defoamer usage response to concentrate sulfide mineralogy (Oct 2023 – Mar 2024).

However, it is also clear that significant defoamer is still required across the entire operating range, with an average of 400 g/t even on bornite dominant ores. The use of defoamer has become an integral part of the reagent suite at Carrapateena. Particularly as milling rates continue to increase well past initial design rates, defoamer dosing underpins the lifting of concentrate transport and thickening constraints on concentrator production.

CONCENTRATE THICKENER FEED SYSTEM MODIFICATIONS

The concentrate thickener feed system, as built consisted of several elements presented in Figure 18. First, the final concentrate discharged into a sampler feed box and its subsequent multistage metallurgical sampler. Sampler discharge passed on to vibrating screen deck, designed to remove any rogue coarse material ahead of entering the concentrate thickener and filter. The screen undersize passed into wide, but short, undersize chute. From this chute, the slurry was required to flow under gravity to the in-thickener deaeration tank, and finally into the thickener's feedwell.

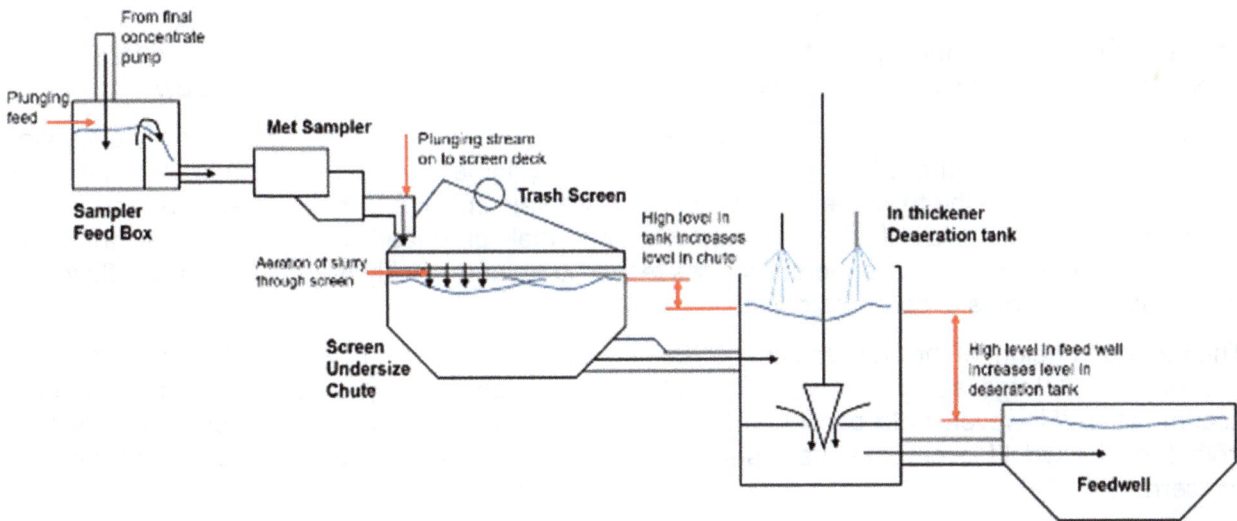

FIG 18 – Thickener feed system – prior to modifications.

This original feed system has since been thoroughly modified with several aims:

- Adress critical circuit flow bottlenecks from initial plant design; and expand flow capacity commensurate with both lower than design concentrate % solids and increased throughput requirements.

- Introduce additional concentrate deaeration stages.

- Minimise additional air entrainment into the concentrate as is passes through the feed system.

Sampler feed box

The sampler feed box accepts the pumped slurry from the final concentrate pump. This slurry enters via top entry nozzle, creating a plunging jet into slurry pooling behind the feed box's internal weir. Observations indicated the vertical drop was reaerating the slurry and froth was again being formed in the first part of the sampler. This feed nozzle was extended such that the inlet was now submerged beneath surface of pooled slurry.

Final concentrate metallurgical sampler

An Outotec MSA 1/40-H multistage sampler collects sample cuts of the final concentrate stream for assaying and accounting purposes. There were several flow mechanisms in the sampler that were identified as likely contribute to slurry aeration. These actions were deemed necessary to complete accurate and robust sampling, thus left unmodified. However, the outlet nozzle of the samplers was modified. High, flooding, slurry level in the sampler discharge chamber was often observed. The outlet nozzle size was considered a capacity constraint, and importantly, also viewed to be transferring slurry at too high velocity into the next stage of the feed system, generating an reaeration risk. This nozzle was removed and replaced with wide sloping outlet launder to both increase capacity, preventing flooding within the sampler and to reduce the speed of slurry entering the next stage of the feed system, which was viewed to contribute to slurry aeration.

Thickener feed screen removal

Two issues were quickly identified with the thickener feed trash screen. Firstly, the undersize chute was a regular source of spillage. When liquor levels rose in the thickener, feedwell and deaeration tank, this would transfer to a higher slurry level in the undersize chute. Even in early 2020 commissioning activities when operating at design rates, this chute was a near constant source of spillage due to both froth accumulation or slurry backing up from the thickener. Further the vibrating nature of the screen was viewed to be source of additional aerating the final concentrate.

After a few months of operation with no material reporting to the trash screen bunker, it was determined that the screen could be removed to reduce the opportunity of re-aeration of the final concentrate slurry at a low risk of equipment damage because of coarse material. With the screen removed, the underflow launder became for all intents and purposes, a concentrate thickener feed hopper. To attend to the spillage of this hopper, the walls of his hopper were increased by approximately 1m, with a dedicated overflow line that was directed to the ground below.

The overflow line was introduced to control the spillage, such that it would report directly a location near a sump pump, and not fall over the surrounding equipment. A trial was made to redirect this overflow directly in to the thickener, in an attempt to contain spillage within vessels, and reduce the excessive housekeeping effort. Unfortunately this trial produced likely the worst thickener performance ever observed by the operations technicians, and quickly reverted to an overflow pipe to ground within hours of commissioning.

The extended height of the hopper did successfully eliminate the overflow of backed-up slurry from this hopper. A large spray bar was also fitted to the hopper in attempt to knock down froth accumulation. However, froth continued to accumulate, and during frothing episodes the extreme froth tenacity would overcome the installed spray bars and froth overflow was an unresolved concern.

Several opportunities to improve the makeshift hopper were identified. As mentioned above, the sampler outlet nozzle created plunging high velocity stream into the hopper, aerating slurry further. The wide rectangular, open hopper design offered little in deaeration capability. Finally, the hopper outlet was able to readily draw entrained air through to the thickener feed slurry. From these observations, engineering work to design a fit for purpose thickener feed deaeration tank commenced.

FIG 19 – Interim concentrate thickener feed hopper.

Thickener feed deaeration tank design

After installation of the interim thickener feed hopper, engineering work commenced in mid-2022 on a permanent thickener feed deaeration tank to replace the removed trash screen. This tank intended to supplement the existing in-thickener deration tank, providing repeated deaeration in the thickener feed system. The design incorporated many of the learnings so far around froth management and resulted in a two stage tank.

The first stage, present in Figure 20 (left), acted a stilling stage. It accepted sampler discharge through a wide sloping inlet launder intended to reduce inlet velocity, plus the froth recycle pump discharge. The first stage outlet was fitted with penstock valve that is adjusted to maintain a high operating level in the tank. The high level aims to reduce froth generation when the slurry from the sampler lands in the box, and to provide some initial froth separation, that can be attended to by froth sprays.

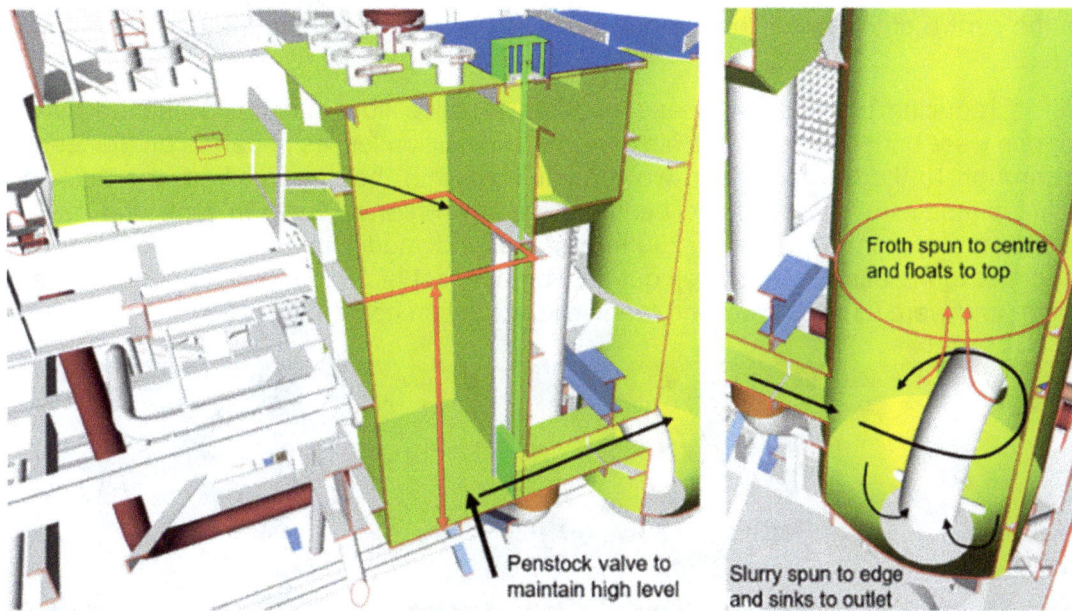

FIG 20 – thickener feed tank operating concept: (left) stage 1 stilling tank; (right) stage 2 deaeration tank.

The second stage, presented in Figure 20 (right), acted as a deaeration stage. The slurry transfer from stage 1 enters through a submerged tangential inlet to this circular tank. The rotation action spins froth to the centre and upwards to the surface, to be attended to by froth sprays. Slurry moves outwards and downwards to down turned outlet. This downturned outlet is designed to reduce the chance of air entrainment, by both forcing a more tortuous route for the leaving slurry to follow and to ensure the outlet is always submerged by a distance set by the downstream thickener.

Both stages are significantly taller than the previous interim feed hopper, with the intent to reduce the likelihood of froth overflowing the tanks. The overflow pipe is directed close to the nearest sump pump to reduce housekeeping effort when this occurs. Further, both are covered with HDPE lids and fitted with easily serviceable spray bars.

The thickener feed deaeration tank was installed in May 2023. Since installation, routine manual clarity measurements of the thickener overflow, taken by the area technicians have markedly improved, refer Figure 21. Further technicians reported that while frothing episode do still occur on the thickener, they appeared to be shorter lived and less likely to cause a dirty overflow.

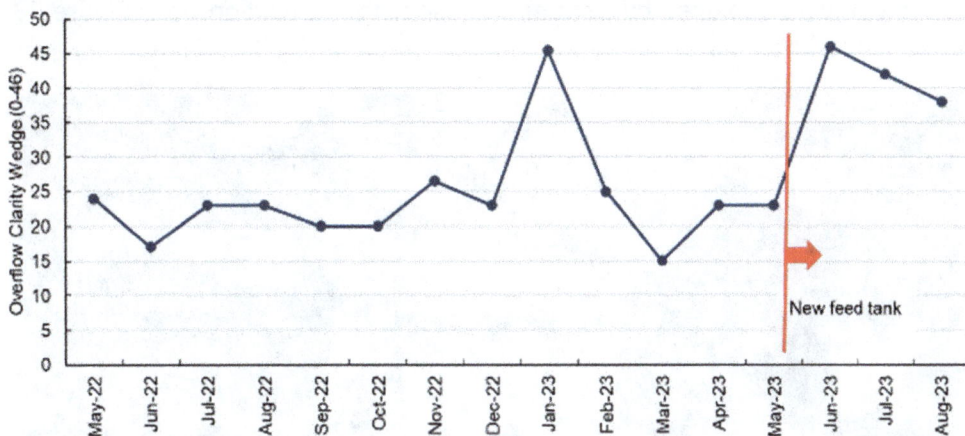

FIG 21 – Thickener overflow clarity measurements – average by month (46 highest clarity).

Feedwell modifications

The concentrate thickener was design with an in thickener deaeration tank between the concentrate trash screen and the feed well of the thickener with a dart valve and additional sprays. Feed flow takes a tangential path of submerged inflow. The rotation of the slurry in the tank promotes confining slurry to the centre of the tank. The dart valve maintains a high level in the tank to keep the inflow submerged.

To assist in reducing the hydraulic head required to drive flow into the thickener, two feedwell modifications were undertaken. These aimed to reduce friction losses imparted on the feed slurry as passing through to the thickener and were completed with the support from the thickener OEM Metso. The diameter of the feedpipe between the in thickener deration tank and feedwell was increased diameter from 150 mm to 200 mm. A second modification was to lower the feedwell deflector cone, to increase the feedwell out flow area. A set of 50 mm spacer blocks were sourced from Metso, and were used to lower the feedwell distribution cone, Figure 22.

FIG 22 – Feedwell outlet spacer installation: (left) before; (right) after.

The thickener feedwell was fitted with an auto dilution port, which supports feedwell dilution to the appropriate density for growth of flocculated solids, by drawing clear supernatant in to the feedwell. However, it was more common than not to observe flow and froth coming backwards out of feedwell through this port. This was believed to be having a negative effect on the distribution of flocculated particles into the thickener, allowing them to short circuit the feedwell. Further this allow froth to escape the feedwell which was fitted with sprays.

As such the port was temporarily blocked with sandbags to determine if any unforeseen consequences resulted. The loss of dilution (during the infrequent times, when the port was flowing forwards) was compensated through earlier dilution, and the low-density nature, typically <15 per cent solids, of the final concentrate stream. As no negative consequences were observed, a permanent solution was deployed by permanently sealing the dilution port, Figure 23, during the next maintenance shutdown.

FIG 23 – Feedwell dilution port sealed: (left) before; (right) after.

Final feed system design

This original feed system has since been thoroughly modified with several aims:

- Adress critical, spillage inducing, circuit flow bottlenecks from initial plant design.

- Expand overall flow capacity commensurate with both lower than design concentrate % solid and increased throughput requirements.

- Introduce additional concentrate deaeration stages to reduce the impact of flotation frothing episodes on the concentrate thickener performance.

- Minimise additional air entrainment into the concentrate as is passes through the feed system.

The present thickener feed system is summarised in Figure 24. Potential future modifications that are under consideration include: replacement of the sampler feed box with a third swirl tank for additional deaeration; removal of turbulence bars from the first stage of the sampler, automation of the level control of the Thickener Feed Stage 1 Stilling Tank.

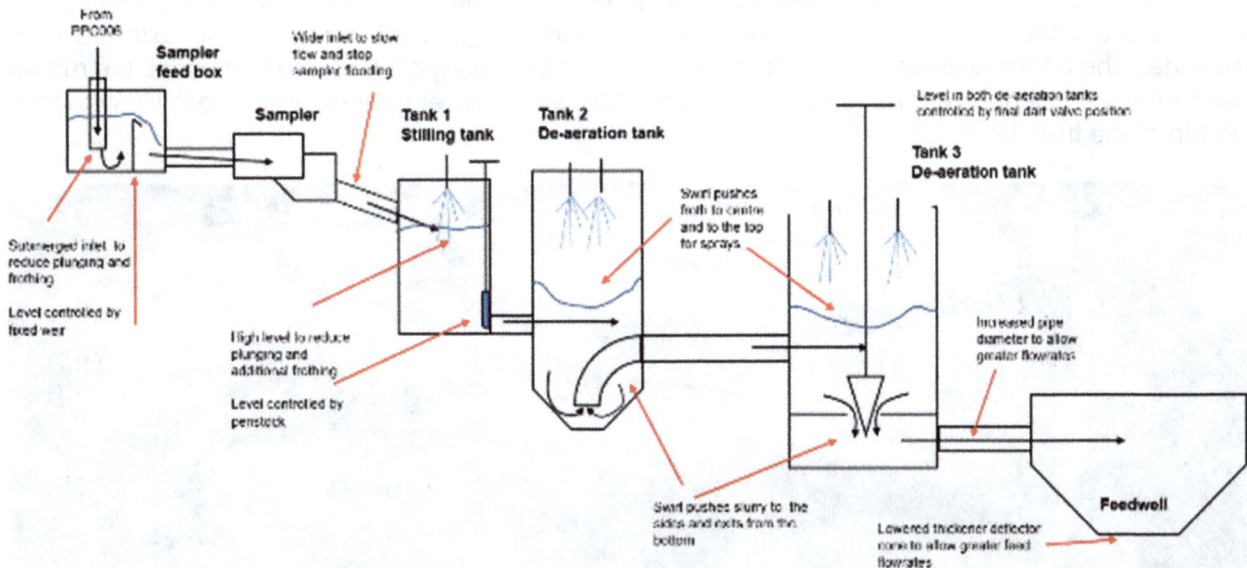

FIG 24 – Thickener feed system – post modifications.

CONCENTRATE THICKENER AND OVERFLOW PIPE MODIFICATIONS

In parallel to the efforts on the thickener feed system to reduce aeration of the feed slurry, several modifications to the thickener itself. These aimed to better accept and manage froth, and to provide capacity to accept larger inflows, particularly due to higher Jameson Cell water recovery, dilution to assist pumping, and spray water additions.

Froth spray upgrades

Whilst sprays were supplied on the concentrate thickener, they were limited to a single line of downward facing sprays along the thickener bridge, spanning only half the diameter. These were inadequate at knocking down any moderate level froth on the thickener. Operations technicians initially installed hose stations on the thickener bridge and surrounding walkways to provide greater spraying effort, which was replaced in June 2021 by a permanent spray ring fitted to the lip of the thickener tank, refer Figure 14 (left). This spray ring encircled the thickener entirely and used horizontal orientated fans sprays that intended not to knock down the froth, but instead to confine it to the centre of the thickener, thereby keeping the froth away from the outer overflow lip.

With later installation froth booms, these sprays were modified to allow froth to encroach the outer launder in the quadrant of the thickener leading into the froth removal beach.

Installation of froth booms with recycle pump

Despite attempts at improving the effectiveness of the concentrate thickener sprays, there were occasions where the sprays were proving to be ineffective. In consultation with the thickener vendor, a design was established for a froth removal system, consisting of booms and a froth beach and froth recycle pump.

These froth booms were attached to the thickener rakes. They rotate around the thickener slowly moving accumulated froth towards a newly installed froth beach. This beach sits just proud of the expected thickener liquor level and captured froth that was being moved by a rubber flap on the outer portion of the froth boom. Froth was re-directed via the beach to vertical shaft pump and then recycled to the concentrate thickener feed deaeration tank.

For the most part, the utilisation of the froth booms, recycle system and sprays together proved effective. On occasion, however the froth can become so stable that despite the froth booms moving it into the path of the sprays, it would not break. After additional time, the froth layer could grow such that it was so dense that the rubber flaps at extremities of the boom no longer push the froth on the beach surface, and instead continually drag across the top of the froth surface. This process became what was anecdotally called 'froth screeding' as seen in Figure 25. During these extreme froth episodes, the booms were largely ineffective. To reduce frequency of the screeding event, the rubber flaps on the thickener were replaced with less flexible, and heavier material that is less likely to float on top of the froth layer.

FIG 25 – Froth boom – 'screeding' of extremely stable froth.

Thickener overflow pipe capacity

The thickener overflow was transferred by a gravity-fed pipe to the process water dam for reuse by the process plant. This pipe was relatively long (150 m), had a shallow overall slope (1.8 m fall over a 100m horizontal length), included several short radius pipe elbows, and included a ~30 m long flat section that was a buried low point. Due to frequency that the concentrate thickener would operate with poor overflow clarity; combined with the low fluid velocities, the solids would settle and accumulate in the lowest dead leg of the buried pipeline. The volumetric capacity of the line progressively decreased, requiring a larger driving head, resulting in a rise of the liquor level of the thickener overflow launder. Eventually the rising launder level would restrict flow across the supernatant weirs; and lift the liquor thickener itself. With the installation of the froth removal beach, this was now the lowest point on the thickener – the liquor would rise above the level of beach start to overflow supernatant through the froth recycle system, ultimately causing the froth recycle pump to spill.

Several pipe modifications were undertaken, presented in Figure 26. Firstly, in May 2022, three breathers were installed in locations where air was assessed to accumulate and potentially air locking the pipe.

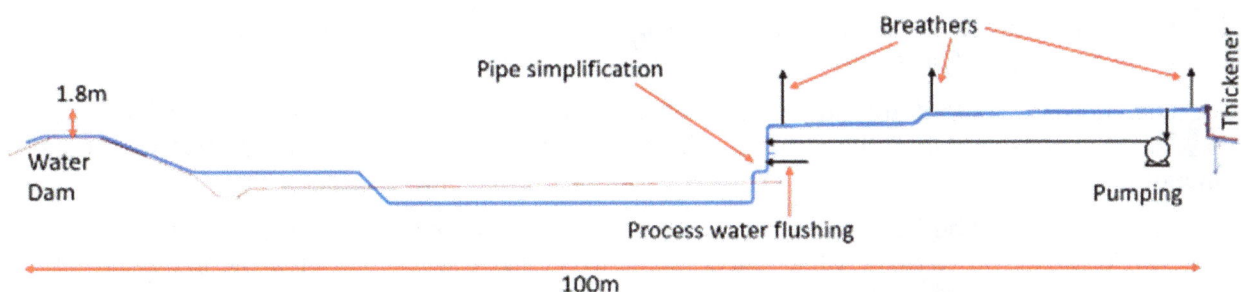

FIG 26 – Thickener overflow piping arrangement and completed modifications.

Secondly, in November 2023, modifications were made to: (a) flush the line with high volumes of pressurised process water to purge settled solids out; and (b) replacement of the most tortuous section of piping path with lesser number of shallower bends. The periodic purging became a routine task for the operational technicians to maintain this pipe capacity as high as possible, by keeping the pipe clear of restrictions. The streamlined piping path was chosen to reduce the head losses of the piping.

Finally, in February 2024, further pipe modifications were completed to install a portable diesel pump, Figure 27. This provided an interim mechanism to control of the thickener liquor level and increased capacity of pipeline above that provided by the gravity fed system. From this point, operation technicians were able to significantly reduce thickener spillage, via the froth beach, during frothing episodes by starting the overflow pump when required to relieve the flooding of the overflow launder which would increase liquor in the thickener proper.

FIG 27 – Thickener overflow pumping.

ADDITIONAL WORKS

Despite the significant modifications and upgrades applied during the first four years of operation at Carrapateena, management of froth tenacity is still an ongoing concern. Many modifications that have bene successful, have often had their improvement margin consumed by a progressively increasing milling throughput. For instance, at time of writing peak achieved milling rate at Carrapateena is 800 t/h compared to the original design rate of 500 t/h; with further confirmed expansion works underway to move to average milling rates of 850 t/h in the near term. Consequently, the effects of a frothing episode are not considered settled issue for the Carrapateena processing team.

Despite repeated laboratory dynamic thickening test work – necessarily completed on 'stale' deaerated plant samples – indicating the existing 15 m concentrate thickener has capacity to accept the current and near future feed rates. It is clear from observation of plant performance that conclusion cannot be made when including considering the impacts of froth tenacity on the thickening performance. As such, engineering work commenced on the expansion of concentrate dewatering capacity at Carrapateena.

These designs are centred around the installation of the second concentrate thickener and storage tank, for the separate thickening of the Cleaner Scalper Jameson and Cleaner 3 Jameson Concentrates.

The new thickener will be 22 m high-rate thickener, and will accept the higher flow Cleaner Scalper Jameson. The existing 15 m thickener will continue to receive the flow from the Cleaner 3 Jameson. Froth mitigation efforts included in the design include:

- The feed system will include the replica of the upgraded external deaeration feed tank.

- The thickener will include, froth spray ring, froth boom, beach and recycle pumping.

- Circumferential walkway on the thickener allowing for regular cleaning and maintenance of the launder and sprays.

- A larger and deeper froth baffle to better separate the froth layer from the supernatant as it exits the thickener and to reduce water rise rates through the area between the baffle and overflow launder.

A new Jameson Scalper Concentrate hopper and pump is included in this design to transfer slurry to the new thickener. Froth mitigation efforts included in the design include:

- selection of a vented froth pump

- design and incorporation of deaeration swirl pot on the hopper inlet

- tangential hopper entry nozzles

- tall hopper to improve suction head under aerated conditions

- hopper fitted with traversable lid and maintainable high-capacity spray bars.

A thickener overflow pumping system will be fitted to both the existing and new thickeners.

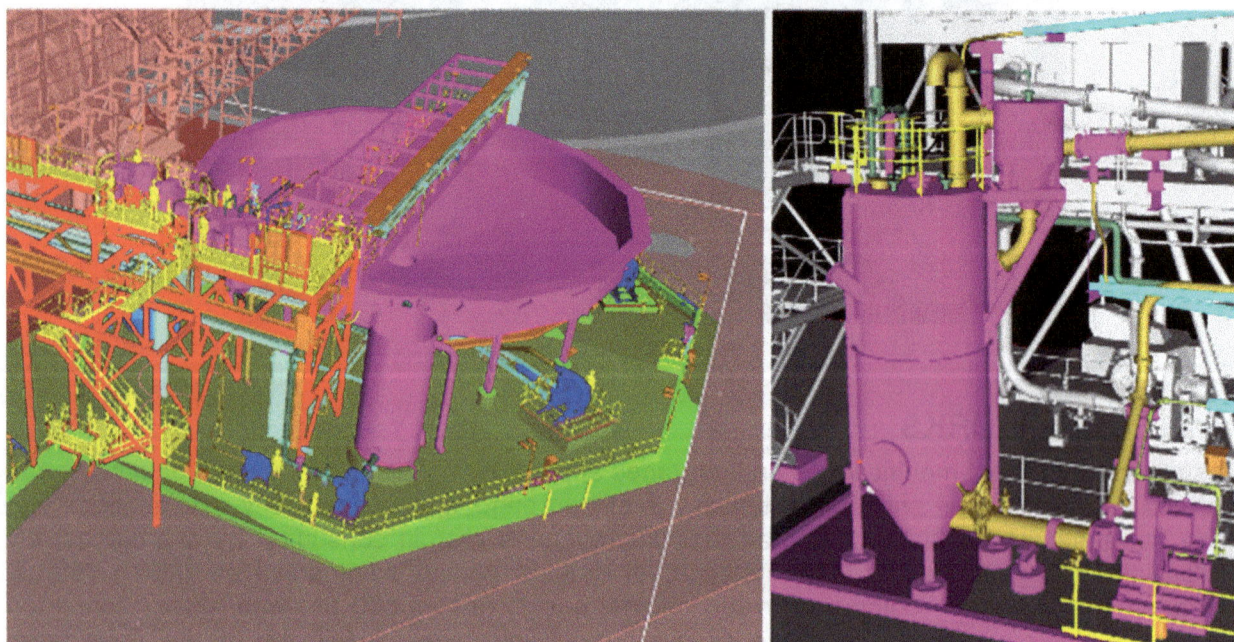

FIG 28 – Concentrate handling expansion: (left) 22 m high rate thickener; (right) cleaner scalper concentrate pump.

EXTENDING PREVIOUS ADVICE FOR OTHERS

Woodal, Smith and Pease (2021) have previously published an extensive guide of advice to operators and designers regarding 'difficult froth' pumping, which is undoubtedly recommended reading and was a helpful reference for Carrapateena. Additional advice authors can provide to that previous guide include:

- Any vessel required the use of sprays to attend to frothing should be fitted with lids. The housekeeping and reliability benefits of keeping overspray contained, can't be overstated.

- Full cone sprays typically work the best for hoppers; however, the distance of the nozzle from the froth surface greatly affects their performance. Consider installing these at multiple heights to keep froth breakage rates high across a range of hopper levels.

- Deaeration is rarely a one and done process:
  - Deaeration of tenacious may take multiple stages to be completed effectively.
  - Can be easily undone in the vessel due a design choice that reaerates slurry.

- Wherever possible all vessel (boxes, chutes and hoppers) inlets should be submerged and tangential to prevent formation of a plunging jets that will entrain air into the slurry.

- Include downfacing outlet of deaeration tanks wherever space allows. Keeping this outlet always submerged reduces re-entrainment of air due to an exposed inlet and, increases the effort entrapped air must go through to reach the outlet.

- Accessible concentrate thickener launders are often considered capital luxury, but the cost of scaffolding to reach the overflow launders for cleaning can be surprisingly just as much.
  - Regular cleaning of these launders will ensure unit operation efficiency.

- The standard concentrate thickener froth baffle is usually too shallow. This allows the bottom of the froth layer to be drawn under the baffle and into the overflow:
  - This may be exacerbated by froth removal booms that can disrupt a thick froth layer as they pass.
  - Work with the manufacturer to supply a deeper baffle, and one that is located further inward from the overflow launder to reduce local upward rise rate between the launder and baffle.

- Install breathers any long gravity flows pipelines, or any lines where a pipe reduction exists, or near just after outlets. With tenacious concentrates be aware that:
  - Air locks can occur where you wouldn't typically expect.
  - Breathers may need to be higher than hydraulics suggest, due to froth build-up and overflow of the breather.

- Encourage your metallurgists to not shy away from tackling what would often be considered 'operational' problem:
  - A technical team that always has time set aside for address items that make the operational technicians life a chore, will be a long way toward holding an engaged relationship critical for a successful concentrator.
  - Don't accept the hopper that always overflows.

- Engage with your reagent and equipment suppliers, they are often just as eager as you to embark on an improvement journey, whether it be equipment reliability or reagent efficacy.

REFERENCES

Reinhold, J, Burns, F, Seppelt, J and Affleck, K, 2023. Plant Modifications and Operating Strategies to Improve Concentrate Quality at OZ Minerals – Carrapateena, in *Proceedings of the MetPlant Conference*, pp 267–289.

Woodal, P, Smith, D and Pease, J, 2021. Advances in Flotation Froth Pumping, in *Proceedings of the 15th Australasian Mill Operators' Conference*, pp 229–240 (The Australasian Institute of Mining and Metallurgy: Melbourne).

CSA mine new SAG mills optimisation using grindcurves

G D Figueroa Salguero[1], B Palmer[2], S Ntamuhanga[3] and J Buckman[4]

1. MAusIMM, Senior Metallurgist, CSA Mine Metals Acquisition Limited, Cobar NSW 2835. Email: german.figueroa@metalsacqcorp.com
2. Senior Process Specialist, JKTech Pty Ltd, Brisbane Qld 4068. Email: b.palmer@jktech.com.au
3. Metallurgy Superintendent, CSA Mine Metals Acquisition Limited, Cobar NSW 2835. Email: stephen.ntamuhanga@metalsacqcorp.com
4. Ore Processing Manager, CSA Mine Metals Acquisition Limited, Cobar NSW 2835. Email: jade.buckman@metalsacqcorp.com

ABSTRACT

Two SAG mills at the CSA Mine Concentrator in New South Wales Australia were recently replaced and optimised using grindcurves as an operational tool. The mill upgrade included the exchange of the motor with a higher-power motor capacity including variable speed drives (VSD). The internal design of the shell lining was also changed to a skip row design with increased lifter spacings. After mill commissioning and start-up, grindcurves determined the optimum operating guidelines for throughput and product size for different circuit configurations based on mill filling, bearing pressure and power draw. A simplified 3-point grindcurve was also developed to facilitate the overall process, gain stability of the mill, and minimise production risks during low and high mill filling conditions. This 3-point curve was analysed with basic statistics and robust regressions for throughput and particle size. In overall, Mill 2 was optimised in approximately two months at different mill speeds, whereas Mill 1 took approximately nine months due to the need of readjusting bearing pressures after commissioning. This paper provides an overview of the metallurgical mill optimisation after commissioning.

INTRODUCTION

CSA is an underground copper mine located in Cobar NSW, Australia. The mine has been operating since 1965. Annual production is circa 1.1 Mt/a, processing copper ore feed grades ranging from 3–4 per cent. The two SAG mills (Mill 1 and Mill 2) were recently replaced due to aging with significant shell structural deterioration. Upgrades included the installation of a motor with a higher capacity of 1600 kW (previously 1100 kW) and installed variable speed drive (VSD) up to 11.94 rev/min (73 per cent critical). The shell liners were also changed from a 48- to a 24-lifter 30° face angle Metso skip-row to increase lifter spacing to standards recommended to reduce the probability of packing (Powell, 2019). Mill 2 replacement took place in mid-2022 and Mill 1 in mid-2023, both successfully executed by CSA Projects Team, Contractors and Metso reline team. Commissioning was conducted by Metso and supported by CSA with the mills delivered to the Ore Processing a week later for metallurgical optimisation. Mill 1 took approximately nine months to optimise due to required bearing pressure adjustments and Mill 2 took approximately two months for the specific mill rev/min conditions evaluated. Figure 1 shows the Mill 1 and 2 upgraded in 2023 and 2022, respectively (Metso:Outotec, 2022).

FIG 1 – SAG Mill 1 and 2 (6.7 m diameter × 2.2 m length) replacement in mid-2023 and 2022, respectively (Metso:Outotec, 2022).

The mills were optimised for throughput and product size by employing grindcurves. The operation of SAG mills (particularly when run as a single-stage SAG) is sensitive to volumetric mill filling. The throughput, power draw, and product size have been shown to peak at different mill filling levels and mill speeds (Powell, van der Westhuizen and Mainza, 2009). To generate a mill grindcurve, the mill fill is progressively stepped up throughout its range at different mill speeds and the resulting power draw, throughput and indicative grind size is measured (Powell, van der Westhuizen and Mainza, 2009; Powell *et al*, 2011). Therefore, developing grindcurves to relate mill filling to other performance indicators such as throughput, power draw, bearing pressure and product size can assist in achieving optimal mill operations (Powell, Morrell and Latchireddi, 2001; Powell and Aubrey, 2006; Powell, van der Westhuizen and Mainza, 2009; Powell, Perkins and Mainza, 2011; Bueno *et al*, 2013). The optimisation was conducted in closed circuit (CC) and open circuit (OC) for Mill 2 using traditional grindcurves. Whereas for Mill 1, at one mill speed condition only, a simplified 3-point grindcurve was used to stabilise and minimise any production risk associated with member set points. In addition, mill discharge samples (D/C) were also collected to estimate the production of fines at different bearing pressure levels.

This study presents the metallurgical optimisation for each SAG mill using grindcurves. A 3-point grindcurve model development, operating guidelines, and recommendations.

Grinding circuit and mill configurations

The CSA grinding circuit has two SAG mills (Mill 1 and Mill 2) and one ball mill (Mill 3) with high-low shell lining, all Ø6.7 m diameter (D) × 2.2 m length (L) dimensions with an effective grinding length of ~1.7 m (Metso:Outotec, 2022; REV-01-FR Report, 2015). Each SAG mill has a cluster of two Krebs hydrocyclones (spigot 87 mm, 150 mm vortex finder) and the ball Mill 3 has a cluster of six Cavex hydrocyclones (spigot 90 mm, 170 mm vortex finder). The SAG can operate in OC or CC defining different mill configurations which are selected depending on feed grade and milled tonnes required.

For low throughput, a standard SAB configuration consisting of one SAG in OC with mill discharge (D/C) to the ball Mill 3 D/C hopper can be selected (Figure 2a). For high throughput, one SAG in OC and the other in CC with cyclone overflow (O/F) discharge to Mill 3 D/C can be used (Figure 2b). Otherwise, the two SAG mills can be set in OC. Also, Mill 2 hydrocyclones underflow (U/F) can bleed one cyclone U/F into Mill 1 feed to increase feeder rate into Mill 1 (Figure 2b).

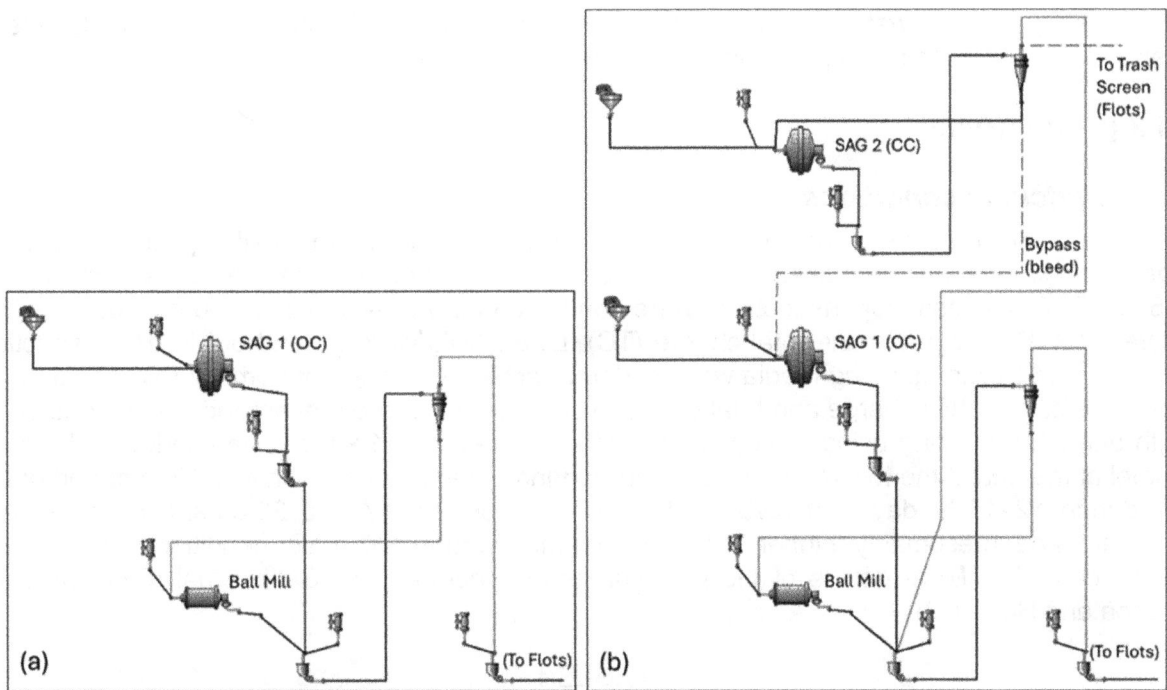

FIG 2 – Standard SAB configuration, SAG in OC (a); SAGs in OC and CC with ball mill (b).

In terms of mill loading, Mill 1 and 2 are charged with 94 mm forged steel balls media, whereas Mill 3 is charged with 50 mm Each SAG mill is fed from two coarse ore storage bins, with each bin having two apron feeders. Individual feeders can be selected to run in manual or automatic at different rates. A constraint limited PID controls the feeders rate in auto to maintain the bearing pressure and power input set points to avoid mill overloading (it is important to mention that CSA does not have feed weightometers on the mill feed conveyors). All mills are also installed with a trommel of 10 × 25 mm apertures having a scatreturn spiral system. The particle size classification is carried out using hydrocyclones. The feed pipes to the SAG mill cyclones also have flow metres (L/min) and density gauges (SG) to obtain mass flow tonnage (t/h). The product grind size P_{80} required from grinding and classification circuit is 106 µm (REV-01-FR Report, 2015).

SAG mill optimisation using grindcurves

To optimise the operation of the new mills, grindcurves were generated to assist and provide understanding on the impact of ball charge, mill speed and mill load on performance. The grindcurves were set-up at different mill speed (rev/min) and bearing pressure (kPa) set points. Mill load was increased by increasing feeder rates (per cent) until the desired bearing pressure (kPa) was achieved. Ball charge and crash stop inspections were conducted, whenever possible, to maintain 11–12 per cent fill nominal prior to the next mill speed condition.

For Mill 2, typical grindcurves following the literature were conducted (Powell and Aubrey, 2006; Powell, Perkins and Mainza, 2011). While for Mill 1, a nominated 3-point grindcurve was performed to avoid several set point changes minimising test work time expecting the mill with minimum variance or better stability (Powell and Aubrey, 2006). On the other hand, this 3-point method was expected to minimise any risk of production losses by closely monitoring the low- and high-end mill load conditions bearing pressure selected based on historical data. These end-member conditions can drive the mill to unloading or overloading causing downtimes or surges affecting downstream flotation which should be avoided. The 3-point selection has been reported by Napier-Munn (2014) and well-applied in flotation studies at the JKMRC, The University of Queensland (Figueroa, 2019; Wang, 2017). These studies provided similar level of information using basic statistics (averages and standard deviations) enabling the application of regression models to better support operating guidelines.

For both grindcurve types, mill D/C samples were collected using a 1 L Marcy cup to obtain -150 µm proportions to determine fines production at the different conditions. Per cent solids, specific densities (SG) and feed water were also adjusted during the tests. Mill noise measurements outside the shell were conducted using a smartphone to gain insights of mill loading conditions.

Mill 2 grindcurves

Mill 2 grindcurve conditions

For Mill 2, grindcurves were conducted at 10.50, 11.0, 11.50, and 11.94 mill rev/min at the target 11 per cent nominal ball charge (BC) (Metso:Outotec, 2022) and bearing pressures from 1650 to 1925 kPa in OC. Crash stop inspections were conducted before each mill speed change condition (before tests 'BT') to determine total charge (TC). Later, Mills were ground out for BC inspection. Four drums of 94 mm grinding media were added to achieve 11 percent nominal BC (pending mill inspection results). Ball charge and total charge were determined from averaged void height, chord length and lifter counting using a measurement tape, except at 11.94 rev/min conducted from the trommel at the end of the test (AT). Table 1 shows grindcurve conducted on Mill 2 in a period of five days during 12–14 hr day shift (D/S) roster. TC ranged from 17.2 to 20.8 per cent for 10 and 11 rev/min and presumably higher for 11.94 rev/min before unloaded conditions (<1900 kPa, 37.9 per cent TC). More details of the grindcurves are reported in CSA internal report (Palmer, Figueroa and Ho, 2022).

TABLE 1

Grindcurve plan executed on Mill 2, including ball charge and total charge for OC.

Day	Mill config	Mill speed (rev/min)	Bearing pressure (kPa)	Total charge (avg%)	Actual BC (avg%)	Nominal BC expected after media addition
15/09/2022	OC	10.50	1650–1800	17.24% (BT)	8.83% (BT)	~11%
20/09/2022	OC	11.00	1700–1825	20.77% (BT)	9.40% (BT)	~11%
20/09/2022	OC	11.50	1740–1850			Not drums added
22/09/2022	OC	11.94	1730–1925	37.9% (AT)		~11%
23/11/2022	CC	11.94	1675–1900			~11%
24/11/2022	CC	11.50	1675–1885			~11%

Notes: * BT = before tests, AT = after test.

Mill 2 grindcurve results

Mill 2 in OC grindcurves show a peak of throughput and a gradual increase of -150 µm following a polynomial fit. The power draw increased with bearing pressure until reaching a maximum point at the highest VSD 11.50 and 11.94 rev/min. The best throughput is observed at 11.50 rev/min considering that the data are scattered between 900 and 1100 t/h. However, it is also clear that at 11.94 rev/min, throughput holds at maximum for a wide range of bearing pressures with a drop at the highest bearing pressure conditions >1900 kPa, potentially indicating mill overloading (Palmer, Figueroa and Ho, 2022). This effect is aligned with the increase of power draw maintained at 1250 kW. In terms of particle size, increasing mill filling increased the proportion of fines in the mill discharge, particularly at 11.94 rev/min.

Figure 3 illustrates the grindcurves conducted on Mill 2 in OC at actuals TC and BC before the addition of grinding media to achieve nominated 11 per cent ball charge at different mill VSD (rev/mins). Note that there was no crash stop at the beginning of the 11.94 rev/min test due to time constrains, instead, at the latest kPa condition to confirm suspected overloading conditions.

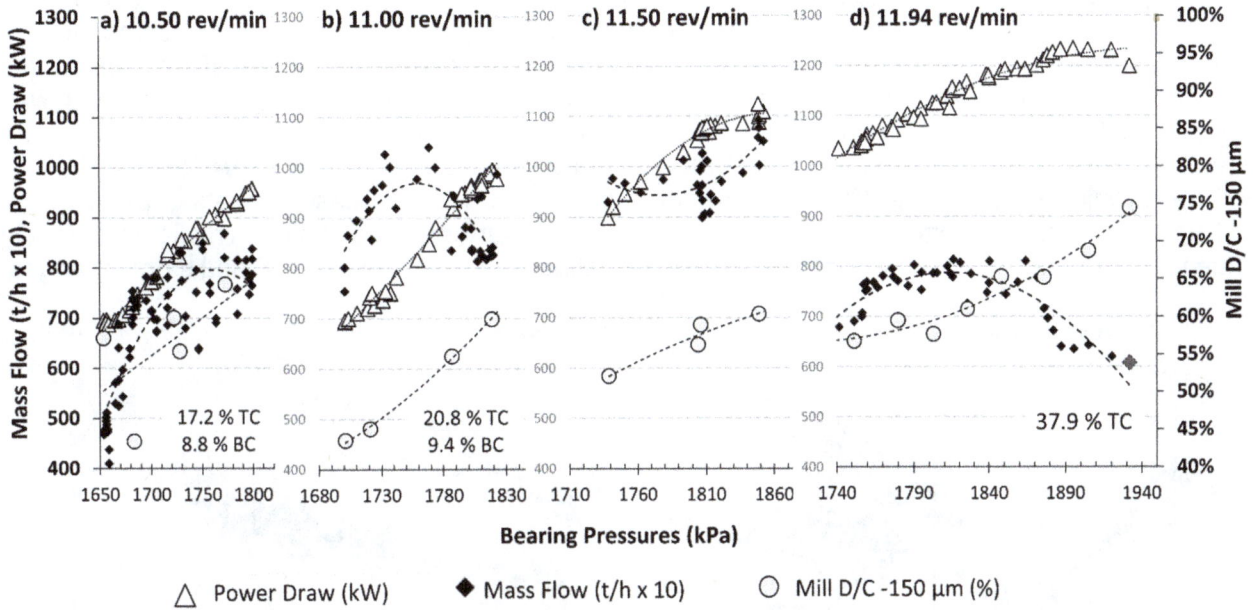

FIG 3 – Mill 2 grindcurves conducted in open circuit at estimated 11 per cent nominal ball charge, TC 17–20 per cent prior to mill filling and different mill rev/min (Mod from Palmer, Figueroa and Ho, 2022).

Figure 4a shows the crash stop inspection conducted at 11 rev/min reporting 20.8 per cent TC on Mill 2. Figure 4c-d illustrate the crash stop performed at the latest condition (1925 kPa) for the 11.94 rev/min to confirm mill overload. The mill was inspected from inside the trommel, reporting about 37.90 per cent TC. In both cases, milled rocks were noted with uniform size, no signs of liner damage packing, or other issues (Figure 4). At overloading conditions, the mill discharge density was about 2.2 SG (~80 per cent solids at 2.8 SG solids) using the Marcy scale. The mill was ball charged with four drums of 94 mm and ground out for post-routine operations.

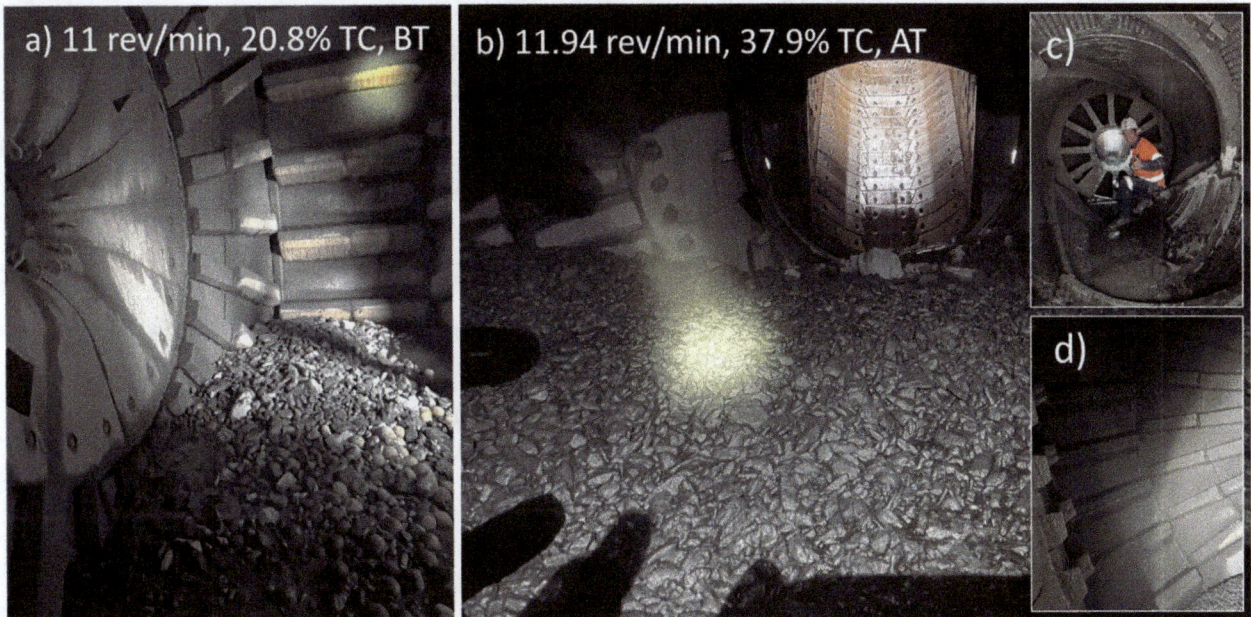

FIG 4 – a) Mill 2 crush stop inspection at 11 rev/min before the test (20.8 per cent TC 10:30 am); b) – Mill 2 crush stop at 11.94 rev/min after the test at high 1925 kPa conditions on 22/09/2022

4:12 PM confirming set point limits (TC 37.9 per cent (v/v)), c) and d) ball charge and lifter inspection from inside trommel.

Grindcurves for Mill 2 in closed circuit (CC) were conducted at 11.50 and 11.94 rev/min (VSD) from 1675 to 1885 kPa as well as media addition before each test. However, neither crash stops, nor BC inspections were performed due to time constrains. The peak and overall shape of the grindcurves in Figure 5 is similar at the same mill speeds as open circuit, except for grind size. This was unexpected considering how different the effective mill feed particle size distribution (PSD) would be. It is noted that the throughput at 1200 kW appears to be higher at 11.94 rev/min than at 11.50 rev/min. Particle size (% -150 μm) also increased with increasing bearing pressure at 11.94 rev/min but is scattered at 11.50 rev/min, requiring further investigation. This is presumably due to other factors such as the influence of recirculating load from cyclone U/F and stope changes.

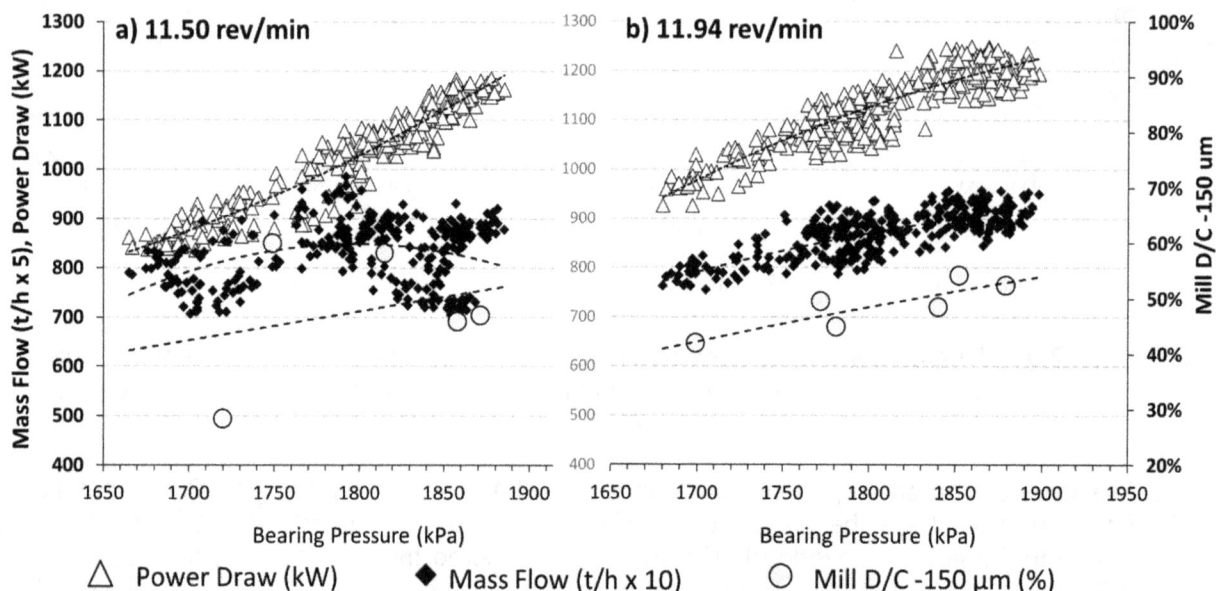

FIG 5 – Grindcurves at selected rev/min on Mill 2 closed circuit (Mod from Palmer, Figueroa and Ho, 2022).

Mill 2 guideline recommendations

Table 2 shows the recommended parameters for Mill 2 in both OC and CC. The key changes in the recommended parameters between configurations are slower speed in OC and higher mill load when in CC, allowing for a finer mill product size. The power draw is well above the predicted maximum of ~1100 kW for both cases to ensure the mill constraint controller is controlling on bearing pressure only (targeting optimum volumetric mill fill), while allowing the constraint controller to prevent the maximum mill power being exceeded.

TABLE 2

Ultimate operating guideline recommendations for Mill 2 (mod from Palmer, Figueroa and Ho, 2022).

Mill 2 recommended parameters	Open circuit (OC)	Closed circuit (CC)
Bearing pressure (kPa)	1825	1850
Power draw (kW) unconstrained	1300	1300
Mill speed (rev/min)	11.0–11.50	11.50–11.94
Discharge specific density (SG)	1.90	1.95

Mill 1 grindcurves

Mill 1 grindcurves

Mill 1 installation (start-up) date was on May 27, 2023 (Wilkinson, 2024). Since commissioning, the mill reported higher bearing pressures than usual. A 3-point grindcurve trial was conducted in July 2023 but no clear differences were observed in throughput at these pressures. The bearing pressure set points were then reduced by operators in September, obtaining better throughputs. To optimise throughput further, on 23 Feb 2024, a 3-point grindcurve was conducted on Mill 1 in OC with the ball Mill 3 cyclones O/F reporting to flotation (basic SAB circuit operation, Mill 2 was shutdown, Figure 2a). A nominal ball charge of 11 per cent was assumed similar to the shutdown mill inspection earlier on 13 Feb 2024 (Figure 6). The ball charge was based on chord length: 4.9 m, void height: 5.4 m measurements, and lifter counting:17.5. After the mill inspection, the mill was charged with three drums of 94 mm ball size daily to maintain the nominal ball charge (Figure 6a and 6b). In addition, part of the optimisation process is to protect the mill shell from rapid wear in the section near the discharge ends (Figure 6c).

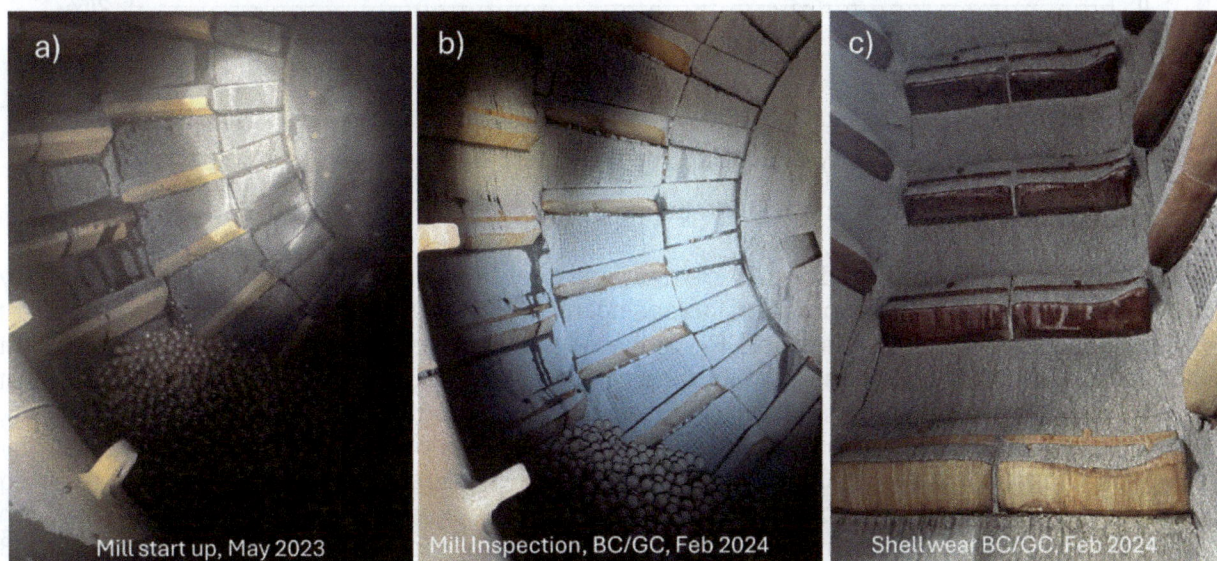

FIG 6 – a) Mill 1 inspection, ball charge (BC) at nominal 11.3 per cent in the start-up, new Mill 1, (May 2023); b) ball charge inspection and maintained for the 3-point grindcurve (GC), Feb 2024; and c) general view of the shell wear – under investigation.

The shell wear is still under investigation with Metso in terms of lifters design. However, it is important to mention this mill usually operates in OC with Mill 2 in CC with cyclone underflow to Mill 1 feed. On the other hand, Mill 1 also operated at relatively high bearing pressures for about four months after installation (up to 2400 kPa). It is possible that a high mill filling and maximum VSD accelerated the wear of the lifters. The shell and head ends will be relined in July 2024 to investigate this issue further after employing optimised bearing pressure and mill speed guidelines from these studies.

The VSD condition selected for the 3-point grindcurve in this study was 11.50 rev/min taking Mill 2 as a reference in OC. The set points selected were 2100, 2300 and 2500 kPa. Feed water was increased from 500 L/min to 650 L/min to maintain SG of the slurry at about 2.00 (since it was 2.15 SG, 76 per cent solids @2.8 Solids SG). The 1 L D/C SG and solids percentage were also collected as usual and taken to the laboratory for -150 µm sizing (per cent). Figure 7 shows the actual t/h, kW, Cu feed grade %/8, and P_{80} data from 410 points (8:00 am to 12:05pm on 23 Feb 2024) exported from PI DataLink, 36 s resolution in the period evaluated.

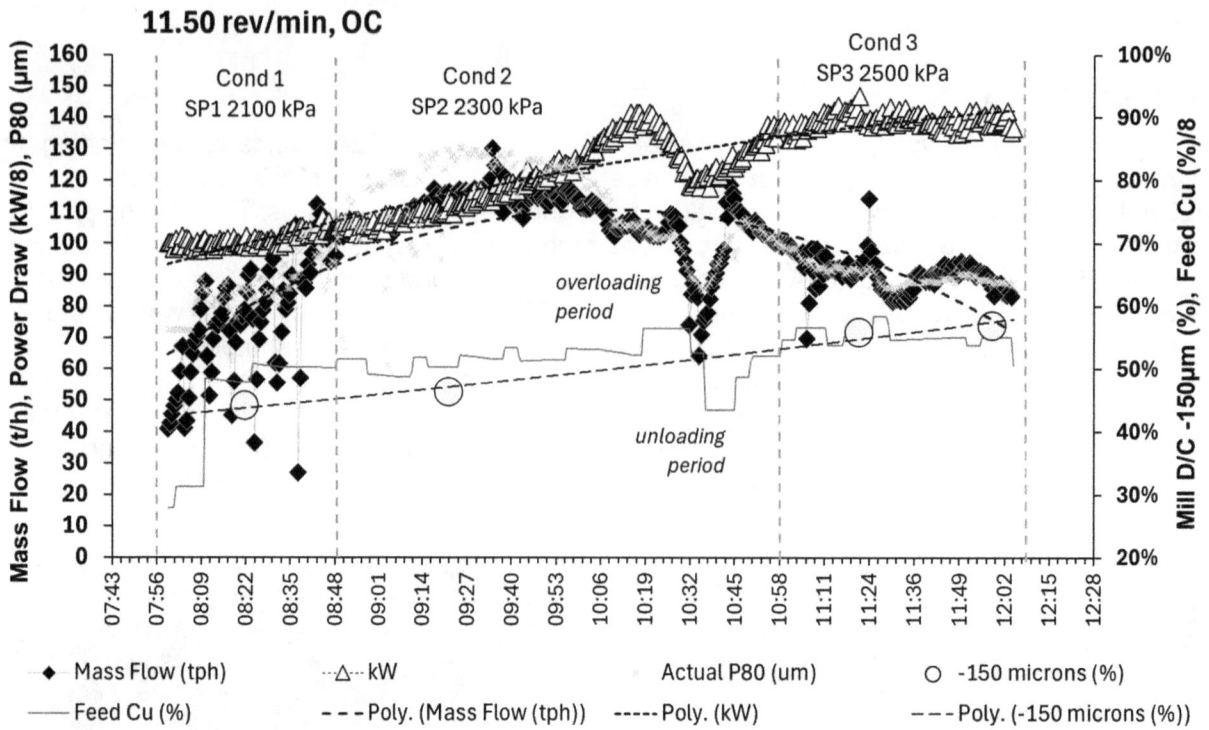

FIG 7 – Actual circuit t/h and PSI/P_{80} during the test work, Feb 23rd, 2024; 410 points.

The instability of throughput and P_{80} is clear during the first condition at 8:00 am. The stability improved in the second condition after increasing mill load and adding 50 L/min feed water at 8:45 am. Of high interest is the unstable P_{80} at the beginning, eventually aligning to the throughput on the second and third conditions. The third bearing pressure set point condition was attempted at 9:50 am. However, power draw suddenly ramped up and throughput decreased about 10:06 am, indicating a mill overloading event (SG Marcy 2.15). This overload was reduced at 10:33 am by adding 100 L/min feed water (650 L/min total) and reducing mill feed rate. The throughput restored prior to the third condition applied again at 11:00 am. This data set of 22 points (from 10:28:48 am to 10:42:36 am, *unloading event*) were excluded in further statistical analysis (averages in Figure 8 and Table 3). Power draw stabilised around 1100 kW in the third condition while throughput decreased. Feed grade was relatively stable during the test work. Table 3 shows the bearing pressure set point conditions and a summary of calculated parameters investigated.

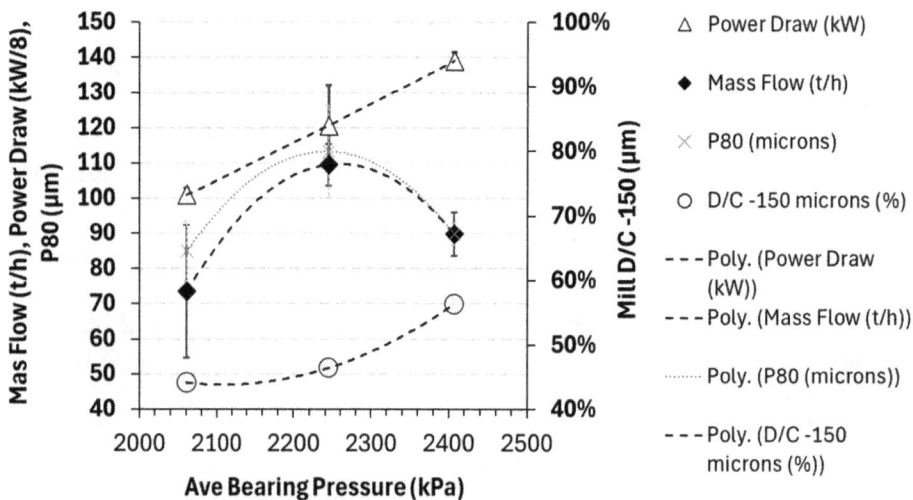

FIG 8 – 3-point grindcurve, Mill 1 open circuit, 11.50 rev/min, estimated 11 per cent ball charge.

TABLE 3

Average and standard deviation summary for Mill 1 OC, 3-point grindcurve, 11.50 rev/min.

3-point grind-curve	Set point bearing pressure	Set point changed	Avg bearing press	Avg throughput	Avg P_{80}	Mill D/C (size)	Avg feed grade
Conditions	(kPa)	(hr:mm)	(kPa)	(t/h)	(µm)	(%-150µm)	(Cu %)
1	2100	8:00am	2061	74	85	44	3.59
2	2300	8:45am	2244	110	113	46	4.13
3	2500	11:00am	2407	90	90	56	4.42
Mill delivered to Operators		12:05pm	STD	STD	STD	STD	STD
1	2100	8:00am	127	18.8	8.5	0.0	0.66
2	2300	8:45am	162	5.9	12.9		0.19
3	2500	11:00am	130	6.2	3.1		0.09

3-point grind-curve	Set point bearing pressure	Set point changed	Feed water	Power draw	Calc avg throughput	Calc avg P_{80}
Conditions	(kPa)	(hr:mm)	(L/min)	(kW)	(t/h)	(µm)
1	2100	8:00am	513	807	76	86
2	2300	8:45am	563	965	108	112
3	2500	11:00am	657	1113	92	92
Mill delivered to Operators		12:05pm	STD	STD	STD	STD
1	2100	8:00am	78.4	15.9	14.9	10.8
2	2300	8:45am	34.9	92.3	7.6	8.3
3	2500	11:00am	28.7	19.8	3.0	4.1

Mill 1 grindcurve results

Figure 8 shows 3-point grindcurve results for Mill 1 in OC. The maximum throughput occurred in the middle condition with a low SD, 2250 kPa, decreasing in the lowest and highest conditions. P_{80} also aligns in the second and third condition, as expected. The high SD in the first condition for throughput and particle size is attributed to mill filling instabilities (low mill filling). This negative effect was noted at the beginning of the grindcurve set-up when the mill rapidly unloaded, with the pressure dropping to 1900 kPa. Increments of 50 kPa were set-up until 2100 kPa was reached to gain better stability.

Similar to Mill 2, the power draw rose to 1113 kW with the increase of feed rate (1113 kW/8). The production of fines is similar in Condition 1 and 2 but increased in the latest condition to 56 per cent. This is probably due to the increase of specific power input when the mill gets loaded, similarly than Mill 2.

Mill 1 sound analysis

No mill inspections or crash stops were conducted after finishing the test work to estimate total charge during the 3-point grindcurve due to time constraints. To gain some insights on mill filling, the outside shell noise was recorded using the Memo voice App in a smartphone (iPhone12) for about 30–40 sec for each condition.

The dB noise average and SD's show a clear step down (noisier mill) from the first condition to the second one (Figure 9). The dB magnitude, however, did not drop any more on the third condition (or the smartphone is not sensitive enough to distinguish any difference). However, in the second and third conditions, there is a distinctive tumbling noise frequency which is reflecting large rock impacts on the shell (Pax, 2001). These peaks frequency occurred three times on Condition 1, and five times on Condition 2 and 3 (see Appendix).

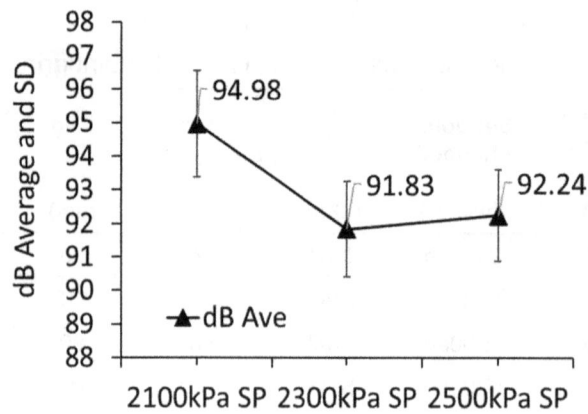

FIG 9 – Mill 1 noise levels (dB) average and SD taken next to the outside shell at three distinct set point (SP) conditions.

These results support the hypothesis that Mill 1 was partially loaded on the first condition, reporting high variance for throughput and particle size which improved in further conditions.

Mill 1 guideline recommendations at 11.50 rev/min

The preliminary operating guidelines recommended to the operators after the test work were as follows: 1) VSD: 11.50 rev/min, 2) bearing pressure: 2250 kPa, 3) power draw: open set point, and 4) feed water: 600–650 L/min. In addition, to maintain ball charge (11–12 per cent), two drums of balls were added on night shift. The mill was observed to be doing well with these conditions as attested by the operators. Further statistical analysis of test work data was conducted to compare with actual day and night shifts to support such recommendations.

As a summary, the simplified grindcurve method prove to facilitate the mill optimisation process in a practical manner. However, due to the simplicity of the process, it would be highly recommended to conduct several grindcurve tests during the life of the lifters, particularly at known ball/mill filling conditions (after shutdowns), different mill speeds (increase % critical), and circuit configurations. Feed water, feed tonnage, and feed grades should also be considered, the latter being highly variable at CSA (Figure 7). It is also recommended that before conducting the test work, gaining mill experience is critical to identify the minimum and maximum loading conditions. The inclusion of all these features will enable a better understanding of the mill to maximise performance.

Mill 1 grindcurve robust regression models

The throughput and particle size (P_{80} from the online Particle Size Analyser or PSI) were regressed in Excel using three predictors: 1) feed water (L/min), 2) bearing pressure (kPa) and power draw (kW). Each term was transformed to the natural logarithm including squared and cubic predictors, example: LN(Water)^2 and LN(Water)^3 (Figueroa, 2019). The model was assessed with 388 observations (excluding those 22 points unloading event). The model resulted with 0.76 R^2 and 0.73 R^2 adjusted for throughput (t/h) and particle size (Outotec PSI data, P_{80}), respectively, with p-values of significance (95 per cent confidence) for water and power related terms. Surprisingly, no bearing pressure is reported which requires further investigation. Details of the model are presented in Appendix 1.

Figure 10 shows a robust regression model for throughput and particle size, respectively. As observed, Condition 1 data did not align well with the intercept due to a high variability (grey circles) previously mentioned. However, better predictions are reported in the second and third conditions due to the increase of mill loading.

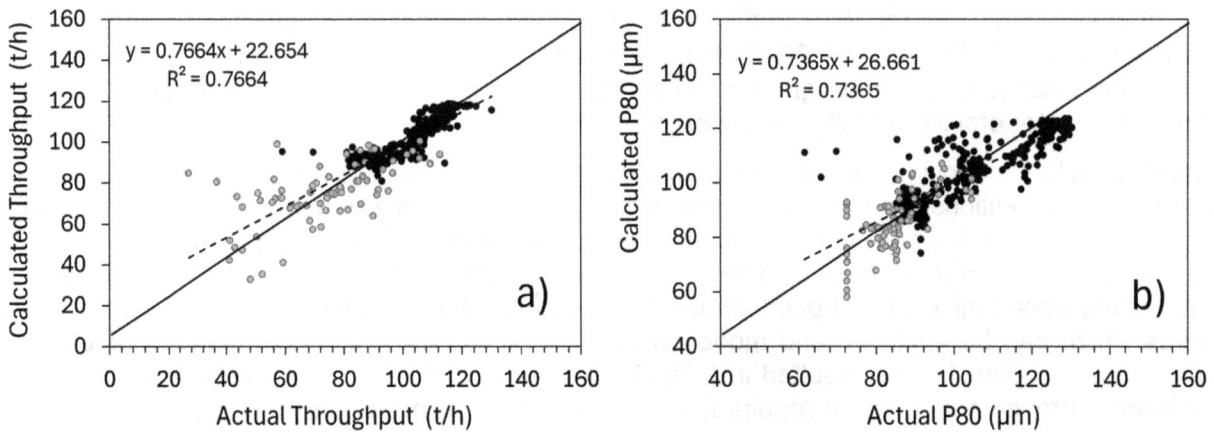

FIG 10 – Regression models for throughput (a) and particle size (b). Condition 1 values in grey colour to visualise their variance.

Nevertheless, future work is required on enhancing model predictors to calculate throughput lower than 80 t/h and overestimated particle size from 6 to 33 μm (data not shown). The P_{80} is somehow expected due to the high variance in the first condition and ultimate PSI-P_{80} is from Mill 3 regrinding and hydrocyclones separation not considered in the model (P_{80} differences from hydrocyclones Mill 2 to Mill 3 are usually observed higher in previous grinding surveys). However, the robustness of the models with three essential predictors can serve as departing point to provide model-supported operating guidelines for future mill optimisation.

Mill 1 model-supported operating guidelines

The preliminary operating guidelines in Mill 1 (Table 4) were used to calculate throughput (115 t/h) and particle size (113 μm) using the regression models. These results are in good agreement with the actual average data during day and night shifts: 111 t/h and 107 μm, 111 t/h and 112 μm, respectively. Similar standard deviations of these values between shifts provided good insights of mill stability.

TABLE 4

Calculated throughput and particle size from the regression model versus actual values.

Shift crew (day/night shift)		VSD (rev/min)	Feed water (L/min)	Power draw (kW)	Bearing pressure (kPa)	Throughput (t/h)	P_{80} (μm)
Calculated t/h, P_{80}		11.5	**600**	**950**	**2250**	115	113
Actual (Avg)	D/S	11.5	619	894	2187	111	107
Actual (Avg)	N/S	11.5	600	896	2200	111	112
SD	D/S	0	24	43	135	10	19
SD	N/S	0	5	52	143	10	14

Note: D/S: 12:30 PM – 6:00 PM; N/S: 6:00 PM – 2:00 PM.

CONCLUSIONS AND RECOMMENDATIONS

CSA replaced both SAG mills that required prompt understanding and optimisation after commissioning. Grindcurves resulted an effective way to assess the new mills performance by providing quick operating guidelines from actual data: mill filling, bearing pressure, power draw and mill speed for different circuit configurations. Mill 2 was optimised in less than two months and Mill 1 in about nine months after reducing bearing pressures. For Mill 2, 11.50 and 11.94 rev/min VSD operation at 11–12 per cent nominal ball filling resulted the optimum throughput of 1825 kPa in OC and 1850 kPa in CC. As for Mill 1 in OC at 11.50 rev/min, the optimum condition for maximum t/h and P_{80} were 2250 kPa, 950 kW, 600 L/min feed water and 11–12 per cent ball charge. However, it is recommended to repeat more grind curve configurations (OC, CC, bypass) at varying conditions (VSD, feed water etc) as the mill liners wear over time.

Though detailed grindcurves are relatively simple to apply, mill stability is critical, including identifying end-member conditions, if unknown (ie new mills). This is because underloading and overloading the mill can quickly happen, causing downtime and affecting milling production and downstream processes. To minimise this risk and reduce test work times (expecting minimum ore variances), grindcurves were simplified to 3-point analysis. This approach provided comparable results with full curves which can be analysed and modelled with robust regression. In this study, predictions of throughput and particle size resulted in 0.76 R^2Adj and 0.73 R^2Adj, respectively. Although model predictor improvements are still required, overall, this approach facilitated mill optimisation using model-supported operating guidelines.

ACKNOWLEDGEMENTS

We wish to express our sincere gratitude to CSA Mine, Metals Acquisition Limited for approval publishing this article. To CSA Operations, Crews and Metallurgists for the support received. As well as to Dr Maria Cristina Vegafria from MSU-IIT on first edits, and Dr. Bianca Newcombe reviewing the pre-final version of this document.

REFERENCES

Bueno, M P, Kojovic, T, Powell, M S and Shi, F, 2013. Multi-component AG/SAG mill model, *Minerals Engineering*, 43–44:12–21.

Figueroa, G, 2019. Investigation of the Effects of Turbulence in Flotation, PhD thesis, The University of Queensland, Brisbane.

Metso:Outotec, 2022. SAG Mill Process Commissioning, Metso:Outotec Australia Ltd, Report.

Napier-Munn, T J, 2014. Statistical Methods for Mineral Engineers – How to Design Experiments and Analyse Data, Julius Kruttschnitt Mineral Research Centre, Queensland, Australia.

Palmer, B, Figueroa, G and Ho, D, 2022. CSA_TN_22_12 Mill 2 Process Optimisation Analysis, Internal Report, CSA Mine Glencore: Glencore.

Pax, R A, 2001. Non contact acoustic measurement in mill variables of SAG mills, in *Proceedings of an International Conference on Autogenous and Semi-Autogenous Grinding Technology (SAG 2001)*, pp 386–391.

Powell, M and Aubrey, M, 2006. Extended grinding curves are essential to the comparison of milling performance, *Minerals Engineering*, 19:1487–1494.

Powell, M S, 2019. Cobar Mine SAG Mill Liner and Discharge Design Review, Internal Reports 1 and 3, CSA Mine Glencore: Glencore.

Powell, M S, Morrell, S and Latchireddi, S, 2001. Developments in the understanding of South African style SAG mills, *Minerals Engineering*, 14:1143–1153.

Powell, M S, van der Westhuizen, A P and Mainza, A N, 2009. Applying grindcurves to mill operation and optimisation, *Minerals Engineering*, 22:625–632.

Powell, M, Perkins, T and Mainza, A, 2011. Grindcurves applied to a range of SAG and AG mills, in *Proceedings of the International Autogenous Grinding, Semiautogenous Grinding and High Pressure Grinding Rolls Technology*, 20 p.

REV-01-FR Report, 2015. Process Description-2015 Rev-01-FR, Internal Report, Cobar management Pty Ltd CSA Glencore CSA Mine Glencore.

Wang, L, 2017. Entrainment of fine particles in froth flotation, PhD thesis, The University of Queensland, Brisbane.

Wilkinson, M, 2024. Mill Lining Wear Trend Report, CSA Cobar SAG Mill 1, Report File 31927, Mar 14 2024.

APPENDIX

Regression models

Figure 12 shows the regression models for throughput (at top) and particle size P_{80} (µm).

SUMMARY OUTPUT

Throughput (tph)

Regression Statistics	
Multiple R	0.875
R Square	0.766
Adjusted R Square	0.761
Standard Error	8.47
Observations	388

ANOVA

	df	SS	MS	F	Significance F
Regression	9	88916	9880	138	1.46E-113
Residual	378	27103	72		
Total	387	116019			

	Coefficients	Standard Error	t Stat	P-value	Lower 95%	Upper 95%	Lower 95.0%	Upper 95.0%
Intercept	-1146233	294102	-4	0.00012	-1724514	-567953	-1724514	-567953
LN(Water)	22358	7047	3	0.00163	8501	36215	8501	36215
LN(kW)	411051	59204	7	1.68E-11	294641	527461	294641	527461
LN(kPa)	54805	110271	0	0.61948	-162016	271626	-162016	271626
LN(W)^2	-3490	1134	-3	0.00225	-5720	-1259	-5720	-1259
LN(kW)^2	-58878	8638	-7	3.71E-11	-75862	-41894	-75862	-41894
LN(kPa)^2	-7028	14307	0	0.62357	-35159	21104	-35159	21104
LN(W)^3	181	61	3	0.00306	62	301	62	301
LN(kW)^3	2810	420	7	8.06E-11	1984	3636	1984	3636
LN(kPa)^3	300	619	0	0.62757	-916	1517	-916	1517

SUMMARY OUTPUT

Particle size (p80)

Regression Statistics	
Multiple R	0.858
R Square	0.736
Adjusted R Square	0.730
Standard Error	8.50
Observations	388

ANOVA

	df	SS	MS	F	Significance F
Regression	9	76335	8482	117	9.89E-104
Residual	378	27315	72		
Total	387	103649			

	Coefficients	Standard Error	t Stat	P-value	Lower 95%	Upper 95%	Lower 95.0%	Upper 95.0%
Intercept	-1019429	295247	-3	0.00062	-1599962	-438897	-1599962	-438897
LN(Water)	22970	7075	3	0.00127	9059	36881	9059	36881
LN(kW)	314878	59435	5	1.99E-07	198014	431741	198014	431741
LN(kPa)	90903	110700	1	0.41207	-126763	308569	-126763	308569
LN(W)^2	-3605	1139	-3	0.00167	-5844	-1366	-5844	-1366
LN(kW)^2	-44848	8672	-5	3.77E-07	-61898	-27797	-61898	-27797
LN(kPa)^2	-11750	14363	-1	0.41383	-39990	16491	-39990	16491
LN(W)^3	188	61	3	0.00220	68	308	68	308
LN(kW)^3	2128	422	5	7.02E-07	1299	2957	1299	2957
LN(kPa)^3	506	621	1	0.41557	-715	1727	-715	1727

FIG 12 – Regression models for throughput (at top) and particle size (at bottom).

Mill sound data and method

The sound outside the mill was recorded using a smartphone iPhone12 located exactly at the same position next to the rotating shell. Each condition (1 to 3) was recorded for 30, 35 and 41 secs, respectively, using the default Memo Voice App. The files were transferred to a Microsoft Clipchamp software in a PC and all re-played loud in a home speaker. The iPhone12 was located next to the speaker to record the sound of the three file one after the other using a dB Meter App v2.8.3. The data was exported containing dB peak and Peak Frequency (dimensionless) into Excel as.txt for averages and SD data analysis. Figure 13 shows the over trends. Condition 1 showing the highest noise, and Condition 2 and 3 more noise peak frequency interpreted as tumbling rock impacts inside the mill. The time is during data transfer, not the test work time.

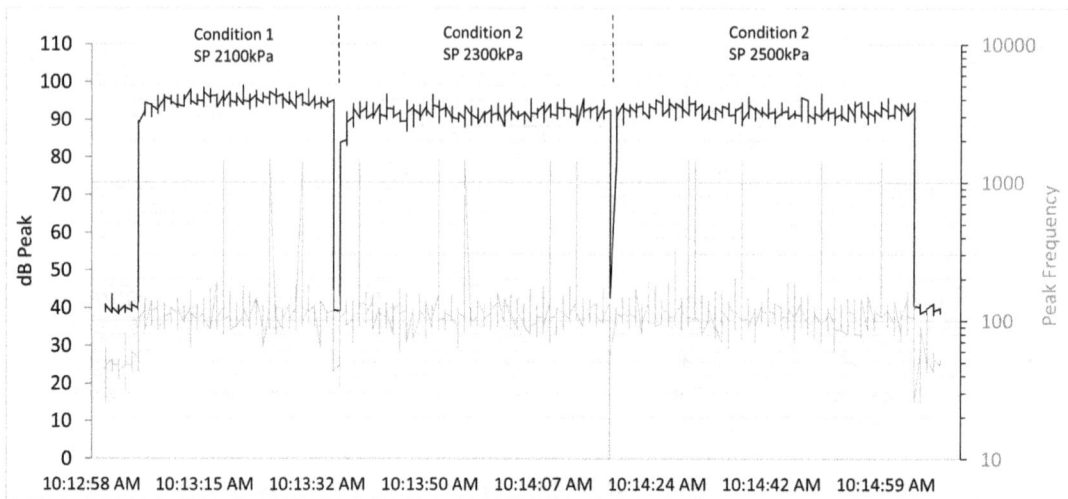

FIG 13 – Decibel (dB) and Peak frequency trends taken from outside the mill shell using a smartphone for each condition (30, 35, and 41 secs recording, respectively).

Regrinding – a subtle mix of liberation and chemistry

C J Greet[1] and B Shean[2]

1. FAusIMM(CP), Manager Mineral Processing Research, Magotteaux Australia Pty Ltd, Wingfield SA 5013. Email: christopher.greet@magotteaux.com
2. Principal Metallurgist – Group, 29 Metals, Brisbane Qld 4000. Email: barry.shean@29metals.com

ABSTRACT

As orebodies become leaner and the mineralogy more complex the need to grind finer to achieve liberation, and therefore produce a saleable concentrate grade has led to the development of several fine grinding technologies. The IsaMill™, HIGmill™ and stirred media detritor (SMD) have all been developed for this purpose. In the first instance, a simple laboratory test can be performed to develop a signature plot to determine the kWh/t required to grind to the desired particle size. This is generally trialled at pilot scale to confirm the energy requirement. However, these tests do not examine changes in pulp chemistry and impact on downstream processing.

Invariably at plant scale fine grinding, whilst achieving the target particle size (ie liberation), rarely achieves the optimum pulp chemistry for flotation. It was observed in the plant at Capricorn Copper Mine (CCM) that the regrind mill discharge typically had a reducing Eh, near zero ppm of dissolved oxygen and a high oxygen demand. This tended to yield poor metallurgical performance in the plant. In the laboratory, regrinding the regrind cyclone underflow produced a pulp that had an oxidising Eh, high dissolved oxygen content and low oxygen demand. Upon further investigation it was also noted that there was a marked difference in pulp temperature between the laboratory and the plant, and produced a superior copper grade/recovery curve.

This paper describes how to conduct a laboratory test that produces a regrind mill discharge pulp chemistry similar to that observed in the plant, by using ceramic media that had been preheated and extending the grind time. These changes to the laboratory procedure produced pulp chemistry readings that were close to those noted in the plant, as well as simulating the poor metallurgical performance. Further work was completed to show how to restore the conditions needed to achieve improved metallurgical results.

INTRODUCTION

The objective of regrinding is to liberate the valuable sulfide minerals from gangue sulfides (for example, pyrite) so that a saleable concentrate can be produced. The modern regrind circuit (eg the IsaMill™, stirred media detritor (SMD) or HIGmill™) generally involves cycloning all or a portion of the rougher/scavenger concentrate such that the fine fraction (the cyclone overflow) bypasses the regrind mill. The coarse fraction (the cyclone underflow) feeds the regrind mill, which is operated at a given power to grind the mineral particles by attrition to achieve the target liberation size. The regrind mill discharge is generally combined with the regrind mill cyclone overflow before reporting to the cleaner circuit.

An examination of the fresh feed to the regrind circuit on a liberation-by-size and -mineral class basis will provide an indication of the particle size to achieve at least 80 per cent liberation of the valuable sulfide mineral. This particle size generally becomes the target cut size for the cyclone, as well as the target regrind product size. For example, if the liberation analysis indicated that the particle size corresponding to 80 per cent chalcopyrite liberation is 20 microns, then the regrind cyclone should be configured such that the P_{80} of the cyclone overflow is nominally 20 microns. Further, the regrind mill should be operated so that the regrind mill discharge has a P_{80} of nominally 20 microns. Therefore, the cleaner feed, a combination of the regrind cyclone overflow and the regrind mill discharge should have a P_{80} of about 20 microns which should yield at least 80 per cent liberated chalcopyrite in the cleaner feed.

This circuit design dominates the modern regrind circuit where the complexity of the ore demands fine grinding employing an IsaMill™, SMD or HIGmill™ to achieve the product size effectively.

It is immediately apparent that liberation studies yield valuable information about the target regrind size, and the locking characteristics of the ore. Equally, development of a signature plot (Figure 1), the relationship between product size (P_{80}) and energy input for the regrind mill feed (ie regrind cyclone underflow) for the expected operating conditions (the ore, pulp conditions and media) is vitally important as this data is used for sizing the mill. However, at no point in these studies are the pulp and surface chemical properties of the particles being treated considered. And why would they be? Invariably, the products from small scale laboratory or pilot plant tests usually exhibit excellent flotation response proving that using these mills are able to achieve the fine particle sizes to obtain liberation, and once liberated significantly better concentrate grades and recoveries are possible. Unfortunately, at industrial scale this is not the case.

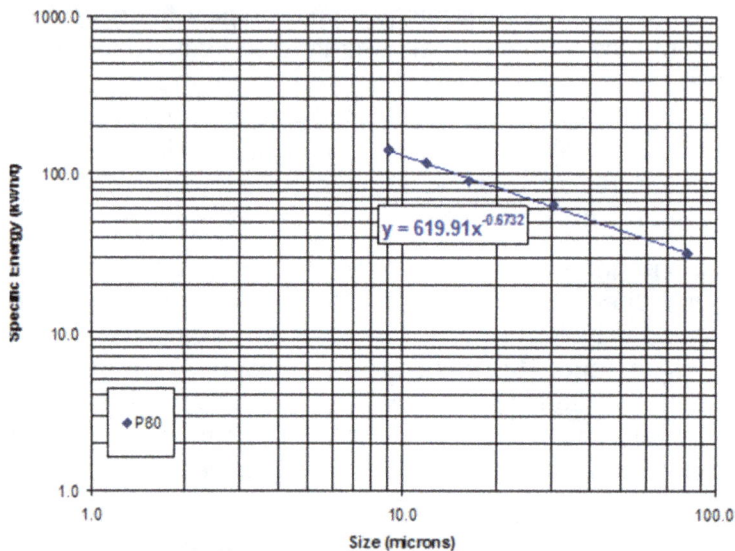

$$y = 619.91x^{-0.6732}$$

FIG 1 – Typical signature plot.

CAPRICORN COPPER MINE

Capricorn Copper Mine (CCM) in far north-western Queensland installed a regrind mill to regrind the coarse fraction of the rougher cells 2 to 4 concentrate to nominally 80 per cent passing 20 microns. After commissioning, it was noted that when the regrind mill was operating the metallurgical performance was below expectation, producing lower copper concentrate grades due to poor selectivity against pyrite. At the time, it was thought that the increased recovery of pyrite may be related to 'copper activation' during grinding. Undoubtedly, some of the pyrite was activated as there is a considerable amount of copper in solution in the process water, but it is unlikely that this cause adequately accounts for the significantly higher pyrite recoveries when the regrind mill was operating.

A pulp chemistry survey around the regrind/cleaner-scavenger circuit indicated that the regrind mill discharge pulp had a lower pH, more reducing Eh, very low dissolved oxygen content and a higher pulp temperature than the rest of the circuit (Figure 2). Further, the oxygen demand (or rate of oxygen consumption or reactivity) of the regrind mill discharge was around 10.5 min^{-1} which is considered to be very high. The data suggested that combining the regrind mill discharge with the regrind cyclone overflow restored the pulp chemistry as measured in the cleaner-scavenger feed, but the anecdotal evidence was that this may be misleading. Undoubtedly, the pulp chemistry of the regrind mill discharge was not ideal for copper sulfide flotation, particularly when considering the near zero oxygen concentration and very high oxygen demand of this process stream. In fact, these conditions tend to favour pyrite flotation (Waterhouse *et al*, 2023; Martin *et al*, 1989). This behaviour has been noted at a number of other site, and appears to be a common condition.

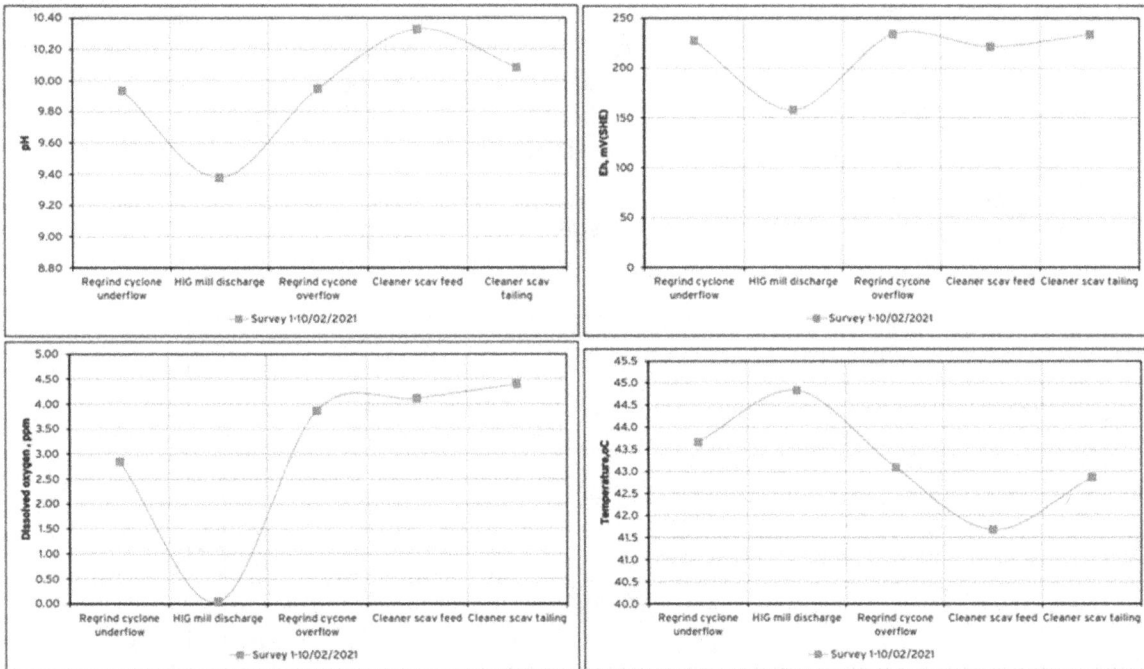

FIG 2 – Pulp chemistry survey of the CCM regrind/cleaner-scavenger circuit.

FINE GRINDING IN THE LABORATORY

Initial laboratory tests were conducted in a small half-litre Netzsch mill, where a 500 g sample of the regrind cyclone underflow was ground to a P_{80} of 20 microns using 2 mm diameter ceramic beads. The ore sample was placed in the Netzsch mill along with the ceramic beads and about 100 ml of process water. The slurry was stirred at 2000 rev/min for nominally 60 sec to reach the desired P_{80}. The regrind product was washed through an 800 micron screen to separate the ceramic beads from the pulp. The pulp was then transferred to a flotation cell where the pH was adjusted to 10.5 and frother was added. Four timed concentrates were collected at 0.5, 1.5, 3.0 and 5.0 mins for a total flotation time of 10 mins. These results were compared to a test conducted on unground regrind cyclone underflow.

Table 1 provides the pulp chemistry of the regrind mill discharge and unground cyclone underflow in the laboratory test. It is apparent that after regrinding the regrind cyclone underflow in the laboratory Netzsch mill that the Eh is oxidising (175 mV (SHE), the dissolved oxygen is moderate (3.9 ppm) and the pulp temperature is close to ambient (23.7°C). That is, the laboratory regrind mill produces a pulp with a pulp chemistry that is suitable for copper flotation. The pulp chemistry of the laboratory flotation feed for the two conditions (Table 1) were similar, meaning that the flotation results should be about the same.

TABLE 1

Pulp chemistry for laboratory tests completed on regrind mill discharge cyclone underflow with and without regrinding.

Test	pH	Eh, mV (SHE)	DO, ppm	Temp, °C
Regrind mill discharge				
No regrinding	-	-	-	-
Regrinding	8.10	175	3.90	23.7
Flotation feed				
No regrinding	10.61	199	1.42	23.8
Regrinding	10.73	187	3.72	23.9

However, when the laboratory regrind mill discharge pulp chemistry numbers (Table 1) are compared with those observed for the plant regrind mill discharge (Figure 2), it is quickly realised that while the Eh is within the same range, the dissolved oxygen content (near zero) and the pulp temperature (at least 20° higher) are markedly different. It is highly likely that these differences are significant and will have an impact on flotation performance.

The copper grade/recovery and copper/pyrite selectivity curves comparing no regrinding with regrinding in the laboratory Netzsch mill are presented in Figures 3 and 4, respectively. The copper concentrate grade and diluent recoveries, at 80 per cent copper recovery are listed in Table 2.

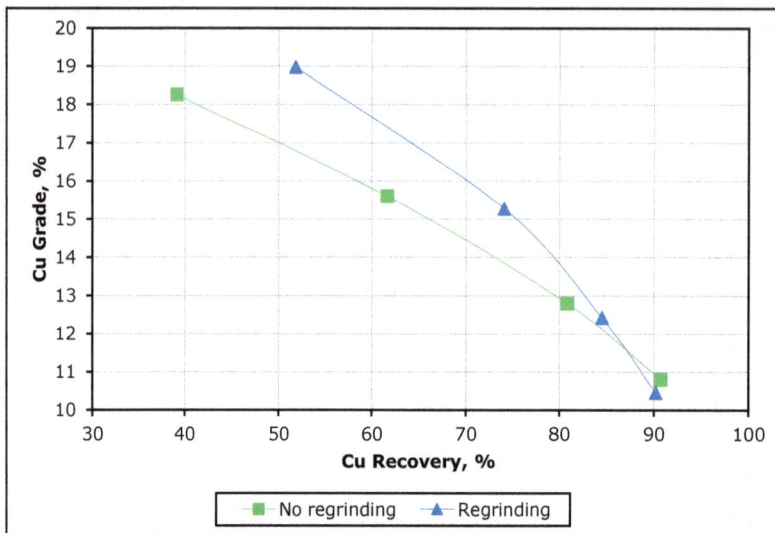

FIG 3 – Copper grade/recovery curves for tests completed on regrind cyclone underflow with and without regrinding.

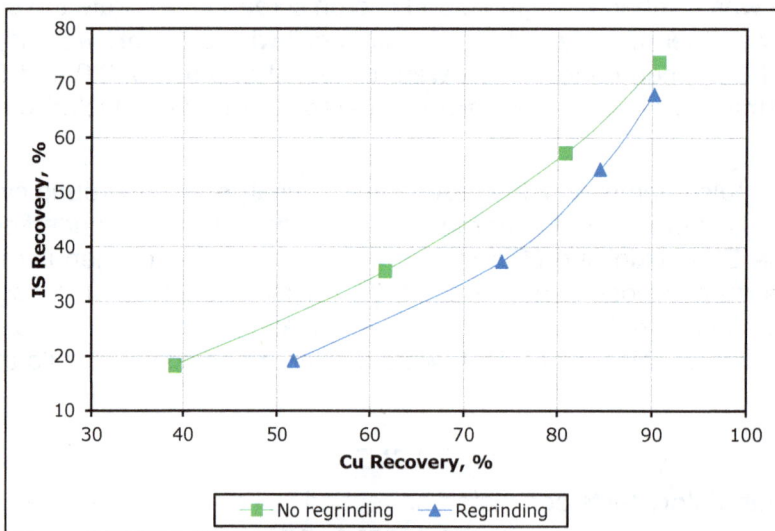

FIG 4 – Copper/pyrite selectivity curves for tests completed on regrind cyclone underflow with and without regrinding.

TABLE 2

Copper concentrate grade and diluent recoveries at 80 per cent copper recovery for tests completed on regrind cyclone underflow with and without regrinding in the laboratory.

Test	Cu grade, %	Diluent recovery, %	
		Pyrite	Non-sulfide gangue
No regrinding	12.9	56.2	19.8
Regrinding	13.7	46.9	22.8

The data shows that after regrinding the cyclone underflow in the laboratory the copper grade/recovery curve is superior to that produced without regrinding (Figure 3). That is, at 80 per cent copper recovery the copper concentrate grade increased by 0.8 per cent (Table 2). The improved copper concentrate grade can be attributed to better selectivity for the copper sulfides against pyrite (Table 2 and Figure 4). As there was little difference in the flotation feed pulp chemistry it is reasonable to assume that the improvement in the grade/recovery curve is related to improved liberation of the copper sulfides from the pyrite.

However, regrinding the regrind cyclone underflow in the laboratory Netzsch mill obviously does not replicate the pulp chemical conditions noted in the plant. So, the flotation results achieved do not resemble the behaviour noted in the plant. The questions posed by CCM were: can a laboratory regrind procedure be developed that replicates the pulp chemistry noted in the plant regrind mill discharge; and how can the flotation response be restored?

MODIFYING LABORATORY GRINDING CONDITIONS TO MATCH THE PLANT

What parameters must change in the laboratory to produce the same pulp chemistry as the plant?

Examining the laboratory regrinding procedure provides some clues as to what can be changed in the endeavour to make the laboratory regrind mill discharge have a pulp chemistry that is similar to that observed in the plant. As noted above, the main differences were that the laboratory regrind mill discharge had more oxidising Eh values and higher dissolved oxygen content (and by default a lower oxygen demand), and a lower pulp temperature.

Working backwards the first parameter to be tackled is temperature. To achieve pulp temperatures similar to those observed in the plant the beads were heated under drying lamps for three to four hours. It was anticipated that the ceramic beads would retain the heat and transfer it to the pulp during grinding thereby raising the pulp temperature to something close to that observed in the plant.

The second strategy adopted was to slow down the impeller speed and extend the grind time. Rather than grinding at 2000 rev/min for 60 secs, the test was initially completed at 250 rev/min for 40 mins.

Ultimately, a procedure was developed whereby 500 g of cleaner feed and 600 ml of process water were added with 1 litre of heated ceramic beads to the laboratory stirred mill. The pulp was homogenised for two mins at 150 rev/min before increasing the impeller speed to 250 rev/min and grinding for a further 38 mins to achieve the target P_{80}. The pulp chemistry was measured throughout the grinding process. Figure 5 shows the plot of the Eh, dissolved oxygen and temperature of the pulp through the grinding.

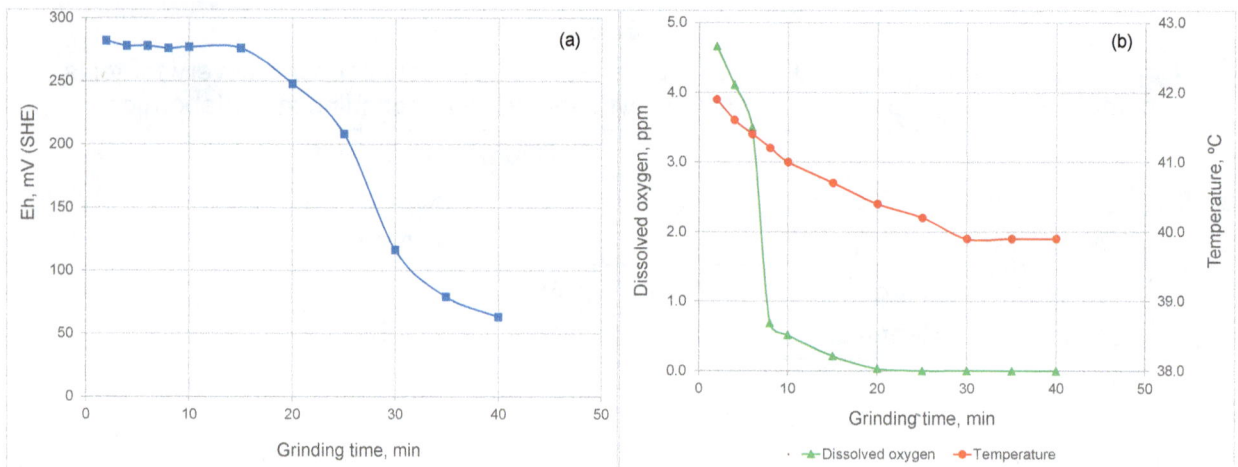

FIG 5 – Laboratory regrind pump chemistry: (a) Eh and (b) dissolved oxygen and temperature.

It is apparent that following this procedure produced Eh, dissolved oxygen and temperature values that were comparable to those observed in the plant (Figure 2). The residual differences are related to variations in the mineralogy of the ore.

To determine the impact on copper flotation, tests were completed with and without regrinding. Table 3 lists the pulp chemistry of the mill discharge and unground regrind cyclone underflow. Firstly, the pulp temperature of the regrind mill discharge was markedly higher. It is also evident that the flotation feed for the test with regrinding had a significantly lower Eh, marginally lower dissolved oxygen concentration, with a markedly higher oxygen demand. This signified that the pulp was hungry for oxygen and one would expect that copper flotation would be poor.

TABLE 3

Pulp chemistry of the regrind mill discharge and flotation feed.

Test	pH	Eh, mV (SHE)	DO, ppm	Temp, °C	OD, min^{-1}
Regrind mill discharge					
No regrinding	-	-	-	-	-
Regrinding	8.01	-38	0.00	38.3	-
Flotation feed					
No regrinding	11.02	162	0.31	27.3	2.64
Regrinding	11.03	0	0.00	30.3	18.12

The copper grade/recovery curves for these tests are provided in Figure 6, and the corresponding iron sulfide selectivity curves are given in Figure 7. As the grade/recovery curves are markedly different the flotation rate constants and maximum recoveries have been determined and are available in Table 4.

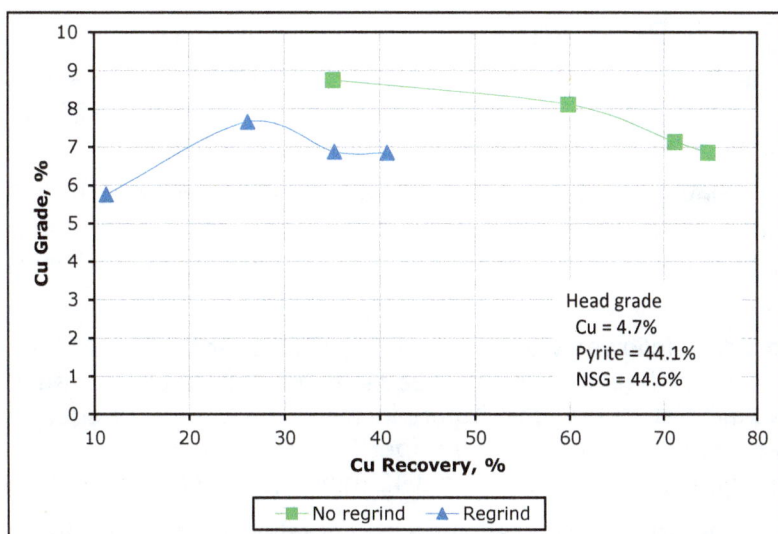

FIG 6 – Copper grade/recovery curves for tests completed on regrind cyclone underflow with and without regrinding (note the high pyrite feed grade).

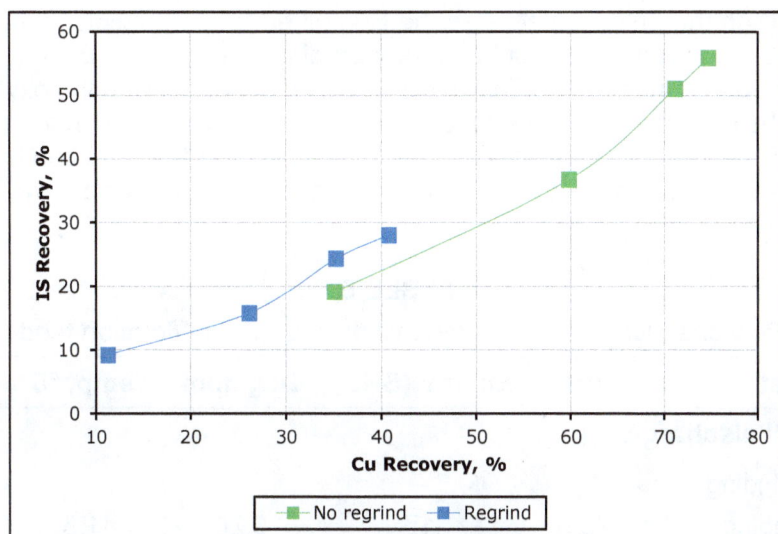

FIG 7 – Iron sulfide selectivity curves for tests completed on regrind cyclone underflow with and without regrinding.

TABLE 4

Flotation kinetic data for tests with and without regrinding.

Test	Cu		Iron sulfide (IS)		Non-sulfide gangue (NSG)	
	k, min^{-1}	R_{max}, %	k, min^{-1}	R_{max}, %	k, min^{-1}	R_{max}, %
No regrinding	1.03	74.7	0.57	57.7	0.58	43.5
Regrinding	0.48	43.2	0.42	30.5	0.43	24.8

An examination of Figure 6 immediately shows that copper grade/recovery curve for the test with regrinding is inferior to the test without regrinding. The copper flotation rate constant and maximum copper recovery for the test with regrinding were nominally half that of the test without regrinding. The flotation rate constants and maximum recoveries for the iron sulfide (Figure 7) and non-sulfide gangue are approximately the same as that reported for copper. As there is very little difference in the copper flotation kinetics and that of the gangue species selectivity is very poor in the test with regrinding, which yields lower copper concentrate grades and recoveries, even when the copper sulfide minerals are liberated.

OVERCOMING POOR REGRIND MILL PULP CHEMISTRY

A second series of tests were completed to firstly validate the procedure employed above, and secondly to determine a methodology to mitigate the poor pulp chemistry of the regrind mill discharge and improve the copper flotation response.

In this instance, to achieve the desired P_{80} a 30 min regrind was required. Further, an oxygen demand test was completed on the regrind product to determine the aeration time required to obtain a more satisfactory pulp reactivity prior to flotation. In this instance a 12 min aeration time was required.

To reiterate, the procedure followed was to add 500 g of cleaner feed and 600 ml of process water were added with 1 litre of heated ceramic beads to the laboratory stirred mill. The pulp was homogenised for two mins at 150 rev/min before increasing the impeller speed to 250 rev/min and grinding for a further 28 mins to achieve the target P_{80}. Three tests were completed: one without regrinding, one with regrinding and a third with a 12 min aeration step following regrinding.

The pulp chemistry of the regrind mill discharge and flotation feed for the three tests are given in Table 5. The regrind mill discharge pulp chemistry was similar for the two regrind tests, having a reducing Eh, low dissolved oxygen content and high pulp temperature. The flotation feed on the other hand varied considerably. That is, the Eh after regrinding was more reducing than the test with no regrinding, but when the pulp was aerated the Eh shifted to more oxidising pulp potentials. The dissolved oxygen concentration in the pulp decreased significantly with regrinding, but rebounded when the aeration step was introduced. This trend was also observed with the oxygen demand. That is, the reactivity of the pulp increased dramatically with regrinding, but add in the aeration stage and the oxygen demand was reduced significantly. So, it would appear that aeration has a positive effect on the pulp chemistry. Does this additional conditioning step have a positive impact on copper flotation?

TABLE 5
Pulp chemistry of the regrind mill discharge and flotation feed.

Test	pH	Eh, mV (SHE)	DO, ppm	Temp, °C	OD, min^{-1}
Regrind mill discharge					
No regrinding	-	-	-	-	-
Regrinding	7.26	-17	0.20	41.8	-
Regrinding + aeration	7.49	-28	0.36	40.3	-
Flotation feed					
No regrinding	10.53	211	4.67	21.7	1.16
Regrinding	10.34	128	0.06	29.2	13.54
Regrinding + aeration	10.22	197	1.04	29.9	3.47

In terms of flotation response, the copper grade/recovery curves for this series of tests are shown in Figure 8, and the corresponding iron sulfide selectivity curves can be found in Figure 9. The flotation kinetic data for copper, irons sulfides and non-sulfide gangue are available in Table 6.

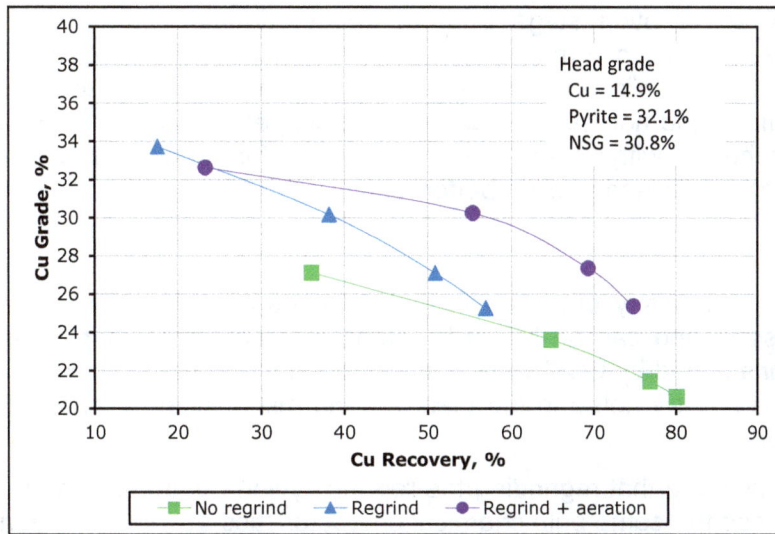

FIG 8 – Copper grade/recovery curves for tests completed on regrind cyclone underflow with and without regrinding, and regrinding with aeration (note the lower pyrite feed grade).

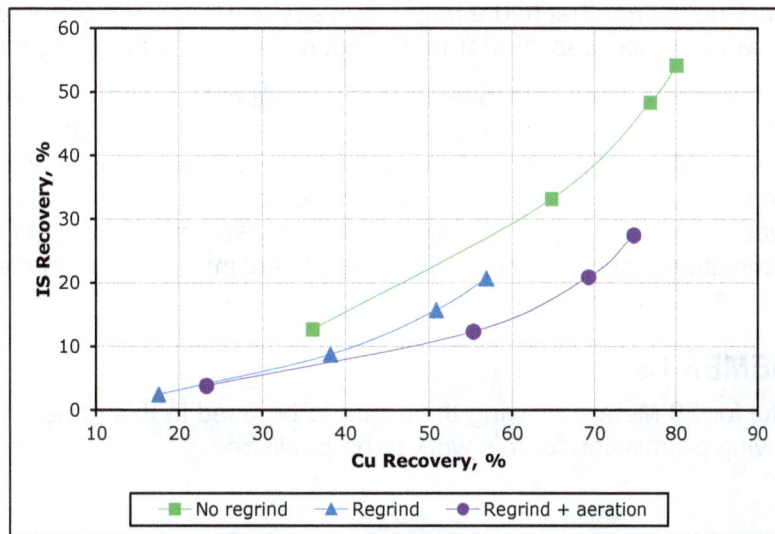

FIG 9 – Pyrite selectivity curves for tests completed on regrind cyclone underflow with and without regrinding, and regrinding with aeration.

TABLE 6

Flotation kinetic data for tests with and without regrinding as well as regrinding with aeration.

Test	Cu		Iron sulfide (IS)		Non-sulfide gangue (NSG)	
	k, min^{-1}	R_{max}, %	k, min^{-1}	R_{max}, %	k, min^{-1}	R_{max}, %
No regrinding	1.00	80.2	0.48	55.6	0.53	46.8
Regrinding	0.55	59.1	0.18	27.5	0.26	25.8
Regrinding + aeration	0.66	76.2	0.25	32.0	0.33	31.4

In this particular case, the copper feed grade is considerably higher and the pyrite feed grade is only two thirds that of the previous set of tests. Despite this regrinding produced an inferior copper grade/recovery curve compared with the test without regrinding (Figure 8 and Table 6). Regrinding saw a 50 per cent reduction in the copper flotation rate constant and the maximum copper recovery decreasing by 25 per cent. The flotation kinetics of the iron sulfides and non-sulfide gangue were also severely affected as well.

The introduction of an aeration stage prior to copper flotation partially restored the copper grade/recovery curve producing a markedly higher copper concentrate grade for a given copper recovery. The improvement in copper concentrate grade can be attributed to improved selectivity against the iron sulfides and non-sulfide gangue. The copper flotation kinetics have increased but were still lower than the test without regrinding. It is felt that further optimisation of the aeration time and reagent additions would lead to even better results.

CONCLUSIONS

Firstly, the poor pulp chemistry observed in the plant regrind mill discharge at CCM has been observed at other sites, and can be linked to inferior flotation behaviour in subsequent flotation stages. The questions posed by CCM were: can a laboratory regrind procedure be developed that replicates the pulp chemistry noted in the plant regrind mill discharge; and how can the flotation response be restored?

The test work demonstrated that regrinding the regrind cyclone underflow in the laboratory did not produce a pulp that had the same pulp chemical characteristics as the plant regrind mill discharge. This potentially produced flotation test results that were better than those achieved in the plant. To replicate the plant pulp chemistry in a laboratory regrind mill the ceramic beads were heated for three to four hours, the mill impellor speed was reduced and the grind time extended. These actions produced a regrind mill discharge that had a pulp chemistry similar to that observed in the plant and further, the flotation performance was inferior to the test completed without regrinding.

Following this procedure made it possible to replicate the plant pulp chemical conditions and flotation response. The next step in the test work was to develop a methodology that would shift the pulp chemistry to more favourable conditions, and improve flotation. This was achieved by completing an oxygen demand test to determine the aeration time of the regrind mill discharge. With the aeration time determined it was shown that applying this conditioning step not only brought the pulp chemistry to more favourable conditions, but also resulted in a significant improvement in metallurgy, with high concentrate grades for a given recovery.

ACKNOWLEDGEMENTS

The author is grateful for 29 Metals allowing their data to be used in this paper. Magotteaux is also acknowledged for giving permission for this work to be published.

REFERENCES

Martin, C J, Rao, S R, Finch, J A and Leroux, M, 1989. Complex sulphide or processing with pyrite flotation by nitrogen, *International Journal of Mineral Processing*, 26(1–2):95–110.

Waterhouse, P, Greet, C J, Munro, P D and Bennett, D W, 2023. Mineralogy, Chemistry and Recovery: The New Century Story – Venturing into extremes, in *Proceedings of MetPlant 2023*, pp 207–230 (MetVal Consulting Pty Ltd: Fulham Gardens).

Improved classification with the Cavex® DE hydrocyclone for mill circuit, coarse particle flotation, and tailings applications

J J Hanhiniemi[1,2], J Heo[3], N Weerasekara[4], E Wang[5], H Thanasekaran[6] and A Kilcullen[7]

1. MAusIMM(CP), Director of Engineering, Weir Minerals, Brisbane Qld 4000.
   Email: jeremy.hanhiniemi@mail.weir
2. Adjunct Fellow, Julius Kruttschnitt Mineral Research Centre, Brisbane Qld 4000.
   Email: jeremy.hanhiniemi@uqconnect.edu.au
3. Process Engineer, Weir Minerals, Brisbane Qld 4000. Email: jun.heo@mail.weir
4. Principal Engineer, Weir Minerals, Sydney NSW 2000. Email: nirmal.weerasekara@mail.weir
5. Process Engineer, Weir Minerals, Perth WA 6000. Email: echo.wang@mail.weir
6. Flotation Process Engineer, Eriez Flotation Division, Melbourne Vic 3000.
   Email: homiet@eriez.com
7. Weir Minerals Technical Centre Manager, Weir Minerals, Melbourne Vic 3000.
   Email: adam.kilcullen@mail.weir

ABSTRACT

The Cavex® DE hydrocyclone is presented, which incorporates two stages of classification in a single device with no intermediate pumping. Experimental and pilot test campaign results show the Cavex® DE hydrocyclone can replace a two hydrocyclone cluster system for coarse particle flotation (CPF) feed preparation. Comparisons of classification performance between the Cavex® DE hydrocyclone and a single stage hydrocyclone at a copper-gold operation in Queensland are presented. Advancements in modelling technique to support selection and optimisation of the Cavex® DE hydrocyclone is also presented, including new methods. Advanced techniques used in optimising the design are described, including multi-phase computational fluid dynamics (CFD) with Lagrangian and Eulerian approaches for coarse and fine particle tracing respectively. CFD modelling was first validated using industrial scale tests, and then used in optimising designs and selection methodology. Some details on example historic application of Cavex® DE technology, as well as new industrial scale test results and optimisation, is provided. This includes example tailings applications, and an application in a Chilean grinding circuit with CPF. The effective generation of separate slimes and sand streams with high recovery is presented, supporting the beneficial use of tailings in impoundment construction. The flexibility inherent in Cavex® DE technology, with its array of interchangeable options and configurations, provides the industry with an adaptable, high performing classifier in the various applications listed above. The test results and operational data presented show the benefits in performance available to the industry.

INTRODUCTION

The Cavex® DE hydrocyclone integrates a two-stage separation process within a single unit. The design incorporates a primary hydrocyclone with a cylindrical section transitioning into an inverted conical section (Figure 1). The bottom chamber has wash water injection feeding a secondary hydrocyclone. An optional air core booster (ACB) can be installed atop each hydrocyclone to support the vortex and air core.

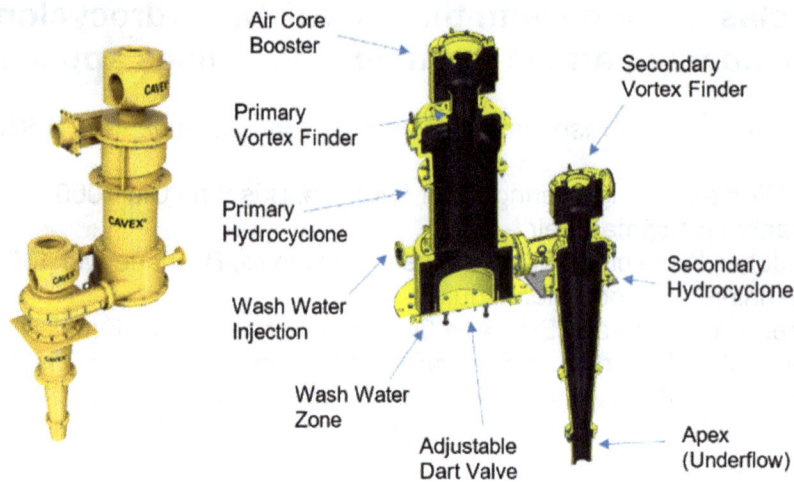

FIG 1 – Cavex® DE hydrocyclone.

The Cavex® DE hydrocyclone's design simplifies operations by eliminating the need for additional pumps, sumps, or civil structures, reducing both capital and operational expenses. The adjustable conical valve influences the distribution of mass between the two stages, optimising cut size and underflow fine and coarse particle size distribution. It acts like an adjustable spigot for the first stage of classification. The wash water zone and chamber smoothly delivers the slurry to the secondary classification stage. It does this in a smooth manner, reducing turbulence, pressure loss, and wear. The addition of wash water is instrumental in determining the cut size of the hydrocyclone, impacting the proportion of underflow fines and the performance of the secondary stage. Distinctly, the Cavex® DE hydrocyclones wash chamber is segregated following the initial classification stage, thus ensuring a controlled, two-stage separation process. Both stages of the Cavex® DE hydrocyclone employ the inlet geometry that was shown to significantly improve both wear life, as well as classification performance, in site measurements taken in trials against older designs (Warman International, 1998). Note the Cavex® DE model number is designated by the first and second stage hydrocyclone diameters in millimetres (mm), respectively, eg 250/150 Cavex® DE.

A comparative analysis of a single-stage hydrocyclone and a Cavex® DE hydrocyclone under identical feed and slurry conditions (to ensure an unbiased assessment) is presented in this paper. The 250/150 Cavex® DE hydrocyclone's secondary stage hydrocyclone, employed in the comparative test work, was configured to mirror the single-stage Cavex® hydrocyclone, providing a consistent basis for evaluating performance differences. The effects of water injection are significant to the performance of the Cavex® DE hydrocyclone, as it influences the slurry density, cut size, and the mass recovery of the final hydrocyclone stage. The effects of water injection are detailed in this paper.

This paper also details several current installations of the Cavex® DE hydrocyclone, especially in tailings and in Latin America, offering insights into their operation. In the construction of tailings storage facilities, it is possible to separate and utilise a free draining sands portion of a tailings stream to support the construction of embankments (depending on the feed particle size distribution). This reduces the overall requirements of tailings management. This is often achieved via hydrocyclones. According to Kujawa (2011) the 'free-draining characteristic of the sand is defined by the mass fraction of particles of less than 75 µm. This is a concept borrowed from the soil sciences where by definition free drainage occurs at a particle mass fraction of 5 per cent passing 75 µm. The definition has been amended for tailing dam construction to slightly higher percentages by mass passing 75 µm, with 15 per cent being common, 18 per cent considered on the high end'. The Cavex® DE hydrocyclone will later be shown to be an excellent desliming solution for such applications.

Another application well suited for Cavex® DE technology described in this paper is in the feed preparation for coarse particle flotation (CPF). Eriez HydroFloat™ CPF is a technology that offers the ability to highly effectively recover valuables at coarse sizes, offering the ability to achieve coarse gangue rejection and thus pre-concentration. This therefore also results in lower energy consumption in comminution. The HydroFloat™ cell 'works on combining the principals of flotation with hindered settling... [and] the key characteristic of the HydroFloat™ cell is the presence of an aerated fluidised bed' (Vollert, Akerstrom and Seaman, 2019). As CPF is less effective at separating fine material, and since fine material below -75 μm can disrupt the function of the HydroFloat™ cell, fines need to be removed from the feed to CPF; target limits for the percentage of solids mass below -75 μm in the feed can often be circa 10 per cent. Therefore, similarly to tailings, the Cavex® DE hydrocyclone is examined in this paper as a solution to CPF feed preparation requirements.

In what follows, the selection and optimisation of the Cavex® DE will be first introduced, including modelling approaches, computational fluid dynamics (CFD) results, and also a summary of recent industrial scale tests that have been used to further improve selection and optimisation methods. Next, comparisons between Cavex® DE and a single stage hydrocyclone are made. Finally, some details are provided for different applications, including in CPF and in tailings processing.

SELECTION AND OPTIMISATION OF CAVEX® DE HYDROCYCLONES

In this section the methodology developed for selecting and optimising the performance of Cavex® DE hydrocyclones is discussed. First the methodology is described, and then some guidance is provided for optimisation.

Modelling and selection methodology

Recent further advancements in the methodology used to predict and optimise the performance of the DE hydrocyclones involved several steps, including extensive test work at the Weir Technical Centre (WTC) in Melbourne Australia, full surveys of the hydrocyclones four slurry streams, and model development. The methodology was also informed through CFD analysis. Slurry samples collected from the hydrocyclone feed sump at a nickel mine concentrator were utilised (primary semi-autogenous grinding (SAG) grinding circuit product) for the portion of the test work reported in this section, as well as other sites and historic data. This test regime and model development methodology considered variables available for selection (eg hydrocyclone geometries and wash water addition). The methods developed further support optimisation and scaling of the hydrocyclone, and provides further guidance for greenfield selection and brownfield optimisation.

CFD analysis was also carried out and indicated no upstream backflows originating from the DE hydrocyclones lower chamber into the primary stage in the conditions reviewed; a full and controlled two stage classification is achieved. This observation aligns with the water mass balance derived from WTC test results. Based on these observations, a two-stage modelling approach has been adopted for performance prediction and optimisation. In lieu of Weir's internal selection and optimisation tools, a two-stage modelling approach could be performed in an application such as JKSimMet® using the following two steps, as an initial estimate of performance. The listed process models could also be applied, which are further described in Napier-Munn (1996):

- First Stage: Employing the 'Single Component Efficiency' model.
- Secondary Stage: Employing the 'Nageswararao' model.

Additionally, the model incorporates a water addition, representing the primary lower chamber. This addition aids in regulating the pressure control and solid concentration for the secondary hydrocyclone feed. It achieves this by enabling precise control over the water volumetric flow rate and secondary feed percent solids. External calculations are used to inform the model parameters of the first stage, as the geometry is dissimilar to conventional hydrocyclones. Full surveys, including of both overflows separately, and pressure measurement at the feed to the second stage, allow for model fitting in both stages. This approach is illustrated in Figure 2.

DE SELECTION, SCALE-UP & OPTIMIZATION

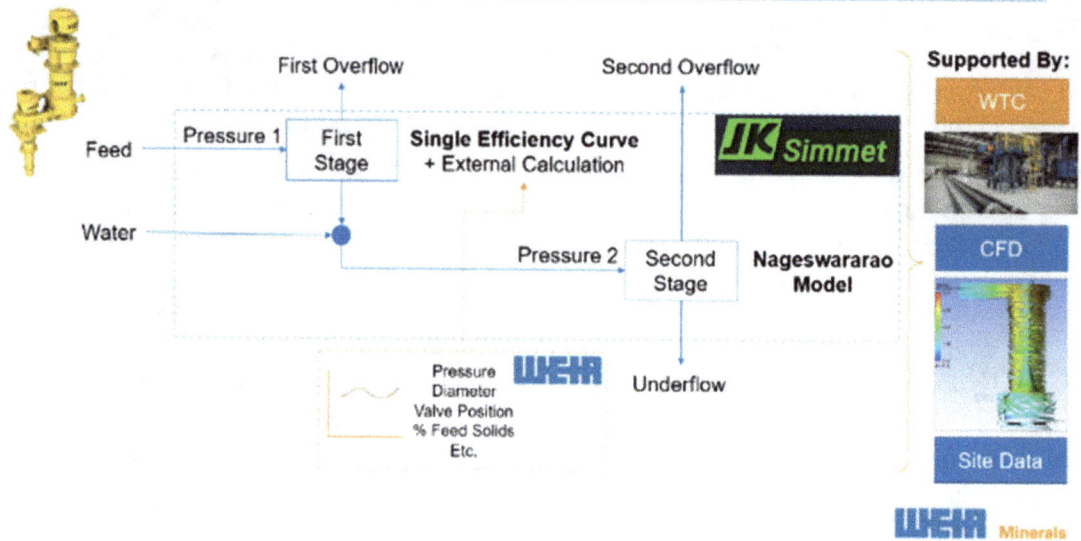

FIG 2 – Cavex® DE selection, scale-up and optimisation modelling approach.

CFD model description

CFD was used to analyse the hydrocyclones performance. As described above, it was used to establish the validity of a two-stage modelling approach, and to improve design and modelling approaches. Slurry flows in hydrocyclones can be described as a multiphase flow, consisting of various sizes of ore particles in a slurry media. This mixture of coarse and fine particle slurry travels inside hydrocyclone system creating a turbulent flow. It is also necessary to capture the air core. These complex flows can be solved by a number of CFD techniques. These include the full Eulerian multiphase approach, simplified Eulerian approaches such as the mixture and Volume of Fluid (VOF) models, and the Lagrangian approach. For this work multiphase Eulerian and Lagrangian approaches were used for fine and coarse particle tracing respectively.

The Eulerian model is the most complex of the multiphase models in Ansys Fluent CFD, the package used in this work. It solves a set of n momentum and continuity equations for each phase. Coupling is achieved through the pressure and interphase exchange coefficients. The manner in which this coupling is handled depends upon the type of phases involved; granular (fluid-solid) flows are handled differently than nongranular (fluid-fluid) flows. For granular flows, the properties are obtained from application of kinetic theory. Momentum exchange between the phases is also dependent upon the type of mixture being modelled.

Given that in a complex slurry structure like the one modelled, there is a range of particle sizes from 'fines' to 'coarse'. Therefore, for relative simplicity, a number of different particle classes are selected to reasonably represent the size distribution of the slurry stream. These selected size classes were represented as multiple granular phases along with water as primary phase.

Example illustrations of CFD results are shown in Figure 3, showing the primary stage and wash chamber, and these mirrored the test conditions at the WTC. The streamline plot (a) shows the outside coarse materials travel, and the internal upward vortex that carries the fine material to the overflow. The lower figures (c) and (d) show the smooth transition to the secondary stage that reduces turbulence, wear, and pressure loss to the next stage. It can be seen from the figures that this is supported by the design and action of the wash water injection.

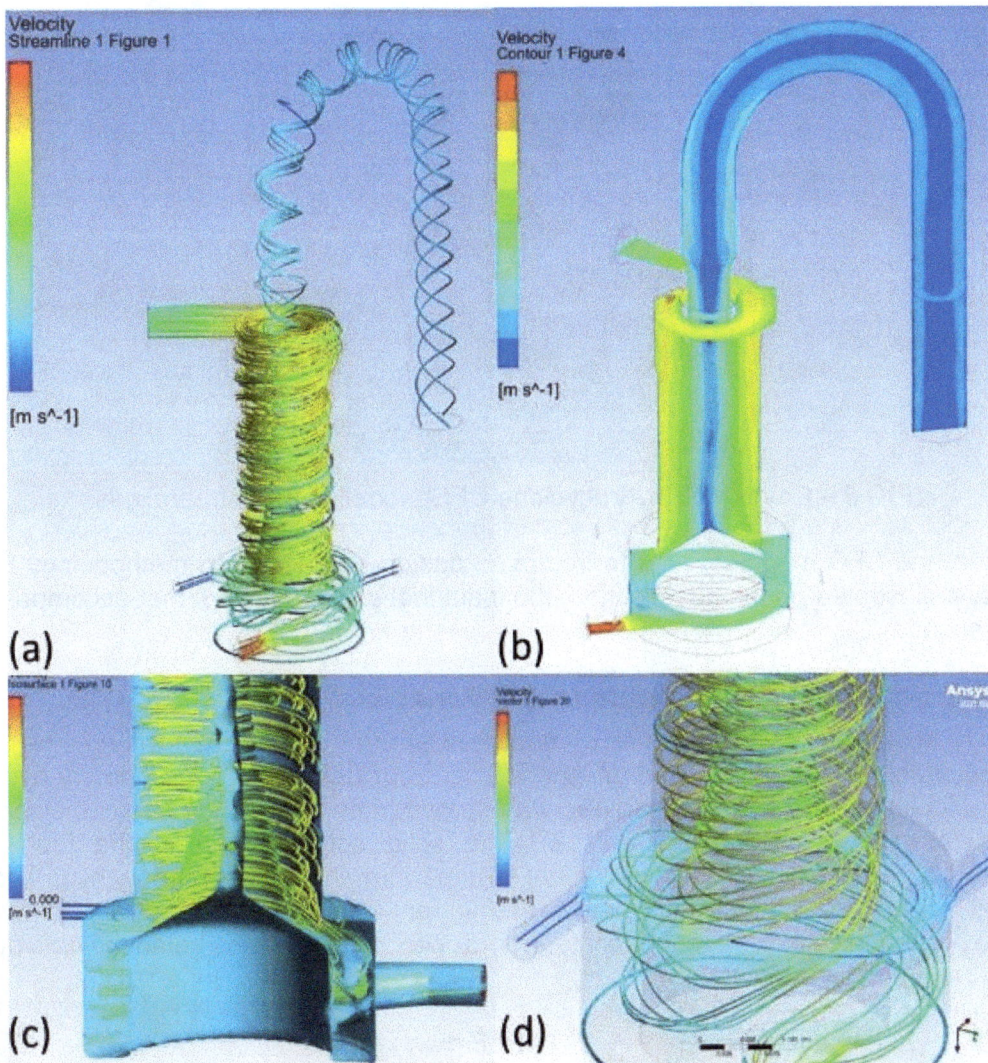

FIG 3 – Cavex® DE hydrocyclone CFD modelled flow velocity field and stream lines showing the primary stage and lower chamber.

Figure 4 shows the modelled flow velocity vector near the dart valve between the first stage and the wash chamber. Flow is out of primary stage, into the wash chamber, and supports the two-stage modelling approach (described above). The streamlines in Figure 5 also supports this observation, with flush water directed to the next stage.

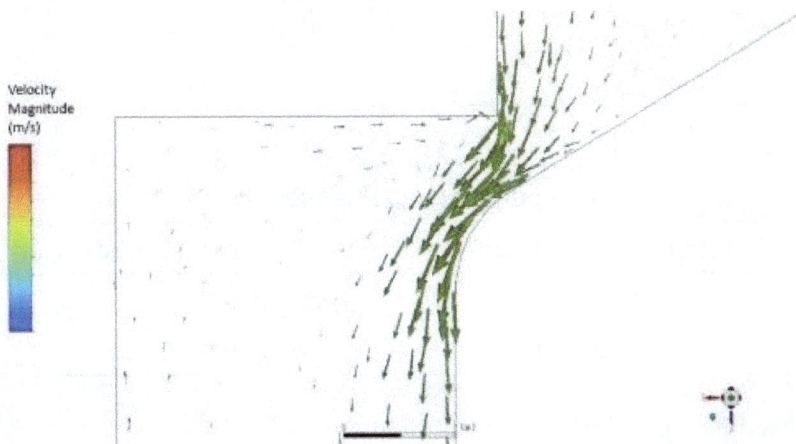

FIG 4 – Cavex® DE hydrocyclone CFD modelled flow velocity vector shown mid plane near the gap.

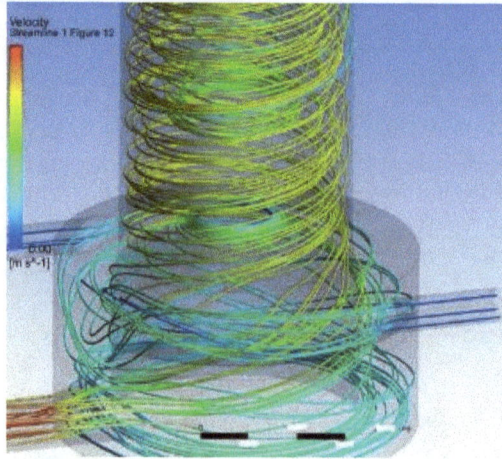

FIG 5 – Cavex® DE hydrocyclone CFD modelled flow streamlines.

In conclusion, the CFD methods used to improved design and selection methodology have been described and illustrated. In the next section, the industrial scale test work that accompanied this is also summarised.

Industrial scale tests – feed conditions and variables

The industrial scale test work and surveys conducted at the WTC in 2022 will now be described. Initially two variables, namely the feed dry tonnes per hour (tph) rate to the primary hydrocyclone and water injection rate to the wash chamber, were tested while keeping the geometric configuration unchanged (as shown in Tables 1 and 2). The solid concentration in the feed remained approximately constant at 55 per cent by weight. The performance of the primary hydrocyclone was primarily influenced by the feed rate. Whereas the percent solids and the feed rate both varied and affected the performance of the secondary stage hydrocyclone. The mass balanced data of the WTC test results can be seen in Table 3.

TABLE 1

Hydrocyclone geometries used in the WTC tests.

	Primary hydrocyclone	Secondary hydrocyclone
Cyclone diameter	250 mm	150 mm
Vortex finder	100 mm	60 mm
Gap size	30 mm	N/A
Spigot size	N/A	25 mm
Inlet diameter	73 mm	41 mm

TABLE 2

Independent variables tested.

Independent variable	Values (mass-balanced)
Water injection*	2, 4, 6, 11%
Feed t/h (dry)	46.4–81.0

TABLE 3

Feed condition and mass-balanced data.

Data set		L2 P2	L2 P3	L2 P4	L2 P5	L2 P6	L2 P7	L2 P8	L2 P9	L2 P10
Feed	t/hr	46.37	66.28	76.79	46.82	80.82	78.32	81.04	79.17	75.45
Primary operating pressure	kPa	78	151	209	84	210	212	210	209	205
Secondary operating pressure	kPa	56	95	127	58	128	128	127	128	123
% Solids primary feed	%	52.9	54.8	55.8	55.3	55.5	55.8	55.1	55.4	58.2
% Solid secondary feed (calculated)	%	61.7	65.7	67.3	60.9	65.7	64.7	64.3	60.4	56.7
Total feed volumetric flow rate	m³/hr	58.2	78.8	88.8	54.9	96.1	94.2	101.2	102.5	81.6
Water injection % to secondary hydrocyclone*	%	0	0	0	0	2%	4%	6%	11%	0

*% of DE slurry feed volumetric rate.

Greenfield modelling guidelines based on above test work

Observations from the test work above provide some insight into Greenfields modelling of the DE hydrocyclone, and these will now be described. Note that these observations relate to the test work described in Table 1 through Table 3, which is limited to a single application rather than a stipulation on general performance, so only the general trends will be described. As described above, to simulate a Cavex® DE hydrocyclone, a two-stage modelling approach in JKSimMet can be applied. The subsequent sections will now cover the major model parameters required for this modelling approach.

Primary hydrocyclone parameters (single component efficiency model)

The primary hydrocyclone geometry is not the same as a conventional hydrocyclone, and so the single component efficiency model in JKSimMet is used rather than other models based specifically on conventional geometry. All model parameters were fitted using this approach for the primary hydrocyclone, and alpha, water split, and D50c were calculated from the test data described in this paper. Responses to each independent variable (eg water injection rate, vortex finder size etc) were tested.

Alpha selection

In the primary stage, the feed rate fluctuated without displaying any significant trend in relation to alpha. The alpha values in the primary stage were observed to vary between magnitudes similar to conventional single stage hydrocyclones. For greenfield projects, similar alpha values for the first stage could be utilised to what is seen in conventional single stage hydrocyclones as a preliminary guideline when more precise estimates are not available.

Water split to O/F selection

For the primary hydrocyclone, the water split to the overflow (O/F) ratio was observed to be within the 63–70 per cent range when operating pressures varied between 100 kPa and 210 kPa in the tests per Table 3. This was based on a consistent solid concentration of 55 per cent in the feed during tests conducted at the WTC. An upward trend in the water split ratio to O/F was noted with increasing pressure. Other data shows that as feed solid concentration reduces, the water split increases (eg in examples with the feed in the 35–45 per cent range, water split increases to 80 per cent). For applications in greenfield projects, this commentary can provide some guidance, and water splits in the first stage of the Cavex® DE are not dissimilar to those obtained in conventional single stage hydrocyclones.

D50c selection

In lieu of Weir's internal toolset, as a first estimate, it might be assumed the 'Nageswararao' model can be employed to provide an indication of the D50c response for the primary stage of the DE hydrocyclone for various sizes and feed pressures. This prediction assumes scaling from the same feed slurry, and the same primary hydrocyclone shape. It also assumes that the nongeometric terms in Nageswararao's empirical model will have the same exponents and terms, despite them being developed for quite a different geometry. For operation with different feeds and geometries (eg vortex finder to hydrocyclone diameter ratios) judgement is required for D50c calculation. With the assumptions above, much of the empirical formula of the Nageswararao model D50c calculation (Napier-Munn, 1996) can be simplified to a single new constant defined in this paper, C_J, specific to like geometry ratios and the same feed characteristics. The ore characteristics and feed conditions need to be considered with care.

$$D50c = C_J \left(\frac{D_o}{D_c}\right)^{0.52} D_c^{0.57} \lambda^{0.93} \left(\frac{P}{\rho_p g}\right)^{-0.22} \tag{1}$$

$$\lambda = 10^{1.82 C_v}/(8.05[1 - C_v]^2) \tag{2}$$

where:

C_J	= A constant determined from experiments for the primary stage
λ	= Hindered settling correction term
Cv	= Volumetric fraction of solids in feed slurry
g	= 9.81 (m/s^2)
D50c	= in m
P	= Operating pressure (kPa)
ρ_p	= Slurry density (t/m^3)
D_o	= Vortex finder (m)
D_c	= Cylinder diameter (m)

This modelling approach was applied to the test work of Table 3, and appears capable of predicting the D50c using the equation given for operating pressures within a ranging of where it was fitted (from 120 to 210 kPa in this study). However, it is important to proceed with caution when predicting performance at operating pressures well outside of the fitted range as it becomes more difficult to predict the D50c.

Secondary hydrocyclone simulation (Nageswararao Model)

As the CFD analysis showed, no flow back into the primary hydrocyclone appears to occur past the dart valve, thus typical Cavex® hydrocyclone behaviour is expected in the secondary stage hydrocyclone. The 'Nageswararao' model can be chosen to simulate the second stage as a first estimation, by including the wash water and the first stage underflow as combined feed, and it provides the ability to simulate performance changes due to adjustments in operating pressure and hydrocyclone dimensions, as well as scaling.

Water split and wash water injection

It was observed that increasing water injection improves the water split to O/F ratio of the secondary hydrocyclone, which in turn improves the classification of fines. However, as the water injection rate was further increased, the gradient of this improvement in fines classification appears to diminish.

Cyclone optimisation

The following sections provides supporting detail on key selections for the Cavex® DE hydrocyclone.

Guideline for primary/secondary Vortex Finder (VF) and spigot size selection

The selection of these parameters is critical to achieving the best performance, and should be tailored to the specific characteristics of the feed, ore properties, and the desired process objective. Weir has completed extensive testing and can support the optimisation of these parameters based on the measured results obtained.

Cyclone pressure optimisation

Primary hydrocyclone pressure recommendation

Results from the WTC tests indicate that the most effective hydrocyclone classification efficiency occurs when operating pressures at the primary hydrocyclone inlet in a similar range to those typical of conventional hydrocyclones, and in the tests conducted in Table 3, improved classification was seen at pressure ranging between 120–160 kPa. If there's a need to make the P_{80} of the global underflow coarser, elevating the pressure in the primary hydrocyclone is a suitable approach. For greenfield, using tools such as Weir CySelect, flow and pressure curves, or any other available resources is advised. These instruments assist in determining the operating pressures aligned with specific flow rates, providing valuable insights into the most fitting hydrocyclone geometries.

Secondary hydrocyclone pressure recommendation

For the secondary hydrocyclone, optimal fines removal in the underflow has been linked to the use of the largest VF sizes coupled with higher operating pressures (100–130 kPa). This set-up exploits the complex relationship between the two hydrocyclone stages, where the secondary hydrocyclone's pressure is typically 70–90 per cent of the primary stage feed pressure (in the tests run). A suitable pressure range in the secondary hydrocyclone has been identified as key to efficiently separate fines, based on WTC test outcomes.

SINGLE VERSUS CAVEX® DE HYDROCYCLONE PERFORMANCE

A comparative study of single stage 150CVX hydrocyclone versus 250/150 Cavex® DE hydrocyclones was carried out. This work was based on test work conducted at WTC under similar conditions. Another key focus of this study was evaluating the hydrocyclones' performance based on variations in the water injection rate into the secondary hydrocyclone. Insights derived from this comprehensive analysis were subsequently leveraged to meet the prerequisites for CPF preparation, details of which are discussed further in the following sections.

The analysis included a survey of each hydrocyclone arrangement utilising identical slurry samples (see Figure 6 for the particle size distribution (PSD) comparison) from a tailings feed from a Copper mine. This method ensured the data sets were comparable, with each maintained at a uniform pressure of 150 kPa. A data set from the 150CVX was chosen to specifically match the configuration of the secondary hydrocyclone of the DE system (also a 150CVX model), as detailed in Table 4.

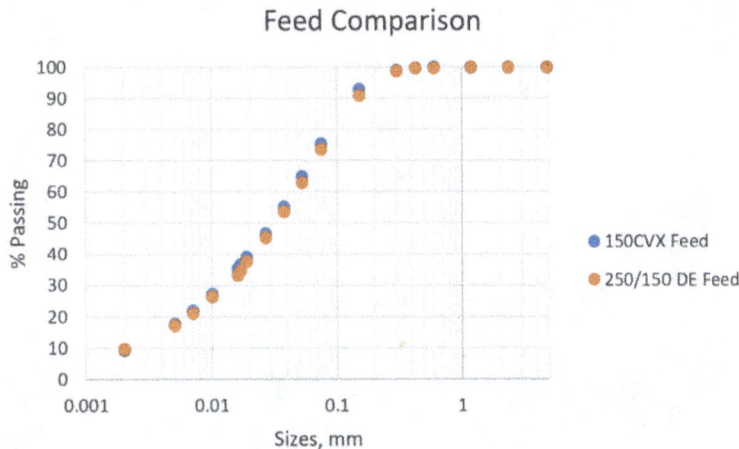

FIG 6 – Feed comparison.

TABLE 4
Single stage hydrocyclone and DE hydrocyclone configuration.

	Single-stage hydrocyclone	Cavex® DE hydrocyclone
Configuration	150CVX	Secondary Stage of the DE (150CVX)
VF (mm)	60	60
Spigot (mm)	25	25

The analysis focused on the underflow and overflow PSDs, represented in Figures 7 and 8, which indicate that the 250/150 Cavex® DE hydrocyclone achieved a lower percentage of sub 75 μm sized particles in the underflow (29 per cent) compared to the 150CVX hydrocyclone (50 per cent). This was achieved while maintaining similar overflow PSDs across both data sets. This can also be seen in Table 5. This result indicates the 250/150 Cavex® DE hydrocyclone's improved performance in fines desliming.

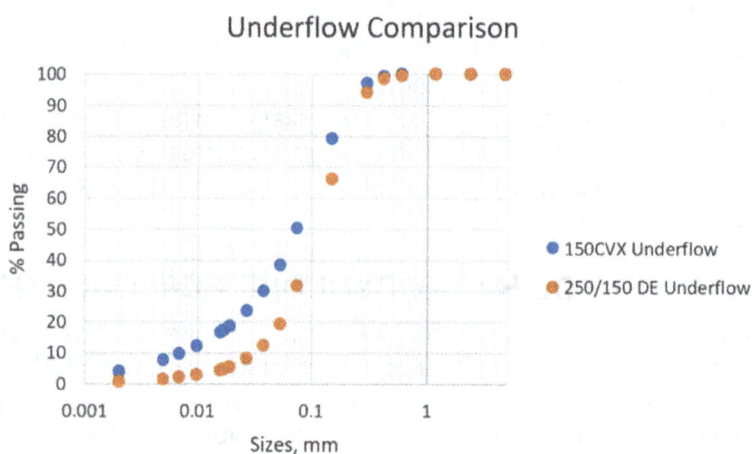

FIG 7 – Underflow comparison.

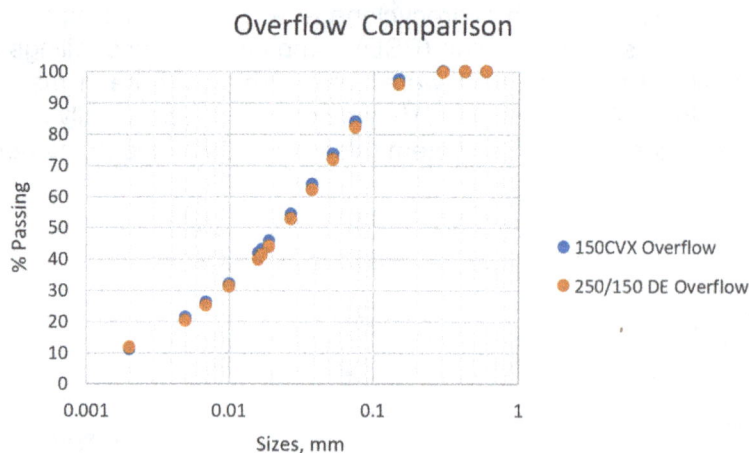

FIG 8 – Overflow stream comparison.

TABLE 5

Cumulative passing of sub 75 μm.

Cumulative passing of -75 μm	F_{80} (μm)	U/F (%)
250/150 Cavex® DE	87	29
150CVX	87	50

The recovery of fines (particles smaller than 75 μm) in both the overflow and underflow streams were also examined (Table 6). The Cavex® 250/150 DE hydrocyclone demonstrated a solid recovery of sub 75 μm particles of approximately 93 per cent from feed to O/F stream, compared to 84 per cent for the 150CVX. This indicates a notably better recovery of fines for the Cavex® 250/150 DE hydrocyclone.

TABLE 6

Mass split/recovery of solids sub 75 μm from feed.

Mass split/recovery of sub 75 μm from feed	To O/F (%)	To U/F (%)
250/150 Cavex® DE	93.0	7.05
150CVX	83.5	16.5

A comparison of the recovery of coarse particles larger than 75 μm indicated that the Cavex® 250/150 DE hydrocyclone had a 62 per cent recovery for coarse materials, compared to a higher rate of 72.4 per cent for the 150CVX (Table 7). This suggests that while the DE hydrocyclone demonstrated superior performance in fine particle classification, attention needs to be placed on maintaining coarse material recovery to the U/F stream (perhaps given the larger diameter of the primary stage in this example) relative to the single stage hydrocyclones; note that the DE hydrocyclone primary VF size can also be adjusted.

TABLE 7

Mass split/recovery of solids larger than 75 μm from feed.

Mass split/recovery of +75 μm from feed	To O/F (%)	To U/F (%)
250/150 Cavex® DE	37.8	62.2
150CVX	27.7	72.4

The observed data indicates a decrease in the overall percentage of mass recovery of solids from feed to U/F stream when comparing the Cavex® 250/150 DE hydrocyclone versus 150CVX as indicated in Table 8. This decrease in the mass recovery aligns with expectations given the improved fine particles classification. Consistent with predictions, there was a marked increase in the quantity of fines reporting to the overflow.

TABLE 8

Total solid mass recovery % from feed to U/F stream comparison.

Overall percentage of solid mass split/recovery from feed	To U/F (%)
Cavex® 250/150 DE	16.4%
150CVX	24.8%

The water balance will now be considered. The Cavex® DE hydrocyclone demonstrates a significant advantage over single-stage hydrocyclones in terms of reduced water bypass to the underflow. In the scenario presented in Table 9, the Cavex® DE hydrocyclone achieved an underflow solids content of 86 per cent, compared to 78 per cent for the 150CVX (Table 10). While this high solids concentration may indicate a condition close to roping, this can be mitigated by adjusting the spigot and vortex finder configurations. Notably only 3 per cent of the total process water reported to the underflow for the 250 Cavex® DE hydrocyclone, compared to approximately 10 per cent for the 150CVX. Moreover, under suitable feed conditions, the water balance of the Cavex® DE hydrocyclone allows its combined overflow to be suitable for conventional flotation, with a solids content of 48 per cent in this example (Table 9).

TABLE 9

Water balance.

	250/150 Cavex® DE	150CVX
Total water volumetric flow rate (m³/hr)	67	20
% Solids U/F	86	78
% Solids O/F	48	55
Water bypass to U/F (% of total)	3	10

TABLE 10

Cumulative passing % of sub 75 µm – Site B Test.

Lowest cumulative passing % of sub 75 µm in the underflow stream	
Site	**Cumulative passing % of sub 75 µm in the underflow**
Site B (F_{80} of 253 µm)	12.9%

In summary, the Cavex® 250/150 DE hydrocyclone demonstrated superior performance in fine particle removal, with 29 per cent of cumulative passing of sub 75 µm particles in the underflow, compared to the 150CVX's 50 per cent. It achieves performances that would otherwise require multiple stages of conventional hydrocyclones with associated increased CAPEX and OPEX costs associated with the intermediate pumping. However, a slight increase in coarse material loss was observed with the 250/150 Cavex® DE hydrocyclone compared to a single-stage hydrocyclone (likely due to the large primary stage and coarser cut). The Cavex® DE hydrocyclones improved capability in fines removal significantly contributes to the overall quality and classification efficiency improvement, producing a coarser product P_{80} of 199 µm versus 153 µm with the single-stage 150CVX hydrocyclone.

The effects of water injection to the secondary hydrocyclone

Figure 9 illustrates the relationship between the wash water injection rate, expressed as a percentage of the total feed volume rate to the primary stage, and the cumulative % mass passing of sub 75 µm in the underflow of each sample. The data indicates that an increase in the water injection rate correlates with a reduction in the fine particle content within the underflow. Consequently, the water injection rate can act as a controllable variable to vary the fines content in the underflow to meet specific targets, such as CPF feed preparation or free drainage for tailings impoundment considerations. However, the overall mass recovery of solids may be affected by variation in the water injection to the secondary hydrocyclone, as seen in Figure 10. The wash injection influences the slurry density and feed rate to the secondary cyclone and hence influences its cut size. In another operation, increasing wash water addition was observed to produce a finer overflow at a given feed pressure, while keeping other operating conditions unchanged. Increasing wash water was also seen to reduce the solids concentration of the underflow and increases the mass recovery to underflow (Banerjee *et al,* 2023). These observations are then consistent across these different operations.

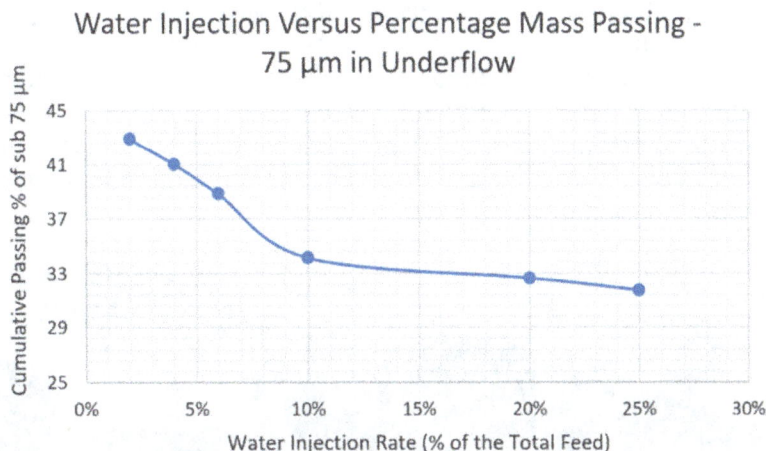

FIG 9 – Water injection rate versus sub 75 µm passing in underflow.

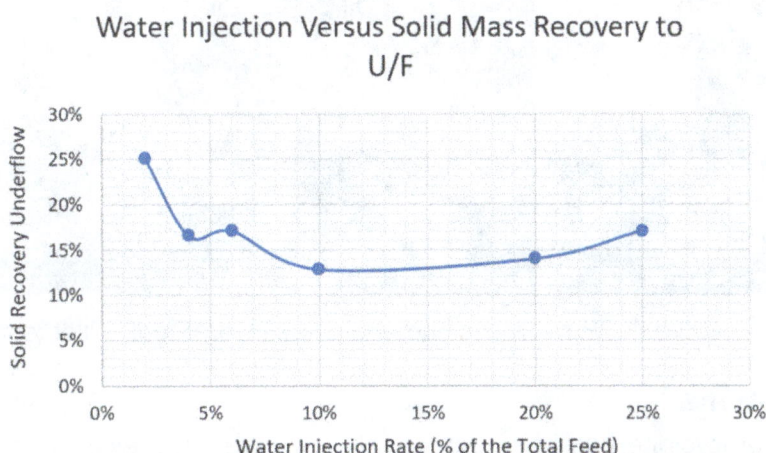

FIG 10 – Solid mass recovery to U/F stream.

CAVEX® DE APPLICATIONS

Having introduced the Cavex® DE hydrocyclone, described its modelling, selection and optimisation methods, and having compared the Cavex® DE against a single stage hydrocyclone, we will now review some example applications in the following part of the paper. This will include CPF feed preparation first, followed by examples in tailings. Finally, an example list of installations is provided.

CPF feed preparation

Weir is targeting sub 10 per cent mass passing of -75 µm in the Cavex® DE hydrocyclones underflow stream for feed preparation to CPF. Achieving this could eliminate the necessity for the Eriez CrossFlow classifier in CPF plants, and added classification stages, improving CAPEX and OPEX performance. Through testing at the WTC, application of the Cavex® DE hydrocyclone has been specifically studied for CPF feed applications, and an example further set of test work from another site (Site B) is now provided. This application was from a relatively fine grinding circuit product, which was fed to the Cavex® DE hydrocyclone. An underflow cumulative mass passing -75 µm of 12.9 per cent was achieved, while the P₈₀ of the feed was 253 µm, per Table 10. This was achieved at relatively low wash water percentage (12 per cent). From subsequent test results showing the continued improvement with increasing water injection rate (refer to Figure 9), the target of below

10 per cent passing of sub 75 µm is expected to be achieved with increased water injection and cyclone configuration improvements. Cavex® DE hydrocyclones have also been employed in CPF feed preparation after semi-autogenous grinding (SAG), and prior to regrinding, at a Chilean Copper mine.

Weir Minerals Australia (WMA) designed and manufactured a Cavex® DE classification pilot test rig that can be integrated with Eriez Australia's HydroFloat CPF pilot test rig. This combined classification and CPF pilot test rig can be transported to sites across the region for piloting. Most recently, Weir and Eriez Australia conducted a pilot plant study to evaluate the Cavex® DE pre-classification and HydroFloat CPF at Northparkes Mine (NPM); a copper and gold mine in central New South Wales. Survey data was collected from December 2023 to May 2024, and assessed the performance of the hydrocyclone to prepare CPF feed and CPF performance itself from both 'coarse' (primary ball mill hydrocyclone overflow), and 'fine' (scavenger tails) feed sources. Some photos of the pilot test rig are shown in Figure 11.

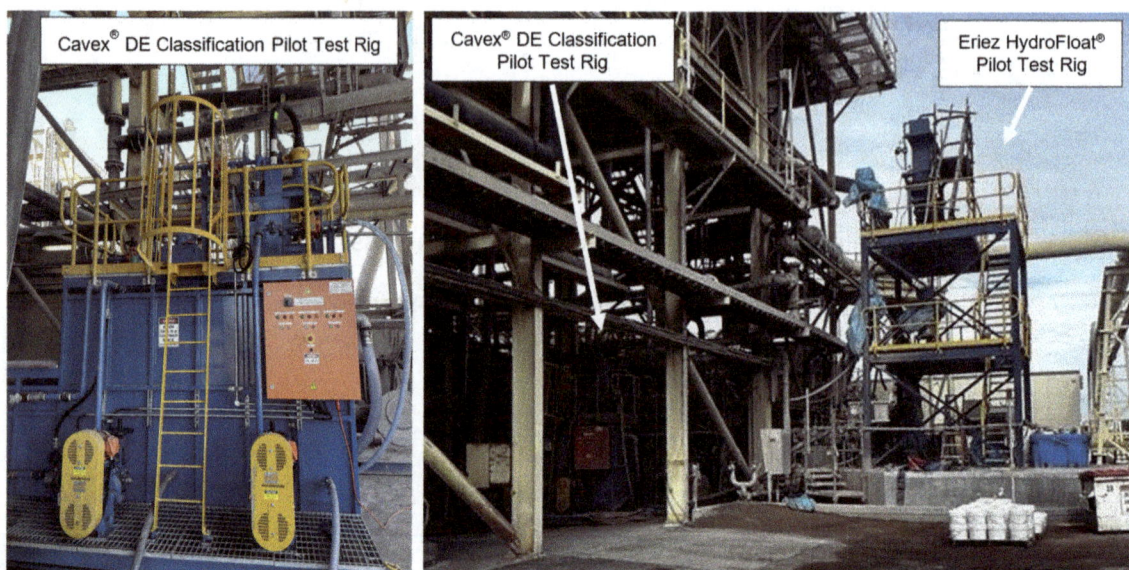

FIG 11 – Cavex® DE hydrocyclone and Eriez HydroFloat® Pilot test rigs.

Tailings applications

The Cavex® DE hydrocyclone has been extensively employed successfully, especially in tailings applications, and especially in Latin America, for particle size classification and dewatering. Many of these applications employ the DE hydrocyclone in desliming of tailings to generate sand for tailings wall construction. It is also used as a classifier in regrind applications, and for CPF feed preparation. Figure 12 shows an example image of a Cavex® DE hydrocyclone cluster installation. The 'Example installation list' section shows just some examples of the extensive list of successful applications in tailings is provided. The next section also provides two more detailed examples of Cavex® DE hydrocyclones operating on tailings feeds (Case Study A and B).

FIG 12 – Cavex® DE hydrocyclone cluster installation examples (bottom) and in a tailings facility in South America (top).

Detailed Case Study A

Hydrocyclone test work was carried out with a tailings sample from a gold-copper mine, 'Site A', at the WTC. The use of the Cavex® DE hydrocyclone resulted in an enhanced water split (93 per cent) to the overflow stream compared to traditional hydrocyclones (70–90 per cent). A fine classification improvement was observed, attributable to an improved water split within the hydrocyclone, which led to a significant reduction in the cumulative percentage of particles smaller than 75 μm in the underflow stream, achieving 14 per cent. The measured data from this application was modelled fitted as a single stage hydrocyclone by combining the overflow streams, and by using the 'Single Component Efficiency Curve Model', which characterises the hydrocyclones performance by three overall performance parameters: sharpness of the cut (alpha), water split % to overflow stream, and D50c cut size. This modelling approach is further described in Napier-Munn (1996).

The DE hydrocyclone showed superior separation efficiency, as indicated by the alpha value in the efficiency curve. The calculated alpha value for the Cavex® DE hydrocyclone was 4.7, markedly higher than the typical range of 2 to 2.5, when model fitted with the single efficiency curve model. A variety of slurry samples with varying feed conditions were also trialled. The volumetric flow rates of the hydrocyclone feed and underflow streams are recorded and measured using the flow metres installed. The density and the PSD data of each stream was measured. Table 11 includes an example feed condition used.

TABLE 11

Feed conditions.

	Site A Cavex® DE hydrocyclone
Feed t/h (dry)	40.2
% Solids, feed	43
Volumetric flow rate, m³/hr	68
Pressure, kPa	100
SG (specific gravity)	2.71

The survey data from the WTC was mass-balanced and then model-fitted in the JKSimMet process modelling environment. The Cavex® DE hydrocyclone modelled in JKSimMet as a single-stage hydrocyclone, to compare its performance parameters with those of the conventional hydrocyclone, is shown in Figure 13. The Single Component Efficiency Curve Model performance parameters can be seen in Table 12.

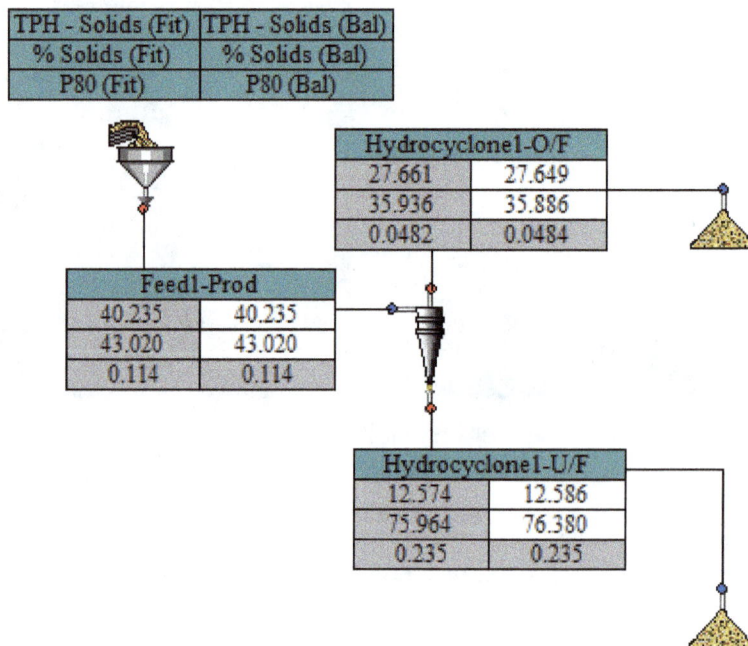

TPH - Solids (Fit)	TPH - Solids (Bal)
% Solids (Fit)	% Solids (Bal)
P80 (Fit)	P80 (Bal)

Hydrocyclone1-O/F	
27.661	27.649
35.936	35.886
0.0482	0.0484

Feed1-Prod	
40.235	40.235
43.020	43.020
0.114	0.114

Hydrocyclone1-U/F	
12.574	12.586
75.964	76.380
0.235	0.235

FIG 13 – Tailings Cavex® DE survey model-fitted as a single stage hydrocyclone.

TABLE 12

Cavex® DE performance parameters.

	Site A
Combined alpha	4.7
D50c (µm)	98
Water split % to O/F	92.5
Combined P_{80} (µm) O/F	48
Feed P_{80} (µm) to Primary Cyclone	114

The industrial standard performance parameters of typical single stage hydrocyclones are alpha's of 2–2.5, and water splits of 70–90 per cent, with D50c varying depending on the dimensions and the slurry conditions. Therefore, based on the survey results, the Cavex® DE hydrocyclone shows excellent performance in these surveys, and for tailings desliming duties.

Detailed Case Study B

The second tailings case study of a Cavex® DE application is taken from literature (Banerjee *et al*, 2023) and is summarised here. This case study related to the pilot scale operation of a Cavex® 500/400 DE (500 mm diameter primary stage and 400 mm diameter secondary stage) processing a sulfide copper tailings feed with a high proportion of slimes (F_{80} of 200 µm, 50 per cent solids concentration by weight, 47 per cent solids mass below -75 µm, at a feed pressure of 103 kPa). The objective was to reclaim a sand product for the purpose of supporting the construction of a tailings dam embankment, and to recover as much water as possible. An underflow solids of sub 75 µm of 17 per cent by mass was achieved in the underflow. The same operation achieved 12 per cent sub 75 µm by increasing feed pressure to 138 kPa. This represents an efficient result on tailings desliming. Table 13 shows the Cavex® 500/400 DE geometry used in this application, and Figure 14 shows the PSDs achieved.

TABLE 13

Cavex® DE hydrocyclone geometry in Case Study B (after Banerjee *et al*, 2023).

Table 3. DE hydrocyclone geometry used for pilot scale testing.

Dimension	Primary	Secondary
Hydrocyclone diameter (mm)	500	185
Inlet diameter (mm)	211	145
Vortex finder diameter (mm)	185	130
Spigot diameter (mm)		70
Cone angle		10

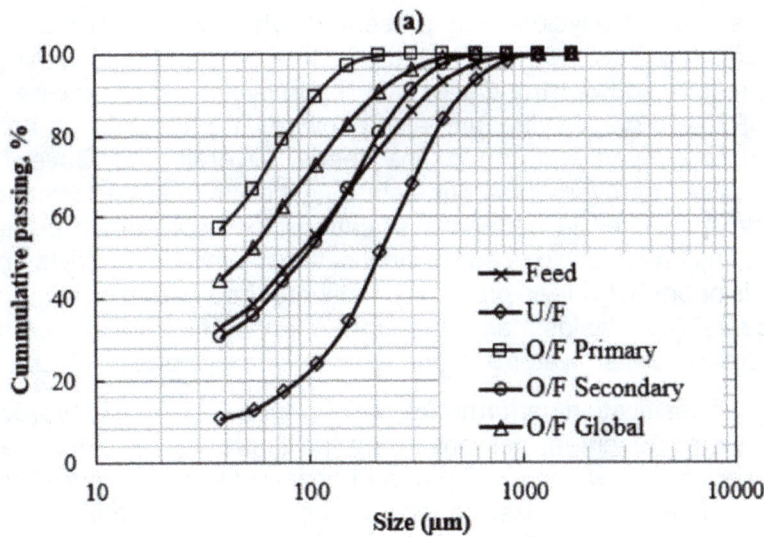

FIG 14 – Cavex® DE hydrocyclone PSDs in Case Study B (after Banerjee *et al*, 2023).

Example installation list

Table 14 shows an example list of installations of Cavex® DE hydrocyclones, in various sizes and cluster configurations.

TABLE 14

Example installations.

Location	Application	Additional information	Model	Cluster quantity	Hydrocyclones per cluster
	Tailings	Wall construction	500/400CVX-DE	1	12
	Tailings	Wall construction	500/400CVX-DE	3	6
	Tailings	Wall construction	500/400CVX-DE	1	20
	Tailings	Sand Deposit	500/400CVX-DE	2	18
Latin American Mines	Tailings	Wall construction	500/400CVX-DE	2	22
	Tailings	Sand Deposit	400/250CVX-DE	1	5
	Tailings	Sand Deposit	400/250CVX-DE	1	5
	Tailings	Sand Deposit	650/500CVX-DE	1	9
	Tailings	Sand Deposit	500/400CVX-DE	1	4
	Tailings	Sand Deposit	500/400CVX-DE	1	8

CONCLUSIONS

The Cavex® DE two stage hydrocyclone was presented, which was shown to achieve the efficiency of two stage classification without the CAPEX and OPEX of intermediate pumping. Experimental and pilot test campaign results showed the Cavex® DE hydrocyclone can replace a two hydrocyclone cluster system for CPF feed preparation, and also achieves the desliming required for free draining tailings sand production, even with quite fine feeds. Overall, the Cavex® DE hydrocyclone demonstrated substantial improvements in fine classification, higher performance parameters, and separation efficiency. It resulted in better classifications for especially the fine material which is critical in CPF and tailings hydrocyclone sands applications. While in the tests and feeds included at the time of writing this paper, CPF feed preparation still required further configuration optimisation to reach the <10 per cent -75 µm solids mass goal for fine feeds, the data presented supports this can be achieved with increased wash water and changes in hydrocyclone configuration.

Direct comparisons of classification performance between the Cavex® DE hydrocyclone and a single stage hydrocyclone were presented, demonstrating the improved performance of the Cavex® DE which resulted in fewer underflow fines (-75 µm) and an increased recovery of coarse product to the underflow, thereby improving fine classifications, compared to a single stage hydrocyclone. The single stage hydrocyclone demonstrated less efficient separation by comparison, although the total mass recovery to the underflow was higher, which could be due to more misclassifications of fine materials to the underflow.

Advancements in modelling technique to support selection and optimisation of the Cavex® DE hydrocyclone were described, which have employed CFD, pilot testing, and industrial scale laboratory testing. Guidance on making first estimations of Cavex® DE performance was also provided, as were some details on selection and optimisation. Furthermore, the influence of significant variables such as cyclone geometries and wash water injection rate etc, on performance were explored.

Example historic application of Cavex® DE technology, as well as new industrial scale test results and optimisation, was provided. This included some detailed examples in tailings applications, and a list of example installations. The flexibility inherent in Cavex® DE technology, with its array of interchangeable options and configurations, provides the industry with an adaptable, high performing classifier in the various applications listed above. The test results and operational data presented show the benefits in performance available to the industry. The Cavex® DE hydrocyclone was demonstrated in this paper to offer excellent classification performance in the various applications reviewed.

REFERENCES

Banerjee, C, Cepeda, E, Switzer, D and Hunter, S, 2023. Application Potential of Cavex® Double Effect Hydrocyclone for the Classification of Mine Tailings – A Pilot Scale Study, *Mineral Processing and Extractive Metallurgy Review*. https://doi.org/10.1080/08827508.2023.2298376

Kujawa, C, 2011. Cycloning of Tailing for the Production of Sand as TSF Construction Material, in *Proceedings of Tailings and Mine Waste 2011*, 11 p.

Napier-Munn, T J, Morrell, S, Morrison, R D and Kojovic, T, 1996. Mineral comminution circuits: their operation and optimisation, Julius Kruttschnitt Mineral Research Centre.

Vollert, L, Akerstrom, B and Seaman, B, 2019. Newcrest's industry first application of Eriez Hydrofloat technology for copper recovery from tailings at Cadia Valley Operations, Copper 2019 Conference, Vancouver.

Warman International, 1998. The Effect of a New Cyclone Shape on Wearlife and Separation Performance, Comminution 1998.

Improved cone crusher chamber design and operation using design and simulation tools

J J Hanhiniemi[1], N S Weerasekara[2] and S Buaseng[3]

1. MAusIMM(CP), Director of Engineering, Weir Minerals, Brisbane Queensland.
   Email: jeremy.hanhiniemi@mail.weir
2. Principal Engineer, Weir Minerals, Sydney New South Wales.
   Email: nirmal.weerasekara@mail.weir
3. Senior Engineer, Weir Minerals, Sydney New South Wales. Email: santy.buaseng@mail.weir

ABSTRACT

This paper presents the use of design tools and simulation methods, including desktop tools, and computational tools like discrete element method (DEM), in the improvement of cone crusher chamber designs.

Custom chamber designs can be completed based on the given feed size distribution, target closed side setting, equipment parameters, and the process objective. Weir has developed an expedient methodology for routine assessment of chamber designs, based on well researched models and techniques, combined with operating data it has gathered. These methodologies and tools were informed by more computationally intensive DEM simulation. It combines equipment design rules with crusher process models.

The DEM simulations employed a full Whiten particle breakage model that provides throughput capacity, indication of power, while giving insight into chamber design and showing relative differences in pressure profile across different liner designs.

An example application is presented, which had a narrow feed size distribution. To prevent localised wear, this application required a progressive liner design to even out applied pressure along the depth of the liner. The example application shown had a very high ore competency and a narrow feed size distribution; it was shown that methods like Bond's crushability index were poor at predicting the required power in such applications, and more sophisticated methods like the Whiten crusher model in JKSimMet® were required. It was shown that the DEM analysis, while providing detailed insight into the behaviour of chamber geometry design, predicted similar comminution power to JKSimMet®.

INTRODUCTION

This paper summarises design and simulation tools and their use in a challenging pebble recycle cone crusher operation from an open pit copper-gold mine. In the design and operation of the upstream circuit (Figure 1), hard ore components from the run-of-mine ore which survive the upstream semi-autogenous grinding (SAG) mill, and subsequent cone crushing application, are being strongly preferentially selected on their hardness/competence, and fed to this cone crusher (shown in the dashed box). Further, the application is intended to achieve a fine closed side setting (CSS) of 7 mm, in a large 2.1 m diameter mantle crusher. The feed distribution is also very narrow, requiring a customised chamber design. For these reasons, the application is challenging, and advanced engineering tools and approaches were applied to improve operation in the field, which are described in this paper.

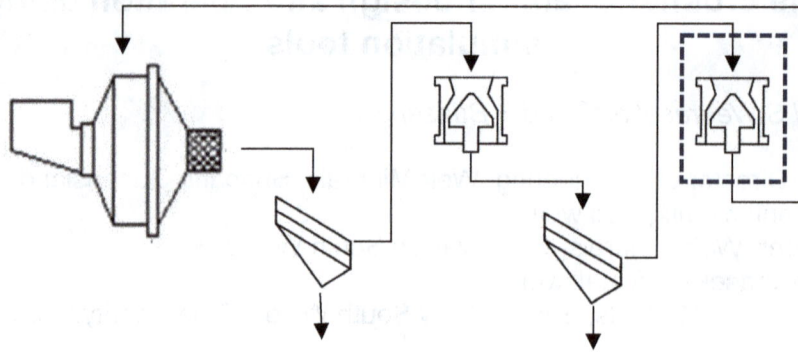

FIG 1 – Upstream circuit.

Feed particle size distribution (PSD) and ore competence testing was conducted on a survey of the ore from the cone crusher feed conveyor. It had a P_{80} of 17.5 mm, with a top size of circa 30 mm. The feed PSD is shown in Figure 2.

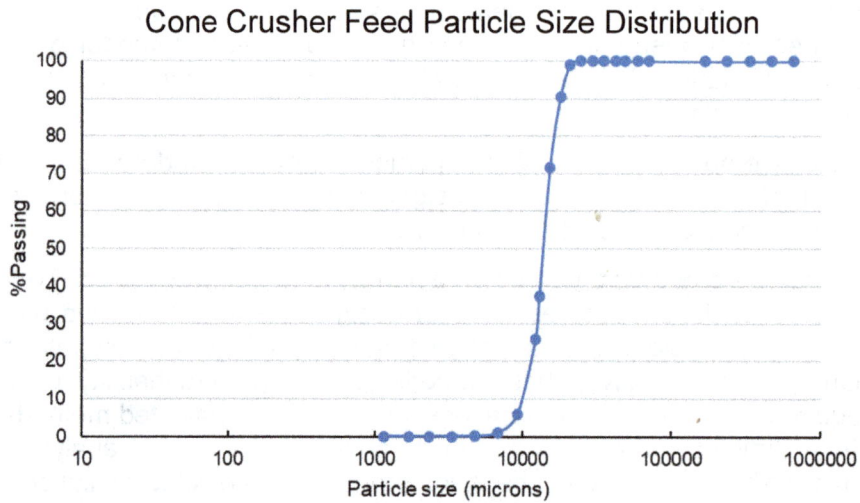

FIG 2 – Feed particle size distribution.

The Julius Kruttschnitt Mineral Research Centre (JKMRC) *Axb* is a measure of ore hardness, with lower numbers being more competent (Napier-Munn *et al*, 1996). The cone crusher feed has an *Axb* of 23.1. Experience indicates that for a target CSS of 7 mm, when operating a 2.1 m sized crusher, a JKMRC Drop Weight Test *Axb* this low (very highly competent) would be challenging the practical limits of cone crusher operation, with crushing forces on ores of this competence being very high. Per Figure 3, this measured ore is harder than any of the sites listed.

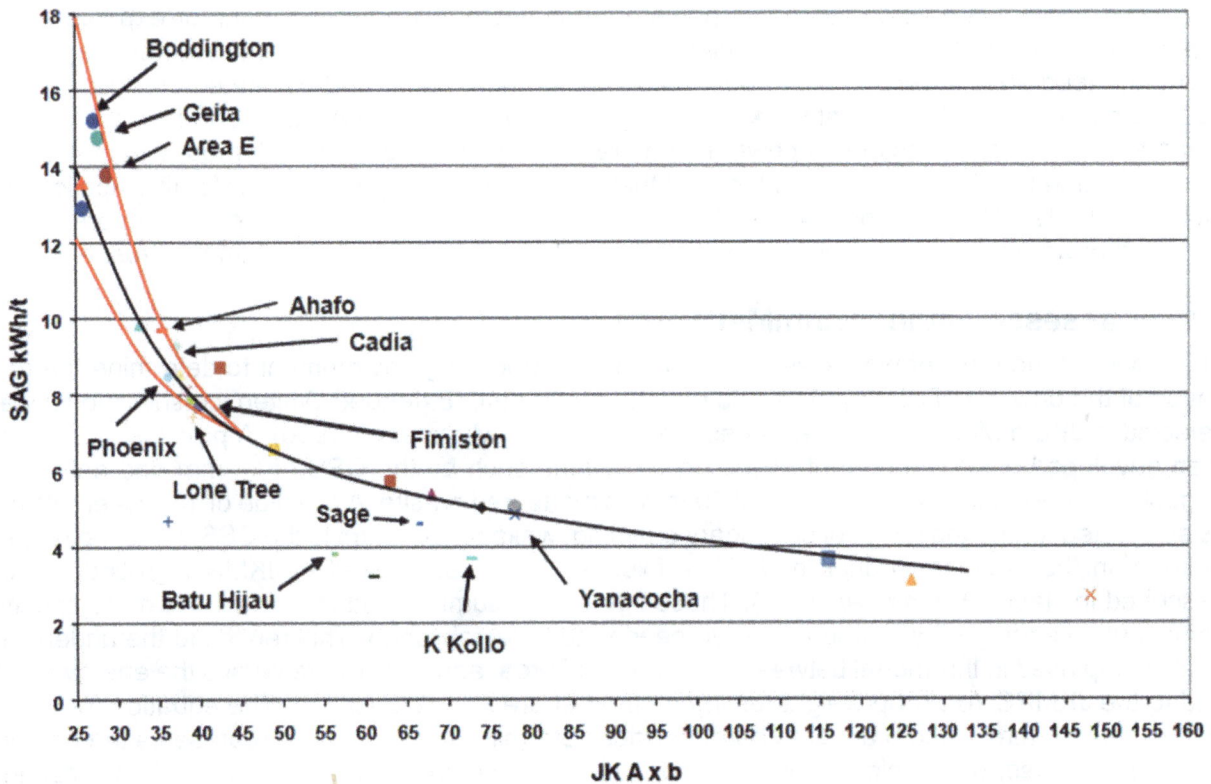

FIG 3 – JKMRC Drop Weight test *Axb*'s for various mines (Bailey *et al*, 2009).

INITIAL CONDITION AND SITE SURVEY

During the first survey, the crusher was operating at a much higher than target CSS, at 15 mm. Furthermore, the feed rate had to be limited to 60 per cent of the targeted tonnage. When these were exceeded, the crusher would start drawing more power, causing electrical current trips to occur, well below what should be required by the duty. It appeared the crusher was at times absorbing far higher power than it was designed for and what should have been required for the duty. It was also observed that the electrical current drawn was highly erratic. In addition to this power issue, there was also an indication that bowl rotation was occurring in the crusher. In the next section an assessment was made to determine what power should be required of the unit using the Whiten approach.

POWER CONSUMPTION

The following model for the energy consumption, E, in kilowatt-hours, of Cone Crushers is used in this paper (Napier-Munn *et al*, 1996):

$$E = (AP_p + P_n)T \tag{1}$$

Where P_p and P_n are the calculated 'pendulum power' and the power draw of the crusher under no load respectively, and A is a constant for the crusher determined by model fitting (Napier-Munn *et al*, 1996). For simplicity P_n is set to the idle power. Industry benchmarks for A, which can be thought of as an efficiency indicator of that application (with smaller numbers being more efficient), are between 1.25 and 1.43 (Napier-Munn *et al*, 1996). P_p is the 'pendulum power' of the ore, or the power that should be required if all the ore was broken in a drop weight test. It can be calculated using the Whiten model in JKSimMet®.

It was reported that electrical current (amperage) trips were experienced on-site. The full load amperage (FLA) is set at the rated electrical current of the motor of 100 A; examination of the control panel confirmed trips occurred at a factor above the FLA. This was occurring when the feed rate was 60 per cent of the targeted throughput of 280 t/h. The crusher motor is rated for 450 kW at 100 A, thus the mechanical power consumed by the crusher during the trip events was well in excess of this power rating. This will now be compared to the power requirement expected based on the Bond relation.

The Bond relation is an old method for power determination but was first used for expedience. A Bond crusher work index of 23 kWh/t was estimated from the ore characterisation testing above. Using the feed size, eccentric throw, operating CSS of the crusher and estimated product P_{80}, the specific power was calculated at 0.2 kW/t, using a work index of 23 kWh/t. Due to the lack of fines in the feed a further correction on power is required due to an increased work requirement, but this was neglected for simplicity. It was identified that the estimated required power for this application, including the target throughput, was well below the installed power. Therefore, the apparent high electrical power measurement on-site indicated highly abnormal operation within the crusher.

Power assessment in JKSimMet®

The power of the crusher was assessed in a process modelling environment to determine the root cause of the power inefficiency observed on-site. A standard 'Extended Whiten' crusher model was selected in JKSimMet®. This model is described in Napier-Munn *et al* (1996). A process model was then developed using the current operating conditions such as the CSS of 14 mm and estimated largest material in the crusher product of 20 mm as measured on-site. A t_{10} value of 15 was employed as suggested for the top range of secondary crushing, while K1 constant is the CSS, K2 is the largest material in the crusher product, and K3 is fixed at 2.3 as suggested in JKMRC procedures as described in Napier-Munn *et al* (1996). These figures are summarised in Table 1. Next the specific comminution energy (Ecs) versus t_{10} curve needed to be established. This relates to the underlying relationship used in this model between the amount of breakage that occurs versus the energy input, and to the JKMRC Axb drop-weight testing method of ore characterisation. The equation for this is shown below. Here A and b are ore characteristics determined by testing, t_{10} represents the percent passing (by mass) of particles smaller than one tenth of their initial size in a drop weight breakage test, and Ecs is the specific comminution energy in kWh/t:

$$t_{10} = A[1 - e^{-b \times Ecs}] \tag{2}$$

TABLE 1

JKSimMet® cone crusher parameter values used.

Parameter	Value
K1	14.00
K2	20.00
K3	2.30
t_{10}	15.00

Some assumptions needed to be made such as the JKSimMet® 'A' value, and variation by particle size was neglected; A is known to be reasonably consistent across various ores, and was assumed to be 60 in this case. Axb value was provided to Weir, which is 23.1. Using these assumed parameters, the b value and the specific comminution energy versus t_{10} table were back-calculated (Table 2). The resulting ores t_{10} is shown in Figure 4. Concurrently, a standard appearance function for a cone crusher, as documented in JKMRC procedures, was employed (Table 3). The no-load power was recorded on-site to be within a range of 110–117 kW. A mean value of 114 kW was used.

TABLE 2

Back calculated Ecs data.

Value of t_{10}	14.50	20.63	28.89	30.00	40.00	Initial particle size (mm)
t_{10} = 10	0.474	0.474	0.474	0.474	0.474	
t_{20} = 20	1.053	1.053	1.053	1.053	1.053	Ecs (kW-hr/t)
t_{30} = 30	1.800	1.800	1.800	1.800	1.800	

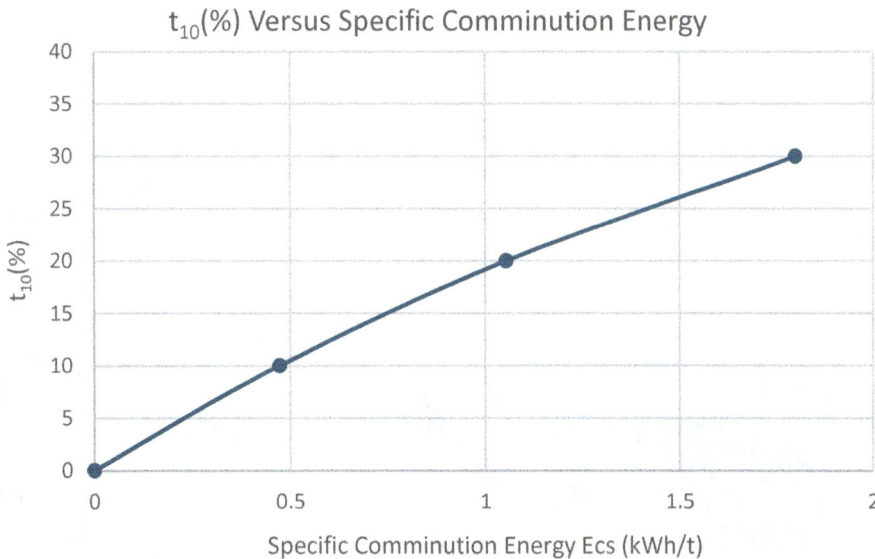

FIG 4 – Ore t_{10}(%) versus specific comminution energy.

TABLE 3

Standard appearance function used for cone crusher, from JKMRC procedures.

Value of t_{10}	T75	T50	T25	T4	T2
$t_{10} = 10$	2.8	3.3	5.4	21.1	49.6
$t_{20} = 20$	5.7	7.2	10.3	45.0	74.9
$t_{30} = 30$	8.1	10.8	15.7	61.4	85.2

The site-measured operating conditions were then determined for the power simulation, based on typical industry performance, at the measured CSS of 14 mm. The resulting power was 250 kW, when utilising a typical power prediction factor of 1.35. This estimated power was then compared with the measured power draw, which was recorded as 330 kW. While closer than the Bond prediction, the industry benchmark predicted power remained much lower than the measured power. This result suggests that there was a power inefficiency within the machine not explained by the feed size distribution or high ore competence. It was postulated that this might be explained by a 'packing' issue within the crusher.

Following this observation, the CSS was iteratively adjusted and then simulated until a power draw of 450 kW was achieved. This process was undertaken to estimate the minimal achievable CSS, based on industry benchmark performance on power efficiency. The model indicated that a CSS of 7 mm could be achieved within the installed power of 450 kW. This was with an industrial standard power prediction factor of 1.35. However, the crusher was not currently achieving this. The recalibrated power factor was found to be 2.12.

In summary, the standard Bond relation was not suitable to estimate power in this application, and the improved power modelling using JKSimMet® provided a more accurate result, but still suggested there might be some machine inefficiencies causing a spike in power draw when adjusting for the target CSS of 7 mm. This led to a review of the cavity design and possible packing addressed in the next section.

POSSIBLE PACKING AND INITIAL CAVITY DESIGN

The measured high electrical current and tripping events, combined with the above power assessments, indicated that 'packing' may be present at high throughputs. If packing was causing high internal forces and power trips, then the liner design may be modified to relieve it. The narrow feed size distribution was being crushed in a narrow area within the chamber (Figure 5) causing high forces and 'packing' in that location. The figure shows top and bottom size particles in the feed versus the location where they are nipped. This is shown in the Figure 5b with F_{100}(30 mm) and F_{80}(18 mm) spheres in the current crushing chamber. The full feed size distribution is nipped in a very concentrated area.

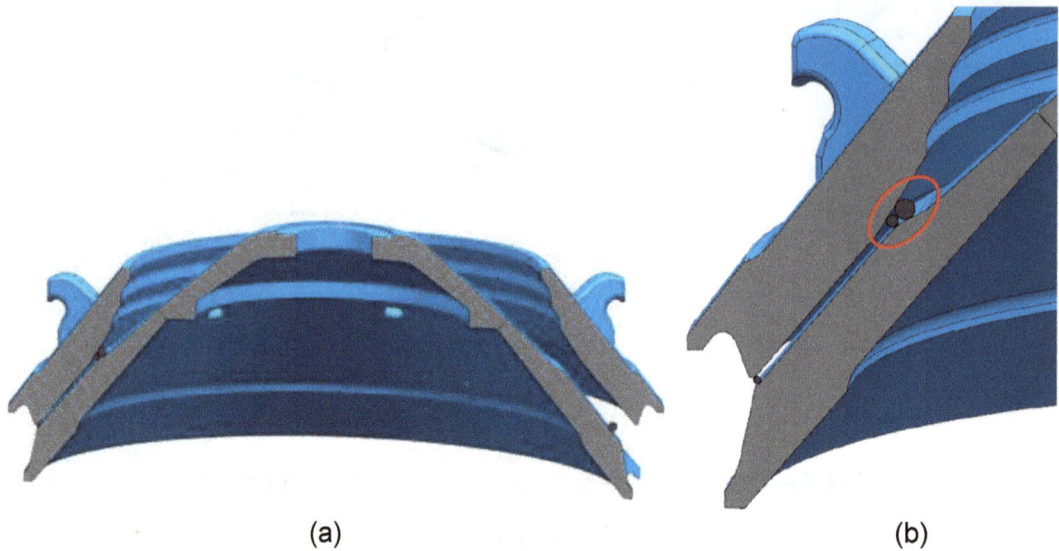

(a)   (b)

FIG 5 – Original crushing chamber.

Such concentration and packing can also potentially lead to bowl rotation. The feed is contacting at the step of the mantle, which can further impede the flow of material into the crushing zone. As the parts wear during operation, the concentrated feed may cause uneven wear patterns which can worsen the existing problems.

REVISED DESIGN AND MODELLING

Several crushing chamber options were reviewed to help reduce or eliminate the issues identified above. The final version that was subsequently trialled was a new ESCO® fine crushing chamber with a shorter parallel zone near the discharge of the crusher. This design concept removes material from the discharge area quicker and was also able to maintain many of the other existing wear parts unchanged. Deciding which option to evaluate considered several factors. The revised crushing chamber was predicted to produce a finer product and has a gradual taper that distributes the feed material over a much larger area; as shown in the Figure 6b. Overall, the proposed finer crushing chambers should be better suited to the crushing application and produce a finer product.

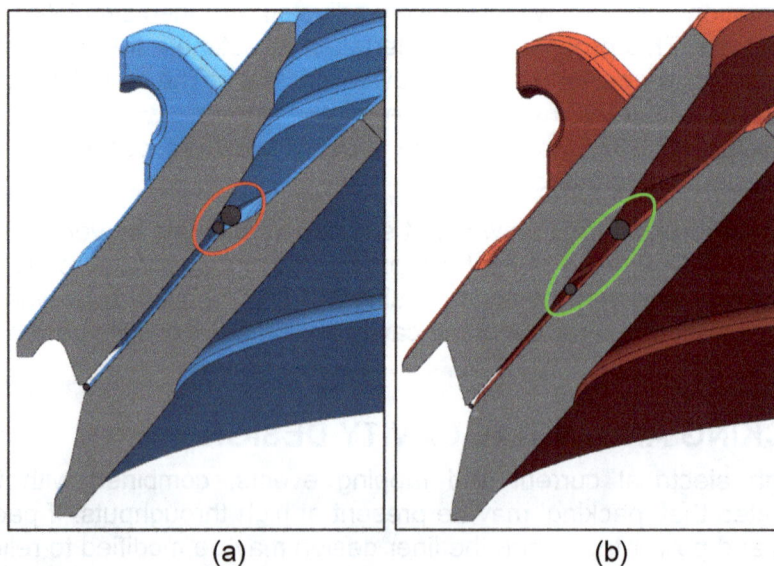

(a)   (b)

FIG 6 – Crushing chambers: (a) original (b) revised.

Discrete element method model of the cone crusher

A discrete element method (DEM) model of the cone crusher, including its feed PSD, chamber geometry, surveyed CSS, and the offset in the feed position on-site was modelled using Rocky DEM. Within this model, particle breakage modelling was employed at the particle level using the t_{10} curves and an appearance function approach from the JKMRC (Napier-Munn *et al*, 1996). An initial simulation of the current chamber design, operating speed (rev/min) of 235, feed rate (t/h) of 280 at CSS of 14 mm were undertaken (Figure 7).

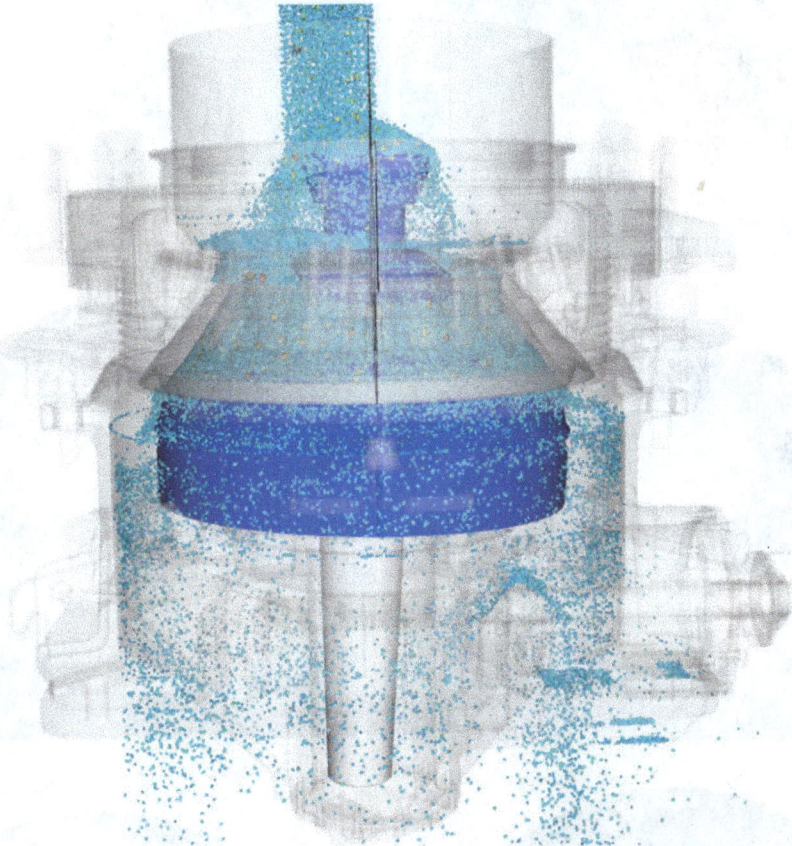

FIG 7 – DEM model of the original crushing chamber.

The revised liner design was also modelled in the same way, to simulate in advance of site trials the likely success and performance of the new chamber design. The results of this simulation are now detailed, beginning with a section of material along the CSS shown together in both chamber designs in Figure 8.

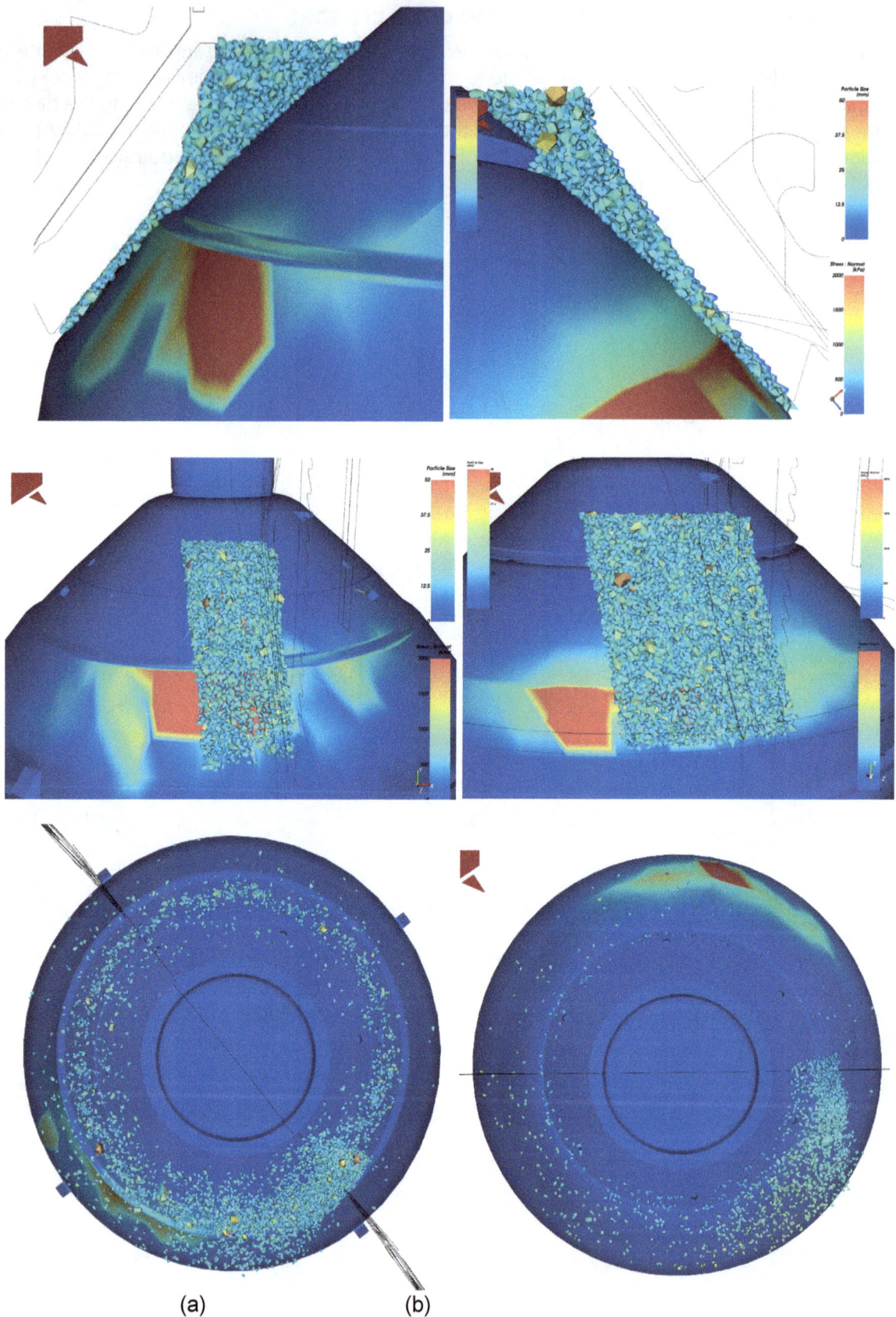

FIG 8 – DEM model of the original crushing and revised design at 280 t/h and 14 mm CSS: (a) original design; (b) revised design.

In this set of simulations, the two designs are run at a higher feed rate which results in a close to choked fed or a choked fed scenario. As can be seen in Figures 7 and 8, the chambers are almost fully filled with rocks. For easy visualisation in Figure 8 only the cross-section of dense region of rocks is shown, along with the level of stress on the mantle liner. The high and sudden stressing of the original chamber design is suspected to be evidence of the crushers 'packing' issues witnessed on-site (Figure 8). When comparing the surface stresses between the original and revised liner design (Figure 9), there is a relative reduction in stress intensity in the revised design, and a wider distribution of work being applied across more of the mantle. This is supported by the revised chamber design matching the feed size distribution and then a gradual reduction in size across more of the length of the mantle. In this way, as several compression cycles are experienced as the material falls through the chamber, the reduction ratio of individual compression cycles is reduced, but more direct reductions are experienced by the top size material.

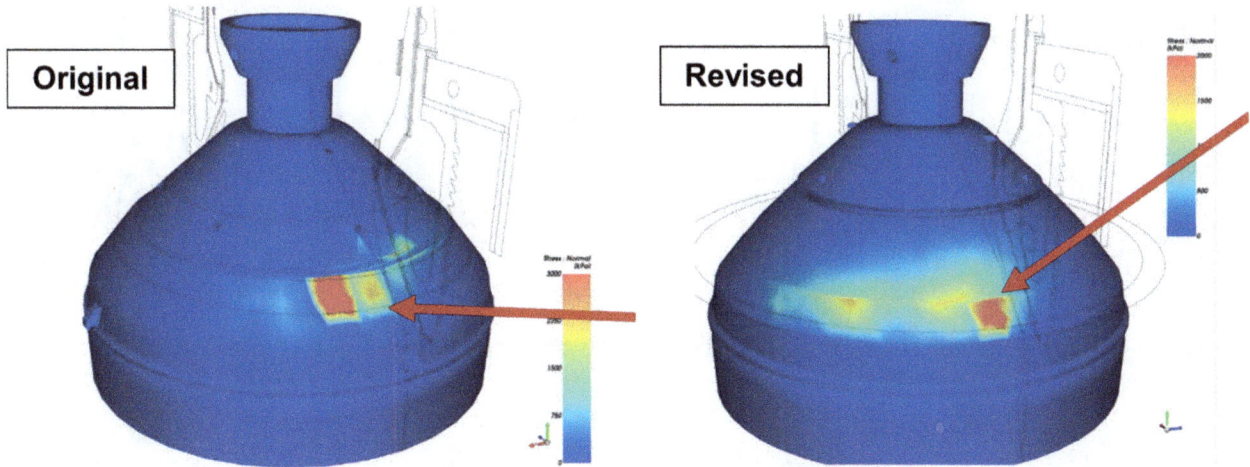

FIG 9 – DEM modelled stress on original and revised mantle.

When the stress versus frequency plots of the two simulations are plotted together in Figure 10 below, it is clear that the revised liner has a lower number of high stress events, whereas the original liner has a high frequency of high stress events, as shown in figure (Figures 10 and 11). This is shown further in Figure 12 with higher power peaking events in the original liners, again indicative of the improved liner design and suspected packing of the previous chamber.

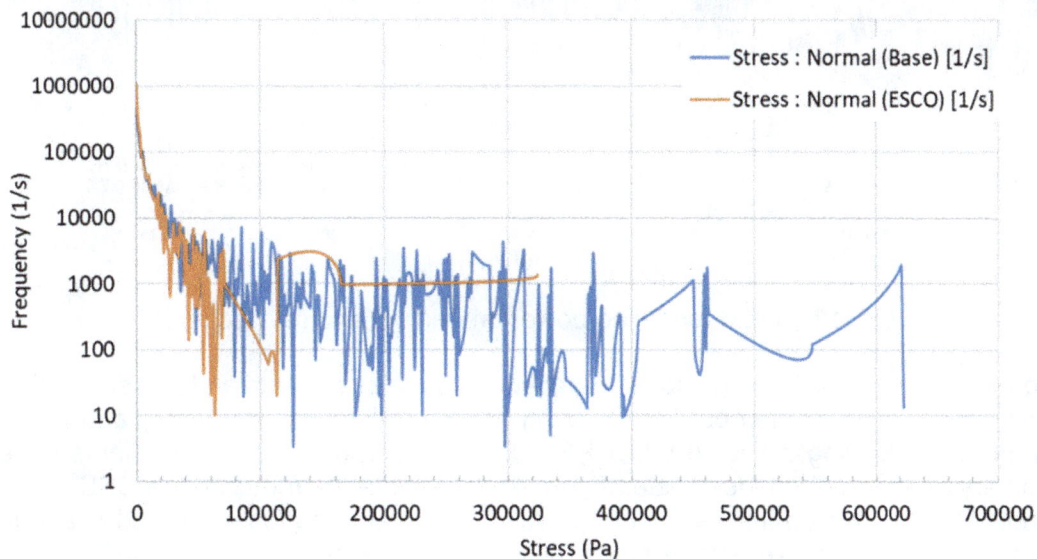

FIG 10 – DEM modelled stress versus frequency of stress events: original (blue) and revised (orange) designs.

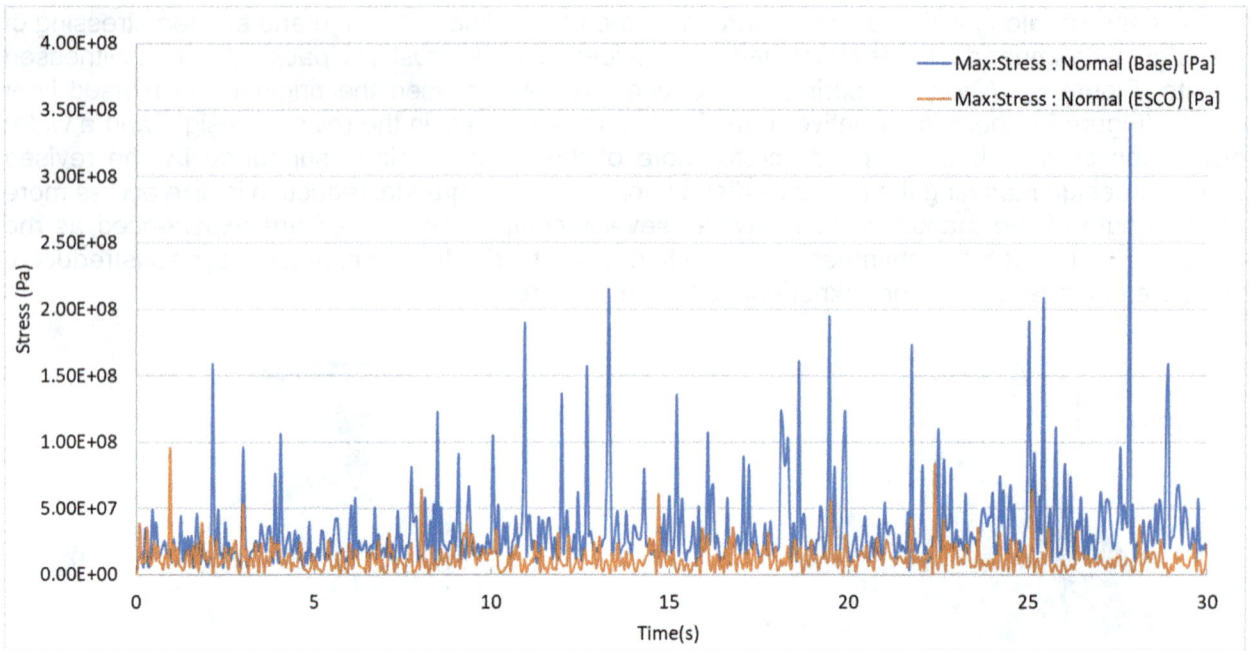

FIG 11 – DEM modelled stress magnitude over time on original (blue) and revised (orange) liner.

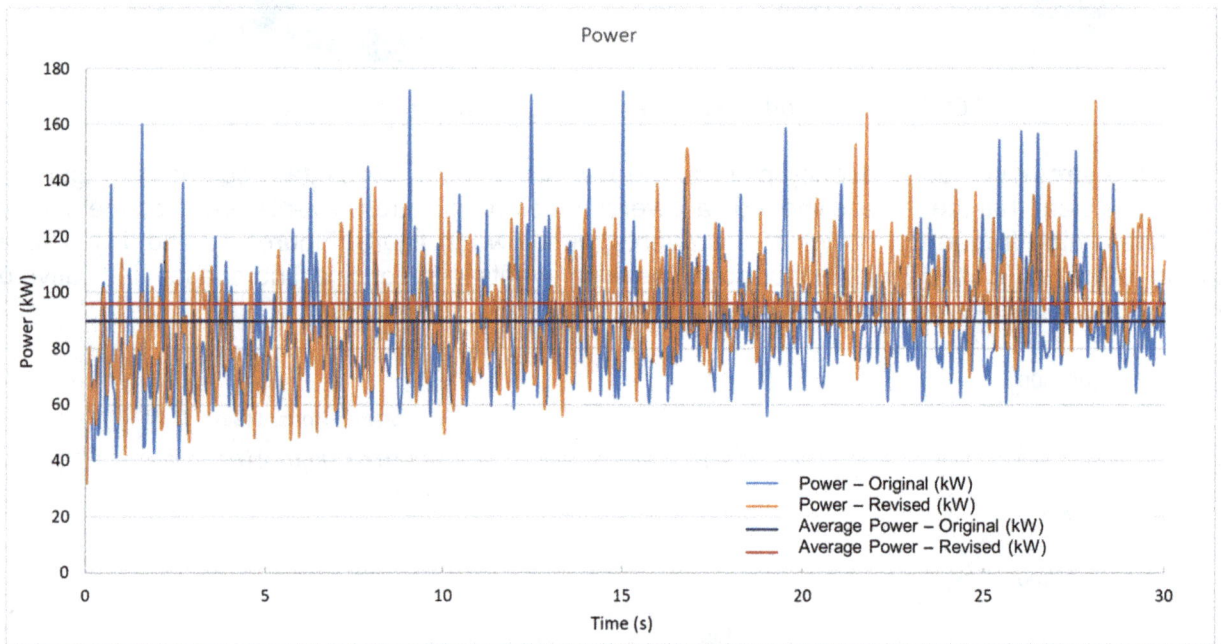

FIG 12 – DEM modelled power on original and revised liner.

Following the success of achieving a lower number of high stress events and power peaking events, further simulations were carried out investigating how to reduce the frequency and magnitude of these events by achieving a finer product PSD. For this purpose the revised liners were further simulated varying the CSS and feed rate. Figure 13 shows that by reducing the CSS up to 8 mm a finer product PSD could be achieved, but to do so the feed rate had to be lowered to about 130 t/h, so that higher stress and power events could be suppressed (Figure 14).

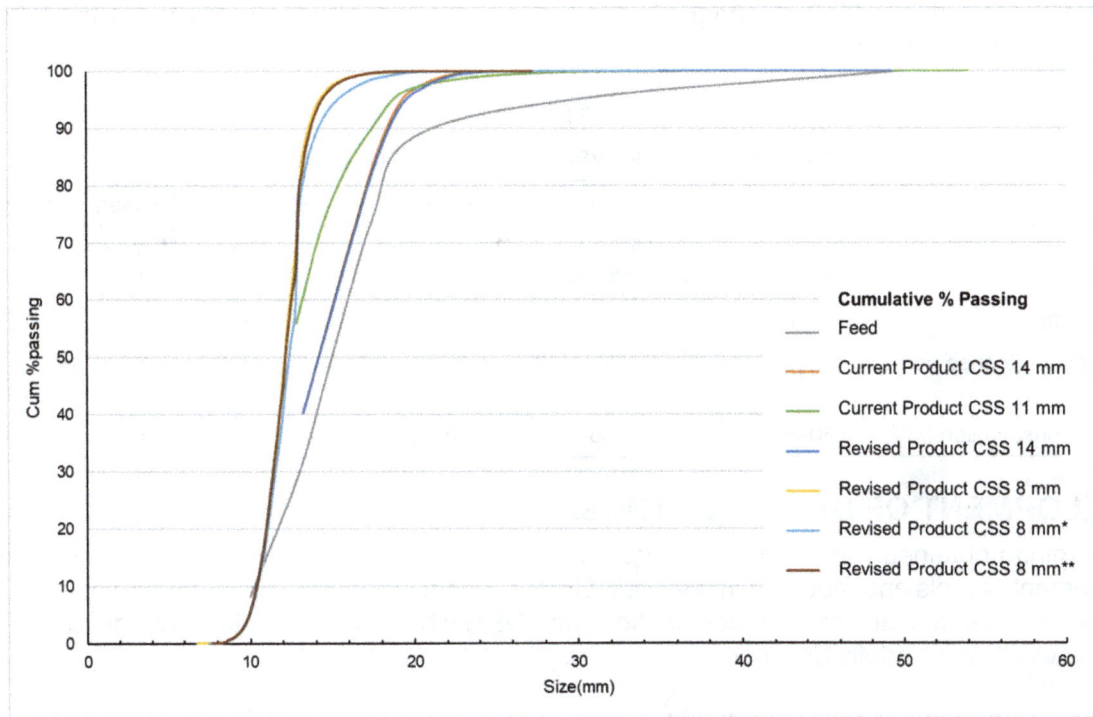

* 10% Reduced Speed ** A Modification on the Revised Liner Discharge Section (not used)

FIG 13 – DEM modelled product PSD from original and revised (ESCO®) liner.

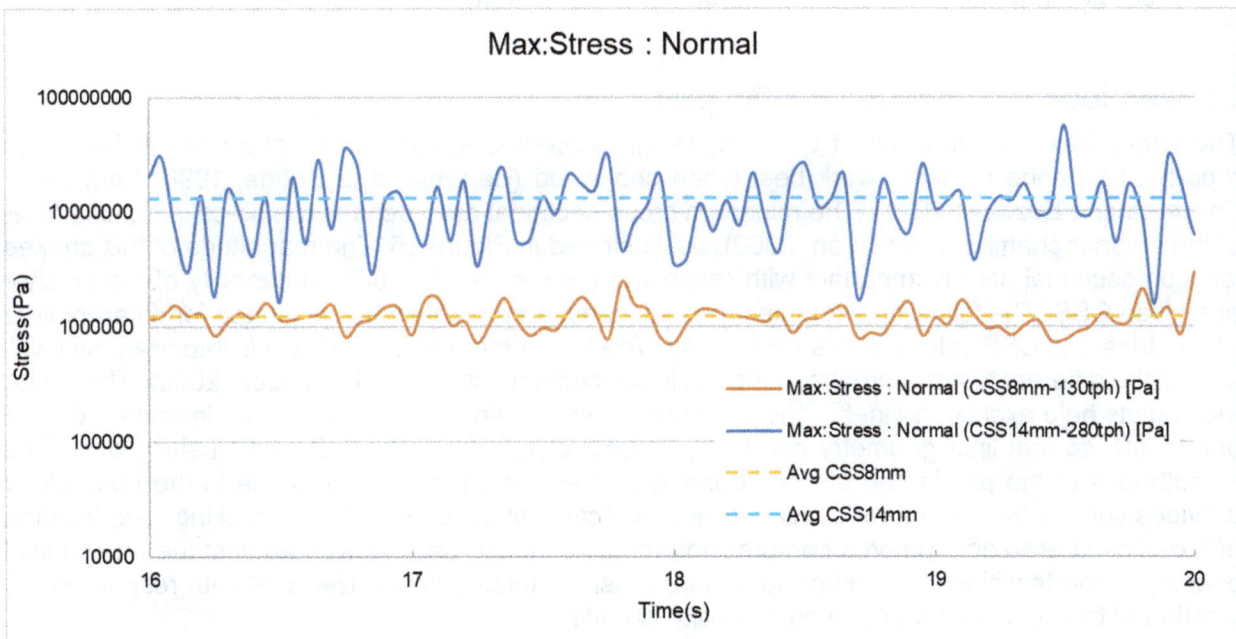

FIG 14 – DEM modelled stress on revised liner at two operating conditions.

NEW DESIGN INSTALLATION AND SITE TRIAL

The new ESCO® liners were successfully installed into the TC84X cone crusher. The new liner set was run and proved capable of achieving higher throughputs despite the lower choked cross-sectional area (consistent with DEM results obtained prior to the trial). The motor current trips were also successfully resolved. The power consumption of the crusher also appeared to be significantly lower. Therefore, based on these observations, it appears that the suspected 'packing' issue was indeed present (based on the two points above) in the original chamber design, and was successfully resolved with the revised cavity profile and design. This was also supported by the DEM results. It was also found to be possible to successfully reduce the CSS to 8 mm, greatly reduced from the

15 mm of the last site visit with the original liner design. Table 4 provides details on measured results of the different liner designs trialled.

TABLE 4

Site trial – original versus revised liner profile.

	Original	Original	Revised	Revised	Revised	Revised
Operation	Uncentred Feed	Trickle Fed Uncentred Feed	Trickle Fed Uncentred Feed	Centred Feed	Centred Feed	Centred Feed
CSS (mm)	14	14	16	15	12	8
Throughput (tonnes per hour)	340	280	200	390–400	300	210
Power Consumption (kW)	480–500	330–355	240–280	400–500	260	160–250

DEVELOPMENT OF DESIGN OPTIMISATION TOOLS

The learnings obtained from applying these advanced DEM methods were then captured into the development of tools and methods in the Weir 'Cone Crusher Chamber Optimisation Toolset' (COT). COT can be used in place of, or in conjunction with, DEM which is a time intensive approach. These methods will now be briefly described.

Feed size distribution chamber layout plots

An initial step in the Weir COT optimisation of cone crusher chamber design involves laying out the feed size distribution against the closed and open side cross-sectional profile of the chamber. This ensures that the chamber design allows for even and well distributed crushing along the length of the chamber.

Choked location and area profile plots

The choke level is a function of the liner design, eccentric speed, stroke and CSS. It has been reported that cone crushers work best when choke fed (Bearman and Briggs, 1998; Yamashita, Thivierge and Euzebio, 2021). The minimum cross-sectional area defines the so-called choke level of the crusher chamber (Evertsson, 2000), as illustrated in Figure 15. The magnitude of this choked location sectional area is important with respect to the required throughput capacity of the crusher at a given CSS. The following example cross-sectional area profile was calculated for different liner geometries and CSS values in this case study. The location of the choke level is important as it will control the different types of breakage modes in the crushing chamber (Evertsson, 2000). Therefore, these plots help explain the inefficiencies in the original design, and serves as an important design parameter, so that liner geometry can be optimised to better suit the required crushing duty. The smoothness of the profile can also be observed. Reviewing the original profile in the case study provides some indication of the sudden area contraction which contributed to packing. The location of the choked area and the ores compression forces can also be reviewed against the mechanical bearing supports within the crusher, to ensure crushing forcing is resolved well with respect to the location of the bearings for improved reliability and life.

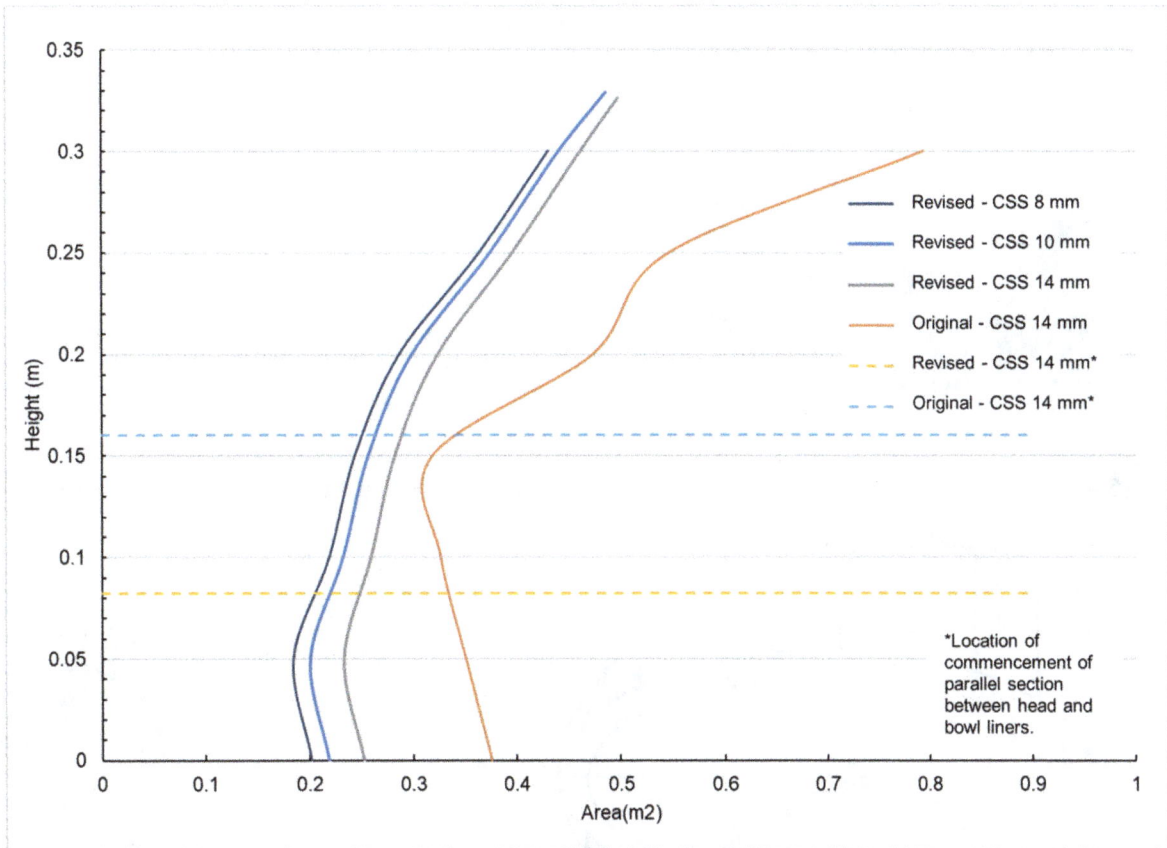

FIG 15 – Area profile and choke location plot example – original and revised.

Throughput and breakage modelling

The generation of the product size distribution from the feed distribution is achieved by the rock breakage methods and results, called 't$_{10}$ curves', and the 'Whiten Crushing Model' described in literature (Whiten, 1972; Napier-Munn *et al*, 1996). Either JKSimMet® or DEM is employed.

For throughput modelling, the maximum capacity of a cone crusher is reached when it is choke fed and is considered to be a function of bulk density, average velocity in the cross-section, and the minimum (choked) cross-sectional area (Briggs, 1997). Weir's COT approach involves first identifying the choked location (per the Choked Location and Area Profile Plots above). The cross-section of this location is then used in the tool, including its varying area at angles around the circumference. A bulk density profile is then application, which varies around the circumference of the cavity. This is determined by the applications ore density, along with a Filling Factor, that describes how volumetrically full the chamber is at angles around the chamber. This Filling Factor is based on a DEM results of filling in an area approximating the choked zone. The DEM results used to determine this filling is illustrated in Figure 16, and the resulting Filling Factor varying around the chamber is also shown. As expected, the filling is higher at the CSS, and drops towards the open side setting (OSS). Then a simplification of the velocity profile of the crushed ore falling around the circumference of the chamber developed by Evertsson (2000) (Figure 17) is employed. Together, these are used to estimate the throughput capacity of any chamber design and CSS.

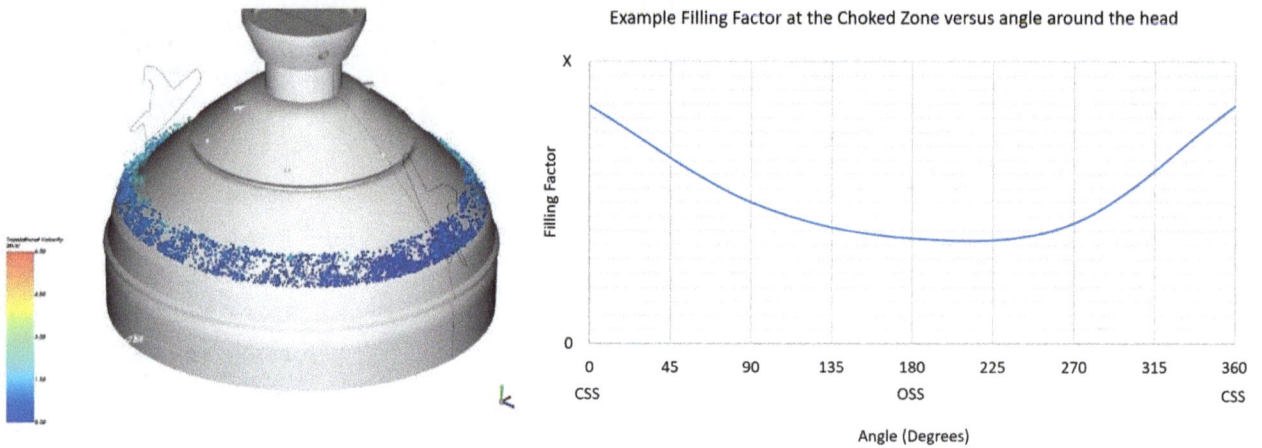

FIG 16 – DEM results of material filling around choked zone (left) and resulting filling factor (right).

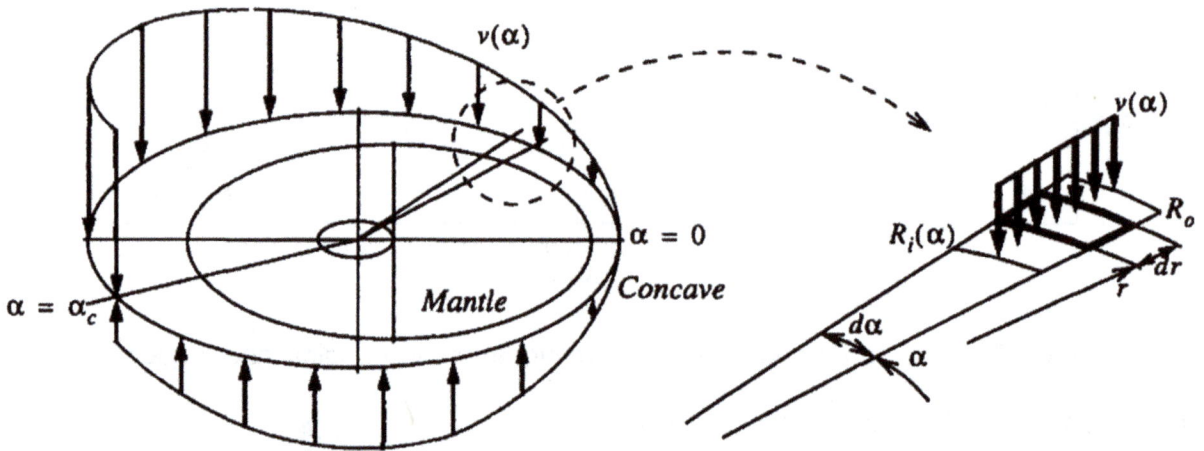

Fig.10 Horizontal cross-section of the crushing chamber.

FIG 17 – Material flow (free fall/sliding down and upwards compression) in a cone crusher (Evertsson, 1999).

Having developed this throughput modelling tool in Weir COT, it was used to predict the throughput capacity of the revised liner. It predicted a throughput of up to 194 tonnes per hour (t/h) at a CSS of 8 mm for the revised chamber, and the highest throughput measured at site at these conditions was 210 t/h, an error of only 8 per cent. It was also used to retrospectively estimate the throughput of the original liner, and had an error of only 7 per cent between measured and predicted values. Figure 18 shows the measured versus predicted and DEM simulation throughputs. It is also noteworthy that despite the lower choked zone cross-sectional area of the revised design, it has a higher throughput capacity, and this was also captured and predicted accurately by the DEM results.

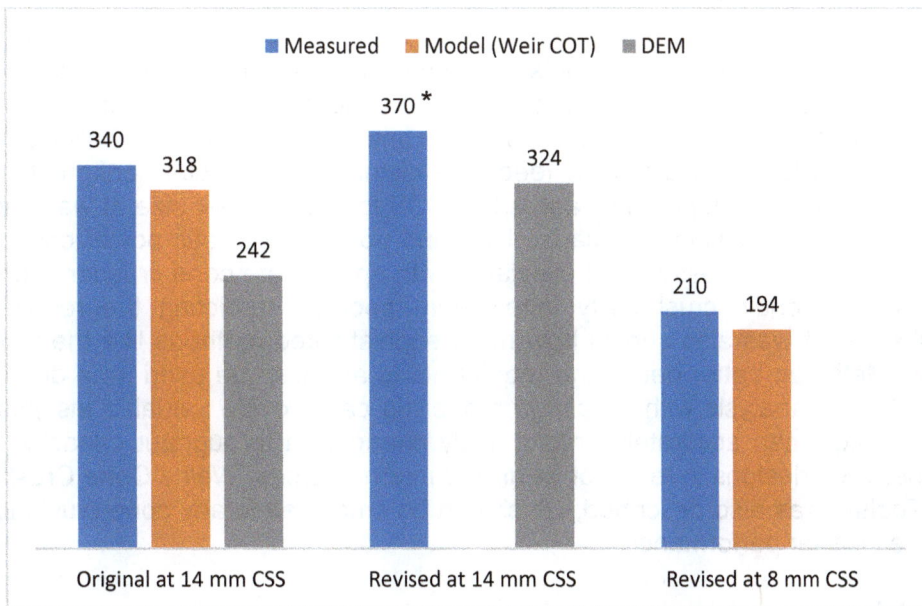

FIG 18 – Simulated full chamber and highest measured throughput (t/h) (* interpolated between measured values at 12 mm and 15 mm CSS).

Product top size control

Another feature of Weir's COT includes an assessment of the adequacy of the parallel discharge section of the cone crusher chamber profile to control final product top size. Dynamic calculations of the crushed material are made to ensure that a final compression cycle of the material is experienced along the parallel section, ensuring that the final product size is controlled. This dynamic calculation considers important variables including the head speed, the liner profile, and the open and closed side settings.

Wear

Liner wear is an important variable in the overall crushing operation as it is intimately related to throughput, size, and cost (Rosario, Hall and Maijer, 2004). Bond developed empirical wear rate approximations for crushers. However, these models are considered 'far from acceptable' as the paddle wheel tests upon which they are based primarily capture abrasive wear (Bearman and Briggs, 1998) and a higher stress wear mode is thought to predominate. Weir uses advanced DEM results to evaluate the distribution and relative magnitude of surface stresses to compare and optimise liner designs. This is supported by Lindqvist and Evertsson (2003, 2004) who state 'squeezing' wear predominates in cone crushing, and model wear, Δw (measured in mm per stoke), with sliding wear considered negligible:

$$\Delta w = \frac{p}{W} \tag{3}$$

where:

p is pressure (Pa)

W   is a constant (N/mm³)

Thus, Weir's use of relative surface stress intensity in DEM is a good proxy for wear performance evaluation between designs. Weir also employs site 3D scanning of worn profiles, which are overlayed with as-new profiles. Relative surface stress can then be combined with these actual wear rates to estimate wear improvements in new designs and in trials.

Power

For power modelling in Weir's COT approach, the Whiten crusher model is used as employed in JKSimMet®, or otherwise the Bond relation is applied where the required data for the Whiten approach is not available.

CONCLUSIONS

This paper presented the use of design tools and simulation methods, including Weir's COT, in the diagnostics of site issues and for the improvement of cone crusher chamber designs. The DEM approaches used to inform this toolset have also been briefly summarised. A challenging case study was also presented, which had a narrow feed size distribution, a circuit configuration selectively generating a very high ore competency, and a low CSS for the crusher size. It was shown that the phenomenon of crusher 'packing' can cause high and potentially erratic power consumption, and that these issues can be addressed and alleviated with appropriate cone crusher chamber design. While methods like Bond's crushability index were poor at predicting the required power in applications like this, it was also shown how more sophisticated methods like the Whiten crusher model in JKSimMet® can better determine requirements and also be used as a diagnostic tool. It was shown that DEM analysis with breakage modelling can provide valuable insights on crusher operation and design, can accurately predict likely changes in throughput capacity, and relative performance between designs in terms of wear and internal forces. Weir's Cone Crusher Chamber Optimisation Toolset was also described, which can be employed in any cone crushing application to improve cone crusher performance.

REFERENCES

Bailey, C, Lane, G, Morrell, S and Staples, P, 2009. What Can Go Wrong in Comminution Circuit Design?, in Tenth Mill Operators' Conference (The Australasian Institute of Mining and Metallurgy: Melbourne).

Bearman, R A and Briggs, C A, 1998. The active use of crushers to control product requirements, *Minerals Engineering,* 11:849–859.

Briggs, C A, 1997. A fundamental model of a cone crusher, The University of Queensland, School of Engineering.

Evertsson, C M, 1999. Modelling of flow in cone crushers, *Minerals Engineering,* 12:1479–1499.

Evertsson, C M, 2000. *Cone Crusher Performance,* PhD thesis, Chalmers University of Technology.

Lindqvist, M and Evertsson, C M, 2003. Prediction of worn geometry in cone crushers, *Minerals Engineering,* 16:1355–1361.

Lindqvist, M and Evertsson, C M, 2004. Improved flow and pressure model for cone crushers, *Minerals Engineering,* 17:1217–1225.

Napier-Munn, T J, Morrell, S, Morrison, R D and Kojovic, T, 1996. Mineral comminution circuits: their operation and optimisation, Julius Kruttschnitt Mineral Research Centre.

Rosario, P P, Hall, R A and Maijer, D M, 2004. Liner wear and performance investigation of primary gyratory crushers, *Minerals Engineering,* 17:1241–1254.

Whiten, W J, 1972. Simulation and model building for mineral processing, PhD thesis, The University of Queensland.

Yamashita, A S, Thivierge, A and Euzebio, T A M, 2021. A review of modelling and control strategies for cone crushers in the mineral processing and quarrying industries, *Minerals Engineering,* 170.

Grinding circuit expansion and optimisation at the B2Gold Masbate Project, Philippines

A Insalada[1], K Bartholomew[2], A J Marcera[3], E Occena[4], M Anghag[5], D Torres[6], N Avenido[7], J Rajala[8] and R E McIvor[9]

1. Metallurgical Superintendent, Phil: Gold Processing and Refining Corp., Puro, Aroroy, Masbate 5414, Philippines. Email: aileen.insalada@pgprc.com.ph
2. Chief Metallurgist, Metcom Technologies, Inc., Grand Rapids MN 55744, USA. Email: kyle@metcomtech.com
3. Senior Metallurgist, Phil. Gold Processing and Refining Corp., Puro, Aroroy, Masbate 5414, Philippines. Email: alexanderjethro.marcera@pgprc.com.ph
4. AVP – Process Plant, Phil. Gold Processing and Refining Corp., Puro, Aroroy, Masbate 5414, Philippines. Email: eugene.occena@pgprc.com.ph
5. Mill Operations Superintendent, Phil. Gold Processing and Refining Corp., Puro, Aroroy, Masbate 5414, Philippines. Email: mario.anghag@pgprc.com.ph
6. Mill Operations Manager, Phil. Gold Processing and Refining Corp., Puro, Aroroy, Masbate 5414, Philippines. Email: dennis.torres@pgprc.com.ph
7. Chief Metallurgist, B2Gold Corporation, Vancouver BC V6C 2X8, Canada. Email: navenido@b2gold.com
8. VP Metallurgy, B2Gold Corporation, Vancouver BC V6C 2X8, Canada. Email: jrajala@b2gold.com
9. Principal Metallurgist, Metcom Technologies, Inc., Grand Rapids, MN 55744, USA. Email: rob@metcomtech.com

ABSTRACT

B2Gold acquired the Masbate Mine, located on Masbate Island, Philippines, through a merger of CGA Mining Limited in January 2013. The deposit is mined by open cut, and current ore processed is a combination of historically stockpiled, low-grade, oxidised material and freshly mined sulfide ore. The comminution circuit consists of a 9.75 m diameter by 3.62 m, 8.5 MW SAG mill followed by 9.6 MW of secondary ball milling. SAG pebble crushing and a three-stage supplementary crushing plant are also used in the grinding circuit. Gold recovery is achieved via a conventional leach/CIP plant. A plant expansion, which included the addition of a third ball mill and crushing circuit upgrades, was commissioned in 2019 to enable an increase from 6.5 Mt/a to 8.0 Mt/a. In 2023 a detailed grinding circuit survey was conducted by the site metallurgical team, with guidance from Metcom Technologies, Inc., to benchmark circuit performance and identify process optimisation opportunities. Bond Work Index analysis was used to benchmark circuit efficiency, and Functional Performance Analysis was used to identify specific performance improvement opportunities in Classification System Efficiency (CSE) and Ball Mill Grinding Efficiency. Drop-weight and rotary mill-based comminution tests were conducted on the 2.4 metric ton SAG feed sample which was collected as a composite before and after the slurry samples were collected. Optimal grind size targeting was achieved via grind versus gold recovery laboratory tests on the same SAG feed composite in combination with the results of the grinding survey. This paper describes the detailed, systematic approach utilised in the comminution circuit optimisation process.

INTRODUCTION

The Masbate Gold Project (MGP) is located at Barangay Puro, Municipality of Aroroy, Province of Masbate, Masbate Island, Region V (Bicol), Republic of the Philippines. The Masbate gold deposit is located near the northern tip of the island of Masbate, 360 km south-east of the country's capital of Manila, and is centred at latitude 12° 28' N and longitude 123° 24' E.

The Masbate Operation has undergone a series of process improvements since start of operation in year 2009 to 2018. A comminution circuit survey in 2013 was completed and an update on the throughput modelling in year 2017 was instigated to identify the process improvements to achieve the 8 Mt/a expansion project. The Masbate plant was able to achieve the production target last year 2023 at 8.3 Mt/a. The company is allowed to operate at 9 Mt/a of gold ores through conventional cyanidation process. A full circuit survey was requested by B2Gold corporate through Metcom Technologies Inc. (2023) to measure circuit performance following the plant expansion to benchmark performance and identify optimisation opportunities.

A grinding circuit survey was conducted on 11th of July 2023 aiming to acquire samples for Bond Work Index analysis to benchmark circuit efficiency, Functional Performance Analysis to identify specific performance improvement opportunities in Classification System Efficiency (CSE) and Ball Mill Grinding Efficiency, along with the Grind-Recovery laboratory test works for optimal grind size targeting and the oxygen uptake test to confirm the oxygen for the Masbate samples.

MASBATE GOLD PROJECT BACKGROUND

The Masbate Gold Project is an open pit mining operation located on Masbate Island in the Republic of the Philippines. The mine operations were acquired by B2Gold in a merger with CGA Mining Limited in January 2013. The indicated resource as of December 31, 2023 was 109.6 Mt at 0.81 g/t Au with 2.87 M ounces of contained gold

Masbate mines gold ore present within a low sulfidation epithermal deposit. Dore produced by Masbate typically contains 60 per cent Au and 40 per cent Ag which may vary dependent on ore source. The gold mineralisation of Masbate occurs within a massive multi-phase quartz vein, quartz calcite veins, and quartz vein stock-works and en échelon veining between major structures. The principal host rock to the gold mineralisation is a fractured andesitic – dacitic, tuffaceous agglomerate. Mineralisation occurs within quartz veins and associated altered and quartz-stockwork wall rocks and breccias. Gold is typically hosted in grey to white crystalline to chalcedonic quartz and is frequently associated with pyrite, marcasite, and minor amounts of chalcopyrite and sphalerite. High-grade veins are generally narrow (<1 m) but some may reach 20 m in width, while sheeted zones with stockworks can reach as much 75 m in width.

The mill throughput in 2023 was 22,745 t per day, which represented 8,302,076 t ore processed at 0.97 g/t Au and with a mill availability of 90.32 per cent. The target grind size P_{80} (leach feed) was 150 µm with an average P_{80} acquired at 128 µm. The CIL circuit used sodium cyanide, lead nitrate (one time addition at start), and quicklime. The dissolved oxygen concentration maintained in the circuit was 4–9 mg/L dependent on the ore type treated. Activated carbon (12 g/t) was used in the CIL circuit and hydrochloric acid was used for acid washing loaded carbon prior to elution. The final CIL tailings underwent cyanide destruction using the Caro's acid process. The detoxified tails were sent to the tailing's storage facility. The average circuit retention time was approximately 24 hrs.

The total plant gold recovery averaged 74.48 per cent in 2023. A simplified block flow diagram is shown in Figure 1.

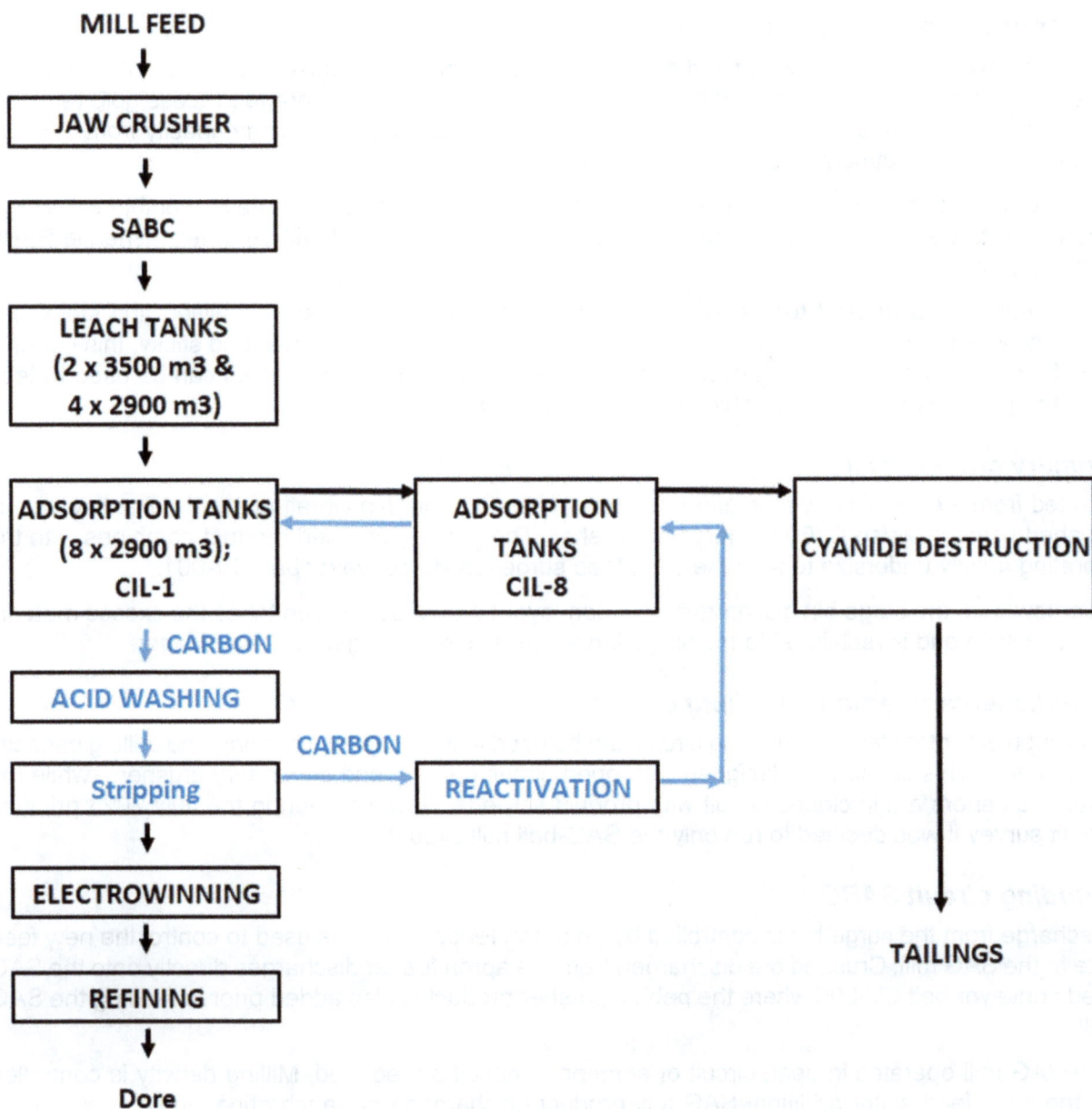

FIG 1 – Masbate mill block flow diagram (2023).

MASBATE PROCESS DESCRIPTION

The Masbate circuit comprises of the following, crushing, grinding, leaching and adsorption, elution, electrowinning and smelting gold recovery stages, tailings detoxification stage prior disposal into the tailings storage facility (TSF), and a reclaim water treatment plant that treats a portion of the decant water from the TSF prior to discharging to the ocean to maintain the integrity and stability of the tailing's storage facility.

The Masbate process flow sheet (in 2023) consisted of the following:

- Primary jaw crushing (C160)
- 8.5 MW SAG mill and pebble crusher
- 2 × 3.6 MW ball mills and 1 × 6 MW ball mill
- 2 × 3500 m³ and 4 × 2900 m³ leach tanks
- 8 × 2900 m³ adsorption tanks
- AARL elution, recovery, and regeneration
- Caro's acid detoxification treatment plant

Comminution circuit description

The original grinding circuit comprised of a SAG and two ball mills in closed-circuit with the cyclones. The circuit targeted a cyclone overflow grind size of 150 µm and an average process gold recovery of 75 per cent. In 2019, a third ball was commissioned increasing the milling capacity from 6.5 Mt/a to 8 Mt/a for the treatment of oxide and transitional ores.

The comminution circuit comprises a primary crushing circuit, optionally combined with a supplementary tertiary crushing circuit, followed by a SAG and ball mill with recycle crusher, ie SABC configuration.

Ore is mined and trucked to the ROM pad. On the ROM pad, ores are classified and stockpiled separately as fresh, transition and oxide ores. As oxide material is often wet and sticky, this material only feeds the primary crushing circuit while the fresh and transition materials can be used to feed both the primary and tertiary crushing circuit as required.

Primary crushing circuit

Ore fed from the ROM bin via an apron feeder is passed through a vibrating grizzly. The oversize is crushed using a Metso C160 primary jaw crusher. The primary crushed product combines with the vibrating grizzly undersize to feed the SAG feed surge bin via conveyor belt CV-001.

Overflow from the surge bin discharges onto conveyor belt CV-009, which takes the excess material to a stockpile and is reclaimed to the surge bin when required using a front-end loader.

Supplementary tertiary crushing circuit

This supplementary tertiary crushing circuit can be used to increase the crushing and milling capacity as required. This circuit was designed with open circuit primary and secondary crushers, while the tertiary crushers are in closed circuit with product screens. However, during the July 2023 grinding circuit survey it was decided to run only the SAG-ball mill circuits.

Grinding circuit SABC

Discharge from the surge bin is controlled by an apron feeder, which is used to control the new feed rate to the SAG mill. Crushed ore discharged from the apron feeder discharges directly onto the SAG feed conveyor belt CV-006 where the pebble crusher product is also added prior to feeding the SAG mill.

The SAG mill operates in open circuit or semi open circuit as required. Milling density is controlled by the SAG feed water addition. SAG mill product discharges onto a vibrating single deck screen with oversize (or pebbles) discharging onto conveyor belt CV-007, which is installed with a belt magnet to remove steel balls or scrap metals ahead of a Metso HP6 pebble crusher.

Pebbles from CV-007 are discharged onto conveyor CV-008, which is the pebble crusher feed conveyor belt. This belt has also been installed with a belt magnet to remove any remaining scrap metal. It should be noted that the pebble crusher can be bypassed when required to maintain the SAG mill load. The pebble crusher product is discharged directly onto the SAG feed conveyor belt CV-006.

The SAG discharge screen undersize gravitates to the cyclone feed hopper, where it is combined with the two-ball mill discharges and dilution water prior to being pumped to the cyclone cluster for classification. The cyclone overflow, at the target grind size, reports to downstream processes, while the coarse cyclone underflow is split into two equivalent streams to feed the two ball mills. It should be noted that no water is added to the ball mill feed. An option is also available to split cyclone underflow back to the SAG mill if required.

A portion of the feed from the cyclone hopper is split into the new milling circuit (ball mill 3) via the mill slurry transfer pumps where the transfer is controlled by mass flow rate as required as required by the new ball mill. Feed mass flow is calculated using the flow and density measurements. Cyclone overflow transfer pumps were made available for pumping to the Leach Trash Screen Feed box. The grinding circuit operates 24 hrs per day, 7 days per week. The grinding and classification plant is designed based on 92 per cent availability.

2023 GRINDING CIRCUIT SURVEY

Survey preparation

A pre-survey site visit to Masbate was made by Metcom Technologies in May 2023 to plan and prepare for the grinding circuit survey. The purpose of the pre-survey site visit was to collect needed information about the process and the plant, prepare the survey team, and prepare the site for the survey. Standard comminution equipment data sheets were filled out by the site team so that the site team and Metcom had a common set of agreed-upon process information for the mills, motors, pumps, and cyclones. A majority of the site metallurgists had already completed Metcom training, and survey procedures were discussed and agreed upon during the pre-survey site visit. The ore blend to be processed during the survey (planned for July 2023) was chosen to best represent the long-term ore blend, a 60/40 per cent blend of 'transition' material which is fresh feed (harder) and oxide material (softer) from a large, aged stockpile.

The instrument data which was to be collected during the survey was discussed during the pre-survey visit and a template for SCADA data collection was constructed and tested. A checklist of instrument calibrations (eg SAG feed and pebble crusher conveyor weightometers) was generated and calibrations were conducted prior to the July 2023 survey. Non-functional analogue pressure gauges on cyclone feed clusters were identified and were replaced prior to the survey. The survey buckets required were identified and a labelling scheme was agreed upon. Approximately 105 × 20 L buckets were identified as being needed for the survey along with 20 × 200 L drums for the bulk SAG feed sample. All buckets were labelled, and tare weights written on buckets and lids prior to the survey.

Sampling protocol was discussed and the sampling teams for each sample point were identified during the pre-survey visit as well.

Sample point identification, inspection and modifications

During the initial site visit the sample points required for the survey were identified as shown in flow sheet (Figure 2).

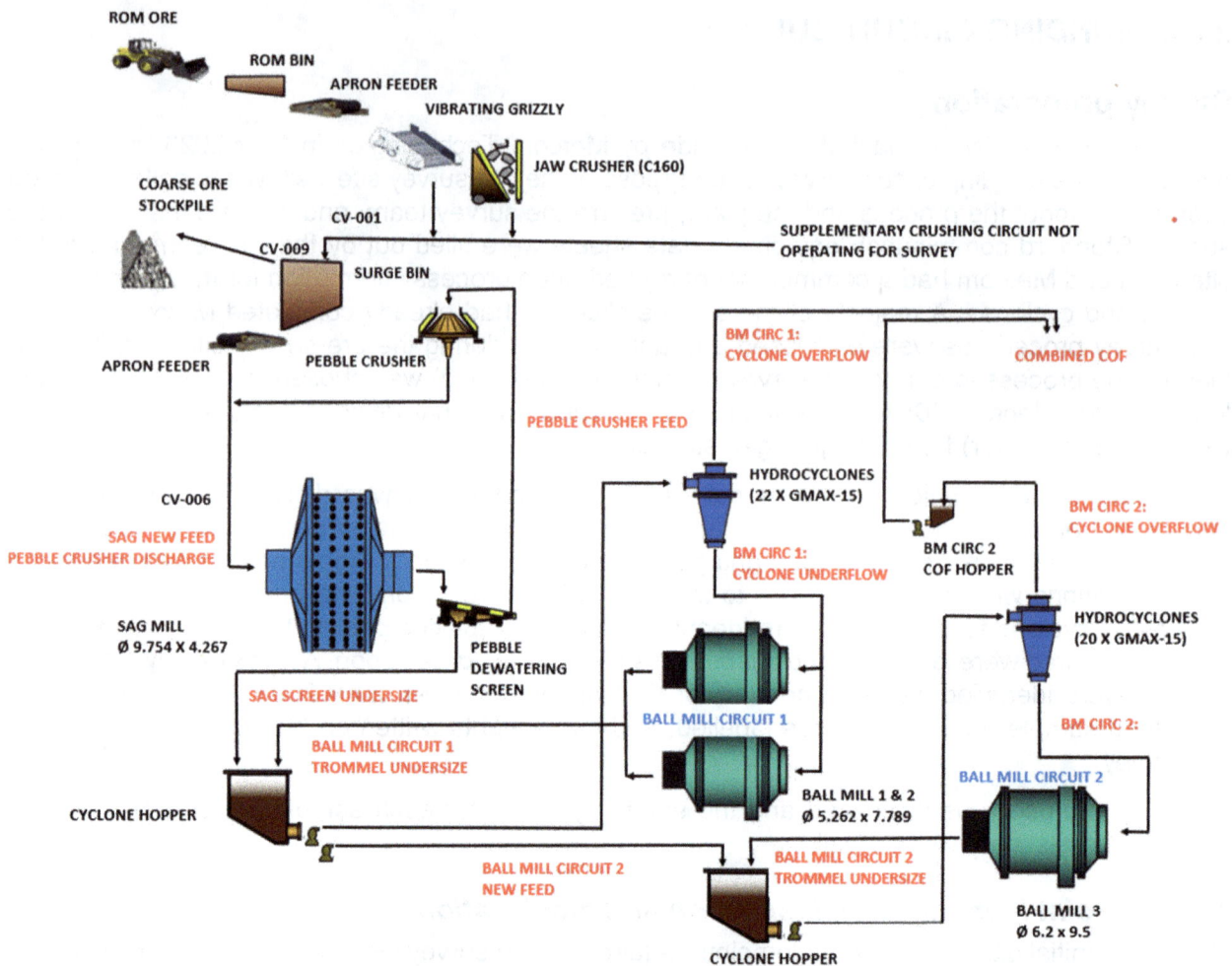

FIG 2 – Masbate flow sheet with grinding circuit sample points.

There is insufficient space to sample the SAG circuit new feed prior to the location where pebble crusher discharge is loaded onto the SAG feed belt (CV-06). To obtain an unaltered SAG circuit new feed sample, pebble crusher discharge was diverted to the ground until CV-06 contained only new feed. Similarly, new feed was stopped and crushed pebbles were loaded onto CV-06 to allow for a pebble crusher discharge sample. Pebble crusher feed was sampled by belt cut.

The original two ball mills are in parallel and for the purpose of the survey were called 'ball mill circuit 1' and for survey purposes are treated as a single unit operation. In 2019 a third ball mill was added to the grinding circuit, and this is called 'ball mill circuit 2' and it is fed from the cyclone feed pump box of ball mill circuit 1. This circuit arrangement requires an independent measure of mass flow into ball mill circuit 2 for the mass balance. The methodology of this method is described later in this paper.

All remaining samples were collected as slurries using full stream cuts with custom fabricated sample cutters for each sample. For ball mill #2 (one of the two parallel mills in ball mill circuit 1) the ball mill discharge sample access was opposite of the 'uphill' side of the trommel screen, meaning that the majority of the discharge slurry would be far from the sample port (Figure 3). A new sample port was added, and a staircase was moved, to provide better access to the ball mill discharge slurry (Figure 4).

FIG 3 – Ball mill 2, access port not ideal, clockwise mill rotation.

FIG 4 – Location of new access port, ball mill #2 (and stairs to move).

SAG screen undersize is typically a difficult sample to collect properly during grinding circuit surveys. While the sample was still challenging at Masbate, fortunately the SAG screen undersize did gather in an individual launder prior to the cyclone feed pump box. A cross-stream cut was possible (with careful cutter design, taking the cut in slices) at this sample point.

Ball mill circuit 2, newly added in 2019, had excellent access to both the circuit new feed slurry (Figure 5), and the ball mill screen undersize. Similar to the SAG screen undersize, the ball mill circuit 2 screen undersize reports to a launder which allows for a good sample at the launder lip (Figure 6). Cutters were designed, fabricated, and tested prior to the survey for these sample points.

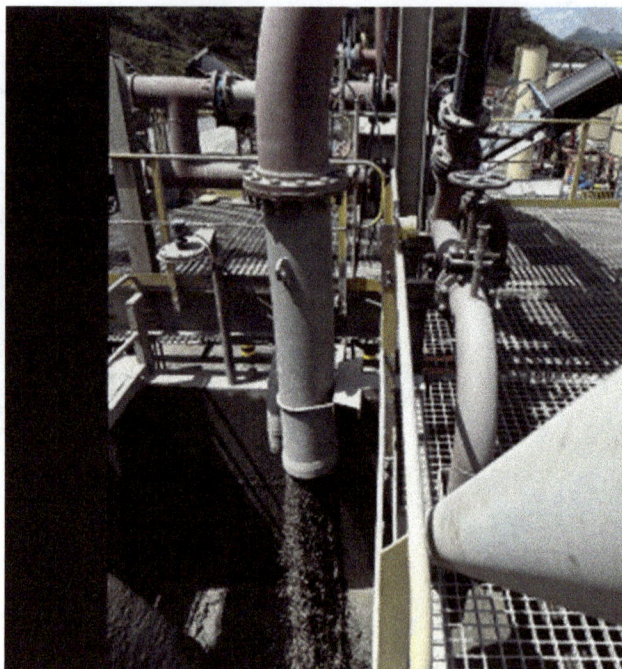

FIG 5 – Ball mill circuit 2 new feed.

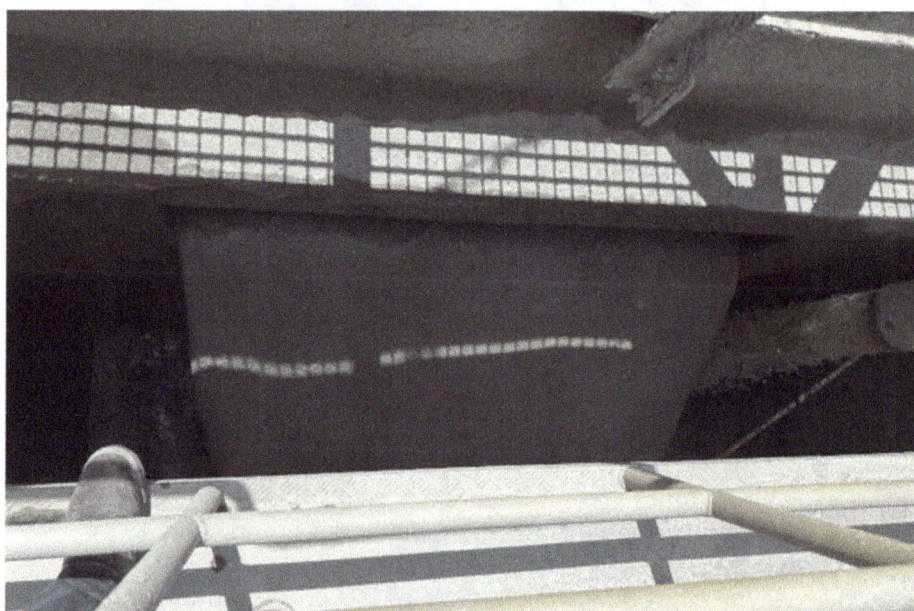

FIG 6 – 1.8 m wide ball mill screen undersize stream.

Cyclone overflow and underflow samples were to be collected at each operating cyclone for each circuit. Sample access was good for each cyclone cluster.

A combined cyclone overflow sample point was also available using the cross-stream cutter of combined leach feed. The automatic sampler discharge line was modified to allow collection of the primary cut into buckets (Figure 7). During the survey, manual samples of leach feed were taken for shift sample purposes to allow the survey team to actuate the cross-stream leach feed cutter as needed.

FIG 7 – Modification of sampling line going to secondary sample cutter of COF sample cutter. Combined COF sample using cross-stream leach feed cutter.

Notably, cyclone feed was not sampled as part of the survey. While it is common practice in many plant surveys to use a blocked cyclone to collect a 'sample' of cyclone feed, it is clear that this is not a full stream cut which is problematic in highly segregated slurry streams such as cyclone feed. Instead, Metcom designs surveys such that the cyclone feed can be back calculated from the other circuit streams. This method does require high-quality samples of mill discharge streams.

Ball mill circuit 2 new feed mass flow measurement

To measure the mass flow of new feed into ball mill circuit 2, a transfer hopper was used as a volumetric flow measurement tank. The transfer hopper collects cyclone overflow from ball mill circuit 2 and transfers it to the leaching system. This location of this hopper in the process can be seen in by referring back to Figure 2 to see 'BM Circ 2 COF Hopper' in the grinding circuit flow diagram.

The transfer hopper has a variable speed pump on its discharge. This allowed the hopper to be pumped down to level below a visible transition in the hopper. Then the discharge pump was stopped and time to fill the hopper was measured. This timing, coupled with the hopper dimensions, was then used to calculate the volumetric flow rate of ball mill circuit 2 cyclone overflow. This volumetric flow, along with slurry % solids can be used to calculate the dry tonnage leaving ball mill circuit 2 which at steady state would be equal to the ball mill circuit 2 new feed rate. Photos of the hopper along with hopper dimensions are shown in Figures 8 and 9.

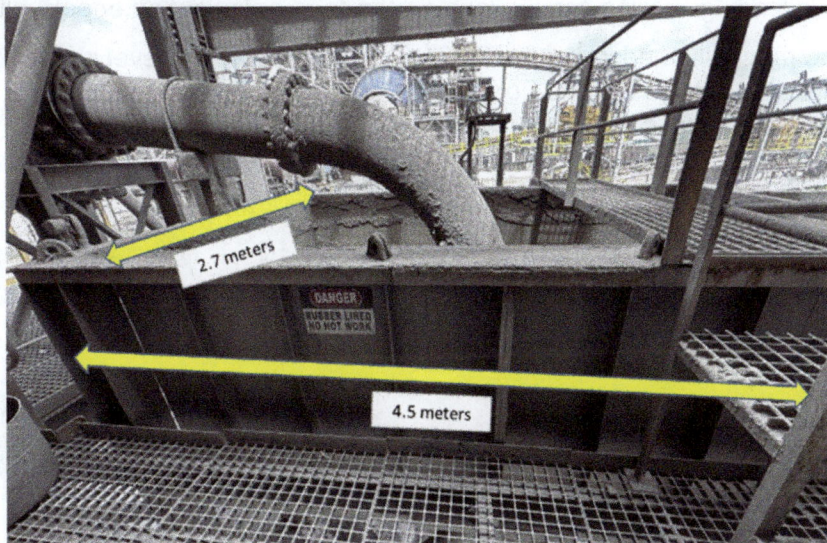

FIG 8 – Length and width of ball mill circuit 2 COF hopper.

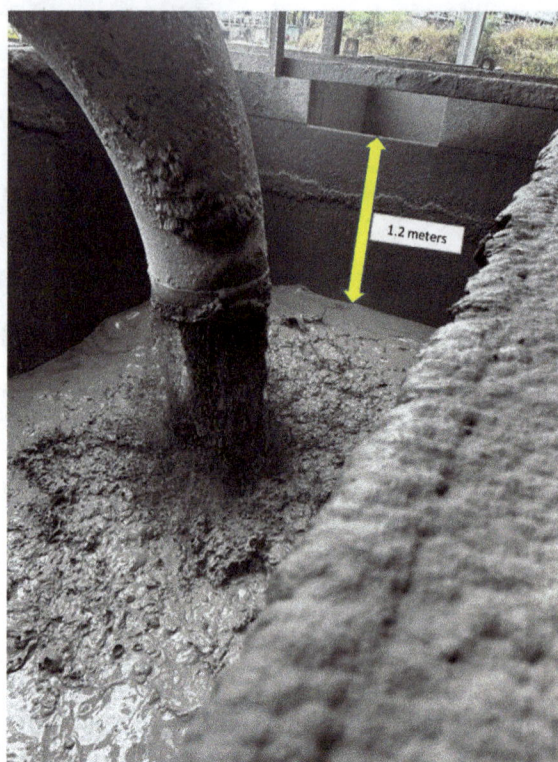

FIG 9 – Height of slurry fill from transition to overflow.

This method was tested during the pre-survey site visit. During testing an interlock which needed to be bypassed to keep ball mill circuit 2 from stopping when the COF hopper discharge pump was stopped for the measurement was identified. While this was easily bypassed to allow for the measurement, if the method hadn't been tested prior to the survey, the survey itself could have been spoilt by this interlock, thus proving the vital need for pre-survey preparations and testing of procedures.

Once the interlock was bypassed a test was run to compare the hopper fill measurement to the mass flow estimated by the flow metre and density gauge in the ball mill circuit 2 new feed line.

Calculated fill volume was 4.5 × 2.7 × 1.2 = 14.58 cubic metres. Fill time for the first trial was 52 secs equating to 1009 m³/h. Slurry density was measured at 44.0 per cent solids (2.7 SG solids) which yields a dry tonnage of 614 t/h. The instruments were reading 703 m³/h at a slurry SG of 1.646 which equates to 438 t/h dry. Clearly the instruments were reading significantly low and if they had been used to calculate the survey mass balance the measured efficiencies would have been highly affected, potentially leading to false conclusions and missed optimisation opportunities.

Note, in addition to manually measuring the fill time by stopwatch, the level sensor data collected in the SCADA system was used to plot the hopper fill to give a second measure of fill time as shown in Figure 10. The fill time was 51 secs by stopwatch and 52 secs by SCADA.

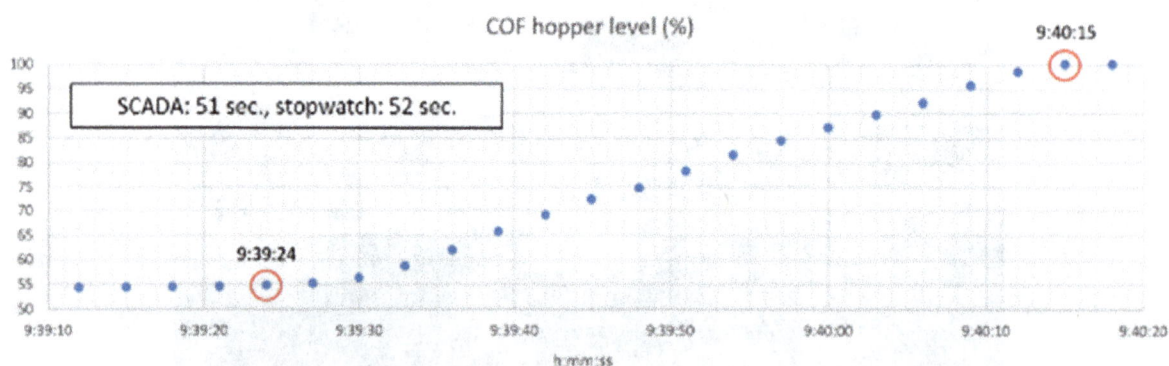

FIG 10 – SCADA level sensor reading of COF hopper fill.

Additional trials of this method were conducted by the site team between the pre-survey visit and the survey which all indicated a similar offset (about 35 per cent higher dry t/h for tank-fill method) between instrument readings and hopper fill tests. Based on this, it was decided to conduct a hopper fill test during the survey and use that as the measurement of mass flow into ball mill circuit 2. Additionally, with the hopper level validation shown above it was possible to conduct additional repeats of the tank-fill tests without creating the mess of the overflow. Instead, the fill time and fill level was measured by instrument and volumetric flow was calculated.

Survey execution

The grinding circuit survey was conducted on July 10, 2023. Overnight process trends were checked between 6:30 and 7:30 am and the survey plan was reviewed with the sampling teams. At 8:00 am the first SAG new feed sample was collected.

SAG new feed sample collection

SAG new feed, and the pebble crusher feed and discharge belt cuts were taken before the survey with sufficient time given to restabilise the circuit for the survey. Then the belt cuts were repeated between the two sets of slurry samples, again with time given to restabilise the circuit before collecting the second set of composites. A total of approximately 2.5 t of SAG new feed were collected during the two 5-meter belt cuts (Figure 11).

FIG 11 – A 5 m belt cut, SAG new feed.

Since the jaw crusher at Masbate is fed by loaders, and the feed is from two distinct sources, it was decided to blend a large sample of material in the correct ratio on the ROM pad the shift before the survey such that the SAG feed belt samples would be reasonably well mixed when sampled. Approximately 300 t of material were blended as shown in Figure 12.

FIG 12 – A 60/40 per cent blend of transition/oxide ore for survey SAG feed cuts.

While the 300 t of material wouldn't be sufficient to run the whole survey (1000 t/h for several hours), the material was selectively fed to the circuit prior to each of the two SAG new feed belt cuts.

Slurry samples

After the first SAG new feed and pebble crusher feed and discharge samples were collected in the morning, the circuit was restarted and allowed to stabilise. Sample buckets and cutters were distributed while the circuit was stabilised. At 11:00 am slurry samples were cut at each of the points noted in Figure 2. Five cuts were taken at each point during the first half of the survey to make a single composite at each sample point. However, a process disturbance caused the SAG feed tonnage set point to drop from 1150 t/h to 500 t/h at the end of the first sampling period ('half-sample 1') and it was decided to not include the fifth cut in the half-sample 1 composite.

After collecting half-sample 1, a second set of SAG new feed and pebble crusher samples was collected between 1:45–2:05 pm. The mills remained in operation during the collection of the second set of belt cuts to minimise the amount of time required to return the circuits to steady state for half-sample 2 collection. Half-sample 2 collection started at 3:30 pm and continued until 4:15 pm when all five cuts for half-sample 2 had been collected as a composite sample at each sample point. Following completion of half-sample 2 the SAG mill was crash stopped to measure total charge level. Grind-out of all three mills was conducted the following day to measure ball charge levels.

Instrument data collection

Instrument data was collected by the SCADA system during the survey and was exported once the survey was completed. Manual readings were also being taken periodically during the survey as a backup, and as a means of watching process trends to see if the process was stable during sampling.

The key instrument readings of SAG new feed tonnage, SAG power, ball mill powers, cyclone feed densities, and cyclone feed hopper levels are shown in Figures 13–17. Half sample 1 and half sample 2 collection periods are shown in yellow on each plot. The process disturbance near 12:00 pm and the second set of belt cuts at 1:45–2:05 pm are also visible in the figures.

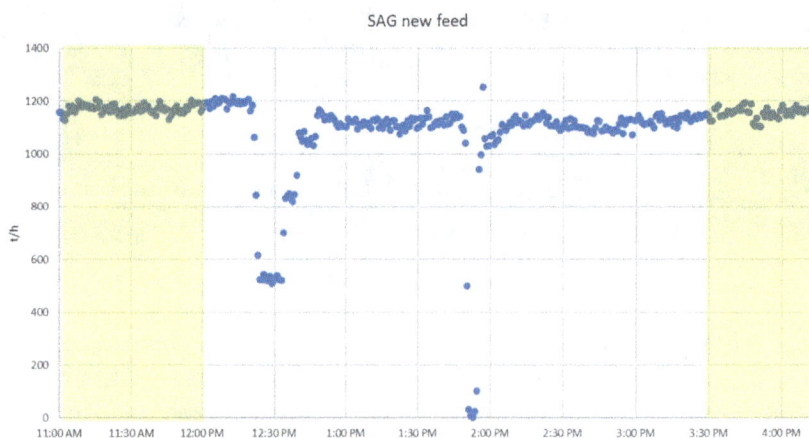

FIG 13 – SAG new feed (weightometer reading).

FIG 14 – SAG power (kW).

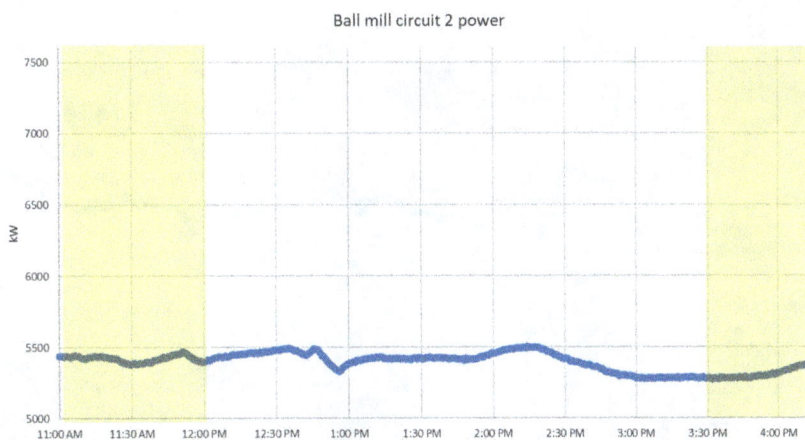

FIG 15 – Ball mill circuit 1 power (total, both mills in parallel) and ball mill circuit 2 power.

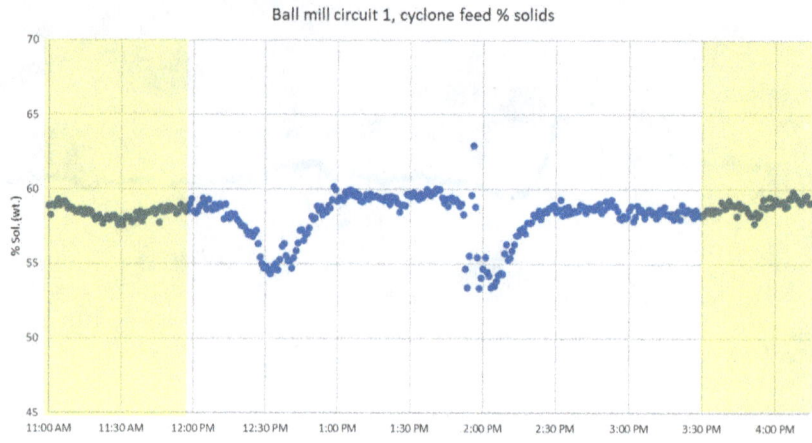

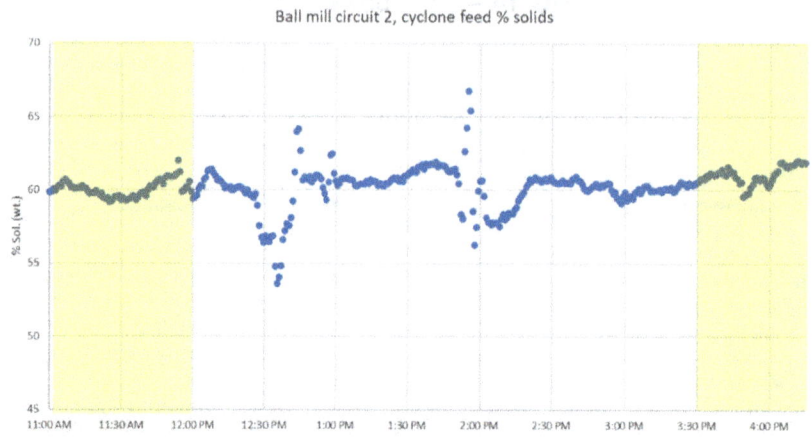

FIG 16 – Ball mill circuit 1 and 2, cyclone feed % solids (wt), densometer readings.

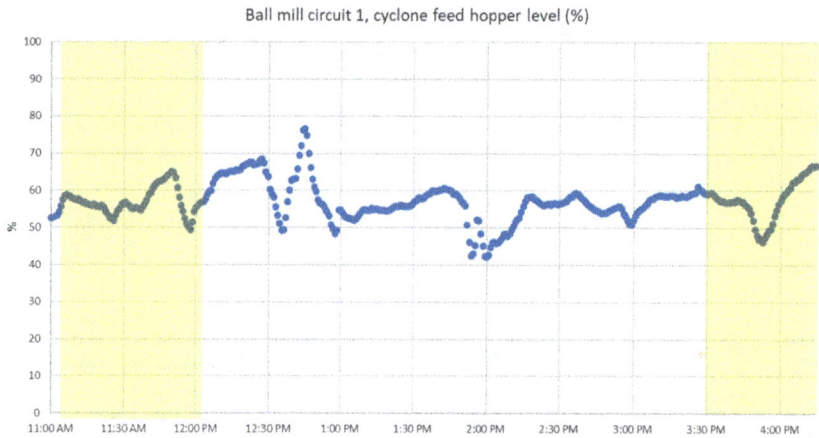

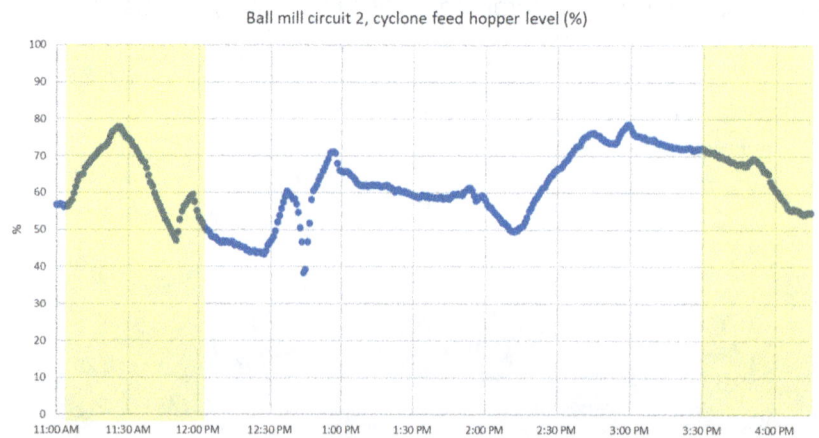

FIG 17 – Ball mill circuit 1 and 2 cyclone feed hopper level.

Cyclone feed pump speeds were locked in manual at a fixed speed for the duration of the survey (during active sample periods) to minimise cyclone feed disturbances. In general, the conditions during half-sample 1 and half-sample 2 were reasonably close to each other, and process variability appeared to be more cyclical than trending in a single direction, so the survey was deemed to be reasonably representative of average process conditions.

Ball mill circuit 2 mass flow was measured, as described earlier in this paper, with results shown in Table 1.

TABLE 1
Ball mill circuit 2 COF hopper tests.

	15-May	6-Jul			11-Jul		
	trial 1	trial 1	trial 2	trial 3	HS1	HS2	
Hopper width	2.73	2.73	2.73	2.73	2.73	2.73	m
Hopper length	4.50	4.50	4.50	4.50	4.50	4.50	m
Fill height	1.20	1.20	1.25	1.26	1.04	1.04	m
Fill volume	14.73	14.73	15.34	15.50	12.78	12.76	m3
Fill time	52.0	52.0	54.0	49.0	46.0	45.0	seconds
Slurry vol. flow rate	1020	1020	1023	1139	1001	1020	m3/h
COF Marcy scale reading	44.0	50.0	49.0	51.5	49.0	49.0	% sol. (wt.), 2.7 SG scale
Dry solids SG	2.70	2.70	2.70	2.70	2.70	2.70	
COF % Sol. By volume	22.5	27.0	26.2	28.2	26.2	26.2	% (vol.)
Dry solids volumetric flow	230	276	268	321	263	268	m3/h dry solids
BM Circuit #2 COF dry tonnage	**621**	**744**	**725**	**868**	**709**	**723**	mt/h dry solids

BM Circuit #2 new feed from instruments

	15-May	6-Jul			11-Jul		
BM Circuit #2 new feed flow	427	552	427	427	465	516	m3/h
BM Circuit #2 new feed density	1646	1573	1646	1646	1597	1626	kg/m3
BM Circuit #2 new feed flow	703	868	703	703	743	839	mt slurry / hr
BM Circuit #2 new feed slurry SG	1.646	1.573	1.646	1.646	1.597	1.626	
BM Circuit #2 new feed % sol.	62.3	57.9	62.3	62.3	59.4	61.1	% sol. (wt.)
BM Circuit #2 new feed tonnage	**438**	**502**	**438**	**438**	**441**	**513**	mt/h dry solids

The two tests conducted during the July 11, 2023 survey (rightmost two columns) resulted in an average mass flow of 716 t/h (dry) leaving ball mill circuit 2. Previous method trials are also shown in the table for reference.

Survey results

The survey belt cut samples were packaged in lined drums, weighed, and sealed before shipment to SGS Lakefield. Slurry samples were weighed on-site to arrive at net wet slurry weight. Partial slurry sample dewatering (decanting/filtering) was also conducted on-site prior to shipment and moist filter cake weights were recorded before final packaging. SGS weighed all received samples and all weights corroborated. Final sample drying was done at SGS Lakefield to arrive at a % solids for each sample. All samples were measured for dry solids SG and particle size distribution.

The full SAG feed composite was also dried, measured for moisture, particle size distribution, and representative subsamples were collected and processed for Bond, JK, and SMC ore hardness tests.

Per cent solids summary

The SAG feed sample was measured to contain 3.5 per cent moisture by weight. This is in line with typical moisture readings at the plant.

As can be seen in Table 2, most of the % solids values from half-sample 1 to half-sample 2 were within about 1 per cent of each other with the exception of the ball mill circuit 1 and ball mill circuit 2 cyclone overflow samples and SAG screen undersize. The two ball mill circuit cyclone overflow samples shifted in % solids between HS1 and HS2 with ball mill circuit 1 COF becoming lower in % solids, and ball mill circuit 2 COF becoming higher in % solids. However, the composite cyclone overflow sample was only about 1 per cent different between HS1 and HS2 so it was decided to average the data from the half-samples to best represent the composite circuit performance during the survey.

TABLE 2

Survey % solids summary.

Half Sample 1

	SAG Screen U/S	Pebble crusher feed	Pebble crusher discharge	Ball mill trommel U/S, circuit 1	Ball mill trommel U/S, circuit 1	Cyclone U/F circuit 1	Cyclone O/F circuit 1	Ball mill circuit 2 new feed	Ball mill circuit 2 trommel	Ball mill circuit 2 cyclone U/F	Ball mill circuit 2 cyclone O/F	Combined cyclone O/F
Specific gravity	2.67	2.70	2.69	2.68	2.70	2.68	2.67	2.67	2.70	2.70	2.68	2.68
% Solids, wt.	64.1	97.9	96.4	70.7	70.9	71.1	44.5	62.7	74.2	74.4	45.0	43.3

Half Sample 2

	SAG Screen U/S	Pebble crusher feed	Pebble crusher discharge	Ball mill trommel U/S, circuit 1	Ball mill trommel U/S, circuit 1	Cyclone U/F circuit 1	Cyclone O/F circuit 1	Ball mill circuit 2 new feed	Ball mill circuit 2 trommel	Ball mill circuit 2 cyclone U/F	Ball mill circuit 2 cyclone O/F	Combined cyclone O/F
Specific gravity	2.68	2.68	2.68	2.68	2.70	2.68	2.68	2.68	2.70	2.68	2.68	2.68
% Solids, wt.	61.5	97.8	97.2	71.4	71.6	72.4	38.9	63.6	74.7	75.7	50.8	44.5

Average

	SAG Screen U/S	Pebble crusher feed	Pebble crusher discharge	Ball mill trommel U/S, circuit 1	Ball mill trommel U/S, circuit 1	Cyclone U/F circuit 1	Cyclone O/F circuit 1	Ball mill circuit 2 new feed	Ball mill circuit 2 trommel	Ball mill circuit 2 cyclone U/F	Ball mill circuit 2 cyclone O/F	Combined cyclone O/F
Specific gravity	2.68	2.69	2.69	2.68	2.70	2.68	2.68	2.68	2.70	2.69	2.68	2.68
% Solids, wt.	62.8	97.8	96.8	71.0	71.3	71.8	41.7	63.2	74.5	75.0	47.9	43.9
HS1-HS2 (% Sol.)	2.56	0.16	-0.80	-0.70	-0.75	-1.30	5.57	-0.93	-0.53	-1.26	-5.87	-1.13

SAG feed: 3.5% moisture

Ore grindability characterisation

The SAG new feed sample was measured for grindability as shown in Table 3.

TABLE 3

SAG new feed ore grindability.

	SG	JK	SMC	Work indices (kWh/t)			AI
	g/cm³	A × b	A × b	CWi	RWi	BWi	(g)
SAG new feed	2.67	54.5	47.8	11.9	14.5	14.1	0.277

Size distributions and mass balance

The SAG new feed sample had a top size of 200 mm and a K80 of 120 mm. A plot of the SAG new feed PSD is shown in Figure 18.

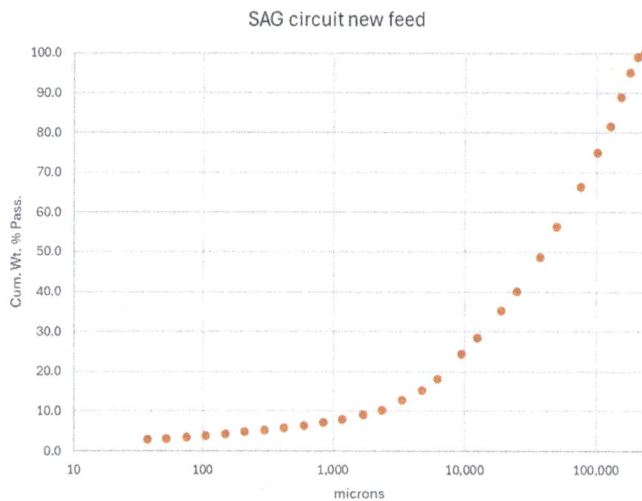

FIG 18 – SAG circuit new feed PSD.

Pebble crusher feed and discharge PSDs are shown in Figure 19.

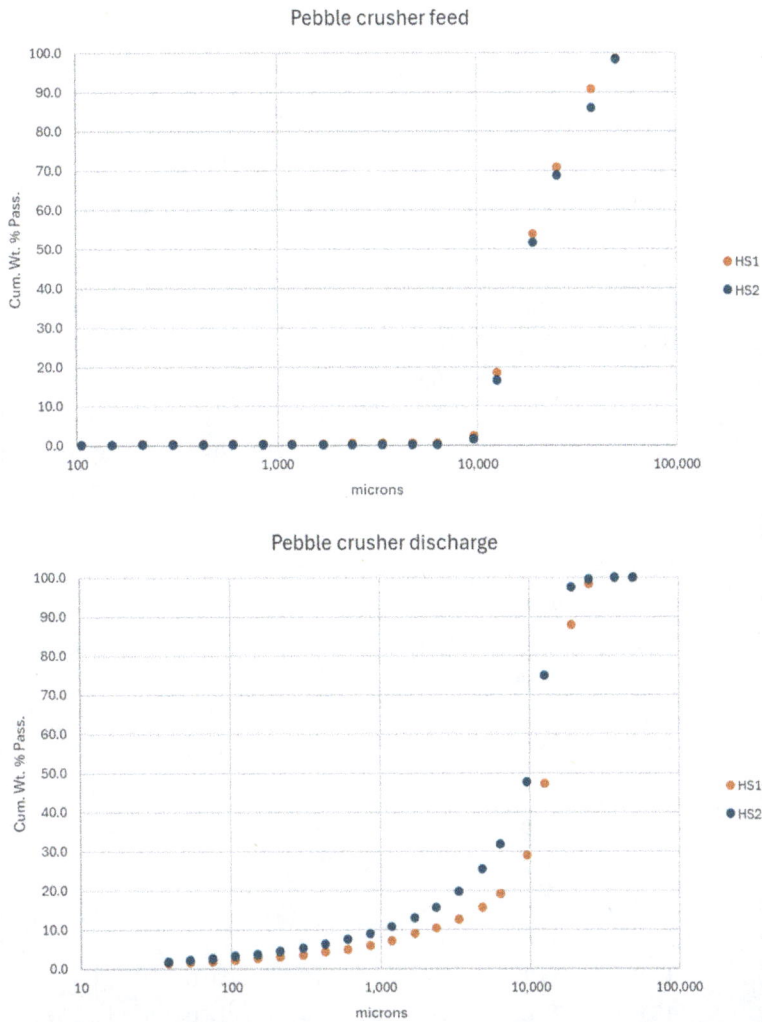

FIG 19 – Pebble crusher feed and discharge PSDs.

The pebble crusher feed F_{80} for HS1 was 31 mm and 33 mm for HS2. The pebble crusher discharge P_{80} was HS1 was 18 mm and 14 mm for HS2. On average this gave a pebble crusher F_{80} of 32 mm and a P_{80} of 16 mm.

The raw PSD values for the ball mill circuit samples is shown in Table 4. These are the average values for half-sample 1 and half-sample 2.

TABLE 4

TABLE 4
Raw cumulative wt % Passing, ball mill circuit survey samples.

Micron	SAG Screen U/S	Cyclone U/F, circuit 1	Ball mill trommel U/S,	COF BM Circ 1	Ball mill circuit 2 new feed	Ball mill circuit 2 trommel	Ball mill circuit 2 cyclone	COF BM Circ 2	Combined COF
50,000	100.0	100.0	100.0	100.0	100.0	100.0	100.0	100.0	100.0
37,500	100.0	100.0	100.0	100.0	100.0	100.0	100.0	100.0	100.0
25,000	100.0	100.0	100.0	100.0	100.0	100.0	100.0	100.0	100.0
19,050	100.0	100.0	100.0	100.0	100.0	100.0	100.0	100.0	100.0
12,500	100.0	100.0	100.0	100.0	100.0	100.0	100.0	100.0	100.0
9,500	98.9	99.4	99.9	100.0	99.1	99.6	99.4	100.0	100.0
6,300	93.5	96.6	99.4	100.0	95.5	98.3	97.0	100.0	100.0
4,750	89.7	94.4	99.0	100.0	93.5	97.6	95.6	100.0	100.0
3,350	85.1	91.4	98.4	100.0	90.7	96.8	94.0	100.0	100.0
2,360	80.5	88.3	97.6	100.0	87.8	96.0	92.4	100.0	100.0
1,700	75.1	86.6	97.0	100.0	86.0	95.5	91.6	100.0	100.0
1,180	65.7	82.2	95.2	100.0	81.2	93.9	88.4	100.0	100.0
850	59.8	77.9	92.9	100.0	77.2	91.9	85.1	100.0	100.0
600	54.1	71.9	88.9	99.9	72.1	88.3	79.9	99.8	99.9
425	48.4	63.6	82.3	99.4	65.6	81.3	71.4	99.1	99.3
300	43.1	53.0	72.6	96.8	58.0	70.0	58.7	96.1	96.6
212	38.2	42.1	61.9	91.8	49.8	55.8	44.8	90.4	91.4
150	33.6	32.6	50.2	83.9	41.3	41.6	31.7	81.2	82.8
106	29.5	26.2	41.3	75.2	34.4	31.4	22.8	70.8	72.6
75	25.7	21.0	34.6	64.5	29.6	24.8	17.7	61.5	63.2
53	22.6	17.8	29.3	56.4	25.9	20.6	14.3	53.1	54.5
38	20.2	15.6	25.6	49.9	23.4	18.0	12.2	46.9	48.1

This data was then mass balanced using the measured SAG circuit new feed rate and measured ball mill circuit 2 feed rates. The balanced data is shown in Table 5.

TABLE 5
Balanced size distributions.

Balanced Data, Cum. Wt. % Pass

Micron	SAG Screen U/S	Cyclone U/F, circuit 1	Circuit 1 trommel U/S	COF BM Circ 1	Circuit 2 new feed	Circuit 2 trommel U/S	Cyclone U/F, circuit 2	COF BM Circ 2	Combined COF	Circuit 1 Cyclone feed	Circuit 2 Cyclone feed
50,000	100.0	100.0	100.0	100.0	100.0	100.0	100.0	100.0	100.0	100.0	100.0
37,500	100.0	100.0	100.0	100.0	100.0	100.0	100.0	100.0	100.0	100.0	100.0
25,000	100.0	100.0	100.0	100.0	100.0	100.0	100.0	100.0	100.0	100.0	100.0
19,050	100.0	100.0	100.0	100.0	100.0	100.0	100.0	100.0	100.0	100.0	100.0
12,500	100.0	100.0	100.0	100.0	100.0	100.0	100.0	100.0	100.0	100.0	100.0
9,500	98.8	99.5	99.9	100.0	99.1	99.6	99.4	100.0	100.0	99.6	99.5
6,300	93.4	96.6	99.4	100.0	95.5	98.3	97.0	100.0	100.0	97.3	97.7
4,750	89.6	94.4	99.0	100.0	93.5	97.5	95.7	100.0	100.0	95.6	96.7
3,350	84.7	91.6	98.4	100.0	90.7	96.8	94.2	100.0	100.0	93.3	95.4
2,360	79.9	88.7	97.6	100.0	87.8	96.0	92.6	100.0	100.0	91.0	94.2
1,700	75.4	85.9	97.1	100.0	85.9	95.6	91.6	100.0	100.0	88.7	93.4
1,180	67.0	80.6	95.5	100.0	80.6	93.9	88.5	100.0	100.0	84.5	91.0
850	61.4	76.2	93.2	100.0	76.5	91.9	85.3	100.0	100.0	81.0	88.5
600	55.6	70.3	89.1	99.9	71.4	88.2	80.2	99.8	99.9	76.3	84.5
425	49.7	62.5	82.4	99.4	65.0	81.3	71.7	99.1	99.3	70.0	77.7
300	44.3	52.6	72.2	96.9	57.5	70.1	59.3	96.1	96.6	61.5	67.3
212	39.3	42.6	60.8	91.9	49.5	56.2	44.7	90.4	91.4	52.5	54.8
150	34.6	33.3	48.9	84.0	41.2	42.7	31.4	81.2	82.8	43.5	42.3
106	30.5	26.9	40.1	75.3	34.3	32.7	22.5	70.7	72.6	36.7	33.1
75	26.6	21.9	33.1	64.9	29.6	26.3	17.4	61.4	63.2	30.6	27.0
53	23.5	18.7	28.1	56.5	25.8	21.9	14.1	53.3	54.5	26.3	22.7
38	21.2	16.3	24.6	50.0	23.2	19.0	12.2	47.5	48.1	23.1	19.9
K80	2359	1143	392	128	1146	410	591	144	137	783	479

The mass balance required no more than a 1.0 per cent adjustment to any given size class indicating good quality survey data.

Ore water balance

Using the balance PSD information, slurry % solids data, and tonnage readings determined by the SAG feed weightometer and the ball mill circuit 2 COF hopper fill tests, the following ore-water balance was calculated for the survey. Figure 20 shows the ore-water balance assembled from the survey data for the SAG and pebble crushing circuit.

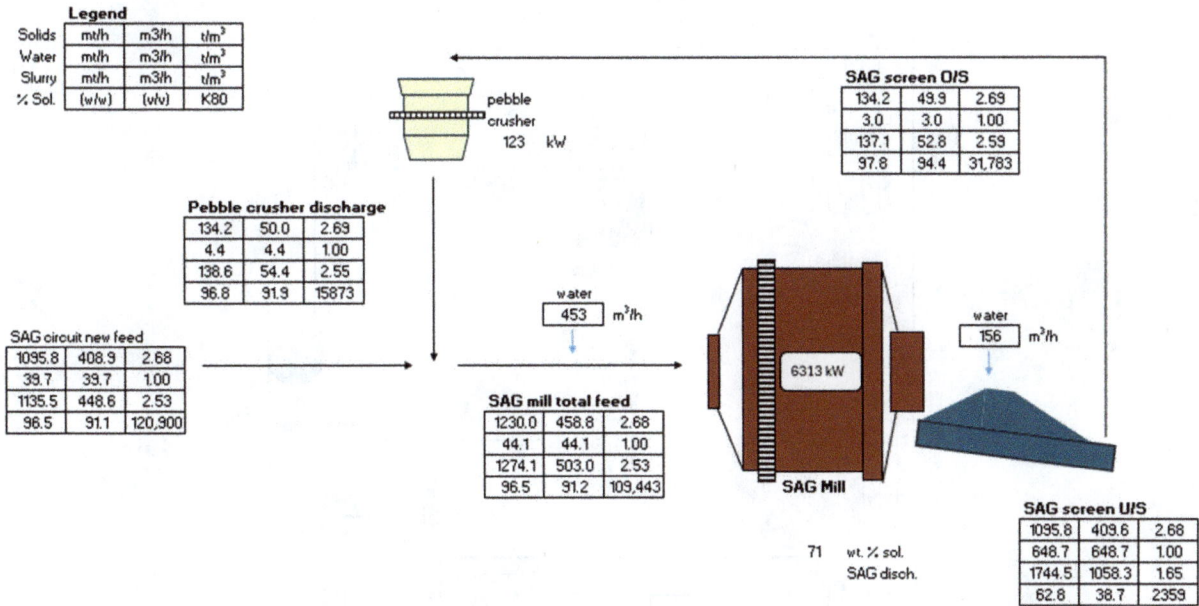

Legend

	mt/h	m3/h	t/m³
Solids	mt/h	m3/h	t/m³
Water	mt/h	m3/h	t/m³
Slurry	mt/h	m3/h	t/m³
% Sol.	(w/w)	(v/v)	K80

pebble crusher 123 kW

Pebble crusher discharge

134.2	50.0	2.69
4.4	4.4	1.00
138.6	54.4	2.55
96.8	91.9	15873

SAG circuit new feed

1095.8	408.9	2.68
39.7	39.7	1.00
1135.5	448.6	2.53
96.5	91.1	120,900

water 453 m³/h

SAG mill total feed

1230.0	458.8	2.68
44.1	44.1	1.00
1274.1	503.0	2.53
96.5	91.2	109,443

SAG screen O/S

134.2	49.3	2.69
3.0	3.0	1.00
137.1	52.8	2.59
97.8	94.4	31,783

6313 kW

water 156 m³/h

SAG Mill

71 wt. % sol. SAG disch.

SAG screen U/S

1095.8	409.6	2.68
648.7	648.7	1.00
1744.5	1058.3	1.65
62.8	38.7	2359

FIG 20 – Survey ore-water balance (SAG circuit).

Figure 21 shows the continued ore-water balance for the ball mill circuits. Legend is the same as shown in Figure 20.

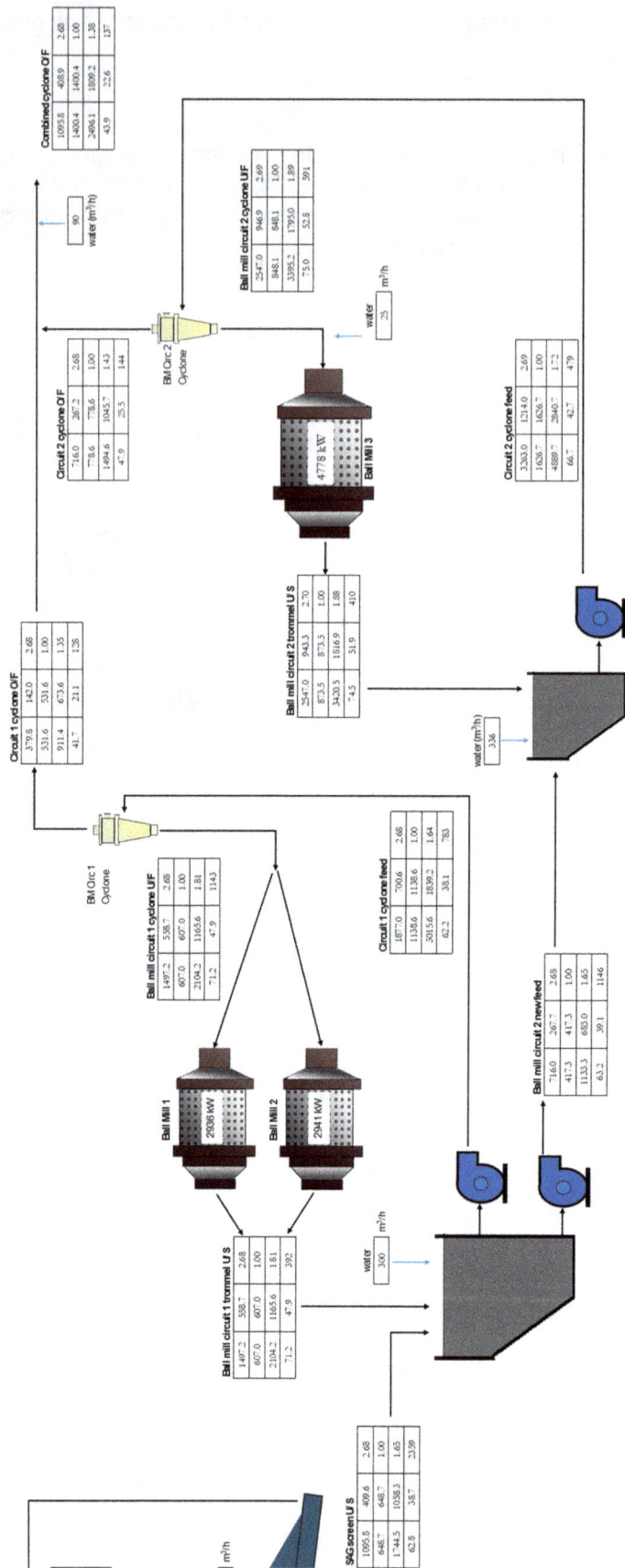

FIG 21 – Survey ore-water balance (ball mill circuits).

Efficiency benchmarking

With the survey data reconciled, it was possible to use Bond Work index methods (Global Mining Guidelines Group, 2021) to benchmark the efficiency of the grinding circuit as a whole, the SAG circuit, and the ball mill circuits. Table 6 is a summary of the Bond Work Index analysis.

TABLE 6
Bond Work Index analysis.

	SAG Circuit	BM Circ 1	BM Circ 2	BM Circ combined	Overall SAG-ball	
Circuit tonnage	1,096	380	716	1,096	1,096	mt/h
Circuit power	6,438	5,877	4,778	10,655	17,093	kW
Circuit F80	120,900	-	1,146	2,359	120,900	µm
Circuit P80	2,359	-	144	137	137	µm
Circuit W	5.9	-	6.7	9.7	15.6	kWh/mt
Circuit Wio	33.2	-	12.4	15.0	18.9	kWh/mt
Circuit Bond WI Eff. (base)	42	78	114	94	75	%
Circuit Bond WI Eff. (fines-corrected)	87	55	82	67	N/A	%

The Bond Work Index analysis was conducted for the overall circuit, the combined ball milling circuit (all three mills), ball mill circuit 2, and the SAG circuit using directly measured data. The work index efficiency of ball mill circuit 1 was calculated by a weighted average method using the directly measured work index values and circuit power draws. This was done because an unknown amount of SAG screen undersize material has the potential to bypass ball mill circuit 1 directly into ball mill circuit 2 so F_{80} and P_{80} of ball mill circuit 1 are not directly measurable.

The methods used to calculate the 'fines-adjustment', which account for the different shape of SAG discharge size distribution curves compared to rod mill discharge PSD cures, are described in Metcom's training system (McIvor, 1990–2024). Essentially, the method gives credit to the SAG circuit for 'extra' fines produced, in excess of a typical rod mill.

The overall work index efficiency of 75 per cent (SAG + all ball mills) shows that overall circuit efficiency can be improved since 100 per cent is considered 'average.' The fines-adjusted SAG circuit WI efficiency of 87 per cent could be improved but is reasonable for an unoptimised circuit. The total ball mill circuit fines-corrected WI efficiency of 67 per cent shows considerable room for improvement, with ball mill circuit 2 outperforming ball mill circuit 1 substantially.

While Bond methods are able to identify where there are opportunities for efficiency improvement, the methods do not give particular direction as to how to improve. For this, Functional Performance Analysis was conducted.

Functional performance analysis

Developed by McIvor (1988) Functional Performance Analysis is able to assign separate efficiencies to the classification system (pumps and cyclones) (Bartholomew, McIvor and Arafat, 2019) and the ball mill environment itself (McIvor, 2006).

A brief summary of the Functional Performance Equation, and relevant process parameters, is given in Figure 22.

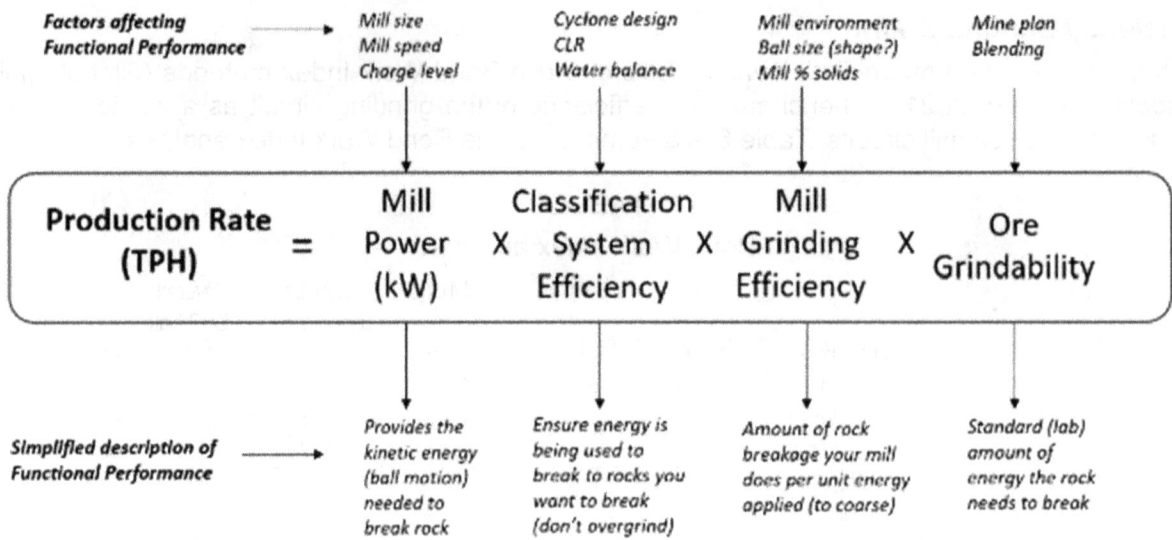

FIG 22 – Summary of the Functional Performance Equation.

Using the survey data the Functional Performance Equation was populated for both ball mill circuits at Masbate as shown in Table 7.

TABLE 7

Functional performance summary.

BM Circ 1	BM Circ 2	All BMs Combined	Functional Performance
242	286	528	prod. rate new -150 µm, mt/h
58.9	63.0	61.5	CSE (%) @150 µm
3461	3009	6548	effective BM power, kW
0.0698	0.0952	0.0807	ball mill gr. rate, mt/kWh @150 µm
1.79	1.79	1.79	grindability (g/rev); SAG feed Bond test
0.0390	0.0532	0.0451	ball mill gr. eff, mt/kWh per g/rev
394	356	369	Circulating load ratio (%)
62.2	66.7	65.0	Cyclone feed % solids (wt.)
0.337	0.312	0.331	Selectivity Index (S.I.), d25c/d75c
55.0	47.0	50.0	Bypass, est. (%)
71.0	75.0	73.5	Cyclone U/F % Solids (wt.)
50/50% 80/65 mm	100% 65 mm	-	Ball sizes & charge ratios

Table 7 also shows the values for relevant operating conditions affecting Functional Performance as indicated in Figure 22 (the simplified FPE).

The following observations about the Functional Performance Summary can be made:

- The production rate of new -150 µm material is reasonably well split between the two ball mill circuits. Even though ball mill circuit 2 is processing more tons than circuit 1, it is also starting with a finer feed, already partially ground by circuit 1.

- Classification system efficiency of ball mill circuit 2 is 7 per cent higher relative to circuit 1. This is primarily due to a lower bypass of fines. This appears to be primarily related to a lower cyclone underflow % solids. This can be addressed with a spigot change.

- Ball mill circuit 1 cyclones have a sharper separation (selectivity index) which is due to the lower feed % solids.

- Ball mill grinding efficiency in ball mill circuit 2 is 36 per cent higher relative to circuit 1. This is most likely related to the better match between ball size and grinding duty in circuit 2, but also the mill density is higher in circuit 2.

The difference in bypass fraction and cyclone selectivity can be observed in the cyclone recovery curves plotted in Figure 23.

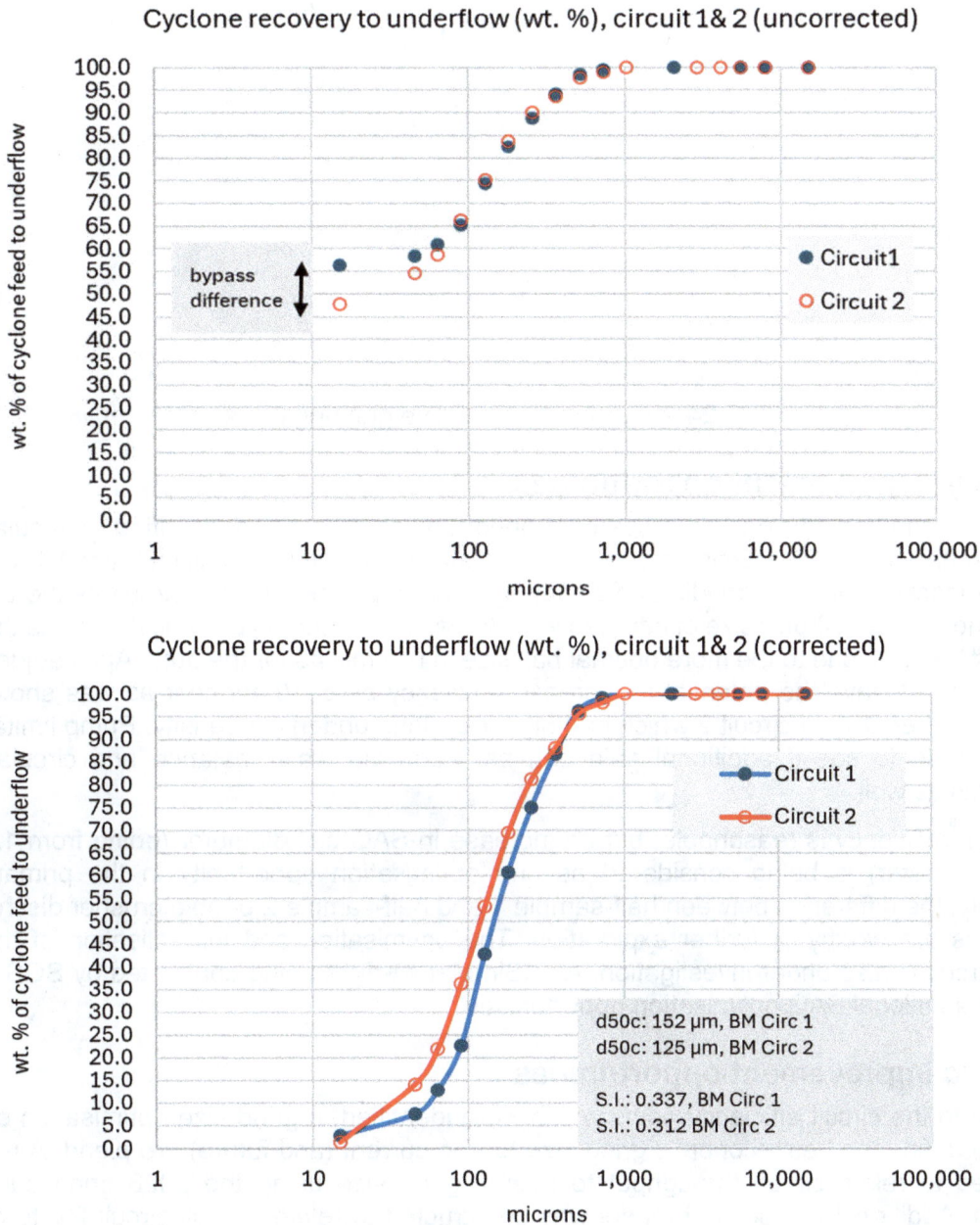

FIG 23 – Cyclone recovery curves.

Ball mill cumulative grinding rates

Apparent ball mill cumulative grinding rates can be used to assess ball mill media size performance (Bartholomew, McIvor and Arafat, 2019). These rates are plotted in Figure 24. At the size of interest for Masbate, 150 µm, the grinding rate was 0.0681 t/kWh for ball mill circuit 1 whereas it was 0.0955 for circuit 2. Ball mill circuit 2 is charged with 100 per cent 65 mm balls while circuit 1 is charged with a 50/50 per cent blend of 80/65 mm balls. It is expected that ball mill circuit 1 will need a somewhat larger average ball size to handle coarse material from the SAG, but it appears from the ball mill grinding efficiency and cumulative grinding rates that the match between ball size and particles size distribution is better in circuit 2. This presents a good opportunity for efficiency improvement through media size optimisation in ball mill circuit 1.

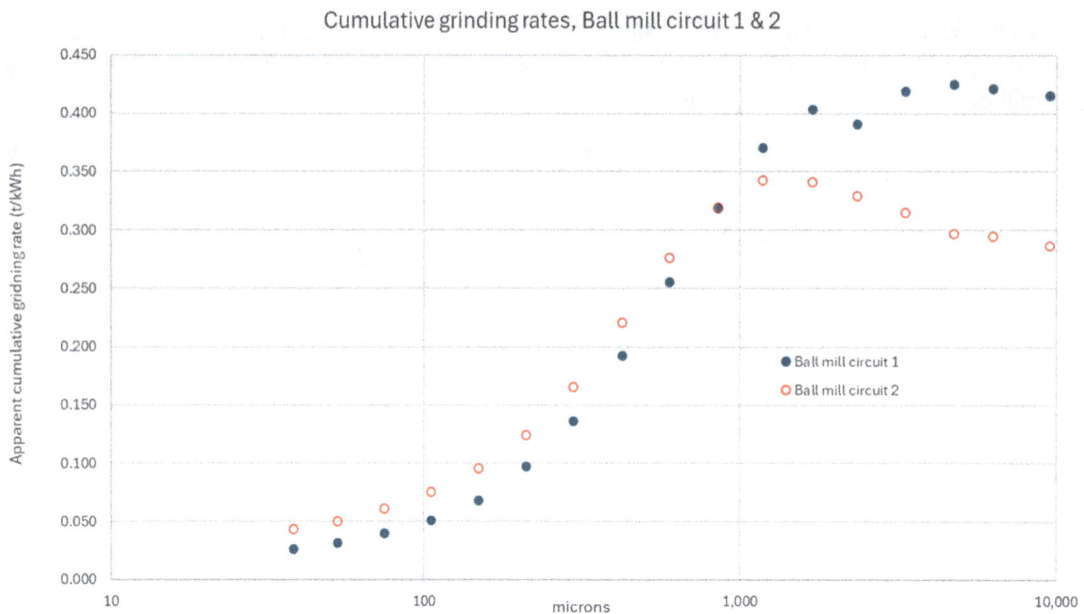

FIG 24 – Apparent cumulative grinding rates.

Efficiency improvement opportunities

A number of grinding efficiency improvement opportunities have been identified, particularly in the ball mill circuits. The most promising of these is ball size optimisation in ball mill circuit 1 which could potentially increase ball mill grinding efficiency by over 20 per cent in that circuit as the cumulative mill grinding rate at 150 µm (size of interest) was 36 per cent higher in ball mill #3 versus the parallel ball mills #1 and #2 due to the more optimal ball size in ball mill #3 for the duty. Apex adjustments in ball mill circuit 1 have the potential to increase efficiency by 5–10 per cent as was shown by the 7 per cent higher CSE in circuit 2 which had higher cyclone underflow density. Pump limitations are being explored to see if additional gain can be made via water balance and circulating load adjustments as well.

SAG circuit efficiency is reasonable, but an increase in SAG ball diameter (going from 125 mm to 140 mm diameter) is being considered as one optimisation opportunity in the primary circuit. Additionally, the difference between half-sample 1 and half-sample 2 pebble crusher discharge size distributions are worthy of further exploration. The optimisation and full utilisation of the pebble crusher circuit is also under investigation. A JKSimMet study is being conducted by SGS Lakefield to explore additional SAG optimisation opportunities.

Economic improvement opportunities

In addition to the circuit efficiency gains which were identified, a grind size optimisation study was conducted to find the best economic grind size for the current (and future) ore blend. A model was constructed to relate circuit throughput to overall grind size using the 2023 grind survey as a calibration. Additionally, a second model was constructed to relate overall circuit P_{80} to combined ball mill power to be used as the most conservative estimate of savings if a coarsening of the grind was enacted at design tonnage (993 t/h).

A series of grind versus gold recovery tests were conducted at SGS Lakefield to better understand the effect of grind size versus recovery on current and future ore blends (Figure 25). This data will be coupled with the grind models, and known plant operating costs, to arrive at an optimal economic grind size. At the time this paper was written the throughput limitations of the SAG mill were being explored by modelling.

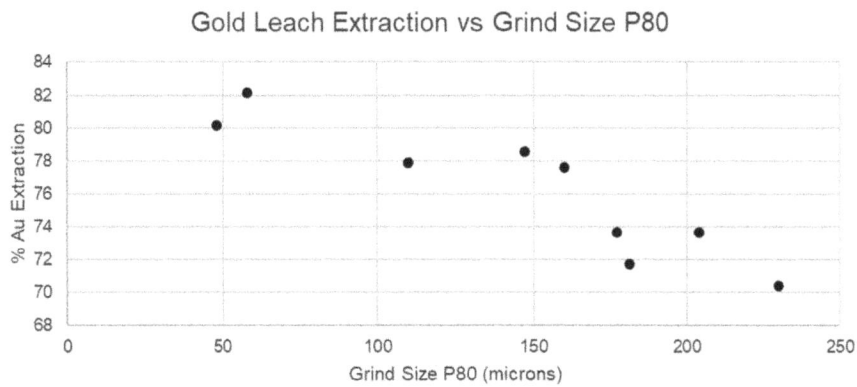

FIG 25 – Grind versus recovery test data.

CONCLUSIONS

A grinding circuit survey was conducted in 2023 at the Masbate gold mine following the 2019 plant expansion for the purpose of benchmarking performance and identification of optimisation opportunities. The survey was carefully planned and executed and produced good quality data for efficiency benchmarking and identification of circuit optimisation opportunities. The flow sheet complexity required an additional mass flow measurement to allow for the mass balance to be closed. This was accomplished using a cyclone overflow hopper timed fill method which showed to be much more accurate than process instruments. The most promising efficiency improvement opportunities appear to be ball mill media size adjustment along with minor adjustments to the cyclone circuits. Bond Work Index Analysis clearly showed the different circuit efficiencies, and Functional Performance Analysis gave direction as to how to best approach practical optimisation projects. SAG circuit opportunities being explored included a ball size increase and better utilisation of the pebble crushing circuit. Finally, a grind size versus gold recovery study using the expected life-of-mine ore is also being conducted to find the optimum economic grind size at Masbate.

ACKNOWLEDGEMENTS

The authors from Metcom would like to thank B2Gold for permission to publish this paper. The authors would also like to acknowledge the efforts put in by the Masbate operations and maintenance teams in support of the survey.

REFERENCES

Bartholomew, K M, McIvor, R E and Arafat, O, 2019. A Method for Ball Mill Media Sizing for Different Upstream Processes, in Proceedings of the 2019 SAG Conference, Canada.

Global Mining Guidelines Group, 2021. Determining the Bond Efficiency of Industrial Grinding Circuits.

McIvor, R E, 1988. Technoeconomic Analysis of Plant Grinding Operations, PhD Thesis, McGill University, Montreal.

McIvor, R E, 1990–2024. The Metcom Instructional Program for Measuring and Improving the Processing Performance of Plant Grinding Operations, Metcom Technologies, Inc., Grand Rapids MN.

McIvor, R E, 2006. Industrial Validation of the Functional Performance Equation, *Mining Engineering*, 58(11):47–51.

Red Chris flotation circuit expansion – from piloting to full scale

K Li[1], D R Seaman[2], B A Seaman[3] and J Baldock[4]

1. Senior Process Engineer, Newmont Red Chris, Vancouver BC V6E 3X1, Canada.
   Email: kevin.li@newmont.com
2. Manager-Directional Studies (Metallurgy), Newmont Corporation, Subiaco WA 6008.
   Email: david.seaman@newmont.com
3. Head of Minerals Processing, Newmont Corporation, Subiaco WA 6008.
   Email: brigitte.seaman@newmont.com
4. Chief Metallurgist, Newmont Red Chris, Vancouver BC V6E 3X1, Canada.
   Email: joshua.baldock2@newmont.com

ABSTRACT

The original flotation circuit at Red Chris Mine included bulk rougher flotation using conventional tank cells, followed by cleaner flotation consisting of two columns that were operated in series: The concentrate from the first column was sent to the second, smaller column to further upgrade to final concentrate-grade. As mining operations advanced, ore mineralogy has exhibited increased variability, necessitating a more robust cleaning circuit, especially to accommodate higher head grades. To address this requirement, a third column was piloted on-site to verify metallurgical performance across various configuration options. Following successful piloting, a third column was integrated into the circuit, and commissioning was completed in June 2021.

In parallel with the cleaning circuit expansion, efforts were directed towards improving the performance of the rougher flotation circuit. A benchmarking exercise revealed the potential for improved recovery of sulfide minerals through increasing rougher flotation residence time. This finding was further validated by the lab scale rougher tailings flotation tests, which confirmed enhanced recovery rates on a laboratory scale. In addition to inherent limitations in residence time, the rougher circuit often needs to be reconfigured, converting the last two flotation cells from bulk rougher flotation to pyrite flotation duty to generate non-acid generating (NAG) tails. This reconfiguration further reduces available residence time and negatively impacts rougher recovery, as evidenced by plant data. To extend residence time and alleviate capacity constraints, the implementation of Eriez StackCell® technology was explored through piloting, owing to its compact design that aligns with the available plant space. Based on the piloting results, full-scale StackCells were installed as pre-roughers and commissioned in July 2022.

This paper discusses the journey of the flotation circuit expansion at Red Chris, detailing the progress from piloting programs to full-scale implementation.

INTRODUCTION

Red Chris Mine is currently operated as an open pit mine, located in north-west British Columbia, approximately 80 km south of Dease Lake, and 1000 km north of Vancouver. The processing plant was commissioned in 2015 with 10 Mt/a nameplate throughput, producing copper concentrate with gold and silver as by-products. The comminution circuit consists of a Semi-Autogenous (SAG) Mill, Ball Mill and Pebble Crushing Circuit (SABC), with target product grind size of 80 per cent passing 150 to 160 µm. The original flotation circuit consisted of six Outotec TC-200 tank cells for rougher flotation. Rougher concentrate is directed to a two-stage regrind circuit comprised of a regrind Ball Mill and Metso Verti-Mill producing a target regrind size of 80 per cent passing 30 µm. The regrind product feeds to two Eriez Columns for cleaning flotation and a bank of five Outotec TC-50 tank cells for cleaner scavenger flotation.

Shortly after commissioning, a seventh tank cell (TC-160) was added to the back end of the rougher circuit to expand the rougher capacity and was configured to have mixed duty along with tank cell #6 ie both cells may be operated in a sulfide flotation duty to maintain the rougher tailings neutralisation potential ratio (NPR) target. When in sulfide flotation mode, the concentrates collected from tank cell #6 and #7 are combined with the Cleaner Scavenger Tails and discharged to the

Tailings Pond as potential acid generating (PAG) Tails. The flow sheet of the Red Chris Process Plant prior to 2020 is shown in Figure 1.

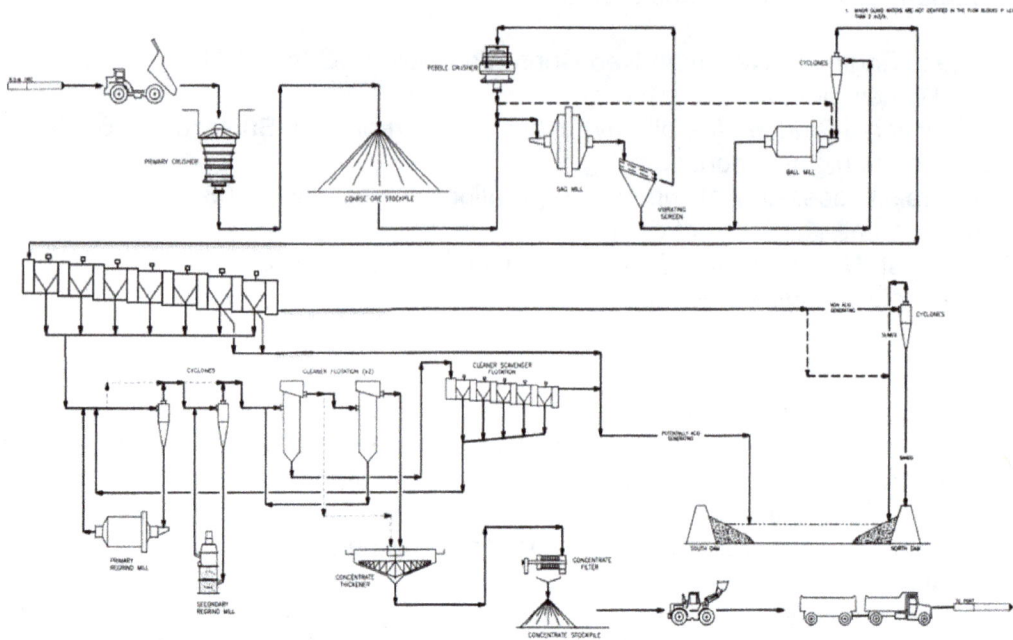

FIG 1 – Red Chris process flow sheet prior to 2020.

CLEANER CIRCUIT EXPANSION

Historical plant data indicates that rougher flotation recovery can be improved by increasing mass recovery. Figure 2 (left) shows the increasing relationship of rougher recovery with increasing mass pull. A series of optimisation projects have been implemented to improve the rougher circuits performance including the modification from radial launders to concentric (donut) launders to improve froth transportation, the implementation of a flow controller to improve consistency in the rougher pulling rate and the implementation of a short interval control (SIC) program to standardise operational discipline and provide tools to gauge circuit performance. An evident improvement in rougher recovery driven by the increased mass pull was observed post the implementation of these projects (Seaman *et al*, 2021).

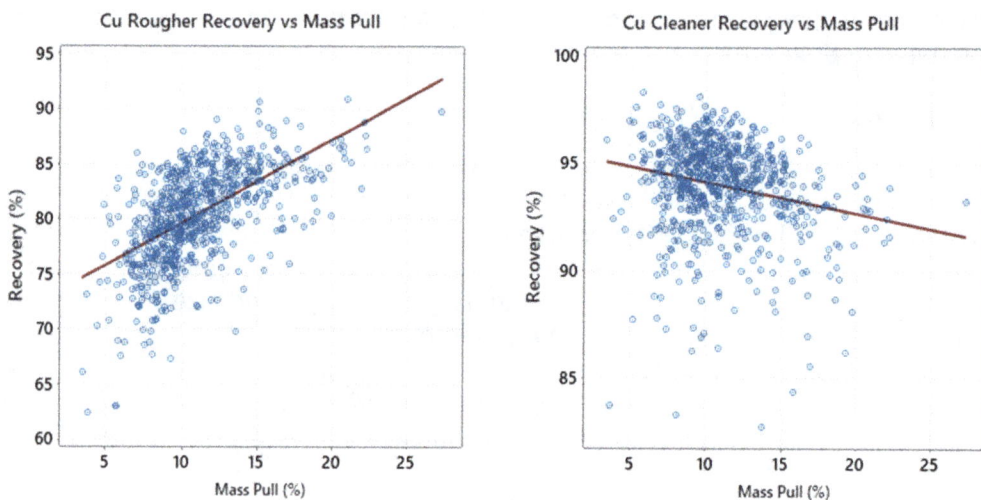

FIG 2 – Copper rougher recovery versus mass pull (left) and copper cleaner recovery versus mass pull (right).

However, increasing rougher mass pull adversely affects the cleaner recovery by effectively reducing the residence time and often created operational issues mostly due to limitations in downstream materials handling capacity. The impact of mass pull on cleaner recovery is also shown in Figure 2 (right), illustrating a decreasing trend of cleaner recovery with increased mass pull. Up to an optimum point, maximising mass pull results in higher overall recovery.

The original Red Chris cleaner circuit consisted of two columns: a larger column with a diameter of 4.5 m and a height of 14 m, and a smaller column with a diameter of 3.6 m and a height of 8 m. Initially designed to operate in series, the two columns were later adapted to also operate in parallel to better handle higher mass and metal loading during periods of elevated head grade and throughput. While this flexible configuration provided tools to manage increased loading from the upstream circuit, it came at the cost of cleaner recovery due to the reduced residence time in the parallel configuration. To address the challenge of the increased mass pull and maintain cleaner recovery, expanding the capacity of the cleaner circuit was necessary.

Both existing columns at Red Chris are Eriez Column Cells equipped with the Cavitation Tube systems. Columns are often used in cleaning duties (Harbort and Clarke, 2017). Compared to traditional mechanical cells, the column provides the benefits of improved grade and recovery performance due to its long cleaning zone and the addition of wash water to the froth phase for the purposes of rejecting un-selectively entrained material. The vertical configuration also requires smaller floor space. Lower operational and maintenance costs are expected due to the absence of internal moving parts. It is effective in flotation of fine particles due to its ability to produce fine bubbles through the sparging system using compressed air (Moon and Sirois, 1987). Eriez's CavTube® sparging system further enhances fine bubble generation and maximises bubble-particle collision rates by inducing hydrodynamic cavitation caused by fast liquid flow-through external pumping (Wang *et al*, 2023).

The Eriez CavTube® column was selected for evaluation in expanding the cleaner circuit, primarily because of the operations team's familiarity with the technology and its demonstrated performance at Red Chris. Additionally, the availability of common spare parts with the existing column would reduce the holding costs of spares and facilitate integration into the existing maintenance programs. Given the constraint in available floor space within the existing plant, but relative openness in vertical height of the mill building, the selection of the column cell was considered an even more attractive option than alternative machine options.

To validate the benefit of adding a new column and evaluate the best circuit configuration to incorporate the new column, a piloting program using a 20-inch diameter column was carried out at Red Chris. The grade-recovery performance of the pilot column can be used to predict the industrial column up to 30 times the column diameter (Knoblauch, Thanasekaran and Wasmund, 2016).

Column piloting method

Liberation analysis on monthly rougher concentrate samples revealed high concentrations of liberated copper sulfides in rougher concentrate (50 per cent), and that the degree of liberation increase at finer grinds size with the -38 micron fractions exceeding 80 per cent liberation.

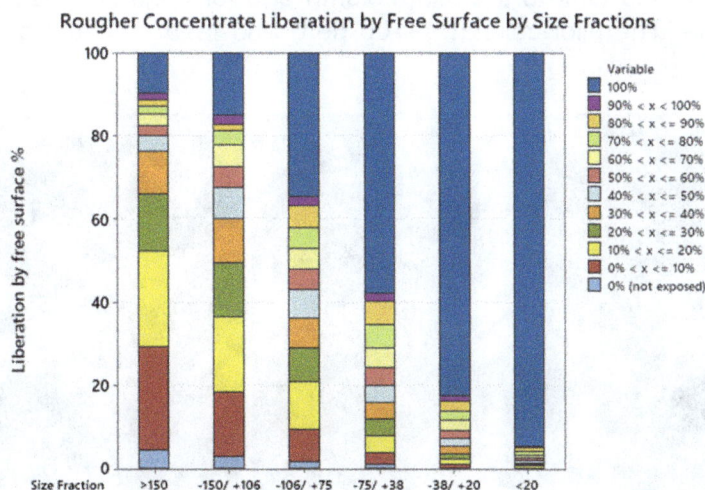

FIG 3 – Red Chris rougher concentrate liberation by free surface by size.

The high degree of liberation in rougher concentrate presented opportunities to explore the options for a cleaner scalping column that may be either directly fed by the rougher concentrate or by partially reground rougher concentrate as illustrated in Figure 4.

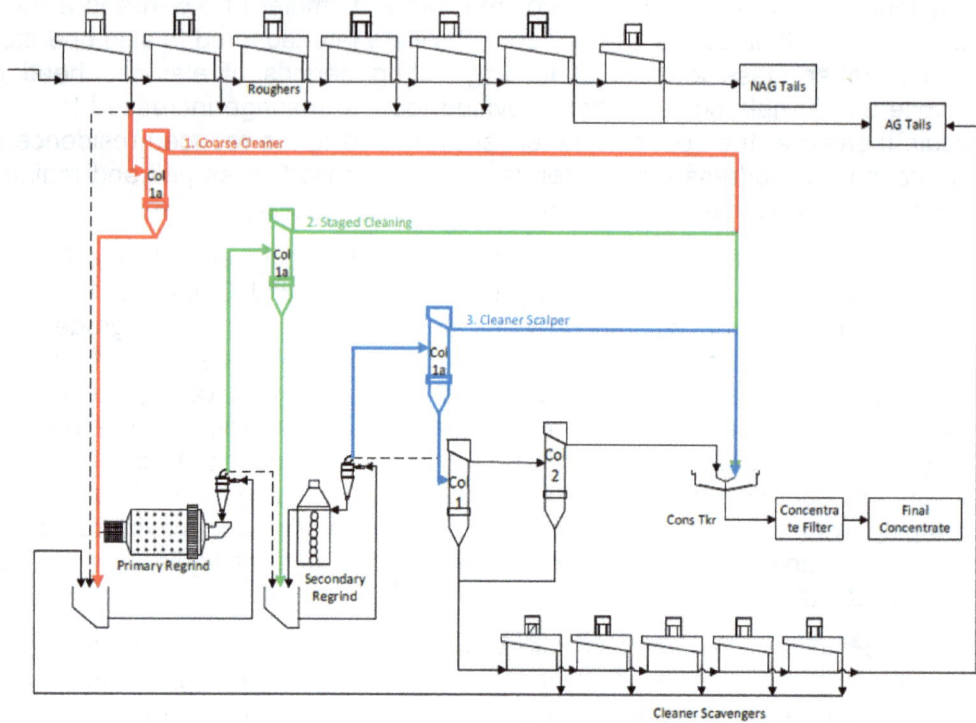

FIG 4 – Potential cleaning circuit configuration options with the new column.

As a cleaner scalper, the concentrate produced may be directed to final concentrate. In this configuration, the ability to recover liberated copper sulfides 'early' has the advantage of avoiding overgrinding in the regrind circuit and reduces downstream treatment and alleviates the material handling restrictions in the dewatering circuit.

To evaluate the configuration options, a flexible feed arrangement to the pilot column was implemented by utilising the multiplexer sample return lines of the On-Stream Analyser (OSA) sampling system. Five of the multiplexer sample return lines were re-routed to a manifold and then diverted to a feed tank by switching valve positions on the manifold. The streams being tests on the pilot column include rougher concentrate (Ro Con), primary regrind cyclone overflow (PRCOF), secondary regrind cyclone overflow (SRCOF or Col Fd), column #1 tails and cleaner scavenger tails (PAG tails). A hose pump (Apex 35) equipped with variable speed drive (VSD) was used to transfer feed material from a mixing tank to the pilot column and for controlling feed flow rate to provide desired residence time. The pilot column set-up (left) and feed manifold arrangement (right) are shown in Figure 5.

FIG 5 – The pilot column plant installation (left), the pilot column in operations (middle) and feed manifold arrangement (right).

Column piloting results and discussions

A total of 78 tests were completed on the five mill streams. The tests performed on column 1 tails and PAG tails were mostly exploratory to investigate the recovery improvement potential (col1 T tests) with increased residence time and the ratio of flotation recoverable minerals remaining in the cleaner circuit (PAG) tails. The effect of particle size and liberation on column performance is best demonstrated comparing the following streams: Ro Concentrate→ PRCOF (Primary Regrind Cyclone Overflow) → SRCOF (Secondary Regrind Cyclone Overflow). As shown in the grade and recovery chart in Figure 6 (left), the enhancement in grade-recovery performance is clear as the feed stream moves downstream in the cleaning circuit. This improvement results from materials passing through primary and secondary regrind mills, providing additional particle size reduction and further mineral liberation.

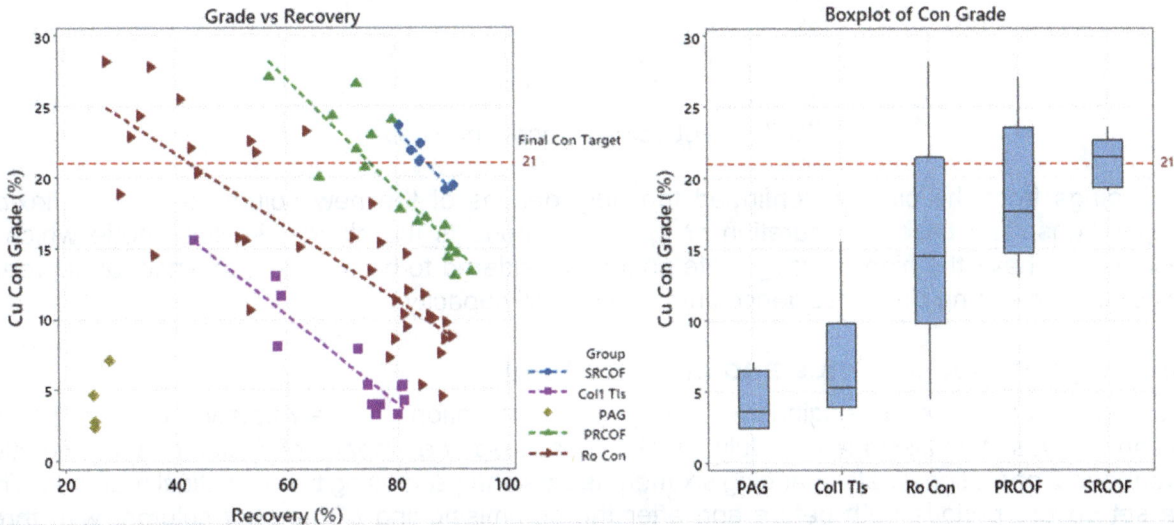

FIG 6 – Pilot column grade-recovery scattered plot (left) and box plot of pilot column concentrate grade (right).

The achieved copper concentrate grade for each stream is shown in a boxplot in Figure 6 (right). The median value of the concentrate grade increases as the pilot column feed source moves downstream within the cleaner circuit: Ro Concentrate→ PRCOF → SRCOF. The achieved concentrate grade values for tests with rougher concentrate as feed have the widest distribution, while SRCOF as the test feed generates the most consistent upgrading performance with a narrow grade distribution. It is hypothesised that the large performance variations when treating the rougher concentrate stream is due to the lower degree of liberation of the rougher concentrate, making the piloted cleaner outcome more susceptible to ore changes due to the variations in mineral grain sizes.

Lime addition to control pH in the cleaners at Red Chris is added to the rougher concentrate hopper, as such the tests were all performed at similar and relatively high pH of around 11.

Several pilot sample sets were selected for size-by-size assay analysis and results are shown in Figure 7. The recovery losses occur at the ultra-fine fraction (-11 μm) and at the coarser fractions (+53 μm). In the case of the ultra-fine fraction, these losses are possibly due to having a lower probability of collision with air bubbles and being more susceptible to hydrophilic surface coatings. In the case of the coarser fractions, these losses are likely due to the lower degree of liberation. The optimal recovery is achieved at around 30 μm which is the target regrind product size. The size-by-size recovery observations are in line with expectations and current operating strategies.

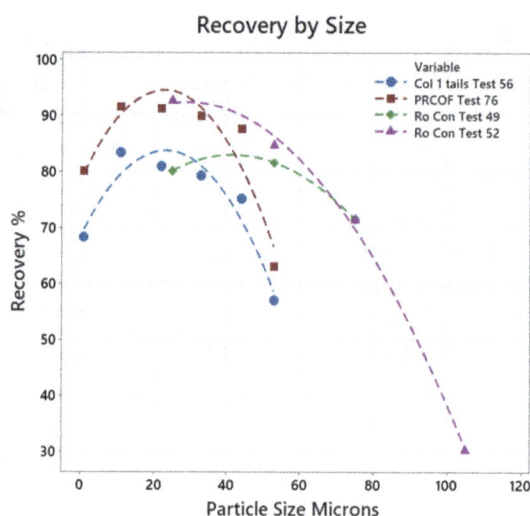

FIG 7 – Pilot column recovery by size.

The findings from the piloting confirmed the duty options of the new column to enable flexible configurations. The base configuration or operation mode is the cleaner scalper mode which is anticipated to have the highest usage rate as it is considered to be a more conventional approach to extend the cleaning circuit residence time and overall capacity.

Column plant performance and improvements

Following the completion of engineering design and installation, the new flotation column, named Column 3, was commissioned in July 2021. To assess the impact of Column 3 on flotation performance, operational data spanning six months was analysed using two statistical methods. This data set covers periods both before and after the commissioning of the new column, with three months preceding and three months succeeding its installation. The ore processed during this time remained relatively consistent in terms of geological origins sourced from the open pit, thereby minimising the influence of ore variability on the results.

The evaluation of recovery performance was conducted by regression modelling employing two statistical evaluation methods.

Method 1 – comparison of recovery model residuals

In method 1, recovery models were developed by multilinear regression based on shift production data. The following input variables were included:

- Mill throughput (tonnes per operating hour (tpoh))
- Head grade (gold or copper respectively)
- Head sulfur grade
- Head carbon grade
- Rougher concentrate mass flow (tpoh)
- Fault zone ore (binary input)
- Final concentrate grade or P_{80} (copper or gold respectively)

The secondary regrind Vertimill was out of commission from February 2021 to June 14th, only one week after the commissioning of Column 3. During the Vertimill outage, a lower copper concentrate grade was targeted which is expected to have offset the recovery impact attributed to the Vertimill. Vertimill operational status was excluded as a regression input due to its high correlation with Column 3 operational status. Figure 8 shows the comparisons between the modelled and the actual achieved recoveries of copper and gold.

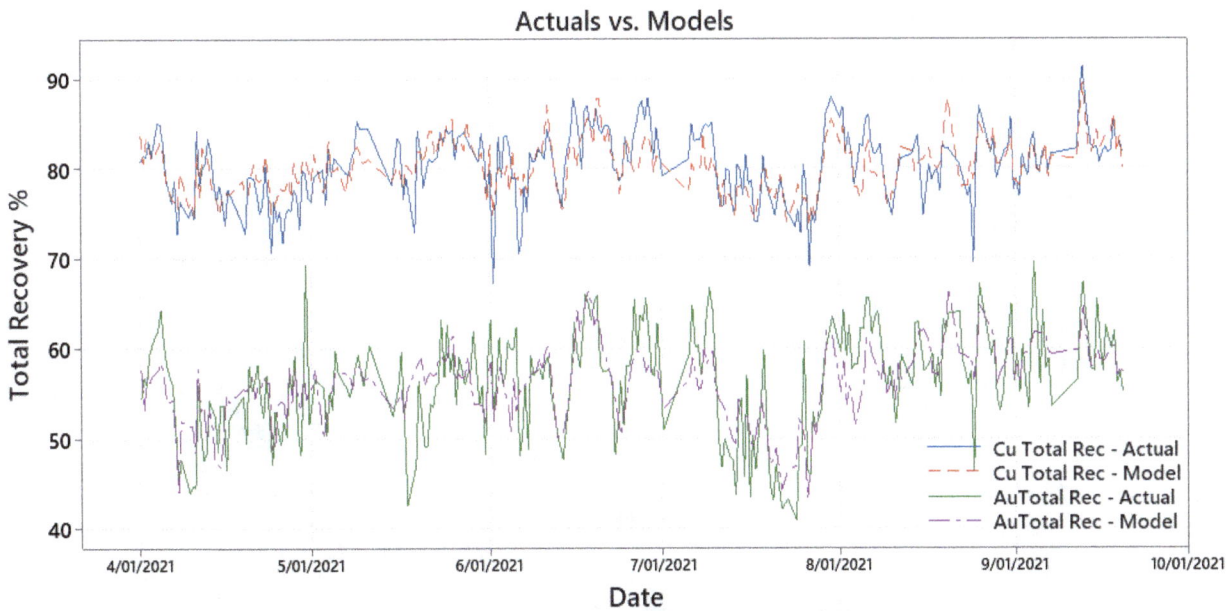

FIG 8 – Time series model performance comparisons.

The residuals of the regression models can be used to assess the performance of any particular shift. Shifts with negative residuals indicate that the actual plant recovery is underperforming and is below its predicted value, and *vice versa* when the residuals are positive. The mean difference of recoveries for shifts with and without Column 3 are compared using Two-Sample T-Tests to assess the impact of Column 3 on recoveries.

The descriptive statistics of the Two-Sample T-Tests are summarised in Table 1.

TABLE 1

Descriptive statistics of the Two-Sample T-Tests.

	Col 3 status	# of data	Mean	StDev	SE Mean	Mean diff (on – off)	T-value	DF	P-Value
Cleaner copper recovery	ON	170	0.9	1.89	0.15	2.1	-7.47	188	0.000
	OFF	109	-1.17	2.46	0.24				
Total copper recovery	ON	170	0.53	2.46	0.19	1.1	-3.55	218	0.000
	OFF	109	-0.59	2.64	0.25				
Cleaner gold recovery	ON	155	0.48	3.8	0.31	0.9	-1.81	209	0.072
	OFF	106	-0.44	4.23	0.41				
Total gold recovery	ON	155	0.29	3.67	0.29	0.5	-1.07	201	0.288
	OFF	106	-0.26	4.3	0.42				

The mean values of the model residuals, with and without Column 3 are also presented graphically in boxplots and histograms in Figure 9, where 1.1 per cent copper recovery uplifts and 0.5 per cent gold recovery uplifts can be observed.

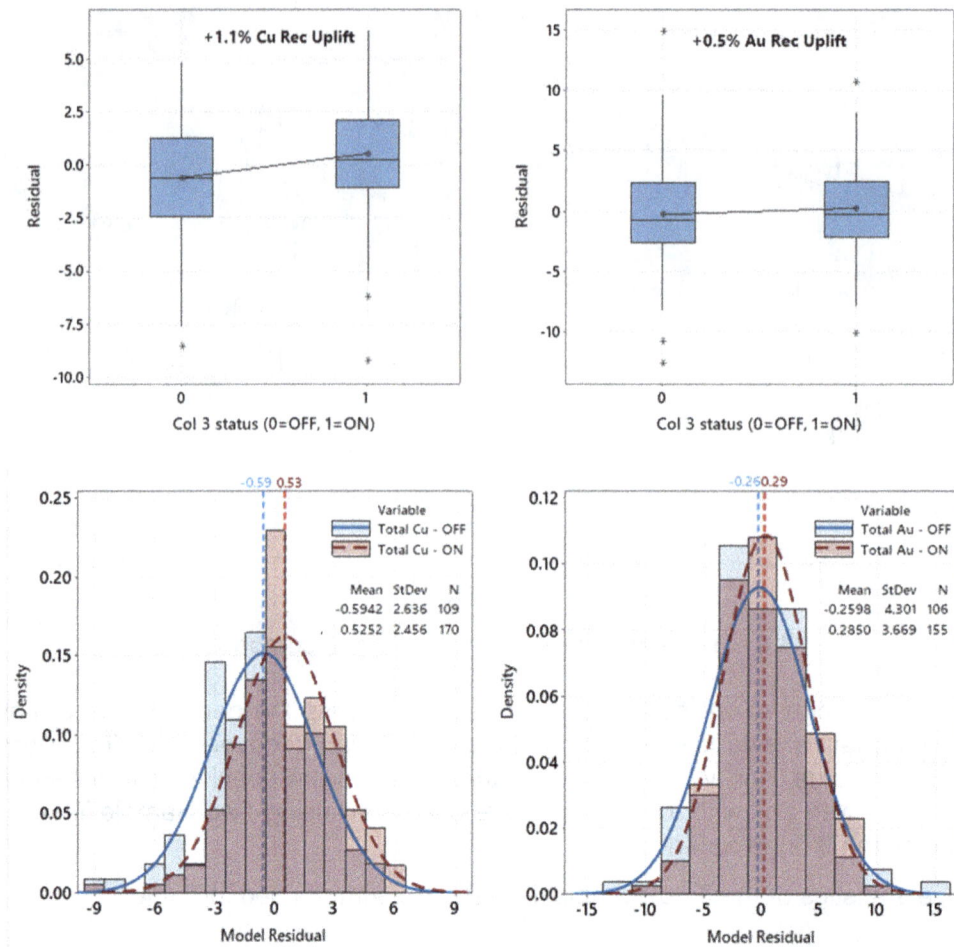

FIG 9 – Boxplots and histograms of model residuals.

Method 2 – regression with column 3 status as categorical input

Another series of regression analyses were conducted using Column 3 status as categorical inputs in binary form (0=OFF, 1=ON). Final concentrate grade and P_{80} were included to reflect the effect of regrinding for copper and gold respectively.

The fitted model versus actual for copper and gold recoveries are plotted in Figure 10. Both models have a moderate level of R squared values at 0.58 and 0.52 for copper and gold respectively.

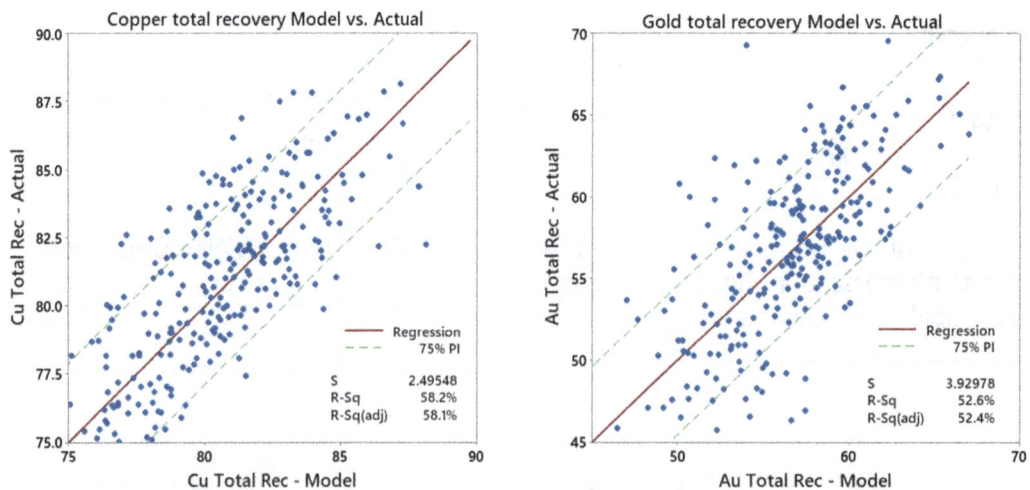

FIG 10 – Actual versus model for total copper and gold recovery.

To examine how each input variables influences the recovery responses, particularly Column 3 status, graphs for these variables are plotted with the coefficients and P-values of each variable shown in Figure 11.

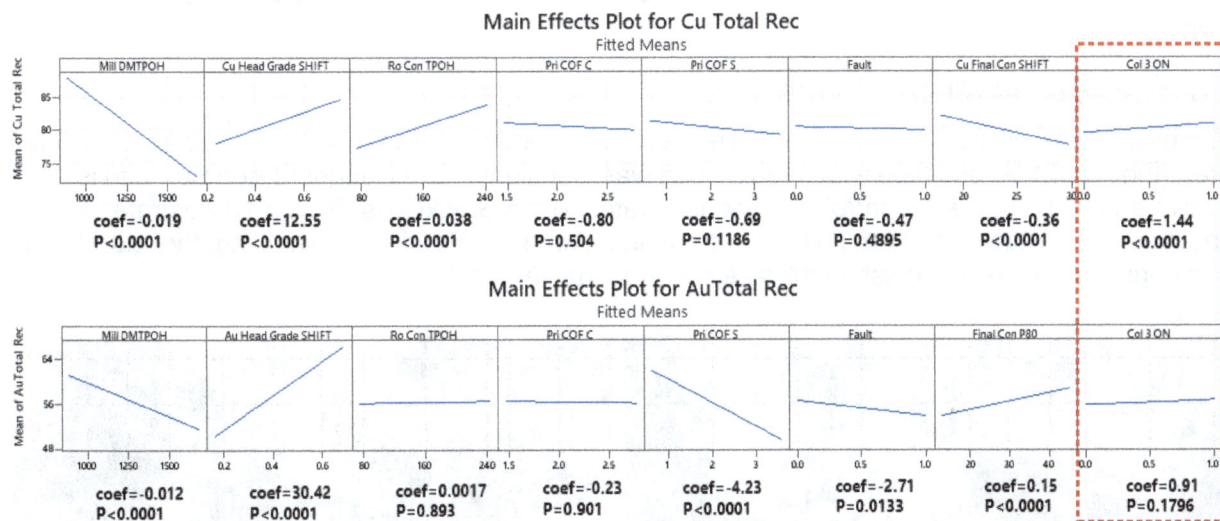

FIG 11 – Multilinear regression outputs, coefficients and P values denoted below each main effect plot.

As seen in Figure 11, Column 3 status ON shows a positive influence on both copper and gold total recoveries. When Column 3 is online, 1.44 per cent total copper recovery uplift is observed with greater than 95 per cent confidence and 0.91 per cent total gold recovery uplift is observed with 82 per cent confidence.

The outcomes of the two statistical analyses are consistent in confirming the recovery uplift achieved by installing the new column, despite only the cleaner scalper mode being enabled due to flow control issues in the other operating modes. An upgrade project to install a flow control valve is underway to enable the coarse cleaning and stage cleaning modes, which is expected to provide greater operational flexibility, particularly in handling higher-grade ore.

Learnings from operating and maintaining the existing column informed the design improvements for the new column. The number of Cavitation Tubes was reduced from 14 to 8 by using larger CavTube units, and the Cavitation Tube isolation pinch valve was replaced with knife gates, which reduces sanding at the valve location and facilitate easier inspections and change-outs. These upgrades significantly improved the wear performance of the Cavitation Tubes, reducing both change-out frequency and CavTube blockages during operations.

ROUGHER CIRCUIT EXPANSION

The residence time of the rougher circuit at Red Chris was benchmarked against other Newmont sites, revealing that the rougher circuit had insufficient residence time. This finding was supported by both the lab Denver kinetic tests and similar tests conducted on the flotation of rougher tailings. The requirement to produce non-acid generating (NAG) tailings from rougher flotation further exacerbated the issues with insufficient residence time (Seaman *et al*, 2021), as the last two rougher tank cells are required to switch to non-copper rougher duty which further reduces the available residence time by over 25 per cent. The estimated impact on copper recovery when switching cell 6 and both cell 6 and 7 is estimated at 1.4 per cent and 2.2 per cent respectively based on multivariate statistical analysis using plant operating and production data.

Due to the spatial limitations within the existing processing plant at Red Chris, newer mechanical cell technologies with smaller footprints were considered, namely the Woodgrove Staged/Direct Flotation Reactor (SFR/DFR) and the Eriez StackCell. Both technologies feature high intensity mixing zones isolated from the separation zone, allowing the process of particle and bubble collisions and the subsequent froth recovery process to be optimised independently (Araya and Lipiec, 2024). This feature is achieved without external pumping requirement like those in the Glencore Jameson

Cells and Metso Concorde Cells (Araya and Lipiec, 2024). Six StackCell pilot units from Eriez were made available to Red Chris to collaboratively investigate their performance. The advantage of piloting six units in series allowed for an in-depth examination of StackCell with extended residence time, that is sufficient for a side-by-side comparison to the existing plant conventional mechanical cells.

Eriez StackCell piloting method

The pilot StackCell units were installed in the mill basement within the rougher tank cells containment area (Figure 12). A feed pipe with trash screen was installed on the Rougher Feed Box with isolation valves and pinch valve to control the feed flow rates to the StackCells. Figure 13 illustrates the flow sheet for the StackCell pilot units fed from the Rougher Feed Box enabling the side-by-side comparisons against the existing full size conventional tank cells.

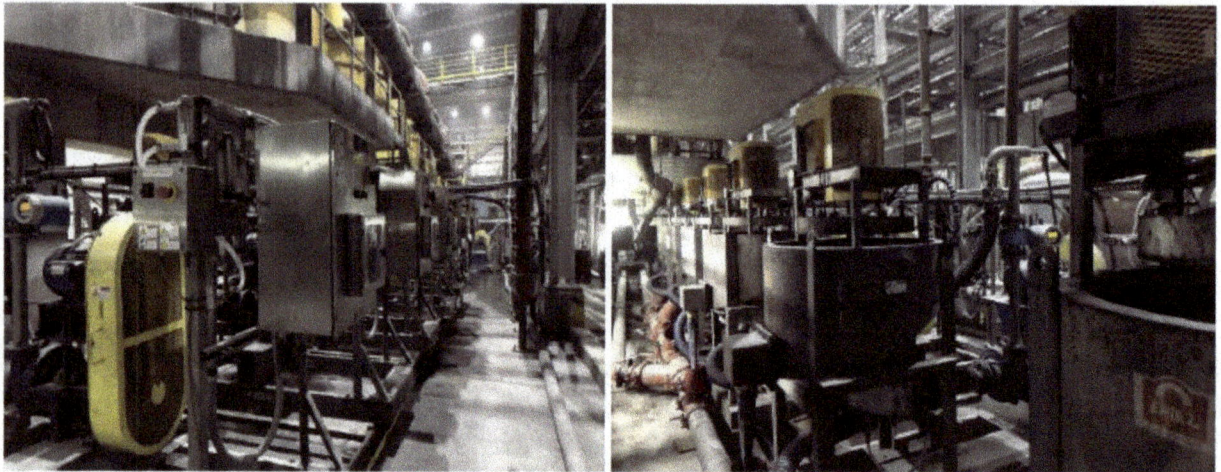

FIG 12 – Installed StackCell pilot units in series.

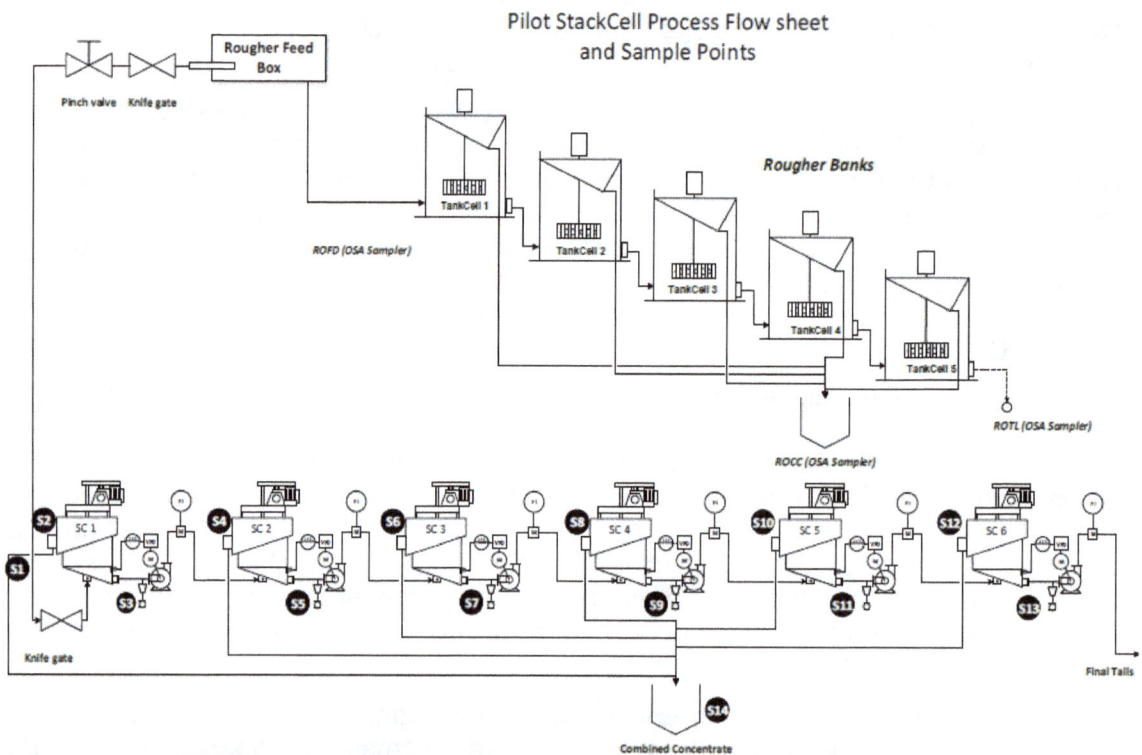

FIG 13 – Pilot StackCell units flow sheet and sampling points

Eriez StackCell piloting results and discussions

A total of 40 sets of StackCell surveys were conducted during the piloting. A full plant down the bank (DTB) survey of the rougher tank cells was also carried out, with results compared against several pairing StackCell surveys from Test 36–38. The paired DTB surveys were assayed for both copper and gold. It was found that the StackCell kinetics were at least four times faster than the plant tank cells while producing an equivalent copper rougher recovery and higher gold rougher recovery (Seaman *et al*, 2021). The respective kinetic curves for copper and gold minerals are shown in Figure 14.

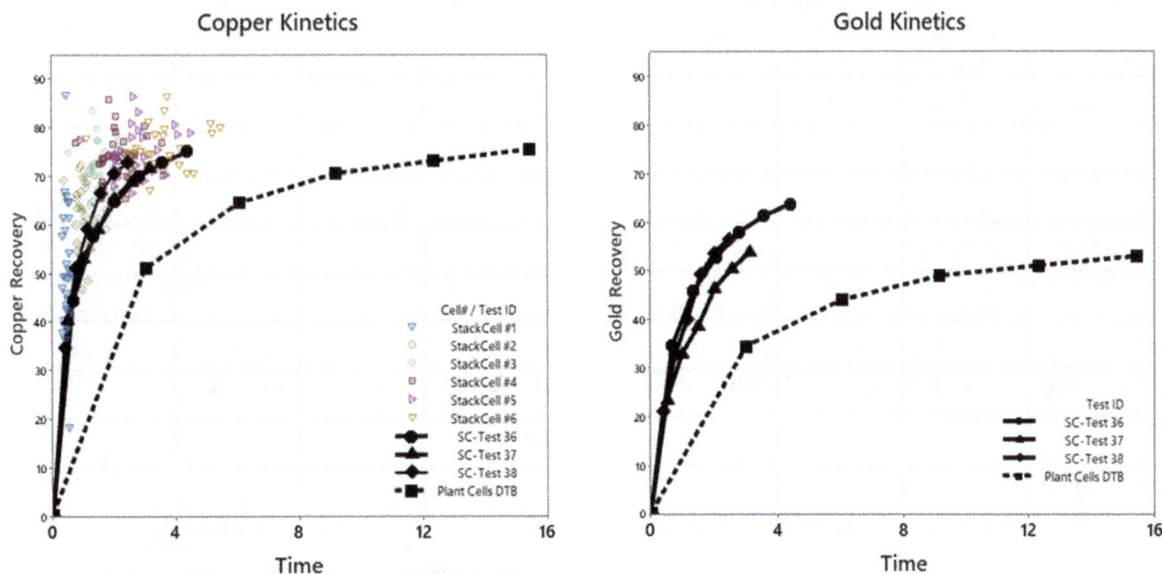

FIG 14 – StackCell kinetics and comparisons against plant survey for copper (left) and gold (right).

The recovery-by-size performance for a selected survey are shown in Figure 15. There is no significant difference in the recovery-by-size performance between plant and the StackCell Pilot results for copper. Recovery improvements on mid-size size fractions were observed for gold with no significant difference in recovery performance for the fines and the very coarse fraction. Recovery-by-size results for sulfur showed a significant increase in the -20 microns size fraction as well as improvements across the mid size fractions (Seaman *et al*, 2021).

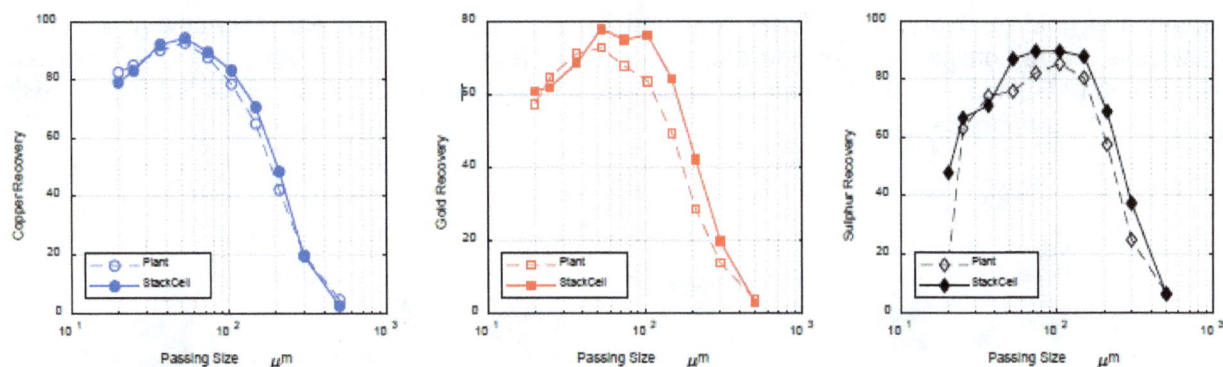

FIG 15 – Size by size recovery comparisons between StackCells and plant for copper (left), gold (middle) and sulfur (right).

The piloting of StackCells in rougher duty confirmed that they can generate much faster kinetics compared to conventional tank cells without compromising flotation performance in terms of total recovery and size-by-size recovery. The fast kinetics achieved were thought to be attributed to the unique features of StackCell technology: The bubble-particle collision occurs in the contacting chamber with intensive mixing, promoting higher degrees of bubble attachments with recoverable mineral particles. The loaded froth then enters the quiescent separation chamber, thus limiting drop back of heavier particles due to turbulence that is often experienced in conventional mechanical cells (Wasmund *et al*, 2018). The expected improvement in coarse particle recovery, however, was not observed based on the size-by-size recovery analysis. The hypothesis for this unexpected observation was that it could be attributed to the lack of recycle. Since particles only have the opportunity for a single pass through the mixing chamber of the StackCell, any particles that become detached in the collection chamber do not get another chance to be recovered in that cell (Seaman *et al*, 2021).

In expansion scenarios, adding StackCell in front of the conventional tank cells provides the benefit of recovering fast-floating, higher liberated fine particles upfront. Subsequently, the tank cells may then be set-up to recover poorly liberated coarse particles that float slowly and require longer residence time. This arrangement also aligns with the available floor space at Red Chris, leading to the decision to install the StackCells as pre-roughers.

Eriez StackCells plant performance and improvements

Two SC-200 StackCells, each with 65 m^3 volumetric capacity were commissioned in June 2022. A series of plant surveys were conducted to evaluate the StackCells performance and compare against the piloting results. The plant surveys followed on and off patterns with StackCells online (ON) and bypassed (OFF) to assess their impact on flotation performance, particularly on the roughers. Paired T-tests using five sets of ON and OFF tests were carried out to evaluate the recovery impact on the rougher circuit with (ON) and without (OFF) the StackCells. The statistical analysis outputs are summarised in Table 2.

TABLE 2
Summary of statistical analysis output for StackCells recovery impact.

	Mean difference (ON – OFF)	T-test confidence (% confidence difference is real)	95% confidence interval for mean difference
Cu rougher recovery	+1.9	76.9	-1.9 ~ +5.7
Au rougher recovery	+3.5	66.9	-5.3 ~ +12.3
Sulfide rougher recovery	+4.8	92.4	-0.8 ~ +10.3
Rougher tails NPR	+2.4	96.6	+0.3 ~ +4.4

Copper and gold recovery saw 1.9 per cent and 3.5 per cent improvement respectively when the StackCells were online with moderate confidence levels (76.9 per cent and 66.7 per cent respectively). Sulfide recovery also saw a higher degree of increase at 4.8 per cent with 92.4 per cent confidence.

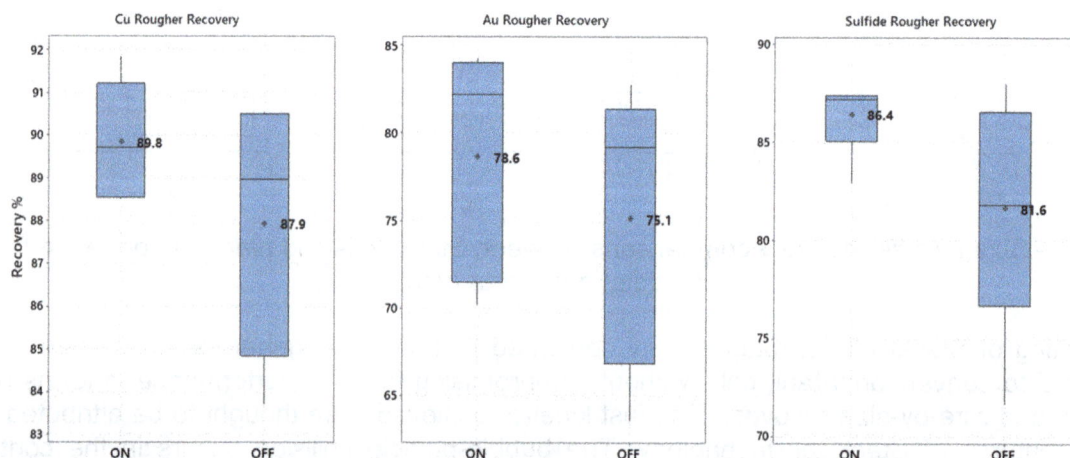

FIG 16 – Boxplots of rougher flotation recoveries with (ON) and without (OFF) StackCells.

To assess the kinetic performance of the full scale StackCells, plant survey data are overlayed with pilot data on the same chart for comparison in Figure 17. For copper kinetics, StackCell plant data matched the pilot performance. Survey data also showed that more than 4.5 per cent mass pull is required to achieve pilot equivalent copper kinetic performance. For gold, pilot level kinetics were matched while several plant data sets exceeded pilot performance. Full down the bank surveys were also conducted with and without the StackCells running. The kinetic curves of the plant surveys are shown in Figure 17. The ore being processed during the time of the survey had feed grades of approximately 0.6 per cent copper and 0.6 g/t of gold, higher than average run-of-mine head grade around 0.4 per cent of copper and 0.3 g/t of gold.

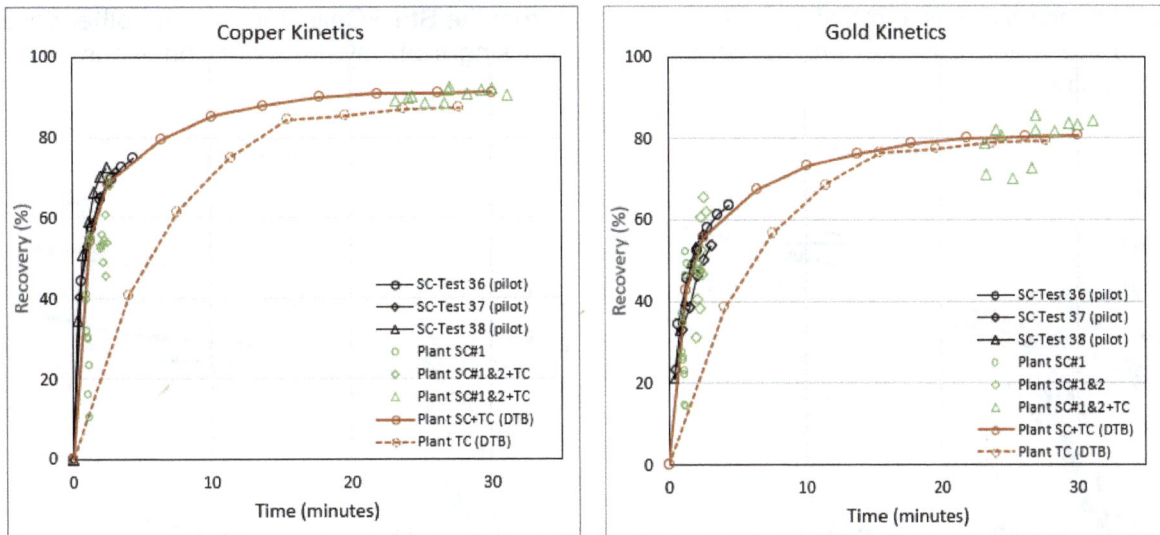

FIG 17 – Copper and gold StackCell flotation kinetics plant (SC200) versus pilot.

To better understand the factors driving the recovery improvements, size-by-size recovery and liberation analyses were conducted on same survey samples.

Copper recovery-by-size distribution charts are depicted in Figure 18 (left), illustrating the recovery performance comparisons across each size fraction. Moderate improvements were noted for the fine to intermediate size fractions, while a more significant degree of improvement was observed for the coarser fractions. This observation is supported by the liberation analysis conducted on selected size fractions, as shown in Figure 18 (right). The analysis revealed that the improvements in fine fractions were predominantly derived from improved recovery in liberated particles, whereas the improvements in coarse fractions were driven by improvements in the poorly liberated class (0–20 per cent). Interestingly, such improvements in coarse particle recoveries were not observed during the piloting program when the StackCell pilot units were operated alone. It was concluded that the combination of StackCells followed by tank cells allowed the tank cells at the back end of the bank to be operated in a manner more favourable to coarse particle flotation.

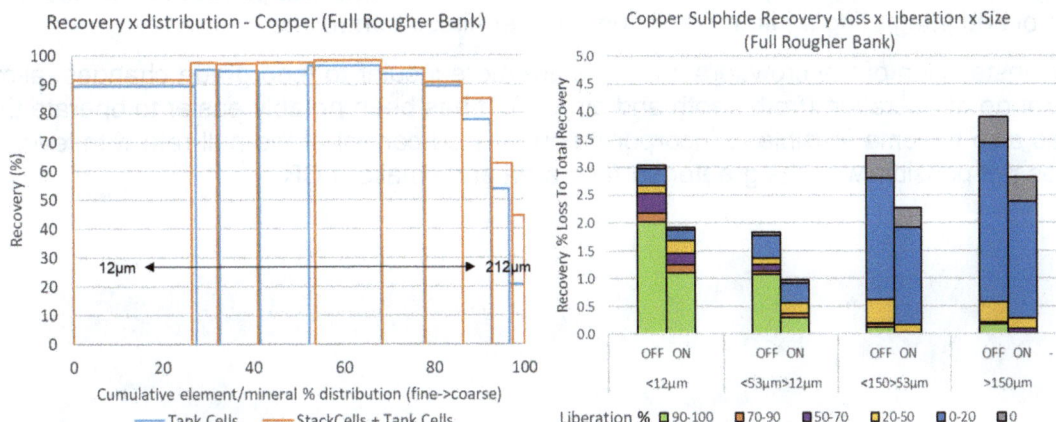

FIG 18 – Recovery by size by metal distribution for first cell (left) and full rougher bank (right).

In complement to the higher grade survey results, two sets of subsequent ON and OFF down-the-bank (DTB) surveys were conducted to evaluate the performance of StackCell at lower-grade ranges. The first set of surveys were conducted in December 2022 with a head grade of 0.4 per cent copper and 0.5 g/t gold, while the second survey set were conducted in July 2023 with a head grade of 0.2 per cent copper and 0.2 g/t gold. Kinetic curves for copper and gold for both surveys are plotted in Figure 19. Overall, recovery improvements are evident for both copper and gold. However, the accelerated kinetics at the front end, attributable to the StackCells, seems to diminish as the head grade decreases. Similar observations were noted during daily operations, where the StackCells encountered challenges in maintaining a stable froth layer and consistent froth pulling into the concentrate launders at lower-grade ranges. Consequently, operators often intervened to make necessary adjustments to maintain froth pulling rates from the StackCells. Increasing frother dosage rates to the StackCells also proved beneficial in stabilising froth layers under conditions of lower metal loading.

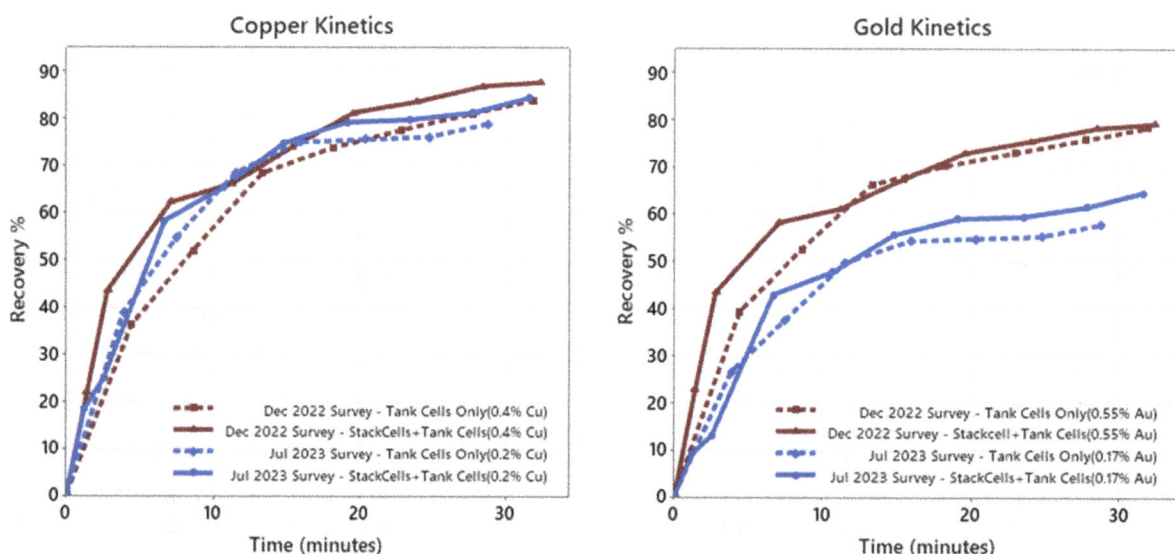

FIG 19 – Kinetic curves for copper (left) and gold (right) at lower flotation feed grades.

To improve the performance of the StackCells when treating lower grade ore, efforts were focused on increasing the froth carry rate (FRC). FCR is a measure of the froth stability and the transportation efficiency of froth. Insufficient mineral particles in the froth to form and maintain a stable froth are symptoms of low FCR (Murphy and Heath, 2013). Froth crowding can be employed to reduce the froth surface area which will increase FCR thus improving froth stability and overall performance. Accordingly, detachable crowders were designed jointly by Newmont and Eriez and installed during a shutdown in October 2023.

In addition to installing crowders, several other modifications were made to the StackCells. These modifications include redesigning the spin-lids from a two-piece design to a single piece, aimed at improving wear life. Additionally, soft sensors have been implemented to facilitate concentrate pull rate control and integrating it into the existing rougher circuit controllers.

Since the installation of the crowders, the cells are more insular to feed grade changes, allowing a broader range of operation (froth depth and air rate). It has been notably easier to operate the cells and it has also become feasible to incorporate automated control of the pull-rate with level and air which was not possible with the greater surface area and smaller FCR.

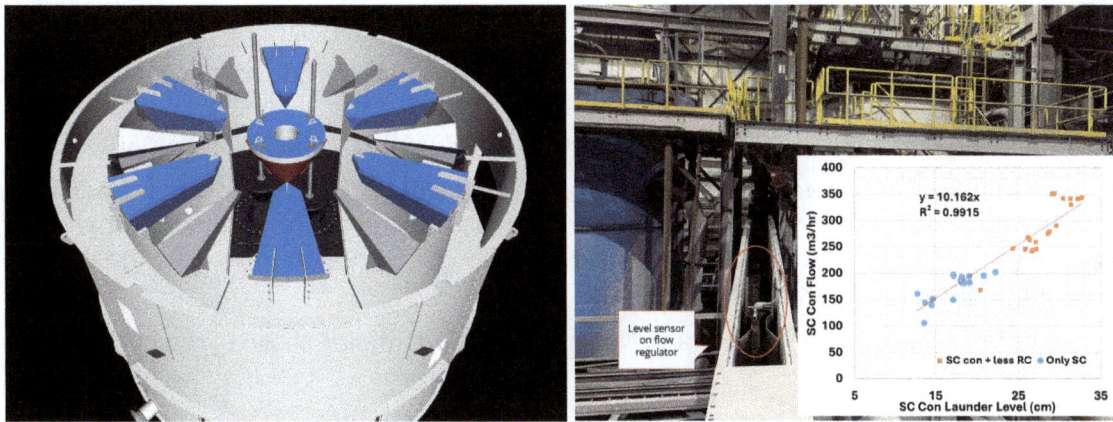

FIG 20 – Screenshot of StackCell 3D model with retrofitted crowders (left) and stackcell flow control using level senor in open launder (right).

CONCLUSIONS

The expansion of the Red Chris flotation circuit, which involved adding a new flotation column to the cleaner circuit and the incorporation of two Eriez StackCells as pre-roughers, has improved the overall performance of the flotation circuit. Piloting programs played a crucial role in evaluating the potential benefits of these modifications and informing the design of the flow sheet. The updated flow sheet is shown in Figure 21. The new cleaner column and StackCells together with their configuration options are shown in red.

Piloting serves as a valuable tool for assessing equipment performance before committing significant capital to full-scale implementation. It helps build confidence in the effectiveness of these implementations, particularly for emerging technologies. In brownfield projects, piloting offers an additional advantage by allowing for comparisons against existing equipment, aiding in the evaluation of alternative technology and flow sheet configuration options.

Close collaboration with vendors during piloting programs is essential to incorporate learnings into the full-scale design. This ensures that the designs are fit-for-purpose and enables troubleshooting of any operational issues encountered post full-scale commissioning, facilitating necessary upgrades and modifications to the design.

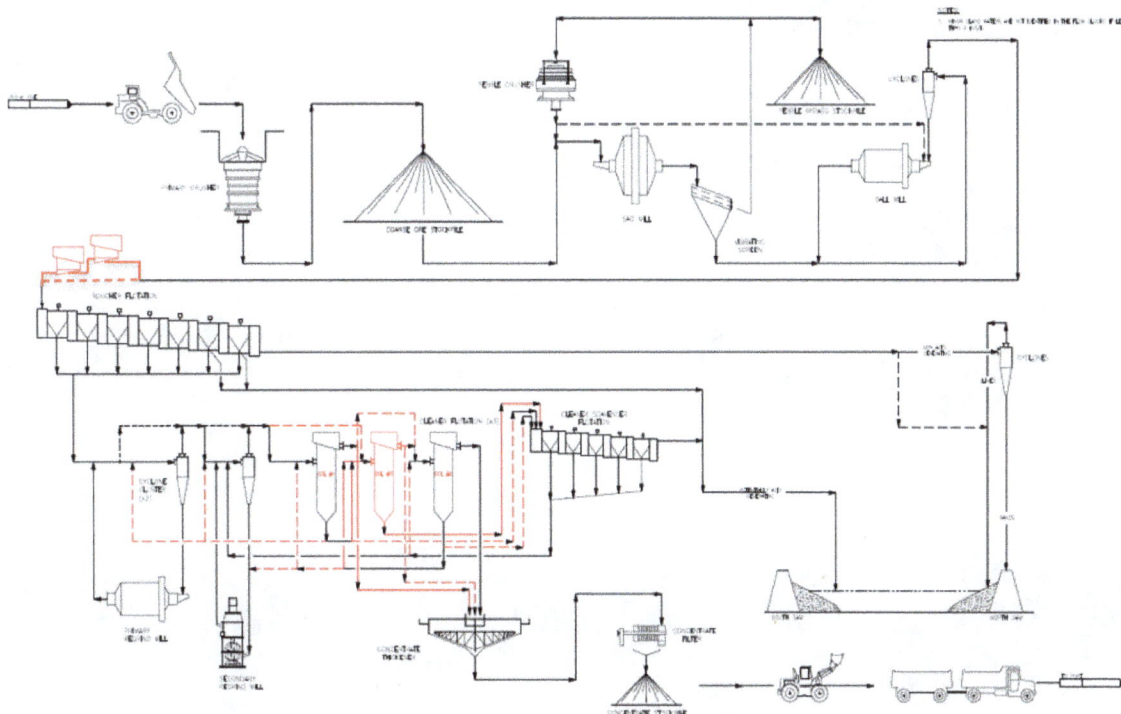

FIG 21 – Updated Red Chris process flow sheet with StackCells and new column.

ACKNOWLEDGEMENTS

The authors would like to acknowledge the Red Chris Metallurgy team and many other site personnel who provided support for these projects from piloting to post-commissioning and optimisation, as well as the guidance and support from the Corporate Innovation and Technology team.

The authors would also like to acknowledge the Eriez Flotation Division for their dedication and collaborative approach, which greatly contributed to the successful implementation of these projects. Ray Van Wagoner of Rio Tinto Kenecott Copper was also kind enough to loan us three of the pilot cells needed to make the bank of six.

Finally, the authors would also like to thank Newmont Corporation for allowing the publication of this paper.

REFERENCES

Araya, R and Lipiec, T, 2024. Review of Cutting-edge Flotation Devices, *Canadian Mineral Processors Operators Conference*, pp 20–30.

Harbort, G and Clarke, D, 2017. Fluctuations in the popularity and usage of flotation columns – An overview, *Minerals Engineering*, 100:17–30.

Knoblauch, J, Thanasekaran, H and Wasmund, E, 2016. Improved Cleaner Circuit Performance At the Degrussa Copper Mine with An In-Situ Column Sparging System, Eriez Flotation Division.

Moon, K S and Sirois, L L, 1987. *Theory and Application of Column Flotation* (Ottawa: Canadian Government Publishing Centre).

Murphy, B and Heath, J L, 2013. Selection of Mechanical Flotation Equipment, METCON, Quezon City.

Seaman, D, Li, K, Lamson, G, Seaman, B and Adams, M, 2021. Overcoming rougher residence time limitations in the rougher flotation bank at Red Chris Mine, in *Proceedings of the Mill Operators Conference 2021,* pp 193–207 (The Australasian Institute of Mining and Metallurgy: Melbourne).

Wang, J, Wang, L, Cheng, H and Runge, K, 2023. A comprehensive review on aeration methods used in flotation machines: Classification, mechanisms and technical perspectives, *Journal of Cleaner Production*, 435:1–24.

Wasmund, E, Christodoulou, L, Mankosa, M and Yan, E, 2018. Benchmarking Performance of the Two-Stage StackCell™ with Conventional Flotation for Copper Sulfide Applications, Canadian Mineral Processors Operators Conference, Ottawa.

The optimisation of a vertical regrind mill at Renaissance Minerals, Okvau Mine

A Paz[1], A Siphanya[2] and K Vansana[3]

1. MAusIMM, Sales and Process Manager, Swiss Tower Mills Minerals, Perth WA 6000. Email: andres.paz@stmminerals.com
2. Plant Metallurgist, Renaissance Minerals (Cambodia) Limited, Khan Chamkar Mon, Phnom Penh 12302, Cambodia. Email: abay.siphanya@rnscambodia.com
3. Senior Metallurgist, Renaissance Minerals (Cambodia) Limited, Khan Chamkar Mon, Phnom Penh 12302, Cambodia. Email: khonesavanh.vansana@rnscambodia.com

ABSTRACT

The Okvau Gold Mine is located in Mondulkiri province of eastern Cambodia. The 2MPTA concentrator consists of a primary crusher, ball mill, flotation, vertical regrind mill (1800 kW 9000 L), leaching circuit, gold elution and electrowinning.

The vertical regrind mill had high wear on the rotors and grinding media. After investigation, it was identified that multiple improvement measures could be implemented and trialled, such as rotor configuration, media selection, operational philosophy. The rotor configuration and ceramic grinding media change allows for 86 per cent increase in vertical regrind mill wear life and larger reline intervals. The trial on new ceramic grinding media indicates a large 59 per cent reduction in media wear rate. Overall, the reduction in Operational Cost Expenditure (OPEX) and Greenhouse Gas (GHG) emission is substantial.

This paper will focus on the details of the improvements mentioned above, and the analysing the results which lead to more stable vertical regrind mill performance, and lower vertical regrind mill OPEX and GHG emissions.

INTRODUCTION

The Okvau Gold Mine is located in Mondulkiri province of eastern Cambodia, approximately 275 km north-east of the capital city, Phnom Penh. The Okvau Gold Mine is owned and operated by Renaissance Minerals (Cambodia) Limited which is 100 per cent owned by Emerald Resources. In 2017 Emerald's team commenced the journey to bring the Okvau Gold Project into production. Emerald's highly experienced exploration geologists and management group completed a definitive feasibility study that demonstrated the potential of the project. The mine was further developed and commissioned on time and in budget in 2021.

The 2MPTA concentrator consists of a primary crusher, SAG mill (8500 kW) with pebble crusher, flotation (6x TC-260), vertical regrind mill (1800 kW 9000 L), thickener, leaching circuit, gold elution and electrowinning. After assessing the baseline operation of the plant, the optimisation of various units occurred, one of which was the vertical regrind mill. The focus of this optimisation project was to increase the reline period and reduce Operational Costs Expenditure (OPEX), which naturally reduces GHG emissions.

The vertical regrind mill technology description and operation its well established and is described in many papers such as Lehto *et al* (2013), Mezquita *et al* (2022), Paz *et al* (2019, 2021). The optimisation of the vertical regrind mill is dependent on the process conditions and selection of grinding media. The major steps in this optimisation are: evaluation (or identify the required improvement), modelling the changes, implementing the change, and assessing the result.

GEOLOGY AND MINERALOGY

The Okvau Deposit is largely hosted in a Cretaceous diorite intrusion emplaced within an upper Triassic metasedimentary host rock package.

Gold mineralisation is contained in a north-east trending fracture set in a narrow off-shoot or apophyses from a larger diorite intrusion however extends beyond the diorite contact into the metasediments.

Gold mineralisation is concentrated along a network of brittle/ductile shears and arsenopyrite-rich sulfide veins. The mineralised shears typically comprise 10 to 50 m wide core of strongly altered, fractured, and/or sheared rock locally with a weak planar fabric, surrounded by 0.5 to 2 m wide less intensely altered halos which retain relict diorite texture. Variably deformed pyrrhotite, arsenopyrite and/or pyrite-rich layers up to 10 m wide also commonly occur in the core of the shears.

CIRCUIT DESCRIPTION

In the Okvau processing plant ore is ground in jaw crusher and SAG mill closed with pebble crusher, ground ore at F_{80} = 90–106 µm is treated by flotation to produce a concentrate with a solids Specific Gravity (SG) of ~3.94 t/m^3 at P_{80} = 30–35 µm which is thickened and sent to the vertical regrind mill for grinding to P_{80} = 10–11µm. The ground concentrate is sent to leach tank leach tank 1 and the flotation tails is designed to be sent to the leach tank 5. The gold is associated with pyrite and arsenopyrite and hence the requirement for ultrafine grinding in the vertical regrind mill.

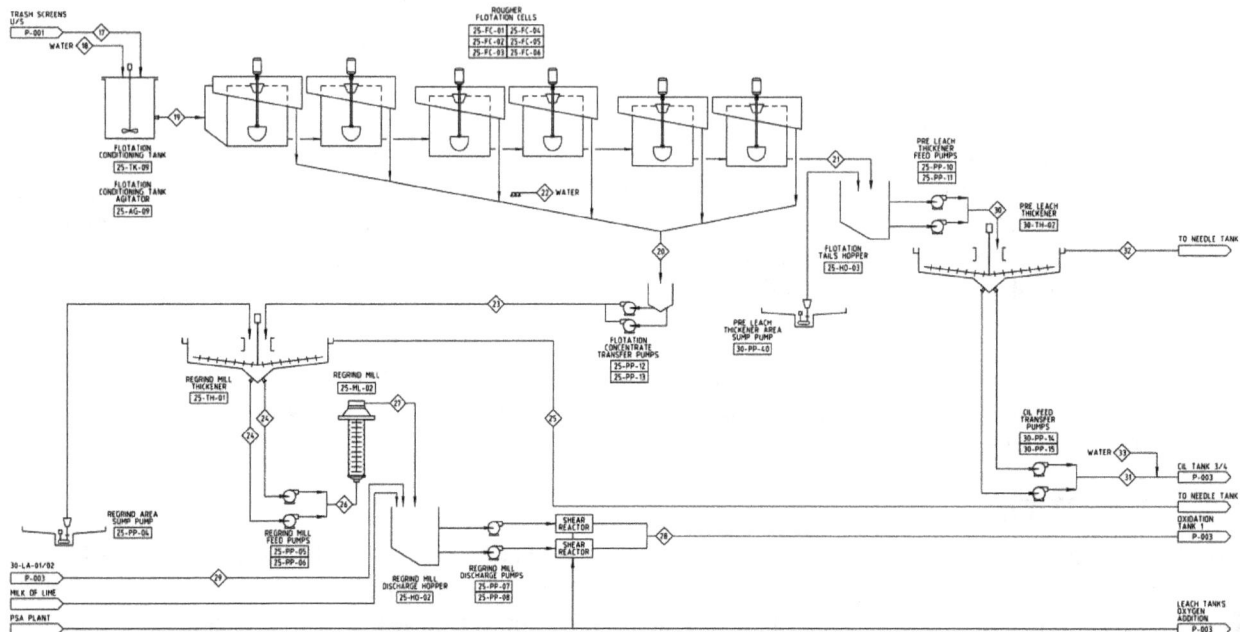

FIG 1 – Process flow diagram (concentrate regrind).

MILL DESIGN AND SIZING

STM's technology partner conducted small sample test on the arsenopyrite sample in 2017 (Lehto *et al*, 2017). Sample quantity was limited to 5kgs, so a Small Sample Test (SST) was conducted. The F_{80} was 65.5 µm and the target grind size was down to P_{80} = 10 µm. The regrind slurry density was 50 per cent w/w with a solids specific gravity of 4.2 t/m^3. The media selected for test work was a 4 mm topsize bead with specific gravity of 3.9 t/m^3. The Small Sample Test method is described in the paper Paz *et al* (2021).

The test results gave a specific grinding energy result 59.3 kWh/t (Figure 2) and together with extra contingency were used for sizing the vertical regrind mill 9000 L 1800 kW.

During the project execution there were concerns that slurry viscosity was going to be extremely high due to the slurry density 45–50 per cent w/w range coupled with ultrafine particles below 10 µm, which may have resulted in excessive media washout and loss of grinding efficiency. In order to reduce the ramp up period and achieve the grind size, the media selected was SG 4.7. The vertical regrind mill is extremely flexible and can operate with a range of media SGs with a typical range from 3.6 to 4.7 t/m^3.

Performance curve

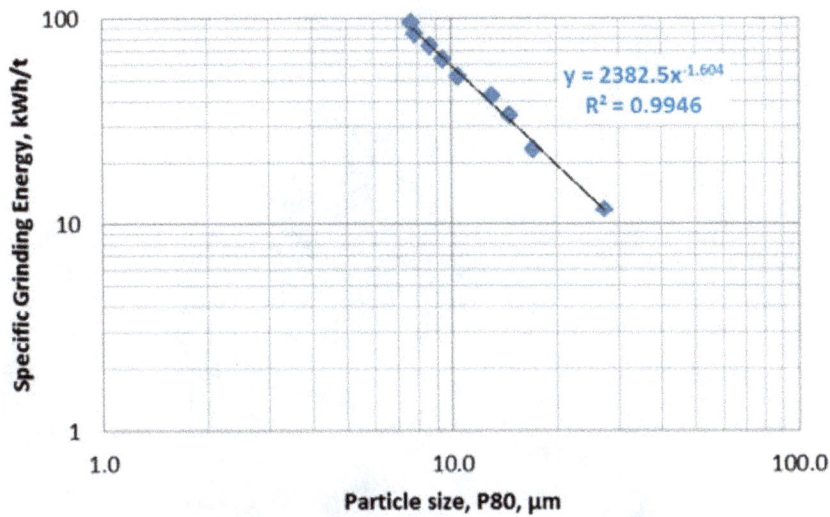

FIG 2 – Performance curve (P_{80} versus specific grinding energy).

EVALUATION

An evaluation was conducted by STM Minerals in Nov 2022 to provide an initial review of the operational performance of the vertical regrind mill.

The evaluation required a complete set of mill operating data, including media additions, confirmation of media Specific Gravity (SG) and photos of the wear componentry.

The issues experienced with the mill were:

- The rotor wear resulting in a reline periods ranging from 12–13.9 weeks.

- Higher specific grinding energy at 68.6 kWh/t versus the expected 59.3 kWh/t.

The logical approach to any optimisation of wear parts in mills is to analyse the wear components and study the wear mechanisms. Simply by looking at Figures 3 and 4, we see that the following observations can be made:

- Rotors are worn from position 1 to 7 with minor wear in rotors 8 to10.

- The flat rotors in positions 11 to 15 have no wear.

- Wear on stators from positions 1 to 8.

- Delaminating (pealing) of the shell liner at position 8 (opposite rotor #8).

- Loss of the polyurethane liner on stator position 11.

- Evidence of media packing on the stators below position 8.

- Evidence of the worn ceramic media having irregular shape with angular edges.

FIG 3 – Reline Nov 2022.

FIG 4 – Sample of worn SG 4.7 media.

The rotor wear pattern shows highest wear on the bottom rotor and lowest wear on the top rotors. This pattern is caused by high pressure of the grinding media combined with high angular sharp edges of the worn ceramic media. Essentially the media is less fluidised and more weight is pushed into the rotors causing high wear rates, and the wear profile is in accordance to the pressure exerted on the rotors.

The starting media level is 50 per cent v/v with SG 4.7 media. During the first week of operation with a new rotor set the operating torque range was 75–80 per cent. It is possible to operate the mill at slightly higher media levels allowing for a lower rotor speed, and higher operational torque around 80–85 per cent. This may bring slightly longer wear life due to the power draw being shared by more rotors.

The media wear level at the reline was 66.7 per cent v/v, which covered rotor position 10. The low rotor wear life was essentially spread over ten rotors and other rotor positions above are unutilised.

It is possible that the delamination of the shell liner was caused by the high operating temperature of the process combined with holding the media level relatively constant.

The vertical regrind mill was performing well achieving the target 10–11 μm grind size, however the specific grinding energy was about 17 per cent higher than the design value of 59.3 kWh/t. After internal review the most probable cause for lower efficiency was low slurry density.

The design slurry density was 45–50 per cent w/w, whereas the operation was set at 35 per cent w/w. Higher grinding density could allow for higher residence times, higher breakage rates and higher grinding efficiencies. However if the grinding density is too high the viscosity increases and there is a loss in grinding efficiency. For this ultrafine grinding application, the grinding density is dependent on the exit temperature of the mill in combination with the power draw, and the slurry discharge temperature high alarm of 70°C and downstream optimum leach density requirements.

Keikkala *et al* (2018) reported a relationship between mixing rotor design (flat rotors versus castellated rotors) and the grinding energy efficiency, which could also indicate that as rotors severely wear they become less effective at mixing and contribute less to optimal grinding efficiency. This can be offset by increasing the media level as the rotors wear.

MODELLING

During the evaluation several opportunities for improvement were identified. Firstly a change of media SG from 4.7 to 3.7 t/m³ and secondly a change in rotor configuration to include two more Patented castellated rotors. The rotor configuration changes are shown in Figure 5.

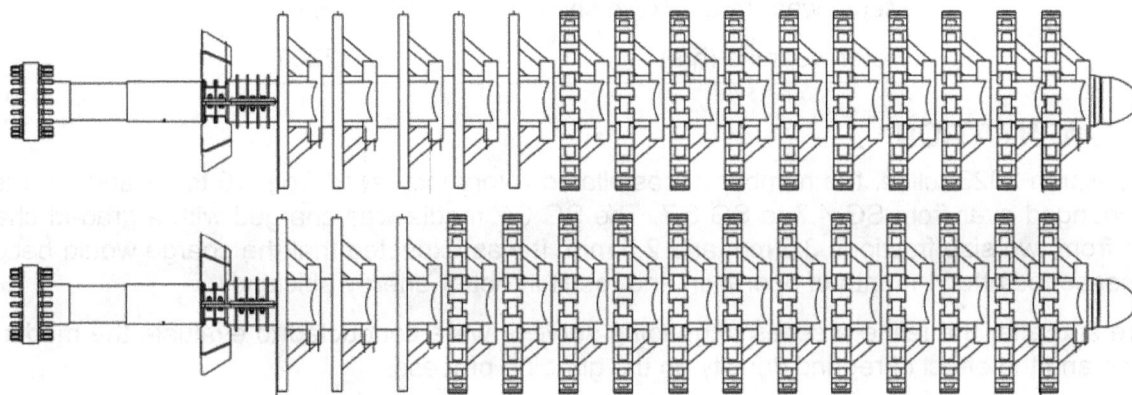

FIG 5 – Rotor configuration before (top) and after (bottom).

Modelling was performed to simulate the solution and provide an indication of the modelled outcomes. A power model was constructed based on the one proposed by Heath *et al* (2016) and fitted to the commissioning power calibration data and operating data. The power model takes into account a variety of factors including the geometric properties such as rotor diameter, speed and process factors such as slurry density, slurry viscosity, media size, media SG, media filling level and fluidisation velocity. The Heath power model was selected as it has been demonstrated to provide good correlation for predicted power against measured power. The power model together with a rotor wear model was used to predict the outcome of changes to the mill operation.

Due to time constraints and difficulty with exporting sample out of the country, test work was not conducted to validate grinding efficiency and the optimum grinding parameters.

The SG 4.7 t/m³ media in service had a wear rate of ~6 g/kWh. The media wear rate was not a focus for testing or modelling in this case as the expected wear rate of the 3.7 t/m³ media SG was expected to be lower.

The operational data used for the modelling is in Table 1. The results of the modelling indicated:

- The selection of SG 3.7 t/m³ and 3.5 mm top size would not excessively fluidise the media, so no media loss is expected via elutriation.

- The same power can be drawn at a lower speed and higher 70–80 per cent v/v media filling level.

- Increase in rotor wear life from 2300 hrs to ~3500 hrs, a 51 per cent increase.

TABLE 1

Six months operation with SG 4.7 t/m^3 media.

Description	2022 avg data
F_{80} (um)	31.7
P_{80} (um)	10–12
Specific grinding energy (shaft) (kWh/t)	68.6
Motor power draw (kW)	1246
Mill feed (per mill) (tph)	17.1
Slurry density (%w/w)	35.4
Flow rate (m^3/h)	35.1
Media density (t/m^3)	4.7
Mill speed (%)	82.4
Media wear rate (g/kWh shaft)	6.0
Rotor wear life (hrs)	**2300**

IMPLEMENTATION

At the March 2023 reline, the number of castellated rotors increased from 10 to 12 and the media was changed over from SG 4.7 to SG 3.7. The SG 3.7 media was charged with a graded charge made from two size fractions 3.5 mm and 2.5 mm. It was expected that the charge would become fully seasoned after 1 month of operation and result in better energy efficiency.

Before and after the reline a series of grinding surveys were conducted to evaluate the media SG change and the effect of regrind density on the grinding process.

The key observations were:

- The grinding efficiency of the SG 3.7 media could be slightly better than the SG 4.7, however there are points intermixed and firm conclusions can't be made (Figure 6).

- Increasing slurry density to 41 per cent w/w with SG 3.7, the SGE reduces from 75 to 65 kWh/t, and a finer grind was observed (Figure 7). No media elutriation was observed.

- No detrimental effect was observed on the ratio of P_{98}/P_{80} between the media SG types. The ratio of P_{98}/P_{80} improves for the product against the feed material. The scatter was less for the SG 3.7 media, which could be considered a benefit (Figure 8).

- Figure 9 shows that the product Particle Size Distribution (PSD) shape of the full scale survey matches the 2023 laboratory test work despite the significant difference in feed PSD due to sieve sizing versus laser sizing. The ratio of P_{98}/P_{80} for the lab test work was 3.52 (sieve) for the feed and 2.81 (laser) for the product. The ratio of P_{98}/P_{80} for the full scale 20230311S2 was 3.10 (laser) for the feed and 2.72 (laser) for the product.

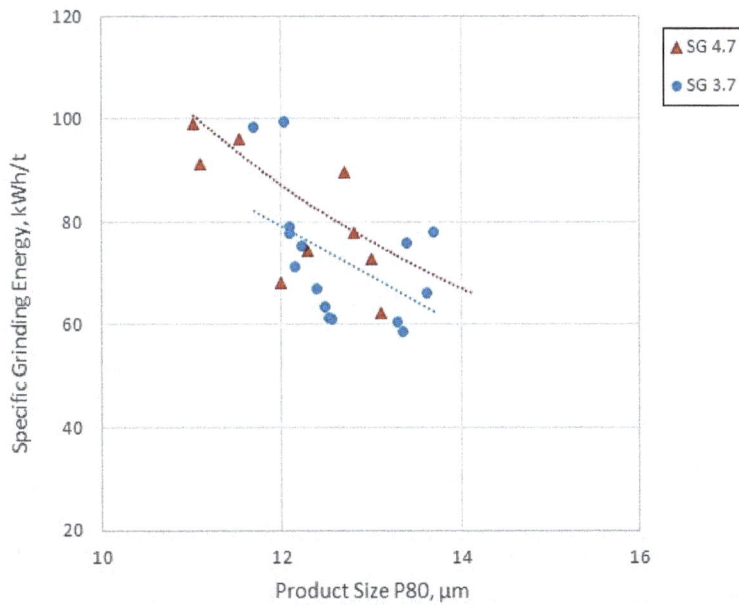

FIG 6 – Performance graph P_{80} versus SGE for SG 4.7 and SG 3.7 media.

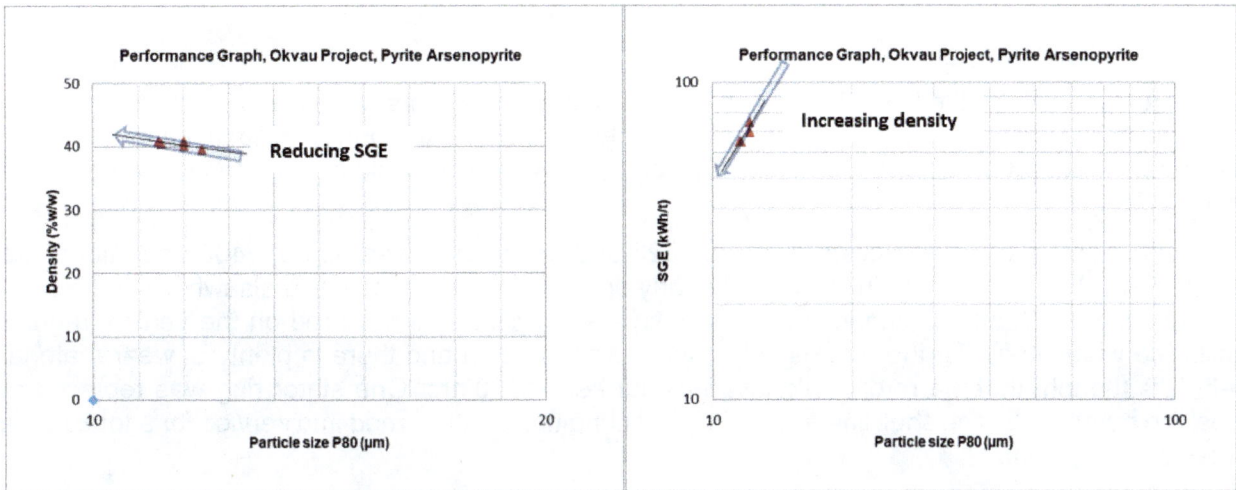

FIG 7 – Performance graph P_{80} versus SGE versus slurry density for SG 3.7 media.

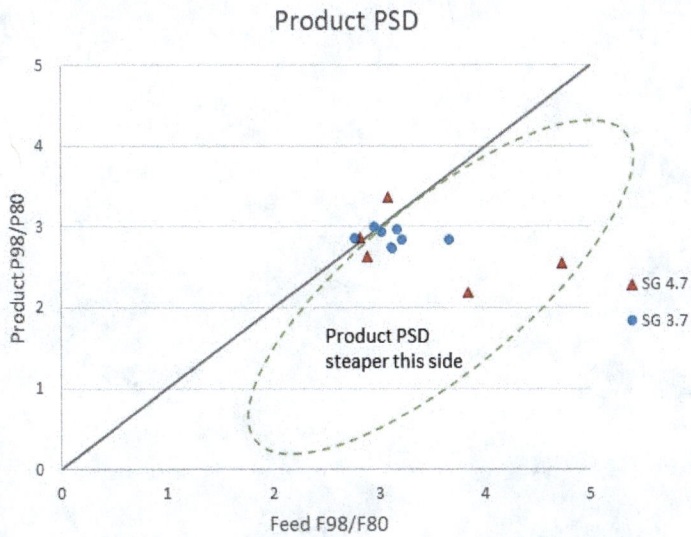

FIG 8 – Feed F_{98}/F_{80} versus P_{98}/P_{80} for SG 4.7 and SG 3.7.

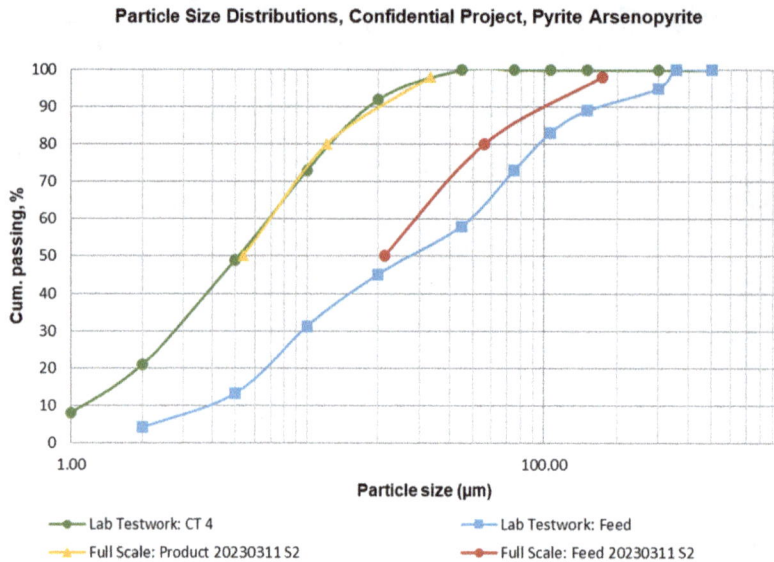

FIG 9 – Feed and product PSDs for test work versus full scale.

OPERATIONAL ASSESSMENT

Wear performance

The rotor wear life for the SG 4.7 media in March 2023 was 2900 hrs, which was up on the previous interval of 2300 hrs. After analysing available process variables, the only significant change was an increase in regrind density from 33.3 to 35.7 per cent w/w, which could be the reason for increased in wear life.

The SG 3.7 media's reline occurred in Oct 2023 and achieved a wear life of 5400 operational hrs (Figure 10). The Oct 2023 reline was made early due to a maintenance issue elsewhere in the plant determining the plant maintenance schedule, so a strategic reline occurred on the vertical regrind mill. The wear profile for the rotor is very even from top down and there is plenty of wear material left. It is thought that the rotor could last for a further ~1000 hrs. One stator ring was replaced at position number 12. The shell liners were in good condition and returned into service for a total reline interval of 70 weeks.

FIG 10 – (Left) Reline March 2023 at 2900 hrs; and (right) reline Oct 2023 at 5400 hrs.

It should be noted, that a full (shell and rotor) reline from slurry off to slurry on, takes 13 hrs.

Operational performance

The operational philosophy was an important factor for consideration with operating with media SG 3.7. The starting media level was 60 per cent v/v and the final target at the end of the rotor life was 80 per cent v/v, therefore during this period the media level was slowly raised to compensate for rotor wear, maximising rotor life. The media level was monitored every four days for the first two weeks then every 2–4 weeks there after until the grinding media wear rate stabilised.

The particle size is measured several times a day with a Malvern Master Sizer, the mill speed is adjusted to achieve the grind size in fixed speed mode. Automatic control of the mill speed using SGE control is not used due to several reasons associated with this application. The feed density varies in accordance with the thickener density output, as such if the grinding density increases beyond 40 per cent w/w combined with grinding less than 8µm, the viscosity would increase and cause the mill speed to ramp up, with a reduced response in motor power. As the rotors wear the viscosity effect becomes more pronounced. The mill is very capable to run at higher slurry densities >40 per cent w/w and further work into this area will be conducted. Expert control would be required to implement the automatic SGE control, where the SGE set point varies with slurry density.

Density control is important for obtaining the optimal regrind efficiency. It has been observed when the regrind density reaches >38 per cent w/w it easier to achieve <10 µm grind sizes with SGEs <50 kWh/t. Conversely when the slurry density drops to the 20 to 25 per cent w/w range, the specific grinding energy increases to the 80–100 kWh/t range. This can be further shown if we plot the averages for the time against the original Small Sample Test work used for the regrind mill design (Figure 11), it's clearly shown that the points above the test work line are lower slurry density (20–34 per cent w/w) and the points below the line are higher slurry density (38–45 per cent w/w). Figure 11 also indicates that the scaleup is 1:1 which is in line with expectations observed at other sites such as Cracow and Sunrise Dam (Paz *et al*, 2019, 2021).

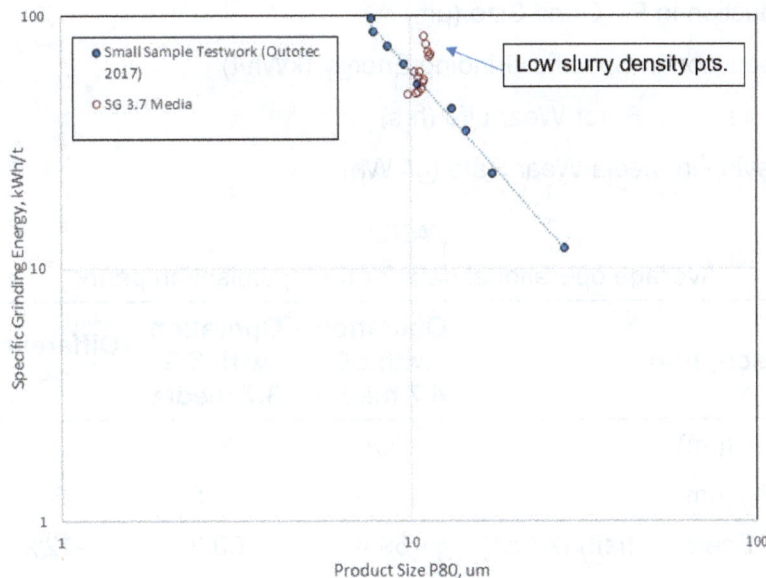

FIG 11 – Performance plot (P$_{80}$ versus specific grinding energy) for small sample test and SG 3.7 media.

The average operational density is 34–35 per cent w/w which is similar for media types trialled. Having the slurry density stable for the trial reduces the effect of this variable on the results. In the future there is an opportunity to further optimise the density higher.

The operation period for the trial is presented in Figure 12. We can see that the throughput varies depending on the concentrate mass pull and the density varies depending on the thickener performance. There is no intermediate tank between the thickener and the mill to allow dampening of the throughput and control of slurry density. However the current slurry dilution in the pipeline works, and the overall grind size is achieved.

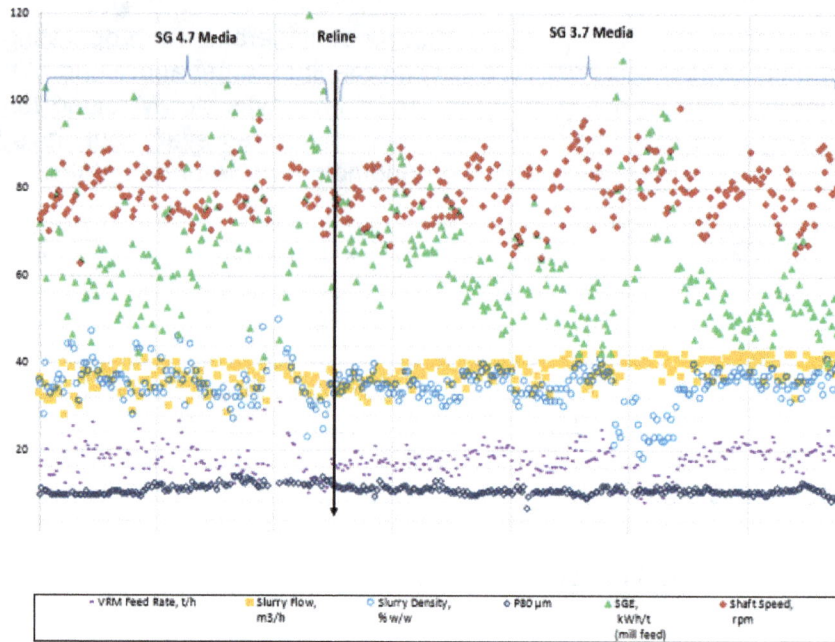

FIG 12 – Operational data for the assessment period.

The average results of the optimisation program are presented in Table 2. A T-Test was performed with a 95 per cent confidence interval to some of the tested variables. With confidence the results are:

- 5 per cent reduction in P_{80} Grind Size (μm)

- 12 per cent reduction in Specific Grinding Energy (kWh/t)

- 86 per cent increase in Rotor Wear Life (hrs)

- 59 per cent saving in Media Wear Rate (g/kWh)

TABLE 2

Average operational data for the optimisation period.

Description	Operation with SG 4.7 media	Operation with SG 3.7 media	Difference %	T-Test P(T<=t) two-tail
F_{80} (um)	NA	NA		
P_{80} (um)	11.4	10.8	-5%	2.60×10^{-7}
Specific Grinding Energy (shaft) (kWh/t)	69.4	60.8	-12%	2.95×10^{-3}
Motor Power draw (kW)	1305	1160	-11%	5.28×10^{-4}
Mill Feed (per mill) (tph)	17.7	18.0	2%	3.96×10^{-1}
Slurry Density (%w/w)	35.7	34.4	-4%	1.37×10^{-2}
Flow rate (m³/h)	35.9	38.5	8%	5.49×10^{-17}
Media Density (t/m³)	4.7	3.7	-	-
Mill Speed (%)	82.8	78.8	5%	3.26×10^{-1}
Media Wear Rate (g/kWh shaft)	6.0	2.5	-59%	-
Rotor Wear Life (hrs)	2900	5400	86%	-

In addition, there is a large Media Cost ($/t) saving for switching to lower SG media.

GHG emissions

More recently, Environmental, Social, and Governance (ESG) considerations are playing a more significant role in the mining industry (Foggiatto *et al*, 2023). Greenhouse gas emissions are an important indicator in assessing the environmental impact of mining activities, and in this case stirred milling.

The effect of changing the media type was assessed via calculating the GHG emissions of the regrind mill. The GHG emissions were analysed as scope 2 indirect GHG emissions. At Okvau electrical power is available from the local grid, so scope 1 emissions were not considered, and scope 3 emissions were not considered due to having a minor role in the overall GHG emissions for this project.

The estimated input parameters used when calculating the GHG emissions are:

Electrical Energy   0.77 GHG $kgCO_2$/kWh   (Australian Government, 2023)

Ceramic Media (SG 4.7)   5.78 GHG tCO_2/t   (Media supplier, 2023)

Ceramic Media (SG 3.7)   3.59 GHG tCO_2/t   (Media supplier, 2023)

Rubber Wear Parts   6.49 GHG tCO_2/t   (Yulex, 2023)

The total reduction in greenhouse gas emissions was 8851 tCO_2/a for the SG 4.7 scenario and 7651 tCO_2/a for the SG 3.7 scenario, which is a 14 per cent GHG emission saving.

Operating cost

Operational costs (OPEX) are an important factor when assessing the optimisation outcomes. The OPEX for ceramic media, wear parts, and electricity, were estimated for the trial periods (Figure 13). The costs for labour were not considered due to being a small portion of the total. The total reduction in OPEX was estimated as 35.7 per cent.

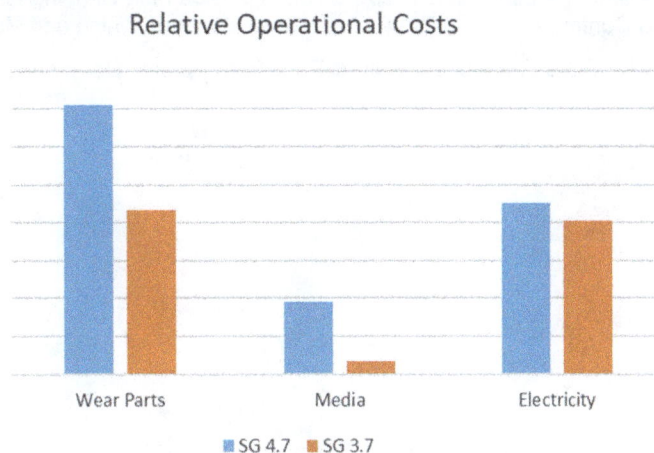

Relative Operational Costs

SG 4.7   SG 3.7

FIG 13 – Relative OPEX for vertical regrind mill.

CONCLUSION

In summary, the optimisation of the vertical regrind mill is dependent on the process conditions and selection of grinding media to enable fluidised conditions and maximise operational media levels, which in this case has allowed Okvau to fully realise the potential of the vertical regrind mill.

The Okvau vertical regrind mill was optimised and achieved a significant 35.7 per cent reduction in operating cost and 14 per cent reduction in GHG emissions. The increase in rotor wear life exceeded expectations with an 86 per cent increase in wear life to 5400 hrs. The shell liner went on to last another reline interval for a total of 70 weeks. Another advantage was the 59 per cent saving in Media Wear Rate (g/kWh).

The vertical regrind mill achieves its required performance with average grind sizes are in the P_{80} range 10–11 µm and is operating very well. The energy efficiency and product PSD matches the SST test work with indicated scaleup 1:1.

ACKNOWLEDGEMENTS

The Authors would like to thank Renaissance Minerals (Emerald Resources) for permission to publish this work. Thank you to the suppliers of the vertical regrind mill: Swiss Tower Mills Minerals together with Metso (formally Outotec) for your support.

REFERENCES

Australian Government, Feb 2023. Australian National Greenhouse Accounts Factors, Table 1 (National Scope 2 and Scope 2), p 8.

Foggiatto, B, Quinteros, J, Ballantyne, G, Crane, C and Lagos, S, 2023. The Impact of Media Embodied Carbon Emissions on Regrind Technology Selection, in Procemin Geomet 2023.

Heath, A, Keikkala, V, Paz, A and Lehto, H, 2016. A Power Model for Fine Grinding HIGmills with castellated rotors, *Miner Eng*.

Keikkala, V, Paz, A, Komminaho, T and Lehto, H, 2018. Energy Efficient Rotor Design for HIGmills, paper presented in Comminution 2019.

Lehto, H, Roitto, I, Paz, A and Åstholm, M, 2013. Outotec HIGmills; Fine Grinding Technology, paper presented in IMCET2013.

Lehto, H, Paz, A and Jamieson, E, 2017. HIGmill Testwork Report – HIG5 for the Okvau Gold Project, Outotec.

Media Supplier, 2023. Internal communication regarding benchmarking GHG for different media SG.

Mezquita, H, Wright, A, Wang, Fisher, W and Rosario, P, 2022. Implementation of Fine Grinding and Dual Circuit Concept at Santa Elena Mine, in 54th Canadian Mineral Processors Conference.

Paz, A, Ghattas, G, Loro, S and Belke, B, 2019. Fine Grinding Implementation at the Cracow Gold Processing Plant, paper presented in Metplant 2019.

Paz, A, Ypelaan, C T, Ryan, M and McInness, 2021. The Application of Ultra Fine Grinding for Sunrise Dam Gold Mine, in Proceedings of Mill Ops Conference 2021 (The Australasian Institute of Mining and Metallurgy: Melbourne).

Yulex, 2023. Rubber Chronicle 19: CO_2e Emissions of Natural Rubber, Neoprene, Geoprene and SBR, Rubber Chronicles, table 2, p 8.

Whiskey, puddings, shampoo and insanity – a taxonomy of circulating loads

J D Pease[1] and P D Munro[2]

1. FAusIMM, Principal Consultant, Mineralis Consultants Pty Ltd, Brisbane Qld 4068.
   Email: jpease@mineralis.com.au
2. Principal Consultant, Mineralis Consultants Pty Ltd, Brisbane Qld 4068.
   Email: pmunro@mineralis.com.au

ABSTRACT

While the attribution to Albert Einstein is incorrect, it is often repeated that insanity is doing the same thing over and over again and expecting different results.

After an argument about whether children preferred to read about fairies or food and fighting, Norman Lindsay wrote 'The Magic Pudding' featuring Albert the 'cut an' come again' pudding who maintained his full capacity no matter how many slices were cut.

The most successful three words in marketing history are renowned to be 'rinse and repeat'.

Whiskey is distilled at least twice to improve its quality. The finest whiskeys are distilled three or more times.

These conflicting observations mirror the industry's confusion over circulating loads. Sometimes they are good, sometimes they are fanciful but harmless, sometimes they are bad. How can we tell the difference?

This is a field guide to help you identify and assess the value of your circulating loads.

INTRODUCTION

Circulating loads can be a valuable tool in any processing duty, including mineral processing. When used properly, they significantly increase the effectiveness of separation. But like all good tools, sometimes they are over-used, and sometimes they are used inappropriately and make our plants less efficient. How can we know the difference?

CORE PRINCIPLES

This discussion is built on two core principles that are so obvious they are rarely stated:

- *First*, that if <u>no</u> payable product can be made from a stream, then it isn't ore. It is tailings.

- *Second*, that the net effect of a circulating load can't be assessed from <u>*within*</u> the load, since the load is a product of its own circulation. In MS-Excel terminology, a sample or calculation would be disallowed as a 'circular reference'.

Ore or tailings?

Plants seek to maximise margin at target product grade. The 'easy' minerals are recovered first at higher-than-target grade. Then more difficult middlings are recovered at progressively lower grade until the final addition 'breaks even' on incremental payment and operating cost. Lower-grade components get lower payment for values, higher penalties for impurities, and higher transport and treatment charges.

If the combined product target grade is T per cent, then the incremental payment may turn negative for additions below (for example) 0.75 times T per cent. In this example, if processing an intermediate stream <u>can't</u> produce at least some product above 0.75 × T per cent, then that stream is tailings since no product can be profitably recovered from it.

It is more complex if some intermediate-grade product can be made from the stream. In the above example, additional product at 0.8 × T per cent <u>might</u> be profitable – <u>*but*</u>, if the final product already averages target T per cent, then we can only add 0.8 × T per cent to it if we first remove an equivalent amount of low-grade component. Figure 1 illustrates this for a simple two-component system. For example, to recover more from flotation cleaner tailings we would have to 'make room' in concentrate by sacrificing recovery from (e.g.,) scavenger concentrate. It is impossible to quantify the trade-offs in real time (or within a load). That is why we build high circulating loads. We _want_ to recover from the cleaner tailings, but we _don't want_ to sacrifice recovery from the scavengers, but we _must not_ reduce overall product quality. We can't solve this trilemma with the available data.

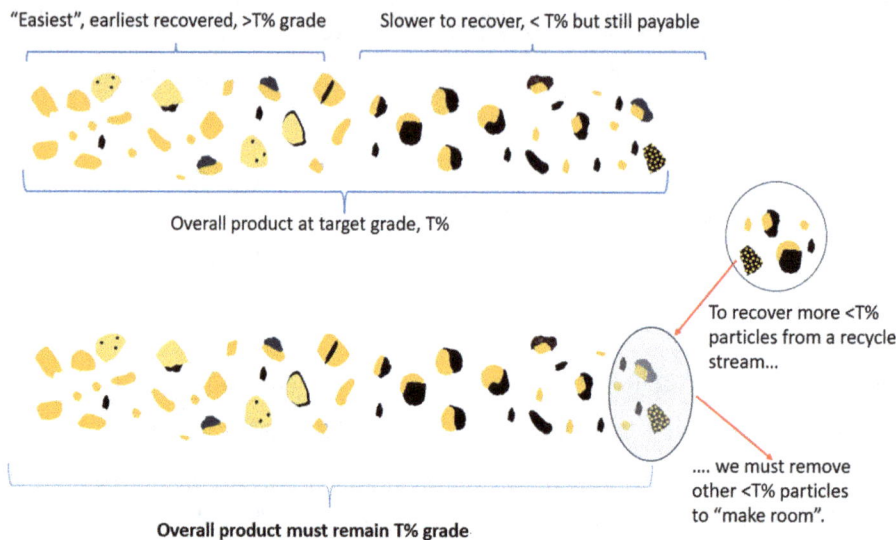

FIG 1 – Making room in concentrate for a lower-than-target grade increment.

A high circulating load demonstrates that we tried anyway.

The circular argument

In the 'middlings trilemma' described above, the particles considered too high-grade for tailings but too low-grade for product will be circulated multiple times until a physical constraint is reached (eg pumping capacity). The most 'indecisive' particles, such as those in Figure 2, may circulate five times or more, so mass flow of the stream is meaningless to predict what might happen in open-circuit. However, laboratory test work is still informative. Recirculation concentrates the 'indecisive' particles, so test work on this stream will indicate whether any profitable concentrate can be made from them. If tests in laboratory conditions on a concentrated source of middlings can't make a profitable addition to concentrate, then we can be sure that the plant can't either.

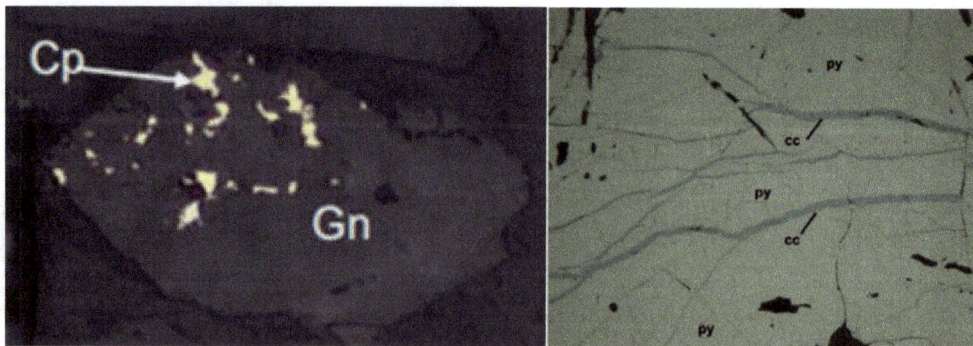

FIG 2 – Recirculating these ~37 µm particles is unlikely to recover more product at target grade.

SIMPLE RECIRCULATION RULES

Taggart's rule

Practitioners have sought to concisely capture this complex topic. Taggart (1945) espoused a simple rule-of-thumb:

Return the stream to the point where its assay value and water composition are nearest those of the stream that it joins.

When applied to flotation, this sometimes works – for example, returning cleaner 2 tailings to the feed of cleaner 1. But it is more complex for returning cleaner tailings to roughing – that may be beneficial, unhelpful, or harmful. Taggart's rule was reasonable at the time – practitioners did not have ready access to quantitative mineralogy or particle surface analysis, and the ores processed were coarser grained and higher grade than those processed now. Since he was a proponent of separating *size* classes in 'sands/slimes' flotation (Taggart, 1945), it seems likely he would have applied the same principle to the different *mineral* classes in Figure 3 if he had been able to easily 'see' and quantify them.

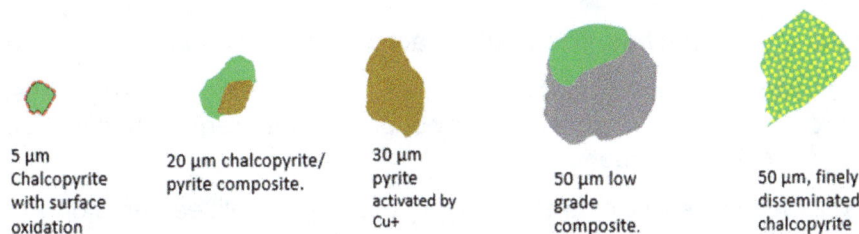

5 µm
Chalcopyrite with surface oxidation

20 µm chalcopyrite/ pyrite composite.

30 µm pyrite activated by Cu+

50 µm low grade composite.

50 µm, finely disseminated chalcopyrite

FIG 3 – These particles in cleaner tailing have similar flotation rate. The stream assay is higher than new feed. Should it be mixed with fresh feed, or do these particles need specialist treatment?

Johnson's rule

A more widely applicable rule – we will call it 'Johnson's Rule' after its enthusiastic exponent, Dr N W Johnson – recognises the difference between particles:

> *'Don't recirculate a particle unless you are going to do something different to it next time'.*

This is a much better rule informed by modern techniques to observe the behaviour of different classes of particles like those in Figure 3. Unless we change the particles or the separation conditions, it is illogical to expect they will perform differently next time.

Even this logical rule can be misinterpreted, and doesn't fully articulate some situations:

- It doesn't explicitly explain why recirculating cleaner 2 tailings to cleaner 1 is helpful. Because if that is helpful, then why doesn't the same apply to returning cleaner 1 tailings to roughing?

- The most common mis-use is to say: *'but we send it back to regrind, so we do change it'.*

EXTENDED RULES

To address the misinterpretation of 'Johnson's rule', we suggest a set of three wider rules:

Rule 1: When not changing particles

> *Recirculating without changing particles is helpful when there is a genuine rate constant differential between the particles classes you want to separate, and this differential is not changed by recycling, and it does not harm the fresh particles in the receiving stream.*

Rule 2: When changing particles

> *Recirculating particles that you intend to change next time is helpful when you can selectively recycle the right particles, and you genuinely change them, and this does not harm the performance of fresh feed by more than it helps the recycled particles.*

Rule 3: The exceptions

> *There are practical exceptions to improve operability or for organisational expediency.*

Further explanation

Rule 1 supports sending cleaner 2 tailings to cleaner 1. Because there is a sustainable and repeatable difference in rate constant, the more times the recycle is repeated the better the separation. Three stages of cleaning (or whiskey distilling) is better than two. Four or more stages is even better, so long as the diminishing returns justify the additional cost and circuit complexity. But the provisos show that care is needed – in some situations the benefits may be offset or outweighed:

- If the behaviour of the target minerals changes with time (eg surface oxidation from excessive residence time).

- If the pulp chemistry of the recycle stream alters the pulp chemistry of the fresh feed in the receiving stream.

- If the recirculating volume has a harmful impact on the residence time of fresh feed.

Rule 2 (when changing particles). As for Johnson's Rule, this does not support sending cleaner 1 tailings directly to rougher feed. The particles will not behave differently when you do the same thing – but they will reduce residence time for fresh feed.

Similarly, sending a cleaning recycle stream back to the regrind mill earlier is only helpful if:

- *The correct particles are recycled* – genuine coarse composites may benefit from another chance to get to the regrind mill, but complex composites and fines with difficult surface chemistry won't. Such recalcitrant particles usually dominate cleaner recycle streams, not the simple composites we like to imagine.

- *There is a **net** liberation benefit* – most recycled particles won't get to the regrind mill; they will go straight back to the hydrocyclone overflow where they came from the last time. A small proportion will get to the mill and may be reground. Even so, the liberation benefit isn't clear – a minor size reduction will not significantly increase liberation for complex composites like those shown in Figure 2. Further, the additional mill feed consumes residence time that may be better applied to simpler composites in fresh feed. The recycle is only helpful if we can prove there is a _net_ liberation benefit, allowing for the effect on both fresh feed and the recycle stream.

Rule 3 – practical exceptions include managing head grade and feed rate changes (eg partial recycle of Jameson Cell tailings to cell feed) or improving froth mobility at the end of scavenging. And sometimes an 'Augustinian' compromise is justified, eg if there is relentless pressure to 'just do something' about the high assay of a mineralogically recalcitrant stream.

The following discussion further illustrates these rules in common mineral processing situations.

COMMINUTION AND CLASSIFICATION

In mineral processing we first encounter circulating loads in comminution. We usually use screens to 'close' a crushing stage, and hydrocyclones to 'close' the grinding mill. In grinding we often employ the highest circulating load that the equipment can bear to maximising grinding efficiency.

This satisfies Rule 2: by preferentially recycling coarse particles we increase their residence time in the mill (or crusher) compared with finer particles. This improves comminution efficiency.

Even more efficient classification would be even better – eg two stage hydrocyclones, with the first underflow being diluted and sent to another stage, minimising the 'water recovery' of some fines to the mill. It would be even better if we could close the mill with a 'sharp' cut from a vibrating screen. But usually the law of diminishing returns – and the needs of operability and maintainability – mean that we accept one stage of imperfect classification.

Classification acts on the whole-of-particle properties of size, specific gravity and shape. While grinding may affect surface chemistry, this does not affect the behaviour of particles during classification or grinding. Recent developments in coarse separation offer a tantalising prospect. If a 'hybrid' classifier separated on both particle properties _and_ surface properties – (eg combining hydrocyclones with coarse flotation) then it might selectively return those coarse particles with small amounts of target mineral to grinding, while rejecting barren coarse particles. This would more efficiently direct grinding energy to the particles that may benefit from it, though with additional circuit and operational complexity.

COARSE SEPARATION METHODS

Coarse separation is governed by whole-of-particle properties. Therefore, circulating a particle will not adversely affect the separation characteristics of either the particle or the receiving stream. This is much simpler than flotation, where circulating a particle may harm the separation response by unintentionally changing surface oxidation, surface deposits, or pulp chemistry.

Even with this advantage, coarse separation technologies rarely employ significant circulating loads (with the exception of size classification in comminution). In many circuits new feed is split into narrower size fractions for separate processing. If middlings of one stage are sent to size reduction, they are more likely to be directed to a different stream rather than returning to their original feed point. This 'stage' processing recognises both the need for additional liberation and the processing differences between size fractions.

Gravity separation

Gravity separators recover coarse gold from a grinding circuit before leaching or flotation. The gravity separator itself does not have a circulating load, but it exploits the high circulating load of high-SG particles in hydrocyclone underflow during size classification.

Many dense medium separation (DMS) circuits consist of a single stage of separation, with the aim to reject low-grade gangue in a single step and a wide size range – eg to separate 'heavies' from lower-SG coal, or to discard siliceous gangue before downstream grinding and flotation of base metal sulfides. For these relatively low value products, a small loss to rejects is acceptable.

However, minimising losses is important for high-value products like diamonds. Here, middlings should be stage-processed so that the conditions can be specifically tailored to their needs, rather than returning to feed.

Spodumene sits on the product value chain between coal and diamonds. DMS recovery circuits may be split into different size streams (Figure 4) where middlings are stage-processed. In the simpler Figure 5, separation is in a single size range and middlings are crushed and returned to feed. This is a circulating load that satisfies our 'Rule 2':

- The middlings have dedicated crushing capacity, the do not compete for it with new feed.

- Returned crushed middlings won't adversely affect the behaviour of new feed (so long as the DMS cyclones have adequate capacity.

- DMS is a more binary 'sink or swim' separation than flotation. It is unlikely to build a large circulating load of 'indecisive' particles.

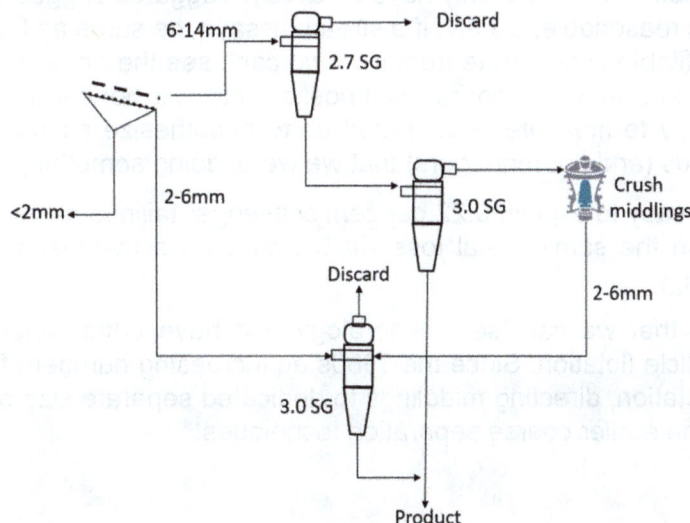

FIG 4 – Dense medium recovery with two size fractions and middlings recycle.

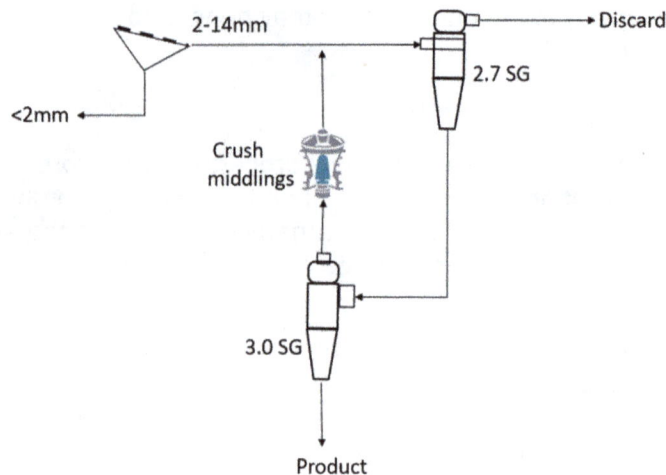

FIG 5 – Simpler DMS circuit, single size fraction with middlings recycle.

Magnetic separation

Compared with gravity separation, a relatively small content of a magnetic mineral can cause a particle to be recovered by magnetic separation. Whereas for gravity separation the rejects of one stage are the middlings, for magnetic separation the middlings will be in a concentrate. Yet the principle is the same – once the middlings are concentrated in a stream, grind that stream to liberate gangue, and then reprocess (preferably in a dedicated separator rather than returned to feed).

Ore sorting

Depending on the sensor, ore sorting can rely on whole-of-particle properties or surface characteristics (eg colour). It is practiced in relatively narrow size distributions – typically the largest particle to a separator should be no more than three times the size of the smallest. There is little scope for circulating loads to the feed of the same unit. Crushed middlings would need processing on a different unit suitable (with a lower processing rate) to suit their finer particle size. Staged crushing, screening and sorting would be justified for high value products, including diamonds.

FLOTATION

In contrast to established coarse separation practice, early flotation practitioners often employed large circulating loads to whole-of-feed. This is ironic since flotation may be the least suited technology to mixing streams – separation is profoundly affected by both particle size and surface chemistry; *and* the surface chemistry can change over time or with mixing.

This is probably because, unlike coarse separation, we can't 'see' what the minerals are doing in real time in a fine-grained slurry. If we only have an assay, Taggart's suggestion to return to a point of similar assay seems reasonable. Surely, if a stream assays the same as feed, if we just try again then we can make profitable concentrate from it? If we can't see the particles in cleaner scavenger concentrate, then perhaps they are coarse composites that will benefit from another turn in the regrind mill. The inability to measure encouraged us to hypothesize scenarios and then propose solutions that assured us (and our managers) that we were 'doing something'.

Besides, it is relatively easy to explain 0.22 per cent scavenger tailings assay to your manager. It is much harder to explain the same metal loss via 0.2 per cent scavenger tailings plus 2 per cent cleaner-scavenger tailing.

This is changing now that we can 'see' mineralogy and have better understanding of surface chemistry and fine particle flotation. Since the 1980s an increasing number of sites have employed staged grinding and flotation, directing middlings to dedicated separate size reduction and flotation steps – analogous to the earlier coarse separation techniques.

Internal cleaner tailings

Recirculating tailings within cleaning is almost universally employed and is a positive use of circulating loads. It complies with Rule 1 – we aren't trying to change the particles, we are enhancing an established difference in rate constant between the target particles and unwanted particles, which may be:

- Slower floating contaminants like iron sulfides.

- Gangue minerals with a rate constant matching the water recovery.

- Composite particles which are too low-grade for concentrate but too high-grade for tailings.

Because there is a repeatable rate constant differential, the more times we repeat the process, the better the separation, though with diminishing returns for each repeat. Like whiskey distilling, the importance of product quality will determine how many 'repeats' are justified – usually two or three stages for base metals processing, sometimes ten or more for molybdenum where maximum separation from copper is needed.

The flywheel effect

Circulating loads in cleaning serve another practical purpose. By smoothing variations in feed rate and head grade, they 'buffer' cleaner operation and protect the final concentrate from instability. The importance of this to 'operability' can't be calculated or modelled, but is real nevertheless. Early Jameson Cell installations in base metals suffered from the absence of this stabilising buffer (in contrast, coal installations had relatively consistent feed and head grade). That was a significant barrier to operability in this technology with a small residence time compared with conventional cells and columns, and where bubble generation is also affected by feed volume, The development of partial tailings recycle was critical to make them robust and operable enough for base metals (Young et al, 2006).

The flywheel effect is also likely to be an underappreciated benefit of closed-circuit cleaning. As well as enhancing the separation, the recycle between multiple cleaning stages 'buffers' the first cleaners from variations in rougher concentrate flow and assay. Therefore, in the rare cases when only one stage of cleaning is justified, a partial recycle of the first cleaner tailings stream may be justified, making a single cleaner analogous to a Jameson Cell with partial tailings recycle.

Rinse and repeat

These are renowned to be the three most successful words in marketing. For shampoo, they are as profitable as they are chemically unnecessary. But what about froth washing?

The answer – as so often – is, it depends. Usually, one stage of froth washed cleaning – equivalent to around three stages of dilution cleaning *if done well* – is enough. If there is a positive wash water bias, and the water is evenly distributed across the froth surface, then most entrained particles will be eliminated. But correct practice is not always sustained – wash trays and rings partially block, or flow rate is reduced for operational reasons. Even this *reasonably* good practice is often enough, but if the concentrate is close to strict penalty or limits on an impurity in gangue – eg F, U, As – then a second stage of washed froth cleaning might be justified to get closer to full entrainment control. The tailings of the second stage would recirculate to the first stage, in typical closed-circuit internal cleaner configuration.

First cleaner tailings

Dealing with the back end of the first cleaner – both the concentrate of the last cells and the tailings – is where good intentions often become harmful. These streams contain a mix of particle size, composition and liberation class that have similar rate constants (as illustrated for chalcopyrite flotation in Figure 3), as well as rejected entrained gangue (not shown in Figure 3).

These particles are genuinely hydrophobic with a similar intermediate flotation rate. To recover one you must recover them all. But the combination is below target concentrate grade, so cleaning must reject them all.

Circulating these particles directly to rougher feed as in Figure 6 will have several effects:

- It won't change any particle liberation or surface, so _at best_ they will do the same thing next time. The operator still has to decide – can I reduce concentrate grade, or will I send them around again?

- The particles will continue to recirculate, increasing the circulating load until it hits the physical limit of launders, sumps or pumps. Ultimately the particles will exit as either concentrate (reducing grade), tailings, or spillage (deferring the concentrate/tailings decision to another day).

- While recirculation won't help these difficult particles, mixing them with feed may harm fresh particles. As a minimum, it will reduce residence time for new feed in both roughing and cleaning.

- This incurs a metallurgical cost, higher operating cost and more spillage compared with open-circuit. Since particle behaviour can't be changed, there can be no metallurgical benefit compared with open-circuit.

- The cleaner tailings will be higher volume and higher assay than it would be in open circuit, since the recycling particles are overrepresented by several times compared with new feed. This may cause operators to overstate the recovery loss if the cleaner tailings were to be open-circuited to final tailings. Once the circulating load is unwound, the same mass of particles – those that can't be put in concentrate -will be lost in either configuration. Similarly, samples taken in closed circuit, or models built from surveys, cannot represent open circuit operation.

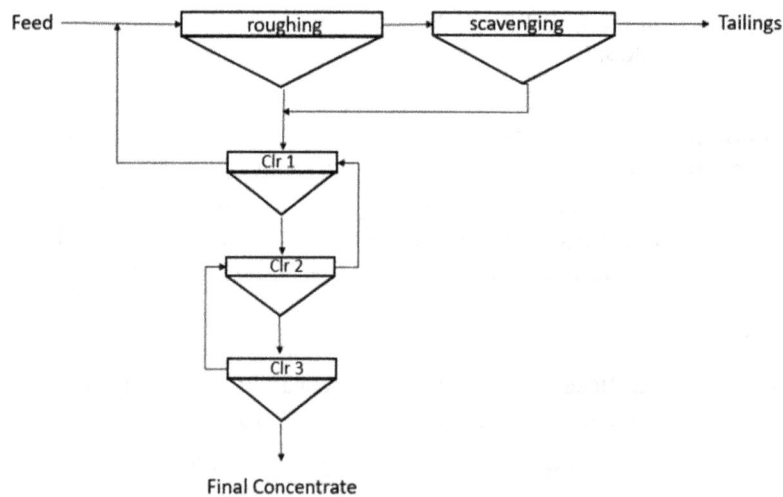

FIG 6 – Simple rougher-scavenger-cleaning circuit with recycle (with 'triple distilled' cleaning).

However, we later discuss exceptions that can justify at least partial circulation of cleaner tailings to improve operability. This includes improving froth stability at the end of scavenging, and the 'flywheel effect' to improve stability for fluctuating feed. If so, it is important to understand the source of improvement – it is not through recovering more particles from the circulating load, rather it is their effect on operating robustness. Once this is recognised, we can find lower cost ways to achieve the operability improvement rather than mixing the entire cleaner tailings with dissimilar fresh feed.

Recycling cleaner scavenger concentrate to regrind

Sometimes cleaner scavenger concentrate is recirculated as shown in Figure 7.

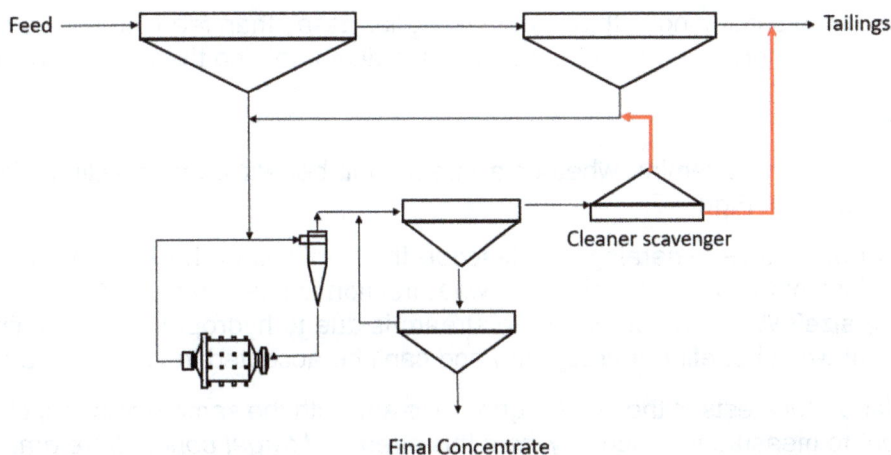

FIG 7 – Sending cleaner scavenger concentrate 'back to regrind' (two stages of cleaning).

The premise is that cleaner scavenger concentrate will contain composites that will be reground by recycling. This simple premise relies on important assumptions, which in our experience are rarely met:

- That this concentrate has a significant component of relatively simple, value-containing coarse composites – as distinct from highly complex composites (like Figure 2), slow-floating fines, and intermediate-floating contaminants.

- If it is dominated by simple composites, that those particles will report to hydrocyclone underflow, instead of going straight back to overflow, where they came from last time.

- That, for the minority that do report to underflow next time, that their size is reduced…

- …and this measurably increases the liberation of valuable mineral (ie they are relatively simple, coarse composites).

- That there is unused capacity in the regrind circuit, so that regrind power can increase to maintain the regrind size and liberation outcome for fresh feed, in spite of the extra mill feed.

This set of assumptions is virtually impossible to test in an operating plant (particularly within a circulating load that is created by the assumptions). But the onus of proof should be to prove why we _should_ recirculate this stream, not why we _shouldn't_, because:

- There is no reason to expect the particles will be significantly reground. We would not send crusher screen undersize back to crusher feed and expect a different result. Why would it work in regrinding?

- We have rarely seen a plant with excess regrind capacity; the correct level at commissioning is soon diminished by higher-than-design plant feed rate and (often) increasing ore complexity.

- The assumed mineralogy – that there are coarse, simple composites in this stream – is rarely shown or quantified.

- It violates the principle of mixing 'like' streams. Examining the mineralogy of these streams usually show they are quite dissimilar to fresh rougher concentrate.

Assuming that regrinding can accept ever-more feed but continue to achieve the design grind size and liberation outcome is a metallurgical Magic Pudding. Sadly, no one has yet designed the '_cut an' come again_' mill, so we have to expect that sending more feed to a full regrind mill has – _at best_ – no net benefit.

We think this unhelpful practice evolved for two reasons:

1. The guidance by Taggart and others to mix streams of similar _assay_. It is very likely that if Taggart had access to quantitative mineralogy and surface analysis, he would instead have said to mix streams with similar mineralogy and behaviour, not similar assay.

2. The desire to 'do something' – this stream is higher-assay than feed, surely we can't just send it to tailings? But there are more effective ways to do something than recycling without benefit.

A simple test

There are simple ways to examine whether a stream will benefit from recycling. For the cleaner scavenger concentrate in Figure 7:

- Use sized quantitative mineralogy to describe the particles in the stream. What is the size-liberation class of the valuable minerals, what fraction is likely to benefit from regrinding, and at what P_{80} size? What proportion of the stream is due to hydrophobic contaminants (eg iron sulfides) that won't benefit from recycling and can't be accepted in concentrate?

- Conduct laboratory tests at the plant regrind size and with the same number of cleaning stages as the plant to measure the recovery from this stream *at target concentrate grade*? This is the *best* the plant could do. (Note the test *must* be at target concentrate grade. Anything lower would reduce average plant concentrate grade, which would need to be compensated by increasing grade/reducing recovery from fresh feed).

Because the sample of cleaner scavenger concentrate is taken from within a circulating load, a sample is not *quantitatively* representative. However, it is *qualitatively* helpful. Complex particles like those in Figure 2 will be over-represented due to multiple recycles. Laboratory tests to determine what conditions are required to make a profitable contribution to concentrate. If applying the plant conditions in the ideal laboratory environment to enriched feed can't make profitable concentrate, then we can be certain that the plant can't either.

Staged regrind and clean

The mineralogy and laboratory demonstration described above will frequently demonstrate:

1. That there can be no benefit from sending the complex minerals in this stream back to the same regrind circuit.

2. However, the stream may benefit from a significantly finer regrind (and/or a significantly different reagent regime).

This is the simple principle behind staged grinding and flotation circuits. The finer regrind/different reagent regime may or may not be economic for this stream:

- If it isn't, then the stream should not be recycled (Figure 8).

- If it is, then use a staged processing circuit like Figure 9.

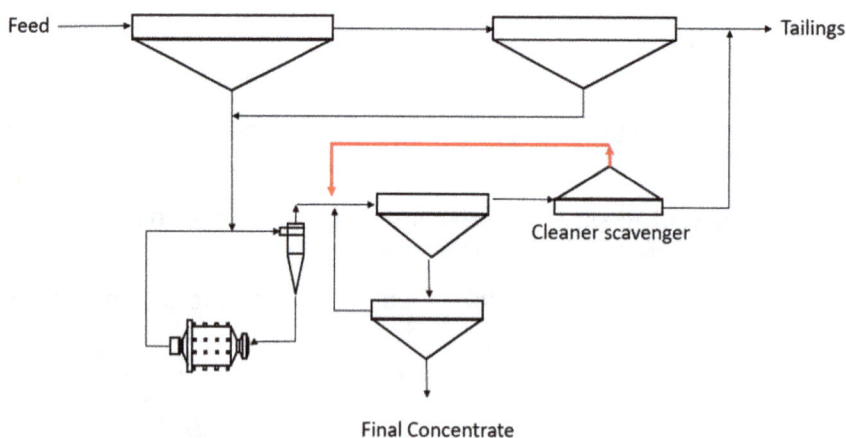

FIG 8 – Single stage regrind, open circuit cleaning.

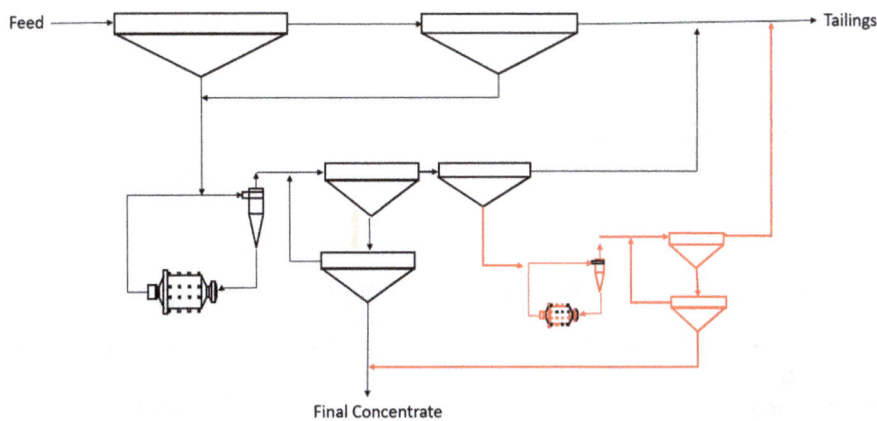

FIG 9 – Staged (finer) regrinding and flotation.

Such approaches have been increasingly adopted for complex ores since quantitative mineralogy was available from around the 1990s. We regard the designs in Figures 8 and 9 to be best practice. They have been widely adopted by operations including Brunswick, Phu Kham, Prominent Hill, Carrapateena, Mount Isa (lead-zinc and copper concentrators), Sentinel and, Escondida, and have been detailed by several authors (Shannon *et al*, 1993; Barns, Colbert and Munro, 2009; Bennett, Crnkovic and Walker, 2012; Lawson *et al*, 2017; Pease, Curry and Young, 2005). The approach satisfies good-practice in several ways:

- Either do something genuinely different to particles, or else don't recycle them.

- If you do something different (eg finer regrinding) only do it for the particles that need it.

- Process similar mineral size fractions and liberation classes together, in conditions that suit them. For example, the narrow size range in the second stage cleaners in Figure 9 means we do not have to separate unwanted 30 µm mildly-hydrophobic iron sulfide particles from 5 µm target minerals with a similar flotation rate.

Otherwise, recycling particles without changing them will – *at best* – use more energy, more reagents, and probably cause more spillage, for no benefit. At worst it will also harm recovery from fresh feed.

Ironically, these good-practice principles are the same as those historically used in coarse processing. Figure 4 (DMS) and Figure 9 (flotation) both show that streams are either stage-processed, or recycled after genuine size reduction. We postulate that flotation designs digressed from this good practice because, unlike coarse particles, we couldn't 'see' the minerals. We could tell ourselves that 'sending to regrind' was the same as 'regrinding'. This also conveniently reduced capital by not requiring additional regrinding. We would never return fines to an earlier stage of crushing; now that we can 'see' fine minerals there is no excuse for making this mistake in flotation.

Wills and Finch (2016) revisited work in the 1980s at Brunswick Mine originally reported by Shannon *et al* (1993):

> *...sending cleaner scavenger tailings to final tailings removed material that actually decreased performance if recycled to the rougher...the circuit modification was a major simplification...During the 1990s the Noranda group moved to adopt (this) circuit across operations.*

Finch and Tan (2021) compared several configurations and concluded that *"especially for complex ores, the open version is favoured over the closed, in addition to isolating roughing from cleaning which aids circuit control"*.

We have worked with many operations that were designed with, or changed to, open-circuit cleaning (sometimes with staged regrinding). We have never seen an operation that wanted to change back.

EXCEPTIONS

Froth stability

Sometimes circulating cleaner tailings to roughing improves froth mobility in scavenging. This can assist the recovery of low-grade coarse composites, thereby improving recovery (so long as the recovered composites can be effectively reground before cleaning). While this is a genuine benefit, there are lower-cost ways to achieve it:

- Using froth crowders in scavenging to improve froth mobility.

- Adding a full or partial recycle stream to the final cells of scavenging, rather than to rougher feed, as shown in Figure 10. This improves froth mobility where it is needed, without unnecessarily consuming capacity for most of roughing and scavenging.

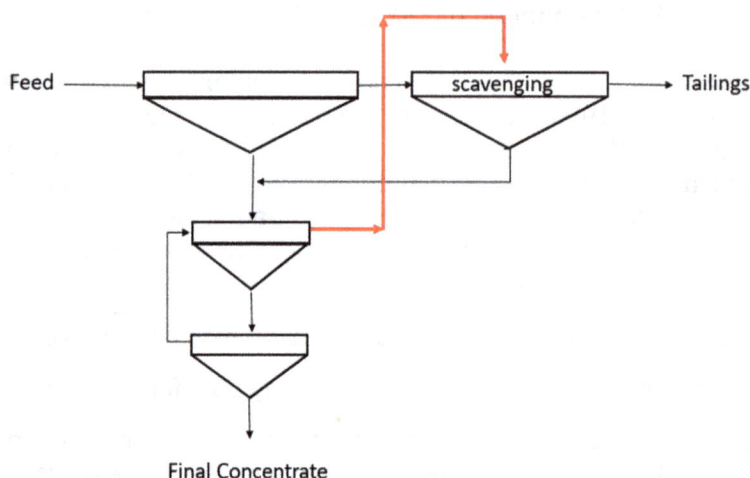

FIG 10 – The 'froth stability' and 'Augustinian' recycle.

This circuit also provides the 'flywheel effect' if it is needed, for example in a plant with highly variable feed but only one stage of cleaning.

The Augustinian recycle

We name this after the Augustinian doctrine of the lesser evil.

Though open-circuit cleaning is best practice in most cases, a visible tailings stream that is higher grade than plant feed will inevitably attract management attention. Ideally, this will create an opportunity to explain mineral size-liberation classes and circuit design principles. To demonstrate how a 0.2 per cent scavenger tailing plus a small mass of 2 per cent cleaner tailings is the same metallurgical outcome as a 0.22 per cent combined closed circuit tailings, but it uses less energy.

Nevertheless, the siren call of the 2 per cent stream may prove irresistible to management. You may be dashed on the rocks of harmful circuit changes, reagent changes, or management changes – silver bullets or golden handshakes. In that case the lesser evil may be to adopt a circuit like Figure 10. The metallurgist will still work to improve the performance of the now less-visible particles in combined final tailing. Such recalcitrant particles may respond to a rigorous technical program; they will not respond to management decree.

MODELS

Lauder (1992) undertook an extensive review of the recycle mechanism in flotation and the use of models to analyse the resulting networks. He concluded that:

> it is important to optimise the network containing a circulating load rather than the stages within that network. The circulating load …is symptomatic of the network … as is the performance (of) that network.

That is, a model can be useful to _analyse_ a circuit, but the underlying mineralogy should be used to _design and optimise_ it. Models will accurately predict the outcomes of circulating loads when rate constants are consistent and are not changed by the recycle. This applies to circulating between cleaner stages (or whiskey distillation).

However, they can _not_ accurately predict the outcomes when:

- mineral liberation or surface chemistry is deliberately changed during recycle

- or surface chemistry is unintentionally changed by the recycle (eg surface oxidation, pulp potential changes during regrinding)

- or the recycle affects fresh feed – eg 'sending to regrind' reduces the power available to fresh feed.

We have not seen a model that can accurately predict the changes in Figure 11 – changes to size, liberation and surface chemistry for both the recycle stream and the receiving stream. The best way to estimate the effect of the physical and chemical changes on all mineral classes is from laboratory tests supported by quantitative mineralogy to 'see' and quantify what actually happens.

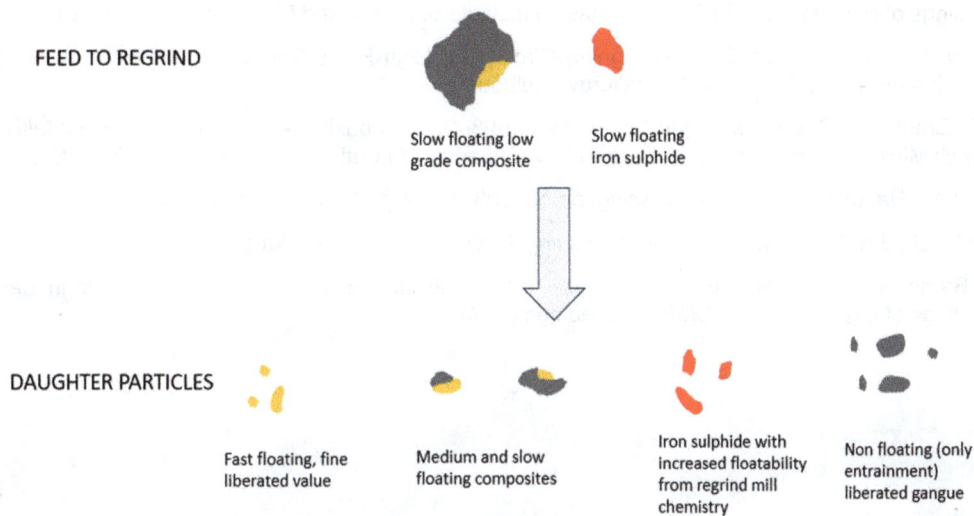

FEED TO REGRIND

Slow floating low grade composite

Slow floating iron sulphide

DAUGHTER PARTICLES

Fast floating, fine liberated value

Medium and slow floating composites

Iron sulphide with increased floatability from regrind mill chemistry

Non floating (only entrainment) liberated gangue

FIG 11 – Model this?

CONCLUSIONS

The most fundamental test before recirculating a stream is: if you can't make some profitable product from it in the laboratory (using plant conditions), then it is tailings. If so, the sooner the better.

Repeating a process without fundamentally changing particles is only helpful when there is a sustainable and repeatable difference in rate constants between the wanted and the unwanted. This applies to multiple stages of cleaning as it does to whiskey distilling.

Circulating a particle to improve its performance should only be done if it is _genuinely_ changed for the better. Classification and coarse separation circuits routinely selectively return the coarser particles for further comminution, or separately crush middlings before reprocessing.

Because of a historic inability to easily 'see' fine particle mineralogy, flotation is the one field in mineral processing that circulates particles without significantly changing them. A particle that is 'sent back to regrind' will probably exit immediately as hydrocyclone overflow (as it did last time). In the smaller chance it gets to the regrind mill, it is unlikely to get significantly better liberation – but it will consume milling and classification capacity that would be better directed to fresh feed particles.

Now that we can more easily 'see' and quantify particle mineralogy, it is clear that recirculating cleaner tailings to roughing, or sending a stream 'back to regrind' is unhelpful at best, and often harmful.

Current best flotation practice uses open circuit cleaning, sometimes enhanced by staged regrinding and cleaning. This echoes the designs for coarse separation technologies – only recycle if you can genuinely change the mineral liberation, and if it recycling doesn't adversely affect fresh feed. Ideally, process different size and mineral classes in separately – coarse and fine DMS cyclones, or staged regrinding and flotation – rather than in one mixed 'soup'.

There are exceptions where at least partial recycle is justified to assist operability and stability – and sometimes to simply placate management. When they are necessary, use these exceptions sparingly, knowingly, and in a way that minimises adverse effect on fresh feed.

REFERENCES

Barns, K E, Colbert, P J and Munro, P D, 2009. Designing the Optimal Flotation Circuit – The Prominent Hill Case, in Proceedings of the Mill Operators Conference (The Australasian Institute of Mining and Metallurgy: Melbourne).

Bennett, D, Crnkovic, I and Walker, P, 2012. Recent Process Developments at the Phu Kham Copper-Gold Concentrator, Laos, in Proceedings of the 11th Mill Operators Conference (The Australasian Institute of Mining and Metallurgy: Melbourne).

Finch, J A and Tan, Y H, 2021. A comparison of Two Flotation Circuits, *Minerals Engineering*, 170:Aug 2021.

Lauder, D W, 1992. The Recycle Mechanism in Recirculating Separation Systems, *Minerals Engineering*, 5(6).

Lawson, V, Muller, M, Radulovic, P and Wallace, J, 2017. Mt Isa Copper Concentrator Cleaner Circuit Redesign, in Proceedings of Met Plant 2017 (The Australasian Institute of Mining and Metallurgy: Melbourne).

Pease, J, Curry, D and Young, M, 2005. Designing Circuits for High Fines Recovery, in Centenary of Flotation (The Australasian Institute of Mining and Metallurgy: Melbourne).

Shannon, E R, Grant, R J, Cooper, M A and Scott, D W, 1993. Back to basics—The road to recovery: Milling practice at Brunswick Mining, in Proceedings of the 25th CIM Canadian Mineral Processors Annual Operators Conference.

Taggart, A F, 1945. *Handbook of Mineral Dressing*, pp 12–103; 12:22; 12:96 (Wiley: New York).

Wills, B A and Finch, J A, 2016. *Wills' Mineral Processing Technology* (Elsevier: Amsterdam).

Young, M F, Barns, K E, Anderson, G S and Pease, J D, 2006. Jameson Cell: The 'Comeback' in Base Metals, in Proceedings of the 38th Annual CMP Proceedings (CIM).

Rockmedia

M S Powell[1], A N Mainza[2], L M Tavares[3] and S Kanchibotla[4]

1. FAusIMM, Liner Design Services, Brisbane Qld 4069; Emeritus Professor, JKMRC, Sustainable Minerals Institute, University of Queensland, St Lucia Qld 4068. Email: malcolm@milltraj.com
2. Professor, CMR – Dept of Chemical Engineering, University of Cape Town, Rondebosch 7700, South Africa. Email: aubrey.mainza@uct.edu.za
3. Professor, Department of Metallurgical and Materials Engineering, Universidade Federal do Rio de Janeiro, COPPE-UFRJ, Rio de Janeiro RJ, Brazil. Email: tavares@metalmat.ufrj.br
4. Seshat Consultants, Pullenvale Qld 4069. Email: sarma@seshatconsultants.com

ABSTRACT

The rock content of semi-autogenous grinding (SAG) mills is dominated by the fraction of coarse and competent rock in their feed with variations in this rock fraction driving fluctuations in mill filling and throughput. This is generally seen as a limitation of SAG milling, with a strong industry focus on minimising coarse rock in the feed. However, overly aggressive size reduction or fluctuating blends of harder and softer ores leave SAG mills operating well below installed power, while overloading ball mills. The 'Rockmedia' technique is proposed to help redress this underperformance by using relatively simple and low-cost mill utilisation and control methods. The technique is to maintain an independent feed source of rock in the media size range 90–250 mm, and to have direct control of this feedrate. Competent feed is diverted to a buffer stockpile or a bin to be direct-fed to the mill feed belt as an independently controlled bleed-in, according to current mill filling. The required coarse rocks can be sourced by processes such as using blasts designed to preserve a coarse fraction; rehandling from the stockpile perimeter; deliberately biasing feed delivery off conveyors; screening out a top size; or diverting recycle pebbles. Rockmedia can significantly increase productivity in existing SAG circuits and be built into future circuits to maximise equipment utilisation while reducing specific milling energy through more efficient and stable operation, and reducing indirect steel grinding media energy usage, both in the SAG mill and downstream ball mills.

MOTIVATION

Semi-autogenous grinding (SAG) mill rock content is dominated by the fraction of coarse and competent rock in the feed. This is generally seen as a limitation of SAG milling which is addressed through finer blasting (Mine to Mill concept) or blending and full pre-crushing to remove or limit this fraction in the feed. However, the authors have repeatedly observed SAG mills operating well below installed power and ball mills suffering overload due to a lack of rock in the SAG mill feed that arose either from aggressive feed preparation or fluctuating blends of harder and softer ores. This was described in the work of Powell et al (2021), where under-utilisation of SAG mill power caused the chronic overload of the downstream ball mills. The consequences of a full pre-crush were also addressed in the work of Powell, Hilden and Mainza (2015). Their review noted that the generic issue of inadequate larger rocks in the feed leads to an inability to build-up a competent rock grinding media. The mill instead builds up a charge of smaller, critical-sized rocks causing it to lose total filling. This creates the need for a new liner design to prevent liner and ball damage, which forces the mill to be operated at a slower speed, losing power draw from both a lower filling and speed. The ball load is then increased to compensate for the lack of coarse rock load which leads to increased ball consumption in the SAG mill and considerable added cost. The work of Guzmán and Sepúlveda (2016) (Figure 1) shows an expected doubling of ball consumption as the fraction of balls in the mill charge increase from 0.43 to 0.71.

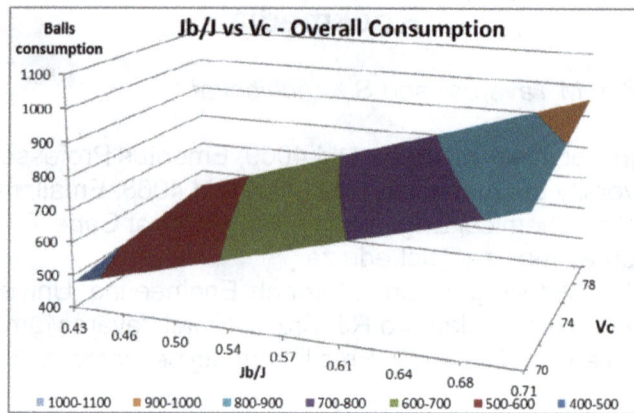

FIG 1 – Modelled relationship of ball consumption with the ball to total charge fraction (Jb/J) (Guzmán and Sepúlveda, 2016).

The reality is that there is a significant dip in SAG mill efficiency in the valley between SAG and ball mill operation. The SAG mill suffers a significant loss of efficiency in the region of too-fine-for-SAG milling and too-coarse-for-ball milling. It is well known that feeding oversize to ball mills reduces capacity, coarsens product, and leads to severe scatting issues (Powell *et al*, 2021; Ballantyne *et al*, 2015; Tavares and de Carvalho, 2009; de Carvalho and Tavares, 2013). In essence, this is the same problem mill operators face when the SAG mill is repurposed for use as a coarse-feed ball mill.

The industry is drifting towards the 1960s practice of three-stage crushing followed by two-stage milling, in an effort to climb out the efficiency dip, which requires ever-more feed-size reduction. Although the massively larger equipment, with throughputs over 1000 tonnes per hour (tph), reduces the number of parallel processing trains required, this still increases stages of reduction and classification, recycles, conveying, and number of comminution units.

The alternative is to return to operating SAG mills in their sweet spot, of a bimodal feed with a rock load fraction of at least 0.5 of the total mill load. This paper explores this option using the 'Rockmedia' approach, which reduces operating cost and CO_2 footprint while making it more robust against fluctuating metal prices and consumable costs. This is primarily achieved through stabilising mill operation, which allows greater utilisation and improved recovery.

The other aspects that compromise SAG mill performance and motivate the shift to ever-finer feed are variability in feed and varying blends of feed ore which both lead to fluctuating performance. Again, the finer-feed, high-ball-load approach offers a solution to this by greatly reducing the impact of feed variation on mill performance. The Rockmedia approach also addresses this.

THE SIMPLE SOLUTION

The process team on any mine is always seeking to open up capacity. The focus usually falls on the SAG mill as the obvious bottleneck. Standard approaches to increase throughput of the SAG mill include:

- higher blast intensity to produce a finer feed
- top-size pre-crush
- full pre-crush through secondary cone crushers
- higher filling and larger top-size of balls
- large pebble port size and open area to increase the contribution of recycle crushing
- increased recycle crushing capacity, potentially with a new recycle crusher station
- larger balls in ball mills
- added ball mills to compensate for coarser grind and higher throughput from the SAG circuit.

Most of the above strategies work adequately when the process plant is SAG-mill constrained, but they are not effective when the circuit is ball mill constrained. Generally, these strategies result in a coarser overall grind and some loss in recovery, but 'throughput is still king', so reduced recovery is usually glossed over when the successes are reported. It was noted from reviewing the application of aggressive blasting and full pre-crush work that the impact on overall grind and recovery barely gains a mention.

An added ball mill and recycle crusher are in the above list, as knock-on needs. These are extremely expensive upgrades that minimise the need for increased SAG mill capacity. Such an outcome is to be expected from incremental upgrades, as the SAG mill is a large and expensive single unit that is unsuited to a staged upgrade in 20 per cent increments.

As noted earlier, the SAG mill builds up a critical size charge that needs to be addressed. The solution to the build-up of critical size rock in the mill appears to be to crush and blast even finer. The use of the largest cone crushers, around 1 MW, is applied to this pre-crush, such as presented by Lee *et al* (2015). However, large cone crushers are limited in the degree of reduction they can achieve from the coarse primary-crusher product, so they are limited to a closed-side setting (CSS) of about 50 mm, although more careful feeding and chamber design can push this lower. The SAG mill then operates more like a ball mill with excessively coarse feed and a low rock load, in the region of 3–8 per cent mill volume. In this state they are commonly known as a barely autogenous grinding (BAG) mill. This coarse feed to the BAG mill exacerbates the build-up of critical-sized rock in the mill—necessitating pebble porting and recycle crushing.

A high-pressure grinding roll (HPGR) can effectively reduce the feed to a fine size suited to ball milling. The upgrade at Cadia mine pioneered the integration of a HPGR into the pre-crush circuit, as presented by Engelhardt *et al* (2015). Originally the HPGR designed to receive a large fraction of the fresh feed crushed to minus 50 mm by two large, secondary cone crushers (MP1250s) in a closed circuit. In essence the circuit is a three-stage crush, two-stage milling process. HPGRs are expensive, so their installation should not be taken lightly. The operation initially suffered from issues of low SAG filling, liner issues, and HPGR underutilisation that compromised the effectiveness of the upgrade (Engelhardt *et al*, 2015). Waters *et al* (2018) documented a concerted plant-wide effort from commissioning the equipment in 2012 to reaching desired equipment utilisation and capacity in 2018.

The alternative of more aggressive blasting leads down a similar route—pushing coarser rocks into the mid-size range, which still require pre-crushing to meet the objective of a far finer feed. However, unlike crushing, there is the added benefit of greater fines (-2 mm) produced by higher blasting energies.

STATUS QUO?

Examples of poor circuit utilisation are provided to illustrate the drawbacks of moving towards a finer feed as the only pathway to increased production.

Increasing ore competence

The example in Figure 2 is compiled from the data of a few large copper-ore milling circuits, with observed data as a realistic example of what can be seen in operation. What really stands out is the underutilisation of circuit power, with 13 MW of available power not used.

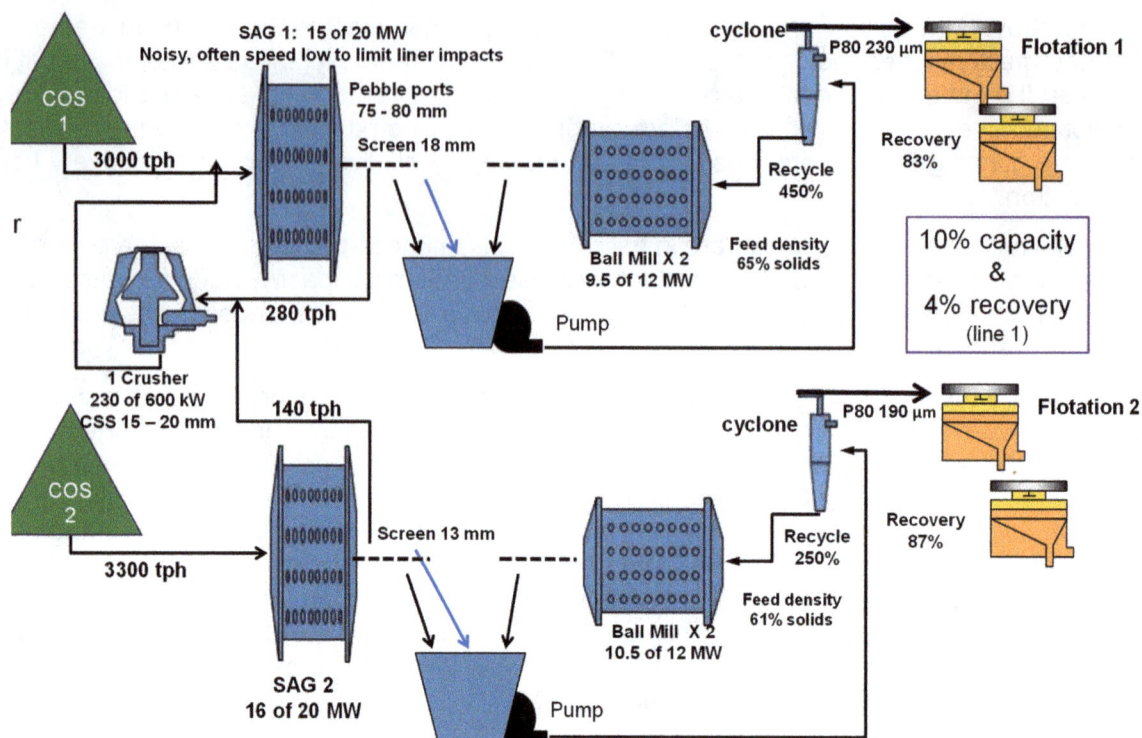

FIG 2 – Underutilised SAG milling circuit: 13 MW not used out of 84 MW installed power.

In the case of the above comminution circuit not achieving the desired throughput, it is not unusual to be asked by the process team to help specify a new ball mill and pebble crusher, when, instead, improved operation should be able to recover the 13 MW equivalent. Utilising the SAG mill can be quantified using the overall equipment effectiveness (OEE) approach (as described in the work of Powell *et al*, 2022) of plotting the operating data distribution from lowest to largest values over a period, as shown in Figure 3. The largest piece of equipment on the mine had 19 per cent unutilised capacity. This should stand out as a critical area to be addressed; however, all too often this underutilised power appears to be completely overlooked as a perceived state of affairs that cannot be altered.

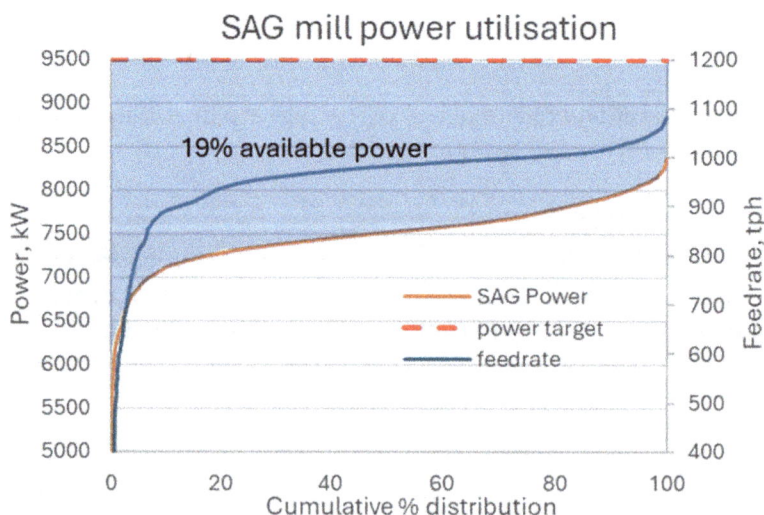

FIG 3 – OEE visualisation of underutilisation of power in a rock-depleted SAG mill.

Figure 4 shows a large SAG mill with 15 per cent available power arising from fluctuating operation and inadequate control. Hidden in this distribution is the widely varying applied grinding energy given by the specific energy (SE) presented in Figure 5. This ranges from 5 to 20 kWh/t, but mainly from 5.5 to 8.5 kWh/t. The knock-on effect is a wide range in circuit product size, as shown in Figure 6.

The 80 per cent passing (P_{80}) varies between the two cyclone clusters (fed from one ball mill by two pumps on the common SAG-ball mill sump – which has inherent segregation issues) and across the range from 200 to 350 μm, which is excessive for a recovery objective and definitely detrimental for recovery.

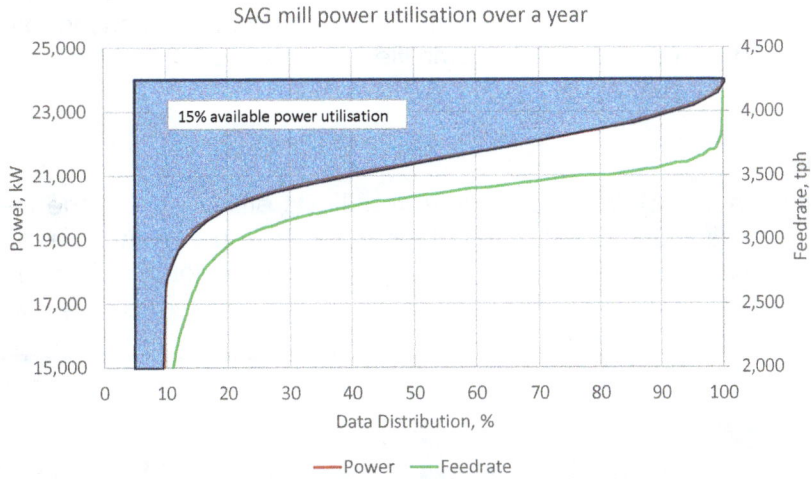

FIG 4 – SAG mill utilisation on 24 MW mill.

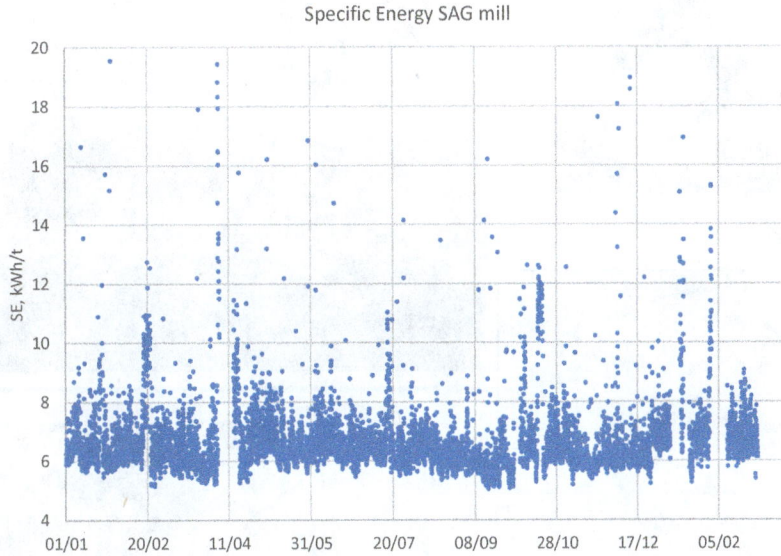

FIG 5 – Variability in specific energy delivered by the SAG mill over a year.

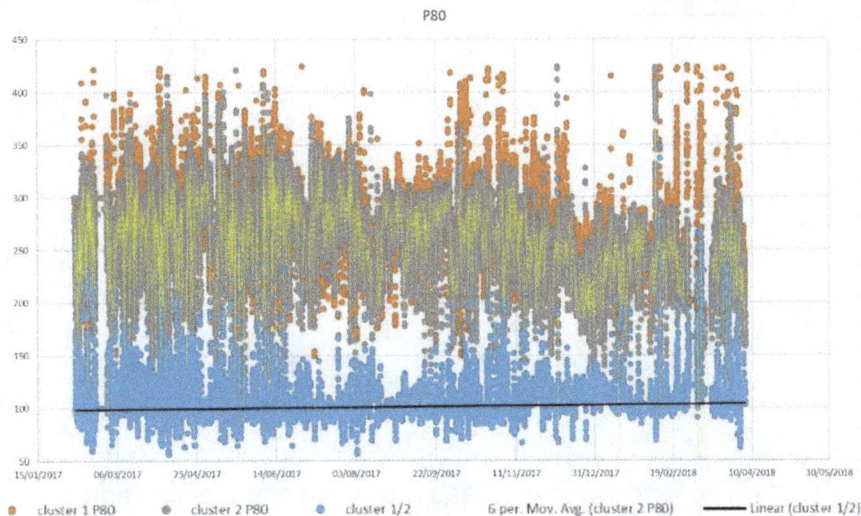

FIG 6 – P_{80} values for SAG mill with two cyclone clusters.

Reduced power draw in a SAG mill can arise from other factors, such as reduced speed with inappropriate new liners, leading to a run-in period and unknown mill filling especially after relining leading to low filling. However, the mills should still achieve target power draw for most of the life of the liner, and as illustrated in the operating variability above, this variation is a clear signature of underloading and variation in mill rock load. The source of variability in product size in this example, and many like it, is the ore variability linked to mill filling variation. The consequent variation in grind size can be dampened by good control or amplified by over-control, but the source remains the variation in the SAG mill performance.

Feed variability

It is not just a depletion of rock in the SAG feed that limits efficiency, but the well-known issue of variability in feed size-distribution and competence. Figure 7 presents an example of the cycle between feed variation and circuit operation, ending in a variable flotation feed size distribution and feed rate.

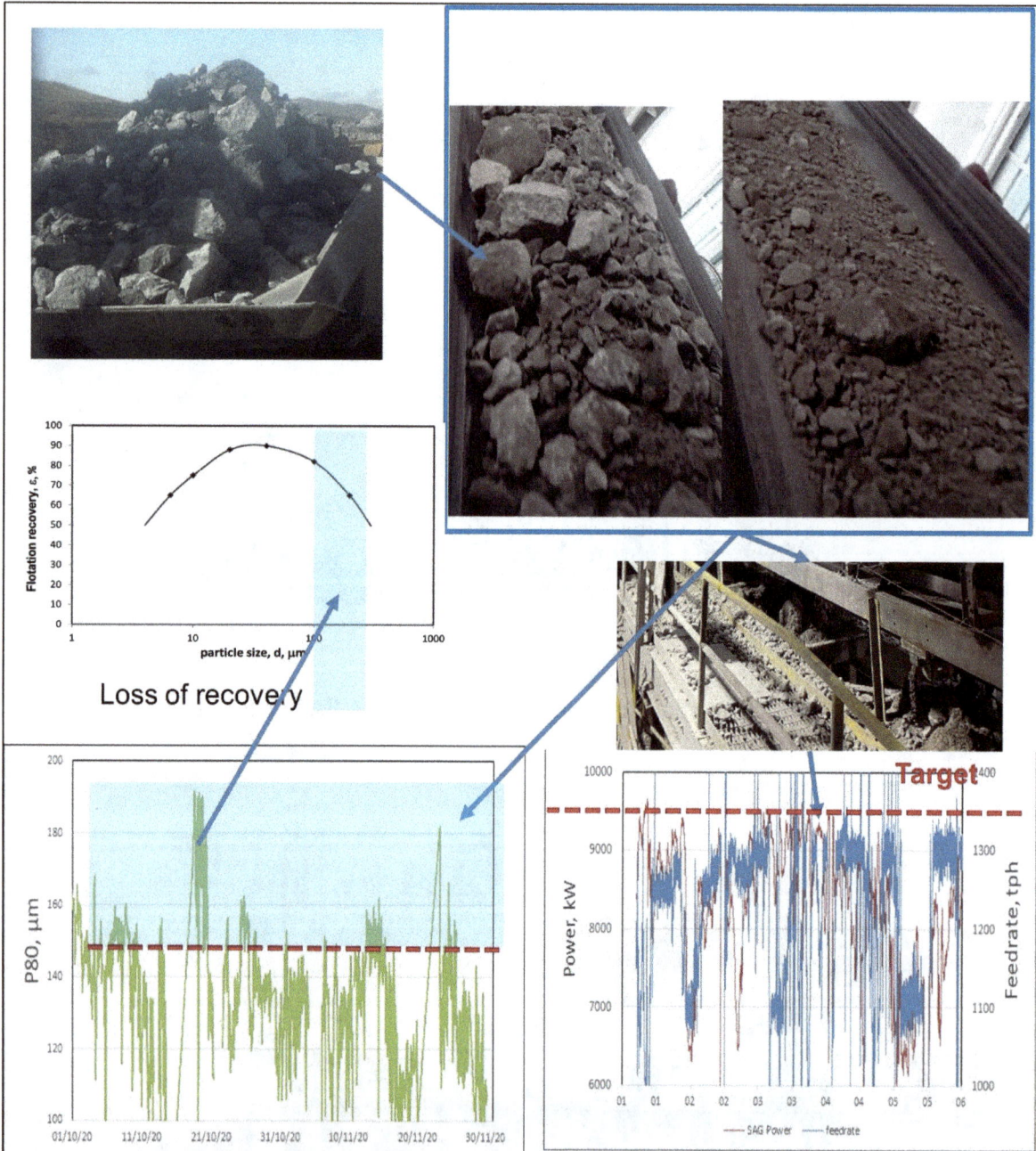

FIG 7 – The vicious cycle of feed size and competence variation.

Figure 7 presents the vicious cycle of:

- cyclic variation in feed size from lack of coarse rock to excess coarse
- wide variation in pebble recycle around the mill, which leads to
- ineffective removal of ball scats during overload periods
- conveyor overflow
- opening of recycle crusher CSS to protect it from damage
- frequent bypass of feed to reject steel not rejected at the magnet and remaining on the feed belt to the crusher
- rapid and ongoing variation in mill filling, feedrate, and power
- large variation in product size
- losses in recovery.

The cyclic feed size is exacerbated by the common tendency to doze stockpiles when they run low. The typical impact is illustrated through the screenshot of a control room monitor in Figure 8 for two parallel SAG mills. Due to feed shortages (arising from primary crusher hold-ups) the larger mill stockpile (upper screen) is being dozed, while the other is not. The difference in mill performance is obvious from the wild fluctuation in one mill while the trends for the parallel mill are stable. The impact of this on grind size and recovery are well understood. It would be far preferable to follow the well-understood practice of stabilising mill load and a balanced SE between the two circuits. It is also likely that in this case, the control system was exacerbating the cyclic filling through over-control.

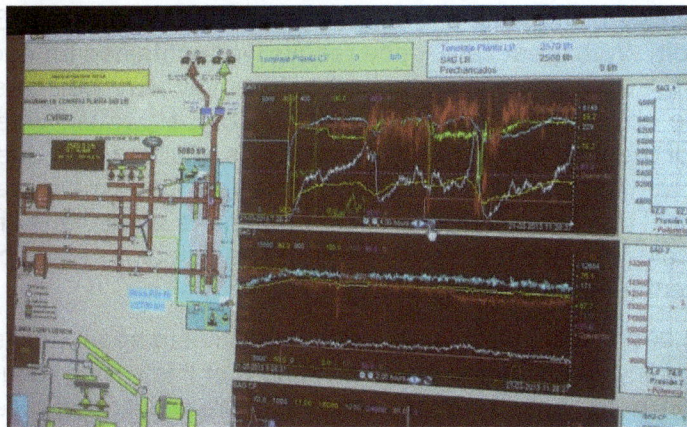

FIG 8 – Illustrating mill response to stockpile dozing.

Blend variation—hard and soft components

Blend variation leads to a high sensitivity to the fraction of competent component, as measured and described in the work of Bueno *et al* (2011), which highlights the 'dominance of the competent' component. The plot in Figure 9 shows how the more-competent silicate built up in laboratory, pilot, and industrial mills leading to situations such as 75 per cent silicate in the mill for 50 per cent in the feed. The authors remember well an unpublished example of an autogenous mill receiving UG2 platinum ore. UG2 is black and friable, while the hanging walls and footwalls are a white quarzitic rock. During some underground development work, additional waste quartz was hauled with the ore, due to shaft limitations. This increase in quartz fraction was around 9 per cent to 15 per cent. The mill manager phoned in a panic, as the mill's throughput had halved. He was advised to crash-stop the mill and inspect the contents. Only white quarzitic rock was evident. The added 6 per cent silica in the feed had tipped the mill content over to being dominated by the competent quartz, and the milling rate dropped to a value close to the throughput of quarzitic rock alone. There is a lot to be learned from such an experience. The mill required competent rock media, but clearly not too much—thus the quartz fraction required control. On another site milling the same ore, the operator could not maintain mill load, as there was insufficient coarse, competent rock in the feed. Faced with a choice of cooperating with the mining group to increase the percentage of hanging wall in the feed by about 5 per cent, they chose to instead transition to run-of-mine (RoM) ball milling, now often referred to as BAG milling.

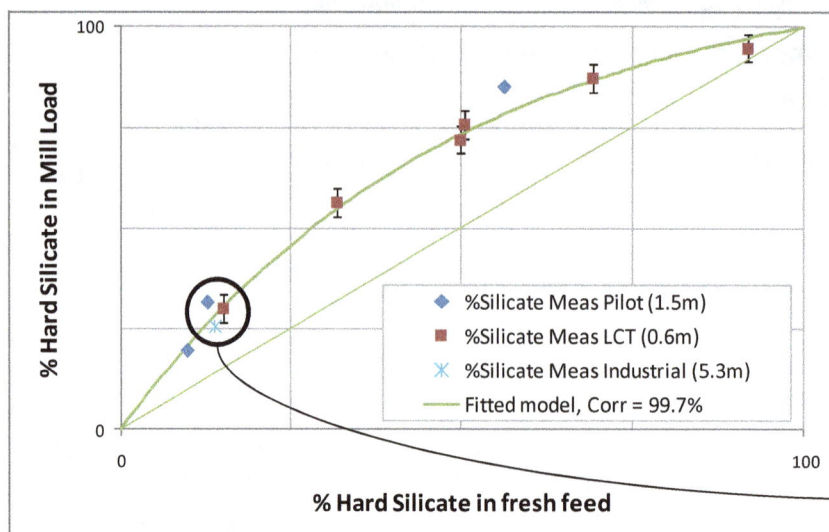

FIG 9 – Build-up of competent rock in a mill with a feed blend (Bueno *et al,* 2011).

This sensitivity of SAG performance to feed blend of competent fraction can be particularly dramatic when conducting pilot-scale tests. Indeed, the authors designed and overlooked a campaign in a 6 foot-diameter mill with iron ores which varied in A*b values from as low as 54 to as high as 370 (Rodrigues *et al*, 2015, 2021). Typical values of A*b of ores range from 35 (competent) to 50 (medium) to over 80 (soft). With great care, good results were obtained for each ore, but in one test with a very soft ore (A*b > 200), the mill performed surprisingly very well, in spite of the fact that it was not expected to be able to generate autogenous media. However, when the mill was finally interrupted at the end of the test, it became evident that its success was due to a few tough large lumps that were mixed with the feed and that stayed for a very long time in the mill, evidenced from its exceptional rounding. This shows that this effect of blend is particularly critical when designing and carrying out pilot-scale tests at the greenfield stage. The outcomes also point to the dependency on a fraction of competent feed rock when treating soft ore.

The authors have often been asked why a mill is over-sensitive to apparently small changes in feed blend, as reported according to the mined ore, indicating that this strong response of a SAG mill to feed blends that contain a competent fraction is a widespread issue. These data and examples outline the challenge and opportunity of ore blending when dealing with distinctive ore types which require operators at all times to closely control the competent rock percentage in the feed over short time intervals, not sporadically based on a long-term average. Batching or surging of a hard or soft component is to be avoided by designing it out of the process.

In another example, a circuit receiving feed from seven separate shafts, the process geologist reported that they attempt to balance the grade, but that they have not managed to concurrently balance feed competence from the widely varying feed ore types. Although it is preferable to include greater control through scheduling, this can be too constraining, or not met with support, from mines that are struggling to achieve production targets. The high fluctuation in these blends leads to an almost uncontrollable mill, as illustrated in the operating trends of Figure 10. The variability in the feed competence and attendant circuit instability is beyond the control system's ability to properly address. Feed instability cannot be fixed or solved in the control system, it can only be dampened to reduce the negative impact; often, instability is amplified rather than attenuated in the milling circuit— due to the sensitivity of the SAG mill to variations in feed ore, and control systems that are over-reactive. This feed variability is handed on to the flotation plant as a wide variation in feed size, slurry percent solids, and feed rate—affecting residence time, viscosity, and particle size-distribution. These factors shift the system away from the ideal range of operation, leading to a continuous loss of recovery relative to what is achievable.

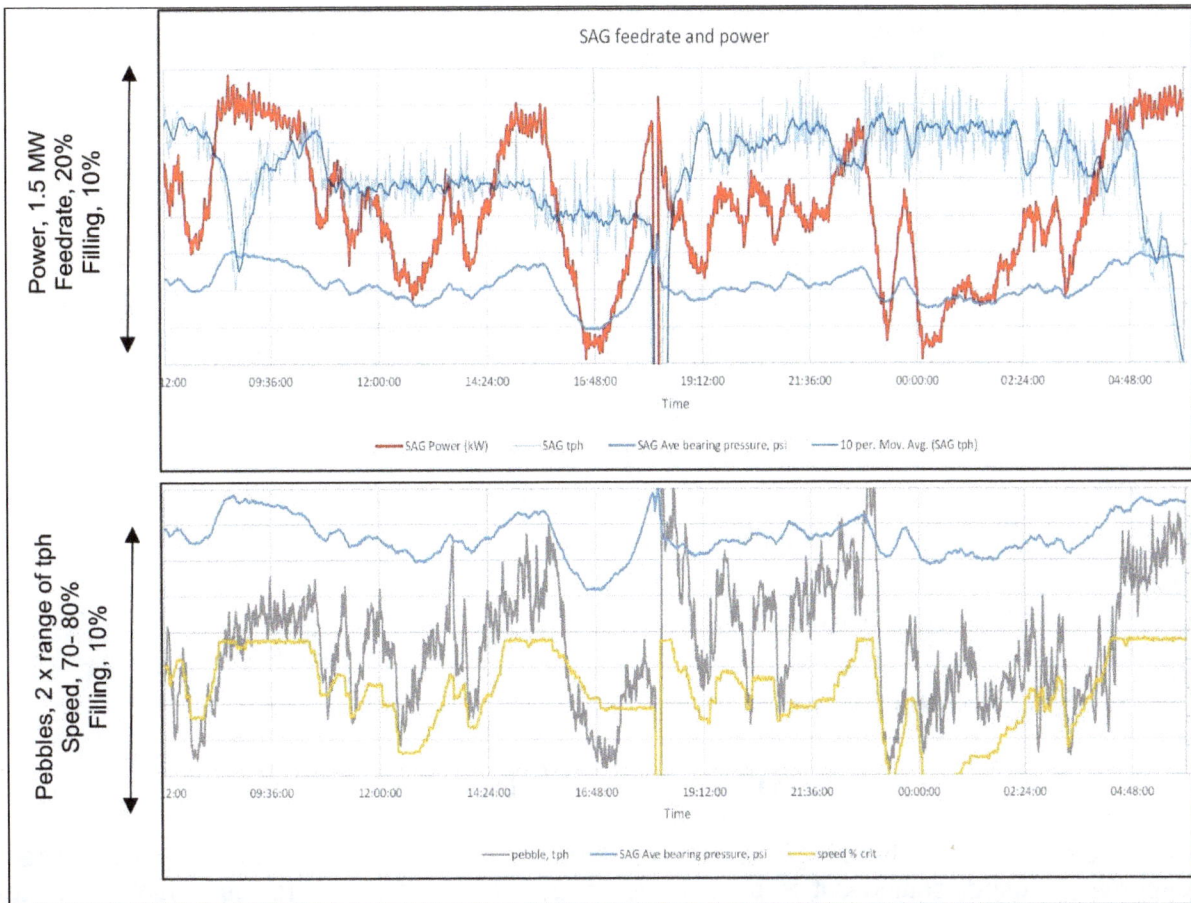

FIG 10 – Daily fluctuations in feed rate, pebble discharge, power, and load.

The operating plots in Figure 10 illustrate the described wide variation in SAG mill performance over hours and even minutes, which results in rapid power cycling up to 2 MW, and commonly cycles of over 0.5 MW, with major changes in feed rate occurring throughout the day. Overloading is common, with sudden step-downs in feed rate (end of trend), and short grind-outs (10 min) an almost daily occurrence. In Figure 10, fluctuations in pebble discharge are shown in the lower trends. The variability in SAG mill operation is passed on to the ball mills and flotation circuit.

POTENTIAL SOLUTIONS

Having highlighted some of the issues facing mill operators with respect to stabilising mill load, we present some techniques that can potentially redress the problems.

Blasting to create optimum feed size for sag mills

The primary crusher and blasting processes influence SAG mill feed-size distribution (McKee, Chitombo and Morrell, 1995). Generally, primary crushers do not produce much fine product (ie −10 mm), and most of the fines in the SAG feed are produced during blasting (Kanchibotla *et al,* 1998). The duty of primary crushers is to reduce the top size from the run-of-mine (RoM) ore. Passing an inappropriate feed-size distribution to the SAG mill reduces its throughput and induces operational instability because the mill charge and grinding media are obtained from the feed coarse fraction, which may vary widely, with what appear to be minor variations in overall RoM distribution. The ideal size distribution for a SAG mill depends on the breakage characteristics of the ore (rock strength) as well as the operating conditions of the mill (Diameter, length, lifter design, grate design, mill speed and rock charge). A well-designed Mine to Mill blast combined with appropriate primary crusher settings and controlled blending can produce a feed size with lots of fines and controlled top size (media), but it has little critical size control, which can sometimes negate the benefits of additional fines. However, if the circuit has enough pebble-crushing capacity to crush the critical size, the negative impact of feeding additional critical-sized material can be negated.

As a general strategy to produce optimum feed for SAG and AG mills, ore blasts should aim to maximise fines (−20 mm) generation, minimise the critical size (20–80 mm), and maintain the media (+100 mm) around 10 per cent–15 per cent (Figure 11a).

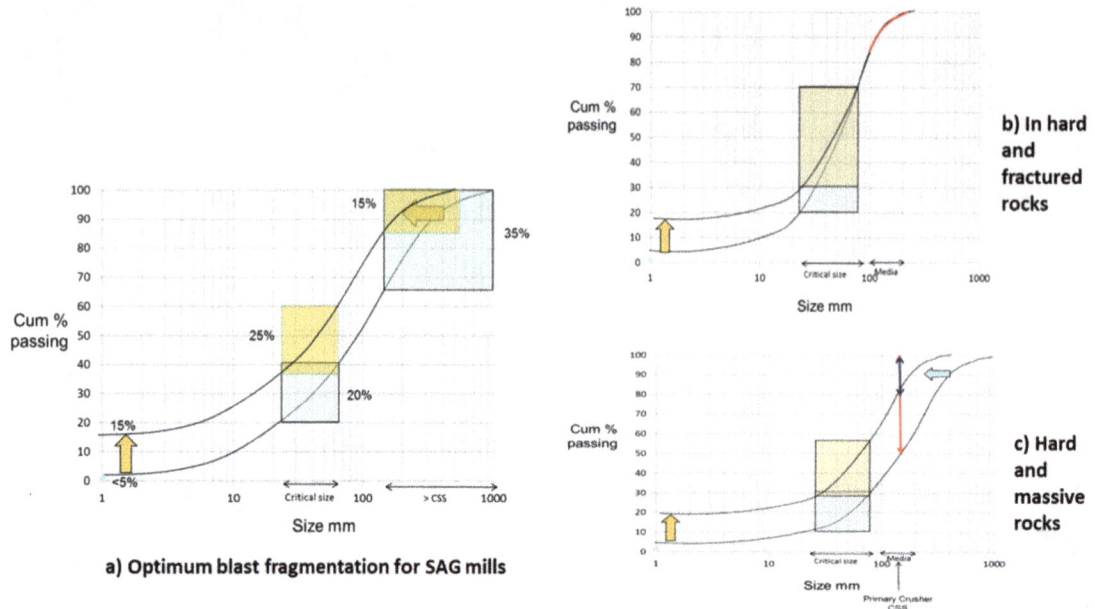

FIG 11 – Ideal blast-size distributions.

In hard and massive ores there is generally enough rock media; therefore, by adjusting blast patterns and primary crusher settings, it is possible to reduce the rock media to less than 15 per cent (Figure 11c). In highly fractured ores, maintaining rock media may be difficult because it is governed by the *in situ* size distribution. Hence, blasting strategy in these ores should aim to increase fines and retain *in situ* grinding media as much as possible by optimising the stemming length (Figure 11b).

In soft and friable ores, it is difficult to preserve rock grinding media because the rock lacks a coarse *in situ* fraction. If the *in situ* rock structure is fine, it will naturally blast to that *in situ* pattern; this cannot be avoided. Hence, rock grinding media should be obtained by controlled blending of coarse, hard ores with the soft, fragmented ore. In such cases, it is necessary to stockpile the rock grinding-media from low-energy trim blasts, blast edges, and carefully designed blasts in hard ores. The stockpile rock grinding-medium can be blended with the fine soft feed in a controlled manner to balance the power between SAG and ball mills (Figure 12).

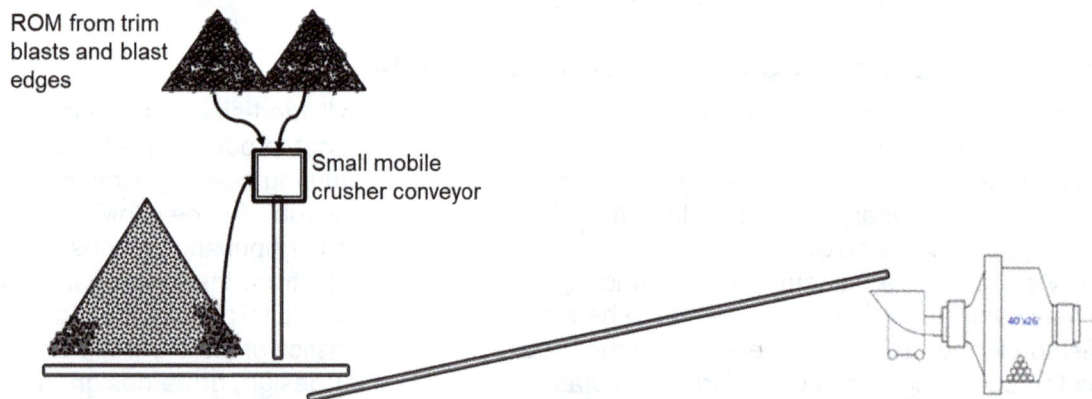

FIG 12 – Utilising preserved competent rock.

Blending

Controlling the fraction of coarse rock fed to a SAG mill is confounded by the variable range of rock competence being mined and delivered to the mill. The required fraction of top-size varies, but with a fixed blast-pattern and energy, the softer rock— needing more coarse material—will be blasted finer and result in a depleted rock grinding-media in the mill when more is in fact required. Thus, there is a need for a variable blast based on ore type. Blending of ore by grade is standard, but by competence is rather rare. Balancing rock grinding-media in the mill by blending and differential blasting provides a significant and low-cost opportunity to increase production. Powell *et al* (2019) demonstrated this through controlled surveys on the Cortez mine site. The mine followed up on the practical implementation of blending with the next phase of ore changes, as the mine shifted to the crossroads pit that hosts two dramatically different ore types; with the modelled outcomes presented by Hanhiniemi and Powell (2022) and practical application presented by Ntiamoah *et al* (2020), as plotted in Figure 13.

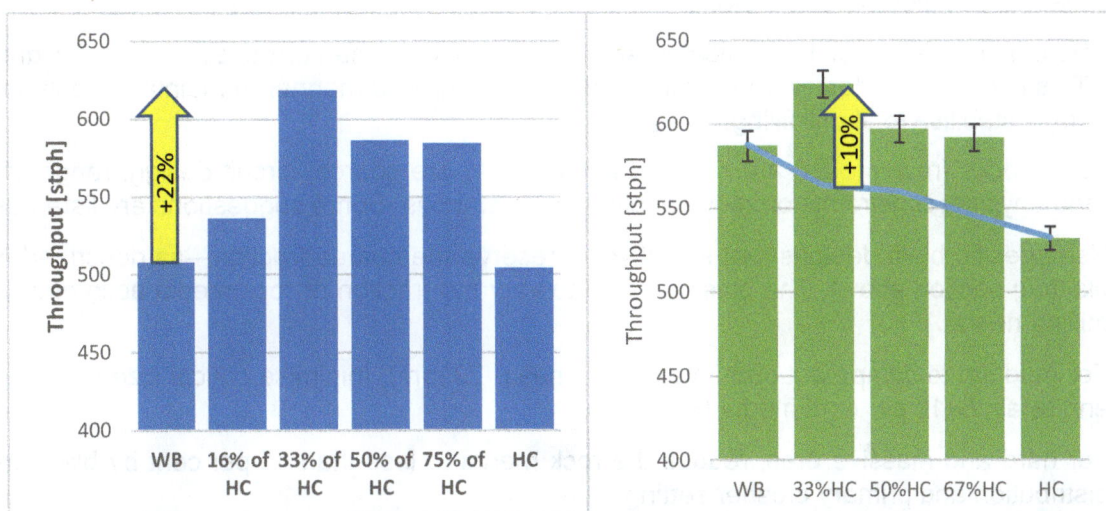

FIG 13 – Surveyed and practical implementation of ore blending at Cortez mine (Ntiamoah *et al,* 2020). NB: stph = short tons per hr; WB = Wenban and HC = Horse canyon ores.

The survey, modelled and production data (from the practical implementation) show that the blend does not need to be precise, merely in a range of 30–70 per cent soft ore to retain throughput. However, the drop-off is steep at the extremes of blends (below 10 per cent of either ore). Blending can add a marked cost to production, due to the need for rehandling. At Cortez, it comes at no extra cost as all RoM ore from the pits is dumped on the RoM pad prior to being fed to the jaw crusher, which cannot handle the large haul trucks. Blending is implemented by keeping the feed consistently within the desirable range and allowing the RoM pad to buffer the erratic delivery of ore by pit area. The site needed only to expand the RoM pad area, thus providing an extra 10 per cent throughput at almost no extra cost, with the standard site operating earth-moving equipment used for the task. For sites that do not need to rehandle, the blending should be on the fly, with: monitoring and controlling of a consistent blend range by truckload; closer cooperation with mine planning; and using a buffer RoM stockpile to provide sufficient capacity to absorb the daily variations in ore delivery by ore type. This could then be implemented most of the time with only a marginal extra cost for rehandling approximately 20 per cent of feed, which should easily be justified by stable and higher mill throughput.

ROCKMEDIA CONCEPT

To help redress the issues of mill instability; throughput loss; low power draw; liner damage; ball spalling and accelerated wear; coarsening and variability of product size; and overloading of ball mills that arise from inadequate rock grinding media, we propose a relatively simple and low-cost mill utilisation and control method. The Rockmedia technique aims to maintain an independent rock feed in the grinding-media size range.

Rock grinding-media is in the +90 mm to −250 mm range. This is a bit of 'back to the future', as it was accepted practice a few decades ago to screen out coarse rock and control its addition to mills for mill stabilisation. However, the proposed method is a far more practical and cost-effective approach, yet easily as effective. In essence, competent feed is diverted to a buffer stockpile that is independently controlled as a bleed-in according to current mill filling. This coarse rock has to be fed directly—'direct-fed'—to the mill feed belt to minimise lag in the control system.

> *'Direct-fed' rock is not fed via any intermediate stockpiles, but controlled with a dedicated feeder at a rate determined by the control system, based on mill filling.*

This direct-feed approach has been dubbed 'Rockmedia' as the control of the rock feed allows the rock to act similarly to the controlled addition of steel grinding media. The Rockmedia approach can significantly increase productivity in existing SAG circuits, and can be built into future circuits to maximise equipment utilisation while reducing direct and indirect energy usage.

Rockmedia implementation would be site-specific, especially in retrofits, but could be standardised in greenfield applications or expansions.

The basic principle is to build a dedicated stockpile of rocks in the correct size range for grinding media. This is fed with a dedicated reclaim feeder in a controlled manner to maintain media load in mill and thus stabilise total mill filling.

The coarse rocks (rock media) are sourced according to ore source, circuit design, range of rock types, and physical constraints on rock handling and distances. Some suggestions are listed below:

- Use specific blast designs that selectively preserve the coarse fraction—as described in the blasting section above. The blast is tuned to vary the fraction of rock media according to the milling needs.

- For medium to competent ores: maximise fines (–20 mm), minimise critical size (20–80 mm), and retain 8–15 per cent media (+90 mm).

- For hard and massive ores, reduce the rock media to less than 15 per cent by blast-energy distribution and primary crusher setting.

- For highly fractured ores, retain *in situ* rock media by optimising the stemming length.

- Separately handle trim-blast product as preferential coarse media.

- Rehandle off the stockpile perimeter. This requires ongoing dozer usage to move the rocks to the independent feed zone; the outcome is markedly different from varying the balance between reclaim feeders.

- Bias feed delivery off conveyors in a deliberate and designed manner (rather than inadvertently). The coarser rocks are projected further on a parabolic trajectory off conveyors (Figure 14); this is often an issue in equipment performance (Powell *et al*, 2011; Powell, Evertsson and Mainza, 2019), but can be turned into an advantage by scalping off the coarse material with a diverter and separate belt. This can be implemented at transfer points, transfer bins and chutes, storage silos, or even at the stockpile feed.

- Use coarse scalping screening to divert the coarse fraction to a feed-stockpile, with excess being directed to a mid-size crusher.

- Integrate mid-size crushing with the coarse scalping. By limiting the top-size fed to the crusher, it can be set to a finer chamber and CSS to effectively eliminate the slow-grind critical-size range (20–80 mm). The crusher can share duty with recycle pebble crushing, as the feed-size range matches, as per concepts presented by Powell (2018).

- Use recycled pebbles as highly effective media to control mill filling. With a continuous flow to the recycle bins, stockpile, and crushers, this flow can be diverted and rapidly drawn upon to control mill filling without the need for an additional storage facility. This requires a continuously variable diverter and transfer system designed for continuous operation that directs pebbles back to the mill, as required, or crushes them as usual.

- Preserve competent rock in a friable orebody. This can be from the blast face, based off geometallurgical zones, or off an in-pit coarse screen. The coarse ore is delivered to the independent competent feed stockpile.

The physical layout can also vary widely depending on local layout and constraints. A few conceptual layouts are presented for illustrative purposes.

FIG 14 – Bias by size of rock being projected off a conveyor.

A standard conical stockpile can be segmented to isolate the front reclaim feeder for rock media feed, as illustrated in Figure 15. This allows coarser and competent components, excess pebbles, and coarse edge material to be taken off the stockpile to be delivered by front-end loader. Additionally, if the divider wall overflows during periods of high stockpile level the overflow will be biased towards the coarser rock that is projected to the far side of the stockpile, as shown in Figure 15.

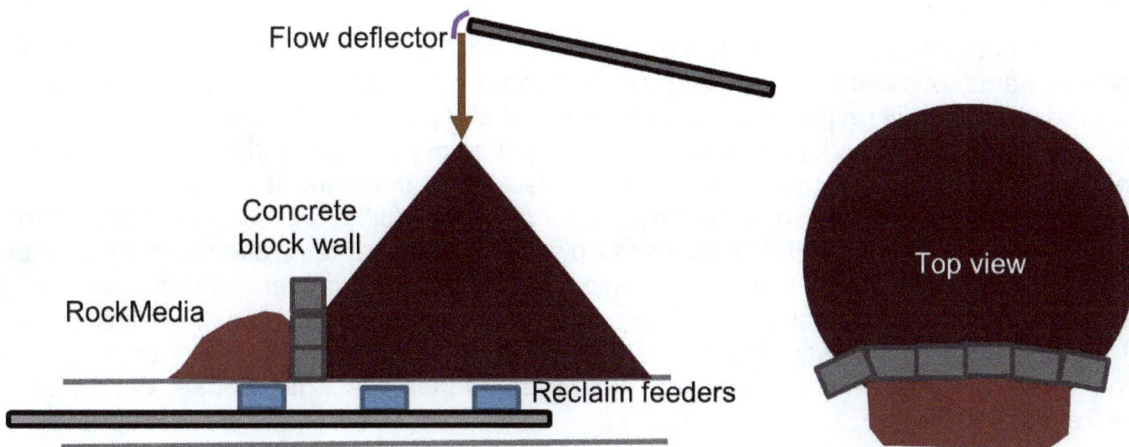

FIG 15 – Conceptual layout of split conical stockpile.

Another alternative is a dedicated stockpile or feed bin linked into an existing mill feed belt. A potential layout for a trial is illustrated in Figure 16, with the orange square showing a potential location of a bin feeder, directed in the direction of the arrow of the mill feed belt. Once trialled a more permanent installation can be constructed.

FIG 16 – Dedicated side stockpile.

Trials may be important in quantifying the impact in order to justify a business case, overcome scepticism or, assess the capacity requirements of the Rockmedia feed—which is strongly dependent on the ore competence and blends. The approximate feed rates are expected to define the size of stockpile or bin.

Being mindful of the potential negative impact of added complexity, the rock media feed should be implemented as a redundant system, so that the circuit reverts to prior operation during any service or downtime.

MODELLING FOR DESIGN

It is highly desirable to predict the Rockmedia concept influence in advance of implementation, as no business wants to invest in guesswork. We face the issue that the current models are woefully poor at predicting the mill response to varying blends of competence. This requires, at minimum, a discrete multi-component model that captures the response of mill filling, throughput, and grind size to a mill feed of independent rock types. This was tackled in the work of Bueno *et al* (2012) and Bueno (2013) tackled this with some success, but it did not result in a predictive model based on laboratory ore characterisation. Hanhiniemi and Powell (2022) presented a pseudo multi-component modelling approach to enable implementation in a single-component model and simulator, but it has to be fitted to controlled plant trials. Additionally, as the issue is one of dynamic variability, a dynamic model is required to adequately assess the opportunity and predict the outcome or realistic feed variation. There have been various attempts at multi-component SAG mill models, but all fall short of being reliably predictive.

Hence, at this stage it is best to conduct a limited trial with temporary feed systems and front-end loaders to prove the most appropriate approach on any particular site.

It is also necessary to understand residence time, segregation, and flow in the stockpiles that are being manipulated to this end. This requires a stockpile and bin model. Discrete element modelling (DEM) can be of use in understanding feed segregation, but it is poor at simulating wide ranges of particle size that must be dealt with in mill feed. Ye *et al* (2022) and Ye, Hilden and Yahyaei (2022) have addressed this. The model runs in real time, so it is suited to both design and online control (Figure 17).

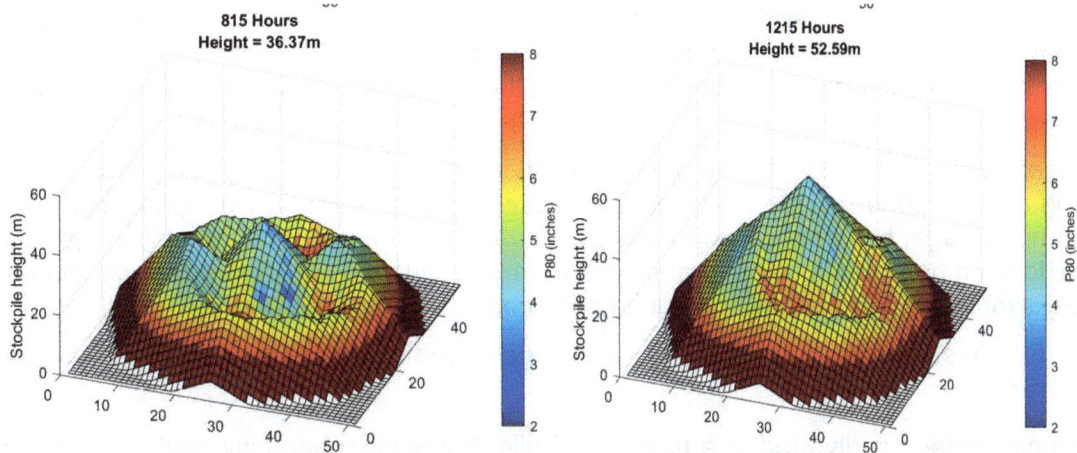

FIG 17 – Stockpile model responding to filling and multiple drawpoints to predict size segregation (Ye *et al*, 2022).

Such a modelling capability points to the opportunity of linking expected feed size to Rockmedia implementation requirements for predictive control. RoM feed size distribution into the stockpile (from standard belt scanners), extraction rate out from each reclaim feeder, and a level sensor provide the required inputs for online prediction of mill feed size distribution. The model can also be used to inform design of segregated stockpiles, biased feed systems, scalping off stockpile perimeters, and so on.

CONTROL

Crucially, the Rockmedia implementation must be integrated into the mill control and overall process control logic. The most direct control loop is the link to mill filling which is indicated by bearing pressure or loadcells, but preferably by more advanced mill filling sensors such as the vibration or audio systems now available or the JK MillFIT calculator (Yahyaei *et al*, 2021). As the mill empties the rock media feedrate should be increased, and as the mill fills the rock media feedrate should decrease. Having a measure of what is being drawn off the coarse ore stockpile will inform the control system, rather than being purely reactive. Model-based feed-forward control from the stockpile model can also be of considerable value in optimising implementation.

The control system is linked to the Rockmedia stockpile to directly control the reclaim rate with the available stock. The control system should also be linked to the mining side of the operation so it can call for more or less rock media. A control system based on process models, especially a mill model that responds realistically to rock filling and dynamics, as discussed in the work of Powell *et al* (2023b).

The more tightly the Rockmedia stockpile is controlled in terms of size distribution and competence, the more predictable its impact on mill filling will be. An unknown blend of reject pebbles and scalped stockpile edge material will result in a varying, and over-reactive, mill response—as their impact on rock build-up in the mill is so different. Thus, the Rockmedia stockpile content should be monitored and controlled.

The fraction of rock media can be used not only to control mill filling, but to shift work between the SAG and ball mill, so as to maximise circuit production through ensuring full utilisation of all milling units.

A FRESH START

Controversially, it can be argued that full-feed pre-crush is the lazy approach to utilisation of existing equipment, process, and installed power for increased throughput. Inefficient use of energy has become accepted as standard practice, so the inefficiencies are not even noted in the application to upgrade brownfield processes that are operating well below potential. The industry appears to be following the path of least resistance; inadequate pre-crush has pushed up SAG throughput but has overloaded ball mills and coarsened product. Incremental fixes have improved the process and now it is recognised that a far finer feed recovers mill efficiency. Is it not time for a reset, to stand back and ask if this is the correct path to be on? A fork in the road has been followed, but this does not mean we should persevere to the conclusion along this path, when a simpler, cheaper, more-efficient one presents itself.

Challenges of varying blends, friable ores, and varying feed-size distribution also lead to fluctuating mill filling and performance, giving rise to issues such as mill instability; throughput loss; low power-draw; liner damage; ball spalling and accelerated wear; coarsening and variability of product size; and overloading ball mills. Many of these arise from inadequate or excessive rock grinding-media, a relatively simple and low-cost mill utilisation and control method is proposed:

> *Rockmedia used to maintain an independent feed of rock in the grinding media size range.*

The competent feed is diverted to a buffer stockpile that is independently controlled as a bleed-in stream according to current mill filling. This coarse rock has to be 'direct-fed' to the mill feed belt to minimise lag in the control system. Direct feeding is not via any intermediate stockpiles; rather is controlled with a dedicated feeder at a rate the control system determines based on mill filling.

This approach, dubbed 'Rockmedia', can significantly increase productivity in existing SAG circuits and be built into future circuits to maximise equipment utilisation while reducing direct and indirect energy usage.

We propose that the size of the prize from this approach bears serious investigation in contrast to high operating cost options that are also working directly against our imperative to reduce indirect energy use and greenhouse gas emissions—inherent in steel grinding media. Rockmedia is the cheapest most environmentally friendly media you will use!

ACKNOWLEDGEMENT

This paper was originally published as a reserve paper in the SAG conference of 2023 (Powell *et al*, 2023a), but was not part of the presentations. It is published here with minor updates.

REFERENCES

Ballantyne, G R, Clarke, N, Elms, P, Anyimadu, A and Powell, M S, 2015. Application of research principles to identify process improvement at Sunrise Dam Gold Mine, in Proceedings of Metplant 2015 (The Australasian Institute of Mining and Metallurgy: Melbourne).

Bueno, M P, Kojovic, T, Powell, M S and Shi, F, 2012. Multi-Component AG/SAG mill model, *Minerals Engineering*, 43–44:12–21.

Bueno, M, 2013. Development of a multi-component model structure for autogenous and semi-autogenous mills, PhD thesis, University of Queensland, Brisbane, Australia.

Bueno, M, Niva, E, Powell, M S, Adolfsson, G, Kojovic, T, Henriksson, M, Worth, J, Partapuoli, Å, Wikström, P, Shi, F, Tano, K and Fredriksson, A, 2011. The dominance of the competent, in International Autogenous Semiautogenous Grinding and High Pressure Grinding Roll Technology 2011 (eds: Major, Flintoff, Klein, McLeod) (CIM: Vancouver).

de Carvalho, R M and Tavares, L M, 2013. Predicting the effect of operating and design variables on breakage rates using the mechanistic ball mill model, *Minerals Engineering*, 43–44:91–101.

Engelhardt, D, Seppelt, J, Waters, T, Apfelt, A, Lane, G, Yahyaei, M and Powell, M S, 2015. The Cadia HPGR-SAG circuit – from design to operation – the commissioning challenge, in International Autogenous Semiautogenous Grinding and High Pressure Grinding Roll Technology 2015 (eds: Klein *et al*) (CIM: Vancouver).

Guzmán, L R and Sepúlveda, J E, 2016. The effect of key operating conditions on grinding media consumption in Semiautogenous mills, in 12th International Mineral Processing Conference, Santiago, Chile.

Hanhiniemi, J J and Powell, M S, 2022. Multicomponent Modelling of Ore Blending in Grinding Circuits: A New Model and Case Study, in IMPC 2022: XXXI International Mineral Processing Congress (The Australasian Institute of Mining and Metallurgy: Melbourne).

Kanchibotla, S S, Morrell, S, Valery, W and O'Loughlin, P, 1998. Exploring the effect of blast design on SAG mill throughput at KCGM, in Mine to Mill Conference (The Australasian Institute of Mining and Metallurgy: Melbourne).

Lee, K, Rosario, P, Schwab, G and Bogadi, A, 2015. An analysis on SAG pre-crush circuits, in International Autogenous Semiautogenous Grinding and High Pressure Grinding Roll Technology 2015 (eds: Klein *et al*) (CIM: Vancouver).

Mckee, D J, Chitombo, G P and Morrell, S, 1995. The relationship between fragmentation in mining and comminution circuit throughput, *Minerals Engineering*, 8:1265–1274.

Ntiamoah, K, Rader, L, Yalcin, E, Powell, M, Jokovic, V, Hilden, M and Hanhiniemi, J, 2020. The influence of ore blending on mill throughput — Cortez mine case study, Nevada Mineral Processing Division Conference, Reno, USA.

Powell, M S, 2018. Integrated performance optimisation of comminution circuits, in Complex Orebodies (The Australasian Institute of Mining and Metallurgy: Melbourne).

Powell, M S, Benzer, H, Mainza, A N, Evertsson, M, Tavares, L M, Potgieter, M, Davis, B, Plint, N and Rule, C, 2011. Transforming the effectiveness of a HPGR circuit at Anglo Platinum Mogalakwena, in International Autogenous Semiautogenous Grinding and High Pressure Grinding Roll Technology 2011 (eds: Major, Flintoff, Klein, McLeod) (CIM: Vancouver).

Powell, M S, Bozbay, C, Kanchibotla, S, Bonfils, B, Musuniri, A, Jokovic, V, Hilden, M, Young, J and Yalcin, E, 2019. Advanced mine-to-mill used to unlock SABC capacity at the Barrick Cortez mine, in International Autogenous Semiautogenous Grinding and High Pressure Grinding Roll Technology 2019 (CIM: Vancouver).

Powell, M S, Evertsson, C M and Mainza, A N, 2019. Redesigning SAG mill recycle crusher operation, in International Autogenous Semiautogenous Grinding and High Pressure Grinding Roll Technology 2019 (CIM: Vancouver).

Powell, M S, Evertsson, C M, Mainza, A N and Ballantyne, G R, 2022. Practical measures of process efficiency and opportunity, in IMPC, 2022: XXXI International Mineral Processing Congress (The Australasian Institute of Mining and Metallurgy: Melbourne).

Powell, M S, Hilden, M M and Mainza, A N, 2015. Full pre-crush to SAG mills – the case for changing this practice, in International Autogenous Semiautogenous Grinding and High Pressure Grinding Roll Technology 2015 (eds: Klein *et al*) (CIM: Vancouver).

Powell, M S, Mainza, A N, Tavares, L M and Kanchibotla, S, 2023. Rockmedia, in International Autogenous Semiautogenous Grinding and High Pressure Grinding Roll Technology 2023 (eds: Simonian *et al*) (CIM: Vancouver).

Powell, M S, Mainza, A N, Tavares, L M, Weatherley, D K, Vien, A, Mular, M and Ballantyne, G R, 2023. Will AG milling make a comeback?, in International Autogenous Semiautogenous Grinding and High Pressure Grinding Roll Technology 2023 (eds: Simonian *et al*) (CIM: Vancouver).

Powell, M S, Yahyaei, M, Mainza, A N and Tavares, L M, 2021. The endemic issue of ball mill overload in SABC circuits, in Proceedings of the 15th Australasian Mill Operators conference 2021 (The Australasian Institute of Mining and Metallurgy: Melbourne).

Rodrigues, A F V, Delboni Jr, H, Powell, M S and Tavares, L M, 2021. Comparing strategies for grinding itabirite iron ores in autogenous and semi-autogenous pilot-scale mills, *Minerals Engineering*. 163:106780.

Rodrigues, A F V, Powell, M S, Tavares, L M, Donda, J D, Pinto, P F, Lima, N and Ferreira, M T S, 2015. Autogenous and Semi-Autogenous Pilot Trials with Itabirite Iron Ore, in International Autogenous Semiautogenous Grinding and High Pressure Grinding Roll Technology 2015 (eds: Klein *et al*) (CIM: Vancouver).

Tavares, L M and de Carvalho, R M, 2009. Modeling breakage rates of coarse particles in ball mills, *Minerals Engineering*. 22:650–659.

Waters, T, Rice, A, Seppelt, J, Bubnich, J and Akerstrom, B, 2018. The evolution of the Cadia 40-foot SAG mill to treat the Cadia East orebody: a case study of incremental change leading to operational stability, in 14th Australasian Mill Operators conference 2018 (The Australasian Institute of Mining and Metallurgy: Melbourne).

Yahyaei, M, Hilden, M, Reyes, F and Forbes, G, 2021. Soft sensors and their application in advanced process control of mineral processing plants, in 17th International Conference on Mineral Processing and Geometallurgy (Procemin-Geomet 2021).

Ye, Z, Hilden, M M and Yahyaei, M, 2022. Development of a three-dimensional model for simulating stockpiles and bins with size segregation, in IMPC 2022: XXXI International Mineral Processing Congress (The Australasian Institute of Mining and Metallurgy: Melbourne).

Ye, Z, Yahyaei, M, Hilden, M M and Powell, M S, 2022. A laboratory-scale characterisation test for quantifying the size segregation of stockpiles, *Minerals Engineering*, 188:107830. https://doi.org/10.1016/j.mineng.2022.107830

Tooling up for future innovation – turning overlooked process opportunities to immediate and long-term benefit

M S Powell[1]

1. FAusIMM, Liner Design Services, Brisbane Qld 4069. Email: malcolm@milltraj.com

ABSTRACT

From decades of researching mill operation and conducting process improvement studies, the author has derived a host of valuable insights into the milling process as a system. The inter-connectedness of the process is partially appreciated in the industry at a high level, but from the author's experience is lacking at an operational level. Furthermore, beneath this there is a gap in solid metallurgical knowledge of standard practice and principles of operation. Graduates are placed on sites and expected to operate complex processing circuits with barely any direct training in the subject. The expectation is that they will learn on the job, but from whom? An understanding built on an insight into the physics of the equipment and processes will payback immediately and provide the upliftment needed for future more advanced processing. Transitioning to high-performance systems with current processing and available technology, will provide the foundation, the springboard for our transformative technologies that will require us to operate the more complex and technically demanding processes of the future. Regardless of progress made in advanced technology, it is surely of value to uplift existing operations, for immediate return, and to provide a platform of technology and operating ability that will be able to effectively operate the more complex and technically demanding processes of the future. This must surely be the pathway to the major transformation in productivity while reducing environmental footprint that the industry is being driven towards. Embracing the opportunities lie firmly, and affordably, in the grasp of the mining industry, through leadership from the top. This paper provides the author's view of the approach that the industry may take in liberating up to 20 per cent upside in production of current operations while tooling up for future innovation.

BACKGROUND

From decades of researching mill operation and conducting process improvement studies, the author has derived a host of valuable insights into the milling process as a system. These have been shared with the industry, both in on-site training and in a number of conference deliveries: Powell, Mainza and Yahyaei (2023), which lists 12 specific published examples; Powell *et al* (2021) addressing the energy balance between SAG and ball mills; Powell, Evertsson and Mainza (2019) presenting opportunities for improved utilisation of recycle crushers; Yahyaei and Powell (2018) providing general production opportunities; Powell *et al* (2018) presenting opportunities of integrating blasting and blending with mill performance; Ballantyne *et al* (2015) demonstrating the use of good process measurement to identify process opportunities; Powell, Hilden and Mainza (2015) focusing around SAG mill issues; and Powell *et al* (2012) showing how an understanding of the inter-connectedness of the crushing-screening-HPGR circuit opened up considerable untapped capacity. There are common themes that run through the reviews over the years. Much of what these authors have uncovered appear prosaic or obvious; are well-understood within specific engineering disciplines; would be defined as mineral processing 101; or are standard process control logic, but these are lacking in the production environment. The reasons are multi-fold, from appropriate training through the ranks; to a mismatch of design to reality of performance; to poor performance or inappropriate product from upstream processes; to the ever-present culprit of ore variability.

Being able to diagnose the root causes of processing issues and plan how to rectify them appears to be poor in the industry. It is postulated that a deeper level of insight into the physics of the equipment and processes is an area of neglect, that can be a pathway to understanding and solving the wide range of issues that a processing plant faces. This should be the role of the metallurgists, but advanced training for advanced technical roles is lacking. Graduates are placed on sites and expected to operate complex processing circuits with barely any direct training in the subject. The expectation is that they will learn on the job, but from whom? Observation indicates that most learn the local lore, as opposed to sound operating principles.

The inter-connectedness of the process is partially appreciated in the industry at a high level but, from the author's experience, is lacking at an operational level. The opportunities highlighted in the prior work are often related to integrated problems and thus solutions. A broader view is presented in the study by Powell (2018) addressing integrated performance optimisation of comminution circuits applied to complex orebodies. Designing independent units and then linking them together will never produce the optimal outcome. The industry uses various simulation packages to assess circuit performance, usually with a few selected variations in ore feed. This certainly helps to map out the required operating range and potential performance, but critically lacks the dynamics of the system. These dynamics roll out into transient behaviour, differences in lag and response time of different equipment along the process and in closed-loop systems. Post-design, and often post-commissioning, a control system is pasted on top of the operating plant. Despite the names applied to these systems, they are in principle rules-based (off operator input) or response-based off empirically plotted response of individual units to specific step changes in operating conditions. The point here being that they lack any basis in the physics of the system.

Underperformance that is appropriately diagnosed, provides opportunity for improvement, as highlighted in the author's prior work. An example is a site being advised to add a fourth recycle crusher, when in fact a simple analysis of crusher utilisation indicated that overall utilisation was below 50 per cent, ie working out how to use the installed crusher capacity was the sensible path forward (Powell, Evertsson and Mainza, 2019). This diagnosis saved the site tens of millions of dollars in unnecessary infrastructure. Careful analysis of the root cause of underperformance opens up the pathway to capture production gains, many of which can be at little or low cost.

Lying hidden under this thick layer of grey opportunity is neglect of the industry in developing high-performing systems. It is the author's view that we are fooling ourselves, for no gain, if we purport to have attained excellence in process performance in mineral processing. There is much that can be improved in existing processing plants.

On the other hand, the mining industry strives to transform itself to meet future needs within the growing constraints imposed on any industry, and especially on resource extraction. There is considerable hype around the targets of reduced energy and water usage, reducing or eliminated tailings dams, mine rehabilitation etc, along with investment in advanced technologies aimed at achieving the required step-changes.

However, while pursuing the vision of ore sorting, dramatically energy-efficient mills, and coarse recovery, the inability to effectively operate existing plants is overlooked. It may be fair to say that these act in complete isolation of one another, reflected in the insular business units addressing each area within mining companies. Regardless of progress made in advanced technology, it is surely of value to uplift existing operations, for immediate return, and to provide a platform of technology and operating ability that will be able to effectively operate the more complex and technically demanding processes of the future. Embracing the opportunities lie firmly and affordably, in the grasp of the mining industry, through leadership from the top.

This paper provides the author's view of the approach that the industry may take in liberating up to 20 per cent upside in production of current operations while tooling up for future innovation. The future may be closer than we think.

OPPORTUNITIES

A distillation of opportunities from the prior work is presented here as a succinct summary of the style of opportunities that are available to the industry. The readers are referred to the reference list for deeper insight and explanation of these. The focus in the paper is on SAG-ball mill circuits. Clearly there are more opportunities available, which experienced readers can add to this list.

It is not by chance or poor design that a mining process, from rock extraction to saleable concentrate, has many stages. The scale from kilometres at the mine to microns at the concentrate, appears unprecedented in industry. The steps are illustrated in Figure 1. It is naïve to ignore how interconnected this process is. Additionally, there would not be a mineralised orebody if it was nice and homogenous, so by the very definition of what our objective is, variability is at the heart of the process. This needs to be embraced, not ignored – in both design and operation.

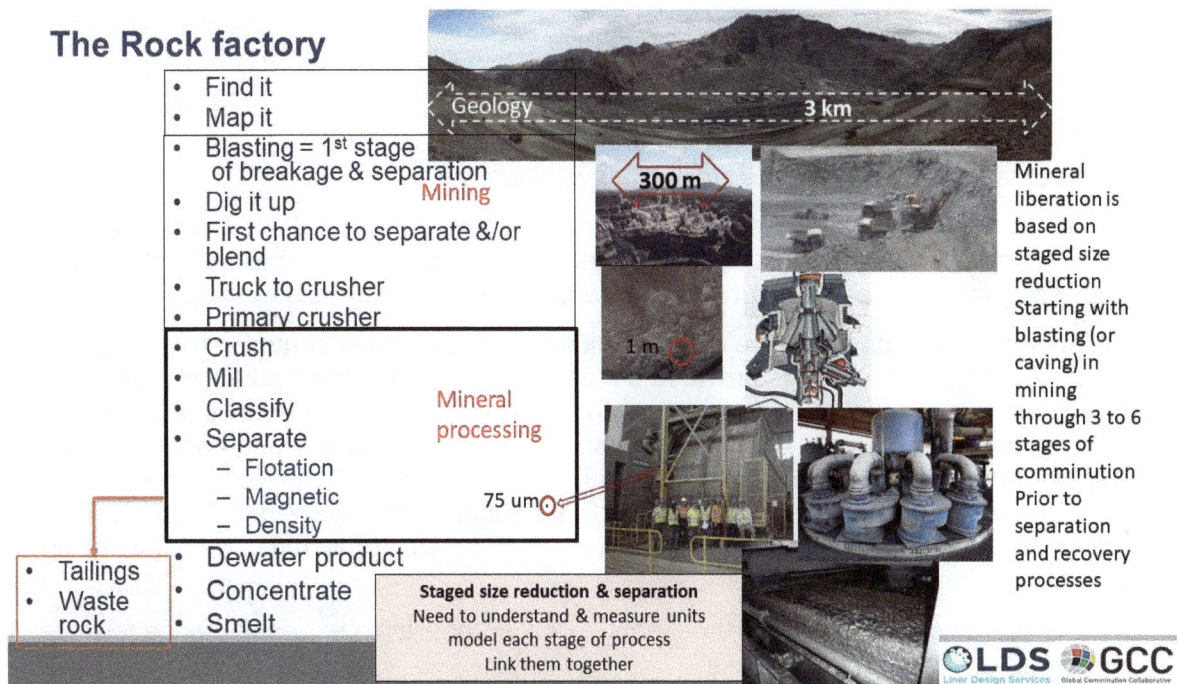

FIG 1 – Stages from mine to concentrate.

Each stage has a distinct objective which supports the overall mine production, however these tend to be focused on individually to the detriment of the essential downstream processes that actually determine the overall production of metal from the mine, and thus productivity and profit. A high-level walk-through of the process and associated issues and opportunities is listed as the basis of this analysis.

1. Blasting: breaks the rock out of the *in situ* orebody. It is known that Mine to Mill application to modify blast energy can be used to increase SAG mill throughput. But blast modification just as importantly:

   o decreases truck loading time

   o increases truck average load

   o increases primary crusher capacity while also improving availability and reliability, and decreasing operating costs

   o can be tuned to differentially blast soft and hard ore to suit the mills

   o can be coarsened to allow simple removal of gangue or sub-grade ore

   o can be carefully timed with electronic explosives to manage throw and dilution of ore

   o or additional energy can be used to provide a finer mill feed but reduce metal recovery through uncontrolled additional movement and resultant dilution (Powell *et al*, 2018).

2. Primary crushing: this is used for top-size control but can be pushed to improve mill throughput:

   o Reducing chunky near-size from the blast allows the crusher to be run tighter and in a slightly bed-breakage mode to reduce top-size to the SAG mill and increase fines production (Dupont, McMullen and Rose, 2017).

   o Improved distribution and flow of rock into the crushing chamber reduces asymmetric stress and thus mechanical stress and failures while also increasing instantaneous capacity.

3. Stockpile(s): these are used as buffers to absorb the mismatch between cyclic batch delivery of ore from mine to primary crusher, including blast cycles, shift changes, relocation and maintenance of shovels, etc. Stockpiles introduce a number of issues and opportunities:

   o Segregation within stockpiles, and inversion as the level reaches a critical low, is well-known and treated as an unavoidable inconvenience.

- They can provide an opportunity to blend different ores, but typical batching of blends onto the stockpile does not result in a neat sequential delivery to the mill.

- The typical conical stockpile with an overhead feed discharge and line of reclaim feeders offers limited ability to control blend and they suffer from shifting centre of mass along the line of reclaim feeders as the stockpile level changes.

- It is not known what part of the earlier mined ore is being drawn out to the mills at any given time, arising from a lack of direct link between dumped ore and arrival on the stockpile, segregation in the pile, and unknown flow and mixing through the stockpile.

4. Ore tracking: this is possibly one of the weakest areas of current control in mineral processing plants. The Metso SmartTags provided the first method to reliably track blocks of ore through the system, from shovel, to truck, crusher stockpile and mill. The experience of those who have used trackers is universally of being surprised how different the transport and residence time was to that expected based off simple plug flow. In the author's experience, a simple painted rock experiment measured the ore arriving 3 hrs earlier than expected for a controlled survey – indicating rat-holing through a silo. Although tracking electronic tags is most valuable, they have distinct limitations:

- They are a consumable that represents an added operating expense, hence few sites use them as a routine operating tool.

- They are a single size. It is well-known that rock flow is not plug-flow. Tracers are reliable for batch truck units, but the ongoing handling through crushers, conveyors and especially stockpiles introduces compounding segregation and flow rate differentiation by size. Thus, a small tracer can potentially be well off the median transport time and almost certainly not provide a correct measure of the distribution of the batch of ore over time into the SAG mill.

- They do not provide a predictive measure ahead of process control. Measuring the arrival of a specific ore type is useful in mill control, but not nearly as useful as predicting some hours in advance what is arriving, and thus allowing proactive rather than reactive control.

5. Feed size measurement: a number of systems have been developed to measure the rock size distribution being fed into SAG mills. These provide a valuable input to mill control as a quantified input to inform operators on control decisions, and allow a short-term feed-forward to control systems in responding to an increase or loss of mill load. Their limitations include:

- Measuring only the visible surface of the rock piled on a conveyor belt. This is invariably the ore from the last reclaim feeder, which is the coarsest feed to the mill. The system is blind to changes in feeder ratios, especially further back along the reclaim feeder belt.

- Inability to measure below a certain cut-off size, around 10 mm for most systems receiving primary crushed run-of-mine (RoM) ore. This is due to pixel resolution, image processing overhead, poor definition of clumped finer material, and shadows. This lack of identification of fines is exacerbated by the natural percolation of fines down through the rocks, effectively hiding them.

6. SAG mill control. Much can be written about the shortfall of SAG mill control systems. In the positive sense, there are many control systems that successfully keep SAG mills out of trouble, maintaining continuous and reliable performance. SAG mills are intrinsically complex systems, with strong non-linear interactions driven by the natural and unavoidable variation in the feed rock, described int eh work of Powell, Hilden and Mainza (2015). Despite the impressive names applied to these control systems: model-based, fuzzy logic, rule-based etc, they are in principle rules-based (off operator input) or response-based off empirically plotted response of individual units to specific step changes in operating conditions. This leads to a number of intrinsic limitations:

- By design, they cannot extend beyond the input range of experience, be it operators (with their inherent bias, limited range of experience and limited process knowledge), or response mapped off prior test conditions. Any system mapped off a window of operating conditions is unpredictable outside that window. Additionally, it will not seek to move outside the historic conditions for just this reason. These systems can thus not seek an optimal condition beyond historic operation, and thus trap an operation into historic performance limits.

For example, in the work of Powell *et al* (2018), the CSS of the primary crusher was opened up and more of a harder, component feed added to the mill feed to increase circuit capacity by 17 per cent – a knowledge-based action entirely beyond the grasp of existing optimising control systems.

o The control systems can keep a mill circuit closer to previous best performance conditions and thus provide a measurable improvement to production, but they are not capable of exceeding these old limits.

o The systems are dictated to by external targets, that can push a control system to cyclically overload a mill in attempting to reach the unobtainable, or to idle when there is actually more head room.

o They do not have any intrinsic means of seeking an optimal operating point. The author's experience is that they can find a local optimum, but have no method to pass through a local minima to find a global optimum.

o Control systems are generally programmed to push to a limit and then react to the response of the mill, repeatedly disturbing the mill operation rather than letting it reach internal load stability, such as the example in Figure 2.

o Having been calibrated and trained on historic data, they invariably respond poorly when mill conditions are changed, such as a change in discharge, higher ball load, different ore type (such as a lot more friable), finer blast or the implementation of pre-crushing. It is regularly noted that SAG mills are operated well below installed power after a finer feed is introduced.

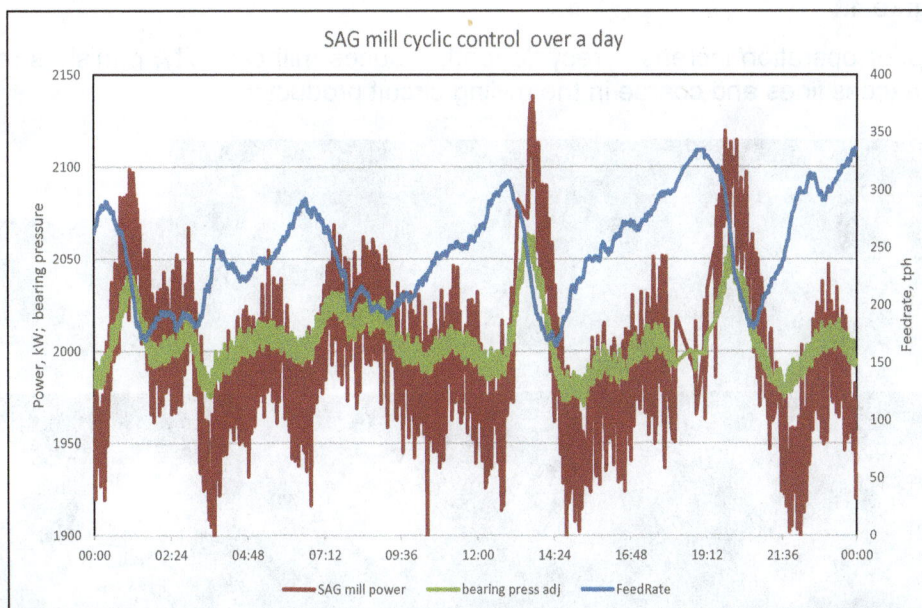

FIG 2 – Cyclic operation of a mill over a day, induced by poor over-control (Powell, Mainza and Yahyaei, 2023).

7. Ball mill selection and control: the duty and performance of the ball mill is inextricably linked to the performance of the upstream SAG mill. The work of Powell *et al* (2021) highlights the issues around this and the significant upside to be derived from balancing the work between SAG and ball mills. Some issues are:

o The SAG mill is pushed to its limit and then the condition of the ball mill considered, which can be well after the ball mill is overloaded far beyond the peak of the power curve. This leads to a loss of grinding power, coarsening of milled product, overflow of rock scats, balls leaking out the mill (especially at start and stop), such as in Figure 3. The installation of ball retaining rings tends to merely exacerbate the issue.

- o Depression of power can lead to a mistaken ongoing addition of balls to recover power, when instead the ball load creeps above the mill discharge, as the reference point is the reduced power brought on by sludge and scats overload.

FIG 3 – Sludge-filled mill, balls above discharge, and scatting and ball discharge from an overloaded ball mill.

8. Cyclones: classification offers a hidden opportunity, which tends to be overlooked, but can be cheap to remedy, or at least improve. Some common issues are:

- o A few large cyclones are operated in a cluster, which limits the ability to control cut-size by maintaining feed density.

- o Uneven wear and lack of synchronisation of maintenance leads to uneven operation of cyclones around a cluster, delivering a wider range of product size and decidedly poor classification with dilute flaring dramatically increasing recycle of fines to the mill, illustrated in Figure 4.

- o The poor operation increases recycle load, reduces mill capacity, punishes recovery with both excess fines and coarse in the milling circuit product.

FIG 4 – Photographs of five cyclones in a cluster with different operation (top) and one cyclone pulsing over 3 secs (bottom) (Powell, Mainza and Yahyaei, 2023).

9. Recycle crushers are almost universal in SAG milling circuits due to the boost in performance that they deliver, with increased SAG mill throughputs of over 20 per cent quoted. The work of Powell, Evertsson and Mainza (2019) highlight the many issues and corresponding opportunities afforded by improved physical set-up and control of recycle crushers.

- o Low power utilisation appears to be universal in pebble crushers.

- o Coarse product, delivering a recycle in the critical sized range, is driven by force overload as the CSS is reduced.
- o Long run-in period of new liners is taken as a fact of operation.
- o Symmetric feed may be the most important factor in unlocking crusher capacity, reducing product size, improving reliability and decreasing maintenance costs.
- o Continuous compensation for chamber wear will keep the crusher at optimum power draw and minimum product size.
- o Installing variable speed drive allows the crusher capacity and product size to be controlled, a factor mysteriously missing in crusher implementation.

10. Milling circuit control: this is the integration of the units. It has been noted over many sites that the control system rarely or only minimally treats the milling circuit as an integrated system with sub-units. Instead, the units are controlled then coupled via some basic outputs to limit the throughput.

- o The control system should primarily be monitoring the circuit response against the production target, then feeding this back into local set points of each unit. It is common to observe the operators totally ignoring grind size and a control system chasing a throughput target without any consideration of how the control could be modified to improve grind – the win-win of reaching target throughput while also improving grind.
- o A limited-horizon view is used that does not consider optimal production of metal over a longer time range. The author has repeatedly observed the control pushing milling targets in the short-term, even though it is known that there will be insufficient feed over the week or month.
- o The system needs improved understanding of the different residence times of units around a circuit: lag through mills is around 20 mins, crusher bins are variable in lag from 5 to 30 mins, and segregation by size in bins depends on filling, eg pushing the SAG mill to its limit, as the pebble bin steadily fills, results in the cyclones being hit by increased flow rate and a coarser feed which is passed on to the ball mills up to 20 mins after a feedrate change was made. This leaves the circuit in an acute overload situation that can then take hours to unload.

11. Recovery of the valuable metals is the objective of the mineral processing plant, so should intrinsically be the operating and control objective.

- o Too few sites have online recovery measures to inform control.
- o Extended recycles and multiple variables to manage, control and play with make a typical flotation circuit tricky to control let alone optimise.
- o Linking flotation circuits direct to SAG milling circuits, that are intrinsically unstable, leads to continuous fluctuation of feed density and grind size into the circuit, which only leads to reduced recovery.

12. Materials handling and distribution is an integral and major task of mineral processing plants handling hundreds of thousands of tonnes of ore per day. Aside from wear, segregation is the most significant problem observed to arise from the design of materials handling systems:

- o Segregation across bins leading to reduced capacity and uneven distribution to the reclaim feeders onto screens or to crushers, noted to be a major source of lost productivity in the work of Powell *et al* (2012).
- o Segregation across belts, such as in Figure 5, leading to unequal feed size to neighbouring crushers, (resulting in under- and over-loading), and segregation within crusher feed chambers.
- o Pumping out of one sump to two or even three ball mills. The feed into a direct-fed sump (under a mill discharge) is always segregated, leading to uneven feed size and slurry density to each ball mill. This results in clearly measurable differences in performance of parallel

ball mills, often with one being overloaded, scatting rocks, having and excessive circulating load, and producing a coarser grind; such as reported in the work of Hanhiniemi and Powell (2023), which was based on a rigorous survey conducted by the JKMRC under the AMIRA P9 project, and presented in Engelhardt et al (2015).

FIG 5 – Segregation across a belt – typical after a 90° transfer point.

Some integrated examples

Some examples of the interaction of the integrated system are provided to illustrate the importance of seeking the source and addressing the observed limits or issues holistically.

The linkage illustrated in Figure 6 shows the outcome of poor ball mill performance linking back to coarse ore in the primary crusher, which leads to oversize feed to the primary crusher and hence coarser feed to the secondary crushers which produce a coarser product, this then results in less efficient HPGR operation, and finally to reduced capacity of the ball mill due to coarser feed and especially due to inappropriate top-size. The site was asking advice on adding larger balls to the ball mill (which would lead to coarser product and reduction in recovery) and a ball retaining ring (which would lead to overload and loss of power), when in fact, the poor blast practice needed to be addressed.

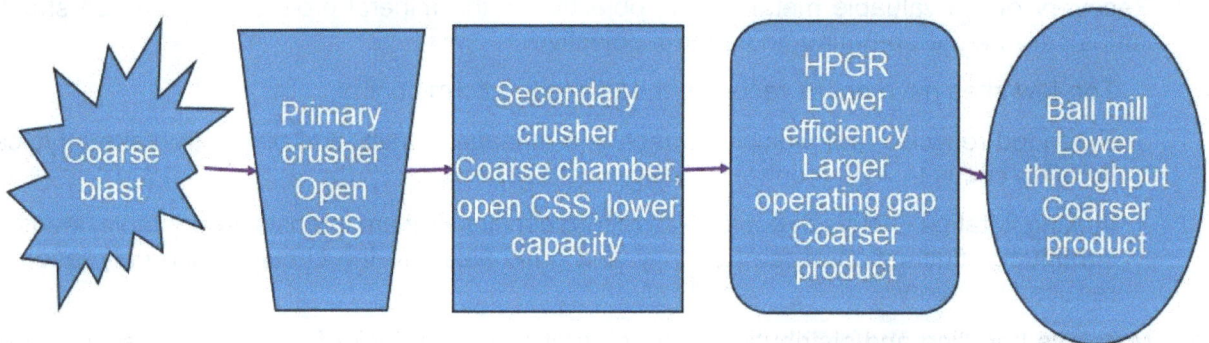

FIG 6 – The linked size reduction across an HPGR circuit.

The cycle of inadequate wash water is a common cause of misallocation of blame. For example, the drenching spray from holes in a pipe shown in Figure 7, pushing the operations to coarser screens, which results in pebbles being fed to the ball mill. This in turn leads to ball mill overload, loss of power and coarser grind size, as demonstrated in the work of Powell et al (2018) and Powell, Mainza and Yahyaei (2023).

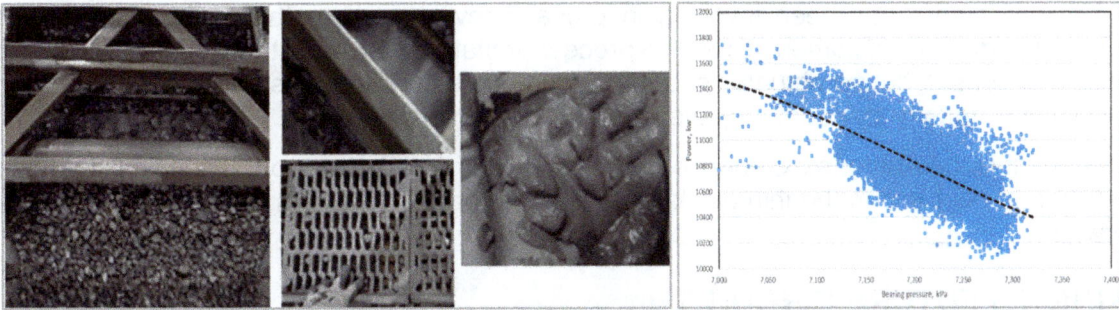

FIG 7 – Cycle of poor wash water, coarser screen, and pebbles in ball mill feed leading to power loss with mill filling.

Figure 8 presents an example of the cycle between feed variation and circuit operation, ending in a variable flotation feed size distribution and feed rate. The issue of sliming and oversize in the flotation circuit, links all the way back to blast practice, addressing each step along the chain is the simple secret to unlocking the capacity on this circuit and overcoming the dominant flotation issues.

FIG 8 – The vicious cycle of feed size and competence variation (Powell *et al*, 2024).

Too often the author has been involved in process reviews where the desired pre-determined outcome is to add new equipment to resolve process limitations. It is held that this is not only wasteful of capital investment, but perpetuates sub-par performance that becomes entrenched as acceptable or even called 'industry standard', or, even worse, 'best practice'.

In summary, it is the Author's experience that most issues and thus opportunities lie in the interface between equipment and in the thin black lines that link them on process drawings (to quote Holger Lieberwirth).

LINKING TO FUTURE CIRCUITS

Considering the points outlined above of common issues in the operation of comminution circuits, let's consider what existing technology can be used to uplift operations across the board.

If the solutions are implemented in a manner that is forward looking, then we can win on at least three fronts:

1. Short-term production benefit.

2. Ability to integrate existing systems with future technology.

3. Building tools and skills needed to operate the future, technology intensive, circuits.

An approach proposed by this author over a decade ago, is to build flexibility into circuits, an approach dubbed FlexiCircuits (Powell, Foggiatto and Hilden, 2014). A concept proposed by the author is illustrated in Figure 9. This lays out an integrated process with multiple potential pathways for the rock to be directed.

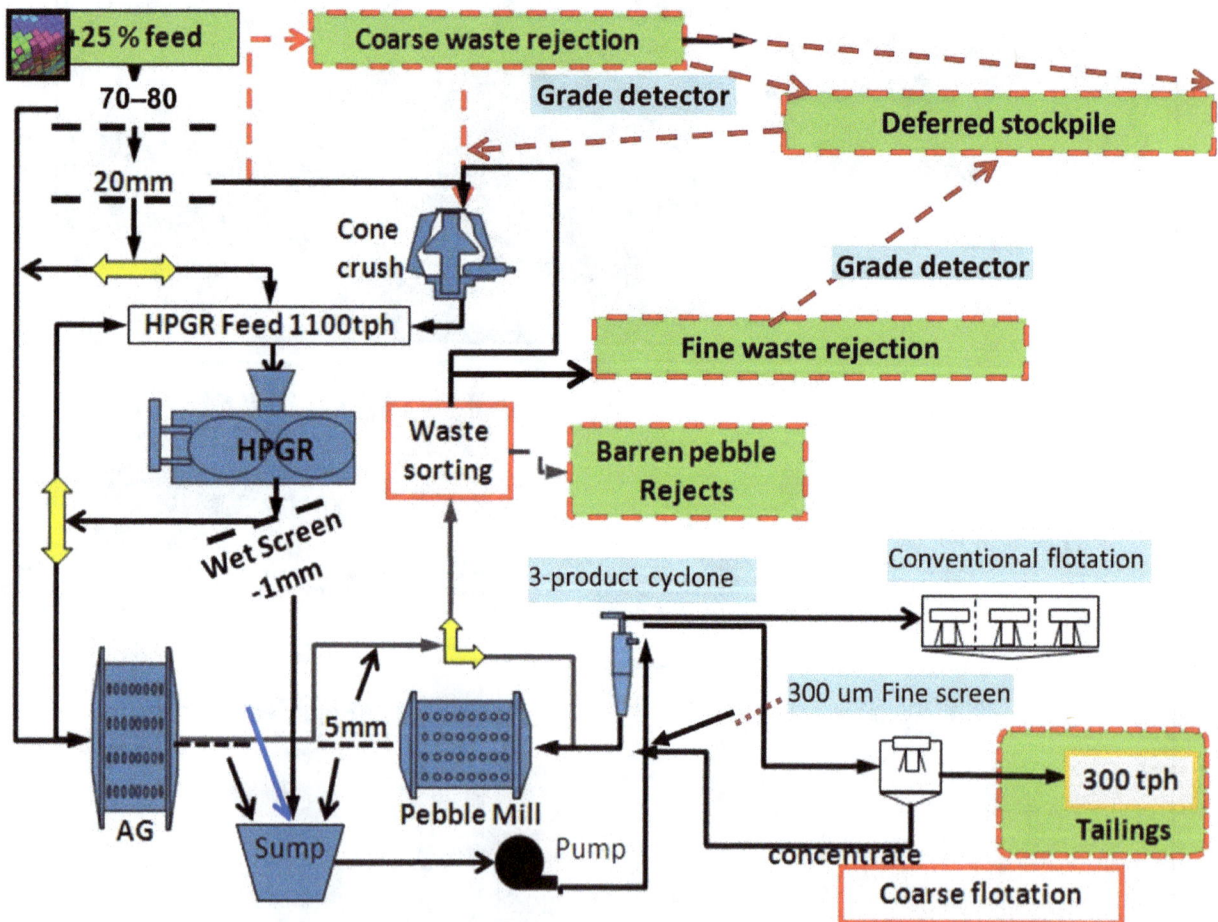

FIG 9 – Conceptual FlexiCircuit.

The first step is to pass the RoM ore over a double-deck primary screen. This relatively simple step removes the mid-size and directs the top-size to a conventional AG or SAG mill, providing the required rock grinding media. The mid-size reports to a large cone crusher, which then feeds an

HPGR with minus 20 mm ore, mixed with a variable portion of the minus 20 mm from the primary screen. The fine feed ensures a high-efficiency in the HPGR operation, with an even and high bed pressure. The HPGR product passes over a wet screen, with the fine undersize reporting as feed to the secondary milling circuit. The screen oversize is ideal feed for the AG/SAG mill but can also be recycled to the HPGR to consolidate the bed and ensure a finer HPGR product. The flows can be varied to balance the work conducted by the HPGR versus the AG/SAG mill, with the objective of maintaining a stable filling in the mill and thus more consistent product size. The load taken by the HPGR varies depending on the feed type, with more work as the ore becomes more competent. The pebbles discharged by the AG mill are directed to the pebble mill as required to draw power, with excess recycled via the pebble crusher. Classification can be via a hybrid cyclone and fine screen or include the three-product cyclone (Mainza, Powell and Knopjes, 2004) to provide three distinct streams: a fine to conventional flotation, a coarser middlings product to coarse flotation and oversize scalped back to the mill via a fine screen.

Layered on top of the circuit is a waste rejection capability that can reject barren rock, or extremely low-grade ore. Potentially viable ore grade can be processed or deferred for later processing when the circuit has excess capacity. This rejection circuit utilises current grade detectors and/or future capability as it is refined. This circuit requires a smaller conventional coarse ore stockpile, instead directing capital to building the deferred stockpiles that can have reclaim feeders or utilise front end loaders to recycle back via installed conveyors.

Ahead of the circuit is ore characterisation, orebody mapping, blast control, ore tracking – all providing a prediction of what is arriving at the comminution circuit. Thus, grade detectors and diversions are activated as required, avoiding the risk of sorting high-grade batches of ore, instead focusing on middling or uncertain grade – ensuring value add not value depletion.

In order to capture the value of such a circuit at design stage, all the features need to be captured. This level of detail is beyond current models and simulators, but has been a long-held vision, presented in the work of Powell, Foggiatto and Hilden (2014) and Foggiatto (2017). Intensive modelling work, calibration off operating circuits, modification of existing models and customising models to respond to the differential feeds within the circuit that are generated, indicated that such flexibility could provide a 25 per cent increase in capacity, with as little as half the comminution energy for metal production and with higher recoveries. This could be predicted at fixed steady-state conditions proposed by the researchers, but the awkwardness of the hand-crafted simulations did not facilitate searching for optimal solutions. Furthermore, a significant advantage of such circuits is the capability to shift the ore flow as the feed competence changes and the target grind for recovery varies, achieved by varying the work between crushing, bed breakage, primary and secondary milling. To adequately simulate this, a genuinely dynamic simulation platform is required, with models that can handle multi-component feeds. Such simulation capability does not currently exist.

Although there is nothing revolutionary about the technology presented in the FlexiCircuit concepts, the fact that we cannot adequately simulate their performance points to where we should be tooling up for the future in terms of designing for future circuits based off current layouts and expansion options. Designing a circuit to provide a more stable feed to recovery will undoubtedly increase recovery and overall circuit value. However, without the necessary tools, the value of the added capital to achieve this cannot be quantified and thus justified, leading our future circuits to merely be copies of our old technology.

FUTURE TOOLS

Some ideas on future tools are presented as a basis for illustrating the potential link of the future with current improved practice.

Critical tools include:

- Significant advances in simulation of mineral processing circuits:
    - dynamic simulation of equipment and circuits
    - model multi-component ores
    - include mineral association and liberation.

- Added in-field sensors:
  - cyclone underflow flare
  - individual cyclone product size (for cluster balancing)
  - mill filling: rock, ball, slurry:
    - physical vibration and acoustic sensors
    - soft sensors based off a range of available input, including redundant data, such at the JK MillFIT.
  - the practical process measures presented by Powell *et al* (2022) provide simple measures that can be linked into online data (such as product size, mill power draw).
- Build and modify circuit layout to enable measurement for ongoing control and optimisation and enhanced with added data collection points, such as mill product. Ideas include:
  - inserting ore-flow transfer points for greatly improved image analysis and quantification of ore feed size distribution
  - utilising an upfront primary screen as a measure of three-point size distribution
  - shifting mill discharge flow to pass over launders with an inbuilt sampling point – enabling the critical measure of mill product
  - installing feed samplers on cyclone clusters
  - size analysis of pebble crusher product.
- Advanced and responsive blast design:
  - From the orebody map which is steadily refined during mine life to final resolution at the blast bench; every blast can be customised to refine explosive loading between and downholes, drill hole spacing, added holes, precise timing modified to provide optimal rock blast movement so as to minimise missing and misplacement of grade.
  - 3D models used to model this ahead of every blast to control and predict the ore location after each blast.
  - Significant improvements in delivered grade and corresponding reduction in dilution and loss have been measured when this is trialled, such as presented in the work of Powell *et al* (2018).
  - Incorporate ore sorting at the face, with systems such as XRF-based ShovelSense (Futcher, Seaman and Klein, 2024), to redirect low or middling grade shovel-loads and recover high-grade ore that can be lost to the waste dump.
- Materials handling designed to prevent segregation.
  - Advanced simulation tools exist to simulate rock flow and enable redesign of all transfer points. Many examples of successful use of DEM modelling indicate that this should be a standard design tool. For example, a recent adjustment of feed flow to a recycle crusher extended crusher chamber life from 3 weeks to over 4 months (Evertsson and Quist, 2022; Quist and Evertsson, 2010).
  - Do not feed multiple mills off one sump. A concept design is illustrated in Figure 10 of the SAG mill feeding a transfer sump that lifts the slurry to a multichannel gravity splitter above two slurry tanks, that are each closed with a ball mill. This offers a number of advantages:
    - lower volume, smaller SAG sump, thus less sump and mill foundation height needed
    - ability to properly split between the ball mills
    - maintenance of even split by the ball mill circuits being separated
    - the splitter can divert all flow to one mill in the case of maintenance.

- o Control ore flow onto stockpiles to prevent them shifting with height, reduce dust issues and minimise segregation across the stockpile. A deflector at the head of the feed conveyor can be used to force vertical flow, which addresses much of this (Powell *et al*, 2024).

- o Deliberately induce segregation to enhance controllability, such as proposed by Powell *et al* (2024) to provide a controlled feed of rock for media to a SAG or AG mill.

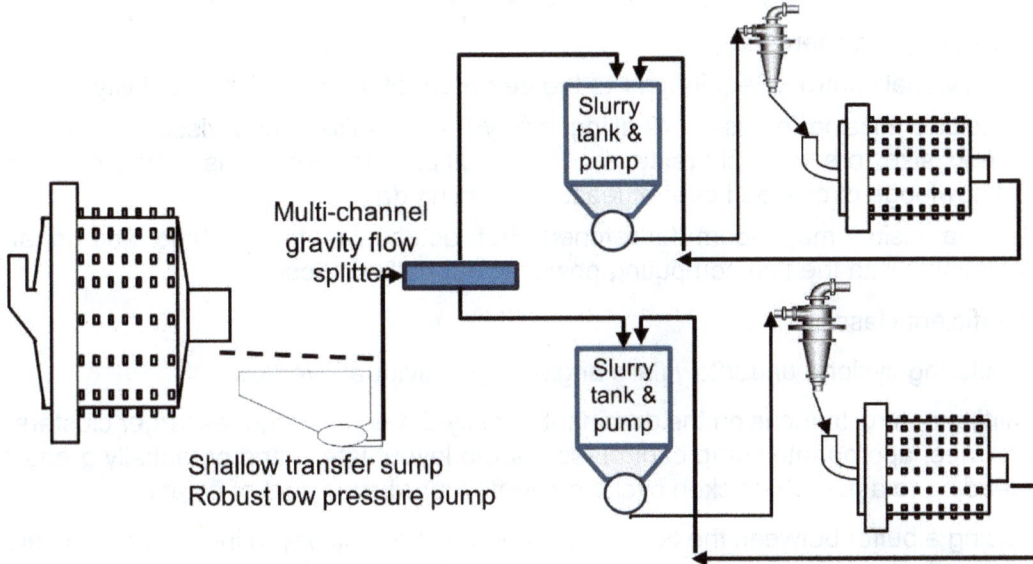

FIG 10 – Conceptual layout of two-stage pump and splitter to ball mills.

- Ore tracking:
  - o Integrate the geomet, blasting, shovel, trucking, crushing, conveying and stockpile models to provide a predictive simulation of ore flow from mined face to mill and on to recovery.
  - o Utilise blast movement modelling to control and track ore deportment from every blast – feeding forward to shovel control.
  - o Include high-precision GPS on all shovels, trucks, front end loaders so that ore movement can be tracked within the resolution for each piece of earth moving equipment, and automatically fed into the tracking system.
  - o Implement stockpile modelling to track ore-storage and movement and incorporate flow and segregation of ore in stockpiles, as developed at the JKMRC (Ye, Hilden and Yahyaei, 2022).
  - o Utilise tracking tags to check and calibrate the tracking system.

- Model-based control:
  - o Upgrading models to be genuinely dynamic, ie realistically track real-time dynamic responses, allows them to span the full range from predictive simulation for design through to online models used in daily control.
  - o Implementing a suitable dynamic simulation platform underpins dynamic simulation, such as the Dyssol system developed for modelling multi-property particle systems.
  - o Utilising machine learning to continuously tune the unknown variables in the models (such as slurry viscosity, varying granular packing in mills, wearing screens) and also highlight poor model predictions indicated by the need for unrealistic tuning of physical parameters (such as unrealistic charge porosity or slurry viscosity).
  - o The models will intrinsically be able to seek new operating modes, as they are predictive well outside any current window of operation.

- The control system based on predictive models will seek stability, as the dynamics and detrimental impact of instability are inherent in the dynamic models.

- This approach allows a coherent, concerted effort on modelling with a seamless interface between design and control models.

- The entire circuit through to recovery should be in the process control model – allowing the process control to target the maximising of metal recovery over the longer term, avoiding:
  - chasing short-term targets
  - individual control of equipment to the detriment of the overall productivity
  - include missing inputs – mill filling, slurry load in mills, slurry viscosity, more complete feed size, ore feed characteristics for multiple blended feeds, content of stockpiles, knowledge of ore feed over at least the coming day.

- Such a vision may seem far-fetched, but actually is far-reaching and is absolutely achievable with the fast computing power at our daily disposal.

- **More efficient classification:**

  - Monitoring cyclone underflow flare angle, and individual overflows.

  - Shifting control to focus on the dominant density driver, this requires larger clusters of many cyclones, appropriate sump control (see sump layout later), and potentially greater dilution – leading to a need to thicken cyclone overflow product ahead of flotation.

  - Placing a buffer between the SAG circuit and flotation will pay in increased recovery. Again, this needs dynamic measures and modelling to demonstrate the value. Adequate thickening to shift an ideal cyclone overflow per cent solids from 25 to the 32 per cent required for flotation, will open up cyclone control to allow more appropriate feed density, escaping the 60 to 65 per cent commonly seen on plants. The thickening also provides a buffer ahead of flotation and ensures a consistent feed density regardless of fluctuations handed down from the cyclones.

  - Hydrocyclones are intrinsically inefficient. Hybrid classification utilising fine screens and cyclones are likely to dominate future circuits in order to increase efficiency and recovery, such as demonstrated by Frausto *et al* (2017) with fine screening, and Mainza, Powell and Knopjes (2006); Becker *et al* (2008) with the three-product cyclone integrated with a fine screen.

- **Incorporate mineral association:**

  - Mineral separation for metal recovery is the purpose of mineral processing, yet our models lack this underlying data.

  - Building toolboxes of ore characterisation, ore testing, and modelling that are based on mineral association, is seen as a critical aspect to dramatically upskilling our abilities and being prepared for the future.

  - The approach of Hilden and Powell (2017) in developing a geometric texture model that can rapidly reproduce mineral association of fragmented ore, leads the way in incorporating mineral liberation into our modelling.

- **Ore characterisation to capture the significant primary ore properties:**

  - Moving away from proxies and measures that are locked into specific equipment and circuits opens the door to innovative circuits and operation.

  - Targeting ore primary properties in measures of orebody characterisation (Powell, 2013) allows the response of different options of equipment and circuits to be derived from an orebody model.

  - Ore characterisation with a consistent and precise measure that can be applied to any equipment, such as that presented by Ali *et al* (2022). The test can be applied to large or small samples, so as to provide an equally precise measure for small samples taken from drill cores as for bulk samples, thus removing the need for proxies of conventional measures. The Geopyörä test being another to provide such a measure for coarser rock samples, but still small sample sizes (1–2 kg) (De Paiva Bueno, Almeida and Powell, 2023; Bueno *et al*, 2023).

All of the suggestions listed above provide avenues to immediate improvement of existing circuits, but also provide a pathway, tools, and measures to future circuits. As presented by Powell and Mainza (2012) there is no giant leap to dramatic circuit changes, instead the leap is made in many steps that build off one another:

- Predictive, mechanistic models on a dynamic platform allow the design with confidence of novel circuits that appear too risky or complex with current design tools.

- More robust online models for control allow more complicated multi-factorial decisions to be made that are beyond even a skilled operator, and facilitate FlexiCircuit implementation driven by value.

- Future responsive circuits will be dependent on relevant and sufficiently high-resolution mapping of the orebody, along with tracking for ore flow to and through the system.

- Growing the workforce and operating expectation to be more technically advanced, builds the workforce and operating platform needed for the uptake of future advanced circuits and equipment.

- The improved tools and confidence in novel technology will allow the industry to tackle big-picture opportunities, such as studied by Powell *et al* (2023) in a team effort that took on the challenge of designing viable technology for a flexible mine and circuit operation powered solely by variable solar energy.

Will the proposed approach liberate up to 20 per cent upside in production of current operations? From the experience of the author and collaborators, any one of the process modifications can represent 10 per cent improvement, when integrated with the multiple steps of improvement available to a site, 20 per cent is a realistic aspiration. The author is also confident that in applying this more considered approach, built off a higher level of people skills and integrated modern technology of high-powered computing and data processing, the companies will be tooling up for future innovation – ready and capable of taking on new technologies with confidence and low risk.

ACKNOWLEDGEMENTS

The many researchers, operators and metallurgists who have allowed us to study their processes and draw on their years of knowledge is acknowledged to underpin the understanding and insights we have developed over the years. The industry funders who have supported the research studies and applied process studies are acknowledged for their foresight in enabling knowledge generation and being generous in allowing the sharing of opportunities across the industry.

REFERENCES

Ali, S, Powell, M S, Yahyaei, M, Weatherley, D K and Ballantyne, G R, 2022. A standardised method for the precise quantification of practical minimum comminution energy, in Proceedings of IMPC 2022: XXXI International Mineral Processing Congress (The Australasian Institute of Mining and Metallurgy: Melbourne).

Ballantyne, G R, Clarke, N, Elms, P, Anyimadu, A and Powell, M S, 2015. Application of research principles to identify process improvement at Sunrise Dam Gold Mine, in Proceedings of Metplant 2015 (The Australasian Institute of Mining and Metallurgy: Melbourne).

Becker, M, Mainza, A N, Powell, M S, Bradshaw, D J and Knopjes, B, 2008. Quantifying the influence of classification with the 3 product cyclone on liberation and recovery of PGMs in UG2 ore, *Minerals Engineering*, 217:549–558.

Bueno, M, Huusansaari, S, Ranta, J P, Rajavuori, L, Veki, L, Almeida, T and Powell, M S, 2023. Applied Geometallurgy at Agnico Eagle's Kittila Operation using the Geopyörä Breakage Test, in Proceedings of International Autogenous and Semi-Autogenous Grinding and High Pressure Grinding Roll Technology 2023 (CIM).

De Paiva Bueno, M, Almeida, T and Powell, M S, 2023. Extensive Validation of a New Rock Breakage Test, *Minerals*, 13(12):1506. https://doi.org/10.3390/min13121506

Dupont, J-F, McMullen, J and Rose, D, 2017. The effect of choke feeding a gyratory crusher on throughput and product size, in *Proceedings of the 49th Annual CMP*, pp 166–175 (CIMM).

Engelhardt, D, Seppelt, J, Waters, T, Apfelt, A, Lane, G, Yahyaei, M and Powell, M S, 2015. The Cadia HPGR-SAG circuit – from design to operation – the commissioning challenge, in Proceedings of International Autogenous and Semi-Autogenous Grinding and High Pressure Grinding Roll Technology 2015 (CIM).

Evertsson, M and Quist, J, 2022. Insights from utilising DEM for re-design of a cone crusher feed system, in Luleå Conference in Minerals Engineering 2022 (ed: T Karlkvist).

Foggiatto, B, 2017. Modelling and simulation approaches for exploiting multi-component characteristics of ores in mineral processing circuits, Thesis submitted in fulfilment of PhD, University of Queensland.

Frausto, J J, Ballantyne, G R, Runge, K, Powell, M S and Cruz, R, 2017. The Impact of Classification Efficiency on Comminution and Flotation Circuit Performance, in Proceedings MetPlant 2017 (The Australasian Institute of Mining and Metallurgy: Melbourne).

Futcher, W, Seaman, D R and Klein, B, 2024. Redefining the battery limits of processing plants – improving Sustainability through the deployment of sensing technologies, in Proceedings 16th Mill Operators Conference 2024 (The Australasian Institute of Mining and Metallurgy: Melbourne).

Hanhiniemi, J and Powell, M S, 2023. Improved Techno-economic Assessment and Modelling of Comminution with Cadia as a Case Study, Proceedings International Autogenous and Semi-Autogenous Grinding and High Pressure Grinding Roll Technology 2023 (CIM).

Hilden, M M and Powell, M S, 2017. A geometrical texture model for multi-mineral liberation prediction, *Minerals Engineering*, 111:25–35. https://doi.org/10.1016/j.mineng.2017.04.020

Mainza, A N, Powell, M S and Knopjes, B, 2006. Enhancing platinum recovery with the three-product cyclone, *Proceedings of the XXIII IMPC*, pp 1578–1583.

Mainza, A N, Powell, M S and Knopjes, B, 2004. Differential classification of dense material in a three-product cyclone, *Minerals Engineering*, 17(10):573–579.

Powell, M S and Mainza, A N, 2012. Step change – A staircase rather than a giant leap, *Proceedings of the XXVI IMPC*, pp 2750–2760.

Powell, M S, Hilden, M M and Mainza, A N, 2015. Common operational issues on SAG mill circuits, Proceedings International International Autogenous and Semi-Autogenous Grinding and High Pressure Grinding Roll Technology 2015 (CIM).

Powell, M S, 2013. Utilising Orebody Knowledge to Improve Comminution Circuit Design and Energy Utilisation, Proceedings of the Second AUSIMM International Geometallurgy Conference (The Australasian Institute of Mining and Metallurgy: Melbourne).

Powell, M S, 2018. Integrated performance optimisation of comminution circuits, in Proceedings of Complex Orebodies (The Australasian Institute of Mining and Metallurgy: Melbourne).

Powell, M S, Evertsson, C M, Mainza, A N and Ballantyne, G R, 2022. Practical measures of process efficiency and opportunity, in Proceedings of IMPC 2022: XXXI International Mineral Processing Congress (The Australasian Institute of Mining and Metallurgy: Melbourne).

Powell, M S, Evertsson, C M and Mainza, A N, 2019. Redesigning SAG mill recycle crusher operation, Proceedings International Autogenous and Semi-Autogenous Grinding and High Pressure Grinding Roll Technology 2019 (CIM).

Powell, M S, Foggiatto, B and Hilden, M M, 2014. Practical simulation of FlexiCircuit processing options, Proceedings XXVII IMPC.

Powell, M S, Hilden, M M, Evertsson, C M, Asbjörnsson, G, Benzer, A H, Mainza, A N, Tavares, L M, Davis, B, Plint, N and Rule, C, 2012. Optimisation opportunities for HPGR circuits, in *Proceedings of the 11th Mill Operators Conference*, pp 81–94 (The Australasian Institute of Mining and Metallurgy: Melbourne).

Powell, M S, Kanchibotla, S S, Jokovic, V, Hilden, M H, Bonfils, B, Musunuri, A, Moyo, P, Yu, S, Young, J, Yaroshak, P, Yalcin, E and Gorain, B, 2018. Advanced Mine to mill application at the Barrick Cortez mine, in Proceedings of the 14th Mill Operators Conference (The Australasian Institute of Mining and Metallurgy: Melbourne).

Powell, M S, Mainza, A N and Yahyaei, M, 2023. Low-cost SAG milling opportunities, Proceedings International Autogenous and Semi-Autogenous Grinding and High Pressure Grinding Roll Technology 2023 (CIM).

Powell, M S, Mainza, A N, Tavares, L M and Kanchibotla, S, 2024. Rockmedia, in *Proceedings of the 16th Mill Operators Conference*, pp 425–442 (The Australasian Institute of Mining and Metallurgy: Melbourne).

Powell, M S, Reynolds, A, McCrae, S, Agnew, J, Bracey, R, Way, D, Phasey, C, Littlechild, T and Manning, D, 2023. Can we shift to a new paradigm of flexibility in mining and processing to build mines powered exclusively by the variable input of renewables?, in Proceedings 26th World Mining Congress (CSIRO).

Powell, M S, Yahyaei, M, Mainza, A N and Tavares, L M, 2021. The endemic issue of ball mill overload in SABC circuits, in Proceedings of the 15th Mill Operators Conference 2021 (The Australasian Institute of Mining and Metallurgy: Melbourne).

Quist, J and Evertsson, C M, 2010. Application of discrete element method for simulating feeding conditions and size reduction in cone crushers, in *Proceedings of the XXV International Mineral Processing Congress (IMPC) 2010*, 5:3337–3347.

Yahyaei, M and Powell, M S, 2018. Production improvement process optimisation opportunities in comminution circuits, Proceedings 14th Mill Operators conference (The Australasian Institute of Mining and Metallurgy: Melbourne).

Ye, Z, Hilden, M M and Yahyaei, M, 2022. Development of a three-dimensional model for simulating stockpiles and bins with size segregation, in IMPC 2022: XXXI International Mineral Processing Congress (The Australasian Institute of Mining and Metallurgy: Melbourne).

Modelling to prepare for and mitigate the impact of increasing ore hardness at Lihir

L Pyle[1], A Rice[2], P Griffin[3], B Seaman[4], W Valery[5], K Duffy[6] and A Jankovic[7]

1. AAusIMM, Process Engineer, Hatch, Brisbane Qld 4000. Email: lindon.pyle@hatch.com
2. Senior Specialist – Metallurgy, Newmont, Milton Qld 4064. Email: amanda.rice@newmont.com
3. Paul Griffin, formerly Group Manager – Processing, Group Technical Services, Newcrest, Melbourne Vic 3004.
4. Head of Mineral Processing, Group Technical Services, Newmont, Subiaco WA 6008. Email: brigitte.seaman@newmont.com
5. Global Director – Consulting and Technology, Hatch, Brisbane Qld 4000. Email: walter.valery@hatch.com
6. Process Consultant, Hatch, Brisbane Qld 4000. Email: kristy.duffy@hatch.com
7. Principal Consultant, Hatch, Brisbane Qld 4000. Email: alex.jankovic@hatch.com

ABSTRACT

Lihir, located on Niolam Island, 900 km from Port Moresby, Papua New Guinea, is one of the world's largest producing gold mines. Future changes in feed blend (less argillic ores and an increased proportion of porphyry and epithermal ores) will increase ore hardness and thus impact comminution circuit throughput. Lihir has experienced harder ores in the past, although more recently, the feed has been predominantly softer. Unusually, the Lihir circuit can swing between semi-autogenous grinding (SAG) or ball mill limiting conditions due to the variable ore hardness in terms of resistance to impact breakage (for SAG milling) versus fine grinding.

To develop effective strategies to maximise performance for the future ore types, it is necessary to understand the ore characteristics, the impact of these on the Lihir circuit, and when the circuit will be SAG or ball mill limited. Then owner Newcrest (now Newmont) engaged Hatch to assist, and the project involved analysing ore characterisation data, historical and recent operating data, previous studies and surveys as well as developing site-specific comminution models to reflect the current conditions. Several modelling and simulation approaches were used. Morrell power-based modelling estimates the specific energy of the comminution circuit as a function of ore hardness and feed size for the required product size and was used to assess the impact of changing feed characteristics and determine when the circuit would be SAG or ball mill limited. However, power-based models do not have the sensitivity required to investigate, in detail, the impact of changing many operating conditions. This was achieved using calibrated process models, such as those used within JKSimMet. Together these models were used evaluate the influence of ore properties on comminution circuit performance and assess strategies to mitigate the impact of increasing ore hardness.

This paper describes the data analysis, modelling, simulation and the resulting recommendations to maximise production for future ores.

INTRODUCTION

The Newmont Lihir Operation is located on the Niolam Island, which is a part of the Lihir Group in the Province of New Ireland. The island is located approximately 900 km north-north-east of the national capital, Port Moresby. The Lihir processing plant is fed from an adjacent open pit mine and stockpiled ore via a primary gyratory crusher, and mineral sizer and two jaw crushers. The primary crusher and sizer products are conveyed jointly to a coarse ore stockpile (COS) and the jaw crushers feed a secondary conveying circuit to the same coarse ore stockpile. The COS feeds three grinding circuits. Two (referred to as the HGO1 and HGO2 mills) of the three circuits are comprised of semi-autogenous grinding (SAG) and ball mills and shared pebble crushing. The third grinding circuit (FGO) is comprised of a SAG and ball mill. FGO scats are transported back to the coarse ore stockpile where they are reclaimed to the HGO milling circuits. Figure 1 shows a schematic of the Lihir crushing and grinding circuit, with the key equipment considered in the throughput modelling summarised in Table 1. The Lihir processing circuit is comprised of three processing paths: two parallel flotation circuits and an autoclave circuit (direct). Direct ore is dewatered and fed to pre-oxidation tanks where it is blended with flotation concentrate before being fed to one of four autoclaves. Autoclave discharge is washed in a counter-current decantation circuit, pH adjusted using lime and then fed to cyanidation leach tanks where gold is extracted from oxidised solids into solution and recovered via carbon adsorption.

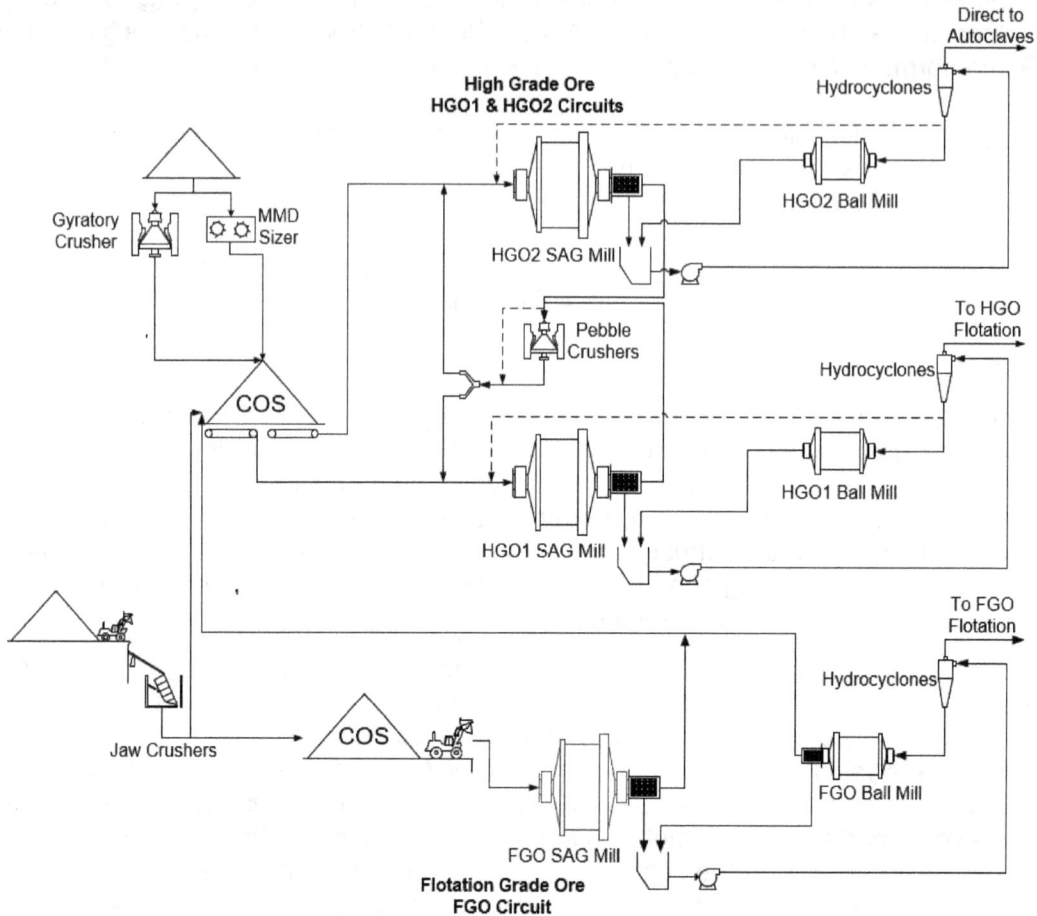

FIG 1 – Lihir comminution flow sheet.

TABLE 1

Grinding circuit key equipment.

Equipment type	HGO1 and HGO2 circuits	FGO circuit
Crushers	Fuller-Traylor 43 inches × 69 inches gyratory, 375 kW	Two × Thyssen Krupp EB 16–12 N single toggle jaw crushers, 200 kW
SAG mills	Two × 8.32 m (inside liners) by 3.75 m, 5.5 MW	7.11 m (inside liners) by 4.55 m, 4.5 MW
Pebble crushers	Two FLS Raptor XL500, 370 kW	-
Ball mills	Two × 5.35 m diameter × 9.51 m long (inside liners), 5.5 MW	5.37 m diameter × 8.58 m long (inside liners), 4.5 MW
Hydrocyclones	8 × Cavex 650 mm, one cluster per mill	8 × Cavex 650 mm

At the time of conducting this evaluation, the mine plan reflected that plant feed would consist of predominantly fresh, ex-pit material from porphyry and epithermal domains which are significantly harder than the ore that was being fed to the plant. With the change in ore from a blend of softer stockpile ore and ex-pit to dominantly ex-pit it is expected that the relatively harder ore will impact comminution circuit throughput rates. Therefore, Hatch was engaged to assess required operational changes in the grinding circuit to make adjustment for the expected changes in ore properties and predict future circuit throughputs. Hatch also investigated benefits of continued pebble crushing or installing secondary crushing.

MODELLING

Different types of models are required depending on the purpose and objectives. Power-based modelling is a common approach to throughput forecasting, as this can estimate the specific energy of the comminution circuit as a function of ore properties (specifically hardness) and feed size for the required product size. However, power-based models do not have the sensitivity required to investigate, in detail, the impact of changing many operating conditions to optimise performance and throughput. This can be achieved by using calibrated (site-specific) process models, such as those used within JKSimMet. Both model types (Morrell power-based and JKSimMet process models) were calibrated and used in this project.

Feed size and ore characteristics, particularly hardness, have a significant impact on comminution circuit capacity and performance. Therefore, these are critical input variables for both modelling approaches and for circuit design and optimisation. Accurate estimates of both are required to ensure reliable models and simulation results.

Ore characterisation

Comminution circuit modelling and optimisation requires measurement of hardness both in terms of resistance to impact breakage (applicable in AG/SAG mills and crushers) and resistance to abrasion breakage at finer size fractions, (ie applicable in ball milling circuits).

The JK Drop Weight Test and the SAG Mill Comminution test (SMC test) measure breakage characteristics to model and evaluate crusher and AG/SAG mill performance. The SMC test requires less sample. The results from these tests include the JK Drop-Weight index (DWi) and Axb parameter which are measures of rock strength when broken under impact as well as comminution parameters M_{ia}, M_{ib}, M_{ih} and M_{ic} which are work indexes for different grinding parameters. M_{ia} is the work index for the grinding of coarser particles (> 750 µm) while M_{ib} is for the grinding of fine particles (< 750 µm) in tumbling mills. M_{ih} is the work index for the grinding in high pressure grinding rolls (HPGR) and M_{ic} for size reduction in conventional crushers.

The Bond Ball Mill Test and Work Index (BBWi) is the industry standard measure of the resistance of the material to fine grinding in a ball mill.

At Lihir, the increasing hardness of future ex-pit ores is expected to have an impact on the throughput of the comminution circuit. Therefore, Hatch reviewed the available ore characterisation data from the Lihir geometallurgical database to understand the changes in ore hardness and identify mitigating strategies.

The ores at Lihir are classified by alteration domains. There are three main groups (porphyry, epithermal and argillic), each of which are divided into three sub-groups for a total of nine domains. The porphyry domains are inner biotite (IB), outer biotite (OB), and chlorite (CHL); the epithermal domains include lower epithermal (LE), upper epithermal (UP), and silica breccia (SB); and the argillic domains are argillic (A), advanced argillic (AA), and upper argillic (UA).

Summary box plots for Axb and BBWi by alteration domain are shown in Figure 2. Note, there was no Axb data and only one data point for BBWi for the upper argillic (UA) domain; however, this domain represents only a small proportion of future ore. The epithermal and porphyry domains are considerably harder than argillics, both in terms of resistance to impact breakage (Axb) and coarse grinding (M_{ia}) for SAG milling and resistance to fine grinding in a ball mill (BBWi). Note that, a lower Axb value indicates harder ore, for all other indices a higher number represents harder ore. The hardness characteristics selected for simulations for each alteration domain were based on the 75th percentile (25th percentile in the case of Axb as lower value is harder) to be conservative.

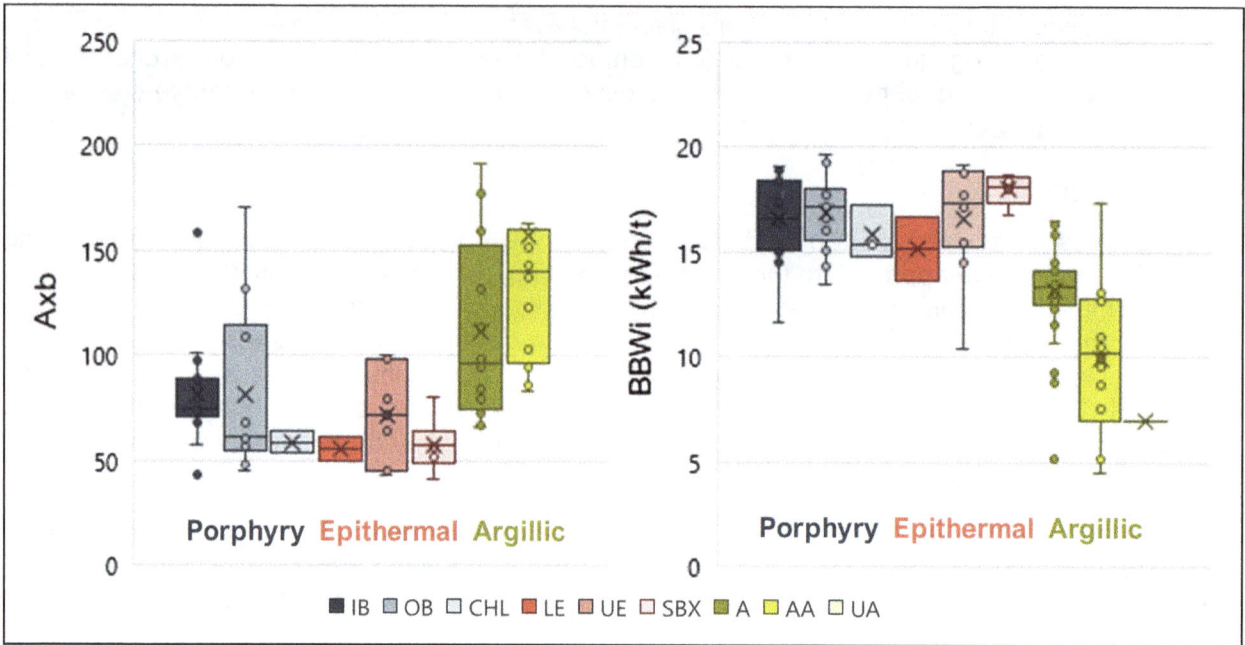

FIG 2 – Summary of ore hardness by alteration domain.

The future feed blend at Lihir would include an increasing proportion of epithermal ore and a decreasing proportion of the softer argillic ores according to the mine plan at the time of this study. However, the mine plan is regularly reviewed and adjusted, so the feed composition and forecast results presented here may not be indicative of the current plans. Rather these results were used for optimisation of strategies over the life-of-mine and also provide feedback to mine planning considering the impact on downstream processes.

Feed size

Comminution performance is also strongly influenced by feed size and the Run-of-Mine (ROM) fragmentation resulting from blasting is strongly influenced by *in situ* ore characteristics, rock structure and strength. Therefore, the feed size will vary for each alteration domain due to the different ore characteristics. For power-based and process modelling to accurately estimate the throughput for different ore types, it is necessary to determine an appropriate feed size for each alteration domain based on the ore characteristics.

Several historical belt-cut size distribution measurements of SAG feed were available for different ore sources with varying hardness. The SAG feed 80 per cent passing size (F_{80}) was estimated for each alteration domain by adjusting historical belt cut results according to the Axb value. The results are presented in Figure 3.

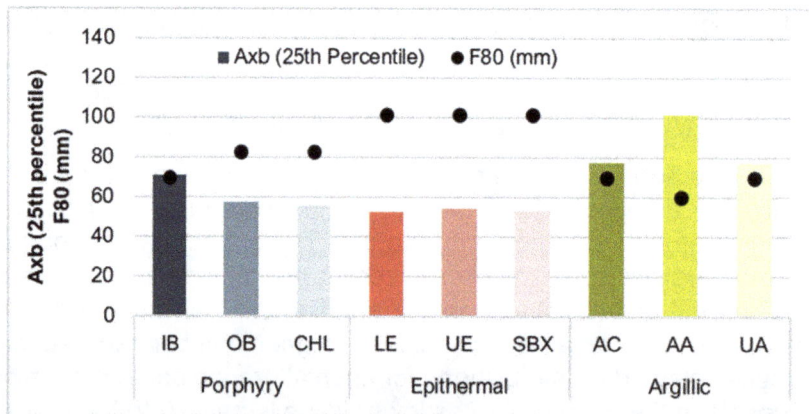

FIG 3 – Summary of feed size by alteration domain.

The variation in feed size due to different *in situ* ore characteristics can be mitigated by optimising blast designs according to the ore characteristics. Increasing fines (-10 mm) generation during blasting can significantly improve the downstream comminution performance, particularly for hard ores. This will impact the accuracy power modelling, which uses only the 80 per cent passing size and does not account for changes in the shape of the size distribution curve. Hatch has a methodology to account for the increase in fines from blast optimisation, as discussed by Rodriguez *et al* (2023), but it was not required in this case.

Morrell power-based model development

Morrell Power-based methodology (Morrell, 2004a, 2004b) can be applied to estimate the achievable throughput over the Life-of-Mine (LOM) considering the available grinding power and feed ore characteristics. The principle is to estimate the specific energy of the comminution circuit as a function of ore properties (specifically hardness) and feed and product sizes.

The total circuit specific energy (Equation 1) is calculated using the approach proposed by Morrell (2004a, 2004b) which is accepted as industry standard (Global Mining Guidelines Group, 2021). The difference between the total circuit and AG/SAG specific energy (calculated from Equation 2) is used to determine if the circuit is SAG mill or ball mill limited.

The key equations for total and SAG/AG specific energy are as follows:

$$Total\ Specific\ Energy\ \left(\frac{kWh}{t}\right) = 4M_{ia}\left(750^{f(750)} - F_{80}^{f(F_{80})}\right) + 4M_{ib}\left(P_{80}^{f(P_{80})} - 750^{f(750)}\right) \quad (1)$$

$$SAG/AG\ Specific\ Energy\ \left(\frac{kWh}{t}\right) = K \times F_{80}^{a} \times DWi^{b} \quad (2)$$

where:

DWi, Axb, M_{ia}, and M_{ib} are ore characteristics as described previously

F_{80} 80 per cent passing size (μm) of the grinding circuit feed

P_{80} 80 per cent passing size (μm) of the grinding circuit product

f(x)   defined functions of F_{80} and P_{80}, as per Morrell (2004b)

K   site-specific factor calibrated to production data

The feed and product sizes of the circuit are incorporated using the 80 per cent passing values (F_{80} and P_{80}) and the site-specific factor was calibrated by aligning results with survey data. The throughput is then calculated from the weighted specific energy based on the blend of nine ore types, allowing for 90 per cent of installed power draw for the SAG mills and historical maximums for the Ball mills. Therefore, a throughput forecast can be determined using the feed ore characteristics from the mine plan. This can be used for long-term strategic planning as this will indicate when changes in ore types will impact plant throughput. In some cases, this may be mitigated by blending or changing operating practices. If not, advance notice is received for when additional comminution equipment would be required to maintain production levels.

The calibrated power-based throughput forecast model for Lihir was used to evaluate the impact of changing ore characteristics on comminution performance over time (Financial Year (FY) 2024 to 2035). This forecast indicates when throughput will be constrained by harder ore, when it will be ball or SAG mill limited, and thus, when and if pebble and/or secondary crushing may be of benefit. This approach was also used to investigate the benefit of feed blending.

Semi-mechanistic process model development

The Morrell power-based modelling discussed above can highlight when changing ore properties may cause issues for comminution circuits, and when the circuit is expected to be either SAG or ball mill limited. However, semi-mechanistic process models, such as those used in JKSimMet, are required to assess the impact of changing operating conditions and evaluate potential mitigating and optimisation strategies.

The comminution models in JKSimMet require both machine and ore parameters. Machine parameters are related to the dimensions and characteristics of the comminution devices. Ore parameters are generally determined from laboratory testing of ore samples. The models are calibrated (using a process called model fitting) for the current operating conditions based on survey data.

Full surveys of the Lihir grinding circuits were carried out in 2017, and at the time, the Hatch team developed and calibrated models for the three grinding circuits: FGO, HGO1 and HGO2. These previous models were updated to reflect current feed characteristics, operating conditions, and circuit changes.

As there have been significant changes in ore hardness and further changes are expected in the future, the models for each circuit were modified to create three versions of each to cover the range of ore hardness of the feed blends: soft (SAG specific energy < 4.5 kWh/t), medium (SAG specific energy of 4.5–6.5 kWh/t and hard ore (SAG specific energy > 6.5 kWh/t), and the SAG feed size was also varied accordingly. These three hardness models allowed assessment of operating conditions for each of the different feed hardness classifications and determination of optimum operating strategies considering changes in feed blend. The models were correlated and validated against operating data.

The site-specific JKSimMet process model was used to investigate the impact of factors such as pebble crushing, SAG grate design and cyclone underflow bleed to make recommendations for circuit optimisation. It is an iterative process, and these results also informed the Morrell throughput forecast model which was used to evaluate the impact of changing ore characteristics on comminution performance over time (FY24 to 35), which in turn indicated areas to investigate with the JKSimMet model to address ball or SAG mill limiting conditions. The results of both models are used together in an integrated approach to determine when and what changes to operating conditions and/or equipment are required to improve performance considering the feed ore characteristics.

RESULTS

HGO circuits

Pebble crushing

One of the focusing questions for this project was if and when the HGO circuits would benefit from pebble crushing. Two FLS Raptor XL500 pebble crushers installed on-site had recently been refurbished, and with a short head fine mantle at closed side setting (CSS) of 13 mm have a capacity in the range of 250 to 320 t/h per unit. Thus, combined HGO mills pebble production with varying ore blends must be reviewed considering this capacity.

Pebble crushing can increase throughput for hard ores, particularly those with high resistance to impact breakage which limit SAG mill capacity. However, many of the alteration domains (ore types) at Lihir have a high resistance to fine grinding in a ball mill (high BBWi). Therefore, the throughput is constrained by the ball mill rather than the SAG mill if the product size must be maintained. Both pebble and secondary crushing have limited benefit when circuits are ball mill limited (unless a coarser product size can be tolerated). Either will result in coarser transfer size to the ball mill, shifting the load from the SAG mill to the ball mill, which is already constrained.

Based on Morrell power-based modelling, the HGO circuits are expected to be predominantly ball mill limited at a required product size P_{80} of 190 µm, as shown Figure 4 (left). The throughput y-axis values in this and subsequent figures have been normalised, whereby X represents a set tonnage, 2X is double this value, and 3X is triple, to indicate the magnitude of throughput variation.

The throughput forecast shows a lower throughput with pebble crushing than without pebble crushing. This is because the Morrell calculations are conducted for at a fixed product size (in this case P_{80} of 190 µm). The transfer size from SAG to ball mill is increased with pebble crushing, shifting load to the ball mill, which in this case is already limited, thus decreasing throughput assuming the product size is constrained. In practice, operators are unlikely to reduce the throughput due to coarsening product size (which is only measured reliably on shift basis) unless there is a significant reduction in downstream recovery. The more likely outcome with pebble crushing is increased throughput, but with coarser product size. This can be shown by comparing with Morrell calculations with coarser product size, as shown in Figure 4 (right).

HGO Circuits - Throughput Forecast FY25 - 35

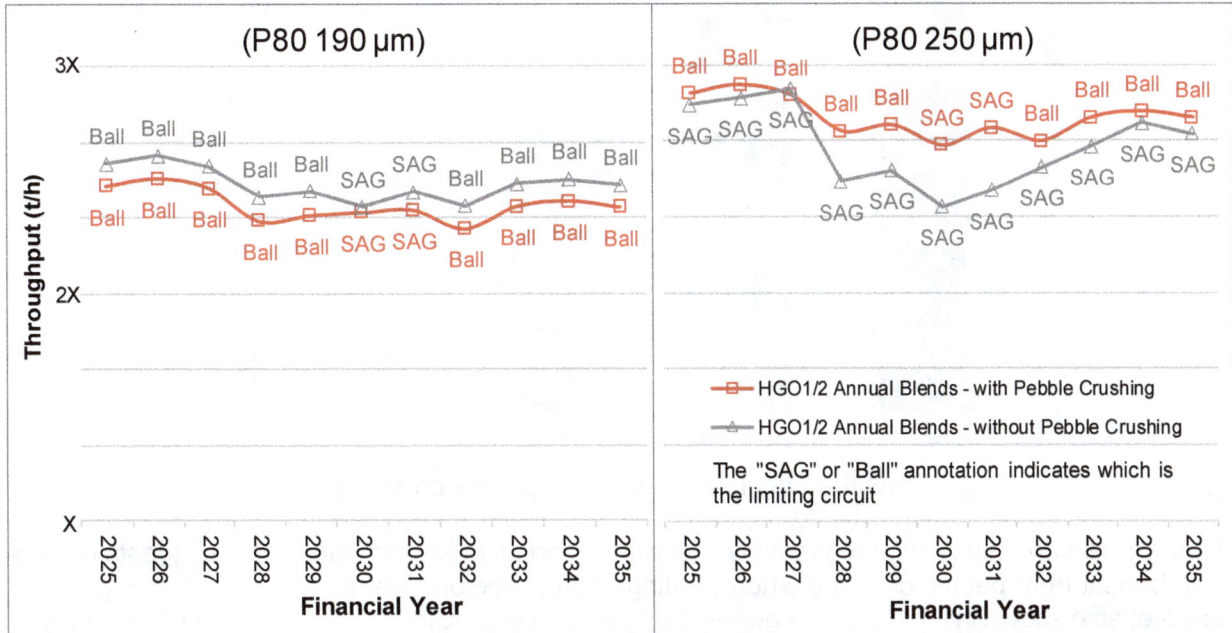

FIG 4 – Throughput forecast for HGO circuits.

Despite the predominance of ball mill limiting conditions, HGO circuit pebble crushing is required to maximise throughput when treating ores with high resistance to impact breakage, ie when the circuits become SAG mill limited. The throughput forecast assumes a constant feed blend is maintained perfectly throughout year. Even with feed blending, the circuit is likely to experience periods of harder (SAG limiting) ore, even in years where the overall average feed blend is softer. In these periods, pebble crushing will be required to maintain high throughput rates, as confirmed by both historical data analysis and JKSimMet simulations. The benefits of feed blending are evaluated and discussed in more detail in a later section.

Simulations with the JKSimMet model indicated minimal benefits from pebble crushing for soft ore, but greater increases in throughput when treating medium (6 per cent increase) and hard ore (10 per cent increase), as shown in Figure 5. In particular, the pebble production increased by over 20 per cent relative to the base case when processing hard ore domains without pebble crushing, indicating the potential for buildup of a recirculating critical size material which may be detrimental to circuit fresh feed rate.

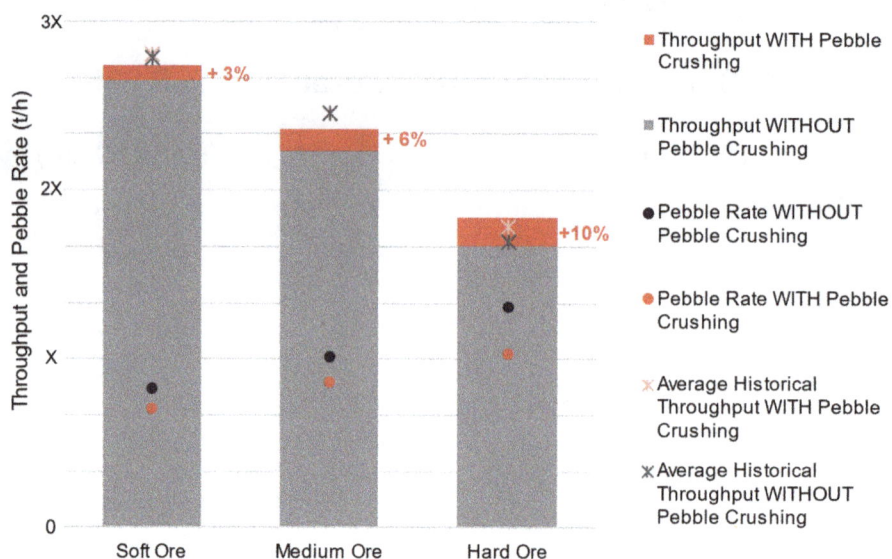

FIG 5 – Simulation results for pebble crushing.

These simulation results align reasonably well with historical operating data which suggests minimal or no benefit from pebble crushing when treating soft or medium ore, but increased throughput for hard ore, also shown in Figure 5. Of course, the operating data is influenced by many other changing conditions and parameters, but overall is consistent with the findings of the simulation results. Considering both modelling approaches and the historical data, it is evident that pebble crushing is of benefit for Lihir HGO circuits when treating hard ores (SAG limiting), but not for soft ores (ball mill limiting). This is expected, as pebble crushing will coarsen the transfer size and thus shift load to the ball mill, so there will be no throughput benefit when treating the softer ball mill limiting ores unless a coarser product size can be tolerated as discussed previously.

Therefore, as a general guide for Lihir, it was recommended to operate pebble crushing when total pebble production from both HGO circuits is greater than 250 t/h. This will ensure choke feeding of the pebble crushers and will typically occur for harder ores when pebble crushing is required. For softer ores the pebble production will be lower and pebble crushing can be bypassed to maintain load in SAG mill.

To ensure sufficient pebble production to achieve choke feeding and avoid ring bounce, it was recommended to install SAG grates with larger pebble ports (a 50/50 mix of 60 and 70 mm apertures). However, larger pebble ports can result in difficulty maintaining SAG mill load for softer ores. Therefore, it was recommended only to change to the larger pebble ports when lower proportions of the softer argillic ore are expected in the feed and this could be aligned with the annual and monthly feed blend schedules.

Secondary crushing

Similarly to pebble crushing, there is limited benefit to secondary crushing when the circuit is ball mill limited, as this too will coarsen the transfer size and shift more load to the ball mill. At Lihir, pebble crushing would be sufficient to increase the SAG mill capacity to match the ball mill capacity when treating SAG limiting ores. Furthermore, the pebble crushing can be operated as required (when pebble rate exceeds 250 t/h) and bypassed, thus maintaining load and finer transfer size from SAG mill, when treating softer ores. Simulations confirmed HGO circuit capacity cannot be increased by addition of secondary crushing at the current required product size.

Cyclone underflow recycle

As the HGO circuits are expected to be predominantly ball mill limited, they may benefit from a partial recycle of the cyclone underflow (U/F) to the SAG mill. This would shift the load from the ball mill back to the SAG mill and help alleviate the ball mill constraint.

Both HGO circuits have the capability to recycle about 20 per cent of the cyclone underflow to the SAG mill, and this has been used intermittently in the past. It is difficult to clearly ascertain the benefits in historical data, due to the large number of other changes; however, small operational improvements were identified when ore was soft and SAG power was low (ie ball mill limiting conditions), as shown in Figure 6. Likewise, it is difficult to quantify the absolute throughput with modelling due to complex interactions; however, simulations with the JKSimMet model confirmed that partial cyclone recycle of about 20 per cent can reduce the ball mill recirculating load and thus alleviate ball mill overloading conditions, also shown in Figure 6. Additionally, the curved pulp lifters currently installed in the SAG mill should handle the extra flow without slurry pooling. Therefore, cyclone underflow bleed is of benefit when treating ball mill limiting (soft) ores, ie when SAG power draw is low. As an easy to implement guide for Lihir, partial cyclone U/F bleed (20 per cent) should be used when SAG Mill power less than 4200 kW.

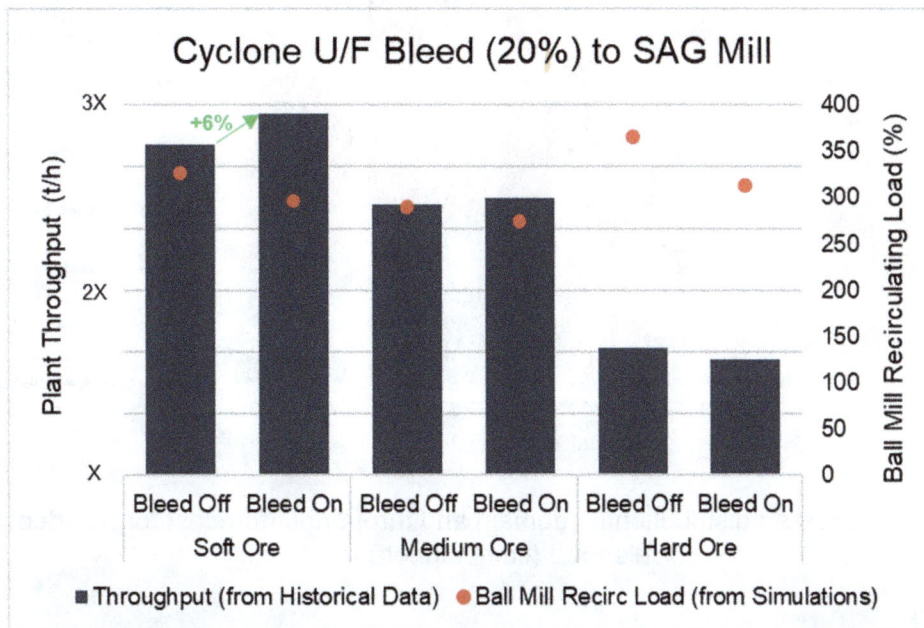

FIG 6 – Impact of 20 per cent bleed of cyclone underflow recycled to SAG mill.

Feed blending

Another of the focusing questions for this project was whether there is any benefit to feed blending? What is the difference between campaigning individual ore types through the plant versus treating a well-blended feed?

Feed blending is a common approach to provide consistent and stable feed to the process plant, to even out extremes of hardness and grade or sometimes to manage penalty elements. In some cases, for very different ore types, it can be more beneficial to campaign these through the plant and allow optimisation of the process for the different ore characteristics. This is not common and incurs logistic and stock challenges (for discharge grates, screen panels etc). However, Batu Hijau is an example where SAG grate, SAG and Ball mill trommels and screen panels were adjusted according to ore characteristics from future mine plan as presented by Wirfiyata, Maclean and Khomaeni (2015).

Power-based modelling (Morrell) was used to evaluate the benefit of treating a well-blended feed compared to campaigning individual ore types through the plant at Lihir. Simulations indicated that a well-blended feed would provide a small increase in throughput of up to 4 per cent for HGO circuits, as shown in Figure 7.

There are several reasons for the increase in throughput. The main reason is that blending ore types that are SAG and ball mill limited provides a better balance of load (breakage work) between the two mills (coarse and fine grinding). This results in greater utilisation of the available milling power and thus greater throughput. Additionally, soft ores theoretically could achieve very high throughput rates, but in practice, the throughput is constrained by other circuit limitations (eg ancillary equipment) and it may also be difficult to maintain load and utilise all the available power in the SAG mill for softer ores. Therefore, there is an advantage to mixing these softer ores with harder ores to fully utilise the available milling power and for more efficient and stable operation.

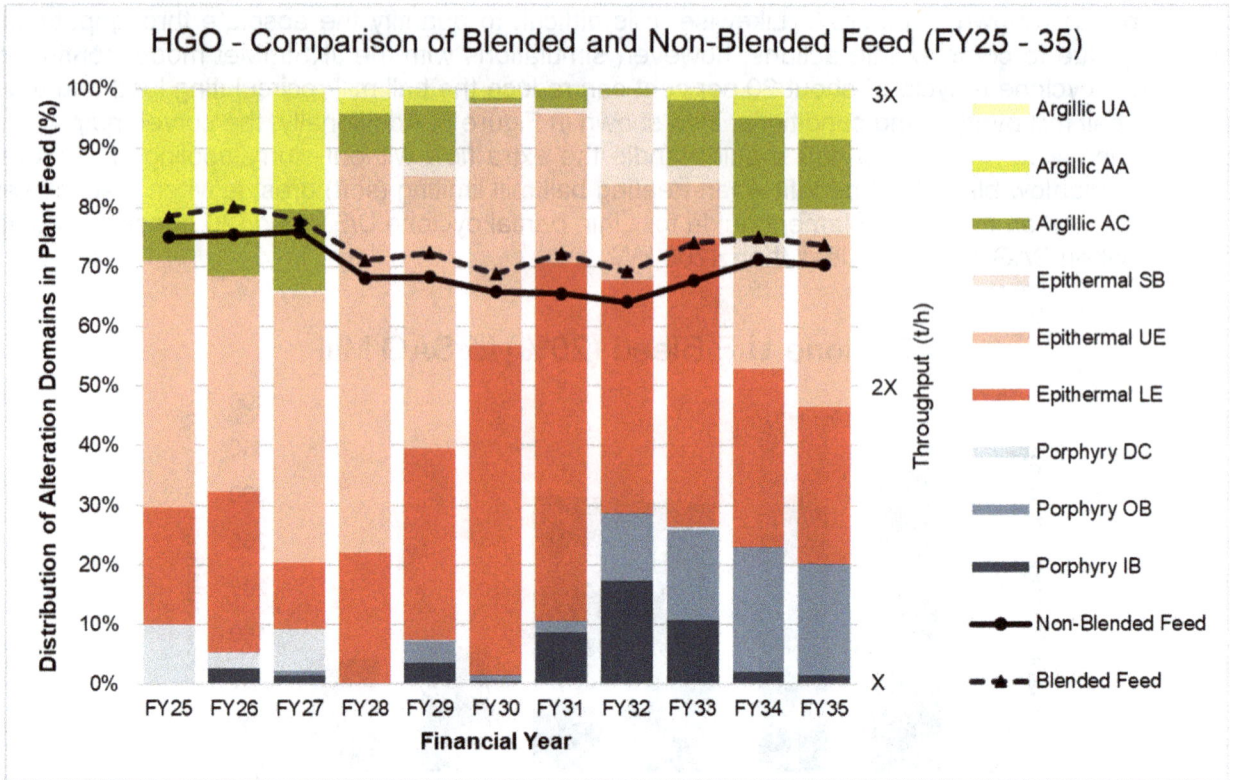

FIG 7 – HGO circuits feed distribution by domain and throughput forecast for blended versus non-blended (campaigned) feed.

FGO circuit

The throughput forecast results for the FGO circuit indicate that it is likely to be SAG mill limited from FY25 to FY30 (based on the provided feed blend plan), as shown in Figure 8. This is due to the high proportion of Lower Epithermal alteration domain which is SAG mill limiting. Therefore, there is minimal benefit from cyclone underflow recycle in this circuit and it is not required at this time.

Pebble or secondary crushing

Due to SAG limitations, it is warranted to reinvestigate the benefit of pebble crushing for the FGO circuit. Currently the FGO pebbles are returned to stockpile and subsequently HGO feed. FGO pebble crushing was investigated previously and shown to be of benefit for harder ores but was not implemented (due to subsequent increase in the proportion of softer ores and other priorities). Given the increasing ore hardness and predicted SAG mill limitations, it would be worthwhile to reconsider FGO pebble crushing. Secondary crushing could also be considered. However, either secondary or pebble crushing may not be viable for the relatively short time frame of SAG limiting conditions (about seven years); although, mobile solutions could be considered.

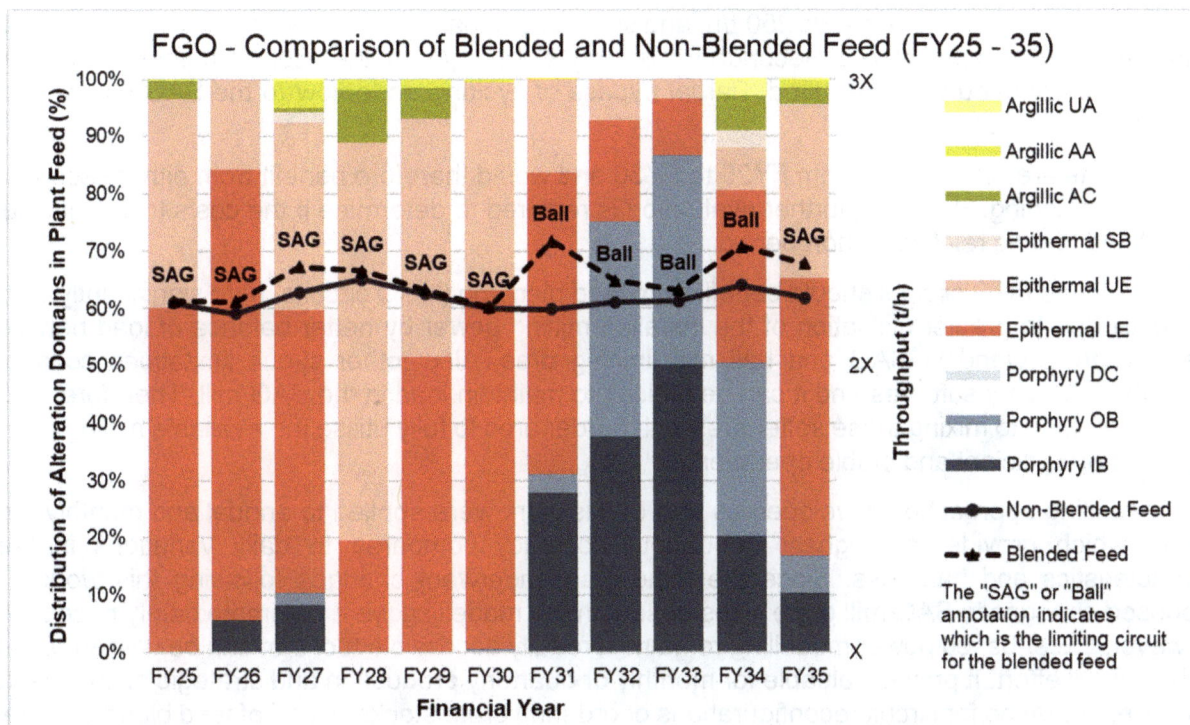

FIG 8 – FGO circuit feed distribution by domain and throughput forecast for blended feed.

Feed blending

The impact of feed blending was also evaluated for the FGO circuit, as shown in Figure 8. The throughput benefits are not as significant for the FGO circuit (compared to HGO circuits) until FY30, as the feed is predominantly epithermal before this. However, the benefits increase after FY31, when the proportion of porphyry increases. This is due to the benefit of mixing SAG and ball mill limiting ores, and thus maximising utilisation of installed power (as discussed for HGO circuits previously).

The estimated throughput benefits assume the annual average feed blend is maintained throughout the entire year. This is not likely to be possible all the time, depending on available ore sources, so the full estimated throughput benefit may not be realised.

However, in addition to the throughput benefits mentioned above, blended feed will improve circuit stability. Sudden changes in ore types and properties introduce instability in the circuit, which require adjustment of operating conditions and may result in lost production through instability and also potentially lost recovery in downstream processes due to changes in feed grade and size. Therefore, overall, in this case, feed blending is recommended for Lihir.

CONCLUSIONS

Several modelling approaches are available to metallurgists, and each have different purposes and limitations. Used together, these are powerful tools to develop optimum strategies to maximise production over the LOM considering expected changes in ore properties. This paper has described how models can be used to assess the impact of multiple ore types with varying comminution indices and feed size distributions on comminution circuit performance and to identify and evaluate optimum operating strategies considering the changing ore properties.

In this case, Morrell power-based modelling, site-specific (JKSimMet) process models, ore characterisation and historical operating data were used together to determine if and when pebble and secondary crushing would benefit the different circuits at Lihir. They were also used to assess the impact of different feed ore types and characteristics, determine when circuits would be ball or SAG mill limited, and develop and evaluate strategies to maximise throughput. These calibrated models can also be used to investigate other changes or opportunities in the future.

At Lihir, the HGO circuits are predominantly ball mill limited, but will still benefit from pebble crushing when treating hard (SAG limiting) ore. Therefore, pebble crushing should only be used for harder ores, ie when pebble rate exceeds 250 t/h, and should be bypassed at lower pebble rates to maintain load in the SAG mill for soft ore. Secondary crushing is not required, the pebble crushing is sufficient and can be operated when required. Partial bypass of cyclone underflow to the SAG mill will help alleviate the load on the ball mill.

FGO circuits are SAG limited from FY25 to FY30 and would therefore benefit from either secondary or pebble crushing. However, further evaluation is required to determine if the costs to install would be justified for this relatively short period.

Both HGO and FGO circuits should benefit from feed blending (HGO circuits more significantly). This is mostly due to greater utilisation of the installed milling power by better balance of load between the mills from blending SAG and ball mill limiting ores. Also, other circuit limitations constrain throughput for very soft ores and it can be difficult to maintain load in the SAG mill. Therefore, there is an advantage to mixing these softer ores with harder ores to fully utilise the available milling power and for more efficient and stable operation.

The modelling approaches developed as part of this work were applied to annual and monthly feed blend, which provide the highest throughput accuracy compared to daily variations in feed characteristics and hardness. Since the mine plan underwent changes following this work, the proposed changes to SAG mill grate sizes based on the modelling were not immediately necessary. However, Excel-based power modelling remains a readily accessible tool that can be utilised on-site with minimal effort. It proves valuable for monthly or quarterly production and strategic planning—for instance, preparing for circuit reconfigurations or ordering grate stocks ahead of feed blend changes. Implementing semi-mechanistic process models presents more significant challenges. The cost of acquiring multiple user specialty software licenses and the time commitment required for developing and maintaining these models can be substantial for site staff. To address these challenges, collaboration between site metallurgy teams, corporate teams and consultants can achieve the most robust outcome from combined modelling approaches.

ACKNOWLEDGEMENTS

Hatch would like to thank Newmont for their collaboration on this project and paper and permission to publish this paper.

REFERENCES

Global Mining Guidelines Group, 2021. The Morrell Method to Determine the Efficiency of Industrial Grinding Circuits (GMG04-MP-2021) [online]. Available from: <https://gmggroup.org/guidelines-and-publications/morrell-method-to-determine-the-efficiency-of-industrial-grinding-circuits/> [Accessed: 22 January 2024].

Morrell, S, 2004a. Predicting the Specific Energy of Autogenous and Semi-autogenous Mills from Small Diameter Drill Core Samples, *Minerals Engineering*, 17(3):447–451.

Morrell, S, 2004b. An Alternative Energy-Size Relationship to That Proposed By Bond For The Design and Optimisation Of Grinding Circuits, *International Journal of Mineral Processing*, 74:133–141.

Rodriguez, L, Morales, M, Valery, W, Valle, R, Hayashida, R, Bonfils, C and Plasencia, C, 2023. Developing an Advanced Throughput Forecast Model for Minera Los Pelambres, in Proceedings of the SAG 2023 Conference.

Wirfiyata, F, Maclean, A and Khomaeni, G, 2015. Batu Hijau Mill Throughput Optimization: Milling Circuit Configuration Strategy based on Ore Characterization, in Proceedings of the SAG 2015 Conference.

Optimisation of Jameson Cells to improve concentrate quality at BHP Carrapateena

J Reinhold[1], J Van Sliedregt[2], F Burns[3], A Price[4] and J Seppelt[5]

1. Senior Metallurgist – Plant, BHP, Pernatty SA 5173. Email: jacqueline.reinhold@bhp.com
2. Metallurgist – Plant, BHP, Pernatty SA 5173. Email: jessica.vansliedregt@bhp.com
3. Superintendent Metallurgy BHP, Pernatty SA 5173. Email: fraser.burns@bhp.com
4. Principal Metallurgist – Jameson Cell, Glencore Technology, Brisbane Qld 4000. Email: adam.price@glencore.com.au
5. Processing Manager – BHP, Pernatty SA 5173. Email: joe.seppelt@bhp.com

ABSTRACT

Carrapateena is an iron oxide copper–gold (IOCG) deposit hosted in a brecciated granite complex located in the Gawler Craton, South Australia. The ore is processed by a conventional sulfide flotation concentrator, producing a copper gold silver (Cu-Au-Ag) concentrate with chalcopyrite and bornite as the main copper-bearing minerals. The concentrate from two Jameson cells, one in Cleaner Scalper duty, the other as a third cleaner, combine to produce the final concentrate.

Since commissioning in 2020, Carrapateena has engaged Glencore Technology to assist with an optimisation program of the two Jameson cells in the Carrapateena concentrator with a focus on concentrate quality. The goal was to collaboratively develop operating and control strategies that aimed to increase copper concentrate grade and reduce non-sulfide gangue recovery by entrainment.

This collaboration continued over the course of the following three years, considering many differing operating parameters including, but not limited to, air addition, froth depth, wash water flow rate, wash water quality and feed density. The work culminated in 2023 with a rapid sprint of improvement activities, cell modifications, challenging of prevailing operating strategies, and installation of alternative instrumentation to improve the reliability of readings and methods of Jameson cell control.

This paper outlines the numerous stages of Jameson cell optimisation conducted, the differing strategies implemented, their success or failure and the future plans for ongoing improvement.

INTRODUCTION

Carrapateena is a copper–gold deposit, located 460 km north of Adelaide, South Australia. The deposit is currently mined using the sub-level cave (SLC) mining method, with future mining to incorporate a block cave footprint beneath the SLC.

The copper mineralogy present at Carrapateena is chalcopyrite and bornite, hosted in an iron oxide breccia; and is relatively typical of iron oxide copper–gold (IOCG) deposits also located in South Australia. A sulfide flotation concentrator produces a Cu-Au-Ag concentrate for sale to copper smelters.

The Carrapateena flotation circuit has been described in detail previously (Reinhold *et al*, 2023); this previous work focused on the reconfiguration, expansion and optimisation of the flotation circuit to improve concentrate quality, particularly during the period from commissioning in 2019 until end of 2021.

A key conclusion of the previous work was that the concentrate quality of the Carrapateena flotation circuit is underpinned by the effective operation of the circuit's two Jameson cells. This paper presents a focused study of Jameson cell operation and performance at Carrapateena through to 2023; noting various challenges particular to Carrapateena, and the modifications and operating strategies that were adopted to address those challenges. This study was completed with the valued support of the Glencore Technology technical team. The study focused on improving cell stability in the context of progressively increasing of concentrator throughput and a changing concentrator feed sulfide mineralogy away from bornite towards chalcopyrite. Coupled with this was the ongoing need to reject ultrafine non-sulfide gangue (NSG) entrainment to maintain deleterious elements below target levels.

Jameson cell duties

Two Jameson cells are currently in operation at Carrapateena. The flow sheet in Figure 1 indicates their position in the flotation circuit.

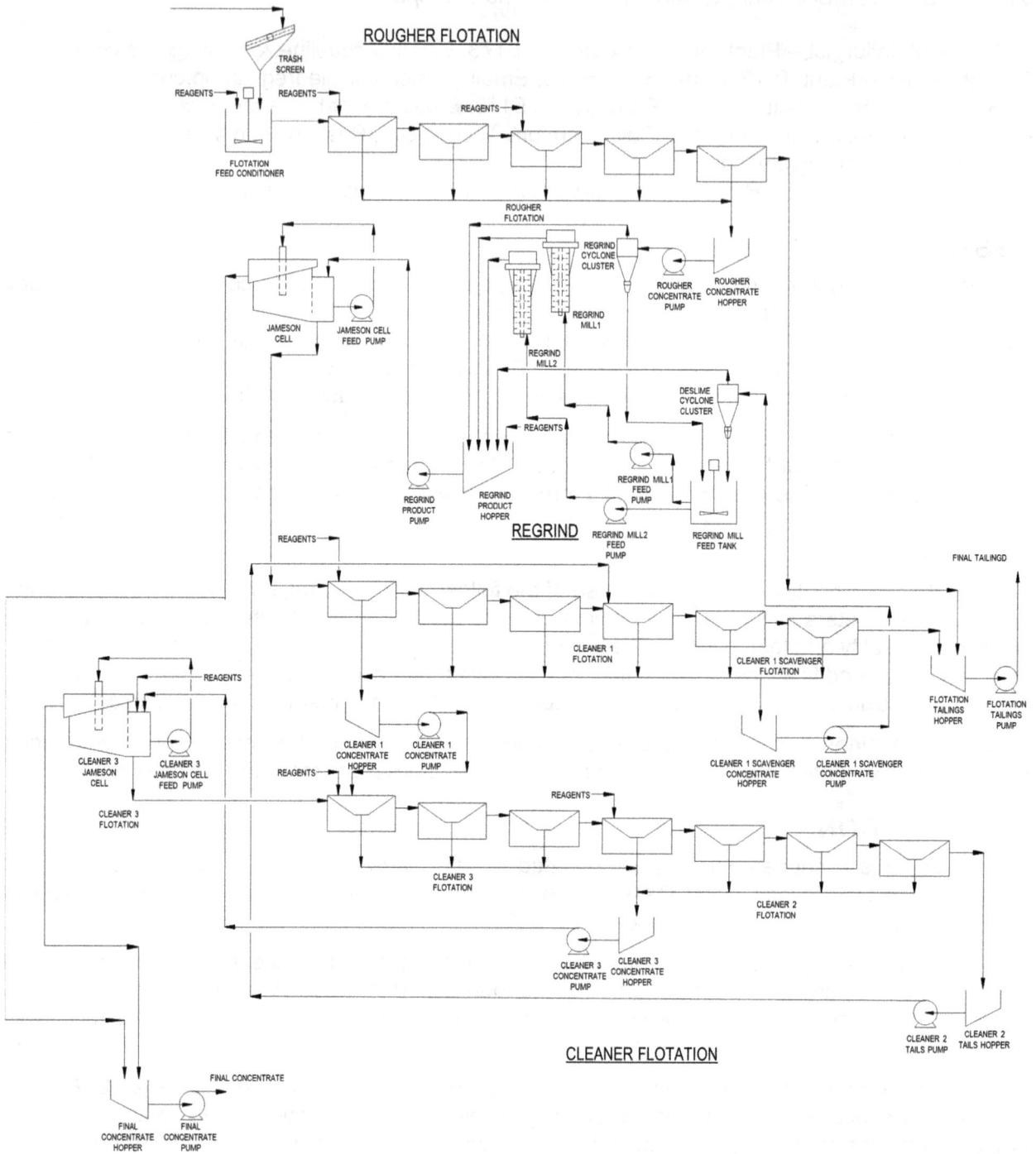

FIG 1 – September 2022 Carrapateena flotation flow sheet.

The first cell – a E4232/10 Mark IV Glencore Technology Jameson Cell – operates in a Cleaner Scalper duty receiving the regrind cyclone overflow, and the product of both regrind mills, typically targeting a discharge P_{80} of 25 µm. The tail of this cell passes to the Cleaner 1 bank of mechanical cell.

The second cell – a E2532/6 Jameson Cell – operates in a third cleaner duty receiving the concentrate from Cleaner 2 bank of mechanical cells. The Cleaner 3 Jameson operates in closed circuit, with its tail reporting back to Cleaner 2 feed.

The concentrate from both Jameson cells combine to form the final concentrate. The mass split between the two cells is typically 75 per cent from the Cleaner 1 Scalper and 25 per cent from Cleaner 3. Given the larger proportion of final concentrate that originates from the Scalper Jameson, it was the primary target of improvements in operating strategy and concentrate quality. Improvements were then applied to the third cleaner cell where relevant.

The flotation flow sheet has seen multiple modifications through the short life of Carrapateena, most modifications have been detailed previously (Reinhold *et al*, 2023). The only further modification relevant to this topic is the addition of a second regrind mill in September of 2022. The second mill operates in parallel to the existing and is run under the same conditions and target product size as the first regrind mill. The intent of this installation was to increase regrind capacity for the higher throughput rate, provide flexibility for finer regrind sizing and to mitigate the risk to concentrate quality of regrind mill downtime. The regrind mills at Carrapateena are Outotec HIG1600.

BASELINE JAMESON OPERATION AND CHALLENGES

As detailed by Reinhold *et al* (2023), prior to the commencement of this review, Jameson operation was characterised by a focus on maintaining Jameson stage copper recovery at a set value, as laid out in the prevailing operating strategy for flotation circuit.

- A higher stage recovery (75–80 per cent) would be targeted if a copper recovery focus was desired.

- A lower stage recovery (65–70 per cent) would be targeted if a NSG rejection focus was required.

- Finally, a moderate stage recovery (70–75 per cent) would be targeted if a balanced approach was appropriate.

Stage recovery was the primary operating target due to variable feed sulfide mineralogy. This prevents the setting of a timely relevant concentrate target for operations technicians as a 'good' clean concentrate which could vary between 30–45 per cent Cu within a shift. This depends on the prevailing chalcopyrite-bornite mix in the concentrator feed. Further, several surveys had shown that the above stage recovery bands were applicable across the various feed conditions experienced.

In a drive to consistently operate at the desired stage recovery, this strategy was encoded to the process plant control system. A live Jameson cell stage recovery was calculated from copper grades measured by the Metso Courier Online Stream Analyser (OSA). A recovery controller manipulated air flow rate and froth depth set points to maintain the measured recovery at the desired value.

Air and depth flow rates were controlled to their cascaded set points, by respective proportional-integral-differential (PID) control loops per standard Jameson operation. Air and depth instruments were the standard supply from Glencore Technology. A thermal mass flowmeter measured air rate and a pair of submerged pressure transmitters, installed at known distance apart, measured a density compensated froth depth.

Wash water, sourced from a raw water distribution system, operated at a constant set flow rate, typically 60 m^3/h. This was value was significantly above the initial design rates (~35 m^3/h) and was selected during commissioning 2020 based on significant survey work to ensure positive water bias was achieved, with wash bias defined as wash water flow rate minus water in concentrate flow rate.

Feed dilution water, sourced from the process water distribution system, was operated at a constant set flow rate into the Jameson feed box. Jameson feed density instrumentation was not fitted during the original plant design and construction. Consequently, feed density was measured by manual sampling (together with a Marcy scale) on the regrind circuit product, as it entered Jameson feedbox separately from the process water. No direct measurement of total feed density was achieved and calculations were used to estimate the feed density after dilution. Typically, a feed density of ~15 per cent solids was targeted and water flow rates of 80 m^3/h (Cleaner Scalper) and 15 m^3/h (Cleaner 3) were used.

Even though significant work had underpinned the prevailing operating strategy, there were several concerns with Jameson cell performance.

- Frothing events where the Jameson cells would race uncontrollably were frequent.

- Several instrument readings were subjected to sudden drift or unexplained variability. It was believed that these, especially depth sensor drift, was contributing to the above froth racing events.

- Feed density was known to be important for NSG rejection but was not measured directly nor successfully controlled via dilution.

- Wash water rates were largely unchanged throughout 2021–2022 despite increases in concentrator milling rates and increased chalcopyrite content. This increased concentrate make due to lower copper concentrate grade and as a consequence impacted wash bias.

- Similarly, with increased concentrator feed rates and the target of low percent solids, whether the Jameson cells were still operating with acceptable limits of recycle.

Of biggest concern were the Cleaner Scalper Jameson racing events. These were characterised by a loss of structure in the cell froth, a very fine bubble size and 'soupy' appearance, a red gangue tinge would be present and the cell water recovery greatly would increase.

Carrapateena's concentrate is known to form particularly tenacious froth requiring various methods of breakage and chemical defoamer to be added downstream to manage froth and slurry transportation. This is likely a consequence of the process water quality used within the plant. Frother addition into the rougher circuit and Scalper Jameson was used sparingly throughout 2020 to 2023, more often not dosed at all. Due to the downstream challenges, its usage was not considered as a part of this review.

Given these concerns, Carrapateena embarked on a full operational review of the Jameson cells and engaged Glencore Technology to provide specialty technical assistance where required.

OPERATIONAL REVIEW

An operational review commenced in December 2022 with an initial focus on benchmarking the performance of the Jameson Scalper cell across a matrix of operating parameters, comparing it against similar surveys completed during the 2020–2021 commissioning efforts. Further, a review into the trend of operating parameters over intervening time was conducted, with consideration being given to the increase in throughput and volumetric feed into the cell over that time.

Matrix survey comparison

The standard approach to mapping out a Jameson cell performance envelope is to complete matrix surveying. During this process, the Jameson cell key operating parameters, such as: air, depth, wash flow and frother dosing are cycled variously through high, med and low, or, on, off values. At each setting, after an appropriate time for stabilisation, a rapid snapshot survey of the cell is performed. For all surveys, scalper feed, concentrate and tails samples were collected for determination of percent solids and assaying. This process is repeated until the matrix of parameter settings is completed. This process is usually performed across the course of a single shift.

The matrix surveys compared copper stage recovery impact on and concentrate quality, as inferred by silica and other NSG content, which represent the major quality challenges at Carrapateena. The surveys completed in December 2022 considered a wider range of process variables than previous, with aim to extend the mapped envelope. Tables 1 and 2 indicate the process variables and ranges relevant for each set matrix surveys.

TABLE 1

Survey 3 conditions – July 2020.

Process variable/level	Low	Medium	High
Air rate (m³/h)	400	500	600
Level (mm)	450	550	650

TABLE 2

December 2022 survey conditions.

Process variable/level	Low	Medium	High
Dilution water (m³/hr)	OFF	NA	ON
Wash water (m³/hr)	0	30	60
Air rate (m³/hr)	150	250	350
Level (mm)	200	400	600

Figures 2 and 3 presents the NSG selectivity curves resulting from the two sets of matrix surveying. The rapid increase in NSG entrainment as the copper recovery increases remains at ~75 per cent recovery; this confirms the selection of operating strategy targets previously identified in 2020 and 2021. The overlapping 2020 and 2022 selectivity curves, importantly indicates that same NSG performance that had previously been capable form the Scalper Jameson, should still be achievable if operating parameters are correctly set.

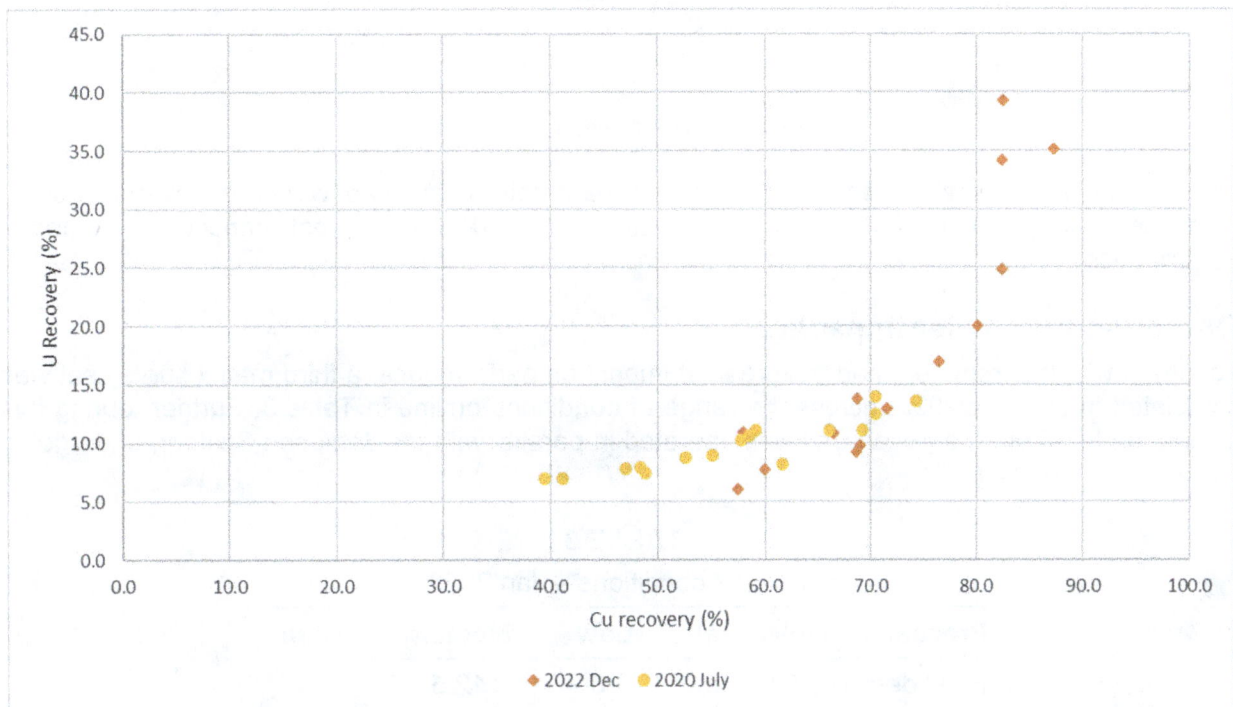

FIG 2 – Copper-Uranium selectivity.

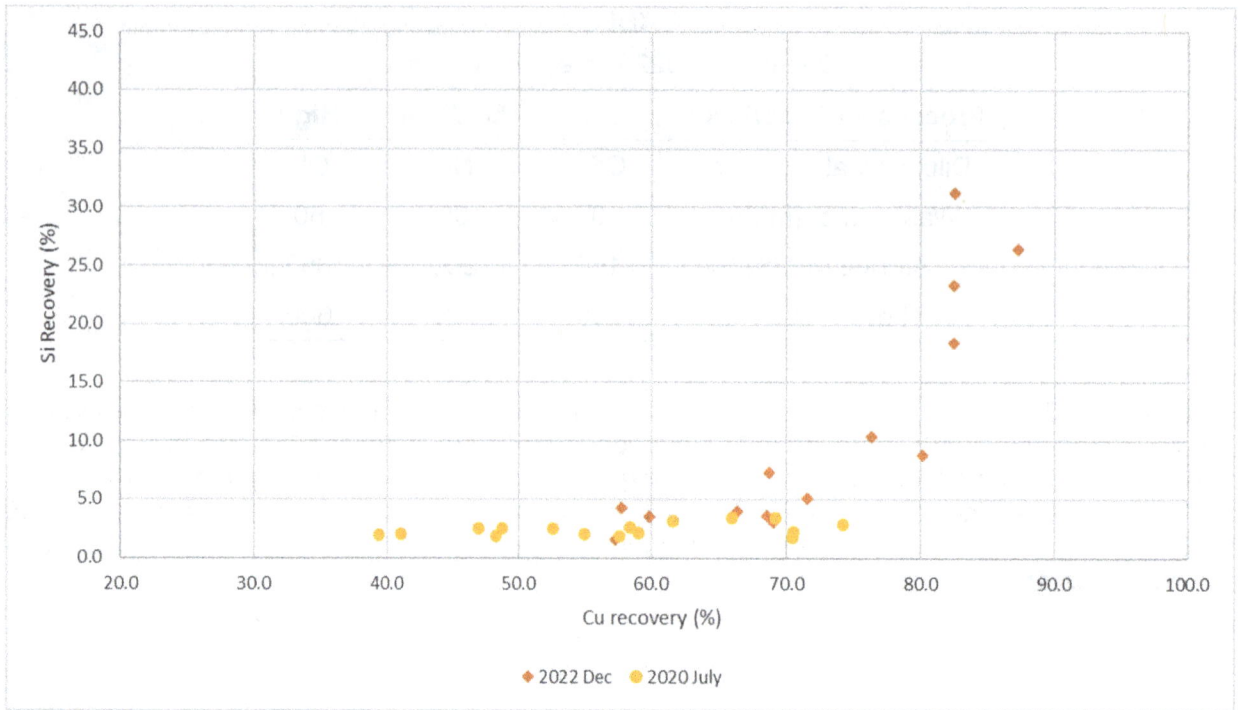

FIG 3 – Copper-Silica selectivity.

This turned the review's attention to operating parameters that had wandered overtime since commissioning in 2020, and which parameters provided the greatest impact on Jameson performance.

Operating parameter impacts

To inform which parameters had the greatest impact on performance, a third matrix survey set was completed in January 2023, across the range of conditions outline in Table 3. Further, during this testing, an L500 pilot Jameson Cell was operated in parallel with the Jameson Scalper, undergoing its own matrix testing.

TABLE 3

Survey 2 conditions – Jan 2023.

Process variable/level	Low	Medium	High
Feed density (% solids)	10	12.5	15
Air rate (m³/h)	200	300	400
Level (mm)	300	500	700

A notable change for this matrix, was a decision to remove wash water. This was held constant at a high setting, as wash water flow is a variable that in reality would not be run at lower levels, to ensure a positive bias is maintained. Instead, Jameson Scalper feed percent solids was introduced into matrix to understand if further dilution cleaning would be prudent. At the time of these surveys, the typical feed density was 14 per cent solids against the design mass balance value of 18 per cent solids.

The Glencore Technology approach to matrix testing often involves the use of Taguchi-Grey Analysis to discern the most influential parameters on performance and concentrate quality. While some caution is used with this analysis, due to a lower than desired number surveys completed, the key outcomes are presented in Rank column of Table 4, where feed percent solids was indicated to be the most impactful on cell performance, followed by air flow rate.

TABLE 4

Taguchi-Grey outcomes – Jan 2023.

Results Analysis And Optimisation Level Selection								
	Parameter	Level 1	Level 2	Level 3	max	min	Delta	Rank
A	Feed Density	0.648	0.625	0.728	0.65	0.63	0.0232	3
B	Air Rate	0.700	0.676	0.632	0.70	0.63	0.0679	2
C	Level	0.619	0.691	0.690	0.69	0.62	0.0716	1

This was supported by review of the grade-recovery results of the surveys, grouped by the various percent solids settings, in Figure 4. Operating at lower Jameson feed percent solids provided a better copper concentrate grade for a given recovery. The lower silica grades are in line with dilution cleaning enabling a reduction of the entrainment of ultra-fine gangue into the concentrate, rather than a change in copper sulfide mineralogy driving the improved concentrate grade.

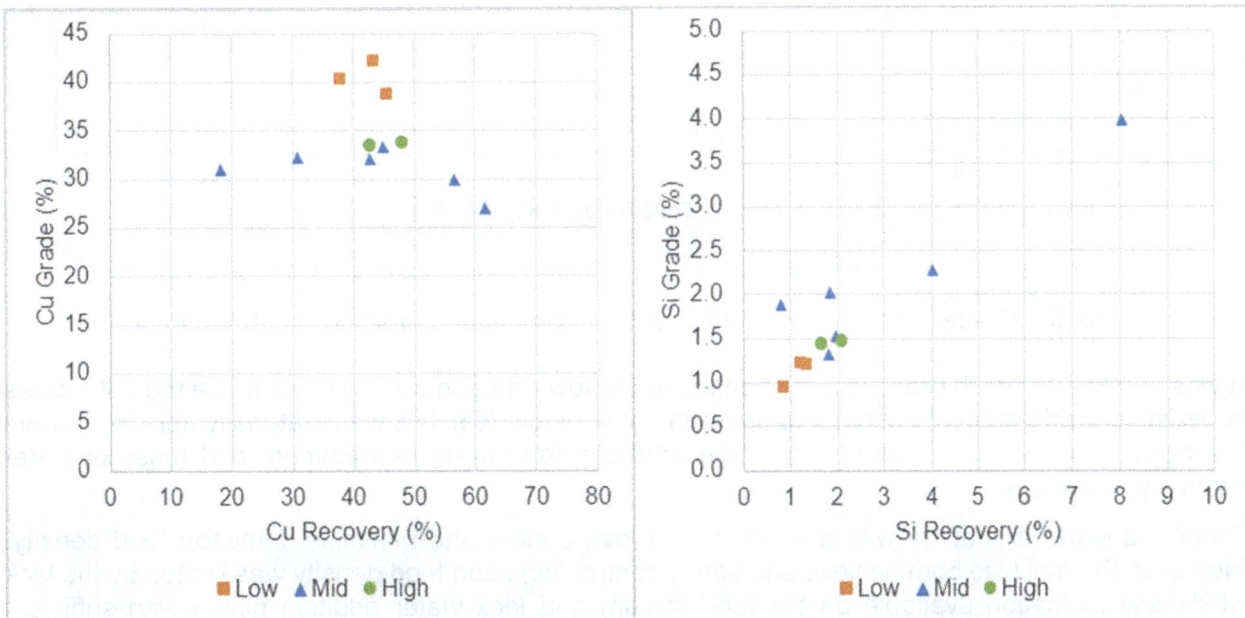

FIG 4 – Grade-recovery by feed % solids for: (left) copper; and (right) silica.

The outcome of air flow rate changes were similar across all matrix survey sets. A higher air flow rate resulted in higher copper and mass recoveries, with a lower air rate producing improved rejection of NSG, to the detriment of copper and mass recovery until a point where there was poor froth formation. Typically, these were found to be more impactful than froth depth changes, as long as a minimum froth depth that avoided pulping was maintained.

Of interest, a significant step change was noted when moving from the prevailing 12.5 per cent solids target to lowest setting of 10.0 per cent solids. This was further investigated by reviewing the Scalper Jameson's performance to that of the L500 pilot Jameson Cell which was being operating in parallel, presented in Figure 5. Two changes were noted: first, the improved grade response was also seen in the pilot Jameson, and second, at the higher percent solids the Scalper Jameson was not operating with the typical 1:1 scale up expected of Jameson cells. The lowest percent solids, appeared to relieve some mechanism in the Scalper Jameson that was holding it back from expected performance, as set out by the pilot cell.

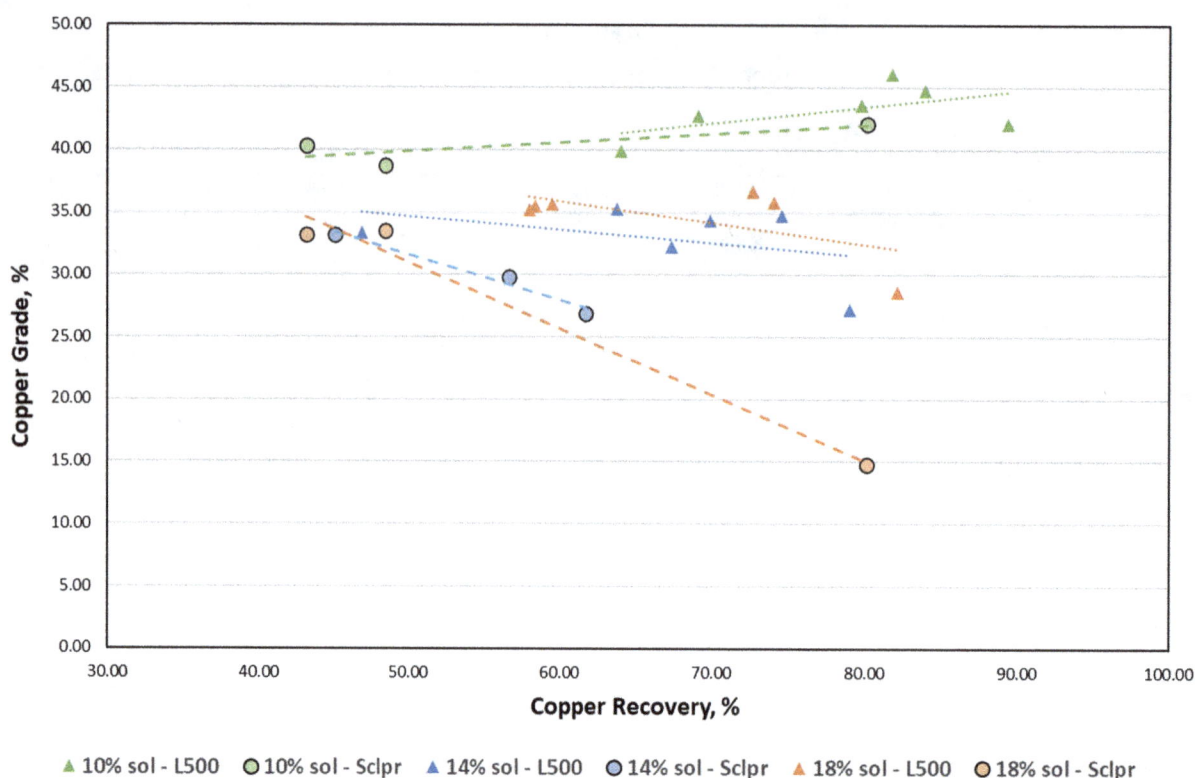

FIG 5 – Grade-recovery by feed % solids – comparison of scalper to L500 Pilot.

It was surmised from the results that changes to air flow rate and depth simply move the cell across an existing grade-recovery curve; but operating at very low % solids was materially able to improve the position of that curve, by changing the relationship between entrainment and mass or water recovery of the cell.

From this work, a strategy was pursued to improve control and minimise Jameson feed density. However, the ability to both achieve and tightly control Jameson feed density was limited by the lack of density indication available on the feed stream and lack water addition piping with sufficient capacity prior to the density indication. A series of upgrades were required to implement the desired control philosophy.

Historic parameter review

A deep dive into the long-term trends of operational parameters since the original commissioning efforts was conducted with due consideration given to the increase in throughput and volumetric feed into the cell over that time.

Wash water rate

Wash water, operated at a constant set flow rate, typically 60 m³/h. This was value was significantly above the initial design rates and was selected during commissioning 2020 based on significant survey and mass balance work to ensure positive wash bias was achieved. The matrix surveying of the Jameson Scalper at that time, identified positive bias as crucial to the quality of the scalper concentrate.

With several years now past since commissioning, inclusive of increased trend of milling rate and reducing trend copper grade of concentrate due to move toward chalcopyrite mineralogy, a review of the wash bias, and required wash rate was triggered. Wash bias was recalculated on a shift basis, using various process instrumentation, and mass balancing shift composite assays. Figure 6 presents quartile box plots of the wash bias for each 6 month period since commissioning. From July 2021 through to December 2022, a steady decline in wash basis was experienced, as increases in Jameson Scalper concentrate make were not paired with wash water flow increases. Further to this, interrogation of the same data set indicated that this reduction in wash bias, clearly related to a poorer outcome in terms of Scalper Jameson silica rejection, shown in Figure 7.

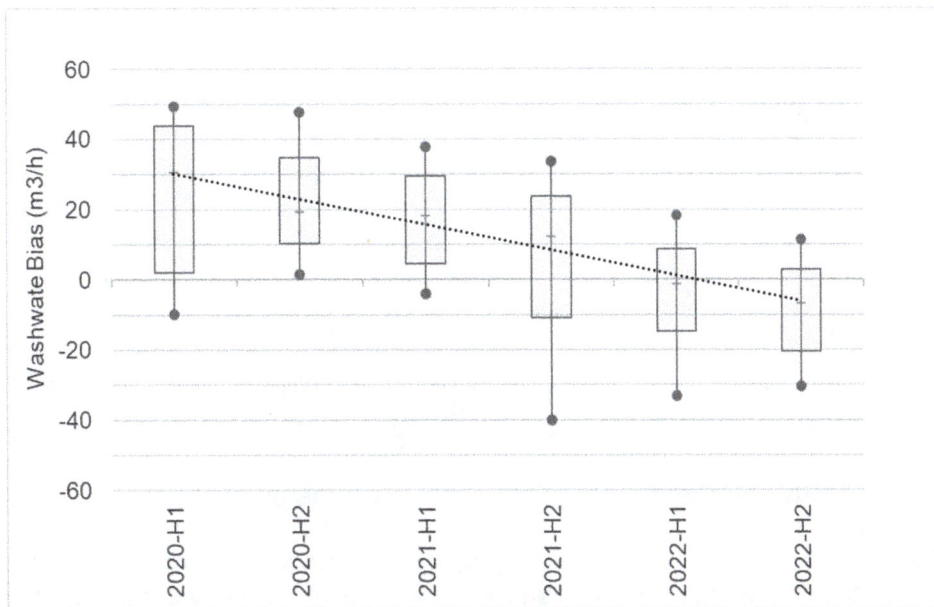

FIG 6 – Cleaner 1 Jameson – box-whisker plot of shiftly wash water bias, grouped by 6 mth period.

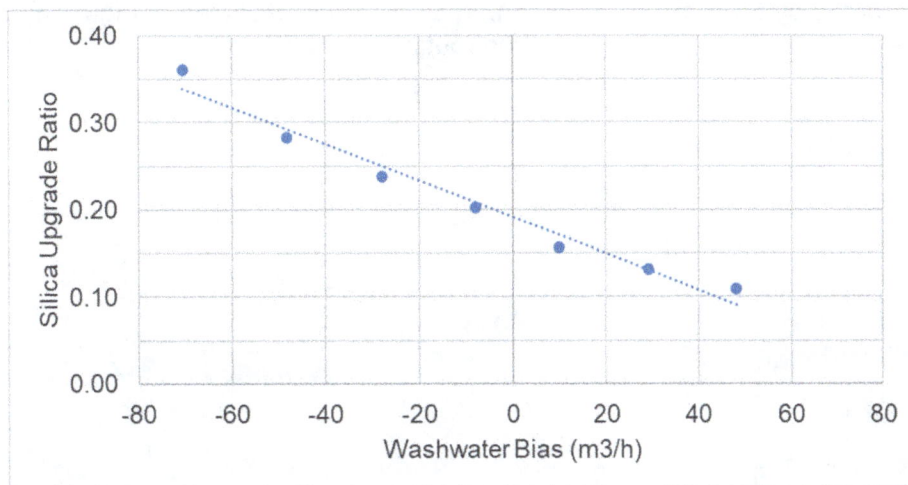

FIG 7 – Cleaner 1 Jameson – wash water bias impact on silica upgrade ratio (2020–2023).

A mass balance assessment of the December 2022 surveys indicated that a wash water flow rate of 130 m³/h would be required to provide a positive water bias and ensure all water in the froth phase was replaced with wash water under all conditions. The wash water flow rate target was rapidly increased in response to this study, however the installed raw water manifold that supplied the wash water, was only able to supply ~85–90 m³/h. While this increase netted an improvement, it saw a positive bias being achieved with only ~50 per cent frequency, up from ~33 per cent. A modification of the wash water supply line was needed to ensure a positive wash bias could be achieved under all conditions.

Air addition, air to pulp ratio and gas hold up

From commissioning in 2020, Jameson air flow rates typically ranged between 400 m³/h to 500 m³/h, this placed the Scalper Jameson towards the lower recommended operating range for the cell. The driver being concentrate quality as the original survey sets indicated that the lower air addition, lowers the entrainment of NSG and impurities. The survey minimum air trialled was 400 m³/h.

However, throughout 2022, air addition rates were notably lower than in 2020 and 2021, Figure 8 below. The drivers here was a strong, overriding operational motivation to run the cell at lower air rates to assist in management of persistent froth tenacity and spillage from concentrate launders, hoppers and thickener. The persistent frothing, manifested at Scalper Jameson as racing episodes that continued even at lowest air rate and greatest froth depths.

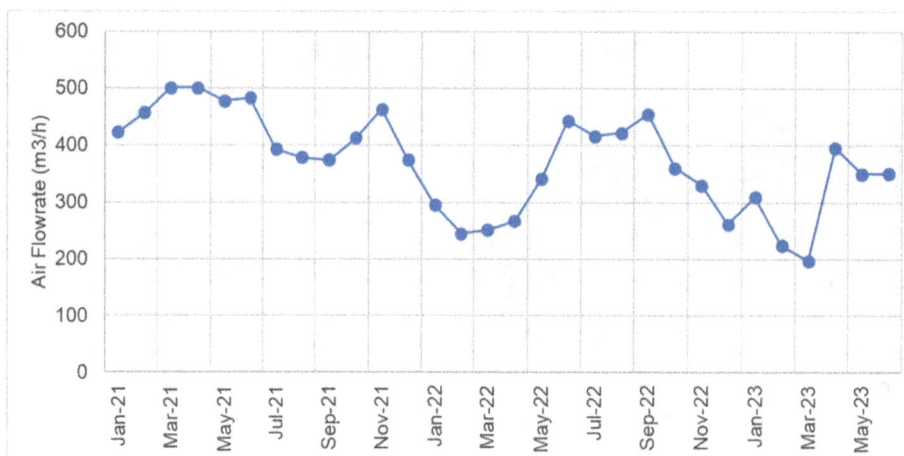

FIG 8 – Cleaner Scalper monthly average air flow rate.

During the 2023 Scalper – L500 Jameson comparisons, Glencore Technology highlighted that the lowest air flow rates in use by the Scalper Jameson were driving the key Jameson operating metric of Air to Pulp Ratio (APR) below their internal minimum thresholds, ~0.4 for base metal applications. For instance, during parallel comparison testing, the lowest APR targeted by the L500 (0.4) was the highest APR targeted by the Scalper Jameson. The air flow rate shown in Figure 8 was converted to APR and presented in Figure 9 displaying the routine drop below 0.4.

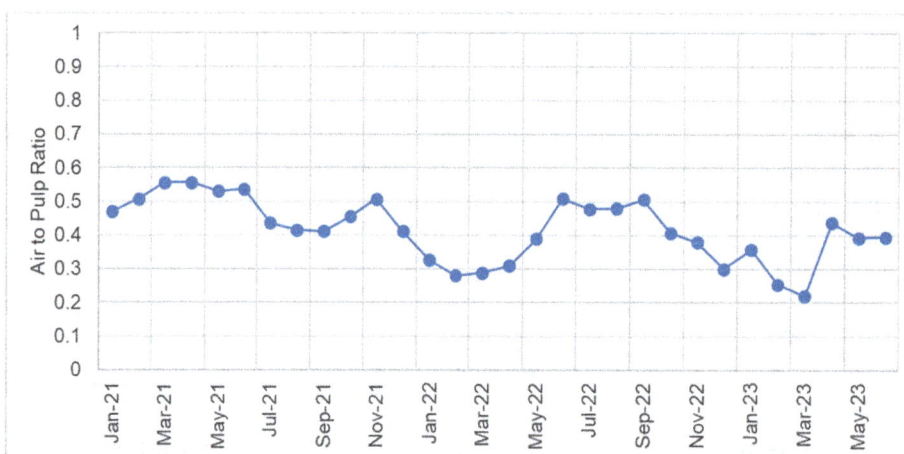

FIG 9 – Cleaner Scalper monthly average air to pulp ratio.

It was theorised that such low air additions were resulting in a reduction of gas hold up, reduction in available air bubble surface area for particle collection and consequentially, poor froth formation (Taşdemir, 2007) (T. Taşdemir, 2007). Further it was posited that low APR and gas hold up was exacerbating, rather than mitigating, the racing frothing episodes.

To confirm, a survey set was completed during a 'racing' episode, comparing gas hold up – as measured with a JKMRC gas hold up device – with the air flow rate that was varied across wide operating range. The surveys confirmed that as air flow rates reduced during a frothing event, gas hold up reduced to below 5 per cent in the extreme case at 150 m^3/h, refer Figure 10. At the time of the survey, visual inspections indicated a poor froth phase that was unable to maintain a stable froth. The bubbles were very fine with poor mineral coverage (Figure 11 (left)) and a higher water recovery reduced opportunity for effective washing with an increased risk of NSG entrainment, and control of the cell was lost as 'racing' commenced.

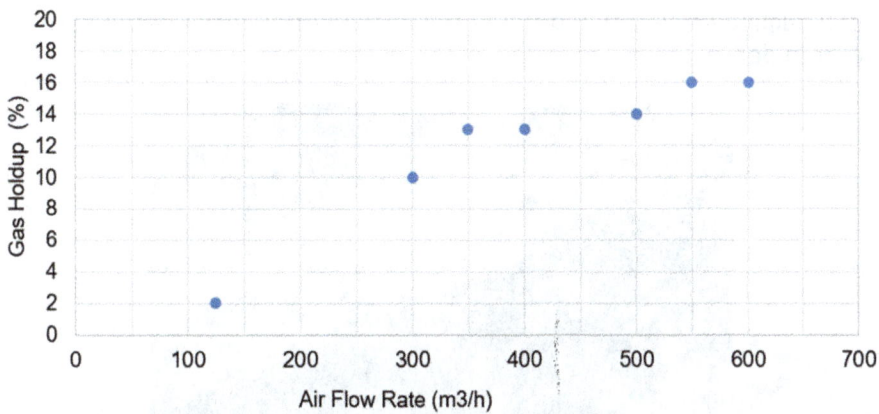

FIG 10 – Cleaner 1 Jameson – gas hold up dependence on air flow rate.

FIG 11 – Cleaner 1 Jameson froth appearance (left) very low gas hold up and APR (right) typical gas hold up and APR.

Further, opposite to prevailing operational wisdom, increasing air flow rate during racing episodes, reduced water recovery, returned froth structure and improved the appearance (colour) of the froth mineralisation (Figure 11 (right)). Tenacious froth remained in downstream hoppers etc, but the volume of froth and concentrate was greatly reduced due to the lower water recovery from cell.

It was recommended that a minimum air flow rate of 350 m³/h be implemented to ensure that gas hold-up remained in the typical operating range for flotation, and during frothing and 'racing' events operation technicians were encouraged to try increasing, rather than decreasing, air flow rates.

Challenges to stable operation and control

Fine control of the scalper froth level is critical to stability and crucial to cell performance for both recovery and quality purposes. Two of the key instruments, for depth and air flow, each posed their own challenges to achieve fine control of these parameters.

Froth depth measurement

The installed level instrument relied on two separate pressure sensors lowered into the cell below the froth depth, with a known height difference between them. This enabled a calculation of an inferred cell slurry density, and thus a further calculation of froth depth. While this level indication is designed to compensate for changes in slurry density, in practice operation technician would find this depth indication would drift with density and aeration changes.

Operations technicians regularly would dip the Jameson cell with a hand-crafted ball float to monitor and correct for sensor drift. This became important during frothing and racing episodes, when the pressure readings were impacted by froth, and subsequent the drift of depth measurement would cause a reaction by the control system that often exacerbated racing events. On occasions, the cell would be found in a pulping condition, Figure 12, when according to the depth sensors it was running a stable sensible froth depth.

FIG 12 – Pulping when operating at stable indicated froth depth.

It was also observed that the readings became increasingly inaccurate as the cell was run deeper. If the upper pressure sensor became exposed to a deep froth phase, and was no longer submerged in the pulp, an erroneous depth reading could result.

Air flowmeter

For the air flowmeter, an issue was identified during 2020 commissioning where the sensor would periodically begin reading erratic flow variation, causing an undesired process control response that impacted froth control. Notably the erratic flows occurred primarily during the day but regained consistency overnight; the flow also remained erratic when the control valve was held in a constant position (which in a Jameson cell should force a stable flow).

Initial lengthy attempts – including recalibrating, confirming all sensors settings, and ultimately replacing the sensor – to the reduce erratic readings from the flowmeter ceased after those attempts then were unsuccessful. Some signal filtering was applied to reduce variation, however it still resulted in erratic movement of the air control valve.

Dart valve

A review of the level control in the Cleaner 3 Jameson identified the tailings dart valves operating in a very small range, between 2 and 10 per cent open but resulting in significant fluctuations in froth depth. At this operating range, the flow characteristic of the valve was non-linear causing sudden changes in flow of the tailings slurry released from the valves. It was likely that the flow rate specified for engineering and design of the valves were significantly higher than those required in operation, leaving the valves oversized for the duty.

Recovery cascade control

Live Jameson cell stage recovery was calculated from copper grades measured by the OSA. In late 2020, a recovery controller was implemented that manipulated air flow rate and froth depth set points to maintain the measured recovery at the desired value as entered by operations technicians, typically 70–75 per cent.

Given the inaccurate and unreliable level instrument on the Jameson Scalper cell, the automated control would from time to time not perform as intended. For instance, as the recovery would decrease, the cell, as expected, would make the cell shallower to increase recovery. On many occasions this would result in the cell pulping as the true level was too shallow for froth flotation despite the level transmitter indicating a deeper level. As a result of the pulping, the recovery would decrease further. Alongside this action, the air flow rate would increase aiming to increase recovery, further exacerbating the poor flotation conditions. The drifting nature of the depth measurement made it extremely challenging to set constant depth set point control limits and to avoid the controllers moving the cells to an adverse location.

Despite the controller working well vast majority of the time; the above situation would occur on a sufficiently semi-regular basis without a clear resolution, that it was determined that an alternative automated control option, or removing the automation, was needed to reinstate stability in the cell to manage and improve recovery and concentrate grade.

CONSOLIDATED ACTIONS FROM OPERATIONAL REVIEW

As an outcome of the operational review, the following of actions were developed to be progressed by the site teams.:

- Operate Jameson Scalper at 10 per cent solids.

- Implement measurement and control of Jameson feed density to assist above action.

- Right-size the Cleaner 3 Jameson dart valves

- Address Scalper Jameson air flow rate accuracy and reliability.

- Set a minimum Scalper Jameson air flow rate set point of 350 m³/h.

- Create 'racing' episode response strategy of increasing air flow rates.

- Install an alternative froth depth measurement method on both Jamesons.

- Increase wash water flow rates to of both Jameson cells to consistently achieve positive bias.

- Expand wash water capacity of the Scalper Jameson to assist above action.

- Cease use of recovery control, and trial alternative automation options for both Jameson cell control, such as utilising froth cameras, or return to operator-managed control.

IMPLEMENTATION OF ACTIONS

Froth depth measurement by ball float

Trials were conducted using the original supplied froth depth sensors to improve the accuracy and reliability, such as setting a minimum inferred slurry density within the level calculation and regular cleaning. However, both proved unsuccessful in resolving the issues of sensor drift that was identified during the operational review. With a hand-crafted ball float becoming a trusted depth measurement tool amongst technicians and metallurgists alike, it was established that a trial ball float would be installed as an alternative level instrument.

Jameson cells rely on comprehensive froth washing to reduce entrainment in the concentrate. As such, there is little froth surface not otherwise covered by a wash tray, limiting the space available for installation. Further, the presence of a work platform and downcomer distributor made location selection more challenging. The ball float was installed so that the shaft sat between two wash water trays and did not impede the washing capability. A hole was cut in the above work platform to accommodate a ball float of sufficient length to cover the desired measurement range from cell lip to ~1200 mm. The depth is measured by a radar level sensor reading a target plate at the top of the ball float shaft, this is in line with other cell ball float instruments in the concentrator, Figure 13.

FIG 13 – Scalper Jameson ball float.

The initial ball float was a modified (lengthened) tank cell float, of stainless-steel construction. During the trials, the longer float was observed to sink occasionally under the greater shaft mass which decreased its buoyancy. A second, reduced-mass, ball float design was developed using a nylon shaft.

After this second design iteration, the ball float level transmitter improved the reliability of the level reading in both Jameson cells and as such the stability of level control, particularly at shallower levels and during frothing episodes. Later modifications were made to the guides and shaft of the ball float to amend issues from sticking and improve reliability at deeper levels.

With a reliable level transmitter, stable controller limits were put in place to ensure reduce the occurrences of boiling (too deep) and racing (too shallow) of the cell.

Resizing of Cleaner 3 dart valves and slurry lens

Operating data, and survey mass balances were provided to Glencore Technology to review equipment sizings of various components of the Cleaner 3 Jameson cell. It was determined that the Jameson was operating at significantly lower feed flow rates than anticipated during engineering and design, driving very high recycle rates (>90 per cent) and the tails dart valve to operate below the preferred operating range. The very high recycle rates were thought to contribute to a build-up of NSG in the recycled downcomer feed slurry, and reducing the recycle rates to normal operating range would ideally improving the concentrate quality recovered by the Cleaner 3 Jameson.

Three changes were proposed and adopted. It was determined that decreasing the dart valve opening would enable a wider operating valve position and finer level control. The dart valve seat bores were reduced from the original diameter of 150 mm to newly sized diameter of 100 mm. To reduce the recycle rate within the cell, a reduction in slurry lens size was implemented from 44 to 36 mm. Thirdly a recycle pump pulley change was implemented, to keep downcomer pressure within guidance with the smaller slurry lens in place.

Scalper Jameson air flowmeter inlet pipe extension

Attempts to resolve the erratic signal from the flowmeter recommenced in 2023 with further trials completed including, but not limited to, replacing the flowmeter with one of the same type, a sunshade over the flowmeter, adjusting insertion depth and re-terminating electrical connections. Lengthy troubleshooting, in conjunction with Glencore Technology, identified that the thermal mass flowmeter, the installed type is most suited to measurement of dry gases. The instrument manufacturer indicated that a minor amount of moisture in the air stream could cause erroneous reading by the flowmeter.

At Carrapateena, the air inlet was located directly above the Jameson cell. Further, the Jameson cells are positioned within walls that protect the cell from high winds. It was entirely possible, that the mist and evaporation from the open-topped wash water trays was trapped within the wind walls and then drawn into the air inlet located directly above.

A trial was completed with a vent bag Figure 14 (left) to extend the flow metre inlet beyond the wind walls, no longer above the cell, allowing the cell to draw in ambient air, free of stray moisture from the froth wash system. An immediate improvement in the flowmeter output was observed, Figure 15. When the valve was held in a single position, the range reduced from ±100 m³/h to what was considered an acceptable variation that could then be managed with process control techniques. After the successful vent bag trial, a permanent hard pipe solution was installed on the Scalper Jameson that extended the inlet of the airline to ensure it received no damp air from the cells, Figure 14 (right).

FIG 14 – Scalper Jameson air inlet extension (left) trial vent bag (right) permanent hard pipe.

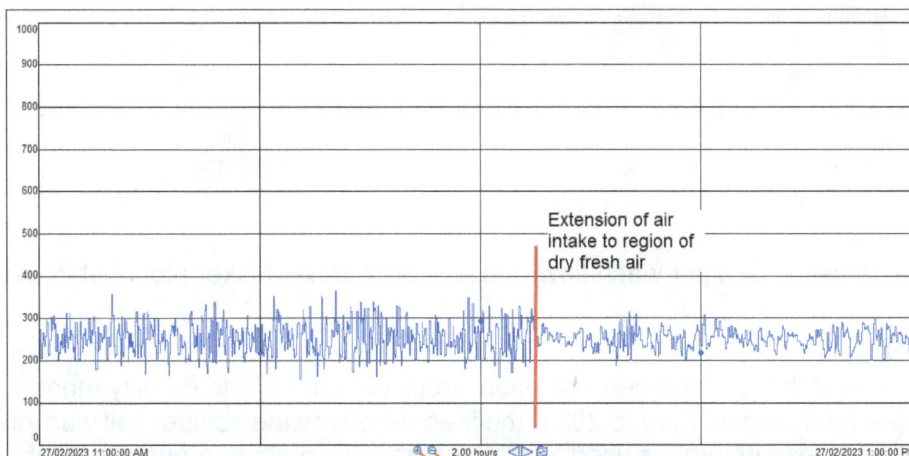

FIG 15 – Jameson Scalper air flow rate – oscillation greatly reduced with extension of air intake.

Wash water manifold

The operational review identified Cleaner Scalper wash water flow rates up to 130 m³/h were required to achieve a positive bias across all conditions. However, the existing raw water manifold (and raw water distribution system) was only able to supply ~85–90 m³/h.

A second manifold was designed and installed in September 2023 to supply up to the required 130 m³/h. This manifold utilised process water due to greater available capacity in the process water distribution system to provide the required flow; and the availability of a nearby process water branch, instrumented with a flowmeter and control valve, that could be quickly repurposed to the wash water duty.

The existing manifold remained in place and in service, with both manifolds able to be operated at the same time. In addition to providing additional capacity for wash water, utilising process water reduced the demand for raw water sourced directly from borefields and provides the operations technicians an additional tool to manage the site water balance. Use of process water for washing minimises the amount being added to the site water balance, which is useful when there is an abundance of process water that needs to be drawn down, particular after large rainfall events. The operations technicians adjust the prevailing mix of process and raw water to suit the prevailing water balance, while ensuring the total wash water flow meets the positive bias requirements. The dual systems also provided some redundancy should there be quality concerns in the process or raw water streams.

Further, mass balance calculations were implemented in the control system to present live estimates of the achieved wash bias, and the required flow to achieve a positive bias to guide the operations technicians. These live calculations were based on the feed flow and density instrumentation, and OSA measurements to provide mass recovery and concentrate percent solids information.

Concurrently, the wash trays were redrilled with addition holes to provide sufficient drainage capacity to accept additional flow.

The increase in wash water flow across both modifications is shown in Figure 16.

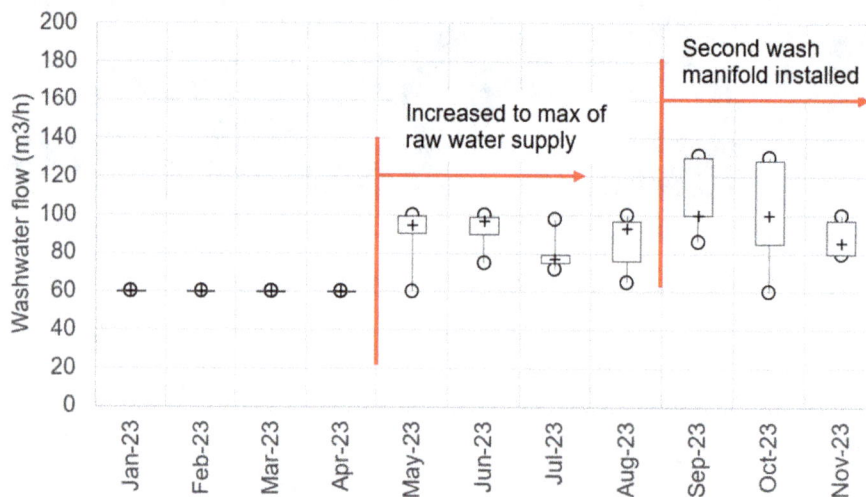

FIG 16 – Jameson Scalper wash flow rate – quartile box-whisker plot of shift averages.

Feed density

The original design of the Carrapateena flotation circuit did not include density measurement on the Jameson Scalper feed stream. Prior to 2021, the feed density to the scalper cell was inferred through the two submerged pressure probes used for froth depth. This method proved unreliable and as such an alternative was sought.

Initially, a calculation was implemented that considered the rougher concentrate and cleaner 1 scavenger concentrate flow and density measurements, along with either a measured or estimated measurement of dilution water in regrinding or in the cell feed box. This enabled the operations team to set a dilution flow rate in response to this calculation. However, manual sampling of the cell's feed box would often expose deviation between real and calculated feed densities.

Consequently, the long-term solution to install a density gauge onto the scalper cell feed line was pursued. This enabled real and reliable density measurements. Together with upgrades to the regrind product hopper dilution water addition, completed in August 2022, these allowed automation of feed density by control of water addition to that hopper upstream of the density gauge.

Initially this control targeted 14 per cent solids. Following the above operational review in January 2023, Jameson Scalper feed density target was progressively reduced to 10 per cent. However, in May 23 the capacity of the dilution water branch in to the regrind product hopper was reached, preventing this target from being achieved. A further upgrade of this piping was completed in November 2023, which then allowing consistent control at the lower percent solids specified in the operational review.

The lower percent solids, resulting in higher volumetric feed, required a review of recycle rate and installed recycle slot. Aiming to maintain downcomer flow and prevent overflow of the scalper feed box, the recycle slot was exchanged from an 80 mm design to a 70 mm design. This reduced the recycle rate as from the scalper tail back to the feed and achieved the desired effect.

The change in density, feed flow and recycle rate as a result of these modifications is given in Figures 17 and 18.

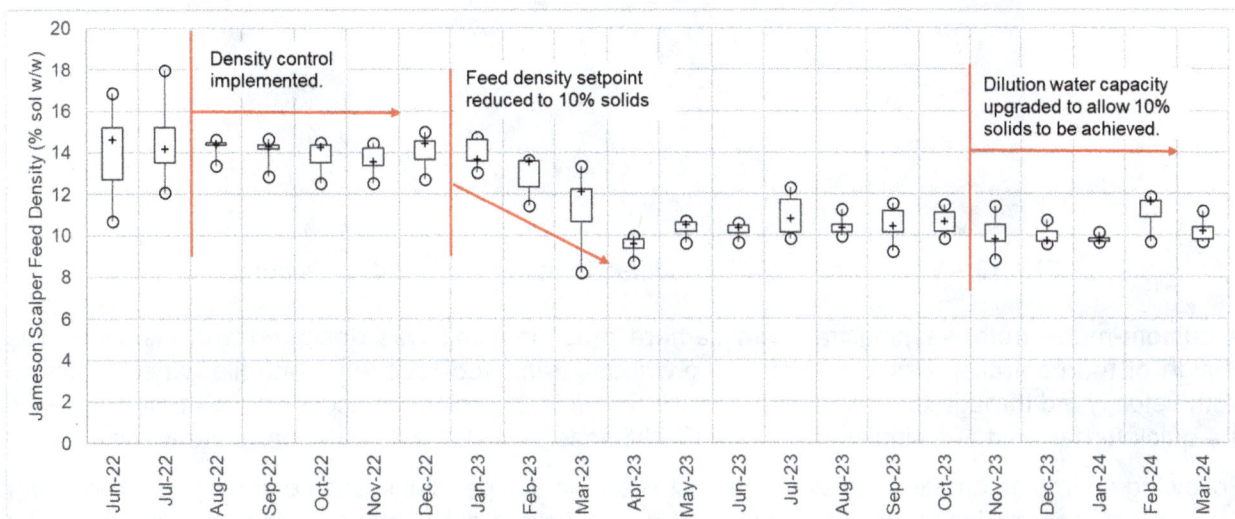

FIG 17 – Jameson Scalper feed density – quartile box-whisker plot of shift averages.

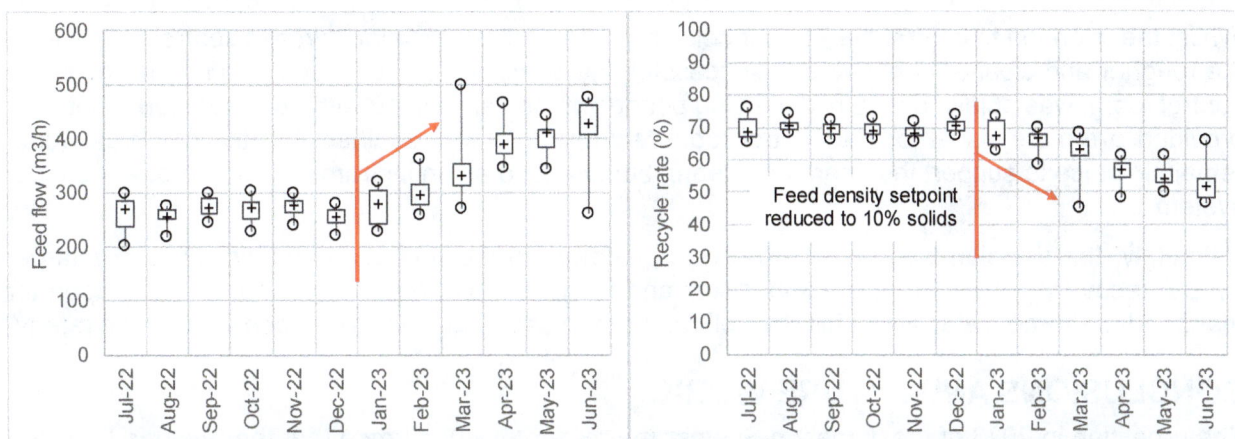

FIG 18 – Lower feed density impact on recycle – quartile box-whisker plot of (left) feed flow (right) recycle rate.

FrothSense camera

The Carrapateena concentrator utilises Metso FrothSense cameras throughout its rougher and cleaner cell banks to manage individual and bank cell mass recoveries with great success (Reinhold *et al*, 2023). A trial was conducted with a camera installed on the Jameson Scalper cell.

The cameras are designed to be installed over the lip of the cells, however given the Jameson cells reliance on wash water for its performance, this would result in a portion of the froth not being adequately washed. Numerous locations and other applications were considered, including whether the monitor the movement of froth off the cell lip, or down the launder as the indication for mass recovery. The decision was made to install it over the cell lip and accept a small portion of froth would not be washed, location shown in Figure 19.

FIG 19 – Wash water tray – modified to fit Metso FrothSense camera.

A custom-made froth washing tray with camera mounting unit was designed and installed. The design of the controller was based off the previously removed recovery controller where a target froth velocity fed through to air and level control. The initial perceived success of this trial resulted in the quick turnaround of a similar install of a FrothSense camera on the Cleaner 3 Jameson.

Following an implementation review of the install on the scalper cell it was seen that the froth under the camera, now no longer impacted by wash water, flowed differently over the lip of the cell. This made the velocity measurement no longer representative of the rest of the cell or relevant for the mass control.

Whilst the intention to control the scalper cell air and level through velocity/mass control as done on the rougher and cleaner banks was unsuccessful; the camera footage, which was on display in the control room, was able to assist in identifying periods of the cell pulping with poor froth formation and provided a remote visual for the control room technician. It was decided that the cameras would remain in place to support the operations team, but not used as an instrument in the process control system.

Ultimately, the decision was made to provide guidance to the operation technicians of the target copper recovery required by the scalper cell; and return the air and level control to being operator managed to promote stability within the cell rather than use of higher level process control strategy.

CONCLUSIONS AND FUTURE WORK

The repetition in 2023 of the Jameson Scalper matrix surveys confirmed that the Scalper Jameson was still capable of performance achieved during commission, even though now operating at significantly higher throughputs and mineral loading. The consistency seen across the three matrix sets of surveys spanning over several years provided certainty in those results.

The review of the operating strategy and historical performance:

- Identification of capacity limitations, that could be addressed.

- Enabled new insight into the cell's operation.

- Ensured previously held assumptions were reconsidered.

- Reset operating windows to suit materially higher milling throughputs and increased dominance of chalcopyrite mineralogy.

Whilst cell performance was only returned to that achieved after the initial commissioning optimisation works; and the trial of froth camera control was unsuccessful, other work proved valuable:

- Implementation of alternative instrumentation such as the ball float and feed density gauge improved the stability of the cell operation.

- Increased wash water and dilution cleaning returned the cell to further reduce the entrainment of fines to the scalper concentrate.

- Removal of the recovery control, enable operation technicians to increase their ownership of the cell performance.

- The development of operator guidance for 'racing' episodes improved the capability of operation technician to keep froth tenacity in control.

The engagement of Glencore Technology to assist with the operational review was instrumental in its success, and significantly compressed the time frames required to execute, as their technical guidance was able to quickly focus the team's energy on the most impactful work fronts.

A conclusion that solidified amongst the Carrapateena team, is that to further improve the concentrate grade and quality, required modifications to the plant external to the Cleaner Scalper cell, as the cell appears to have approached the limit of its capacity. Carrapateena is also currently in the process of developing a block cave expansion project which will see an increase in the ore mined up to 12 Mt/a and may result in expansion of the existing concentrator and/or the addition of a second concentrator.

To address this, engineering designs intended to expand rougher capacity (via inclusion of a Rougher Jameson on flotation feed) are now also incorporating relief of Jameson Scalper capacity. Rougher Jameson concentrate will be cycloned separately from the existing mechanical rougher concentrate, with that cyclone overflowing being cleaned by an additional cleaner Jameson cell to separately from the existing Jameson Scalper cell. A proposed design presented in Figure 20.

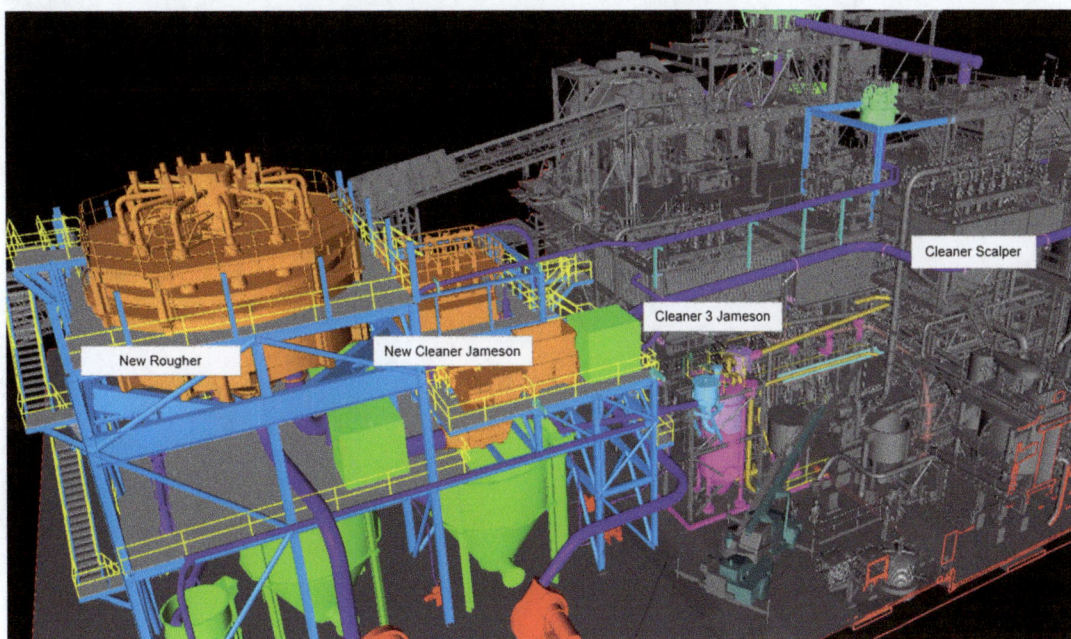

FIG 20 – Proposed rougher scalper expansion.

ACKNOWLEDGEMENTS

The authors would like to acknowledge the Carrapateena Processing team, who have all played a part in the successful implementation of the program and without which, it would not have been successful.

REFERENCES

Reinhold, J, Burns, F, Seppelt, J and Affleck, K, 2023. Plant Modifications and Operating Strategies to Improve Concentrate Quality at OZ Minerals – Carrapateena, in *Proceedings of MetPlant 2023 Conference*, pp 267–289.

Taşdemir, T B Ö, 2007. Air entrainment rate and holdup in the Jameson cell, *Minerals Engineering*, 20(8):761–765.

Optimising flotation pulp conditions to improve recovery at Newmont Red Chris Mine

D R Seaman[1], J Johannson[2], J Baldock[3] and B A Seaman[4]

1. MAusIMM, Manager-Directional Studies (TS – Processing and Metallurgy), Newmont Corporation, Subiaco WA 6008. Email: david.seaman@newmont.com
2. Plant Metallurgist, Newmont Corporation, Red Chris Mine, Dease Lake BC, VOC 1L0, Canada. Email: jennifer.johannson@newmont.com
3. Chief Metallurgist, Newmont Corporation, Red Chris Mine, Dease Lake BC, VOC 1L0. Canada. Email: joshua.baldock2@newmont.com
4. Head of Mineral Processing (TS – Processing and Metallurgy), Newmont Corporation, Subiaco WA 6008. Email: brigitte.seaman@newmont.com

ABSTRACT

The Red Chris orebody is considered a typical copper porphyry system containing chalcopyrite, minor bornite, gold in multiple forms and pyrite. Gold deports between liberated gold, gold associated with copper minerals and gold associated with pyrite in granular and sub-microscopic form. Flotation is used as the primary process to produce a saleable copper concentrate with significant precious metal credits.

Of two open pits, the main pit contains a much higher pyrite content with pyrite to chalcopyrite ratios exceeding ten. This high pyrite content produces a highly reducing environment in the grinding circuit due to galvanic interactions with mild steel grinding media. This also results in a correspondingly high oxygen demand, as well as inadvertent copper activation of pyrite requiring high lime addition to achieve chalcopyrite/pyrite selectivity in the rougher and cleaning circuits.

Pulp chemistry measurements within the grinding and flotation circuits confirmed that conditions were not conducive for collector adsorption. Two laboratory programs identified recovery opportunities in the form of moving from forged steel to high chrome grinding media in the ball mill and adjusting the reagent strategy in terms of addition points and collector synergies.

Plant trials were successful with a combined impact of improving recovery, reducing grinding media consumption and a reduction in lime consumption required.

This paper presents an overview of the plant diagnostics leading up to the trial and presents the trial outcomes which have led to the acceptance of high chrome media and reagent strategy in the operation.

INTRODUCTION

Red Chris operations

The Red Chris Mine is in the Golden Triangle of Northern British Columbia, approximately 80 km south of Dease Lake and 18 km south-east of the community of Iskut. The deposit is a noteworthy copper-gold porphyry deposit with substantial economic value due to its large copper and gold reserves. Its highly deformed structure indicates a past marked by volcanic activity, magmatic intrusions, and hydrothermal alteration, which led to significant copper-gold mineralisation (Rees *et al*, 2015). Newcrest Mining Ltd, and subsequently Newmont Corporation, became the operator of the Red Chris Mine on the 15 August 2019 after purchasing a 70 per cent stake of the operation from Imperial Metals (Seaman *et al*, 2021).

Process flow sheet

The Red Chris processing plant has been designed to treat 10.5 MTPA using a conventional semi-autogenous grinding (SAG) mill, ball mill and pebble crushing (SABC) circuit with a target grind P_{80} of 150 μm. Throughput takes precedence over grind size, resulting in actual grind size P_{80} in the range 130–180 μm depending on crushed ore availability on the coarse ore stockpile and water availability from the tails storage facility. A current flow sheet of the concentrator is shown in Figure 1.

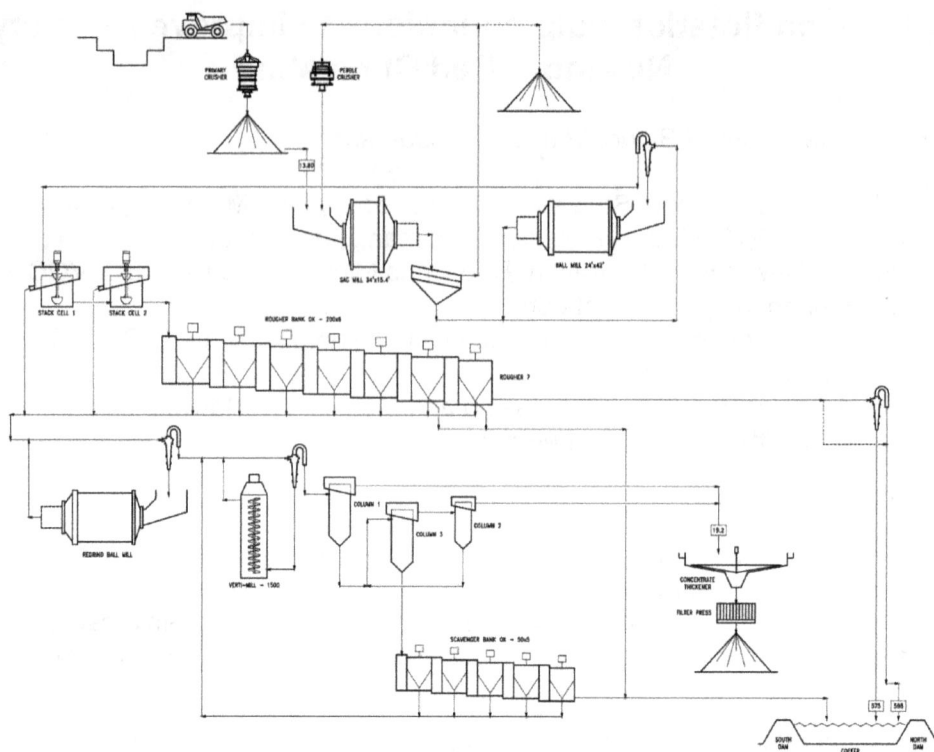

FIG 1 – Process flow sheet of the Red Chris concentrator.

Copper rougher flotation is performed in two Eriez SC200 StackCells, six 200 m³ and one 160 m³ Outotec Tank Cells. At times, the last two rougher flotation cells are directed to sulfide duty to remove excess pyrite from the rougher tailings for separate storage in the tailings impoundment area.

Rougher flotation targets bulk flotation conditions where the recovery of copper minerals (predominantly chalcopyrite) and iron sulfides (pyrite) is maximised by applying relatively low pH conditions in the roughers of 8–9.5. Prior to 2019, the pH in the roughers was operated at higher levels to create selective (copper/pyrite) conditions, and this has been continuously lowered over time with physical and operating strategy changes summarised by Seaman *et al* (2021). This has been a deliberate change towards more bulk sulfide recovery conditions. A selective primary collector, DTP (dibutyl dithiophosphate) is used along with PAX (potassium amyl xanthate) in the roughers and MIBC (metho isobutyl carbinol) as frother. A high mass pull is targeted from the roughers to maximise overall recovery.

The rougher concentrate is directed to the cleaners via a two-stage regrind circuit with a nominal target P_{80} of 30 µm. The cleaner circuit configuration is variable with the flow sheet changing depending on operational requirements (Li *et al,* 2024). The pH in the cleaners is raised to depress pyrite, typically to around 11–12, no additional pyrite depressants have been necessary at Red Chris. It has been noted at times that over-dosing of collector in the rougher stage can be problematic in terms of froth transportation and selectivity in the cleaners. The operation targets collector addition dosing on a total rougher sulfide throughput basis to ensure that overdosing and underdosing is minimised.

Final concentrate with a grade of 22–24 per cent Cu is thickened and pressure filtered ahead of trucking to the Port of Stewart where it is loaded on a ship for delivery to the point of sale. The process plant produces non-acid generating (NAG) and potentially acid generating (PAG) tailings streams. The NAG tailings are further processed by semi mobile cyclones at the tailings dam, producing coarse sand for wall construction. This stream must have a sufficient ratio of neutralising carbonates and acid generating sulfides to achieve a neutralising potential ratio (NPR) of greater than two.

The PAG tailings stream is a pyrite rich product principally composed of cleaner-scavenger tailings and concentrate from the sulfide roughers. The PAG tailings stream is required to be deposited sub-aqueously in the tailing's impoundment area.

In 2019, an intensive recovery diagnosis and improvement program commenced. Other papers (Seaman *et al*, 2021; Li *et al*, 2024) have shown the resulting physical and operational changes made to improve recovery in the Red Chris concentrator which have included:

- The introduction of donut launders in the rougher cells to optimise froth crowding.

- Implementation of a mass-pull controller and associated short-interval-control system.

- Installation of a third Eriez cavitation tube cleaner column and two Eriez StackCell SC200s.

In parallel to these circuit and operational changes, it was observed that laboratory test work on geomet samples often outperformed plant performance after accounting for residence time and grind size effects. A program of measuring and optimising pulp chemistry conditions began in 2020 and ultimately led to the successful implementation of adopting high-chrome media in the ball mill and a change to collector operating strategy.

PULP CHEMISTRY DIAGNOSIS

Mineral flotation systems involve multiple redox (reduction-oxidation) reactions, knowledge of the Eh changes around a circuit allow optimal reagent addition points to be selected (Chander, 2003). Pulp chemistry surveys were initiated at Red Chris late 2019. A set of probes (Eh, DO (dissolved oxygen), pH and temperature) with datalogger was purchased specifically for this task. Procedures for conducting surveys were provided by Chris Greet of Magotteaux Australia (Emery, 2009). Similar procedures are available in Grano (2010). For each survey point around the plant, a 1 litre sample jar is half-filled with slurry, the probes are allowed to come to equilibrium after being rinsed in fresh water, the data logger is started and a few seconds later, the probes are inserted into the sample jar and used to gently stir the sample for 2 mins. Data is logged every second for the two-minute period commencing just before the probes are inserted. The logged data is later used to determine the dissolved oxygen demand (DOD) by fitting a first-order decay curve to the DO reading from point of insertion to the slurry. The remaining parameters (pH, Eh, temperature and DO) are averaged from the last 10 seconds of the log data for consistency.

McMahon *et al* (2016) applied the same procedures to measure DOD at Newmont's Telfer operation to diagnose a significant selectivity deficiency post regrind of pyrite concentrate and demonstrated the criticality of DOD on the flotation recovery of gold and copper as well as its importance in achieving selectivity against pyrite.

The authors transferred this know-how to Red Chris and began making pulp measurements soon after Newmont took over management of operations. Regular pulp chemistry measurements around the circuit are conducted, mostly focused on the ball mill discharge and primary cyclone overflow streams. Figure 2 shows the results from one of the early surveys conducted where several measurements were taken around the circuit.

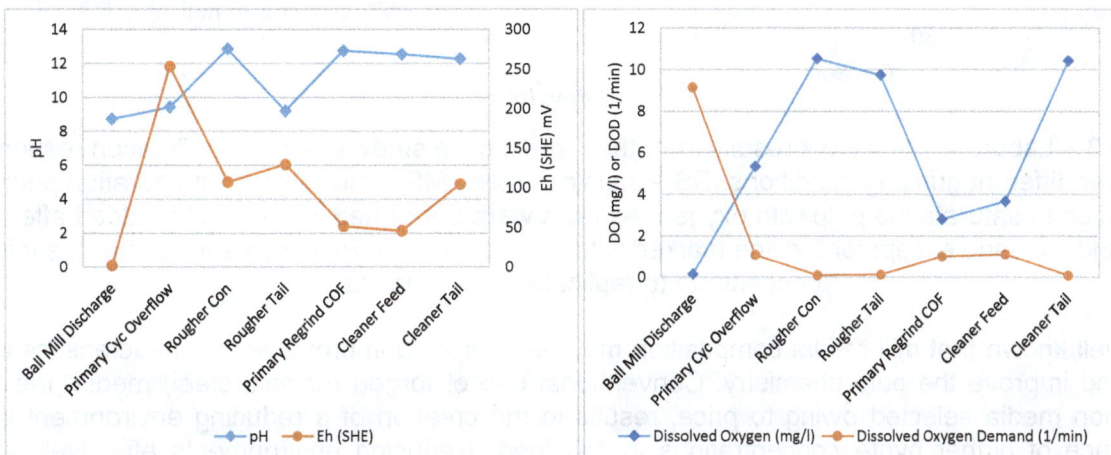

FIG 2 – Initial pulp chemistry surveys conducted around the flotation circuit (2020).

All Eh measurements were made using a Platinum electrode. An adjustment has been made to all readings in this paper to report against equivalent SHE (Standard Hydrogen Electrode) for consistency and easy reference to literature values.

Dissolved oxygen was almost undetectable in the ball mill discharge with a correspondingly high demand. As the survey progresses to the ball mill cyclone overflow, the Eh increases as does the dissolved oxygen. The high pH in the rougher concentrate is a function of lime addition in the concentrate hopper prior to the sampling point. There is another notable drop in dissolved oxygen around the regrind circuit with a similar reduction in Eh and increase in DOD. Collector adsorption requires the presence of oxygen to complete a chemi-sorption reaction and the surveys highlight that the mill environment is therefore not ideal for collector addition.

Subsequent surveys showed the same trend, however there was an apparent change in pulp chemistry as feed grade changed, particularly with more reducing conditions evident at higher pyrite feed grades and more oxidising conditions at lower pyrite grades.

In their paper on the classification of copper minerals into semiconductor types and associated flotation characteristics, Lotter, Bradshaw and Barnes (2016) highlighted that specific Eh-pH regions exists for the favourable collection and optimisation of different copper minerals, and it was found that conditions in Red Chris were too reducing for optimal flotation of chalcopyrite and pyrite.

Based on the preliminary conclusions drawn from the pulp chemistry surveys, ore samples were collected and dispatched to the UniSA (University of South Australia) for characterisation test work as part of the AMIRA P260H project, and a similar sample was dispatched to Magotteaux Australia for vendor test work on evaluating alternative grinding media alloys.

The AMIRA P260H study found that copper recovery could be improved in laboratory tests by using stainless steel (SS) grinding in place of mild steel (MS) and also demonstrate that adding collector to the MS mill rather than after milling reduced the kinetics and copper recovery, a summary of their findings are shown in Figure 3.

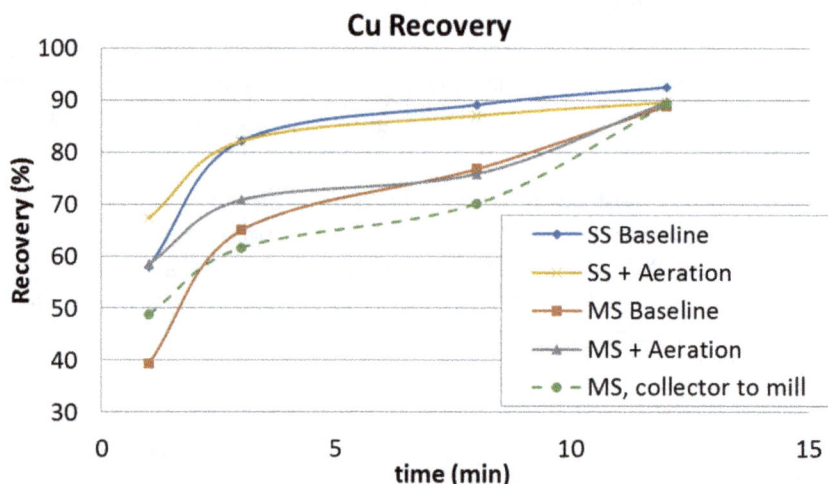

FIG 3 – Laboratory test work results from the P260H case study showing the flotation response under different grinding conditions (SS – stainless steel, MS – mild steel), with aeration prior to flotation to saturate the pulp with oxygen. All tests were performed with collector added after the grinding stage except for the line marked with collector to mill. PAX was added after the third concentrate to replicate the sulfide stage.

It is well known that mill media composition may be altered to improve wear characteristics in the mill and improve the pulp chemistry. Conventional use of forged (or mild steel) media, the most common media selected owing to price, results in the creation of a reducing environment in the presence of higher pyrite concentrations in mill feed. Reducing environments effectively act to corrode the media (through galvanic interaction) and extract metal ions, most notably iron oxy-hydroxides ions. These iron oxy-hydroxides ions can then coat mineral surfaces resulting in the depression of the target valuable minerals (Johnson, 2002). In the case of ores containing copper secondary mineral species, copper ions may be released from the ore during grinding and these ions, when precipitated on the surfaces of pyrite particles can serve to activate the pyrite thus requiring higher lime dosages to improve the efficiency of the separation between copper and pyrite with the outcome of both higher reagent cost and recovery losses at the increased lime dosages.

In his comprehensive paper on grinding chemistry and its impact on copper flotation, Greet (2019) presents five examples where laboratory benefits observed with a change to high chrome media were successfully translated to plant scale. The host orebodies ranged from iron oxide copper-gold deposits, copper-gold porphyry ores and a sedimentary copper silver orebody. In all cases the following benefits were realised:

- Measurable recovery improvements of approximately 1 per cent or more were realised at plant scale.

- Measurable reductions in the EDTA extractable iron concentrations were observed, concluding that there had been a resultant decrease in the media corrosion with the change to the high chrome media. This in turn implies that a decrease in consumption should be reasonably expected owing to the reduced corrosion.

Additional benefits noted regarding reagent usage at Ernest Henry Mine and the Lubambe Concentrator included:

- The move toward no longer requiring lime for pH control where they previously utilised 0.85 gk/t of this reagent in the case of Ernest Henry Mine, and

- A 17 per cent reduction in their Sodium Hydrogen Sulfite consumption at Lubambe Concentrator.

In line with the above case studies, promising copper and gold recoveries were produced from the laboratory test work with statistically significant improvement in copper rougher recovery of greater than 3 per cent and a statistically significant improvement in gold rougher recovery of greater than 4 per cent. Plant expectations for copper and gold recovery were therefore also in line with the industrial findings for these case studies, ie ~1 per cent or greater for copper and gold recovery overall.

Based on this diagnosis and benchmarking exercise, it was agreed to move to a plant trial of high chrome media and an update of the collector addition strategy.

REAGENT ADDITION STRATEGY

Like many operations, the collector addition strategy at Red Chris has evolved since operations began in terms of the types of collectors utilised, the dosage set points and, of course, the addition points. Some of these changes have been based on plant or laboratory experimentation, while others have been influenced by personal experience and opinions of operators and metallurgists either at the operation or from previous experience.

Most metallurgists have heard mythical stories of collectors, such as: *C-crew at A-mine site discovered that moving the collector dosing point to (insert location such as mill discharge) improved recovery. They didn't tell the other crews and basked in the glory of outcompeting the other crews for many years.* While this exact situation was not happening at Red Chris, some practices were certainly based on a limited number of data points or very few ad hoc experiments that would certainly not pass the statistical rigour of Napier-Munn (2014). The mine had also recently switched pits from East Zone (lower pyrite content) to Main Zone (higher pyrite content). Detrimentally, there was little consistency between crews and the leadership was looking for some technical direction to standardise an approach. The authors turned to reference libraries and past Red Chris test work to establish a technically substantiated approach.

As mentioned earlier in this paper, the operating strategy for the roughers is focused on bulk sulfide recovery. Mass pull from the roughers is maximised up to any downstream constraints (typically regrind cyclone feed pumps). Pyrite rejection takes place in the cleaners with increased pH of 10.5–12 through addition of lime. DTP (di-butyl di-thiophosphate) is utilised as the primary collector, with PAX (potassium amyl xanthate as secondary). Trials of alternate collectors have been successful in the past but were excluded from the strategy at this time. Collector dosage in the roughers is normalised to the mass flow rate of sulfur (representing total sulfide load) in feed. This is based on prior experience where overdosing at low sulfur feed grades occurred resulting in over-collected froth in the cleaners and conversely under-dosing (recovery loss) was observed at higher sulfur feed grades.

Collector addition point

The pulp chemistry analysis (Figure 2) and test work conducted by UniSA pointed towards moving the initial collector addition point after grinding where the pulp conditions would be more favourable. This should allow more DO to be present in the slurry and for the Eh to be in a more conducive range at the point of addition (due to water addition and aeration in the hydrocyclones). By adding the collector at sub-optimal conditions, some of the collector may be adsorbed unselectively to non-valuable particles such as clay minerals.

Yoon (1981) demonstrates collectorless flotation of chalcopyrite under certain pulp conditions, this has been proven at Red Chris in the plant (albeit at lower overall recoveries). The laboratory geomet standard test was developed by third party laboratory, BML (Base Metal Laboratories, Kamloops) which employs an optimised strategy of adding collector only after the first minute of flotation has been completed.

Optimising synergistic effects

Many authors cite the synergistic effect of using multiple collectors where the recovery obtained is higher than that achieved by individual components alone (Bradshaw, 1997; Lotter, Bradshaw and Barnes, 2016; Dhar, Thornhill and Roa Kota, 2019).

In copper sulfide and pyrite flotation, these papers suggest combinations of more selective collectors such as IPETC (Isopropyl ethyl thionocarbamate), DTP (dithiophosphate) and Xanthates. Red Chris was at the time (and is currently) utilising a DTP and PAX as sulfide collectors and had already proven the synergy of utilising both on the copper rougher stage but up to now had not formalised the sequencing of addition.

Dhar, Thornhill and Roa Kota (2019) investigated different ratios of collector additions applied to a copper-gold ore and varied the sequence of addition, adding the reagents together or in order of increasing or decreasing selectivity. Their findings on Nussir ore showed that a 3:1 ratio of DTP:SIBX was optimal over the reverse ratio, they also showed that adding the more copper selective collector, DTP, first and conditioning prior to addition of SIBX produced far superior results. This was consistent for copper and gold response for their ore. These authors conducted a wide variety of characterisation techniques including adsorption studies to support hypotheses of the mechanism behind this result. They demonstrated that adding SIBX prior to DTP was sub-optimal since SIBX took up all the available collector sites on the mineral particles leaving fewer sites for DTP to adsorb on. When DTP is added first, it preferentially adsorbs to higher energy (less oxidised) sites, leaving xanthate and its dimer to adsorb on the remaining sites.

Bazin and Proulx (2001) studied the impact of distributing flotation reagents down a flotation bank. These authors reviewed several size-based flotation studies, such as Trahar (1981), and developed an understanding of down the bank addition based the different collector needs and responses for coarse and fine particles. Coarse particles require significantly more collector to be recovered while fine particles need less collector, but they consume the majority of collector due to their higher surface area to volume ratio. In addition, some of the studies reviewed by Bazin and Proulx (2001) showed that fine particle recovery can be reduced at higher collector dosages in opposition to the positive effective observed for coarse particles (Klimpel, 1995).

Bazin and Proulx (2001) completed a laboratory study followed by plant trials at the Brunswick mine where they varied the distribution of collector down the rougher and scavenger banks concluding that by using 50 per cent or less collector at the top of the bank, with larger additions in the downstream cells provides higher recovery of coarse particles at equivalent or lesser reagent consumption.

Collector addition profile

Based on the above review and discussions on-site with operators and metallurgists, a practical strategy was trialled and has been subsequently maintained as shown in Figure 4. The ratio of DTP:PAX has been maintained at around 2:1, with 30 per cent of the DTP added immediately after grinding, and the remainder DTP is added between the two StackCells and the conventional rougher bank. When there is a high pyrite content in the feed, the last two cells are converted to sulfide recovery duty. In this mode, additional PAX is added prior to the cells, and very high mass pulls are targeted on these cells to remove pyrite from rougher tailings.

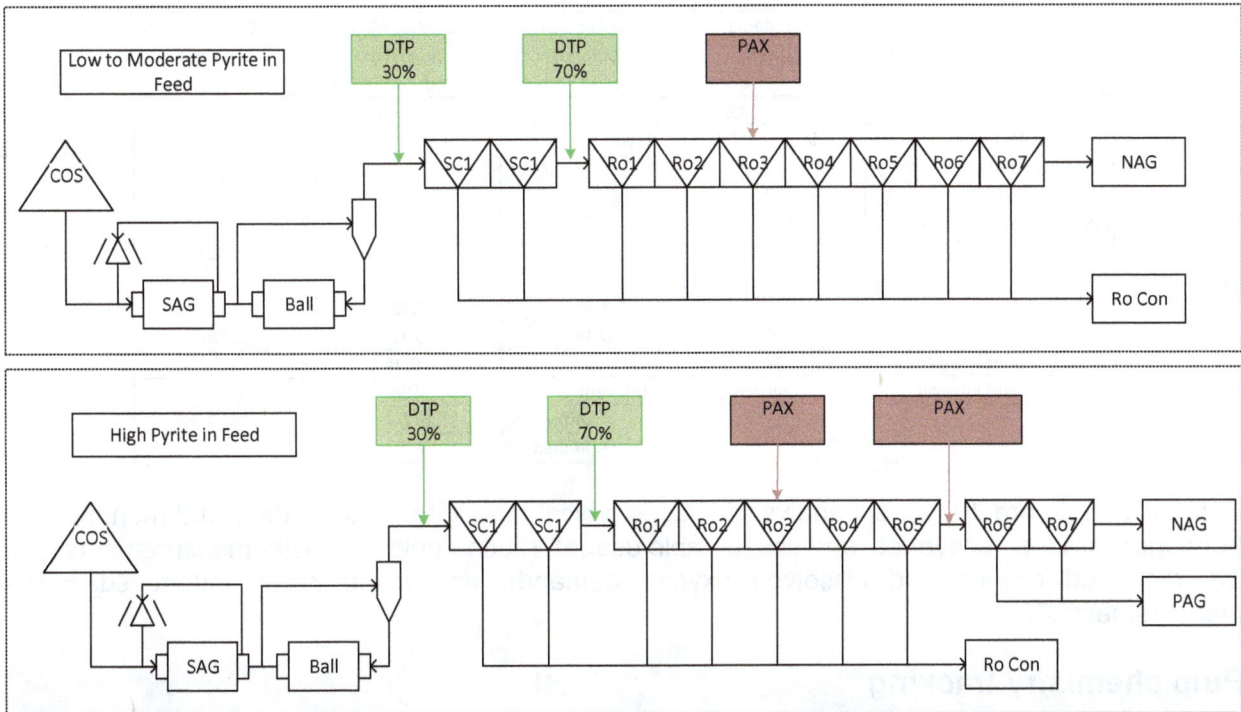

FIG 4 – Collector addition strategy adopted at Red Chris.

When this strategy was implemented on-site, the operation saw an immediate response in terms of stability in the rougher flotation circuit with improved mass pull control and less variation in controller action needed to maintain set points. Operators took to the strategy quite readily as the circuit was easier to run and they could see an immediate uptake in recovery, this strategy has been in place for approximately one year without notable change or deviation.

Further optimisation of this strategy is planned in terms of collector types, sequence and ratio. The addition of extra PAX to aid sulfide recovery is also under review from a cost perspective.

HIGH CHROME IMPLEMENTATION

An ideal approach to high chrome media would have been to drop the forged charge and replace this with high chrome during a mill down. Methods for safely removing the forged charge were explored and discounted by the operation.

Accordingly, a slower purging approach was adopted with high chrome media being used to top-up mill media as the forged media was consumed. It was estimated to take three full replacement charges of ball mill media to achieve full replacement of forged and stable pulp conditions prior to the start of the purge.

The collector strategy was implemented at the same time as the implementation of high chrome charging and pulp chemistry measurements were continued for the duration and beyond.

Prior to the high chrome implementation, SAG mill reject media was added to the ball mill, this practice was and remains discontinued with this reject media being sent away as scrap rather than contaminating the high chrome environment of the ball mill.

For the purposes of conducting the benefit analysis, the data used spans the period of November 1, 2022; to May 2024, excluding the purge period of May 2023 through to the end of September 2023. The data represents a period of seven months prior to and post the trial. Main Zone ore was dominantly fed during this period and feed sulfur grade was considered relatively stable during this period (see Table 1 for confirmation).

TABLE 1

Summary statistics to quantify the significance of the change in ball mill media and lime consumption reduction as a result of using high chrome media over forged mild steel media.

	SAG Media (g/kWh)		Sulphur Feed Grade %Sulphur		BALL MILL Media (g/kWh)		LIME Consumption (kg/tonne)	
	before	after	before	after	Forged	High Cr	Forged	High Cr
mean	*47.3*	*45.5*	*4.1*	*4.2*	*59.4*	*43.8*	*0.6*	*0.3*
standard deviation	6.5	7.4	0.3	0.2	6.7	7.0	0.1	0.1
number of data points	7	7	7	7	7	7	7	7
degrees of freedom	12		12		12		12	
pooled standard deviation	6.95		0.29		6.8		0.08	
t	0.493		0.72		4.3		5.46	
t_95,12	2.18		2.18		2.18		2.18	
tstat	0.63		0.48		0.00055		0.00007	
% prob means are different	37		52		99.94		99.99	
95% conf interval	means are NOT different				7.96		0.09	
				Difference	15.6+/-7.96		0.2+/-0.09	
				% Reduction	26%		43%	

Data utilised toward the benefit analysis includes typical metallurgical shift, daily and monthly data (tonnages, grades, recoveries, and consumable usage), routine pulp chemistry measurements (pH, Eh, dissolved oxygen and dissolved oxygen demand), and some newly introduced EDTA measurements.

Pulp chemistry tracking

Figure 5 shows the results of the pulp potential measurements (mV, SHE) prior to the high chrome trial and through the course of the trial for ball mill discharge and rougher feed streams. The x-axis in these charts shows the amount of new media added to the mill since high chrome was introduced as a ration of a complete steel charge of the mill (~700 t). It may be seen that, as estimated, approximately three replacement charges of high chrome media were required to fully purge the mild steel media and yield a stable chemical environment in both ball mill discharge and the rougher feed.

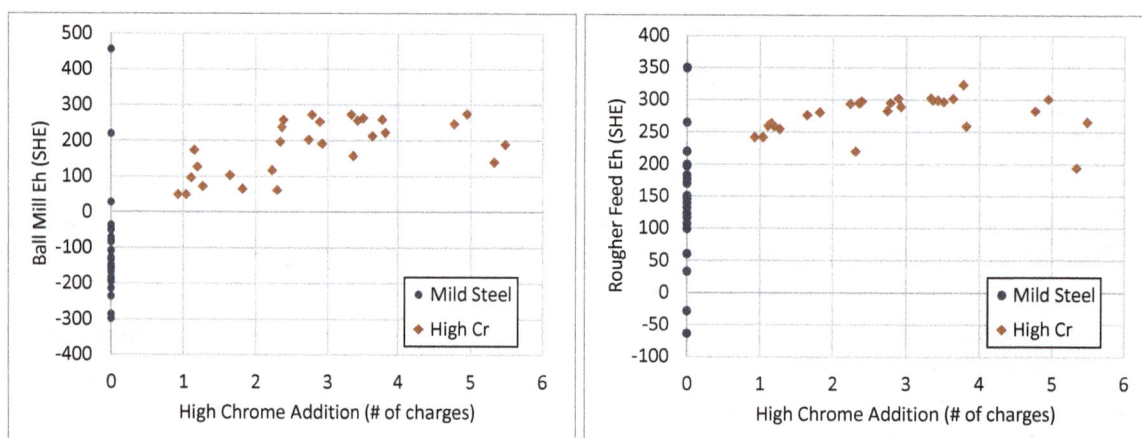

FIG 5 – Comparison of the pulp potential (Eh, SHE) measured for mild steel prior to the high chrome trial and as a function of the high chrome media replacement during the trial for (a) the Ball Mill Discharge and (b) Rougher Feed streams.

The DOD and DO measured in the ball mill discharge stream are shown in Figure 6. It may be seen that the dissolved oxygen demand decreased significantly as purging commenced and continued to decrease throughout and post completion of the purge. The dissolved oxygen measured in the stream only started increasing as purging neared completion thus indicating that there was a threshold oxygen demand that needed to be crossed prior to being able to appreciably improve dissolved oxygen in the ball mill discharge stream.

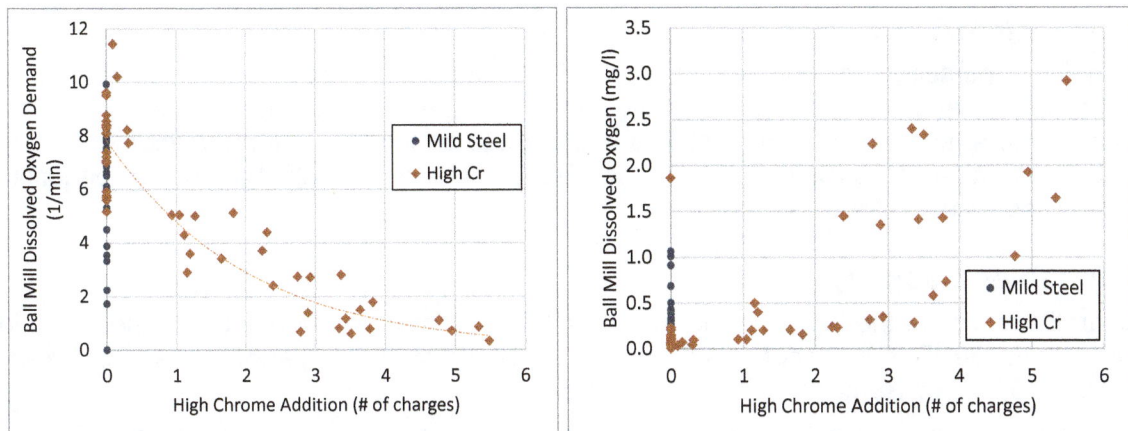

FIG 6 – Comparison of the (a) dissolved oxygen demand and (b) dissolved oxygen as a function as a function of the high chrome media replacement during the trial for the Ball Mill stream.

The change and stabilisation in pulp chemistry measurements highlights the importance in tracking these measures to provide insight into the progress of the media purge as well as additional information with respect to successful realisation of achieving the target objective of moving the flotation feed stream to a more conducive pulp chemistry region for optimising performance.

As mentioned previously, Lotter, Bradshaw and Barnes (2016) presented a graphical representation of the typical flotation regimes in pH-Eh space of different minerals, this has been replicated here in Figure 7 with the plant measurements from the Red Chris ball mill discharge before and after the change in grinding media. The diagram also identifies typical xanthate (BX-) and dithiophosphate (EDTP-) speciation and their dimers (BX2) and (EDTP2). It is apparent that that the pH-Eh regions of operation prior to the high chrome and reagent dosing strategy were not favourable for dixanthogen formation and nor were they generally conducive for optimal chalcopyrite recovery.

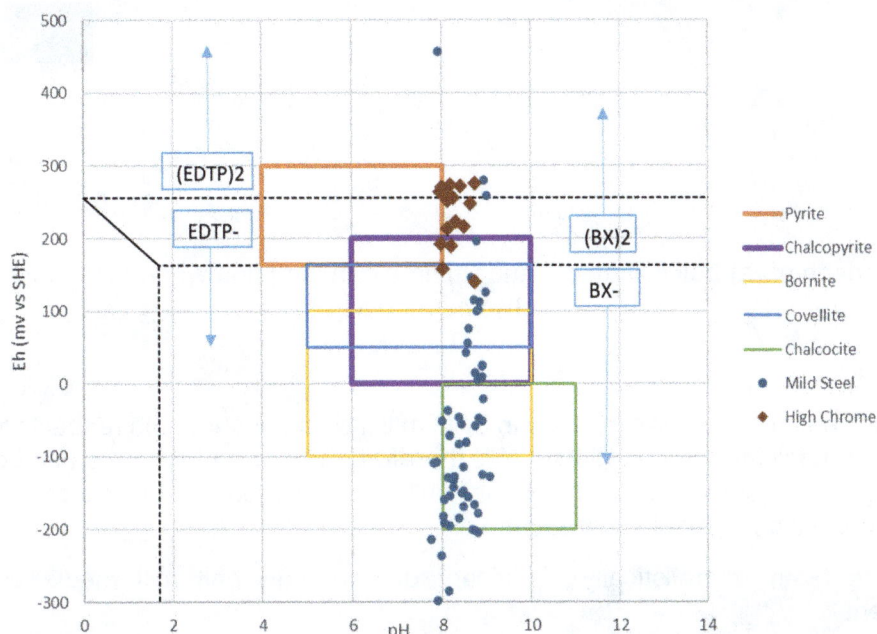

FIG 7 – Empirically arranged flotation domains of chalcopyrite, bornite, chalcocite, and covellite in terms of Eh and pH values, with favoured forms of collectors and their dimers, and Red Chris Ball Mill Discharge stream pulp chemistry measurements overlain (after Lotter, Bradshaw and Barnes, 2016).

In this type of diagnostic measurement, it is also important to consider the pulp chemistry journey a particle goes through before arriving at the point of collector addition and flotation and not just the end point – ie avoiding the highly reactive conditions will produce less surface contaminants at a similar ultimate pH-Eh condition.

BENEFIT ANALYSIS

The benefits of pursuing the change in pulp-chemistry conditions were conducted by looking at the cost implications and the metallurgical outcomes. With the cost of high chrome media being significantly more than mild/forged steel, it is important to understand the net outcome. It is also a lot easier to defend a net cost improvement to the finance department with the recovery benefit being additional icing on the cake.

Consumables tracking

The overall success of the high chrome media trial was ultimately determined by the reduction in consumable usage. This has been readily demonstrated through comparisons made between ball mill media and lime consumption:

The box plot comparisons for ball mill media consumption and lime consumption for forged and high chrome media periods are compared in Figure 8. The decrease in both media and lime consumption are substantial and the assessment of the significance of these changes has been evaluated using single sided t-test analysis as presented in Table 1. Also included in the t-test assessment are two-sided t-test assessments for evaluation whether there has been a change in SAG mill media and feed sulfur grade as a means of assessing whether there has been any reasonable change in feed ore hardness or chemistry that could reasonably explain observed changes in ball mill media and lime consumption beyond the hypothesised change factor.

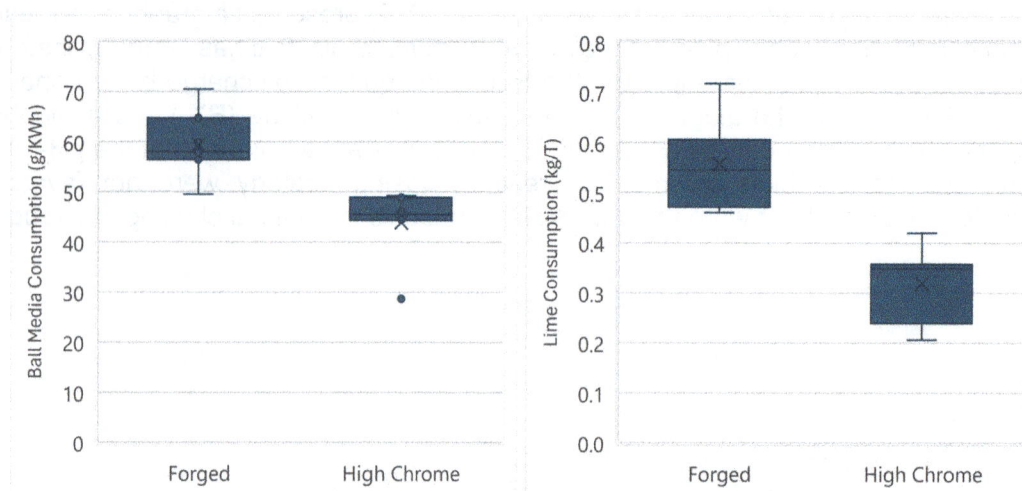

FIG 8 – Comparison of (a) ball mill media and (b) lime consumption when using forged versus high chrome media.

The t-tests confirm:

- There has been no significant change in SAG mill media consumption (expected) or sulfur feed grade as both fail the null hypothesis. It is therefore concluded that there has been no change in feed inputs that could reasonably result in the change in ball mill media or lime consumption other than the type of media introduced.

- There has been a statistically significant decrease in ball mill media consumption of 26 per cent.

- There has been a statistically significant decrease in lime consumption of 43 per cent.

Metallurgical outcomes

Following on from the observations made toward consumable reduction, it is noted that the grinding media consumption rate reduction was clearly a result of reducing the galvanic interaction effects in the ball mill.

The reduction in lime consumption may be related to a change in reactivity in the pulp and, more importantly, a significant change in the operating pH range of the rougher and cleaner banks has been possible (see Figure 9). This reduction in pH has been possible from an improvement in copper selectivity, particularly in the cleaning circuit where it was found that target concentrate grade could be achieved at a lower pH. Similarly, it was found that lowering the rougher pH did not have a deleterious effect on cleaner concentrate grade post the change.

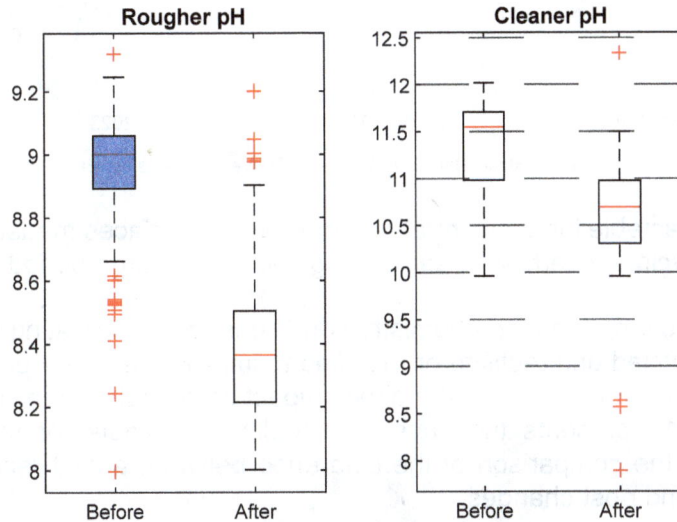

FIG 9 – Circuit pH changes as a result of the modified pulp chemistry conditions before and after the changes.

The grinding media change is expected to have reduced the extent of any inadvertent pyrite activation through a reduction in copper ions dissolved during grinding, while the change in collector addition strategy is likely to have increased the ratio of selective collector (DTP) reaching the cleaning circuit over xanthate since they are not directly competing for adsorption sites in the roughers.

EDTA extraction from particle surfaces (Grano, 2010) were conducted in the on-site laboratory. There were a few challenges in setting up the equipment and conducting the tests, which meant that the period prior to introduction of high chrome and early introduction period were not captured in this analysis. Nevertheless, measurements were carried out on the ball mill discharge stream and the ball mill cyclone overflow during the media purge period. The purge period commenced in May 2023, with first EDTA measurement around two months later. The full purge (three media changes) was complete around the end of September 2023, coinciding with the last of the EDTA measurements. While only limited data was collected, this is a difficult and uncommon procedure carried out in site laboratories.

The results of the EDTA extractions conducted on the primary cyclone overflow are shown in Figure 10. Based on the limited measurements made, the changes in pulp chemistry have significantly reduced the extractable copper, supporting the hypothesis that the observed reduction in required pH to effect chalcopyrite/pyrite selectivity is related to a reduction in EDTA extractable copper ions on the particle surfaces. There was a high degree of variability in the EDTA extractable iron which may be related to the sometimes-high day-to-day variation in pyrite content.

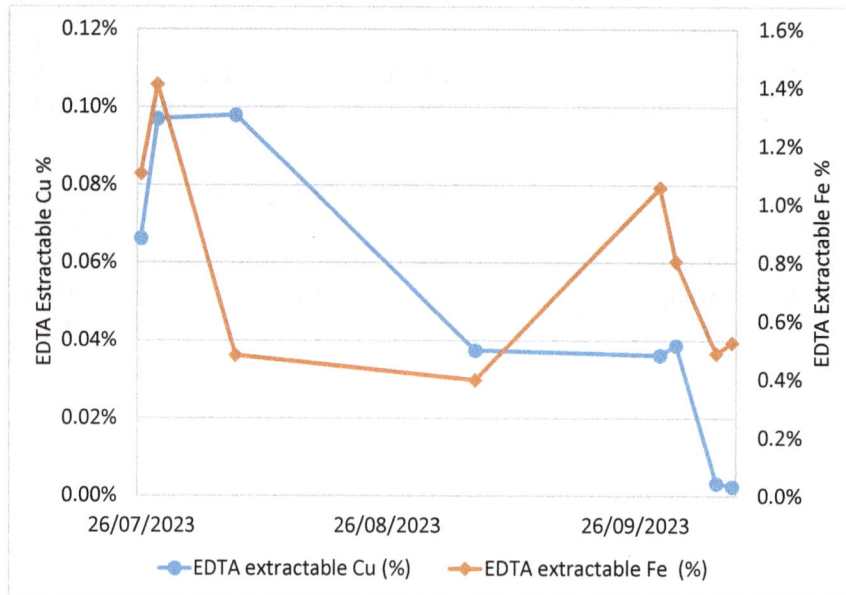

FIG 10 – EDTA extractable ion concentration from particle surfaces measured in the ball mill cyclone overflow stream during the media purge period.

Over the course of the change in pulp chemistry conditions, other operating factors affecting plant recovery were also identified and rectified or modified to improve recovery performance throughout the trial and thus it is difficult to elucidate the direct impact of the pulp chemistry changes on overall recovery itself. Figure 11 presents the substantial uplift observable for the collective changes showing the change in the comparison of the difference between actual recoveries and modelled actual recoveries pre- and post-changes.

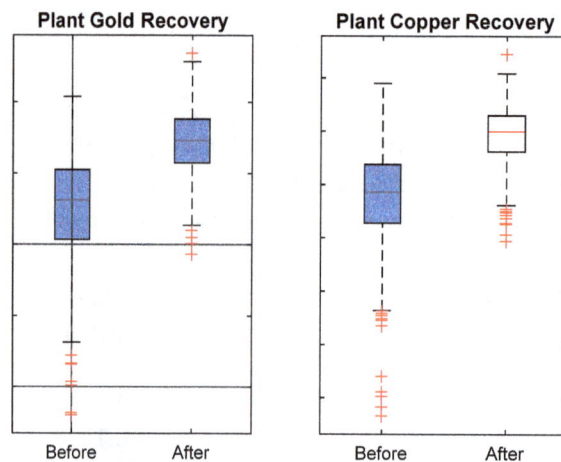

FIG 11 – Gold and copper – Actual minus Forecast Model Plant performance residuals.

Academically, it would have been ideal to step through the changes, measuring the benefit of each as they were progressed. Operationally, recovery was imperative and expedited improvement was more important than academic curiosity. Accordingly, an estimate of recovery impact due to the changes in pulp chemistry of shift-based performance considering only the seven months before change and seven months after change was conducted via linear regression analysis. Other compounding input factors were included in the analysis as best as they could be portrayed, as well as known factors affecting recovery such as mill throughput and head grades. Data was filtered to include only shifts with appreciable runtime of more than 50 per cent. Full details of the analysis cannot be reported for confidentiality reasons, however the overall model fit and coefficients relating to this investigation are summarised in Table 2.

TABLE 2

Regression analysis.

	Gold recovery	Copper recovery
Co-efficient of pulp chemistry change	2.3%	0.7%
p-value of change	0.0001	0.048
Adjusted R^2	0.40	0.57
Number of terms in model	6	5
Degrees of freedom	666	667

CONCLUSIONS

This study highlights several key insights and recommendations for the optimal operation of flotation plants. Firstly, the inclusion of routine pulp chemical analysis is highly recommended as it provides an additional level of insight necessary for optimising plant performance. The absence of this knowledge can lead to significant changes in recovery, both positive and negative. By identifying the underlying pulp chemistry, one can optimise operations through strategic adjustments, such as reagent dosing and replacing forged media with high chrome media in the ball mill.

For high chrome media, achieving stable pulp chemistry conditions required three complete ball mill purges. The shift to high chrome media resulted in a 26 per cent decrease in media consumption and a 43 per cent reduction in lime consumption, both statistically significant changes. These improvements are attributed to a decrease in corrosion due to a less reducing grinding environment, as evidenced by the reduced oxygen demand. Furthermore, the reduction in lime consumption was due to changes in pulp chemistry and the decreased presence of copper ions in solution, which inadvertently activate pyrite-bearing particles.

Relocating reagent dosing from the mill to the mill discharge and down-the-bank promotes a more optimal oxygenated environment, enhancing collector-particle adsorption. The synergy of reagent addition and its profile down the bank are crucial for optimising selectivity and recovery, providing a pathway to improve both fine and coarse recovery.

Using previous studies found in the literature provided a fantastic baseline for the reagent strategy without having to conduct an extensive experimentation program. An opportunity remains to further optimise the strategy and synergy through test work and further trials.

The recovery improvements observed in this study were in line with expectations for copper, showing an increase of 0.7 per cent, and exceeded expectations for gold, with an increase of 2.3 per cent. These findings underscore the importance of routine pulp chemical analysis and strategic operational adjustments in maximising the efficiency and effectiveness of flotation plants.

ACKNOWLEDGEMENTS

The authors wish to thank the Red Chris metallurgy and operations staff for assistance in conducting the trials. Stuart Liu, former plant metallurgist, played a big role in conducting the trial and measurements. Thanks to Newmont for supporting this and giving us permission to publish this work.

The AMIRA P260 project (Bill Skinner and Max Zannin – MZMinerals) provided the authors with the required education (over many years) to identify the opportunity and undertake the necessary programs to complete this work. Chris Greet also provided an invaluable service in assisting with procuring necessary equipment, undertaking vendor test work and providing additional analysis of program results.

The authors also thank Dee Bradshaw for her inspiration and insights on these topics.

REFERENCES

Bazin, C and Proulx, M, 2001. Distribution of reagents down a flotation bank to improve the recovery of coarse particles, *International Journal of Minerals Processing*, 61:1–12.

Bradshaw, D J, 1997. Synergistic Effects Between Thiol Collectors Used in the Flotation of Pyrite, PhD thesis (unpublished), University of Cape Town.

Chander, S, 2003. A brief review of pulp potentials in sulfide flotation, *International Journal of Minerals Processing*, 72:141–150.

Dhar, P, Thornhill, M and Roa Kota, H, 2019. Comparison of single and mixed reagent systems for flotation of copper sulphides from Nussir ore, *Minerals Engineering*, 142:1–12.

Emery, S, 2009. EDTA Surveys, SWP 003.01-EN, Magotteaux Australia Pty Ltd.

Grano, S, 2010. Chemical Measurements During Plant Surveys and Their Interpretation, in *Flotation Plant Optimisation, A Metallurgical Guide to Identifying and Solving Problems in Flotation Plants,* 6:107–122.

Greet, C, 2019. Grinding Chemistry and Its Impact on Copper Flotation, in *Proceedings of Copper 2019, COM2019*, paper 593932.

Johnson, N W, 2002. Practical aspects of the effect of electrochemical conditions in grinding mills on the flotation process, in Proceedings of Flotation and Flocculation: From Fundamentals to Applications (eds: J Ralston, J Miller and J Rubio), University of South Australia, pp 287–294.

Klimpel, R R, 1995. Technology trends in froth flotation chemistry, *Minerals Engineering*, 47(10):933–942.

Li, K, Seaman, D R, Seaman, B A and Baldock, J, 2024. Red Chris Flotation Circuit Expansion – From Piloting to Full Scale, in Proceedings of the 16th Mill Operators Conference (The Australasian Institute of Mining and Metallurgy: Melbourne).

Lotter, N O, Bradshaw, D J and Barnes, A R, 2016. Classification of the Major Copper Sulphides into semiconductor types and associated flotation characteristics, *Minerals Engineering*, 96–97:177–184.

McMahon, M, Seaman, D R, Sharman, P, Adamson, B and Seaman, B A, 2016. Optimising the Telfer Pyrite Regrind Circuit – Effect of Dissolved Oxygen Demand, in Proceedings of the 13th Mill Operators' Conference (The Australasian Institute of Mining and Metallurgy: Melbourne).

Napier-Munn, T J, 2014. Statistical Methods for Mineral Engineers – How to Design Experiments and Analyse Data, Julius Kruttschnitt Mineral Research Centre.

Rees, C, Brock Riedell, K, Proffett, J M, Macpherson, J and Robertson, S, 2015. The Red Chris Porphyry Copper-Gold Deposit, Northern British Columbia, Canada: Igneous Phases, Alteration and Controls of Mineralization, *Economic Geology,* 110:857–888.

Seaman, D R, Li, K, Lamson, G, Seaman, B A and Adams, M H, 2021. Overcoming rougher residence time limitations in the rougher flotation bank at Red Chris Mine, in *Proceedings of the 15th Mill Operators Conference 2021*, pp 193–207 (The Australasian Institute of Mining and Metallurgy: Melbourne).

Trahar, W J, 1981. A rational interpretation of the role of particle size in flotation, *International Journal of Mineral Processing*, 8:280–327.

Yoon, R H, 1981. Collectorless flotation of chalcopyrite and sphalerite ores by using sodium sulfide, *International Journal of Mineral Processing*, 8(1):31–48.

Regrind test work review – analysis of procedures and results

J Thomson[1], B Foggiatto[2], G Ballantyne[3] and G Lane[4]

1. Graduate Process Engineer, Ausenco, South Brisbane Qld 4101.
   Email: julian.thomson@ausenco.com
2. Technical Director, Comminution and Processing, Ausenco, South Brisbane Qld 4101.
   Email: bianca.foggiatto@ausenco.com
3. Director Technical Solutions, Ausenco, South Brisbane Qld 4101.
   Email: grant.ballantyne@ausenco.com
4. Principal Consultant, Ausenco, South Brisbane Qld 4101. Email: greg.lane@ausenco.com

ABSTRACT

There is no generally accepted methodology to determine the energy requirements of regrind circuits that is independent of the milling technology. The Bond ball mill work index test is not suitable for regrind circuit design due to the fineness of the feed materials and the change in log-log exponent when grinding fine enough, so regrind mill vendors have developed their own proprietary methodologies for estimating full scale performance of their specific mills. These disparate methods and the resulting scale-up procedures have led to estimates of required specific energy varying by 100–300 per cent even for similar mills in the same duty (Larson et al, 2011).

To assess the variance between the different vendor methods, the authors compiled test results from Ausenco's in-house database of regrind test work, where more than one stirred mill technology was considered. This paper presents the regrind test work methods used by different regrind mill vendors and compares typical test conditions based on the data set of stirred milling test work results. A review of the test work results is provided, highlighting issues that metallurgists face when analysing test work performed using different technologies such as coarse particle accumulation, differences in power measurement, variable viscosity and different media size.

The objective was to identify a method that allows a direct assessment of test results. Various methods were tested and two are discussed: signature plots (grinding size versus specific energy) and the size specific energy (SSE) (specific energy versus additional fines production) (Mokken, 1978; Ballantyne and Giblett, 2019). Both models were found to produce useful insights for unit performance evaluation as well as identification of deviations from expected performance in the presence of test work issues. The signature plot is a simple model which can be a fitted to the data for estimating the grinding energy of a unit. However, there have been issues observed and discussed regarding this model which require consideration. The SSE model is an effective alternative to the signature plot model, as it avoids issues that stem from utilising the P_{80} to describe the bulk size distribution.

Recommendations for regrind test work planning and analysis are provided to aid in explaining the observed variance between tests in future regrind test work endeavours.

INTRODUCTION

Estimates of energy consumption in mines worldwide indicate that comminution represents 36 per cent of a site's energy consumption on average (Ballantyne, Powell and Tiang, 2012). This is likely to increase if new installations require finer product sizes driven by increased mineralogical complexity, as the most economic orebodies are developed (Taylor et al, 2020).

Fine grinding applications require different grinding mechanisms when compared with primary or secondary tumbling mills such as semi-autogenous (SAG), autogenous (AG) or ball mills. Therefore, established models for scale-up of these technologies cannot be directly applied to scaling test work of regrind milling technologies. Figure 1 describes the change in comminution energy requirements in log-log space for existing models and a representation of the deviation of behaviour outside of the range in which Bond ball mill work index (BWi) is applicable.

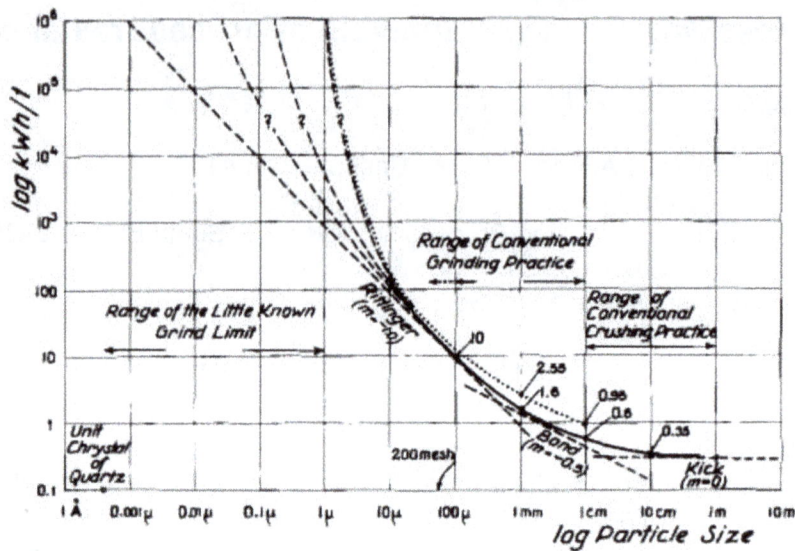

FIG 1 – Example of Comminution Energy Changing Over Size (Hukki, 1961).

Hukki (1961) proposed that the exponent attributed to grinding energy changes with particle size. However, for a sufficiently narrow range of sizes, the exponent can be considered constant, giving the general solution to Charles' equation (Morrell, 2004).

$$E = C(P_{80}^{-\alpha} - F_{80}^{-\alpha}) \tag{1}$$

This is the basis of many grinding models, eg the Bond equation is found if α is set at 0.5 (Doll, 2017).

The use of stirred milling test work results for similar duties using different technologies has proved problematic with variations between 100–300 per cent in predicted regrinding energy requirements (Larson et al, 2011). Most regrind test work methodologies don't require scale-up factors for sizing the industrial mill (ie measured specific energy obtained from the test is expected to be the same for an industrial mill), and can result in significantly different regrind milling power selection. For example, Pease (2010) compared a database of operating work index (OWi) for existing tower milling operations and compared to vendor estimates for similar regrind duties.

Figure 2 indicates that vendor OWi estimates were underestimated. Better understanding of test work conditions and scale up procedures are necessary to reduce the prevalence of poorly sized regrind applications.

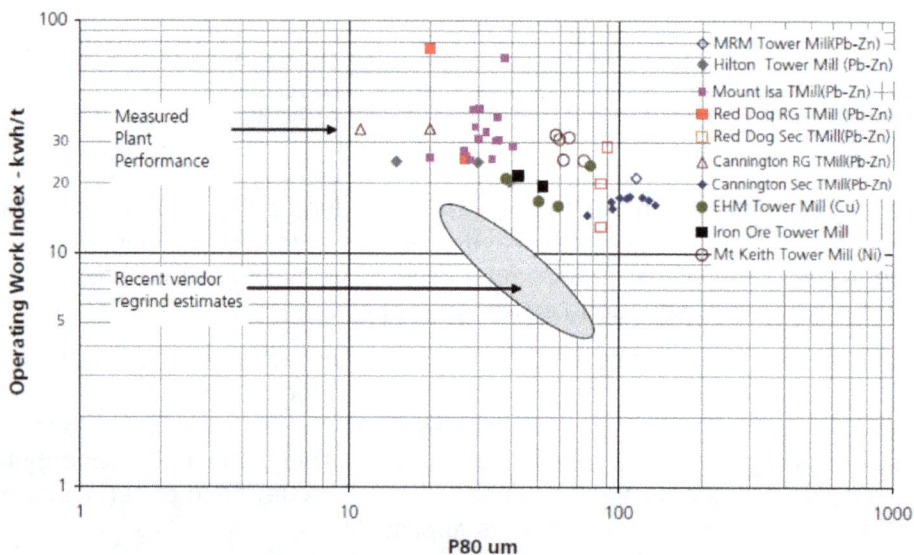

FIG 2 – Tower Mill Operating Work Index Summary (Pease, 2010).

Signature plot

The signature plot equation is a simplification of Charles' equation, providing the means to linearise the P_{80} versus specific energy (SE) without assuming a log-log gradient as prior models do.

$$SE = C(P_{80})^{-\alpha} \tag{2}$$

The simplification follows from the assumption that feed size is usually fixed and does not vary significantly between test work, additionally the energy associated with reducing the top size is of negligible magnitude to the fine size grinding. Therefore, the effects of feed size are negligible and the F_{80} term is removed from Charles' equation (Doll, 2017).

The signature plot equation simplifies the issue of varying gradients in log-log space as the exponent can be easily calculated and applied for SE estimation. However, the signature plot equation still assumes a constant log-log gradient over the size range of the test work and the model should only be applied within the range of data.

The key assumption for the signature plot equation is also the main limitation. Disregarding the feed size means that comparison between different feed size test data requires assumptions to be adopted. Additionally, the calculated SE at feed is non-zero. For example, the SE at feed size predicted by the signature plot for an IsaMill is typically about 5 kWh/t (Larson, 2012).

Size specific energy and fractals

An alternative method to quantifying grinding energy is the size specific energy (SSE) method, which quantifies energy required to generate new material finer than a specific marker size, ie in addition to the fine material present in the feed. The SSE metric is described by the equation:

$$SSE = \frac{SE}{\% \, passing(x) - \% \, passing \, in \, feed(x)} \tag{3}$$

where x is marker size. SSE has the units of kWh/t of new fines generated, where fines are the fraction of material that passes the marker size screen. In the case that marker size is set to P_{80}, the SSE metric will give the same predicted unit SE as the signature plot model. The benefit of this metric arises when comparing tests with different feed distributions.

When applied to a single cycle and utilising an appropriately fine range of marker sizes (typically less than the P_{80} of the cycle particle size distribution), the SSE is linearly related to the marker size in log-log space. Providing the following relationship:

$$SSE = A \cdot x^{-B} \tag{4}$$

B is indicative of the fractal dimension (D) of the fracture surfaces generated over each cycle, where $D = 3 - B$ (Ballantyne, 2023). The fractal dimension can be used to indicate the type of fragmentation, with lower fractal dimensions relating to bulk-splitting (low relative fines production) and higher fractal dimensions relating to catastrophic failure (high relative fines production) (Turcotte, 1986).

Objectives

To assess the variance between various regrind test work methods, a database was compiled for tests performed on the same samples. Four regrind milling technologies were analysed, including:

- Vertimill from Metso.

- Sirred Media Detritor (SMD) from Metso.

- Vertical regrind mill (VRM) from Swiss Tower Mills (STM) which is also licenced to Metso as the high intensity grinding mill (HIGmill) [in this paper VRM is used as the acronym for this technology].

- IsaMill from Glencore Technology.

Each technology has a unique test work method for full scale unit performance estimation. These methodologies differ in both test conditions and inherent unit performance, which results in differences in test work energy requirements for the same grinding duty.

Signature plot and SSE models were used to evaluate the data from the tests to compare results and predicted unit performance. This paper discusses the issues and differences that arose from the review.

TEST WORK DATABASE

The test work database contains results obtained for seven different projects with samples submitted for test work to more than one regrind technology provider, as summarised in Table 1.

TABLE 1

Summary of test work database.

Project	Units tested	Test characteristic, (F_{80}-target P_{80})
A	IsaMill (two sets), SMD	Pb/Ag rougher concentrate, 65–20 μm Zn rougher concentrate, 55–10 μm
B	IsaMill, VRM	Deslimed rougher tailings, 105–35 μm Rougher concentrate, 90–15 μm
C	IsaMill, VRM, Vertimill	Two stage IsaMill and VRM, 140–10 μm Jar mill single stage, 130–25 μm
D	Vertimill, SMD	Single sample, 85–15 μm
E	VRM, Vertimill, SMD	Sample 1, 70–30 μm Sample 2, 48–30 μm Sample 3, 56–45 μm Sample 4, 66–45 μm
F	Vertimill, SMD	Sample 1, two stages: • Vertimill, 208–20 μm • SMD, 26–15 μm Sample 2, two stages: • Vertimill, 212–20 μm • SMD, 26–15 μm

Test work procedures

A brief description of the test work procedures is provided below.

- The Vertimill specific energy is calculated based on the results of a batch Jar mill test that utilises a lab-scale ball mill with 19 mm steel media (single size). Samples are periodically taken from the emptied mill contents at given kWh intervals and power is measured via a torque coupling and tachometer on the mill drive (Neilsen *et al*, 2023). This test requires a sample size of about 1.5 kg and requires a scale up factor, selected depending on the duty, to account for the higher energy efficiency of the Vertimill. The measured specific energy is multiplied by a factor of 0.65 (Mazzinghy *et al*, 2015) for regrind duties. Huang *et al* (2019) demonstrated that the efficiency factor should be proportional to test feed size (F_{80}) and varies between 0.65–0.85.

- The SMD test is conducted under batch conditions in a 1.4 L lab-scale SMD mill and requires a 0.5 kg sample. For a target product size $P_{80} > 20$ μm or a wide feed size distribution, a coarse media is selected (3 to 2 mm), while for finer target grind sizes ($P_{80} < 10$ μm), a finer sand size is used (2 to 1 mm). Samples are taken from the mill contents at incremental kWh milestones for particle size analysis. The torque reaction from the mill is measured by a load cell and converted to an energy reading, the power input is measured continuously and is shown as total kWh. According to Metso test work reports, the SMD specific energy is a direct scale from test operation to equipment sizing.

- There are two different types of VRM (or HIGmill) test work, both conducted in a VRM5 (HIG5) unit: small sample test (5 to 8 kg) and semi-continuous test (30 to 50 kg). In the small sample test, the slurry is continuously pumped at a set flow rate from a small discharge tank to the mill feed. Samples are periodically taken from the discharge of the mill and power is measured via a torque strain gauge and tachometer on the agitator pinion (Neilsen *et al*, 2023). In the semi-continuous method, the mill discharge is captured in the product tank then utilised as the feed for the next cycle, and *vice versa* for the feed tank. Both the small sample and semi-continuous test regimes have a scale-up of 1:1 (Lehto *et al*, 2016).

- IsaMill tests are conducted under semi-continuous conditions in an M4 unit, requiring a sample of 15 kg. The media size and solids content for the small-scale test are selected based on the expected full scale design conditions, with the unit speed (rev/min) set at about one third of the full-scale unit (Taylor *et al*, 2020). The full sample is run through the mill for each cycle, and the sample utilised for product sizing is taken from the mill discharge at the midway point of the pass. Power is measured based on drive energy consumption via a frequency inverter. Due to the similarity in the operating conditions, feed size distribution and grinding mechanisms between the laboratory scale and the full-scale IsaMill, a 1:1 direct scale-up is usually achieved (Larson *et al*, 2011).

SE estimates from signature plots

Power draw estimates for regrind applications have been observed to vary significantly between similar units (Larson *et al*, 2011). The SE estimates for unit performance (per the signature plot model) obtained from the database collated for this paper are no different, with the highest SE estimates falling between 100 per cent and 200 per cent of the lowest estimate on similar samples. The signature plots resulting from the Project C test work are displayed in Figure 3.

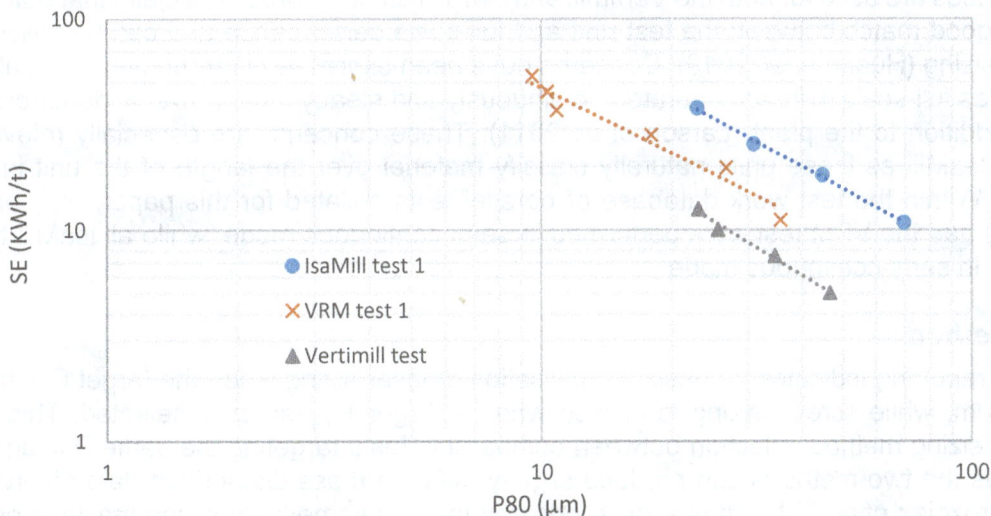

FIG 3 – Project C Signature Plots (adapted from Foggiatto, 2023).

Selecting a P_{80} of 25 µm as a reference point, the IsaMill, VRM and Jar mill SE estimates are 33.4, 18.2 and 11.9 kWh/t, respectively. The IsaMill test SE value is 181 per cent higher than the Jar mill test and 53 per cent higher than the VRM test. Results like these prompted an investigation into the effects of test conditions on the SE measurements and how the models utilised can help identifying potential test work issues.

Initial test work observations

A summary of the operating conditions for each test unit type in the test work database is provided in Table 2.

TABLE 2

Reported operating conditions according to unit type.

Test unit	Test unit volume (L)	Cycles	Media size (mm)*	Media type	Solids density (v/v %)	F_{80} (µm)	Target P_{80} (µm)
Jar (Vertimill)	3.3	1–5	19	Monosize steel	38–44	48–131	15–45
SMD	1.4	1–7	3	Graded ceramic	32–34	26–85	10–45
VRM5 (or HIG5)	6.5	5–8	2–4	Graded Ceramic/Alumina	17–28	52–138	10–45
M4 (IsaMill)	4	4–9	2–4.5	Graded Ceramic/Alumina	16–23	27–158	10–35

* When a media size distribution is provided the top size of the distribution is displayed.

Media size

Selecting the media size based on the application is good practice, as media size is a key consideration for mill performance optimisation and test work should be done utilising a similar charge size distribution to the full-size installation (Larson, 2012). The Jar mill test is the only test that uses fixed media size, while SMD, VRM and IsaMill media size are all tailored for the duty.

In some cases, depending on the required reduction ratio, IsaMill and VRM tests have been conducted in two-stages to optimise media size selection and improve grinding energy efficiency.

Test procedure

Batch methods are used for both the Vertimill and SMD. Test work providers claim that batch regimes provide a good match between the test unit and full-scale performance and can be utilised for full-scale mill sizing (Huang *et al*, 2019). Concerns have been expressed surrounding the scale-up from batch test as full-scale mills are operated continuously and steady state power is dependent on new material addition to the plant (Larson *et al*, 2011). These concerns are especially relevant to the VRM and IsaMill as these units naturally classify material over the length of the unit in full scale operation. Within the test work database of parallel tests collated for this paper, only one project (Project C) had the VRM test work performed in semi-continuous mode, while all IsaMill tests were conducted in semi-continuous mode.

Sizing method

Test work reporting indicates a consistent utilisation of laser sizing when the target P_{80} is less than about 25 µm, while screen sizing is utilised when a larger P_{80} target is selected. This results in consistent sizing method selection between comparable tests targeting the same P_{80}, an important outcome as the two methods can produce slightly differing mass distribution data given the same sample (Drozdiak *et al*, 2011). It is even suggested that the same laser sizing machine be used for comparable tests due to variation in results that can be produced unit to unit (Pease, Young and Curry, 2005).

Slurry viscosity

Viscosity changes or observed effects in test work are not consistently reported, and those that do mention rheology issues do so briefly and respond with addition of water to dilute the solids concentration in later cycles. This serves to decrease the solids concentration over the course of the test and the effects of this on energy consumption are uncertain.

Number of cycles

Jar and SMD test work do not usually only report a single cycle as indicated in Table 2, this is due solely due to the test work conducted for Project E which consistently reached the target P_{80} within one to two cycles for all three units tested. This is due to the unusually low reduction ratio (less than 2) specified for this project, where other projects defined reduction ratios of no less than 2.5. Due to the small number of cycles reported for this project, it was omitted where a regression analysis is conducted over cycles.

Coarse particle accumulation

The build-up of a coarse fraction within the mill chamber due to inadequate top size breakage over the course of a test is the result of undersized media (Larson *et al*, 2011). To quantify the reduction of the top size over the course of a test, the P_{98}/P_{80} ratio is calculated for each cycle, and a significant increase in this metric indicates that the top size is not being effectively reduced relative to the P_{80}.

A regression was conducted over each test to determine how this ratio typically changes between cycles. It was found that on average, the ratio increased by 0.15 per cycle with a 95 per cent confidence interval of [-0.01, 0.31] indicating that the positive correlation is not statistically significant at this confidence level. The average P_{98}/P_{80} across all cycles is 3.03, at 95 per cent confidence the upper limit of the confidence interval is 3.37 (a lower ratio indicates proper top size reduction). This upper limit corresponds to a Rosin-Rammler spread parameter of 0.74.

Figure 4 illustrates the changes in P_{98}/P_{80} ratio between test cycles for selected tests, indicating possible issues occur in all unit types.

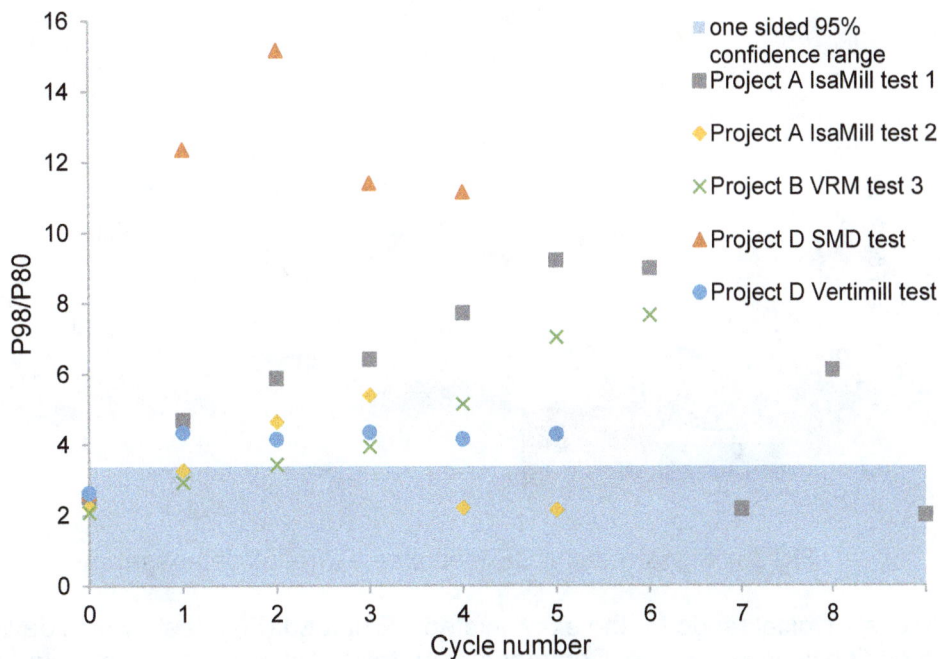

FIG 4 – P_{98}/P_{80} ratio for four outlier tests.

Only one Vertimill test exceeded the ratio of 3.37, this is the Project D test displayed in Figure 4. Despite the increase over the first cycle, the ratio remained consistent at about 4.2 for the remaining cycles, it is likely this is a result of the ore characteristics. The SMD test conducted using the same sample also exhibited a P_{98}/P_{80} ratio of 12.3 after one cycle. The test work provider was aware of this issue, as the report for the Project D SMD test discussed the presence of a 'coarse tail' due to undersized media.

As the P_{98}/P_{80} ratio is typically consistent or increasing between cycles, the dramatic variation between cycles 6–8 in the Project A IsaMill test 1 is concerning, likely indicating sampling or measurement issues and warrants contacting the test work provider. This high variability in P_{98}/P_{80} ratio on later cycles is also observable in Project A IsaMill test 2, which was conducted on a similar sample for the same project, showing a significant drop after cycle 3.

The existence of this issue can affect the energy requirement results in one of two ways, depending on the type of test:

- In the case of a semi-continuous test, the coarse fraction can accumulate within the unit if not effectively reduced or passed to the discharge. This effect cannot be captured by measuring mill discharge if the coarse fraction is retained in the mill, causing the test to underestimate the power required for steady state milling. Larson *et al* (2011) suggests a sample should have at least three to four times the void volume of the test unit to ensure coarse material is displaced to the feed by the end of the cycle.

- In the case of a batch test, where samples are taken from the mill contents in every cycle, coarse particle accumulation is clearly captured regardless of sample size.

MODEL ANALYSIS

Signature plot

The power relationship between specific energy and P_{80} was regressed for all tests. As previously mentioned, one inherent characteristic of the signature plot model is a non-zero SE if the curve is extrapolated to feed size (F_{80}). This means that the test results are only valid if the exact same feed size distribution is experienced in the full-scale plant. This is most significant in cases where the target P_{80} is reached within a single cycle, such as Project E. Figure 5 displays the distributions for the calculated SE values at feed size (F_{80}) for each regrind technology.

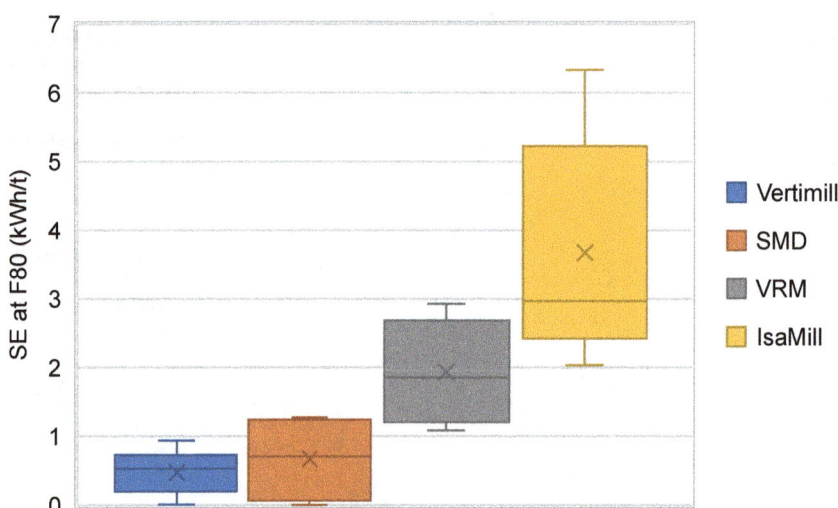

FIG 5 – Signature plot SE feed size (F_{80}) – all database.

Each unit exhibits a typical range for the extrapolated SE at feed (F_{80}), tests which deviate from this range require additional investigation. Commentary on tests that exceed the expected range of this metric is displayed in Table 3.

TABLE 3

SE at feed F_{80} and observations from tests.

	Upper limit of 95% confidence (kWh/t at F_{80})	Tests exceeding upper limit of 95% CI	Observations of test conditions
Vertimill	0.66	Project C Vertimill test	Attributed to sample characteristics as IsaMill and VRM tests required two stages for similar application
		Project F Vertimill test 2	Received a coarse sample for testing with an F_{80} of 312 µm resulting in a high reduction ratio of 15.7
SMD	1.20	Project A SMD test 1	Received an unusually flat feed distribution with a P_{98}/P_{80} ratio of 5 (P_{98} of 361 µm)
VRM	2.34	Project B VRM test 3	Exhibits a shift in log-log gradient over the range of the test
		Project C VRM test 1 and 2	Exhibit high P_{98}/P_{80} ratios and high reduction ratios (12.2 and 13.8), test 2 is a two-stage test with a first stage reduction ratio of 5.5
IsaMill	4.71	Project B IsaMill test 1 and 2	Exhibits a shift in log-log gradient over the range of the test

The observations from each test presented in Table 3 indicate that overall, a high SE at F_{80} is an indication of test conditions such as:

- a high reduction ratio
- small media size (inadequate coarse material breakage)
- feed characteristics such as size distribution and/or coarseness
- change in log-log gradient associated with SE over cycles, discussed further in next section.

An excessively low SE extrapolated to F_{80} is correlated with a high exponent on the signature plot equation. This is exemplified by the Project D test work, where the Vertimill and SMD test work exhibited exponents of -5.1 and -6.0 respectively, with an SE at F_{80} calculated at 4.2×10^{-3} kWh/t and 6.1×10^{-4} kWh/t, orders of magnitude less than any other tests on either unit. Vertimill and SMD tests for Project D also exhibited abnormally high P_{98}/P_{80} ratios as displayed in Figure 4, indicating that the high exponents are likely due to inadequate coarse material breakage.

Deviations from the signature plot model

The signature plot model assumes a constant exponent over the F_{80} to P_{80} range for a given test. As discussed above, a high SE at feed size (F_{80}) can indicate that a data set deviates from this assumption. One example is shown in Figure 6, where one of two IsaMill tests (Project B) on a deslimed rougher tailings sample resulted in a change in slope of the signature plot.

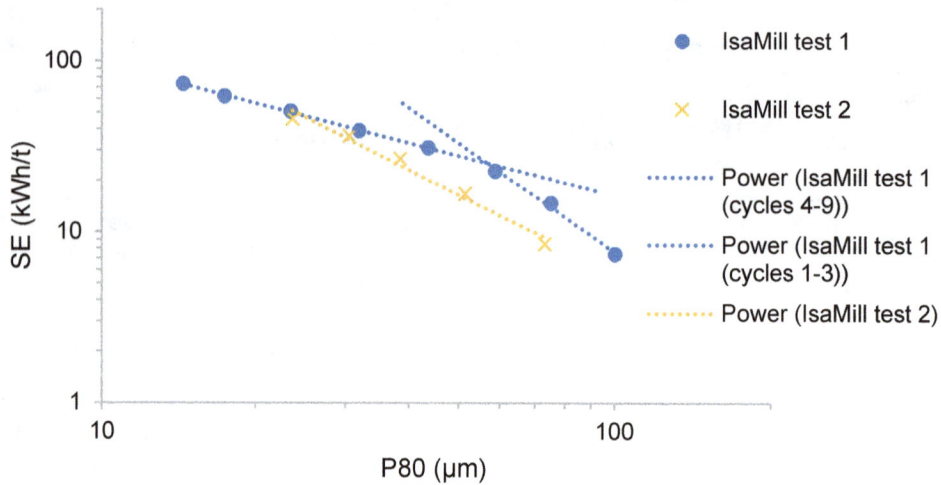

FIG 6 – IsaMill signature plots – Project B, deslimed rougher tailings sample.

The tests displayed in Figure 6 were conducted on similar samples, with the IsaMill test 1 using a 2.8 mm top size charge and targeting a grind size (P_{80}) of 15 µm. Viscosity issues were reported for the IsaMill test 1 and additional water was added to the slurry between cycles 2–5. The IsaMill test 2 was carried out utilising 4.5 mm top size media and targeted a coarser P_{80} of 35 µm. The exponent attributed to the SE curves fitted for test 1 decreases in magnitude after the first three cycles, which can be attributed to the combination of three effects:

- The hold-up of coarser particles within the mill causing a relatively fine sample to be produced earlier cycles prior to the breakage of the coarse fraction.

- The dilution of the slurry feed, which reduced the energy associated with viscosity related losses.

- The higher energy efficiency of the smaller media at finer grind sizes.

A change in signature plot slope was also seen in the comparison between the VRM tests conducted for the same project on rougher flotation concentrate samples displayed in Figure 7. For these two tests, there was no difference in media size (both 3 mm top size charge) and the slurry density was set at 28 per cent v/v in test 1 and 19 per cent v/v in test 3.

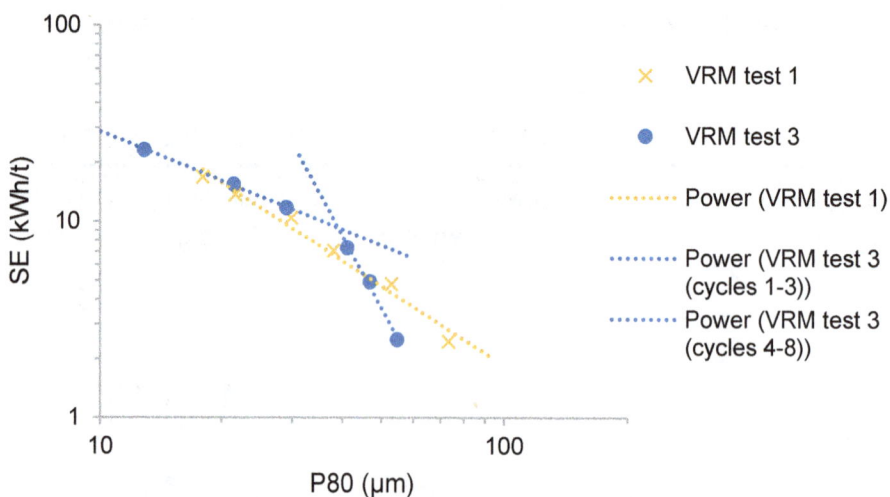

FIG 7 – VRM signature plots – Project B, rougher concentrate sample.

In this case, there was significant coarse particle accumulation during VRM test 3, with the P_{98}/P_{80} ratio increasing to 9.7 by cycle 7 (also shown in Figure 4). The lower slurry density in VRM test 3 caused the 3 mm media to be ineffective at grinding the top size particles. Additional dilution was also required during the test, the slurry feed density was decreased to 16.4 per cent v/v after cycle 5, due to viscosity issues as product became finer.

In both sets of tests for Project B (IsaMill test 1 versus 2 on deslimed rougher tailings sample, and VRM test 1 versus 3 on rougher concentrate sample) the energy required to achieve target grind size was similar for each pair of tests, demonstrating the complexity of unit performance under competing factors. The holdup of a coarse fraction in the mill volume in earlier cycles can artificially depress the measured P_{80}. As grinding occurs the viscosity and measured coarse particle accumulation effects tend to cause SE to increase, while the additional efficiency due to the smaller media at finer applications tends to decrease energy consumption. This serves as a caution to designers, as they should not accept test work data on face value as two cancelling errors do not make the similar outcomes correct.

SSE and fractals

Selecting an appropriate marker size for the SSE methodology application is essential to ensure that the calculated SSE is linear in log/log space. Typically, the SSE selected marker size should be less than the P_{80} of the cycle. The lower limit for marker size is 5 µm, as below this point, mass data become unreliable. Figure 8 shows an example of SSE curves developed for all the cycles of an IsaMill test targeting a product P_{80} of 10 µm, as resulting from the suggested marker size trimming.

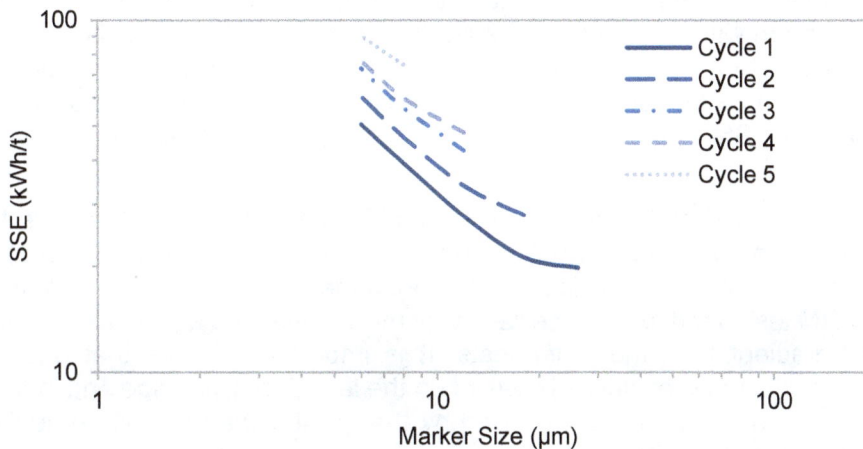

FIG 8 – SSE versus marker size – Project A IsaMill Test 4.

Figure 8 shows that most of the filtered SSE curves are linear in log-log space, except for the SSE curve for the first cycle where a change in slope is observed in the coarser marker sizes. This is commonly observed and is an indication that the selected marker size should be less than the cycle P_{80} in some cases. However, for simplicity this upper limit is maintained for this analysis.

With these filters the SSE curve was fitted for each cycle in each test. The exponents (parameter B in Equation 4) for each cycle in each test were found and utilised to calculate the fractal dimension of the fracture surfaces generated for each test unit. The variability of fractal dimensions for each test unit shown in Figure 9.

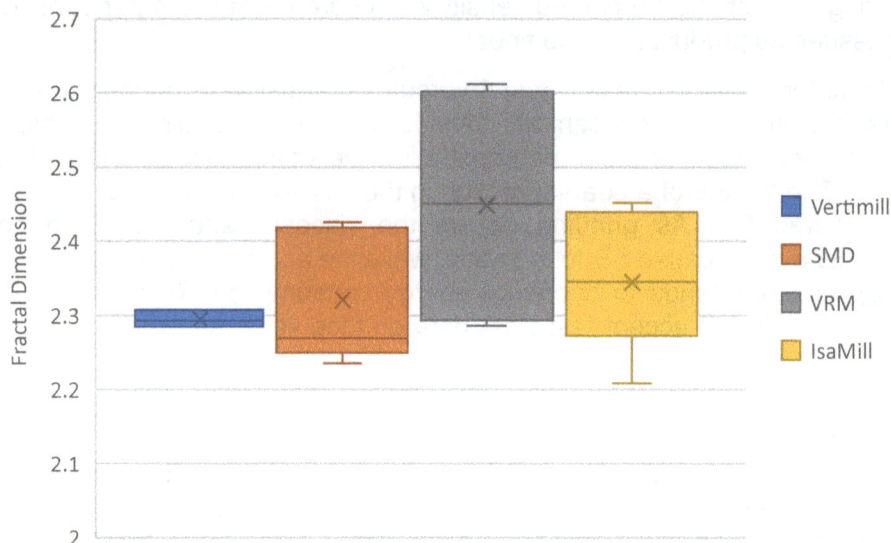

FIG 9 – Fractal dimension per unit – all database.

Figure 9 indicates that out of the four units, the VRM test results exhibited the widest range and highest fractal dimension with an average of 2.46, the IsaMill and SMD resulted in similar average fractal dimension of 2.32 to 2.34. The Vertimill exhibited the smallest fractal dimension at an average of 2.3. It is important to note that only three Vertimill tests are available to display in Figure 9 due to the filtering required for SSE, more are required to provide a valid representation of unit fractal dimension.

As suggested earlier, the fractal dimension of breakage indicates the type of fragmentation, such as bulk splitting (low fractal dimensions) and catastrophic failure (high fractal dimensions). It also relates the change in the gradient of the feed size distribution to the product size distribution. These results suggest that the VRM tests tend to produce relatively more fines, and a product size distribution with a relatively lower gradient than the other tests. It is important to note that due to the filtering implemented, this conclusion is exclusively based on the less than P_{80} range and independent of the coarse particle accumulation, which is indicated by the greater than P_{80} range, ie P_{98}/P_{80}. Further work is required to investigate this effect further.

Technology differences indicated by SSE

Two projects were selected to compare results from tests using different technologies and the SSE methodology. The SSE was calculated at a consistent marker size across two groups of comparable tests:

- The SSE at 10 µm marker size for both Ag/Pb and Zn concentrate regrind duties are shown in Figure 10 (IsaMill and SMD tests).

- The SSE curves at 20 µm marker size for deslimed rougher tailings samples and SSE curves at 15 µm marker size for the rougher concentrate samples are shown in in Figure 11 (IsaMill and VRM tests).

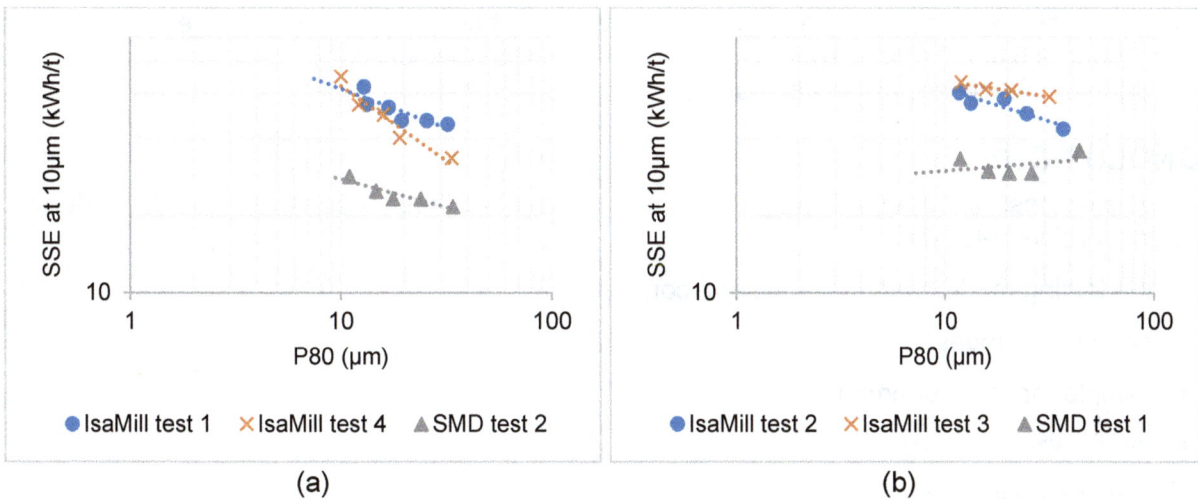

FIG 10 – IsaMill and SMD SSE – Project A: (a) Ag/Pb regrind feed sample; (b) Zn regrind feed sample.

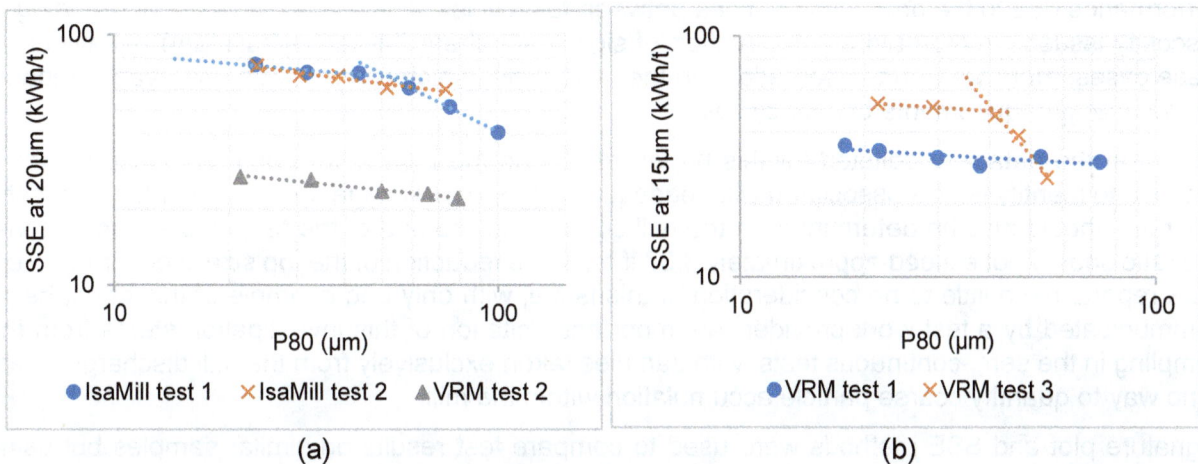

FIG 11 – IsaMill and VRM SSE – Project B: (a) Deslimed rougher tailings sample; (b) Rougher concentrate sample.

The SSE tends to increase as product fineness increases for most tests. The steady increase in SSE for finer grind sizes indicates that the energy required for fine grinding increases at a rate faster than the SSE model expects. Larson (2012) developed a similar model to SSE, in which the difference in percentage passing was squared to account for this increase. On the occasions that SSE decreases between cycles there was an observable issue:

- In Figure 10:

  - The SSE for the Project A IsaMill test 2 (Zn regrind) decreased between cycles 3 and 4. As discussed, the P_{98}/P_{80} ratio registered for both IsaMill test 1 and test 2 suggest a potential sampling issue for these tests.

  - The SSE for the Project A SMD test also decreases from cycle 1 to 2. This test has been mentioned to have a P_{98}/P_{80} ratio of 5 which is unusual. The higher SSE at cycle 1 is therefore attributed to the additional grinding of the top size required by the unit.

- In Figure 11, the SSE for the Project B VRM test 1 decreases after cycle 2. This is caused by sampling issues, as this test resulted in overlapping particle size distributions at coarse sizes (greater than 100 µm), with cycle 1 product particle size distribution intersecting with both the feed distribution and the cycle 2 product.

In the case of Project B, graphing SSE against the P_{80} of a cycle exhibits the same change in exponent in log-log space as the signature plot model. The relatively sharp increase in grinding energy over the first three cycles for IsaMill test 1 and VRM test 3 test are still visible in Figure 11 as seen in Figures 6 and 7 respectively. This indicates that the issue of changing exponent is not solved for these tests via the implementation of SSE.

CONCLUSIONS

This review of test work methodologies highlighted the main differences between test procedures by different vendors related to:

- the testing mode (batch versus semi-continuous)
- number of stages
- sample mass requirement
- media size
- power measurement
- scale-up factor requirement.

There is inconsistent reporting of rheology issues, which is a significant factor in fine grinding test performance due to the production of fines between test cycles. There is an occasional reporting of 'viscosity issues', interpreted as observation of significant shear thickening (dilatant) behaviour. In these cases, more water has been added in the next cycle, which typically has observable impacts on the energy requirements of later cycles.

Based on the database collated for this paper, reviewing the P_{98}/P_{80} for each test cycle makes it possible to identify issues associated with inadequate top size breakage. For semi-continuous tests, this ratio should also be determined on the mill contents at the end of the test. It is suggested that this ratio should not exceed approximately 3.37 if effective reduction of the top size is occurring. Test work reports have little to no consideration of this issue, with only one example of this effect being communicated by a test work provider. An important limitation of this investigation stems from the sampling in the semi-continuous tests, with samples taken exclusively from the mill discharge–there is no way to quantify coarse particle accumulation within the mill.

Signature plot and SSE methods were used to compare test results on similar samples but using different technologies. Each model provides a reasonable analysis of the data but has its limitations.

The signature plot is a simple and effective measure to utilise SE data for scale-up but does not include effects of feed size distribution, as such, its application to tests where target P_{80} is reached within the first cycle is not recommended. The extrapolation of the signature plot to F_{80} tends be an indicator of test conditions and/or a change in exponent as grinding progressed. The latter being a point of concern for this model.

The SSE model serves to include the effect of feed size distribution and provide an indication of fractal dimension of each cycle. As the SSE model utilises fines generation instead of P_{80}, it is sensitive to sampling issues unlike the signature plot, aiding in identifying potential issues. The SSE model also serves to simplify the analysis of test work data, requiring only the marker size (P_{80}) passing % for each cycle, rather than an interpolation of P_{80} at each cycle to predict unit SE.

The available test work has provided some qualitative insights into how test conditions can affect the outcome of test work for regrind units, which will be helpful in the planning of future test work programs. However, future work is required to quantify the effect that test conditions and test work issues can have on regrind power estimation. The authors wish to continue this investigation with fresh test work conducted over samples from existing operations containing stirred milling technologies.

RECOMMENDATIONS

1. When performing stirred milling tests with multiple equipment providers, the feed samples should be prepared independently and distributed as necessary to test work providers, with the objective to ensure that the test work is directly comparable by the signature plot method.

2. While the signature plot is a simple and effective method for estimating SE from test data, it should not be used as a comparison between tests with varying feed sizes.

3. Media should be the same size as a full-size application, including an appropriately graded charge. This is especially true for the Vertimill jar test which currently demonstrates no consideration for media size selection or optimisation, instead relying on a scale-up factor to account for energy efficiency gains for the full-scale application.

4. When viscosity effects are recognised in a test, the process should be restarted with a lower slurry density given sufficient feed sample is available. This will produce results with a consistent slurry density across all cycles. If insufficient sample is available to re-run the test and dilution of the slurry is required for further testing, this should be clearly reported in the test results.

5. In addition to particle size distributions and P_{80}, test work reports should include the P_{98}/P_{80} ratio to clearly display the top size reduction during test work. Coarse material accumulation in the unit volume must also be quantified in semi-continuous testing where samples are taken at the unit discharge. Upon finalisation of any semi-continuous test, a particle size analysis should be conducted over a sample from the mill contents to quantify the potential build-up of coarse material in the mill chamber.

6. To prevent units from reaching the target P_{80} within the first cycle, cycle duration should be sufficiently reduced for low reduction ratio applications. This will ensure that the signature plot method will be valid when estimating SE at P_{80}. Additionally, in cases where the log-log gradient fitting the data appears to change over the course of a test, ensure that the signature plot equation is regressed over the later cycles, filtering out earlier cycles which exhibit a different log-log gradient. This will ensure that the regressed signature plot equation is well fitted to the data around the target P_{80} for SE estimating.

7. The size specific energy metric should also be included in test work reporting as it serves to include the effects of the feed distribution providing a more consistent basis for the comparison of unit energy consumption in the absence of a uniform feed sample. Also providing an indication of how each unit transforms the particle distribution as grinding occurs with the calculation of fractal dimension for each cycle, a result which cannot be achieved by models which estimate SE over P_{80}.

ACKNOWLEDGEMENTS

The authors would like to thank Ausenco for granting permission to publish this work.

REFERENCES

Ballantyne, G, 2023. Fractal Energy: Combining fractal dimension of fracture surfaces and Size Specific Energy to describe comminution, paper presented at MEi Comminution 2023.

Ballantyne, G and Giblett, A, 2019. Benchmarking Comminution Circuit Performance for Sustained Improvement, paper presented at the 7th SAG Conference.

Ballantyne, G, Powell, M and Tiang, M, 2012. Proportion of Energy Attributable to Comminution, paper presented to the 11th AusIMM Mill Operators' Conference.

Doll, A, 2017. Fine grinding, a refresher, presented at the 49th Annual Meeting of the Canadian Mineral Processors.

Drozdiak, J, Klein, B, Nadolski, S and Bamber, A, 2011. A Pilot-scale Examination of a High Pressure Grinding Roll/Stirred Mill Comminution Circuit, presented at SAG Conference 2011.

Foggiatto, B, 2023. The design of fine grinding circuits in an ESG context, presented at the 2023 Series of the JKMRC Friday Seminars.

Huang, M, Chandramohan, R, Foggiatto, B and Lane, G, 2019. Review of Comminution Testwork Requirements for Fine Grinding Mill Sizing and Selection, paper presented to the 15th International Mineral Processing Conference and the 6th International Conference on Geometallurgy.

Hukki, R, 1961. Proposal for a Solomonic Settlement Between the Theories of Von Rittinger, Kick and Bond, *AIME*, 220:403–408.

Larson, M, 2012. Experimental Study of IsaMill Performance Leading to a Preliminary Model, Masters thesis, The University of Queensland, Brisbane.

Larson, M, Anderson, G, Morrison, R and Young, M, 2011. Regrind Mills: Challenges of Scaleup, paper presented to the SME Annual Meeting, Denver.

Lehto, H, Musuku, B, Keikkala, V, Kurki, P and Paz, A, 2016. Developments in stirred media milling testwork and industrial scale performance of Outotec HIGmill, presented at Comminution 16.

Mazzinghy, D, Russo, J, Schneider, C, Sepulveda, J and Videla, A, 2015. The grinding efficiency of the current largest Vertimill installation in the word, presented at SAG Conference 2016.

Mokken, A H, 1978. Progress in run-of-mine (autogenous) milling as originally introduced and subsequently developed in the gold mines of Union Corporation Group, Union Corporation, Ltd, presented at the 11th Commonwealth Mining and Metallurgical Congress, p 49.

Morrell, S, 2004. An Alternative Energy-Size Relationship to That Proposed by Bond for the Design and Optimisation of Grinding Circuits, *International Journal of Mineral Processing*, 74:133–141.

Neilsen, B, Davey, G, Walters, A and Boylston, A, 2023. Stirred Mill Testing and Energy Determination, presented at MEI Comminution 2023.

Pease, J, 2010. Elephant in the Mill, presented at XXV IMPC 2010.

Pease, J, Young, M and Curry, D, 2005. Fine Grinding as Enabling Technology – The IsaMill [online], Glencore Technology, Available from: <https://www.glencoretechnology.com/.rest/api/v1/documents/f2543f5863db5af2e7e3 33d7efffcd45/FineGrindingasEnablingTechnology_TheIsaMill.pdf> [Accessed: 22 July 2024].

Taylor, L, Skuse, D, Blackburn, S and Greenwood, R, 2020. Stirred media mills in the mining industry: Material grindability, energy-size relationships and operating conditions, *Powder Technology*, 369:1–16.

Turcotte, D, 1986. Fractals and Fragmentation, *Journal of Geophysical Research*, 91:1921–1926.

Continuous improvement of the Carrapateena grinding circuit

P Toor[1], J Seppelt[2], F Burns[3], J Reinhold[4], W Valery[5], L Brennan[6] and K Duffy[7]

1. Comminution Consultant, Hatch, Brisbane Qld 4000. Email: paul.toor@hatch.com
2. Processing Manager, BHP Carrapateena, Pernatty SA 5173. Email: joe.seppelt@bhp.com
3. Superintendent – Metallurgy, BHP Carrapateena, Pernatty SA 5173.
   Email: fraser.burns@bhp.com
4. AAusIMM, Senior Metallurgist – Plant, BHP Carrapateena, Pernatty SA 5173.
   Email: jacqueline.reinhold@bhp.com
5. Global Director – Consulting and Technology, Hatch, Brisbane Qld 4000.
   Email: walter.valery@hatch.com
6. Process Engineer, Hatch, Brisbane Qld 4000. Email: lachlan.brennan@hatch.com
7. Process Consultant, Hatch, Brisbane Qld 4000. Email: kristy.duffy@hatch.com

ABSTRACT

Slurry commissioning of the Carrapateena copper concentrator began in Q4 2019, with first concentrate produced in December of that year (Reinhold *et al*, 2023) and nameplate throughput capacity of 500 t/h achieved in Q1, 2020. Having achieved the nameplate capacity, site personnel worked through a comprehensive sequence of operational improvements to debottleneck the plant towards 700 t/h.

Carrapateena engaged Hatch to support the grinding circuit improvement activities, including planning and analysis of a full grinding survey, the subsequent development of a site-specific model of the SABC circuit, and the investigation and validation of opportunities for increasing throughput rates. Hatch were also requested to review the semi-autogenous grinding (SAG) and ball mill lifter and liner designs in consultation with the manufacturer, with a view to optimising the designs for the conditions in the Carrapateena circuit.

The project identified various opportunities to increase throughput, including via increasing power utilisation, optimising SAG parameters such as mill filling, ball load and mill speed, and increasing pebble production rates via optimisation of the grate and pulp lifter designs. Optimisation of the shell lifter and liner designs for both the SAG and ball mill was also identified as an opportunity to provide improved process and maintenance performance.

These initiatives were implemented by the site team, and in combination with the other debottlenecking activities outside of the grinding circuit, the new target peak throughput of 775 t/h (55 per cent increase) has been achieved. The work highlights the synergy of process simulation and detailed lifter-liner studies to provide not only an optimal solution but also allow streamlined implementation.

Following the successful implementation of the recommendations, Hatch was then requested to support the site via analysis and evaluation of existing concepts for possible plant upgrade options to allow throughputs above 1000 t/h.

This paper provides an overview of the journey of the Carrapateena grinding circuit to date – from commissioning, through ramp-up and optimisation – as well as the pathway for future expansion.

BACKGROUND

BHP's Carrapateena operation is located within the South Australia copper-gold province, approximately 160 km north of Port Augusta and 600 km north of Adelaide. The iron oxide copper gold (IOCG) mineralisation of the Carrapateena deposit is typical of other resources in the region, with chalcopyrite, pyrite, and bornite the dominant sulfides.

The Carrapateena concentrator was designed as a conventional SABC circuit fed by primary crusher product, followed by concentration via flotation with regrind. The design feed rate for the comminution circuit was 500 t/h at the target final grind (cyclone overflow) 80 per cent passing size (P_{80}) of 75 µm. A simplified process flow diagram for the comminution circuit is shown in FIG 1.

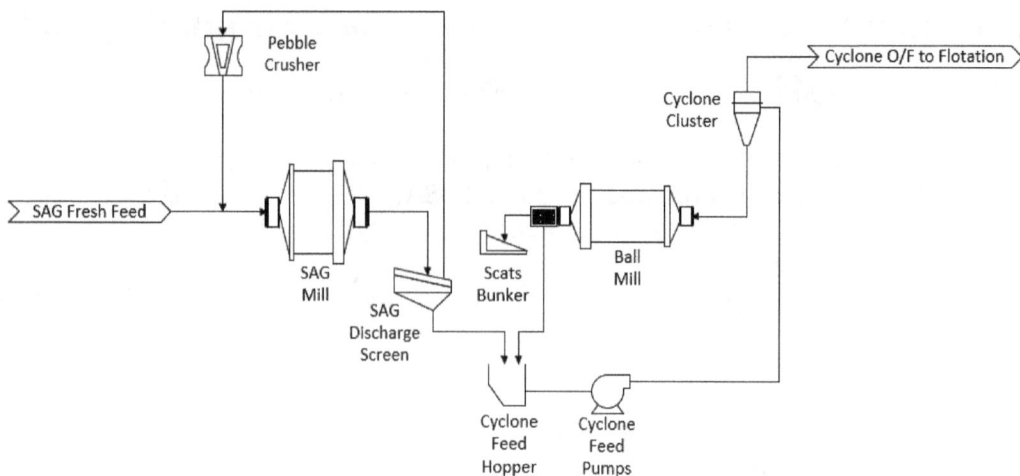

FIG 1 – Simplified flow sheet of the Carrapateena grinding circuit.

The circuit was designed based on an Axb value of 35.6 and Bond Ball Work index (BBWi) of 18.0 kWh/t. Pre-development test work indicated an overall Axb range of approximately 29–95, and BBWi of 7–21 kWh/t. Averages were 47.9 and 16.0 kWh/t respectively, which closely aligned with the feed processed during the commissioning and ramp-up phase; samples collected during that period returned an average Axb of 46.5 and BBWi of 16.2 kWh/t.

The primary semi-autogenous grinding (SAG) mill is a 28 × 15 ft (8.53 × 4.57 m) Outotec mill with a 7 MW single pinion variable speed drive (VSD). Slurry discharges from this mill and reports to a single deck 3.05 × 7.32 m Joest screen with 10 mm apertures. Oversize from the SAG mill discharge screen reports to a Metso HP3 pebble crusher. The secondary mill is an overflow discharge 22 × 36 ft Outotec ball mill (6.69 × 10.97 m), fitted with a 9.5 MW dual pinion VSD and discharge trommel.

COMMISSIONING AND RAMP UP

Commissioning activities were completed in conjunction with the vendor, Outotec. Amongst others, one critical commissioning task within the grinding circuit was the calibration of the bearing pressure derived weight measurements. This was a process that realistically could only be completed once, starting with a mill truly empty of both rock and grinding media. Great care was used to establish reliable SAG and ball bill weight calibrations which could be trusted during operation. An initial calibration measurement was made by filling of the mills with a known amount of water, measured via both the feed flowmeter and by surveying the fill water level in the mill. A second calibration point was established for each of the mills. In the case of the SAG mill, a parcel of ore was inched into the mill and its mass measured by the freshly calibrated SAG feed weightometer. For the ball mill, a mass of balls was added, with the total mass calculated based on a count of bagged media. This careful approach was taken to enable accurate monitoring of the load in each mill, with a view to ensuring the accuracy of set point adjustments and mill power calculations which rely upon those readings.

Carrapateena was commissioned during a time of limited ore supply, with the mine also still in the process of ramping up to design rates. During this period a conscious, deliberate choice was made to operate a campaign milling strategy, rather than turned down continuous running, until October 2020. The purpose of this was to run at design rates or greater for short periods rather than lower rates continuously. This enabled early performance assessments and the identification of bottlenecks, such that timely actions could be put in place prior to those bottlenecks being reached in continuous operation. This strategy also intended to instil a driven, outcome-focused mindset in the processing teams.

The 'as supplied' process control philosophy for the grinding circuit was a focus for updating during the first six months of commissioning. A strong understanding of 'good' grinding circuit control practices was held within the site metallurgy team. This enabled the rapid design and implementation of constraint-based SAG control philosophy, where mill feed rate is increased until one of several physical constraints are reached. Further a water balance control strategy focused around maintaining steady slurry density into each mill and maintaining a ratioed total water addition to the circuit.

A microphone is present on the SAG mill to monitor for overthrow of grinding media into the shell liners. Decibel readings at the standard frequency of 1 kHz were initially included in the control philosophy as a SAG mill speed constraint. While this provided good liner protection, it also unnecessarily limited SAG mill speed on occasion, due to detection of high noise during other transient events, for example a period of coarse, fines deficient feed. As such the vendor, Manta Controls, was engaged with a view towards reducing the occurrence of these 'false positives'. Through an intensive period of collaboration and data sharing, the microphone system was updated to include proprietary signal processing, that ultimately generated a control signal measured in metal impacts per revolution. This new signal was adopted into the SAG mill constraint control and remains in use as the primary protection for the SAG mill from metal on shell impacts.

SAG AND BALL MILL LINER DESIGN

The initial SAG and ball mill lifters and liners were supplied with the mill by the original equipment manufacturer (OEM), with Growth Steel engaged by Carrapateena to design and supply subsequent liner sets. A joint request was made by both Growth Steel and Carrapateena for Hatch to provide a technical review of the liner designs, with a focus on ensuring alignment between the liner designs and the process objectives arising from the grinding optimisation study. No OEM drawings were supplied with the mill, but a summary of the key dimensions of the original SAG and ball mill liners is provided in Table 1 and Table 2 respectively. The SAG mill liners were composed of chrome moly steel whilst the ball mill liners were rubber.

TABLE 1

Summary of OEM SAG mill shell lifter/liner dimensions.

Parameter		Units	Value
Shell lifters/lines			
Inside Diameter (new liners, plate-to-plate)		mm	8320
Plate Thickness		mm	90
Lifter Thickness		mm	310
Lifter Height Above Plate		mm	220
Leading Face Angle		degrees	27.5
Discharge grates			
Grate Plate Thickness		mm	100
Grate Lifter Height Above Plate		mm	160
Grate Lifter Height From Base		mm	260
Pebble Port Aperture Size		mm	70
Grate Open Area		%	8.6
Feed head lifters/liners			
	Inner	mm	80
Plate Thickness	Middle	mm	85
	Outer	mm	85
	Inner	mm	220
Lifter Height Above Plate	Middle	mm	215
	Outer	mm	215
Total Lifter Height		mm	300

TABLE 2

Summary of OEM ball mill shell lifter/liner dimensions.

Parameter	Units	Value
Plate Thickness	mm	70
Total Lifter Height	mm	210
Leading Face Angle	degrees	Not Reported

Another key feature of the original SAG mill liner design included the use of Twin Chamber pulp lifters with square aperture grates, as shown in FIG **2**. The Twin Chamber pulp lifters (also called Turbo pulp lifters) are designed to significantly reduce, or potentially eliminate, flow back of slurry into the mill to prevent slurry pooling. They can provide significant benefit in high throughput, single-stage (closed circuit) SAG/AG circuits, where the high volumetric flow of slurry can result in the mill being prone to slurry pooling and flow back is often a key mechanism.

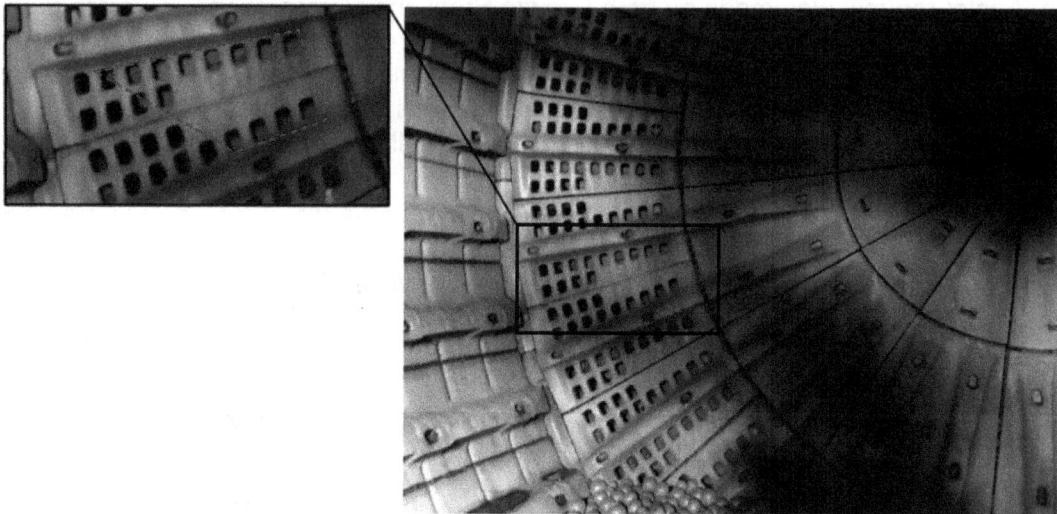

FIG 2 – Mill inspection photo from the grind out of the Carrapateena SAG mill, note the Twin Chamber pulp lifters blocking a number of the grate apertures.

Given the Carrapateena mill operates in open circuit and at moderate throughput rates, there is minimal risk of pooling, meaning that Twin Chamber pulp lifters are likely to provide little or no benefit to overall mill performance. This was confirmed by a pulp lifter capacity assessment conducted as part of Hatch's scope. However, the presence of the Twin Chamber restricts the grate open area, as can be seen in the area outlined with red dashes in FIG **2**. In the case of Carrapateena, the presence of the Twin Chamber was impeding approximately half of the grate apertures. This combination of grate and pulp lifter design was identified as the likely cause for the low pebble production rate, which was typically 10–18 per cent of fresh feed while operating under these conditions.

Growth Steel had recommended the use of a curved pulp lifter and a fully ported, 70 mm aperture grate design. On paper, the new grate design would increase the open area from 8.6 per cent to 10.1 per cent. The recommended grate design is shown in FIG **3**. Apart from the 1.5 per cent increase in open area, the ability of material to discharge through the grates would be increased further through the use of slotted apertures, which are less prone to pegging by near-sized media and rock compared to a square aperture of the same size. Similarly, removing the obstruction caused by the Twin Chamber pulp lifter would see a larger portion of the open area serve as *effective* open area. Hatch endorsed this change as it would facilitate higher pebble rates and therefore greater utilisation of the installed pebble crusher capacity, which had been sized based on a pebble rate equal to 30 per cent of the SAG fresh feed.

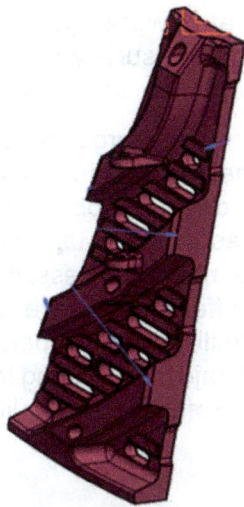

FIG 3 – Growth steel grate design suited to curved pulp lifters.

Another change made to the design of the SAG internals involved increasing the shell lifter face angle – which had originally been 27.5 degrees along the full length of the mill – to 33 degrees on the feed end and 35 degrees on the discharge end. Outermost trajectory simulations conducted by Hatch (FIG **4**) suggested that these lifters will produce suitable trajectories when new and operating at 74 per cent of critical speed (% Nc). The results of the outermost trajectory simulations were in line with the discrete element method (DEM) simulations provided by Growth Steel.

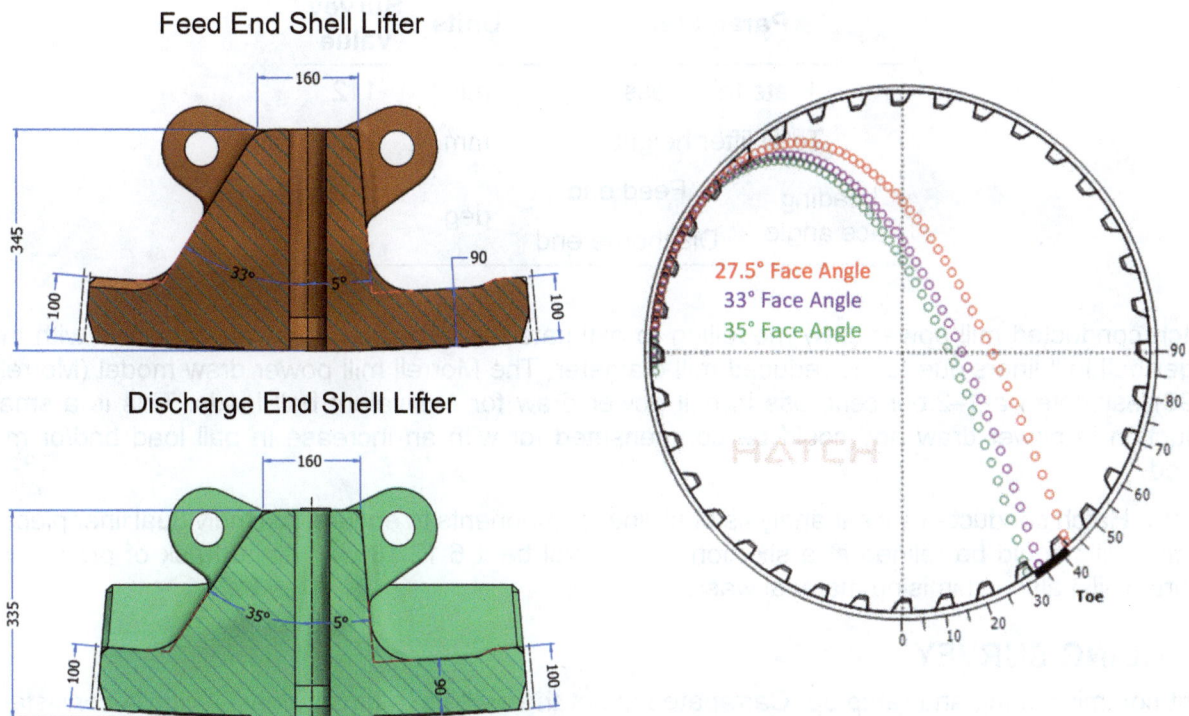

FIG 4 – Cross-sectional profiles of Growth Steel's proposed shell lifters (left), and *MillTraj* outputs (right) for the OEM (red trajectory) and Growth lifter profiles (feed and discharge end being purple and green, respectively).

Based on a benchmarking exercise conducting using Hatch's in-house tools and database – and which considered the power split between the mills, the ore characteristics, and the relatively fine target P_{80} – it was expected that the circuit would become ball mill limited as the throughput ramped up. The revised shell lifter profiles were therefore endorsed by Hatch, on the basis that the more relaxed media trajectory would help to promote attrition and abrasion breakage in the SAG mill. This would yield a finer transfer size and help to shifting some of the grinding workload upstream from the ball mill circuit. It should be noted that power-based modelling and analysis of operating data conducted in association with the subsequent survey and model development also supported the aim of shifting the grinding workload towards the SAG mill.

With regards to the ball mill, Carrapateena was targeting a ball mill reline frequency of 18 months, which would not be possible for rubber liners given the mill size. To achieve this increase in liner life, Growth Steel recommended a composite design (rubber with steel inserts) as well as increasing the plate thickness to 112 mm. Other changes included utilising different feed and discharge end lifter profiles. The feed end lifter features a more aggressive 30 degree face angle, compared to 35 degrees on the discharge end. This difference in face angles was introduced to promote more impact breakage at the feed end of the mill where the particles would be coarser. Conversely the discharge end lifters have a more relaxed trajectory shifting the grinding regime slightly more towards attrition and abrasion. Cross-sectional profiles of both lifters are provided in FIG **5**, with key dimensions summarised in Table 3.

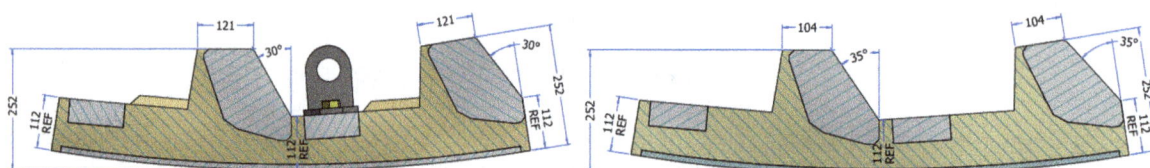

FIG 5 – Proposed feed end (left) and discharge end (right) ball mill shell lifter profiles.

TABLE 3

Summary of proposed ball mill shell lifter/liner dimensions.

Parameter		Units	Survey value
Plate thickness		mm	112
Total lifter height		mm	252
Leading face angle	Feed end	deg	30
	Discharge end		35

Hatch conducted mill power draw modelling to estimate the reduction in power expected with the larger ball mill liners due to the reduced mill diameter. The Morrell mill power draw model (Morrell, 1996) estimated a 1–2 per cent loss in mill power draw for equivalent ball loads. This is a small reduction in power draw and could be compensated for with an increase in ball load and/or mill speed.

Finally, Hatch conducted a wear analysis of all liner components to ensure that individual liner pieces for the mills could be relined at a six-monthly interval be it 6,12,18 with limited risk of premature failure whilst also minimising material waste.

GRINDING SURVEY

Post commissioning and ramp up, Carrapateena set the ambitious target of achieving a consistent throughput of 650 t/h, whilst limiting the deviation from the original target grind size (P_{80} of 75 μm) as much as possible. The Carrapateena technical team understood that development of a site-specific mathematical process model of the grinding circuit would assist in identify process changes and strategies to help in achieving this target. A well-calibrated model allows for simulations of various scenarios to determine the optimum operating parameters and circuit configurations and avoids the need for lengthy and disruptive plant trials.

The Carrapateena team engaged Hatch's Mining and Minerals Processing team to provide support in planning a survey of their grinding circuit and for the subsequent analysis and model development. The development of a high-quality model relies on representative, high-quality survey data. Hatch personnel typically travel to site prior to the survey to check and assess sample points and equipment, train site personnel and develop a detailed survey plan and procedure. However with the survey planned for late 2020 this was not possible, owing to travel restrictions related to the COVID pandemic. Instead, Hatch undertook the survey planning remotely, aided by a 3D model and photos of the concentrator, and discussions with site personnel.

This process identified a few challenges which needed to be considered prior to executing the survey. As an example, a review of the 3D model found that the original inspection port to the ball mill trommel (highlighted blue FIG **6**) would not allow for a representative sample of the trommel undersize to be collected, as the full length of the stream could not be accessed. For this reason, it was recommended that additional ports (marked yellow in FIG **6**) be cut on the adjacent side to allow a representative sample to be collected.

FIG 6 – 3D model showing inspection hatch (blue) and proposed sample points (yellow).

Similar checks were carried out for all sample points, including belt cuts to ensure accessibility. Samples collected during the survey are shown in FIG **7**.

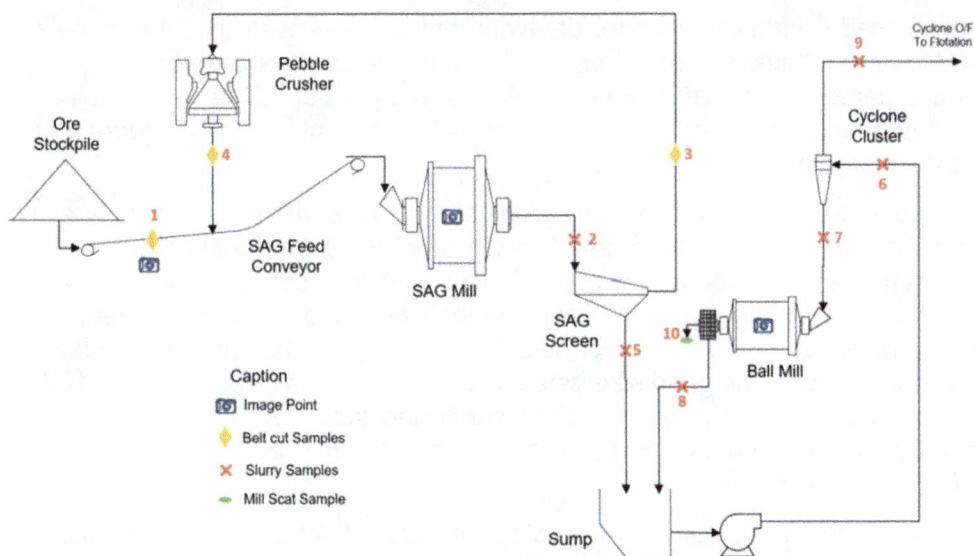

FIG 7 – Sample point summary.

Table 4 summarises key operating parameters at the time of the survey.

TABLE 4

Summary of survey conditions.

Parameter	Units	Survey value
SAG mill feed F_{80}	mm	95.4
SAG mill feed rate	tph	540
Pebble rate	tph	70
Pebble crusher power	kW	85
SAG mill speed	% Nc	67.2
SAG mill power draw	kW	5809
Ball mill speed	% Nc	79.7
Ball mill power draw	kW	6200
Cyclone O/F P_{80}	µm	89
Specific energy	kWh/t	22.4
Operating work index	kWh/t	21.1

MODEL DEVELOPMENT

The Carrapateena grinding circuit model was developed within the *JKSimMet* software package using the standard unit models, fitted (ie calibrated) to the survey data. Ore characterisation test work from the survey returned an Axb value of 39.8, and BBWi of 16.8 kWh/t. These hardness values fall between the design values and the results from recent samples collected prior to the survey, and relatively near to the average values from the data set used during design. The survey ore can therefore be thought of as being broadly 'typical' of the Carrapateena ore in terms of comminution characteristics.

One observation made during the model development process was that the grate open area in *JKSimMet* had to be simulated at about 4 per cent to achieve reasonable results aligning with survey data. This is compared against a measured value of 8.6 per cent. The low simulated open area supports the hypothesis that the Twin Chamber pulp lifter, along with square apertures, was restricting the discharge of coarse material.

Having developed a model based on the conditions at the time of the survey, further adjustment was required to account for the fact that the new Growth Steel SAG liner design was installed following the survey. The changes included a new grate design, and replacement of the Twin Chamber pulp lifter design to a curved pulp lifter design. These changes yielded a significant increase in throughput, largely attributable to an increase in pebble rate (in both absolute and relative terms). A corresponding coarsening of the grind size was observed, with an average P_{80} of ≈ 120 µm following the changes. This step change in performance confirmed that the Twin Chamber pulp lifter was restricting the discharge of coarse particles from the SAG mill and as such was not the best application of this technology.

The model tuned to the survey conditions was updated to reflect the change in mill internals, and hence provide a Baseline case which reflected the new operating conditions. This revised Baseline was validated using power-based methodologies (Morrell, 2022), and by comparison against operational data, and was the benchmark against which the subsequent simulations were compared.

FIG **8** through 10 show the frequency distributions of throughput, pebble rate, and grind size for a period of one week prior to and after the installation of the new mill internals. Indicated on each graph is the corresponding parameters from the Baseline. The Baseline scenario considers a throughput of 640 t/h, at a final product P_{80} of 114 µm which was in line with mill performance post reline with the new liner designs.

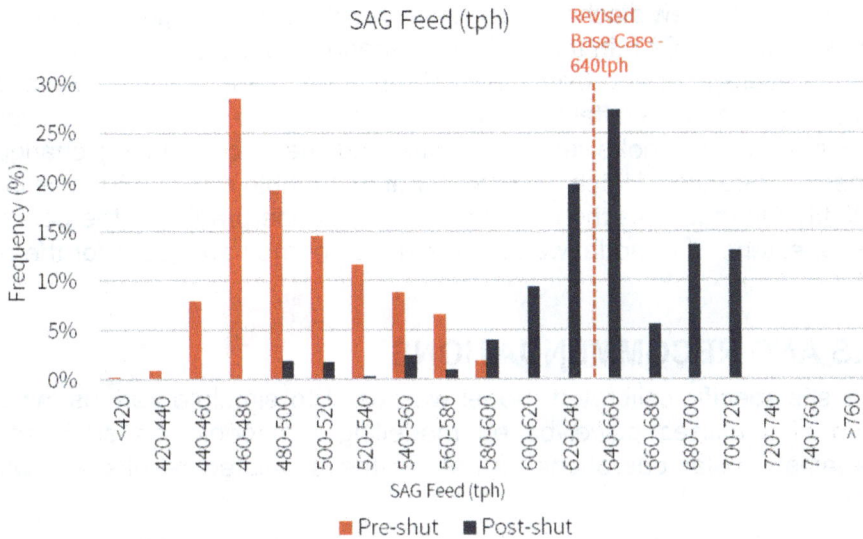

FIG 8 – Comparison of the baseline throughput against the frequency distributions of throughput in the week immediately prior to and after the reline.

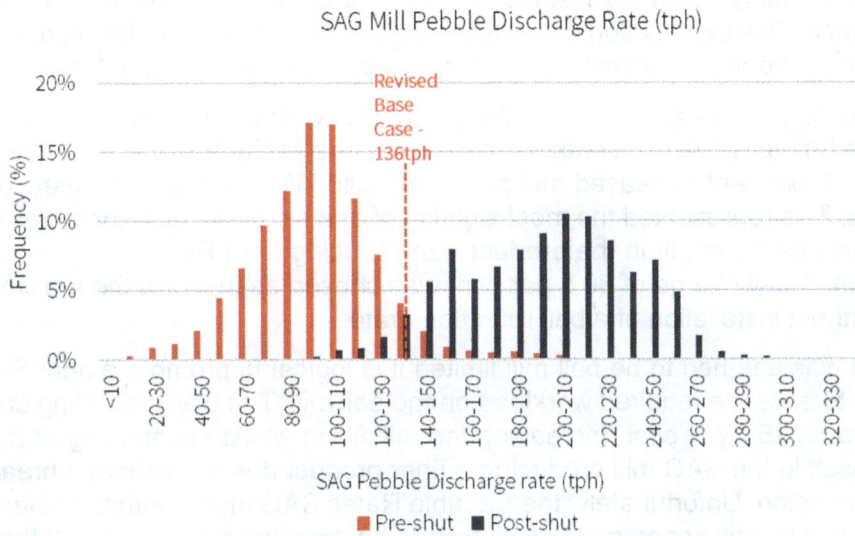

FIG 9 – Comparison of the baseline pebble rate against the frequency distributions of pebble rate in the week immediately prior to and after the reline.

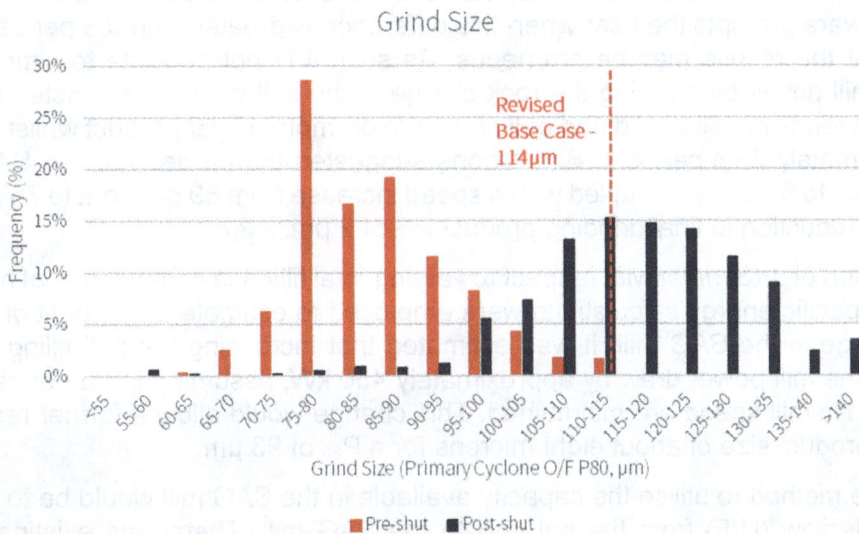

FIG 10 – Comparison of the baseline grind size against the frequency distributions of grind size in the week immediately prior to and after the reline.

While the simulation with the new discharge liners achieved a step-change in pebble rate, from 67 t/h in the original base case to 136 t/h in the revised scenario, this still appeared to underestimate the observed post-reline average of approximately 210 t/h. Two possible factors were identified which could have been exaggerating the difference between the actual and simulated values. The first was the pebble crusher not being choke fed post reline and the second being changes in ore type. Accounting for these differences, Hatch estimated that the pebble rate at 640 t/h is likely to be in the range of 178–193 t/h. On that basis the model may still be underpredicting the pebble recirculation, however to a lesser extent. The model was nonetheless deemed adequate for the purposes of the study.

SIMULATIONS AND RECOMMENDATIONS

The developed a site-specific calibrated model was used to simulate various 'what if' scenarios. Hatch's approach also utilises power-based modelling methodologies such as Morrell's and benchmarking against similar operations to ensure the simulated results are within practically achievable ranges.

The simulation cases were developed with a view towards the likelihood that the circuit will become ball mill limited, based on the relatively fine target P_{80}, and the latent capacity available in the SAG mill. Prior to the grinding survey in December 2020, the SAG mill typically operated below 5 MW power draw – comfortably below the installed 7 MW – including during periods which approximated design performance. This expectation was supported by power-based modelling and an assessment of the power split between the two mills using Hatch's in-house tools and database.

The ball mill was only drawing 6200 kW during the survey, equivalent to about 65 per cent of the 9500 kW installed motor power. Increasing the ball charge of the ball mill from the survey level of 21 per cent to 30.5 per cent increased mill power to 9045 kW with the mill speed kept constant at 78.9 per cent Nc. This represented the most significant and simplest route for increasing total circuit power draw, which would result in the product size reducing from P_{80} of 114 µm to 103 µm at the higher throughput. A ball charge of 30.5 per cent was chosen as this was the maximum ball charge level possible without installation of a ball retention grate.

Given the circuit was deemed to be ball mill limited it is logical to promote a finer SAG mill product (ie transfer size) to ease the required workload on the ball mill. The SAG mill filling during the survey was measured to be 25.4 per cent. Increasing the mill filling whilst maintaining or reducing the ball charge would result in the SAG mill producing a finer product due to increased breakage occurring by attrition and abrasion. Unfortunately, the Variable Rates SAG model within *JKSimMet* only allows for parameters such as mill speed and ball charge to be adjusted and simulated, the total mill filling must remain effectively constant. There are well established empirical correlations that describe the change in the breakage rate curve in response to changes in these and a number of other parameters (hence the 'variable rate' name), however no such correlation exists for the total filling. Indeed, the software prompts the user when this difference is greater than 0.3 per cent absolute as a soft alarm that the results may be erroneous. As such it is not possible to simply simulate an increase SAG mill power by allowing the rock charge in the mill to increase. Instead, the SAG mill was optimised in terms of mill speed and ball charge to promote a finer product whilst retaining a mill filling of approximately 25.4 per cent. Simulations suggested than a decrease in SAG ball charge from 11.6 per cent to 8 per cent coupled with a speed increase from 69 per cent to 75 per cent would result in a small reduction in final grinding product P_{80} of 2 µm.

Given the limitation of *JKSimMet* with respect to varying total filling, a combination of mill power draw modelling and specific energy calculations were employed to estimate the benefit of operating at a higher total charge in the SAG mill. It was estimated that increasing the mill filling to 30 per cent would increase the mill power draw by approximately 450 kW, assuming the 8 per cent ball charge and 75 per cent Nc mill speed are maintained. This change would allow a further reduction in final grinding circuit product size of about eight microns for a P_{80} of 93 µm.

Another possible method to utilise the capacity available in the SAG mill would be to divert some of the cyclone underflow (U/F) from the ball mill to the SAG mill. There was existing facility in the Carrapateena grinding circuit to partially split the cyclone U/F from the ball mill circuit back to the SAG mill, and simulations were conducted assuming a split of 10 per cent to the SAG mill.

Simulations suggest that the final grinding circuit P_{80} may be reduced by about seven microns for a P_{80} of 86 µm. The simulated reduction is at the higher end of the expected range based on similar operations and as such were likely optimistic but nonetheless encouraging.

Cyclone optimisation was also investigated. Analysis of both the survey and historical data suggested the cyclones were typically operating at a pressure of about 90 kPa, which was relatively low for the design cut size, especially considering OEM datasheets reported a maximum operating pressure of 200 kPa. Simulations indicate that increasing the pressure to 130 kPa in combination with reduction of the vortex finder diameter from 200 mm to 185 mm would provide a further reduction of the final P_{80} by 2 µm. This in combination with the cyclone U/F bypass would yield a final grinding circuit P_{80} of 84 µm at a throughput of 640 t/h.

Though simulations suggested that a significant drop in the P_{80} was possible from 114 µm to 84 µm at a throughput of 640 t/h (28 per cent above design), the final P_{80} is still marginally higher than the design value of 75 µm. Simulations suggested that the throughput would have to be reduced to approximately 615 t/h to achieve a final product P_{80} of 75 µm for the survey ore. Conversely, at the new target throughput of 650 t/h post implementation of the recommendations, the final grinding product would have a P_{80} of approximately 88 µm. The cumulative benefits of the recommendations (from simulations) are illustrated in Figure 11

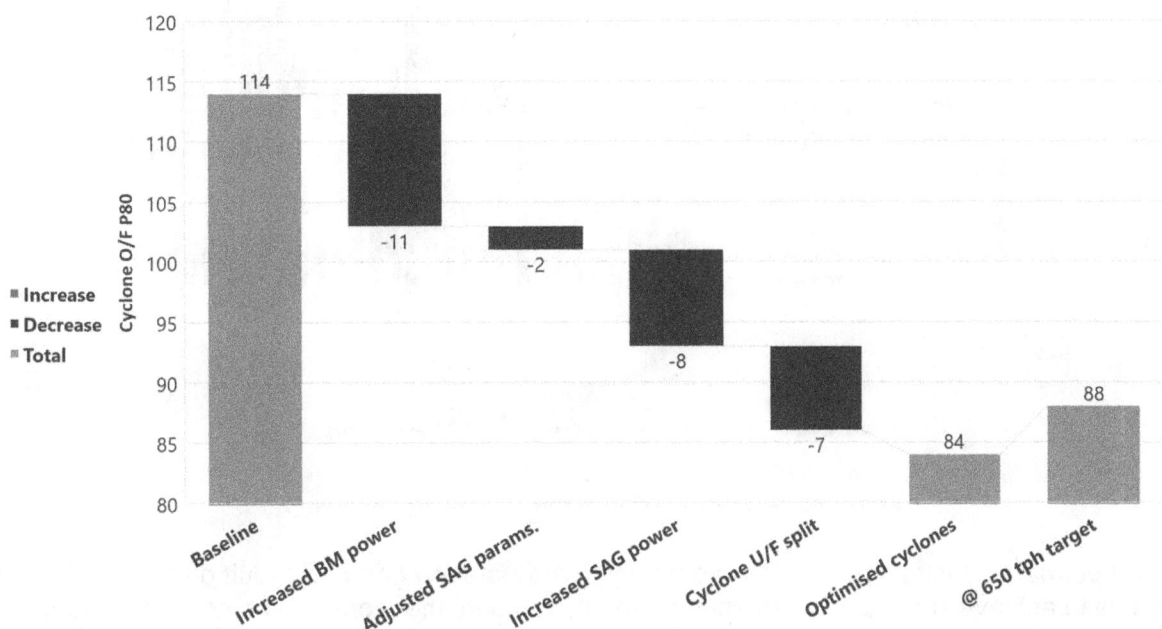

FIG 11 – Summary of simulation results. Note that all simulations except for the last are at 640 t/h, with the change identified by the simulation name building upon the previous scenario.

IMPLEMENTATION

Carrapateena staff began implementing the recommendations almost immediately. The first item implemented was increasing the ball charge level in the ball mill which increased the ball mill power draw from approximately 6200 kW to 9000 kW. The installation of a retaining ring was required to achieve this outcome. Modifications were made to the ring following the initial install to increase the aperture angle and reduce blockages.

Following this, reducing the ball charge levels in the SAG mill whilst increasing the total filling was also employed. Weights were taken of liners before and after a campaign to estimate the steel loss over the time. For a given ball charge, this enabled modelling of the expected total charge volume at a given load value and supported the desire to maintain a stable mill fill over the life of the liners. A SAG ramp-up schedule was put together that reflected optimum maximum speeds during the life of the liners to ensure optimum breakage and reduce the risk of liner damage from ball impacts.

A partial split of the cyclone U/F was also trialled; however, the trial had to be abandoned due to excessive spillage at the pebble screen feed box. Recommendations for changing the cyclone internals were also pursued soon after the optimisation study. Due to the lead times for vortex finders, an interim strategy was required, and this involved reducing the cyclone feed percent solids and allowing the cyclone pressure to increase from 90 kPa to 130 kPa.

Implementation of these strategies allowed for throughputs of 650 t/h and above to be achieved consistently, at a final product size generally ranging between 75–95 μm. There was a decrease in recovery due to the reduced residence time and coarser product; however there was a considerable net benefit in terms of metal produced due to the higher throughput.

FIG **12** provides a summary of mill performance since commissioning. It should be noted that since the operation can be mine limited for extended periods, particularly prior to a planned shutdown, mill performance has been tracked by comparing the most productive day (two entire shifts) for each calendar month and the associated P_{80} and circuit power draw has also been plotted.

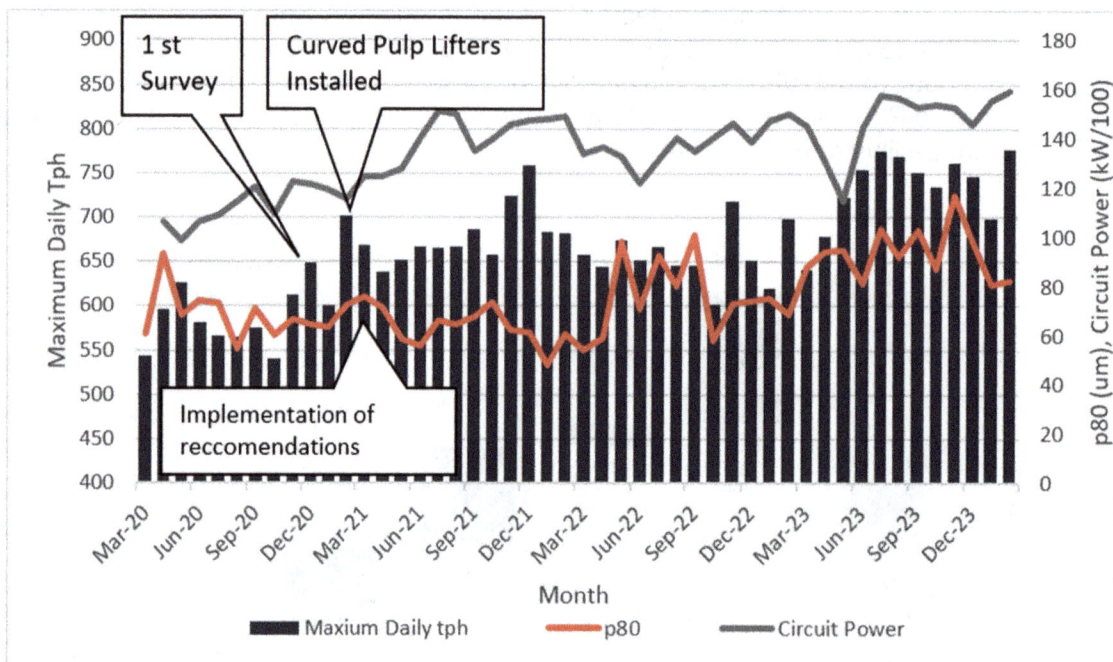

FIG 12 – Summary of grinding circuit performance.

Those upstream constraints notwithstanding, the Carrapateena grinding circuit demonstrated that it was able to achieve and exceed nameplate throughput soon after commissioning. The general trend in throughput since has been a positive one, with peak daily throughputs up to an exceeding 750 t/h achieved in several months during the second half of 2023. Key milestones are identified on the graph in FIG **12**, as follows:

- The first grinding survey being performed in December 2020, less than 12 months after wet commissioning.

- Installation of curved pulp lifters in May 2020 providing a step change in throughput at a coarser product size.

- Implementation of recommendations to reduce P_{80} from the Hatch simulation study.

- Gradual increase in mine output supporting higher more consistent throughput rate.

While the throughput ramp-up coincided with a coarsening of the cyclone overflow P_{80}, it should be noted that the circuit power draw steadily increased over the same period, resulting in more limited coarsening that would otherwise be the case.

EXPANSION OPTIONS

Carrapateena began investigating the possibility of incorporating a coarse flotation stage in their concentrator in 2022. This would allow the target P_{80} to be relaxed significantly – from the original design value of 75 µm to 150 µm – with minimal impact on metallurgical performance. Hatch was requested to investigate the potential to increase mill throughput to 1050 t/h at this coarser P_{80} target, via the inclusion of a secondary crushing circuit and adjustment to operating conditions in the grinding circuit.

Carrapateena conducted another grinding survey with remote support from Hatch with the view to update the *JKSimMet* model, given the various changes made to the liner designs and circuit operation since the first survey. The updated model was used to simulate circuit changes aimed at achieving the target throughput of 1050 t/h, with the model outputs supporting capacity checks for each of the unit operations.

Morrell's power-based methodology was employed to provide an initial estimate of the required reduction in SAG feed 80 per cent passing size (F_{80}) that would allow a throughput of 1050 t/h with final grinding circuit P_{80} of 150 µm. Morrell's method suggested an F_{80} of 50 mm would be required for the new throughput and P_{80} targets. However, to achieve a throughput of 1050 t/h the SAG mill would need to operate at 6500 kW (or 93 per cent power utilisation) consistently. As such, a finer P_{80} of 40 mm was used as the design basis, as in practice it would be easier to maintain a steady load in the mill with a finer feed. Importantly, a P_{80} of 40 mm is still comfortably within the range that can be achieved by open circuit secondary crushing.

Simulations were conducted using Metso's *Bruno* software package to determine the full size distribution of secondary crushed material required for simulation of the SAG circuit as well as for sizing of the secondary crusher. The proposed design involved an open circuit flow sheet, with the crusher receiving a feed scalped at 65 mm. It was determined that a Metso HP6 (or equivalent) fitted with coarse liners would be suitable for this crushing duty, with this arrangement yielding a SAG feed P_{80} of approximately 38 mm. The crushing flow sheet in the Metso *Bruno* software package, and the associated feed and product PSDs, are shown in FIG 13.

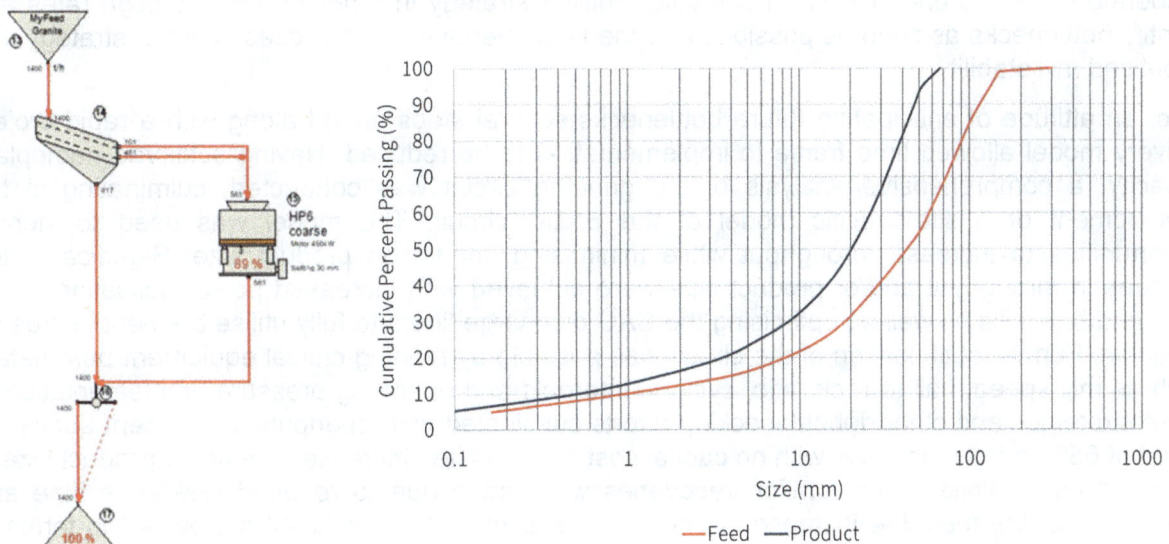

FIG 13 – *Bruno* flow sheet developed for SAG feed size distribution estimation.

In addition to secondary crushing, several grinding circuit changes were identified as being required for the circuit to maintain throughputs of 1050 t/h. Firstly, as already mentioned, the SAG mill power draw needs to be increased to approximately 6500 kW. This can be achieved by increasing the ball charge to 18 per cent and the mill speed to 75.5 per cent Nc. A higher ball load is required as the finer feed results in a lower total filling of about 22 per cent. A ball load of 18 per cent is reported as the allowable maximum due to the mill's structural limit. It was also recommended to reduce the top size of the SAG mill media to 105 mm (to suit the finer feed size).

Changes to the lifters and liners of the SAG mill will also be required, as is always the case when secondary crushing is implemented. It was recommended that the feed shell lifter face angles be increased to 35 degrees to match the discharge end lifters. This will aid in improving the grinding efficiency and liner life given the reduction in mill filling and increased speed. The pebble port size will also need to be reduced from 70/50 mm ports (3:1 split) to 30 mm ports.

An increase in ball mill media top size from 65 mm to 74 mm is recommended to account for the coarsening of the transfer size and promote a coarser cyclone overflow. To operate the Carrapateena grinding circuit at an average throughput rate of 1050 t/h it was recommended that the cyclone cluster be upgraded from the current Cavex 500CVX. Simulations suggested that at 1050 t/h, nine of the original Cavex 500CVX cyclones would need to operate, leaving only one spare and assuming a 66 per cent solids feed density. Operating with only one spare cyclone at very high cyclone feed densities is not considered practical in the long-term. Simulations suggested the circuit will be able to operate with eight 650CVX cyclones open for most of the time, with a feed density of about 60 per cent, yielding an operating pressure of between 80–85 kPa. The simulations suggested that implementing the changes outlined would result in a final P_{80} of approximately 145 µm.

There are several mechanistic limitations within the various equipment units that *JKSimMet* does not address or check to ensure sufficient capacity. In these scenarios Hatch uses its extensive operating database and suite of in-house tools to benchmark performance, as well as liaising with vendors directly. Hatch carried out volumetric capacity checks for the SAG mill pulp lifters, SAG discharge screen, and the ball mill, with no major issues identified.

Carrapateena are currently in the engineering stage of this expansion project, with this concept being one of the potential pathways to increase the concentrator capacity. The Carrapateena block cave expansion is currently planned to increase ore mined and treated to ~12 Mt/a.

CONCLUSIONS

The Carrapateena copper concentrator achieved rapid ramp up following commissioning in late 2019. Key reasons for this success included close engagement with the vendor for support, a deliberate choice to operate with a campaign milling strategy in order to run at design rates and identify bottlenecks as soon as possible, and the implementation of a process control strategy that prioritised mill stability.

Also, an attitude of anticipating future bottlenecks several steps ahead along with a rapid project delivery model allowed time frame to implementation to be reduced. Having achieved nameplate capacity, a comprehensive analysis of the grinding circuit was conducted, culminating in the development of a site-specific model of the SABC circuit. The model was used to identify opportunities to increase throughput while minimising impact on product size. Significant step changes in throughput and/or product size were achieved with increased power utilisation of the SAG and ball mills as well as optimising the SAG discharge liners to fully utilise the pebble crusher available. Further optimisation of the circuit was possible by refining critical equipment parameters such as mill speed, ball to rock ratio, cyclone internals and operating pressure. Implementation of these strategies and other debottlenecking efforts culminated in throughputs 30 per cent above the target of 650 t/h to be achieve, with no capital cost and a limited increase in grinding product size of around 15 µm. While lower flotation recoveries were seen due to reduced residence time and coarsening of the feed, the increase in throughput meant that a significant net benefit in terms of metal production was achieved. The synergistic approach of process simulation and detailed lifter-liner studies not only yielded optimal technical solutions, but also enabled efficient implementation.

An expansion option involving a coarse flotation step and the addition of a secondary crushing circuit was proposed as a potential next step in Carrapateena's journey. This expansion strategy was investigated with the support of Hatch, through comprehensive analysis and modelling to assess the feasibility of increasing mill throughput on the basis of a more relaxed P_{80} target of 150 µm. As part of this work, the *JKSimMet* model was updated using new survey data to reflect changes in liner designs and circuit operation. The updated model – together with power-based methodologies and other capacity checks – was used by Hatch to assess this expansion option. The process feasibility of achieving a grinding circuit throughput of 1050 t/h via this option was confirmed, and the expansion project has since progressed to the subsequent Engineering phase.

ACKNOWLEDGEMENTS

Hatch would like to thank Carrapateena staff for their collaboration on this project and paper and thank BHP for permission to publish this paper.

REFERENCES

Morrell, S, 1996. Power draw of wet tumbling mills and its relationship to charge dynamics – Part 1: A continuum approach to mathematical modelling of mill power draw, *Transactions of the Institutions of Mining and Metallurgy, Section C: Mineral Processing and Extractive Metallurgy*, 105:C43–C51.

Morrell, S, 2022. The Morrell Method to Determine the Efficiency of Industrial Grinding Circuits 2021 Revision, Guideline, Global Mining Guidelines Group (GMG).

Reinhold, K, Assmann, S, Brodie, E, Van Sliedregt, J, Burns, F, Tsatouhas, G and Seppelt, J, 2023. Replacement of Sodium Ethyl Xanthate Collector at Carrapateena, in *Proceedings of the MetPlant 2023 Conference,* pp 124–134.

BRC copper processing – recovery improvements

I A Torok[1] and J Begelhole[2]

1. General Manager Engineering, Wilmar, Townsville Qld 4810.
   Email: istvan.torok@au.wilmar-intl.com
2. Senior Project Metallurgist, Glencore, Mount Isa Qld 4825.
   Email: jason.begelhole@@glencore.com.au

ABSTRACT

Maximising the recovery of secondary copper mineral orebodies is becoming more and more critical. This research reviews the copper recovery of the BRC orebody located at the Mount Isa Mine to maximise copper production and the economics of the operation as feed grades continue to decline. Various reagents and feed blend changes were investigated and referenced to previous research to assist in understanding the complex interactions taking place.

The key results from the study illustrated that the Black Rock Chalcocite (BRC) orebody has four main copper deportments (chalcocite, covellite, chalcopyrite and cuprite) and three main gangue counterparts (quartz, pyrite and kaolinite). The thiocarbamate collector (DSP009) demonstrated significant improvement to copper recoveries of plus 7 per cent compared to sodium isobutyl xanthate (SIBX) alone. Additional BRC copper recovery and concentrate grade improvements may be achieved by dosing the F100 biopolymer dispersant to reduce quartz and kaolinite slimes, negatively impacting copper recovery and concentrate grade. The annualised benefit for the Auxiliary Mill circuit is calculated to be approximately 800 t of additional copper and a reduction in operating costs in the order of A\$1 m.

INTRODUCTION

The north-west mineral province in Queensland contains several world-class operations, namely the Isa Inlier, where Mount Isa Mines owns and operates one of the longest serving copper mining operations in the world. Figure 1 illustrates the location of the Isa operation within the Leichhardt River Trough. One of the orebodies within the Isa Operation is the Black Rock Chalcocite deposit, which is secondary to the primary Chalcopyrite mining operation.

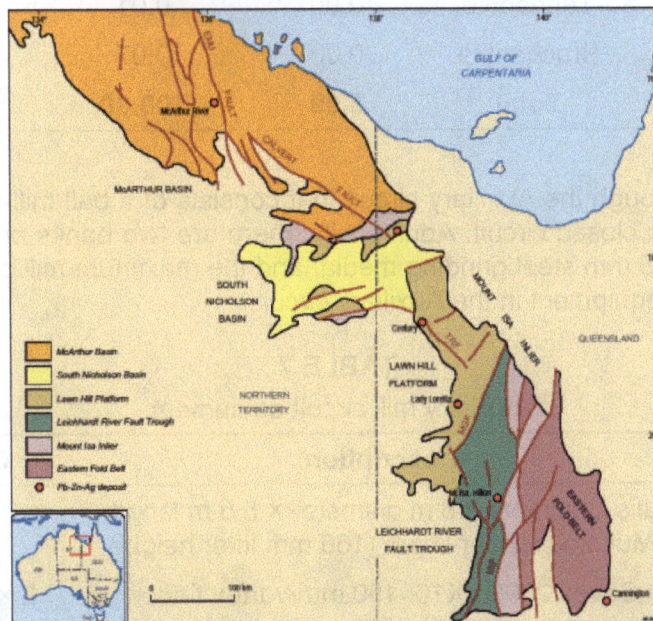

FIG 1 – Location of the Mount Isa Operation in the North-west Mineral Province (Gibson, 2017).

The Copper Concentrator at Mount Isa Mines processed BRC from 1963 to 1966, in what was named the No. 3 Copper Concentrator. The No. 3 Copper Concentrator was replaced in the 1970s with what is now the current Copper Concentrator and is often referred to as the No. 4 Copper Concentrator. A recent evaluation of the BRC orebody in 2017 presented a positive economic position to resume mining and processing the orebody. Mining operations recommenced using sub-level caving and processing through the existing Auxiliary Mill at the Mount Isa Mine No. 4 Copper Concentrator.

The Black Rock sub-level cave (BRC) was evaluated as a 1.4 Mt resource at 4.1 per cent copper with an annual production rate of 500 kt/a. The orebody is under the old Black Rock open cut mine. The blasted rock is crushed to sub-5 mm sizing and stockpiled to feed into the Auxiliary Mill circuit to produce a copper concentrate as feed for the downstream Copper Smelter. The BRC copper deportment mineralogy is captured in Table 1 and details the percentage of copper by mass that contributes to the orebody's feed grade of 5.09 per cent. The results presented in Table 1 are a sub-set of the bulk mineralogy of the mixed ore from numerous core samples used to define the resource, enable laboratory tests and the flow sheet to be developed. Chalcocite is the primary copper species followed by covellite.

TABLE 1

BRC copper mineralogy detailing the copper species (Brindle, 2023).

Copper deportment	% Cu by mass	% deportment of copper
Chalcocite	3.16	62.06
Chalcopyrite	0.36	6.97
Covellite	0.66	13.01
Bornite	0.11	2.13
Native Cu	0.08	1.59
Chrysocolla	0.20	3.85
Dioptase	0.27	5.34
Cuprite	0.25	4.93
Tetrahedrite	0.00	0.00
Tennantite	0.00	0.01
Brochantite	0.00	0.07
Total	**5.09**	**100.00**

BRC was processed through the Auxiliary circuit that consists of a ball mill-rougher flotation circuit with a cyclone bank in a closed circuit. Additionally, there are two banks of Wemco flotation cells. The mill typically uses 78 mm steel grinding media, and the maximum mill power is 1 MW. Table 2 summarises the critical equipment in the Auxiliary circuit.

TABLE 2

Auxiliary mill circuit equipment.

Item	Description	Quantity
Auxiliary ball mill	Allis Chalmers 3.8 m diameter × 5.5 m length, Multotec rubber liners (160 mm lifter height)	1 × Mill
Auxiliary cyclones	Cavex 250CVX10, 150 mm vortex finder, 54 mm rubber spigot	2 × Cyclones
Auxiliary flotation	Wemco 120 flotation cell	2 × Banks 2 × Cells/Bank

The high-level flow sheet for the Auxiliary Mill is illustrated in Figure 2. The discharge from the mill is pumped into the cyclones, cyclone underflow is fed back into the mill, while the overflow is directed to the head of the Wemco Rougher Bank. Concentrate from the rougher is sent to the Jameson Cleaner to produce a 12–15 per cent concentrate for the downstream Copper Smelter. Tailings from the Cleaner Scavenger Cells are combined with the tailings from the Rougher Tank Cells and are directed to the head of the Rougher Scavenger Tank Cells. Final tailings are produced at the last Cleaner Scavenger.

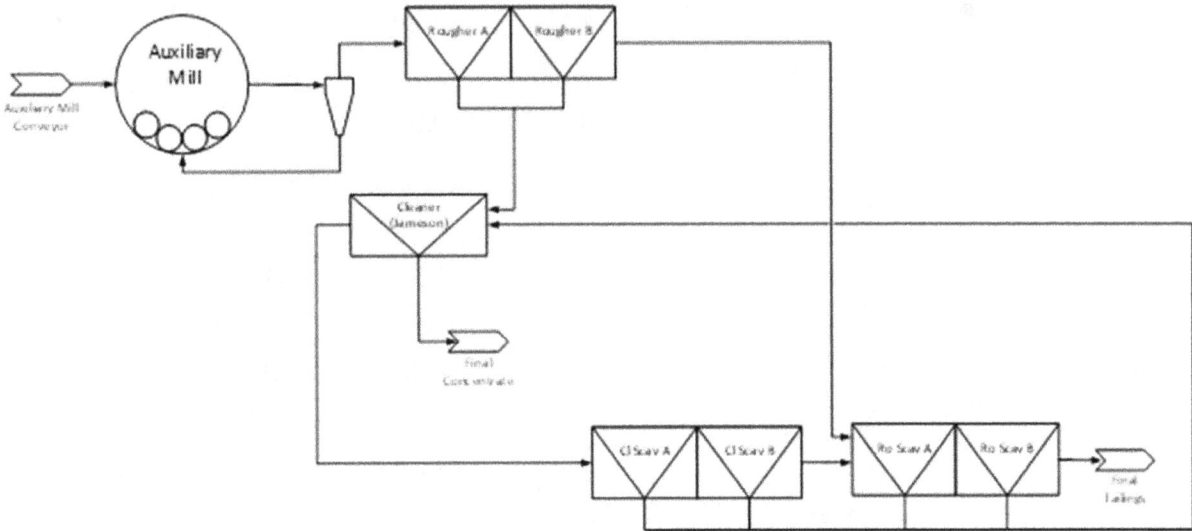

FIG 2 – Simplified auxiliary mill circuit, March 2023.

The circuit typically operates between 50 to 60 wtph, at 30 per cent solids and a corresponding P_{80} of 40-microns. The reagents used in the circuit throughout 2021, 2022 and in the start of 2023 are as follows:

- SIBX – Sodium Isobutyl Xanthate – used as a copper collector.

- MIBC – Methyl Isobutyl Carbinol – used as a frother.

The circuit and reagent profile does not contain any pH modification through lime, blending or any other reagent addition. Nor did the circuit at the time of commencing BRC processing have online pH monitoring.

Shift composite samples of feed, concentrate and tailings from the BRC processed through the Auxiliary Mill were submitted for mineralogical characterisation at the commencement of processing BRC. The results are presented in Table 3 and captures the feed grade being higher, plus 6 per cent versus the composite drill sample mineralogy in Table 1 of above 5 per cent. The feed grade at the commencement of the processing operation presented promising results, however, these feed grades were not sustained and declined to much lower values due to dilution from pit fines and pyrite.

TABLE 3

BRC mineralogy (Brindle, 2023).

Mineral mass in sample	% Mass – head	% Mass – conc	% Mass – tail
Native Cu	0.08	0.42	0.06
Cu sulfides	6.06	28.72	0.66
Cu silicates	1.25	3.89	0.41
Cu oxides	0.28	0.85	0.06
Other Cu minerals	0.01	0.01	0.01
Alloclasite/Cobaltite	0.06	0.10	0.04
Galena	0.93	2.83	0.17
Sphalerite	0.71	4.88	0.16
Pyrite	6.30	25.82	1.55
Pyrrhotite	0.07	0.35	0.05
Fe-oxides	6.81	2.48	7.69
Feldspar	3.67	1.49	4.24
Quartz	54.97	21.17	65.35
Chlorite	1.69	0.44	1.56
Pyroxene	8.39	2.54	9.80
Talc/Kaolinite/Nontronite	6.37	2.98	6.06
Carbonates	0.71	0.27	0.90
Mica	1.16	0.48	1.04
Other	0.47	0.30	0.18
Total	**100.0**	**100.0**	**100.0**

Feed grades became diluted, typically ranging between 2–3 per cent copper from pit fines and reactive pyrite, which presented several processing challenges. The reactive pyrite and pit fines decreased circuit pH, recovery and concentrate grade. The economics of the BRC operation was challenged. To improve recovery and concentrate grade, BRC was blended with a slag product. The slag product used for this blend is referred to as Magnetics or Mags. Mags are the result of the Copper Smelter Converter slag being crushed and through magnetic separation, the magnetic portion of the slag is separated for reprocessing. The blend was approximately 17 per cent Mags to 83 per cent BRC on a mass basis. The average copper content is between 6–8 per cent and the iron oxide content is around 0.6–0.8 per cent. The Mags in the blend assisted with improving recovery due to the high copper content and concentrate grade for the smelter, however, there were several limiting factors, these were:

- The Mags were very abrasive and caused considerable wear through the mill, pumps and process pipework resulting in accelerated wear and higher maintenance requirements. Routine maintenance was previously on a six weekly-cycle and the introduction of Mags dropped this down to two weeks.

- The magnetite in the Mags would concentrate in the silica cyclones preventing the tailings being used as a clean silica flux for the Copper Smelter. More details are captured in 'One plant's trash is another plant's treasure (O'Donnell and Begelhole, 2023).

Maintenance and operational costs increased with the BRC/Mags blend and without Mags, concentrate grade was below the Copper Smelter's requirements. Processing BRC through the Auxiliary Mill presented a significant challenge, and a significant opportunity to enable copper recovery, concentrate grade and BRC silica flux recovery benefits to be achieved.

METHODOLOGY

The methodology adopted for assessing the impact of grinding, flotation and reagent addition on recovery was conducted on the Auxiliary Mill circuit while processing BRC only. The intent of performing the full plant trial was to enable existing analysers and lab assays to be used as part of the data collection process for ease of assessment. The Copper Concentrator Metallurgists coordinated the test work, plant trials, data collection and change management process. Plant modifications were completed by the Copper Concentrator Maintenance and Operations teams. The proposed sequence for improving BRC recovery was based on the following principles:

- Selection of a chalcocite specific collector

- Removal of Mags from the Auxiliary Mill circuit

- Further optimisation of BRC copper recoveries

- Produce suitable concentrate grade for the downstream smelter

- Produce a clean silica flux from the tailings for the smelter

- Reduce copper losses to tailings through improved circuit reliability.

RESULTS

Copper collector

A test work program conducted by IXOM in 2022 demonstrated that DSP009 would improve copper recovery at a suitable smelter concentrate grade for the BRC feed. To further prove this benefit, an on/off plant trial was conducted in February 2023, dosing DSP009 into both Auxiliary Mill rougher banks while only processing BRC feed. The potential recovery benefits captured in the IXOM test work did translate to improved recovery in the on/off plant trial (Vass, 2023). The results from the trial using DSP009 are illustrated in Table 4. When DSP009 was used, copper recovery improved by approximately 8.1 per cent (77.3 per cent to 83.6 per cent) compared to SIBX alone for the BRC feed.

TABLE 4

DSP009 On/Off plant trial recovery and reagent cost assessment (Muhamad, 2023).

Results	Recovery (%)		% Change	Reagent Cost $		% Change
	DSP009 Off	DSP009 On		DSP009 Off	DSP009 On	
No. of Samples	31	20		31	20	
Min	65	78		450	380	
Quartile 1	75	82.75		675	510	
Mean	**77.3**	**83.6**	**8.1%**	**903.9**	**547.0**	**-39.5%**
Quartile 3	81	85		1200	600	
Max	84	86		1550	730	

Table 4 also illustrates the estimated reagent dosing costs for DSP009 versus SIBX. The average DSP009 dosing cost was just under A$550 per shift, whereas SIBX averaged close to A$900 per shift. The results from the on/off plant trial supported a full plant trial, using DSP009 as the collector and Mags being removed from the feed blend. Online pH monitoring was not available at the time of conducting the on/off plant trial.

Feed blend

A full plant trial was conducted without blending the BRC feed with Mags while continuing to dose DSP009. Table 5 illustrates the Auxiliary Mill recovery and rougher pH for BRC feed versus the BRC/Mags blend while dosing DSP009. The results summarised in Table 5 for the Auxiliary Mill circuit recovery used the on-stream analyser (OSA) copper recovery values and the online pH monitor installed at the head of the rougher bank. Online pH monitoring was implemented to understand the relationship between pulp pH and recovery due to the reactive pyrite in the feed.

TABLE 5

Auxiliary Mill copper recovery and rougher bank pH for various feed blends (Muhamad, 2023).

Results	Recovery (%)		% Change	pH		% Change
	BRC only	BRC/mags		BRC only	BRC/mags	
No. of Samples	33	49		33	49	
Min	56	70		4	4.1	
Quartile 1	65	74		4.2	4.8	
Mean	**68.2**	**77.9**	14.3%	**4.5**	**5.2**	15.2%
Quartile 3	71	82		4.6	5.6	
Max	80	85		5.7	6.1	

Recovery for the BRC feed was around 68 per cent using DSP009, much lower than the 83.6 per cent achieved in the on/off trial (Table 4). The BRC/Mags blend illustrated higher recoveries of 77.9 per cent. The rougher bank pulp pH for the BRC feed was 4.5 versus the BRC/Mags blend of 5.2. The addition of Mags to the blend assisted with increasing the pulp pH. For the trials presented in Tables 4 and 5, pulp density was controlled to around 30 per cent solids, P_{80} at 40-microns and feed grade between 2–3 per cent.

The results from the full plant trial provided support to feed the Auxiliary Mill circuit on BRC feed only and remove Mags from the blend. This change took place March 2023 in consultation with the downstream Copper Smelter. The smelter requested these changes to support optimisation of their process (O'Donnell and Begelhole, 2023).

Consequently, circuit reliability significantly improved and BRC tailings were reprocessed and used as a clean silica flux for the downstream Copper Smelter. The copper deportment reporting to tailings was recovered and improved the overall recovery of BRC. The Copper Concentrator Metallurgical team continued to conduct further test work and circuit optimisation.

Dispersant

Various modifiers were trialled in the laboratory, including sodium silicate, lime and F100 against a standard sample, where DSP009 was used as the collector. Figure 2 illustrates BRC copper recovery and concentrate grade for the various reagents.

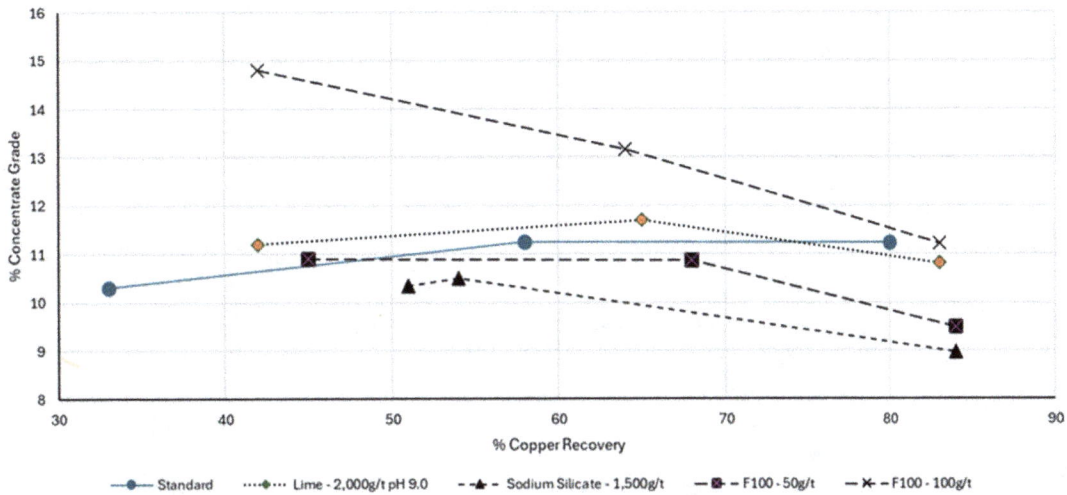

FIG 2 – BRC copper recovery versus concentrate grade with various reagents (Muhamad, 2023).

From the reagents trialled in Figure 2, F100 dosed at 100 g/t illustrated better performance in terms of producing a higher concentrate grade for a similar recovery compared to lime and sodium silicate. Lime, sodium silicate and a lower dosage of F100 at 50 g/t did not appear to improve concentrate grade compared to the standard sample. These samples were conducted with P_{80} of 30-microns and a pulp density of 32 per cent and 2.5 per cent copper feed grade with the reagents dosed at the Auxiliary Mill rougher banks while operating at 50 wtph BRC feed only.

BRC recovery and pH

During the feed blend test work, pH was recorded for both the BRC and BRC/Mag feed during the full plant trials. Figure 3 plots the test work summarised in Table 5 and a linear regression line has been added. A relationship between BRC recovery and rougher pulp pH appears to be present ($R2 = 0.41$) and indicates a higher pH should improve BRC recovery through the Auxiliary Mill circuit.

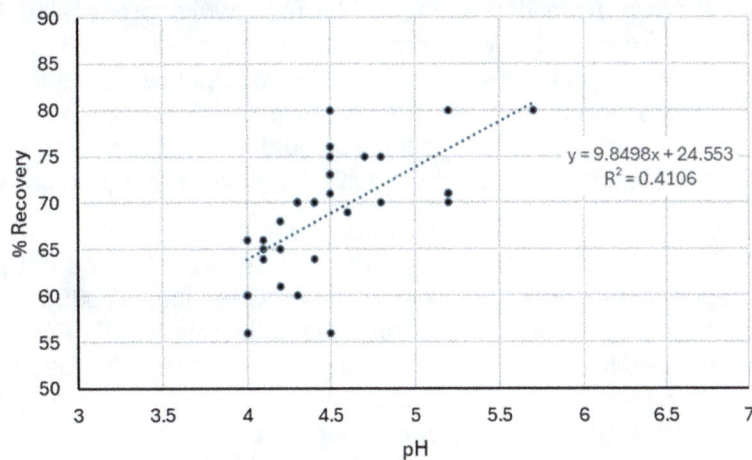

$$y = 9.8498x + 24.553$$
$$R^2 = 0.4106$$

FIG 3 – BRC copper recovery versus rougher pulp pH.

Results summary

In 2023, the Auxiliary Mill processed a BRC/Mags blend from January to March. In March 2023, test work enabled the magnetics from the blend to be removed and DSP009 to replace SIBX as the copper collector. From April onwards, BRC was processed without blending with Mags. Further circuit optimisation was completed by the plant metallurgists, which enabled higher recoveries to be achieved. Table 6 displays a summary of the monthly Auxiliary Mill performance during 2023. BRC tailings were reprocessed and used as a clean silica flux recovering the copper that would have otherwise reported through to the tailings dam. The Auxiliary Mill circuit availability improved and the BRC milled increased as well.

TABLE 6

Auxiliary Mill 2023 processing performance for BRC/Mags and BRC feed.

Enter	BRC/Mags + SIBX			BRC + DSP009							% Change
	Jan-23	Feb-23	Mar-23	Apr-23	May-23	Jun-23	Jul-23	Aug-23	Sep-23	Oct-23	
% Feed grade	2.72%	2.78%	2.51%	2.46%	2.92%	2.56%	2.48%	2.11%	2.02%	1.98%	
% Copper recovery	76.4%	71.5%	73.4%	77.3%	76.3%	80.2%	78.5%	82.1%	85.2%	76.1%	
% Concentrate grade	13.1%	18.0%	16.0%	13.9%	15.8%	17.1%	15.6%	15.6%	16.2%	18.4%	
Milled tonnes DMT	35 592	27 049	28 876	30 200	23 285	33 697	36 554	33 617	32 415	34 256	
Average % feed grade		2.7%					2.4%				-11.7%
Average % recovery		73.8%					79.4%				7.6%
Average milled tonnes DMT		30 506					32 004				4.9%

The results in Table 6 illustrates two periods, firstly BRC/Mags blend with SIBX, followed by BRC with DSP009. Feed grade reduced from the Jan-Mar period when compared to Apr-Oct by 11.7 per cent. Considering the feed grade reduction, recovery increased by 7.6 per cent when the same periods are compared. In addition, circuit throughout increased by 4.9 per cent over the same period, noting May-23 contained the major annual plant shutdown. The overall result enabled an additional 800 t of copper in concentrate to be produced while processing lower feed grade material.

DISCUSSION

The open cut mining operation of BRC back in the 1960s facilitated segregation of the different ores, however, the sub-level caving operation does not. Implementing DSP009 as the copper collector enabled the removal of Mags from the Auxiliary Mill feed while processing BRC. Feed grades continued to decline as illustrated in Table 6 from 2.72 per cent in Jan-2023 to 1.98 per cent in Dec-2023. These feed grades are significantly lower than the results presented in Table 1, which were used for the flow sheet and economic models. When designing flow sheets or re-purposing existing milling infrastructure, a range of process conditions should be modelled and factored into the final flow sheet. Sub-level cave dilution is a known risk with this type of mining technique.

Table 5 presents pulp pH ranged from 4.2 to 5.7 throughout the full plant trial when processing BRC feed only with the DSP009 collector. The Orica Flotation Guidebook does not present a pH range where the DSP009 reagent is most effective (Orica Mining Chemicals, 2009). Castellon *et al* (2022) reported that the most effective range for thiocarbamate collectors like DSP009 is between a pH of 5 to 9.5. Considering changes to feed grade, pH and gangue in the BRC feed while conducting the test work, Figure 2 illustrates a relationship between recovery and rougher pulp pH. Higher recovery may have been achieved with pH modification while using DSP009 as the collector. It is recommended to conduct further test work to assess pH and BRC copper recovery as a higher pulp pH will typically assist recovery when pyrite is present and is the case with the BRC feed (Table 3, approximately 25.85 per cent).

The results illustrated through dosing the F100 dispersant at 100 g/t in Figure 2 may correlate with the clay gangue in the orebody (Talc/Kaolinite/Nontronite). Wang, Lauten and Peng (2016) concluded that F100 performed well in the presence of clays like kaolinite. Basnayaka (2018) proposed this to be due to the biopolymer increasing repulsive forces between metal and gangue particles while reducing pulp viscosity. The relationship reduced fine gangue particles (Slimes) from coating the copper particles resulting in reduced recovery and concentrate grade. Figure 2 illustrates the potential for improved concentrate grade compared or copper recovery, however, requires more test work and plant trials to provide a more substantial understanding.

The Copper Concentrator Metallurgists persistence assisted in the delivery of improved copper recovery with a decreasing feed grade, while increasing production. It is recommended a $/t processing assessment be completed to illustrate the reduction in unit costs and additional revenue. This model may then be used to review the mining cut-off grade.

CONCLUSIONS

The BRC orebody illustrates a complex mix of copper species and gangue, which make copper flotation challenging. Dilution of the sub-level cave orebody impacted feed grade resulting in the addition of Mags to the blend to support the downstream Copper Smelter's requirements, however, impacted the Auxiliary Mill's availability and the ability to reprocess the tailings into a clean silica flux. Test work and plant trials were conducted using a more selective copper collector and dispersants to assess the impact on copper recovery and concentrate grade.

The results from the trials enabled higher BRC copper recovery of plus 7 per cent to be achieved through reduced copper reporting to tailings and producing a clean silica flux for the downstream Copper Smelter. The removal of magnetics from the Auxiliary Mill feed improved circuit availability, resulting in additional BRC processing and copper concentrate being produced. This improvement equated to an additional 800 t of copper per annum based on the forecast feed grade and Auxiliary Mill throughput. The additional metal production has been complemented by an estimated A$1 m reduction in reagent cost and maintenance expenses annually (Torok, 2023). Removing Mags from the BRC feed was critical for enabling these results.

Additional test work is recommended to investigate the benefits of pH modifiers and dispersants. Lab test work and plant trials indicate positive uplift to recovery and concentrate grade. These improvements may present a favourable business case to economically mine and process potentially lower BRC feed grades.

ACKNOWLEDGEMENTS

The Author thanks the support from Mount Isa Mines to research the performance of the Black Rock Cave orebody and the personnel who assisted in gathering information referenced in this report. I must convey my gratitude to the teaching staff at the University of New South Wales and Associate Professor Seher Ata. Finally, this research was only possible with the support from the Mount Isa Mines Copper Concentrator Metallurgical team; Lucian Cloete, Justin Searle, Viv Beehan, Michael Celona, Ben Sullivan, Mai Vu and Nur Muhamad.

REFERENCES

Basnayaka, L R, 2018. Influence of Clay on Mineral Processing Techniques, PhD thesis, Curtin University, Perth.

Brindle, G, 2023. Mt Isa 2021 Surveys – BRC Circuit, Expert Process Solutions.

Castellon, C I, Toro, N, Galvez, E, Robles, P, Leiva, W H and Jeldres, R I, 2022. Froth Flotation of Chalcopyrite/Pyrite Ore: A Critical Review, *Materials*, 2022:15. Available from: https://www.ncbi.nlm.nih.gov/pmc/articles/PMC9572913/pdf/materials-15-06536.pdf [Accessed: 17 May 2023].

Gibson, 2017. Basin architecture and crustal evolution in the Paleoproterozoic Mesoproterozoic sequences of Mt Isa. Available from: https://www.researchgate.net/figure/Principal-tectonic-elements-within-the-Mount-Isa-terrain-and-the-Pb-Zn-mineral-deposits_fig1_320546385 [Accessed: 25 Feb 2023].

Muhamad, N, 2023. DSP009 AUX Mill full plant trial, internal report, May, Mount Isa Mines.

O'Donnell, R and Begelhole, J, 2023. One Plant's Trash is Another Plant's Treasure; a Synergistic Approach to Novel Uses for Tailings Streams, in Proceedings of MetPlant2023.

Orica Mining Chemicals, 2009. *Flotation Guidebook*. Available from: www.orica-miningchemicals.com [Accessed: 17 May 2023].

Torok, I A, 2023. BRC Chalcocite Processing – recovery improvements, MEng Thesis (unpublished), University of New South Wales, Sydney.

Vass, P, 2023. Mt Isa Mines DSP009 Sighter Trials, January, IXOM.

Wang, Y, Lauten, R A and Peng, Y, 2016. The effect of biopolymer dispersants on copper flotation in the presence of kaolinite, *Minerals Engineering*, 96–97:123–129. Available from: https://www.sciencedirect.com/science/article/abs/pii/S0892687516301303 [Accessed: 17 May 2023].

Optimising ball mill grinding circuits – it's not just the mill

J Zela[1] and B Cornish[2]

1. Application Engineer – Grinding Media Division, ME Elecmetal, Arequipa 04000, Perú.
   Email: jzela@me-elecmetal.pe
2. Director of Application Engineering – Grinding Media Division, ME Elecmetal, Port Douglas Qld 4877. Email: bcornish@meglobal.com

ABSTRACT

Ball milling circuit operation is complex due to the interactions between the circuit components; as such, there is little to be gained from attempting to optimise the milling or classification stages independently. Circuit optimisation is best undertaken with a holistic approach, and a well-proven system is the Functional Performance Evaluation as developed by Metcom Technologies, a framework for differentiating efficiencies that are blended in every operating ball milling circuit.

This framework offers metallurgists and operators a more insightful understanding of the mechanisms at work inside a ball mill grinding circuit and allows rapid identification of the causes of poor grinding performance. This analysis involves assessing mill power draw along with grinding and classification efficiencies, including a full review of all mineral characteristics test work and circuit operating data. ME Elecmetal has used this framework to assist customers with grinding circuit optimisation, and this paper presents three case studies where the method has been employed.

INTRODUCTION

Ball milling circuits always include several components: sump(s), pump(s), pipework, a classification system (usually cyclones), and a tumbling mill. The general operating strategy is to generate a healthy circulating load between the cyclones and the mill, so that operating efficiencies of the process stages can be optimised to reliably achieve the product 80 per cent passing grind size (P_{80}) suitable for the downstream plant processes. The circulating load creates complexities due to the interactions between the circuit components; as such, there is little to be gained from attempting to optimise the milling or classification stages independently.

Optimising a ball milling circuit involves evaluating various process stage efficiencies followed by a focus on improving the low efficiency components. These stage efficiencies can be assessed using a specific performance analysis focused on maximising mill power utilisation, classification efficiency, and rock breakage efficiency.

Electromechanical and pumping efficiencies can generally only be improved by upgrading equipment (eg: more efficient motors or optimised pump impellers). Such equipment changes are not included in this discussion as the performance analysis is only about operating conditions.

Circuit optimisation is best undertaken with a holistic approach, and a well-proven system is the Functional Performance Evaluation (FPE) as developed by Metcom Technologies (McIvor *et al*, 2017a, 2017b), a framework for differentiating efficiencies that are blended in every operating ball milling circuit.

ME Elecmetal and Metcom have been partners since 2014, and this has been leveraged at several mineral processing plants globally with ball mill circuit optimisation projects. This paper describes the Metcom FPE process and laboratory test work, along with three case studies where ME Elecmetal customers have established improved ball milling circuit performance.

FUNCTIONAL PERFORMANCE OF BALL MILLING CIRCUITS

The FPE has four elements:

1. **Ball mill power draw**; to be optimised at ~95 per cent of the installed motor capacity. Generally manipulated by the ball charge level (Jb), liner design, and mill speed.

2. **Classification system efficiency (CSE)**; manipulated by the circulating load ratio (CLR) and hydrocyclone (cyclone) operation. This should not be confused with classification efficiency which is specific to cyclone separation curves and cut points.

3. **Mill grinding efficiency (MGE)**; manipulated by the mill operating conditions such as ball size and mill density.

4. **Ore grindability**; a measure of resistance to breakage and although outside of operational control, it will affect the above elements.

These elements are used in effective optimisation of ball milling circuits. The objective is to measure the power draw, ore grindability, as well as the classification and grinding efficiencies. These elements (or factors) are then combined to obtain the overall production rate of the size of interest. More importantly, each element can be assessed to determine the opportunities for improvement, and operating conditions can be manipulated accordingly. Figure 1 summarises the Metcom FPE.

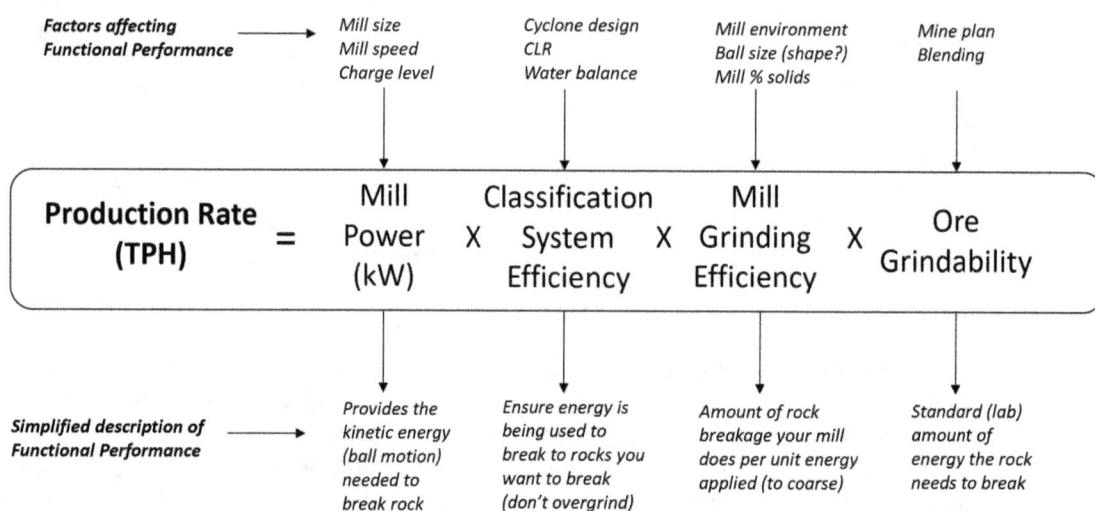

| Factors affecting Functional Performance → | Mill size Mill speed Charge level | Cyclone design CLR Water balance | Mill environment Ball size (shape?) Mill % solids | Mine plan Blending |

$$\text{Production Rate (TPH)} = \text{Mill Power (kW)} \times \text{Classification System Efficiency} \times \text{Mill Grinding Efficiency} \times \text{Ore Grindability}$$

| Simplified description of Functional Performance → | Provides the kinetic energy (ball motion) needed to break rock | Ensure energy is being used to break to rocks you want to break (don't overgrind) | Amount of rock breakage your mill does per unit energy applied (to coarse) | Standard (lab) amount of energy the rock needs to break |

FIG 1 – The Metcom Functional Performance Evaluation (FPE) equation.

The necessary data for the FPE is generally available in most grinding circuit surveys. A circuit mass balance by size fraction gives both the solids throughput of the ball mill circuit streams and the particle size distributions (PSDs). A Bond ball mill work index (BWi) test (or batch grindability test on mill feed), and the mill motor power draw (at the pinion or shell) are also needed.

The FPE elements are critical for any Plant Metallurgist to understand and will be explained in detail in the following sections.

Classification system efficiency

The CSE is the fraction of particles inside the mill that are coarser than the target P_{80}; in other words, the particles that belong in the mill because they require further grinding. This is calculated using the PSD for each of the mill feed and mill discharge samples and observing the % retained of the target P_{80} for each stream; then, for practical simplicity (so long as the circulating load is reasonably high) it is assumed that the mill contents will be the average of the feed and product (McIvor, 1989–2008).

It is critical to understand that *classification efficiency* (without the *system*) is a different efficiency. Classification efficiency describes how well an individual cyclone is classifying (or a nest of cyclones). CSE represents the bigger picture *system* efficiency that captures the net performance of the classification system in terms of achieving its function: focusing milling energy on particles requiring more grinding. For example, it is well-understood that good classifier efficiency with a low circulating load ratio is a sub-optimal operating condition since the extended ball mill residence time leads to overgrinding; further, when over-grinding has occurred, energy has been irreversibly wasted.

The limiting factor on how high classification system efficiency can go, given a cyclone classification system, is typically the density requirement of downstream recovery operations (eg flotation feed density minimum). If the cyclone overflow reports directly to a leach circuit and requires a high density, the grinding circuit CSE will be limited (to 70 per cent or less) because higher cyclone *feed* densities are required for higher *overflow* densities, and higher cyclone feed density reduces cyclone performance. Alternatively, if the cyclone overflow reports to a pre-leach thickener the cyclone circuit can be designed to use a low feed density for excellent cyclone performance. Best in class CSE values of >85 per cent have been measured in plants with low cyclone overflow density, and achievable CSE is in the range of 65–85 per cent depending on cyclone overflow density limitations (McIvor, 2014).

An example survey of a ball mill in a copper porphyry semi-autogenous ball mill crushing circuit (SABC)-A circuit resulted in the following data and Functional Performance metrics:

- Dry solids feed rate to the circuit = 500 t/h
- Dry solids feed rate to the ball mill = 1723 t/h (CLR = 1723/500 = 343 per cent)
- Mill power draw at mill shell = 12 708 kW
- Ball mill feed (cyclone underflow): 22.4 per cent -212 µm (77.6 per cent retained)
- Ball mill discharge: 51.4 per cent -212 µm (48.6 per cent retained)
- 51.4 per cent fines (-212 µm) in mill discharge, 22.4 per cent in the mill feed
- CSE = average (77.6, 48.6) = 63 per cent @ 212 µm (size of interest)
- Effective mill power: 12 708 kW × 63 per cent = 8006 kW.

As indicated above, 63 per cent is a relative low value for CSE so there is likely opportunity to improve. The effective mill power, 8006 kW, is the power being applied to the +212 µm particles, the remaining power is overgrinding fines and is wasted.

Ball mill grinding efficiency

The *grinding rate* at the size of interest (P_{80}) is the production of minus P_{80} material by each kWh of mill energy consumption. The higher the value, the more efficient the mill environment is at utilising energy to grind rock to the target size. Mill environment factors such as ball size and slurry density have strong effects on the mill grinding rate (McIvor, 2006).

The *ball mill grinding rate* at the P_{80}, can be calculated from plant survey data: specifically, ball mill feed and discharge PSDs and the mill power (McIvor, 2006). Clearly, the ball mill grinding rate is a function of not only the internal grinding environment, but also the ore's resistance to breakage (grindability). Ball MGE accounts for ore hardness by dividing the mill grinding rate (kg/kWh) by a standard measure of ore hardness, such as the Bond grindability (g/rev) of the ore. In a similar sense to Bond Work Index Efficiency benchmarking, the MGE is the ratio of the plant grinding rate to a lab mill grinding rate. A practical way of obtaining the lab mill grinding rate is described in the Metcom training program (McIvor, 1989–2008) using a plant-specific batch grindability test on the mill feed (typically cyclone underflow) and using that value in calculation of the MGE. As such, MGE is best used as a relative indication of mill circuit grinding performance at a given plant versus attempting to compare MGE values from plant to plant.

Using the same example survey as given in the CSE calculation example, we can now calculate the MGE:

- Dry solids feed rate to the circuit = 500 t/h
- Effective mill power (total power × CSE) = 8006 kW
- Ball mill (BM) CIRCUIT new feed: 32.5 per cent -212 µm
- BM CIRCUIT product: 80.0 per cent -212 µm
- New fines production rate: 500 t/h × (80.0–32.5) = 237.5 t/h new -212 µm produced by circuit
- Grinding rate: 237.5 t/h/8006 kW = 0.0297 t/kWh

- Batch grindability of mill feed: 1.26 g/rev
- MGE = 0.0297/1.26 = 0.0235 t/kWh per g/rev.

A common ball mill optimisation parameter is the mill density. Figure 2 (McIvor, 2005) shows the effect on grinding efficiency [as kWh/(g/rev)] as a function of the mill density from test work conducted at the Tilden Mine. Higher density generally results in improved grinding efficiency until slurry viscosity causes a sharp drop in grinding efficiency.

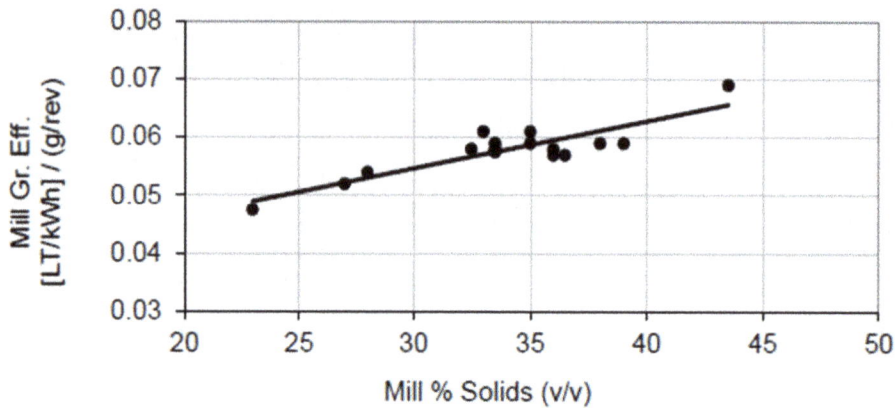

FIG 2 – Mill grinding efficiency versus % solids.

While many ball mills are operated between 45–50 per cent solids by volume, the exceptionally fine grind (P_{80} of 20 μm) at the Tilden Mine (where the Figure 2 data was generated), like most fine grinding or regrind circuits, don't allow operation at 'typical' density values. However, the trend between mill grinding efficiency and density is still clearly shown and this advantage can easily be leveraged. For example, the clustering of data points around 35 per cent solids (by vol) shows the typical operating condition at Tilden. After validating the higher efficiency at higher density, the plant took steps to improve density control and capture the increased MGE.

Cumulative grinding rate analysis, mill grinding efficiency and media size

Figure 3 shows the effect of media size on rock breakage rates (McIvor, 1989–2008) who in turn credits it to Lo and Herbst (1986). Cumulative grinding rate curves, such as those shown in Figure 3, can be generated using a batch laboratory mill with application of a first-order disappearance rate equation (Roberts, 1950). The method of how cumulative grinding rates are calculated is shown below:

$$TPH\ (coarser\ than\ size\ i)in\ mill\ discharge = TPH\ (\ coarser\ than\ size\ i)in\ mill\ feed\ x\ e^{-kt}$$

where:

k: cumulative grinding rate (minus sign due to disappearance rate)

t: specific mill energy per pass (kWh/t)

The value of 'k' for the equation above is evaluated as:

$$k_i = \ln(F_i) - \ln(D_i)t$$

where:

k: cumulative grinding rate at size i

i: size class 'i'

F_i: cumulative tons per hour of material coarser than size 'i' in mill feed

D_i: cumulative tons per hour of material coarser than size 'i' in mill discharge.

t: specific energy per pass {kWh/ton} assuming plug flow-through mill

Mathematically, the cumulative grinding rate at the size of interest is identical to the MGE. The cumulative grinding rate curves show how the ball size affects rock breakage inside a ball mill – too large a ball results in inefficient grinding of the fine fractions, too small a ball results in ineffective grinding of the coarse fraction. For example, in Figure 3 the largest ball size (50 mm) has a high coarse grinding rate (eg 10+ mm), but relatively lower grinding rate at finer sizes (eg 0.2 mm). Conversely, the cumulative grinding rate curve for the 19 mm balls show essentially zero grinding rate at 10 mm, but the highest grinding rate at 0.2 mm. Clearly, the size distribution of the ball mill circuit feed will be an important factor when considering optimal media size, and MGE is a straightforward and effective metric for media size optimisation.

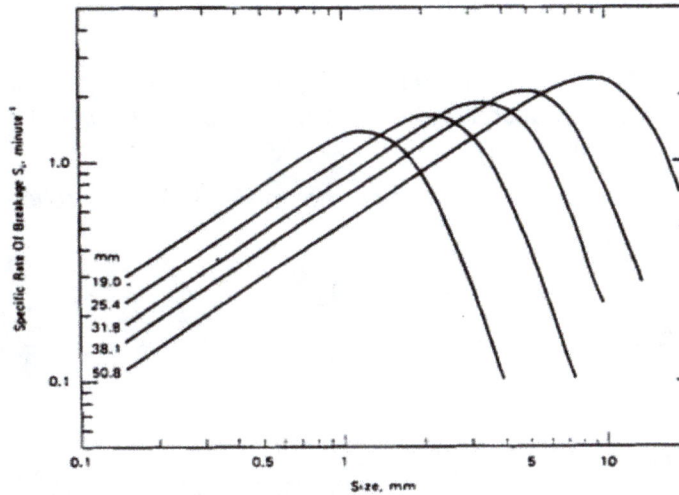

FIG 3 – Effect of media sizing on ball milling efficiency.

From a circuit survey, value of 'k' can be calculated, cumulatively, for all size classes. An example of a cumulative grinding rate curve generated from survey data is shown in Figure 4.

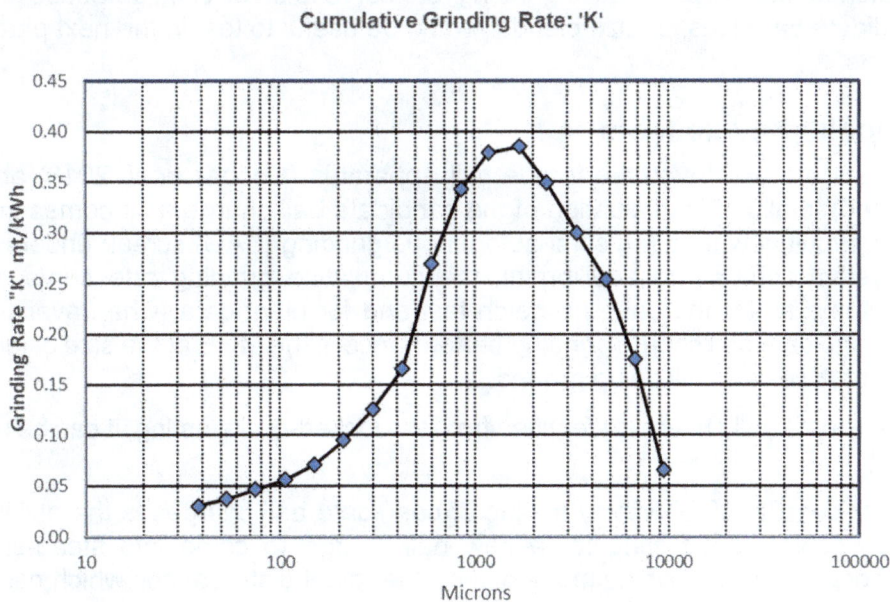

FIG 4 – Ball mill plant grinding rate curve.

The shape of these curves gives some indication about media sizing. When the curve continues upward at the right side (coarse grinding rates are high), oversized media is indicated. The objective of media size testing is to establish a media size or blend of sizes that will increase the grinding rate through the size of interest without resulting in undesirable (near-zero) coarse fraction grinding rates. If the coarse grinding fraction grinding rate is excessively low, there is a risk of accumulation of coarse particles.

METHODOLOGY

Accurately measuring grinding circuit performance at industrial scale is a difficult task requiring a scientific, yet practical, approach. Proper preparation, instrument calibration, plant sampling, sample analysis, quality control, and performance evaluation are necessary.

Metcom ball size optimisation testing uses three metrics to evaluate the performance of a ball charge:

- the FPE including the ball mill grinding rate
- coarse particle counts of pilot torque mill feed and discharge material
- cumulative grinding rate curves (from plant samples and/or the torque mill).

At many mineral process plants, survey data is readily available and can be used for these metrics. In addition, the Bond ball size equation can be used as a reference point along with a review of the global database of similar operating mills.

If a fresh baseline survey is required (always recommended), the following elements will be considered:

- laboratory screening tests
- laboratory confirmation tests
- confirmation plant survey.

Laboratory screening tests

An instrumented torque mill is used to conduct a batch test with the same ball charge as the existing plant ball charge to establish a baseline cumulative grinding rate curve, and the method and apparatus have been described in detail (Conger *et al*, 2018). This is often a more practical way of generating a reasonable cumulative grinding rate curve versus plant sampling due to most plants having inadequate ball mill discharge sample quality for this type of analysis. A comparison of the shape of the batch torque mill cumulative grinding rate curve to an existing database provides insight into which candidate ball sizes, or size blends, would be useful to test in the next phase: laboratory locked-cycle tests.

Laboratory locked-cycle tests

Locked-cycle tests are used with the torque mill apparatus (Conger *et al*, 2018) and a classifier (typically a screen) to allow for evaluation of the candidate ball charge as it comes into equilibrium with the ball mill circuit new feed (eg semi-autogenous grinding (SAG) screen undersize). Once the locked-cycle system comes into equilibrium, the cumulative grinding rate curve and the MGE analysis is the same as for the baseline batch test and for plant data (when available). The final evaluation of a candidate ball charge grinding performance is the MGE at the size of interest, so long as the coarsest particles are still being ground.

Once a good candidate ball charge is identified from the locked-cycle testing, it can be recommended for plant evaluation.

One of the major benefits of laboratory testing of candidate ball charges is the ability to avoid the long times required in plant testing for a new ball charge to come into stead-state (months). Additionally, laboratory testing avoids many of the other plant disturbances which can cloud results when trying to understand the individual effect of media size on MGE.

Plant validation testing

The final step in the media size improvement process is a follow-up survey to establish the new cumulative grinding rate curve and measure the new FPE with the new MGE after a process change (eg cyclone water balance, media size, mill density). Note, as previously mentioned, following a change in media charging practice (ball size or blend of sizes), several months need to pass before conducting the follow-up survey to allow the ball charge in the plant mill to establish a new equilibrium. The exact amount of time required depends on the media consumption rate, and there are calculators available to assess time to equilibrium.

Work index efficiency

$$\text{Work Index Efficiency} = WItest/WIoperating$$

Measured circuit efficiencies cluster around the *standard* of 100 per cent which is typical of many rod and ball milling circuits. However, efficiency values can vary from 50 per cent to 140 per cent (for some well-optimised ball milling circuits). Work Index Efficiency is a valuable metric, both as a global standard for comparison of the performance of any comminution circuit, and also for monitoring overall circuit efficiency increases as plant improvement steps are conducted. However, it cannot isolate the root causes of good or poor efficiency; rather it serves as a global efficiency benchmark. Additionally, due to the errors involved in sampling and laboratory characterisation of the BWi, relatively large changes in efficiency (>10 per cent) are needed for a BWi Efficiency change to be considered measurable.

OPTIMISATION

Ball mill density

Ball mill density has a well-known influence on MGE. In most plants the density can be measured directly at the mill discharge, but occasionally trommel spray water can affect the sample. If the water cannot be stopped during the sampling, then the best estimate of mill density is the cyclone underflow stream so long as there is no water addition to the mill feed.

Classification system

CSE is a net measure of how well the classification system (pumps and cyclones) is at maintaining coarse material in the ball mill. Ball milling energy consumed on grinding particles already at or less than the P_{80} target is wasted, and with an optimised classification system, this energy is now available for grinding coarse particles larger than the P_{80} target.

For example, after a survey, CSE was 69 per cent; meaning that 69 per cent of the mill contents were of target P_{80} size (+300 µm) and thus 31 per cent of the mill contents were already finer than the circuit P_{80} (300 µm). CSE is primarily a function of water addition to the cyclone feed, the CLR, and the classification efficiency of the cyclones. The CLR was high at 510 per cent for this example which means there is no significant opportunity to improve CSE via an increase in CLR. Note that there is often confusion in the industry about the difference between design and operating CLR; there are practitioners who claim that a high CLR is 'bad', especially with the experience of an excursion in the plant well beyond the design CLR. It is also true that when circuit operating changes result in increased efficiency (eg an improved ball size) the CLR will decrease if no other changes are made. This does not mean that a lower CLR is better, but rather is an indication that the circuit is now capable of increasing tonnage or grinding finer with a corresponding return to design CLR. Designing a ball mill circuit for low CLR results in longer than needed mill retention times, overgrinding, wasted energy, and higher operating costs.

In this example, a CSE of 69 per cent could be improved with increased water addition to the cyclone feed and subsequent adjustment to the cyclone geometry to maintain design CLR. This would result in lower density to flotation, but it is well established that lower cyclone feed density increases CSE by increasing the cyclone separation sharpness and reducing the bypass fraction.

CASE STUDIES

The detail of three case studies is presented in this section.

Case 1: 20' and 16' ball mills

This plant has several ball milling circuits operating in parallel with direct feed configuration; two circuits (BM1A and BM2C; 16.5 ft and 20 ft diameter respectively) with an incumbent ball charge consisting of a 50:50 blend of 3:3.5 inch ball size are the focus of this evaluation. A simplified plant flow sheet is shown in Figure 5.

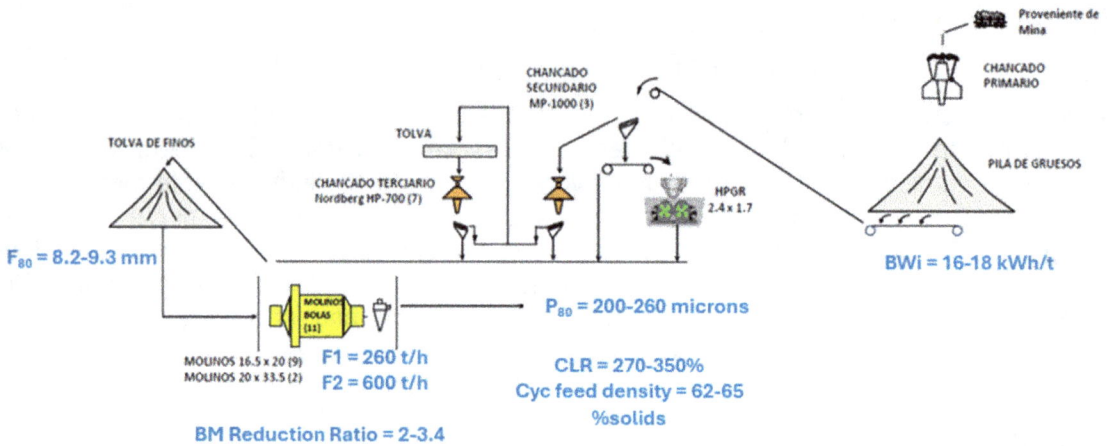

FIG 5 – Plant flow sheet.

The first lab screening test used the incumbent ball blend of 3:3.5 inch on a sample of ball mill feed (cyclone u/f). The torque mill cumulative grinding rate curves have been combined with circuit survey data are shown in Figure 6.

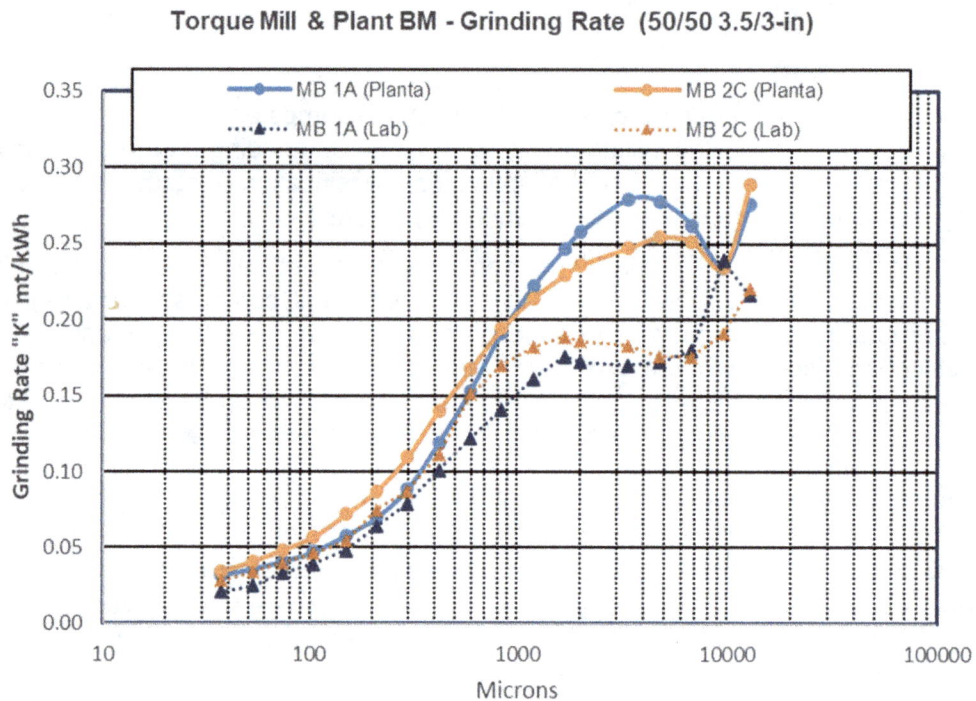

FIG 6 – Cumulative grinding rate in laboratory and plant.

The chart shows the torque mill and plant data cumulative grinding rate curves for circuits 1A and 2C are similar in shape, and that the highest rates are at the coarse end of the curve. While there is often little mass in the very coarsest sieves in this type of laboratory testing, in general the curves are trending toward higher cumulative grinding rates in the coarse region indicating the likelihood that that the average top media size is larger than necessary.

The particle counts on the coarsest sieves of +19 mm and +16 mm was low (<20) but despite this, it is still important to flag that it may be necessary to maintain a small portion of top size (3.5 in) to prevent any accumulation of coarse particles in the circuit. However, it is clear the percentage of top-size media can be reduced which opens up mill volume for smaller balls which are more efficient for the majority of particles in the mill.

The interpretation of these cumulative grinding rates is that the grinding media is likely oversized and mill grinding rates can be likely be optimised by moving to a smaller (average) ball diameter. As such, a plant trial of 3 inch media was tested in the plant. After the purge period was complete, a repeat of the circuit surveys was conducted, and the new plant grinding rate curves can be directly compared to the 3:3.5 in blend. It is fair to expect lower coarse fraction grinding rates and higher rates of the size of interest.

The BWi of the ore was given as 17.9 kWh/t. The operating work index (OWi) of Line 1A was 19.2 kWh/t for a work index efficiency of 94 per cent (17.9/19.2), while Line 2C was 102 per cent (17.9/17.5).

BWi efficiency depends on accurate measurements of F_{80}, P_{80}, mill power, and circuit feed rate. The compounding errors expected in these measurements can make a difference of <10 per cent, so it is fair to say that Lines 1A and 2C are performing equivalently from a BWi efficiency perspective.

From a benchmarking perspective, *average* circuit performance is BWi = OWi, or 100 per cent work index efficiency, and this is essentially where the circuits are operating. This indicates that grinding performance is acceptable, but there is some room for improvement. Bond analysis is limited to pointing out the presence of opportunities, but without specificity. However, the torque mill test gives a strong indication that overall circuit efficiency could be improved with a media size change.

Functional performance analysis (FPA) complements the Bond analysis by identifying the biggest improvement opportunities: the classification system (CLR, fines bypass etc) and the mill grinding environment (ball size and mill density). Table 1 shows the FPA of circuits 1A and 2C from the data provided.

TABLE 1

Functional performance analysis, (Survey data 9-Nov-21).

Parameter	Units	Survey Results	
		BM 1A	BM 2C
New -210µm Production	t/h	399	154
Classification System Efficiency (CSE)	% (at 210µm)	78.2	71.8
Effective Mill Power	kW	4767	1734
Grinding Rate of +210µm	t/kWh	0.084	0.089
Mill Grinding Efficiency (MGE) at 210µm	t/h per g/rev	0.040	0.042
Funtional Performance Efficiency (FPE)	t/h per kW per g/rev	0.031	0.030

These values, especially the MGE, can be used for comparison to any future surveys if media size is changed. CSE is >70 per cent for both circuits which aligns with expectations for the downstream density target.

Line 1A slightly outperformed 2C in CSE with a lower cyclone feed density and higher CLR which are the likely sources of the improvement. Any changes to the circuit which allow for lower cyclone feed density and higher CLR (within reason) would take further advantage of this effect.

Line 1A also had a slightly higher mill grinding rate than 2C which is attributed to the higher ball mill density on Line 1A on the day of sampling. Torque mill testing at the same density did not show any difference in mill grinding rate between Line 1A and 2C samples.

Since CSE and mill density are healthy, and the BWi efficiency is average, the opportunity appears to be in media sizing (moving smaller on average).

Leveraging the Bond media size estimate results in a ball size of 2.7 in (70 mm) based on the plant survey data for both circuits.

The Bond (1958) guideline, one of several similar media sizing estimation tools, is shown below:

$$B = \left(\frac{F}{K}\right)^{0.5} \left(\frac{Sg \; x \; Wi}{100 \; x \; \%CS \; x \; D^{0.5}}\right)^{0.34}$$

where:

F	= 9285 µm
Sg	= 2.69
Wi	= 17.95 kWh/t
%CS	= 0.75
D	= 20 ft
K	= 350
B	= 2.7 in

Based on the FPA, the torque mill test work (cumulative grinding rate curve analysis), and the Bond formulas, a plant trial using the new ball size of 3 in was conducted. Figure 7 shows the improvement of the P_{80} from *262 µm to 244 µm* with the transition from 3.5 in to 3:3.5 in to 3 in ball sizes. Ore hardness is very consistent in this plant and ball mill power maximised.

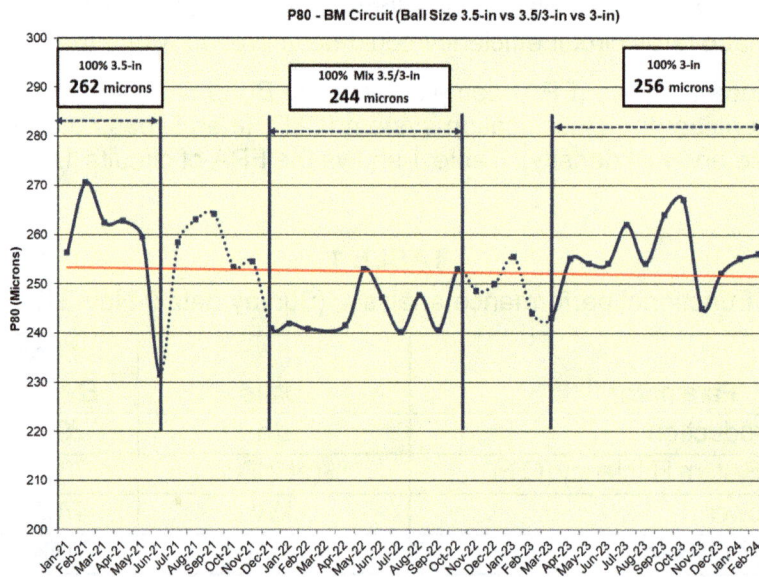

FIG 7 – P_{80} data and trendline.

Figure 8 shows the media consumption trend in terms of g/kWh.

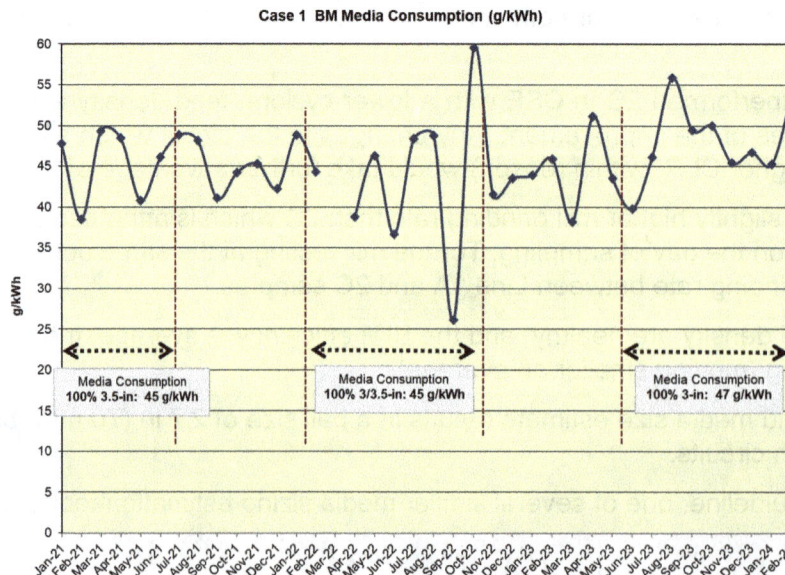

FIG 8 – Media consumption g/kWh.

As can be expected, there is a slight increase in the consumption of grinding media with a reduction in ball size.

Case 2: 28 × 44 ball mills (operation 1)

This plant has a grinding circuit featuring a 40 ft SAG mill and two parallel 28 ft diameter ball mills in standard reverse feed configuration, with an incumbent ball size of 80 mm. The circuit configuration as shown in Figure 9 is SABC-2 with secondary crushed pebbles directed to ball mill feed. The Axb is 42.

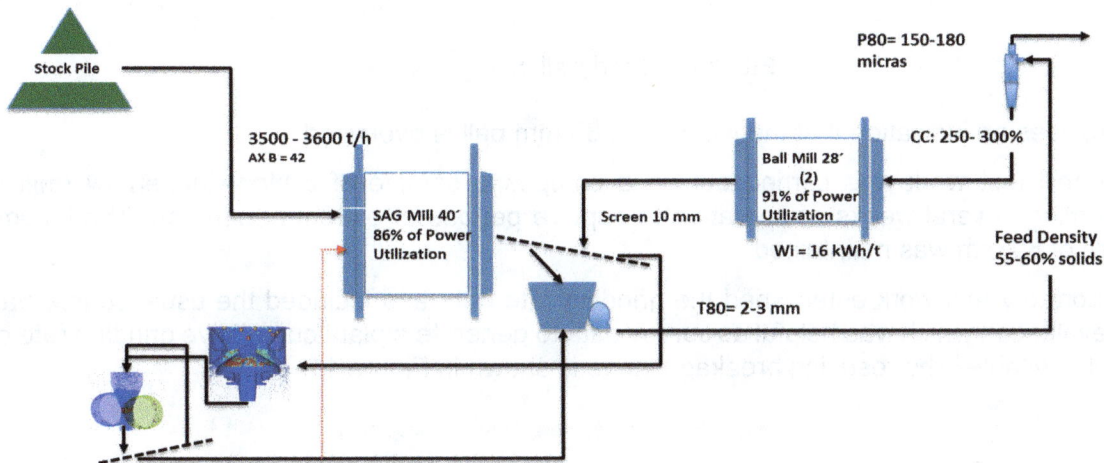

FIG 9 – Grinding circuit flow sheet.

Analysis of the plant operating data is summarised as follows:

- SAG mill feed = 3681 t/h

- BM1 and BM2 power draw = 18.5 + 19.8 = 38.2 MW

- BM combined specific energy consumption (SEC) = 38.2/3681 = 10.4 kWh/t

There was no circuit survey data available for this operation, so operating data was used for the FPE.

Ball Mill Power: The installed combined motor power is 44 MW, and thus 38.2 MW is 87 per cent utilisation. This represents a simple yet significant opportunity to increase to the standard 95 per cent utilisation which will increase the mill grinding rate to enable a higher feed rate or finer grind.

CSE: Operating data indicates that the CLR is in the 250–300 per cent range, and thus there is an opportunity for improvement to target the 350–450 per cent range through a pump and cyclone system adjustments.

Additionally in this case study, the process control strategy was to maintain the pump box level using the cyclone feed pump speed; this introduces unnecessary disturbances into the circuit. A more effective strategy is to control the cyclone feed pressure with water addition to minimise density, and this adjustment has been implemented.

MGE: This is a measure of how effective the mill power is at generating the target P_{80}, and when this improves, the production rate of the size of interest increases for a given mill power.

Industrial benchmarking indicates that ball size can likely be improved, so the Bond ball size guideline and torque mill test work was used to prove that this is the best opportunity for ball mill grinding efficiency improvement.

The Bond ball sizing equation (Bond, 1958) was used as shown in Figure 10 and estimates a 55 mm ball size.

Bond Formula for calculating the largest make-up ball size

$$B = \left(\frac{F}{K}\right)^{0.5} \left(\frac{Sg \quad WI}{100 \quad \%Cs \quad D^{0.5}}\right)^{0.34}$$

F	7500	µm	
Sg	2.72		
WI	15.8	kwh/st	17.4 kWh/mt
%Cs	0.77		11.3-11.4 RPM
D	27.3	ft	assume 4 inch liner thickness
K	350		
B	2.164	inch	
	55.0	mm	

where,

B = Largest make-up ball diameter (inches)
F = Circuit feed 80% passing size, K80 (microns)
Sg = Specific gravity of the ore
WI = Bond ball mill laboratory work index of the ore (kwh/st)
%Cs = Mill speed (as a fraction of the mill critical speed)
D = Mill inside diameter (ft)
K = Constant, 350 for wet overflow mills
330 for wet diaphragm mills (grate discharge)
335 for dry diaphragm mills

FIG 10 – Bond ball size guideline.

This provides an indication that the incumbent 80 mm ball is oversized.

Torque mill test work was carried out on a composite sample of cyclone underflow (mill feed) representing several weeks of operation to capture periods of extreme operation. The incumbent ball size of 80 mm was maintained.

The laboratory test work established the grinding rate data and included the usual coarse particle count evaluation which was helpful as survey data to generate a plant cumulative grinding rate curve was not available. The resulting breakage curve is shown in Figure 11.

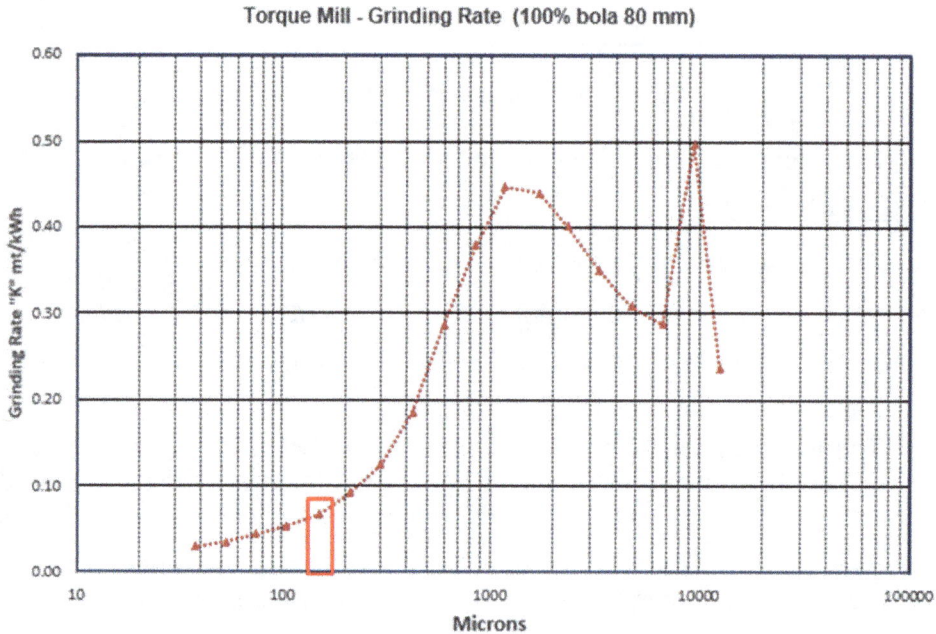

FIG 11 – Cumulative grinding rate.

The coarse particle counts indicated that the 80 mm ball was more than capable of breaking the coarsest particles. Also, the grinding rate at the target P_{80} (150 µm) was 0.068 t/kWh, and this could likely be improved with a smaller ball size without causing an unmanageable reduction in the coarse particle breakage rate.

After establishing the data analysis and the breakage curve of the incumbent 80 mm ball charge using data from the torque mill test, a locked-cycle test with 65 mm balls was recommended, but the plant elected to conduct a plant trial using 65 mm balls instead. The resulting cyclone performance parameters for each ball mill are shown in Figures 12 and 13. Like the previous case, ore hardness, throughput, and ball mill power do not vary significantly at this operation so the measured P_{80} is a good indicator of efficiency changes.

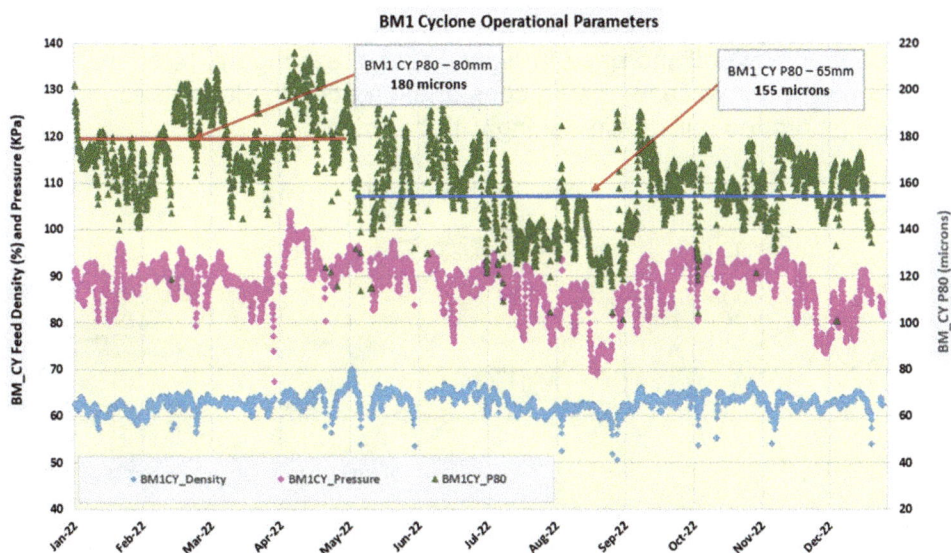

FIG 12 – BM1 cyclone operational parameters.

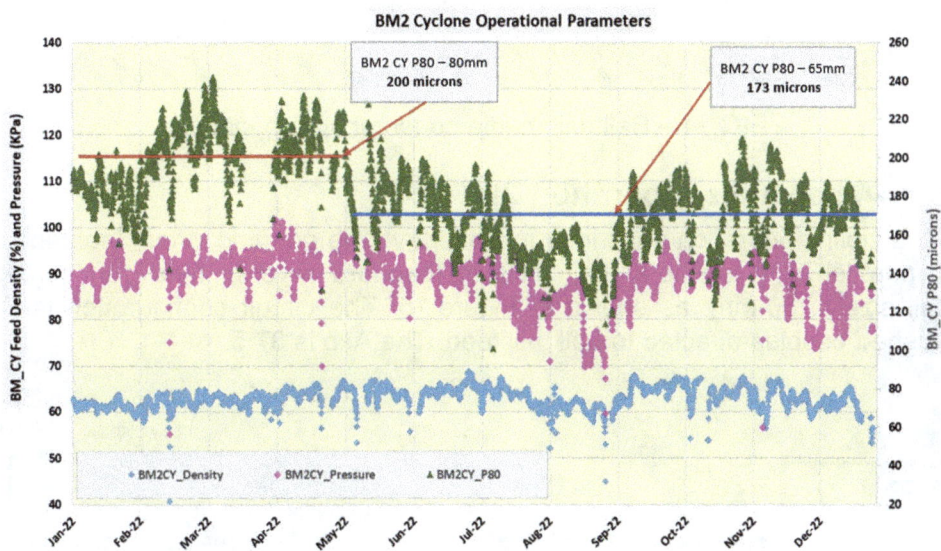

FIG 13 – BM2 cyclone operational parameters.

The average BM1 cyclone overflow P_{80} during the Baseline period is 180 µm and transitions to 155 µm after the introduction of the 65 mm ball size.

The cyclone feed pressure and density are similar between the two periods and these two parameters represent further improvement opportunities to generate an even higher grinding rate of the target P_{80}, as it is known that a lower cyclone feed density and higher pressure may help to reduce the P_{80} further by increasing CSE.

The average BM2 cyclone overflow P_{80} during the baseline period is 200 µm and transitions to 173 µm after introduction of the 65 mm ball size. Further, due to a bias with the introduction of crushed pebbles into the ball mill circuit feed sump, it is known that BM2 cyclone overflow P_{80} tends to be coarser than BM1. While it may seem risky to reduce ball size when dealing with coarse ball mill circuit feeds, so long as the smaller ball (65 mm) is breaking the coarsest rock (as indicated by torque mill particle counts), the greater quantity of balls (impacts) in a 65 mm ball charge versus an 80 mm ball charge is likely to prove up to the task. Operationally, monitoring of ball mill scats gives feedback on the adequacy of the ball size as the ball charge equilibrates toward 65 mm.

As with the BM1 circuit, the BM2 cyclone feed pressure and density are similar between the two periods and these two parameters also represent further improvement opportunities.

It is also important to evaluate the media consumption, as it is expected to increase as the top ball size decreases due to the overall increase in ball surface area; Figure 14 shows the consumption trends in terms of g/kWh. In this case, media consumption actually decreased, but this was due to the supplier change (improved ball quality) along with ball size.

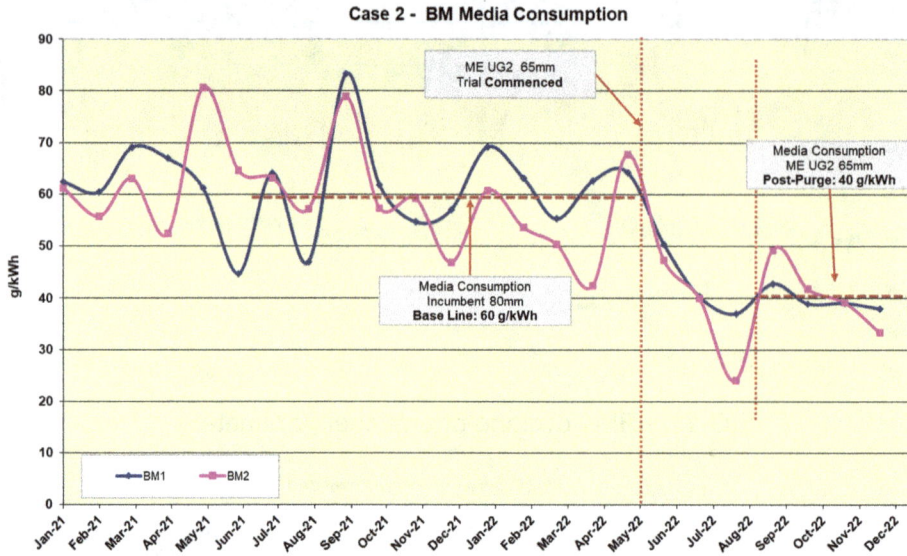

FIG 14 – Ball mill media consumption (g/kWh).

Case 3: 28 × 44 ball mills (operation 2)

This plant is the sibling operation described in Case 2 with a grinding circuit featuring a 40 ft SAG mill and two parallel 28 ft diameter ball mills in standard reverse feed configuration, with an incumbent ball size of 80 mm; as shown in Figure 15. The circuit configuration is SABC-2 with secondary crushed pebbles directed to ball mill feed. The Axb is 37.5.

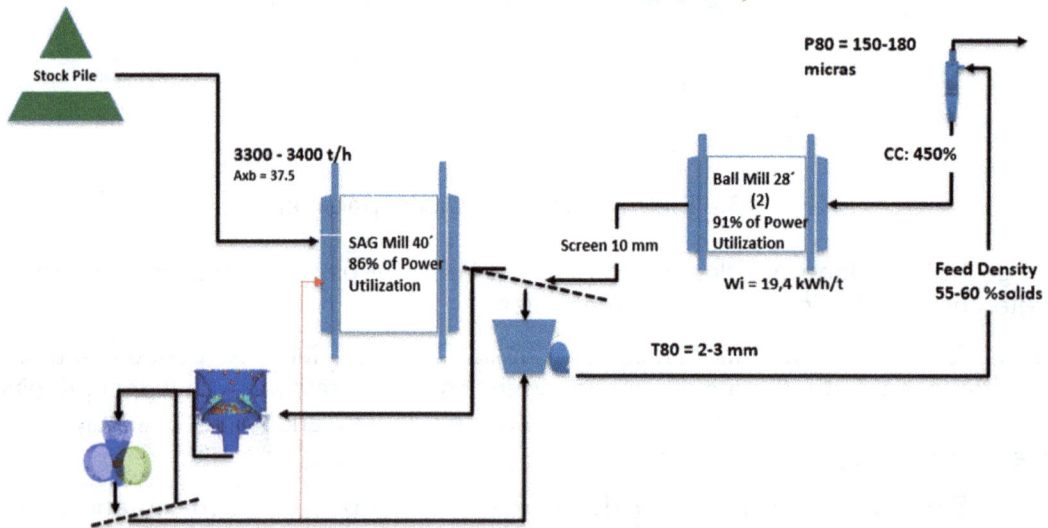

FIG 15 – Grinding circuit flow sheet.

Analysis of the plant operating data is summarised as follows:

- SAG mill feed = 3468 t/h

- BM1 and BM2 power draw = 19.1 + 19.7 = 38.8 MW

- BM combined SEC = 38.8/3468 = 11.2 kWh/t

Ball Mill Power: The installed combined motor power is 44 MW, and thus 38.8 MW is 88 per cent utilisation. As before, maintaining a target of ~21 MW for each ball mill will increase the mill grinding rate to enable a higher feed rate or finer grind.

CSE: Operating data indicates that the CLR is 450 per cent, and at the top end of the standard target range; this tends to minimise mill retention time and the chance of over-grinding. In this case study, the process control strategy was already configured to control the cyclone feed pressure with water addition to minimise density. CSE was estimated at 87 per cent, above the benchmark of 80 per cent.

MGE: This is a measure of how effective the mill power is at generating the target P_{80}, and when this improves, the production rate of the size of interest increases for a given mill power.

Plant survey data was available to use the BWi to estimate the work index efficiency.

$$W = 38\ 800 \text{ kW} / 3468 \text{ t/h} = 11.2 \text{ kWh/t}$$

$$Wio = \frac{11.19}{\left[\dfrac{10}{\sqrt{145}} - \dfrac{10}{\sqrt{2500\ (est)}}\right]} = 17.7\frac{kWh}{mt}$$

$$Work\ Index\ Efficiency = \frac{17.4\dfrac{kWh}{mt}}{17.7\dfrac{kWh}{mt}} = 98\%$$

When using BWi analysis for SAG milling operations it is important to remember that SAG mills, via abrasion, produce a larger a portion of the ball milling circuit P_{80} sized material compared to the rod-ball mill circuit upon which the method was developed. As such, the Work Index Efficiency of the ball mill circuit is lower than 98 per cent because it is necessary to credit the SAG mill for the extra fines it produced. A high performance ball milling circuit in an SABC plant should be >120 per cent, so the Work Index Efficiency represents a significant opportunity for improvement.

The Bond ball size equation also has 55 mm as a recommended ball size.

The same process of torque mill test work was carried out on a composite sample of cyclone underflow (mill feed) representing several weeks of operation to capture periods of extreme operation. The incumbent ball size of 80 mm was maintained.

The test work established the grinding rate data and included the usual coarse particle count evaluation. The resulting breakage curve is shown in Figure 16.

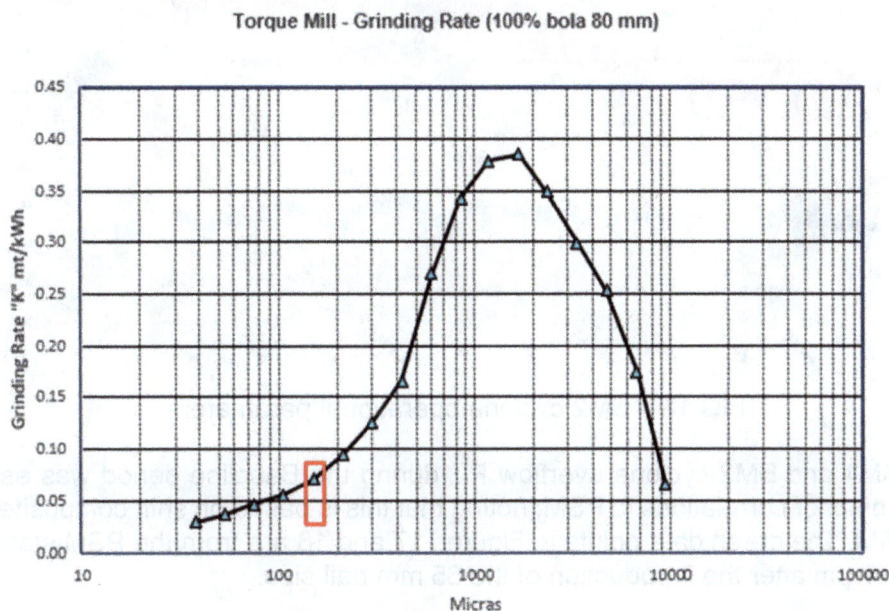

FIG 16 – Torque mill cumulative grinding rate.

The coarse particle counts were even less than Case 2 and the grinding rate at the target P_{80} (150 µm) was 0.071 t/kWh, and this could likely be improved significantly with a smaller ball size without affecting the coarse particle breakage rate.

After establishing the data analysis and the rock breakage curve from the torque mill test, a plant trial using 65 mm balls commenced for both ball mills. While 65 mm may not be optimum (as would be determined by a locked-cycle test program), the plant elected to conduct a plant trial at 65 mm while monitoring for operational indications of undersize media, such as coarse rock scats as previously described. This case study demonstrates the reality that occasionally the most useful operating data is not always available; in this case the online particle size monitor (PSM) was out of service until the last month of analysis. However, other cyclone performance parameters for each ball mill are shown in Figures 17 and 18.

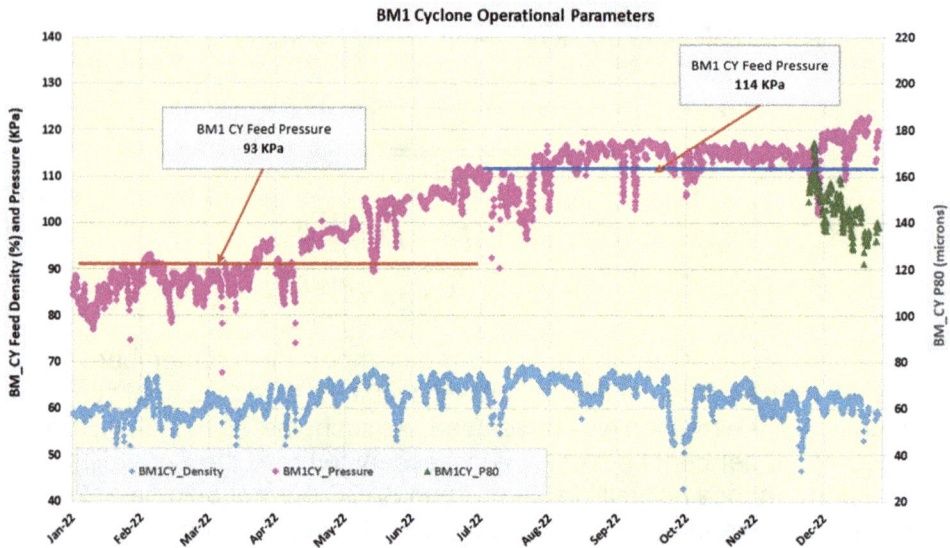

FIG 17 – BM1 cyclone operational parameters.

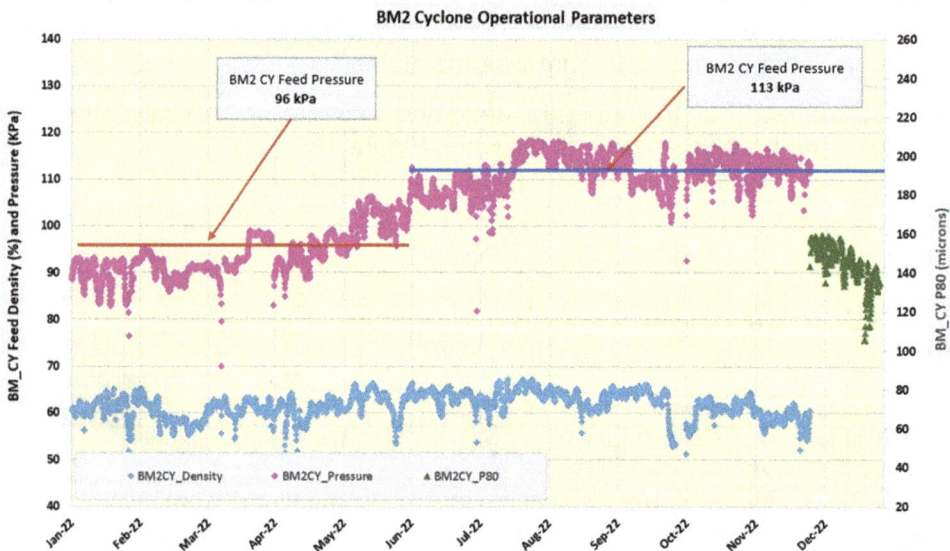

FIG 18 – BM2 cyclone operational parameters.

The average BM1 and BM2 cyclone overflow P_{80} during the Baseline period was estimated to be 180 µm based on prior correlations to PSM; noting that this is based on shift composite samples due to the faulty PSM. The green data points in Figures 17 and 18 are from the PSM starting in Dec-22 and average 147 µm after the introduction of the 65 mm ball size.

The smaller ball size has resulted in an increased grinding rate and an increased MGE; this has in turn enabled a decrease in the CLR and an increase in the cyclone feed pressure from 93 kPa to

114 kPa through using less cyclones. Mill density is consistent at 63 per cent solids between the two periods.

The BM2 cyclone performance trends are very similar to BM1.

In this case study, the transition of ball size and supplier was convoluted due to the variety of ball stocks available on-site, and the subsequent desire to maintain housekeeping standards. As such, despite the steady cyclone operational trends, the media consumption trends were highly variable as shown in Figure 19.

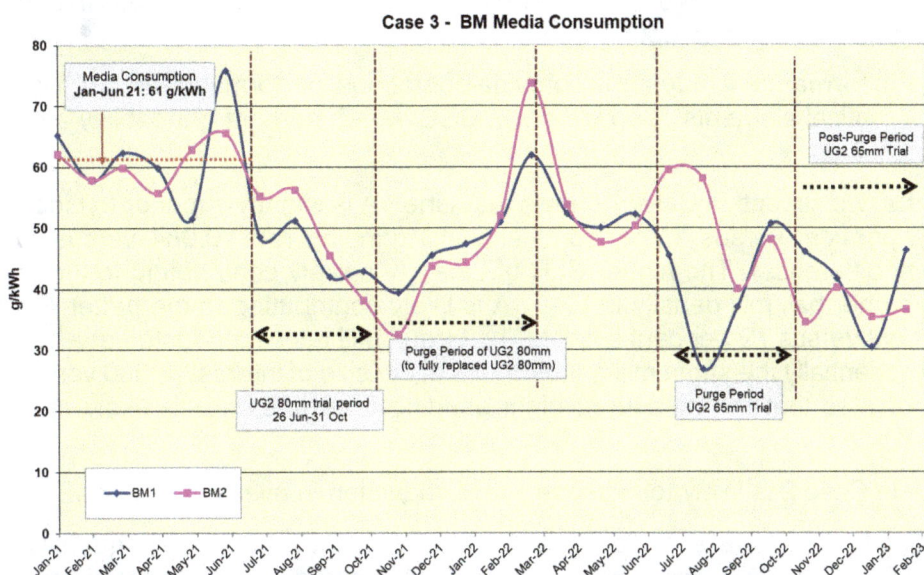

FIG 19 – Ball mill media consumption (g/kWh).

Since making a full transition in 2022, the ball mills in both case studies have maintained supply of 65 mm ball size with excellent operational and media performance.

CONCLUSIONS

The **Bond Work Index Efficiency** provides a good guide for improvement opportunities:

- In Case 1, the two circuits evaluated (1A and 2C) are essentially identical (94 per cent and 102 per cent respectively). This analysis indicates that there is no significant difference in grinding circuit performance between these circuits.

- In Case 2 and Case 3, the BWi efficiency calculations show average performance of ~98 per cent which indicates an opportunity for improvement. Survey data (if conducted) for Case 2 would be beneficial for a similar study as in Case 1.

The **Classification System Efficiency** provides specific insight regarding operation of the cyclones including the CLR and possibility of unnecessary grinding and wasted energy:

- In Case 1, plant data indicates a CSE of 75–78 per cent for the current cyclone feed density. Thus, an improved CSE is possible if the cyclone feed density (and resulting overflow) could be reduced. Such a change requires consideration of the downstream processing constraints.

- In Case 2 there is a potential opportunity for CSE improvement through cyclone specification (vortex finder and apex dimensions) adjustments and an improved control strategy (controlling the cyclone feed density rather than the pump box level).

- In Case 3 the CSE appears to be very high (85 per cent), however the high CLR contributes to this and there are frequent indications of the ball mill circuit becoming overloaded.

The **Cumulative Grinding Rate Curves** as determined from torque mill test work are extremely useful in establishing the performance of the incumbent ball size(s):

- In Case 1 there is a strong indication that the 1:1 blend of 3.5:3.0 inch balls is oversized for the application. An increase in MGE (and thus an increase in throughput rate) is likely with a smaller average ball size blend. This was true for the torque mill tests and the plant data.

- Comparing the Case 2 and Case 3 grinding rate curves to a similar SABC-2 locked-cycle test work program shows promise for 65 mm balls or a blend of 80 and 65 mm balls.

- In any case, it is recommended to conduct follow-up surveys (after completion of the purge period) to allow comparison of the new plant cumulative grinding rate curve to the curves shown in these studies. It is often the case that the best opportunity for ball mill circuit efficiency improvement is in ball size optimisation.

A **Functional Performance Analysis** will provide operational performance insight for any ball milling circuit. The four critical elements of power draw, CSE, MGE, and ore grindability are all included in the analysis.

- FPA of the two circuits in Case 1 shows that Line 1A is slightly superior to Line 2C in terms of CSE (78 per cent versus 75 per cent) and mill grinding rate (0.084 versus 0.080 t/kWh, a 5 per cent difference). The higher CLR of Line 1A is likely contributing to the improved CSE, and the higher ball mill density in Line 1A is likely contributing to the better mill grinding rate (82 per cent versus 79 per cent solids). The torque mill tests (conducted at a constant density) showed essentially the same mill grinding rate at the size of interest (0.060 versus 0.061 t/kWh) which indicates that the measured plant grinding rate difference is likely related to density control.

- The FPA in Case 2 is likely to increase with a reduction in average ball size.

- In a SAG mill circuit (Case 2 and Case 3) it is important to consider the quantity of fines being generated by the SAG mill, and this can be estimated and accounted for based on the throughput rate and ball charge level. It is well known that increased production of fines in the SAG mill by abrasion will reduce the load on the ball mills (assuming an adequate CSE). A complete SAG and ball mill circuit survey would help to understand this and is recommended, although it is always challenging to obtain a representative sample of SAG mill discharge.

- It will generally always be beneficial to the CSE to operate with maximum water addition to the cyclone feed stream. However, care must be taken not to exceed the volume constraints of the downstream processes (flotation, tailings pumping capacity etc).

Overall, Functional Performance Analysis and cumulative grinding rates analysis have provided useful tools for assisting customers in ball mill circuit optimisation of classification systems and media sizing. This is especially valid considering the real-world limitations of quality data needed for reliable computer modelling and subsequent trial-and-error optimisation. The strong linkage and understanding between FPA metrics and conceptual understanding of the interactions in ball mill circuits provides a practical framework for process improvement in a time-efficient manner.

ACKNOWLEDGEMENTS

The authors would like to thank those who collected and published the data used in this analysis. Conversations with Nelver Benavides, Alex Doll and Kyle Bartholomew were integral in the development of some of the ideas expressed in this paper. Finally, thanks to ME Elecmetal for providing the impetus and opportunity to publish this paper.

REFERENCES

Bond, F C, 1958. Grinding Ball Size Selection, The American Institute of Mining, Metallurgical, and Petroleum Engineers (AIME).

Conger, W, DuPont, J-F, McIvor, R E and Weldum, T P, 2018. Advanced in Industrial Ball Mill Media Sizing, SME Annual Meeting, Minneapolis, Minnesota, USA.

Lo, Y C and Herbst, J A, 1986. Considerations of ball size effect in population balance approach to mill scale-up, in *Advances in mineral processing: A Half-century of progress in application of theory to practice* (ed: P Somasundaran), Proceedings of a Symposium Honoring Nathaniel Armb, pp 33–47.

McIvor, R E, 1989–2008. The Complete Metcom Training Program on Improving the Performance of Plant Grinding Operations, Metcom Technologies Inc.

McIvor, R E, 2005. Industrial Validation of the Functional Performance Equation: A Breakthrough Tool for Improving Plant Grinding Performance, preprint no. 05-31, SME Annual Meeting.

McIvor, R E, 2006. Industrial validation of the functional performance equation for ball milling and pebble milling circuits, *Mining Engineering*, Nov, pp 47–51.

McIvor, R E, 2014. Plant performance improvements using grinding circuit 'classification system efficiency', *Mining Engineering*, 66(9):72–76.

McIvor, R E, Bartholomew, K M, Arafat, O A and Finch, J A, 2017a. Use of Functional Performance Models to Increase Plant Grinding Efficiency, in Proceedings of Canadian Mineral Processors Annual Meeting, Ottawa.

McIvor, R E, Bartholomew, K M, Arafat, O A and Finch, J A, 2017b. Ball mill classification system optimization through functional performance modelling, *Mining Engineering*.

Roberts, E J, 1950. The probability theory of wet ball milling and its application, *Transactions of the American Institute of Mining, Metallurgical, and Petroleum Engineers*, 187:1267–1272.

Sustainability

Plant design comparison of dry VRM milling plus magnetic separation with AG and ball milling plus magnetic separation for Grange Resources' Southdown ore

D David[1], D Olwagen[2], C Gerold[3], C Schmitz[4], S Baaken[5], C Stanton[6] and M Everitt[7]

1. FAusIMM(CP), Senior Consultant, Wood, Perth WA 6000. Email: dean.david@woodplc.com
2. Quality and Process Improvement Manager Grange Resources, Burnie Tas 7320.
   Email: dian.olwagen@grangeresources.com.au
3. Senior Manager, Ore and Minerals Technology, Loesche GmbH, 40549 Dusseldorf Germany.
   Email: carsten.gerold@loesche.com
4. Senior Process Engineer, Ore and Minerals Technology, Loesche GmbH, 40549 Dusseldorf
   Germany. Email: christian.schmitz@loesche.com
5. Managing Director, Loesche Australia Pty Ltd, Perth WA 6000.
   Email: stefan.baaken@loesche.com
6. Process Engineer, Grange Resources, Burnie Tas 7320.
   Email: chris.stanton@grangeresources.com.au
7. Geology Manager, Grange Resources, Perth WA 6000.
   Email: michael.everitt@grangeresources.com.au

ABSTRACT

Loesche's Vertical Roller Mill (VRM) has achieved superior pilot plant comminution outcomes when treating Grange Resources' Southdown ore. The Southdown magnetite deposit is a competent and abrasive ore from southern Western Australia, situated in an area with poor access to power and water. From the mid-2000s to the mid-2010s it was extensively evaluated using pilot scale conventional autogenous grinding (AG) and ball milling coupled with sequential tailings rejection by magnetic separation. Recent pilot VRM testing has demonstrated several significant advantages over these wet milling circuits and also over dry milling alternatives, such as the use of high pressure grinding rolls (HPGRs) in series.

An industrial VRM can accept open-circuit secondary crushed feed, make a single-pass P_{80} 20 µm product, operate dry, remove significant moisture from damp feed, selectively liberate hard minerals and produce an ideally classified final product. VRMs are dust free in operation, the throughput and product size are readily controllable, they are simple to maintain, and they utilise the same hydrostatic breakage mechanism that results in HPGR power efficiency enhanced by a shear breakage component. When treating magnetite ores a grit fraction, which is an internal dry classifier oversize stream, in case of the Southdown project, sized -1 mm +85 µm, is intercepted and extracted for dry magnetic separation. Magnetic grits are returned to the VRM for additional grinding and coarse dry non-magnetic grits are rejected to tailings. The final dry product is also suitable for emerging fine dry magnetic separation that would further reduce water consumption. Another major advantage of the technology is that the comminution and classification systems have a 100-year pedigree, having been developed by Loesche in tandem. Industrial design is achieved with high confidence through Loesche's extensive testing and operating database.

Loesche's 'VRM/grit magnetic separation pilot plant' is a novel and unique facility that has demonstrated, in the Southdown test work, to be up to 41 per cent more energy efficient than pilot AG/Mag/Ball milling. In this paper VRM milling is compared with wet milling and, to a lesser extent, HPGR alternatives in areas such as test work, scaleup, capital expenditure (CAPEX), operating expenditure (OPEX), operability, design and plant layout. Special attention is given to the power and water savings, which are 'game-changing' for Southdown.

INTRODUCTION

The Southdown Project is being developed by Grange Resources and involves the construction and operation of an open pit magnetite mine located approximately 90 km east-north-east of Albany. The project location is 10 km south-west of the locality of Wellstead in the Great Southern region of Western Australia. Southdown has been under development since the early 2000s and has been the subject of various feasibility studies. The 2012 Definitive Feasibility Study (DFS) incorporated a processing plant with a capacity to produce 10 Mt/a of high-grade concentrate. The plant was designed by AMEC (now Wood) after a program of bench and pilot testing. The 2012 DFS was revisited by Wood in 2021 and an updated Pre-Feasibility Study (PFS) was completed in early 2022 for a smaller plant, producing 5 Mt/a high-grade magnetite concentrate.

The deposit itself is approximately 12 km in length with 6 km of this included in the current study. It contains over 1.2 billion tons of high-quality mineral resources, including 388 million tons (Mt) of ore reserves (Grange Resources, 2024).

The deposit is estimated to have a life-of-mine in excess of 30 years with a mass recovery to final product in the 34–36 per cent range. The site, illustrated in Figure 1, is located in farmland and requires the complete establishment of all facilities and infrastructure for the mining and on-site processing of the magnetite ore. The project aims to develop an open pit mine and processing plant to treat about 15 Mt/a of run-of-mine (ROM) feed material from which 5 Mt/a of high-grade magnetite concentrate will be produced. The processing plant will be adjacent to the mine and the concentrate will be pumped via an overland slurry pipeline to the filtering, storage and ship loading facility located at the port of Albany. A water pipeline transports recovered filtrate water, together with new water from other sources, back to the concentrator for use in processing.

FIG 1 – Greenfields exploration drill rigs.

The project is challenged by the lack of water and electricity. As there is no large volume water supply in the region, the original DFS (2012) proposed the desalination of sea water. The closest suitable capacity electricity grid is 280 km away in the Collie region, close to Bunbury (Figure 2). In part due to these utility supply and cost restrictions, the target production rate of magnetite concentrate was reduced from 10 Mt/a down to 5 Mt/a. Despite this, power and water remain high-cost commodities and decision drivers for the project.

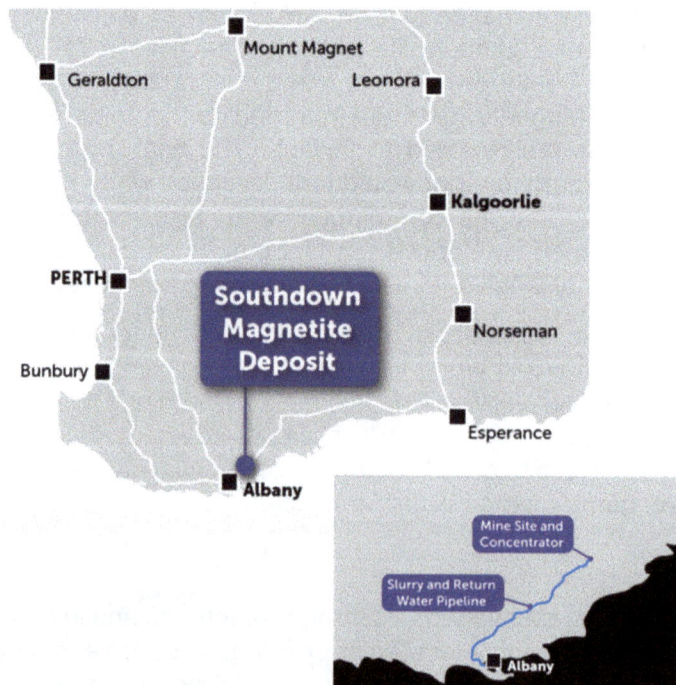

FIG 2 – Project location in Western Australia.

A BRIEF INTRODUCTION TO THE SOUTHDOWN FLOW SHEET

Southdown project design constraints

Until the update of the PFS in 2022, the Southdown flow sheet consisted of a 'traditional' autogenous and ball mill grinding circuit (AG/Ball Circuit) with wet magnetic separation stages between each comminution step. The original production rate of 10 Mt/a of concentrate at P_{80} 45 µm required two parallel AG/Ball grinding circuits. Reducing the production capability to 5 Mt/a removed the need for one of the processing lines.

The AG mill is fed primary crushed ore and the product is screened at 3 mm, with undersize sent to rougher magnetic separation (RMS). Screen oversize is returned to the AG mill feed along with a pebble stream. The -3 mm magnetics is pumped to the ball mill and the -3 mm non-mags discarded as waste. The Ball milled product, at P_{80} 85 µm, is further concentrated within an intermediate magnetic separation stage (IMS). The magnetics from IMS are sent to sulfide flotation to remove pyrrhotite and then on to regrinding and cleaner magnetic separation (CMS). The magnetite concentrate is thickened before being pumped to Albany for filtration and ship loading.

In this circuit there was little opportunity to reduce front end water use without significant additional capital equipment, such as dry stack tailings filtration. The wet RMS stage produced a large volume of dilute and coarse non-magnetics waste that required cyclone classification to produce a coarse slurry and thickening of the overflow to produce a fine slurry. The cyclone underflow was combined with thickener underflow for pumping to a tailings storage facility. Thus, the circuit had a heavy reliance on water due to unacceptable losses to tailings storage. Opportunities to reduce the reliance of the flow sheet on water were proposed and this commenced the investigation into dry milling in 2019.

After encouraging results from an exploratory set of open circuit VRM tests (Stage 1 of the dry milling test work, 2020 and 2021), the 2022 PFS proposed replacing the single line of AG and ball milling (described above) with two lines of dry Loesche Vertical Roller Mills (VRM). Loesche were in the final stages of constructing a new pilot unit, incorporating dry magnetic separation in closed circuit with the VRM, at their facility in Dusseldorf, Germany, at approximately the time the PFS was completed. The pilot plant allowed Loesche to reject dry and coarse non-magnetic waste material and improve the mill efficiency, similar to the RMS stage in the AG/Ball mill circuit. In conjunction with projected power savings offered by the VRM technology, the ability of the new mill configuration to create a coarse dry tailings stream was predicted to provide a material incentive to incorporate it in the Southdown flow sheet. Testing on Loesche's closed circuit test facility in 2022 comprised Phase 2 of the dry milling program.

BRIEF INTRODUCTION TO VERTICAL ROLLER MILLS (VRM)

VRM technology is one of the oldest grinding technologies known. Developed in antiquity, it is a simple technology, and machines could be constructed from common materials. With thousands of VRM installations, dry compressive comminution technology is now a standard in the production of cement, preparing coal for power stations and grinding for various industrial mineral applications. VRM technology is arguably the compressive comminution method with the widest application globally, when all materials needing to be ground are considered.

The VRM in its modern form was invented by Loesche 1928 for the grinding of German hard coal for coal- fired powerplants. In the 1960s VRMs were used for the first time to grind cement raw material, a mixture of mainly limestone with quartz and iron ore, which is the feed to the cement clinker oven. Mill capacities are available from pilot plant scale up to 1200 t/h. Due to their higher energy efficiency compared to Ball Mills, VRMs have become established for milling hard to grind cement clinker since the 1990s. To save costs and minimise CO_2, abrasive blast furnace slags are ground in VRMs and used as a substitute for clinker. Today VRMs are standard for grinding clinker and slags with capacities ranging from 50 to 400 t/h. Loesche alone has sold more than 420 VRMs for this application.

Although limited, there are industrial applications of VRMs for grinding highly abrasive hard rock. Two such Loesche applications are both for VRMs grinding phosphate ores. The first example treats ore with Bond Ball Mill Work index (BBWi) values in excess of 25 kWh/t in South Africa (Jacobs *et al*, 2016) and the second treats BBWi 18.6 kWh/t ore in Kazakhstan (Stapelmann, Gerold and Smith, 2019). The South African application was the first, almost 20 years ago producing up to 800 t/h and provided many valuable lessons for Loesche in regard to managing wear and improving design. Other references for ore grinding, and associated difficult applications, are VRMs treating Manganese ore, copper matte and tin slag.

The first application in the Copper- and Nickel-industry will be the BHP West Musgrave project where two parallel VRM lines will grind 12 Mt/a ROM to flotation feed size. Each line has a capacity of up to 830 t/h (Weidenbach and Gerber, 2023).

Loesche has about 2300 VRM units installed globally at the time of writing.

All Loesche VRM units incorporate a rotating grinding table, grinding rollers which are pulled down on the grinding table via a hydro dynamic spring system and a high efficiency dynamic classifier mounted on top of the mill, as shown in Figure 3. The set-up requires a constant gas flow-through the machine and is a self-contained and compact combination of grinding, classifying, and drying.

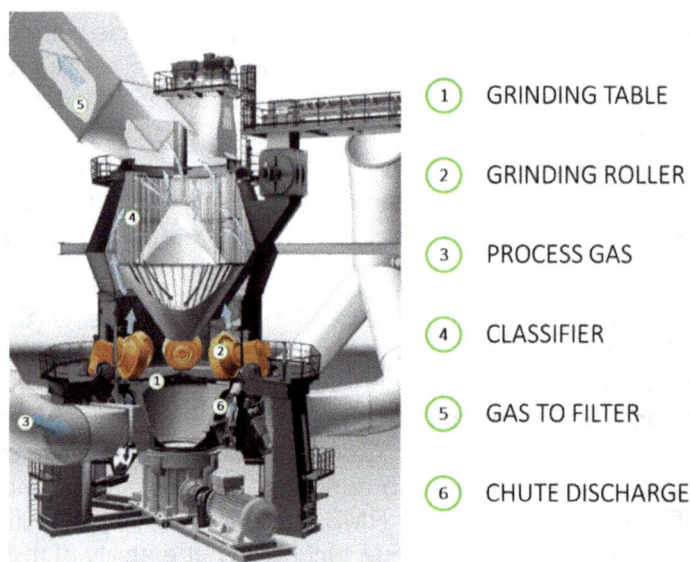

1. GRINDING TABLE
2. GRINDING ROLLER
3. PROCESS GAS
4. CLASSIFIER
5. GAS TO FILTER
6. CHUTE DISCHARGE

FIG 3 – Schematic Loesche vertical roller mill.

Operating principles

Accepting feed sizes up to F_{100} 150 mm and achieving products down to P_{80} 20 µm in a single unit, the Loesche VRM, illustrated in Figure 3, contains all elements necessary to take the feed to the final product.

After tramp-metal removal, material to be ground is fed into the mill and directed to the centre of the grinding table (1). By centrifugal force it moves outwards towards the edge of the rotating table and into the path of the grinding rollers (2). Under compression in the grinding gap the particles are firstly broken and displaced into existing voids within the material bed before they start to break under hydrostatic pressure. This process is known as compressive in-bed comminution. Breakage is induced mainly by compression, supported by small amounts of shear, resulting in optimum size reduction, high energy efficiency, and minimum wear. The in-bed comminution principle is that through the multiple contact points pressure is applied on all sides of the particle and there is a high probability for crack initiation on the mineral boundaries and at other weaknesses. This often results in an improved degree of mineral liberation when compared to crushing and conventional tumbling milling, as has been reported in the literature by Van Drunick, Gerold and Palm (2010), Altun *et al* (2015), Reichert *et al* (2015) and Jacobs *et al* (2016).

The ground material escapes the confines of the compressive breakage region and continues its movement under centrifugal force to the edge of the table. All particles fall into the process gas stream (3) where finer particles are levitated upwards to the classifier (4). The rotational speed of the classifier, delivering centripetal force to the outside, combines with the inward directed speed of the gas flow to define the cut size of the classifier. Product-sized material passes through the classifier vanes and leaves the mill with the gas flow (5) to be collected in the bag-house filter. Coarse material rejected by the classifier can be partly or completely discharged from the mill (henceforth termed as grit), as is done in the Southdown pilot plant test work. However, in most VRM installations the grits are not removed and are guided by the grit cone to the centre of the grinding table for further comminution.

Rapid high energy grinding followed by classification produces a narrow particle size distribution which results in reduced amounts of ultra-fines. A small amount of very coarse material, so-called reject, is not levitated at the edge of the grinding table and it falls through the upward moving gas into a chute discharge (6) where it is removed. The rejects are not lost as they are recycled to the mill feed by the reject conveying system.

A simplified process and air flow for VRM milling (not including Grit extraction) is illustrated in Figure 4.

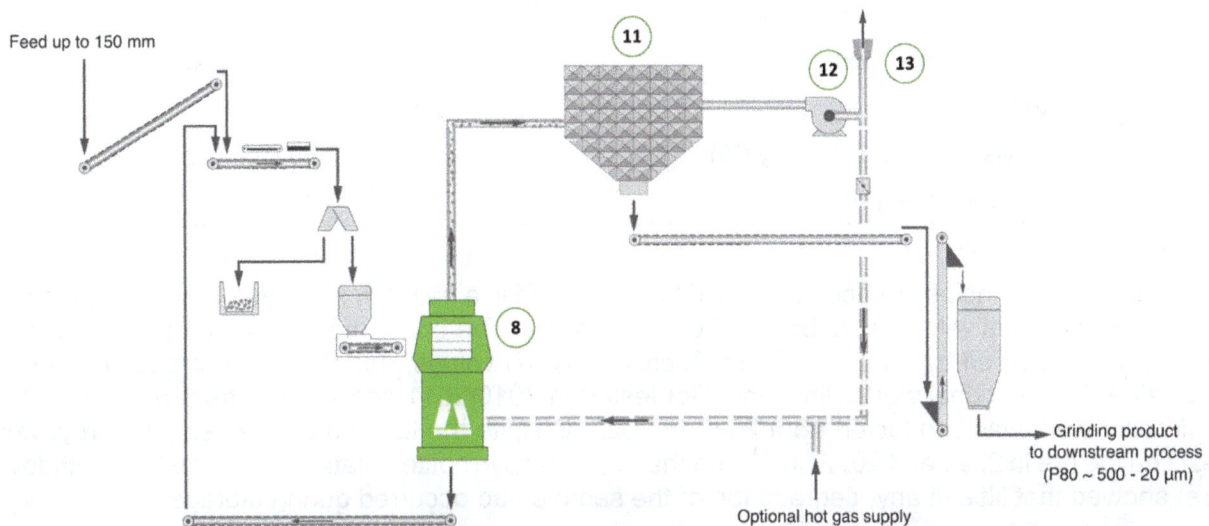

FIG 4 – Simplified VRM flow sheet with main air pathway through mill and classifier (8), filter bag house (11), fan (12) and stack (13).

Although this flowsheet represents the configuration of an operational industrial unit, it is also the functional description of the Loesche open circuit testing unit, used for more than 10 000 tests. The wide application of this testing unit is described elsewhere (Schmitz and Gerold, 2019).

TEST MATERIAL AND GEOMETALLURGY

Drilling was done in 2010 and some cores were drilled at 150 mm diameter and some at 83 mm (PQ) size. These drill samples have been used for all the testing across more than a decade.

Typical core produced by drilling is shown in Figure 5.

FIG 5 – 150 mm diamond core for metallurgical test work.

The competent nature of this core caused comminution difficulty and was part of the reason it remained useful for metallurgical test work over such an extended period. The drilled mass from this program was more than 40 t. The set of properties illustrated in Table 1 are typical of the Southdown ore.

TABLE 1

Comminution properties of Southdown iron ore.

Test	Measurement		
	kWh/t	g/cm³	%
DWi (drop-weight index)	5.8		
CWi (crushing (impact) work index)	12.0		
RWi (rod mill work index)	13.1		
BWi (ball mill work index)	18.2		
Ai (abrasion index)			0.40
SG (g/cm³)		3.52	
Davis Tube Test recovery (%)			36.0
Sulfur content (%)			0.40

Source: David *et al* (2023).

Southdown ore samples are abrasive and difficult to grind in a ball mill. However, the large grain size makes the ore relatively easy to break at crushing sizes. The ore is more than one third magnetite as directly measured by the Davis Tube Recovery (DTR) values, and this is the cause of the high SG value. Conventional wet milling was pilot tested in 2010/2011 and again a few years later. The VRM pilot work was conducted from 2020 to 2022. Despite the long time between pilot programs, tests on the ore in 2021 and 2022 (including the best ever pyrrhotite flotation response for Southdown ore) showed that little, if any, degradation of the sample had occurred during storage.

TEST WORK

Summary results

Within the different feasibility stages of the Southdown Magnetite project, evaluations of conventional and new comminution processes, including circuits incorporating HPGRs and VRMs have been carried out. The wet (AG/Ball) milling circuit and the dry VRM circuit test programs are discussed in detail by David *et al* (2023).

The HPGR test work results were generated before Wood's involvement in Southdown. The results have been assessed but found to be problematic in some areas.

- The tests were performed on a different ore sample to the one used in the work reported here

- Coarse 'dry' magnetic separation is performed on all ore (-32+0 mm) and has been assumed to operate with the same effectiveness in the plant (with naturally damp ore) as it did on perfectly dry ore in the laboratory

- Comminution values reported in the HPGR work were significantly lower than those determined on the ore samples in the work reported here. The BBWi values reported in Metso's HPGR pilot testing and design document for the test composite (Metso Minerals (Australia), 2008) are in the range 13 to 15 kWh/t and the DWi value was measured at 4.9 kWh/m³. However, the variability work conducted by both Metso and Wood indicated the BBWi values are consistently between 17 and 19 kWh/t and in Wood's SMC variability database 53 out of 66 samples have DWi >5 kWh/m³.

While an HPGR comparison would be a strong addition to the report, using the available data is likely to provide a misleading comparison.

AG/Ball mill pilot test work conducted at ALS (then ALS Ammtec) used their 1.6 m diameter pilot AG mill closed with a 1 mm screen. AG product fed a wet low intensity magnetic separation (LIMS) unit then a 7.5 kW ball mill closed with a 150 μm screen was used to grind the LIMS mags stream. The product generated typically had a P_{80} of 78 μm.

Loesche installed a continuously operable closed circuit pilot plant in their test centre in Neuss, Germany. It has the facility to extract and process grits from the VRM then return the magnetic grits to the mill for further comminution. The non-magnetic fraction is a coarse and dry grits waste stream. The final P_{80} 85 μm VRM baghouse product contained the magnetite, as illustrated in Figure 6.

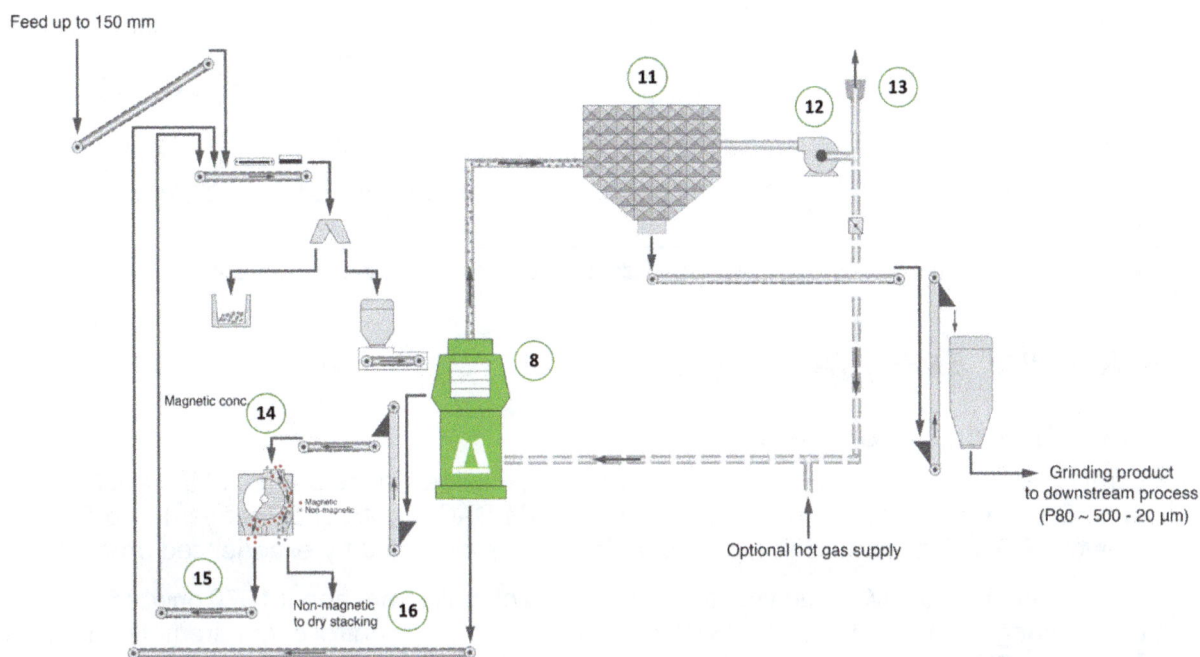

FIG 6 – Continuous VRM and grit magnetic separation pilot plant flow sheet.

This is similar in all ways to Figure 4, except that the grits (14) are magnetically separated, the barren waste (16) disposed of and the magnetics (15) returned to VRM feed.

In grit extraction and processing mode the VRM achieves a 41 per cent reduction in power compared to the AG/Ball mill circuit and reduces the mass to be processed and ground in the final IMS stages of the flow sheet by 5.5 per cent. Each circuit achieves the same Magnetite recovery. After grinding to P_{80} 85 μm the VRM product is slurried, wet magnetically separated to produce IMS concentrate which is subjected to flotation to remove pyrrhotite. The low sulfur flotation product is ground to approximately P_{80} 40 μm before cleaner magnetic separation generates the final magnetite concentrate.

The greater power benefits achieved when the VRM is operated in grit extraction mode confirmed the benefits estimated from open circuit VRM milling, confirmed the promise offered of the grit circuit and confirmed that VRM technology has the attributes to shift project economics in a positive direction.

A general benefit of VRM pilot testing is that the products generated are very close in all respects to the products that will be generated in an industrial circuit. This is a major advantage when conducting downstream test work such as magnetic separation, flotation and solid/liquid separation. In comparison the AG/Ball mill pilot plant did not produce streams consistent with industrial expectations as it was operated with screens where the plant would use cyclones and this leads to non-representative stream size distributions and separation responses. Corrections have been incorporated into the AG/Ball mill circuit design to compensate and allow reasonable scale-up outcomes.

The VRM Pilot results and the AG/Ball pilot results (performed on the same feed sample) have both been scaled to design their respective industrial flow sheets. Key calculated performance measures for these flow sheets are compared in Table 2.

TABLE 2

VRM versus AG and ball mill grinding to P_{80} 85 µm with integrated non-mag waste rejection.

Operating Parameter	AG/Ball		VRM
AG SE* (100% of feed)	9.84 kWh/t	Comminution SE	4.5 kWh/t
Ball mill SE	13.7 kWh/t	Fan, classifier etc	6.8 kWh/t
Mass rejected as coarse non-mag waste	32% of feed	Mass rejected as coarse non-mag waste	35% of feed
Classification and internal conveyors 5% allowance	+1.7 kWh/t	Internal conveyors and mag separation	+1 kWh/t
Total duty power	20.8 kWh/t	Total duty power	12.3 kWh/t
Mass rejected by IMS	27.7% of feed	Mass rejected by IMS	27.1% of feed
Mass of IMS concentrate for regrinding	40.0% of feed	Mass of IMS concentrate for regrinding	37.9% of feed

*SE is Specific Energy, the net power (exclusive of drive losses) required to grind one tonne of ore to the equipment product size.

OPERATIONAL EXPECTATIONS

AG/ball mill circuit in operation

The operating experience of AG/Ball mill circuits is well described in the papers of the AusIMM's Mill Operators' Conference series and especially in the CIM SAG Conference series. In addition, the intricacies of SAG, AG and ball mill circuit operations are understood by experienced practitioners.

SAG and AG circuits took over much of the milling duties globally from about 1970 onward. The main operational change that faced operators was the need to exercise control over parameters that were dependent upon the ore properties and not upon the selected equipment. Crushers and ball mills typically operate with consistent available power and, as long as the ore does not vary in its general nature (as long as it is essentially rock rather than clay, for example) the circuit output is predictable and steady. The power delivered by crushers and ball mills only deteriorates slowly over time as the liners and grinding media wear away.

Conversely, the operation of SAG and AG mills is highly dependent upon the ore properties, especially its competence. Before ore blending was introduced at Grange Resources Savage River mine (Tasmania), the AG mill specific energy would vary from a low of 3 kWh/t to a high of 14 kWh/t as the ore competence changed from shift to shift. AG mill variability directly resulted in ball mill feed rate and size distribution variability. This, in turn led to problematic variability in concentrate size distribution, magnetite production rate and magnetite concentrate quality. It was also demonstrated that both high and low AG mill specific energy periods resulted in significant production capacity losses.

The introduction of in-pit blending resulted in AG mill SE variability being reduced to a range typically between 5 and 8 kWh/t, and this eliminated much of the problematic plant operation. It also allowed an increase in production rate and provided a much more consistent feed to the company-owned pellet plant at Pt Latta.

While the Savage River example is extreme, most AG and SAG circuits experience problematic variations in mill performance resulting from ore competence changes, rock type changes and even ore SG changes. Many of these milling circuits feed flotation banks, which are highly sensitive to grind P_{80} and residence times. Again, flotation feed stream variations are damaging to short-term performance and long-term profitability. Flotation operators that have experienced the stability of crush and ball mill circuits often have difficulty accepting that AG and SAG based grinding has been a positive advancement.

Operating an AG mill circuit in a magnetite plant requires that the mill power be monitored closely and the feed rate varied in response to changes. Normally the aim is to ensure that the AG power is as high as it can be in the circumstances, the pebble stream is within normal range and discharge screening is not problematic. In larger circuits it is also necessary to split the AG discharge screen undersize between ball mills (two, three or even four) and ensure that these mills are drawing similar powers and making similar product size distributions.

When the competence of the feed rises the rock charge in the mill starts to increase, as does the power draw. Left uncontrolled the mill would quickly overload with respect to power or, worse still, overload with rocks and have spillage from the feed and/or discharge. To avoid this outcome the feed rate is reduced, but there is often a lag where the power continues to climb. Some operators will avoid bad consequences by cutting feed rate dramatically. Others will avoid problems by running the feed rate perpetually low so that the mill is always 'safe'. The typical outcome is varying throughput and varying loads on the downstream mills.

In a modern plant much of the expected variation is controlled within limits by machine learning algorithms but given that many of the measurements used infer important mill properties, experienced operators remain indispensable. In many other plants, much of the feed variation is removed by secondary crushing the ore, and this works by creating a BAG (Barely Autogenous Grinding) milling environment. A closer analysis of such a circuit, with secondary crushing and pebble crushing, is that it is another form of the stable three-stage crush and ball mill arrangement.

The introduction of AG and SAG milling has brought with it a seemingly endless search for operational stability for the sake of achieving long lost optimal concentration conditions.

VRM circuit in operation

A detailed list of operational changes, expectations and benefits has been described in an extended abstract (David *et al*, 2024). Many of these aspects are described here in the context of this test work and associated design.

A significant advantage of VRM milling is that it is a return to the relatively stable performance expected of crush and ball mill circuits. In addition, the air classification, grinding force and milling power control system allows product size and throughput targets to be selected and maintained with minimal operator intervention.

The first difference compared to a crush and ball mill circuit is that The VRM can receive a much coarser feed. Three stage crushing for ball milling needs to produce a P_{80} in the range 12 to 15 mm (depending on the ore hardness) and, for high throughput circuits, this requires at least three crushing stages, the last two stages with many units operating in parallel. The VRM can receive up to 150 mm particles in feed (with soft ores) and 85 to 100 mm particles in feed with hard ores. Even hard ores can be prepared to these sizes with a single secondary crusher per VRM line operating in open circuit.

AG and SAG mills (but not BAG mills) only require a primary crusher and will typically supply particles as large as 300 mm to stockpile and mill feed conveyors. The addition of an open circuit secondary crusher to a primary crusher is relatively simple and it provides a significant benefit in that only 150 mm particles (about 13 per cent of the mass of a 300 mm particle) need to be stockpiled, fed, transferred and conveyed.

This means that the first major difference compared to AG or SAG milling is an additional crusher to maintain, but there is less to maintain in the materials handling equipment.

The second major difference is the internal grinding classifying circuit achieved by combining comminution and classification in one machine. 90 per cent of the ore are fed to the mill with up to 150 mm and ground to final product size. A 10 per cent coarse fraction falling against the process gas flow, called reject, is discharged underneath the mill and mechanically recirculated to the mill feed.

Based on the required plant capacity two parallel grinding VRM lines producing at a rate of close to 1000 t/h per VRM line were chosen. In combination with a fine product storage this allows to carry out maintenance on one VRM line without limiting the feed to the downstream process. This is less than can be achieved with large AG/SAG circuits but more than can be accommodated by most crush and ball mill circuits, certainly with a single ball mill. In other industries Loesche already installed bigger mills. Transferring these mill capacities to the Southdown project would result in achieving capacities of up to 2000 t/h per mill.

A VRM line will also process a lower throughput than can be accommodated by a single large HPGR, but in such an arrangement the HPGR does not achieve a fine grinding outcome. Fortescue's Iron Bridge circuit (Fortescue Metals Group (FMG), 2019) is the first commercial approach of dry fine HPGR grinding in Iron Ore. The Iron Bridge circuit is the only one of its kind and, in the most recent quarterly report (FMG, 2024) it was still in commissioning. Multiple fine grind HPGRs appear to be required, each in combination with its own air classifier, to achieve a grinding outcome that competes with VRMs and crush/ball mill circuits.

In the case of the Southdown circuit the VRM plant operator will also be presented with a grit extraction and processing circuit. The VRM classifier grit (in case of the Southdown project 75 μm up to 1 mm) will be extracted from the classifier recirculation, conveyed to dry magnetic separators and separated into mags and non-mags. The mags will be black and up to 80 per cent magnetite. The non-mags will be a whitish brown and less than 0.3 per cent magnetite. Most of the non-mags grit is liberated silicates which can be discarded as waste. As grits are produced by air classification, all the product sized material has been removed from it and it is essentially dust free. So while dry magnetic separation is being conducted it will need minimal, if any dust extraction and will not require continuous heavy-duty cleaning effort.

Mags return to the VRM feed conveyor and >300 t/h of non-mags (35 to 40 per cent of the new feed) is conveyed to tailings management. The only other VRM output seen by the operator is the baghouse product, about 700 t/h of enriched magnetite ore at P_{80} 85 μm.

The VRM product size distribution will be continuously monitored and controlled to the required size target. If, for example, 100 μm is found to be a more appropriate P_{80} target then it is a simple matter to request this of the control system. Adjustments will be made to the airflows, the classifier rotor speed, the roller working pressures (if necessary) the table speed and also to the feed rate, which should increase in this instance. Naturally the ability to benefit from this product size change will be determined by the combined effect on IMS, flotation and CMS separations and the availability of downstream grinding power to manage a larger flow of coarser product. Generally, operation of a VRM is much less problematic than a AG or SAG mill and the effect of control changes in a closed circuit VRM are virtually instant. In a VRM circuit with external grit processing the effect of changes will take a little longer, but the control system will manage the lags and rapidly achieve steady state.

The production rate from the VRM will vary if the magnetite content of the feed varies significantly. A low-grade feed should allow a greater rejection of non-magnetic grits and will produce less final product. Similar issues will arise if the liberation status of the feed changes significantly. There are also expected to be variations in throughput if the ore grindability varies, but currently the dependence on grindability measures, such as Bond Ball Mill Work index (BBWi), has not been explored for Southdown. However, it is known that there is low variability in Southdown BBWi values with the majority being in the range 17 to 20 kWh/t. It will be a function of the mining method and blending practices as to how steady the feed grade can be held and all samples to date have exhibited the coarse grain size shown in Figure 5.

All the variability factors noted above will have a much greater influence on the operation of an AG/SAG plus ball mill circuit than they will have on a VRM. The consistent stream of VRM product will be processed in the IMS, which has high tolerance to feed variability, and then to flotation which has a lower tolerance. It is expected that these parts of the wet circuit will operate with a high degree of stability and consistency of outcome.

As flotation only removes a minor amount of mass it is IMS that determines the loading on the regrind and CMS circuits. Multiple parallel fine grinding mills and CMS banks provide the flexibility necessary to ensure that the quality of final concentrate is consistent.

Added operational benefits of the VRM circuit are fast commissioning (usually only a few weeks) rapid shutdown and restart, high utilisation (>95 per cent is not uncommon), relatively simple inspection of rollers, grinding table and internal wear points and robust baghouse components.

DESIGN CONSIDERATIONS

Technology evaluation

The combination of higher sustainability, better performance data and more flexible and better to control plant operation resulted in the decision to continue the project with VRM grinding classifying circuits over the AG/Ball Mill circuit initially foreseen in the PFS.

Design changes and key project benefits

The change from AG/Ball grinding to VRM brings with it many advantages, other than reduced power consumption, for the Southdown project.

The power benefit of changing from AG/Ball to VRM reduces the maximum and average amount of power that needs to be supplied to the operation. It also means that any given renewable energy installation will account for a greater proportion of the total power.

The ability to reject 35 per cent of the feed mass as dry non-magnetic grits is estimated to reduce the water supply requirements for the project by 24 per cent. Not only is the dry disposal of coarse tailings possible, but it is also possible to mix thickened fine tailings with the coarse material to produce conveyable and compactable tailings with minimal water content (<13 per cent moisture). It may be necessary to dispose of a small proportion of the fine tailings as slurry to ensure that the remainder of the tailings is always conveyable, but the amount of wet tailings disposal required with VRM technology will be at least one order of magnitude below that required with AG/Ball milling. It is also possible to filter minor excess fine tails without a major capital investment in filters.

In the AG/Ball circuit the RMS tails with a top size of 3 mm were cycloned to separate the coarse from fines. The fines feed thickening while the coarse bypasses thickening, allowing mixed coarse and fines to be pumped to a tailings dam or to filtration. In the VRM circuit the lack of wet coarse tailings means that cycloning is no longer required. Instead, a simple drum mixer will be used to combine dry grit tailings with thickener underflow before the mixture is conveyed and trucked to tails stacking or used elsewhere for construction purposes.

Test work showed that while both the VRM and AG mill were able to reject 40 per cent of the feed as non-magnetics, more than half of the AG mill non-magnetics were finer than 85 µm, compared to only 5 per cent finer than 85 µm for the VRM grit reject. Rejection of the same non-mags mass proportion at a much coarser size confirms that the VRM is achieving a much higher degree of selective gangue liberation than AG milling.

Flotation performance after VRM is superior to that achieved in the AG/Ball pilot plant. Very low sulfur levels were achieved in the magnetite concentrate after VRM grinding, wet magnetic separation and pyrrhotite removal by flotation. The levels achieved were 0.02 to 0.03 per cent S, well below the target value required to minimise SO_2 gas production during pelletising. In addition, the pyrrhotite flotation rate for VRM product was very fast and this is likely to result in a smaller flotation section.

The enclosed nature of the VRM means that even though dry milling is employed, the process will not emit dust. This is important given the siliceous nature of the non-magnetic gangue at Southdown. Even the dry magnetic separation section will essentially be dust free due to the complete lack of sub 38 µm in the grit stream. Immediately after VRM grinding the P_{80} 85 µm material is slurried ahead of IMS and flotation.

A secondary crushing stage is needed ahead of VRM compared to AG/Ball. However, as the VRM topsize only needs to be less than 100 mm (specifically for Southdown, coarser sizes are allowed for softer ores) an open circuit secondary cone crusher is sufficient. This compares less favourably with the AG/Ball circuit which only requires an open circuit primary crusher. However, crushing for VRM feed compares favourably to preparation ahead of HPGR, which requires closed circuit secondary crushing to a 50 or 60 mm top size in order to protect the tungsten carbide roll studs.

VRM 3D PLANT LAYOUT DESIGN

Based on the successful test campaigns the flow sheet and hence the plant layout was changed from AG/BM to VRM comminution. The VRM plant layout is illustrated in Figures 7 and 8.

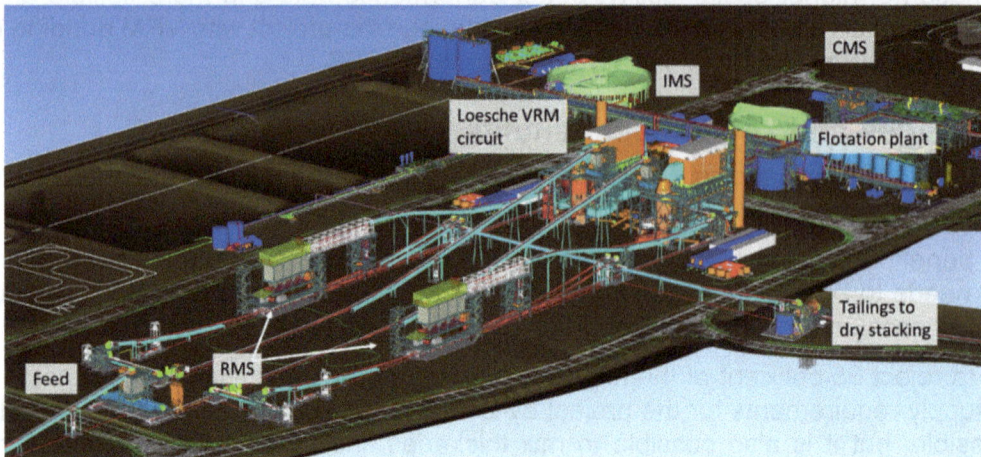

FIG 7 – Southdown plant 3D layout (VRMs and bag houses centre, RMS = rougher magnetic separation, IMS = intermediate magnetic separation, CMS = cleaner magnetic separation).

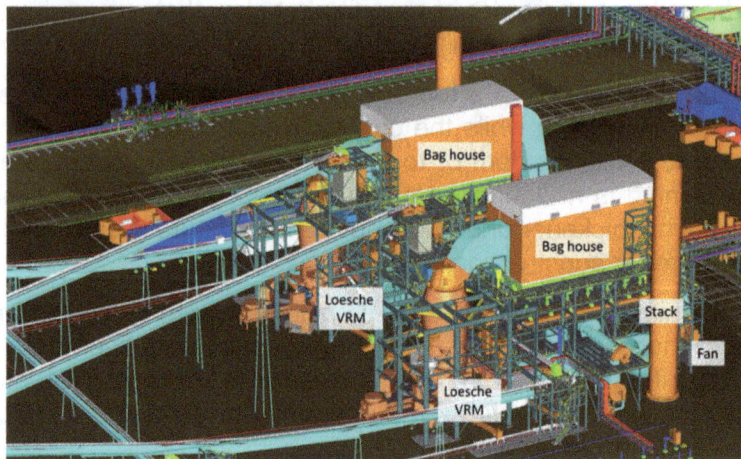

FIG 8 – Closeup of VRMs and bag houses.

Followed by primary and secondary open circuit crushing the ore is fed to two parallel VRM circuits with integrated RMS performed on the grit stream. Each of the VRM circuits is operated individually to allow for maximum flexibility during operation and to ensure separate maintenance schedules for each plant. The RMS plants, shown on the left side of Figure 7 are integrated into the VRM circuit.

The Fe-enriched fine product of the VRM circuits reports to the product pulping tanks (between the two VRM bag houses) ahead of magnetic separation and flotation. The dry (conveyable) non-mag tailings are discharged from the circuit to be stacked or blended with the wet fine tails.

A benefit of the dry grinding is that it provides the capability to easily store the ground product in silos. Dry silos provide much higher capacities than agitated slurry tanks. However, in the Southdown circuit as presented, slurry tanks are incorporated after wet IMS due to the reduced mass flow after magnetic separation. In addition, these tanks will provide stable feed flow and density to the flotation circuit and beyond.

As VRM technology allows 24 hr or more dry buffer ahead of downstream processing they will be considered to replace the wet storage. Large product silos, if installed, substantially increase the decoupling of dry comminution from the wet process circuits. One advantage is that the comminution and wet process can be optimised without interfering with each other. Another advantage is that maintenance on one of the grinding plants can be carried out without affecting the downstream process capacity, as the missing capacity is compensated by the product silos.

TAILINGS CHANGES

The dry circuit prompted some changes in tails handling, the biggest being the ability to dry stack the dry tailings generated from the VRM/Dry magnetic separation circuits given its relatively coarse particle size.

This resulted in a substantial reduction, but not quite an elimination, of wet tailings storage footprint. Some area is needed for emergency storage of fine non-mag thickener underflow when there is no coarse tailings to mix it with for dry disposal. It is also necessary to store the sulfide (pyrrhotite) flotation concentrate in a contained manner to prevent escape of acidic waters into the environment.

Normal disposal of non-mag tails is to mix the dry VRM non-mag tailings with the thickened fine tailings produced in IMS and CMS. After continuous mixing the resulting material with <13 per cent moisture can be conveyed and trucked to tails stacking or be used elsewhere for construction purposes.

SUSTAINABILITY

Energy efficiency and environmental sustainability

The design of the processing plant focuses on maximising process and energy efficiency. This will be achieved through a number of measures, including the use of energy-efficient equipment and the optimisation of the process flow sheet, including incorporation of dry milling utilised prior to the dry magnetic separation stage. Dry tailings disposal also has other environmental benefits and potentially reduces future tailings storage failure risks.

The proposed processing plant will treat about 15 million tons per annum (Mt/a) of ore and have a nominal production capacity of 5 Mt/a of magnetite concentrate. The plant is designed to be modular, so that it can be expanded in the future to meet increased demand, with pipeline capacity design at 10 Mt/a of concentrate. Overall, the change from wet milling and wet separation to dry milling and dry separation for the primary grinding circuit led to an estimated 24 per cent plant water consumption reduction, a substantial step into a more sustainable mineral processing plant design. The final dry product is also suitable for emerging fine dry magnetic separation. That would further reduce water consumption. The main sustainability advantages are summarised in Table 3.

TABLE 3

Industrial design comparison (adjusted to P_{80} 85 µm common basis).

Operating parameter	VRM	WET milling	Relative VRM benefit
Overall SE consumption	12.3 kWh/t	20.8 kWh/t	41%
Mass of non-mag grit rejected	35%	32%	9%
Mass of IMS concentrate for regrinding	37.8% of feed	40% of feed	5.5%
Plant water consumption	3.9 GL/a	5.1 GL/a	24%
Tailings disposal	Dry stacking	Wet pond	

CONCLUSIONS AND RECOMMENDATIONS

The test work and comparisons provide the following conclusions.

Open circuit VRM milling to achieve P_{80} 85 to 100 µm proved to be highly energy efficient for Southdown ore.

Indicative crude grit testing (grit generated by open circuit pilot VRM) showed a high degree of liberation of barren gangue within the grit stream and that magnetic separation of the grit produced barren waste with minimal magnetite losses.

A pilot plant was required to accurately evaluate closed circuit VRM with grit extraction and magnetic separation. Loesche designed, constructed, and successfully commissioned such a pilot plant and integrated it into their test centre in Duesseldorf.

Comparing the VRM and AG/Ball circuits (when both incorporate RMS removal of non-magnetic waste before generating an P_{80} 85 µm product) gives a power benefit of 41 per cent and a water benefit of 24 per cent to the VRM. Between 35 and 40 per cent of the feed is rejected as VRM grits dry non-mag waste whilst Magnetite recovery is kept constant in both cases.

To maximise the extent of ore processing that can be performed dry, the plant design for the Southdown project was changed from AG/Ball Milling to a VRM circuit incorporated RMS on grits.

The benefits of applying VRM technology extend beyond power reduction, process-flexibility, and control to important project factors such as reducing water consumption and simplifying tailings management.

Planning and preparation for the Southdown project has spanned a number of years, during which Grange has established a project closely with key stakeholder organisations and community members. Whilst working on completion of the DFS in 2024, a two year timeline has been set for construction and commissioning (Grange Resources, 2024).

ACKNOWLEDGEMENT

The authors would like to express special thanks to Grange Resources for their permission to publish this paper and to let their employees work on the completion of this paper. Further acknowledgement is given to Wood PLC and Loesche GmbH who supported this work with personal capacities and technical insights.

REFERENCES

Altun, D, Gerold, C, Benzer, H, Altun, O and Aydogan, N, 2015. Copper ore grinding in a mobile vertical roller mill pilot plant, *International Journal of Mineral Processing*, 136:32–36.

David, D, Olwagen, D, Stanton, C, Gerold, C, Schmitz, C, Baaken, S and Everitt, M, 2023. Pilot Testing and Plant Design Comparison of Dry VRM Milling plus Magnetic Separation with AG and Ball Milling plus Magnetic Separation for Grange Resources' Southdown Ore, in Proceedings Comminution 23 (Minerals Engineering International (MEI)).

David, D, Olwagen, D, Stanton, C, Gerold, C, Schmitz, C, Baaken, S and Everitt, M, 2024. Dry VRM Milling for Power and Water Benefits with Magnetite Ore, in Proceedings Mill Circuits 24 (Minerals Engineering International (MEI)).

Fortescue Metals Group (FMG), 2019, 2 April. US$2.6 billion Iron Bridge Magnetite Project approved, media release, FMG. Available from https://cdn.fortescue.com/docs/default-source/announcements-and-reports/iron-bridge-project-approval.pdf

Fortescue Metals Group (FMG), 2024, 24 April. March 2024 Quarterly Production Report, media release, FMG. Available from https://cdn.fortescue.com/docs/default-source/announcements-and-reports/march-2024-quarterly-production-report.pdf

Grange Resources, 2024, 11 July. Grange Resources Southdown website. Available from https://grange.blob.core.windows.net/attachment/a01ba35e-9070-4d08-a5ab-df314ca711f2.pdf

Jacobs, P, Seopa, G, Mofokeng, M, Nienhaus, D, Gerold, C and Mersmann, M, 2016. 16 Years of successful operation of a Loesche Vertical-Roller-Mill Type LM 50.4 in a Hard Rock Application at Foskor Pty (Ltd) in Phalaborwa, in Proceedings of Comminution 16.

Metso Minerals (Australia), 2008, 3 November. Southdown Magnetite Project Interim Process Development Report, internal document.

Reichert, M, Gerold, C, Fredriksson, A, Adolfsson, G and Lieberwirth, H, 2015. Research of iron ore grinding in a vertical-roller-mill, *Minerals Engineering*, 73:109–115.

Schmitz, C and Gerold, C, 2019. A Fundamental Change in Approach - Grinding Ores in Vertical Roller Mills: Presentation of Test Results, IMCET, Turkey.

Stapelmann, M, Gerold, C and Smith, J, 2019. Successful Applications of Vertical-Roller-Mills in Phosphate Processing, in Proceedings of Beneficiation of Phosphates VIII.

Van Drunick, W, Gerold, C and Palm, N, 2010. Implementation of an energy efficient dry grinding technology into an Anglo American zinc beneficiation process, in *Proceedings of XXV International Mineral Processing Congress (IMPC)*, pp 1333–1341.

Weidenbach, M and Gerber, S, 2023. Evolution of the West Musgrave Flowsheet to Maximise Value, in Proceedings 26th World Mining Congress (WMC 2023).

Redefining the battery limits of processing plants – improving sustainability through the deployment of sensing technologies

W Futcher[1], D R Seaman[2] and B Klein[3]

1. Principal Metallurgist, Newmont Mining, St Kilda Vic 3182.
   Email: william.futcher@newmont.com
2. Manager-Directional Studies (Metallurgy), Newmont Mining, Subiaco WA 6008.
   Email: david.seaman@newmont.com
3. Professor, Department of Mining Engineering, University of British Columbia, BC Canada.
   Email: bklein@mining.ubc.ca

ABSTRACT

The role of separating ore from waste in an open pit mine is typically the realm of the production geologist who applies grade control to delineate ore from waste. The resolution of the grade control pattern is much coarser than the mining equipment used to recover and transport blasted ore to the mill or waste dump. This residual heterogeneity presents a unique prospect to initiate separation at the excavation front by means of real-time grade estimation for each shovel scoop. By doing so, shovel scoops or trucks below economic cut-off grade can be diverted to waste-dumps and similarly ore grade material can be directed from waste to the mill.

Once ore is introduced into the process, other sensing systems such as belt scanners are also useful tools in feed-forward control to the process plant. Presently, these innovations in online measurement are yet to become a standard piece of processing equipment, unlike their on-stream analyser counterparts which have gained widespread adoption.

At Red Chris, an X-ray florescence (XRF) based ShovelSense system has been successfully trialled and integrated into the fleet management system to take minerals processing to the mine face. In addition to the shovel mounted sensor, a Prompt Gamma Neutron Activation Analysis (PGNAA) based belt scanner has also been installed post crushing to measure additional ore properties ahead of the mill. This sensor allows the realisation of real-time geometallurgical model outcomes in the form of estimated ore hardness as well as forecasting tailing neutralisation potential.

INTRODUCTION

With the developments in sensor-based sorting techniques, the ability of operations to push the initial processing of material into the mine is becoming more prevalent. The mine-based assessment of material grade facilitates waste rejection as close as possible to the extraction process, maximising operational efficiencies by minimising dilution in mill feed and ore loss to the waste dumps. There are some dependencies however. *In situ* orebody heterogeneity is one requirement, while a sensor, or separation method, capable of detecting grade variability and the separation of ore from waste, is another.

At Red Chris, a process for the assessment of the potential for in-pit sorting combined with belt-based grade sensing was undertaken to determine the potential application of these methods. A theoretical heterogeneity assessment followed by bulk sorting trial using ShovelSense proved positive leading to the full-scale deployment of the technology. During this process, a Prompt Gamma Neutron Activation Analysis (PGNAA) sensor was installed on the conveyor between the primary crusher and the coarse ore stockpile (COS). The combination of these two techniques for the detailed characterisation of mined material has led to the operation's ability to ensure minimal misdirection of trucks. The belt sensor further confirms ShovelSense measured grades and adds detailed rock chemistry, offering the potential for feed forward control to the process plant.

RED CHRIS OPERATIONS

The Red Chris Mine is in the Golden Triangle of Northern British Columbia, approximately 80 km south of Dease Lake and 18 km south-east of the community of Iskut. Newcrest Mining Ltd, subsequently Newmont, became the operator of the Red Chris Mine on the 15 August 2019 after purchasing a 70 per cent stake of the operation from Imperial Metals (Stewart *et al*, 2021).

Geology

The Red Chris deposit is a significant copper-gold porphyry deposit located in British Columbia, Canada, and is economically significant due to its substantial copper and gold reserves. The highly deformed nature of the deposit reflects a history of volcanic activity, magmatic intrusions and hydrothermal alteration, resulting in the formation of valuable copper-gold mineralisation (Rees *et al*, 2015).

The mineralisation is associated with a series of faults and fractures that provided pathways for the movement of mineral rich fluids. This has led to the precipitation of thin wavy and thicker planar quartz veins containing chalcopyrite, bornite and magnetite. Dissemination of sulfide and gold mineralisation is also common locally around these vein structures. Gold predominantly occurs as microscopic inclusions in the copper sulfides, and occasionally as free grains in high-grade zones (Stewart *et al*, 2021). Material is classified in grade (or value) bins of HG (High Grade), MG (Medium Grade), LG (Low Grade), MW (Mineralised Waste), and Waste which is subclassified as NAG (Non Acid Generating) and PAG (Potentially Acid Generating).

Mining and processing

Ore is extracted from the Red Chris deposit using a conventional open pit mining method. Mining is performed by three large shovels: a Komatsu PC7000, P&H 2800 and a Hitachi EX3600, with support from three smaller excavators. The duty of ore and waste haulage is carried out by 18 trucks with a mix of Cat 793s and 785s.

Primary crushing of ore is done by a Metso gyratory crusher linked to a COS via a series of overland conveyors. The concentrator processes ore at a rate of approximately 10.5 Mt/a using a conventional semi-autogenous ball mill crusher (SABC) circuit with a target grind P_{80} of 150 µm. Throughput takes precedence over grind size, often resulting in typical grind P_{80} closer to 170–180 µm.

Rougher flotation is performed in two Eriez StackCells, six 200 m³ and one 160 m³ Outotec TankCells. The rougher concentrate is directed to the cleaners via a two-stage regrind circuit with a target P_{80} of 30 µm. The cleaner circuit configuration is variable with the flow sheet changing depending on operational requirements. Final concentrate with a grade of 23–24 per cent Cu is thickened and pressure filtered ahead of trucking to the Port of Stewart where it is loaded on a ship for delivery to the point of sale (Seaman *et al*, 2021).

The process plant produces non-acid generating (NAG) and potentially acid generating (PAG) tailings streams. The NAG tailings are further processed by semi mobile cyclones at the tailings dam, producing coarse sand for wall construction. This stream must have a sufficient ratio of neutralising carbonates and acid generating sulfides to achieve a neutralising potential ratio (NPR) of greater than 2 (Seaman *et al*, 2021).

The PAG tail is a pyrite rich product principally composed of cleaner-scavenger tailings and is required to be deposited sub-aqueously in the tailing's impoundment area.

OREBODY HETEROGENEITY AND SORTING METHODS

Heterogeneity of a mineral deposits refers to the natural variability in mineral distribution within the deposit. The variability can occur at various scales ranging from the microscopic, exploited by comminution and flotation circuits, to macroscopic differences seen vertically and horizontally within the deposit. The potential for sorting is dependent on the orebodies inherent heterogeneity with grade variability at the scale of the sorting instrument necessary for viability. Alongside this is the sensors capability to detect grade variability with sufficient accuracy and precision to enable confident decision-making using measurement outcomes (Moss, Klein and Nadolski, 2018; Nadolski *et al*, 2018).

Ore sorting, also known as preconcentration, refers to the practice of waste removal as early as possible in the mining system with the goal of increasing efficiency by minimising costs. Ore sorting process can broadly be broken down into two principal techniques: particle-based sorting and bulk ore sorting.

Particle based ore sorting

Particle based sorting refers to the use of systems that assess individual particles using a range of sensors to determine whether they are ore or waste. Typical sensors used by such systems are laser, RGB, X-ray transmission (XRT), X-ray florescence (XRF), electromagnetic, near-infrared (NIR) and Laser Induced Breakdown Spectroscopy (LIBS). This approach to ore sorting is best suited to low throughput operations processing high value commodities as the capital and operating cost are greater than bulk ore sorting techniques.

Bulk ore sorting

Bulk ore sorting, as the name suggests, is a technique used to assess bulk increments of material for determination or material type (Duffy et al, 2015). This approach can further be divided into belt-based sorting and in-pit sorting. Belt based ore sorting can employ the same sensors as particle-based sorting, with the addition of sensors like PGNAA, Gamma Activation Analysis (GAA) and Magnetic Resonance (MR). This approach characterises larger increments of material, typically 1–200 t, for the decision-making process and requires a diversion system to remove waste from the materials handling system.

A typical in-pit bulk ore sorting system can be either sensor based, using XRF or MR sensors, or can exploit the natural characteristics of the ore via mechanisms like screens to produce a waste and ore stream. In-pit sorting is deployed in the mine directly or at the pit crest via a truck-based grade measurement.

Belt based grade sensing using the Prompt Gamma Neutron Activation Analysis technology

PGNAA is a non-destructive analytical technique commonly used on conveyors for grade measurements of crushed ore. The technique uses a Californium 252 source located beneath the conveyor belt to generate neutrons that bombard the material on the moving conveyor. The neutrons interact with the ore on the belt and generate characteristic gamma rays that are registered by detectors mounted above the moving material. This generates a spectrum that is deconvoluted, allowing the elemental composition of the material to be determined (Noble, Thompson and Sanhueza, 2022; Noble, Strombotne and Gordon, 2024).

Newmont has been interested in the PGNAA technology for grade measurement of mined ore for several years. A system was initially installed at Cadia between the mine portal and the COS for the purposes of better understanding the PGNAA's potential in base metal operations and to generate an understanding of the nature of grade heterogeneity of ore in the Materials Handling System (MHS). This installation followed an early trial of a belt-based Magnetic Resonance (MR) sensor on the Ridgeway mine conveyor, later moved onto the Cadia MHS (Miljak, 2016). Cadia operates a block cave with material extracted from drawpoints across an extraction level. Drawpoint sampling indicated grade variability across the extraction level, suggesting that belt based bulk ore sorting may be possible. However, ore is drawn as evenly as possible from across the extraction level to ensure continued and gradual cave propagation, resulting in maximum ore recovery. This, however, leads to the homogenisation of the grade of the ore in the MHS leading to a stable plant feed grade but reducing the potential for belt-based sorting using the installed PGNAA sensor.

PGNAA sensors produce results at a frequency specified by the user. The measurement integration time is influenced by ore grade and a desired level of measurement precision. As part of the assessment of PGNAA usage in a base metal deposit, measurement integration time was evaluated to determine an acceptable level of precision for Cadia. This was done during a planned mine shutdown with an ore sample loaded onto the stationary belt within the sensor. Measurement time for the sample was then varied from 30 secs to 1 hr to generate sufficient data points for the assessment. A mean and a standard deviation were determined from the data to calculate a coefficient of variance as shown in Figure 1. From this analysis, a 2 min measurement integration time was selected as appropriate for Cadia as a reasonable compromise between precision and sampling frequency.

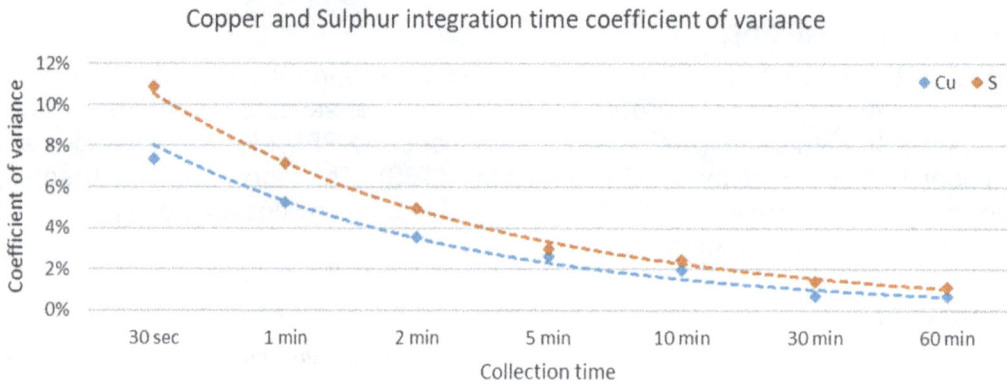

FIG 1 – Cadia PGNAA belt sensor measurement time coefficient of variance for copper and sulfur.

In early 2023, a CB Omni Fusion PGNAA belt sensor, supplied by Thermo Fisher, was installed and commissioned at Red Chris on the conveyor between the primary crusher and the COS. Important for a successful installation is the inclusion of an accurate weightometer proximal to the sensor. The sensor produces spectra that must be de-convoluted and converted into weight percent data. Necessary for this is an accurate input for tonnes in the selected measurement period.

The Thermo Fisher CB Omni Fusion PGNAA belt sensor utilises dual 40 µg Californium sources ensuring even grade data from the full cross-section of the ore on the belt. As with the sensor at Cadia, it measures a comprehensive suite of elements, including copper and sulfur. Analysis of the average daily grades produced by the belt sensor, the mill on-stream-analyser (OSA) for copper and cyclone overflow grab samples for sulfur show excellent agreement, as is seen in Figure 2. This confirms the findings from Cadia that this sensor type works well for base metal deposits.

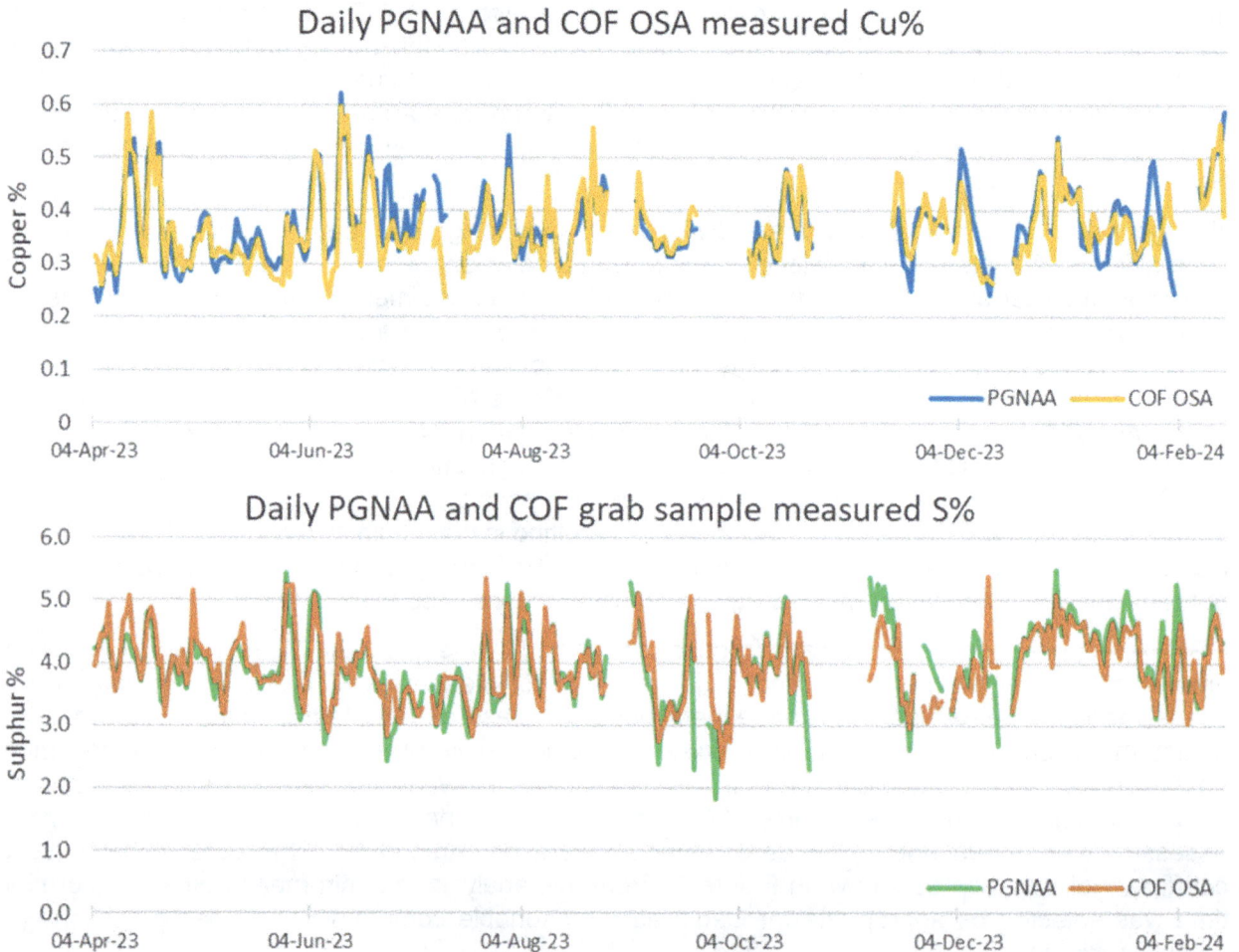

FIG 2 – Average daily copper and sulfur in mill feed as measured by the belt senor and the mill OSA for Cu grab samples.

ASSESSMENT OF BULK ORE SORTING POTENTIAL AT RED CHRIS

Proof-of-concept bulk ore sorting assessment for Red Chris

For the Red Chris deposit, a heterogeneity study was conducted at the University of British Columbia as part of a final year Capstone Project (Acacia *et al*, 2021). A 1 m drill hole assay composite data from across the deposit was provided for the analysis. This was composited to differing lengths to represent the heterogeneity present at the block model, grade control, truck and shovel bucket scale for an assessment of bulk ore sorting potential. Figure 3 graphically shows the four scales typically encountered by a mine.

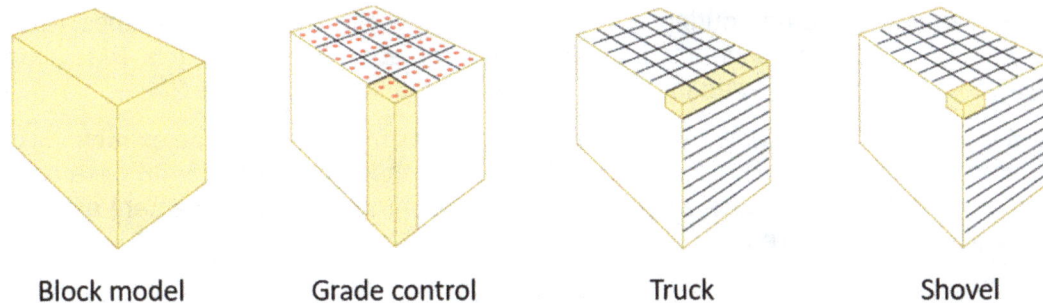

| Block model | Grade control | Truck | Shovel |

FIG 3 – Block model, grade control, truck and shovel bucket typically encountered by a mine, after Redwood (2018).

Re-combining of the assay data at different interval lengths provides an indication of the grade heterogeneity of the deposit (Acacia *et al*, 2021). For the analysis, interval lengths of 2 m and 4 m, representing shovel and truck scale heterogeneity, was compared to a 12 m composite representing grade control heterogeneity. The waste rejection potential from applying grade control on a per shovel scoop or per truck basis compared with the traditional grade control scale are presented in Table 1 for both the East Zone and Mian Zone pits at Red Chris. This provides a theoretical estimation of the *in situ* potential for bulk ore sorting at the shovel and truck scale for the Red Chris deposit.

TABLE 1

Heterogeneity analysis outcomes for bulk ore sorting at Red Chris.

	East zone		Main zone	
	Shovel (2 m)	Truck (4 m)	Shovel (2 m)	Truck (4 m)
Ore from waste	3.0%	2.5%	6.3%	6.2%
Waste from ore	13.5%	12.3%	14.0%	12.2%

SHOVELSENSE AT RED CHRIS

A decision was made to avoid belt based bulk ore sorting techniques to make use of the anticipated orebody heterogeneity as high frequency diversion systems and associated infrastructure are costly and take significant time to design and construct. This approach also sacrifices the waste to ore conversion as grade control defined waste material is not measured and therefore any contained ore is lost. Given that the heterogeneity analysis suggested that waste to ore conversions form a significant proportion of the total benefit, shovel or truck-based sorting techniques were deemed most appropriate.

An assessment of the available technologies that could be deployed at Red Chris to make use of the quantified orebody heterogeneity at the truck or shovel scale resulted in limited options. The gantry mounted magnetic resonance sensor from NextOre was not sufficiently advanced to be considered for deployment. The only commercially available technology deployable at the time was ShovelSense.

The ShovelSense technology is a grade measuring tool mounted in the bucket of a shovel. The system uses XRF to determine the copper grade of the mined material on a bucket-by-bucket basis, allowing a truck average grade to be determined. As material begins to fill the bucket, lasers in the heads detected a level change and activate the XRF emitters. Characteristic X-rays generated by the interaction of primary beam and rock mass are sensed by the detectors and converted to an elemental weight percentage. For Red Chris, this was deemed to be an appropriate technique and a decision was made to advance the project to a trial.

Bulk ore sorting trial using ShovelSense at Red Chris

The smallest shovel at Red Chris, the Hitachi EX3600, was selected for the initial trial with installation of a two head system occurring during a scheduled 72 hr preventative maintenance (PM) period. Fortunately, this shovel was scheduled for a bucket replacement during this PM, allowing the armoured sensor head to be welded into the replacement bucket prior to the PM.

Throughout the course of the trial, 1 092 528 t of material was measured via the extractions of 42 291 buckets of material, resulting in the filling of 5026 trucks. Of the 5026 trucks, 3485 were classified by grade control as ore and 1541 were waste. A high-level summary of ore movement as classified by grade control is shown in Table 2.

TABLE 2

High-level summary of ore movement during ShovelSense trial.

ShovelSense trial	# of trucks	Tonnes	Cu grade
Ore (classified by grade control)	3485	756 761	0.494
Waste (classified by grade control)	1541	335 767	0.120
Total	**5026**	**1 092 528**	**0.379**

Prior to the trial commencing, a general calibration was developed by testing ore samples in a laboratory setting and this was loaded onto the ShovelSense system. As the trial proceeded, a Red Chris specific calibration was developed with a focus on reconciliation at the ore/waste cut-off grade of 0.245 per cent copper. This revised calibration has proved reliable across the grade range and gives confidence that the system is on average, measuring the true copper grade of each shovel bucket with a comparison against grade control data showing excellent agreement as is seen in Figure 4.

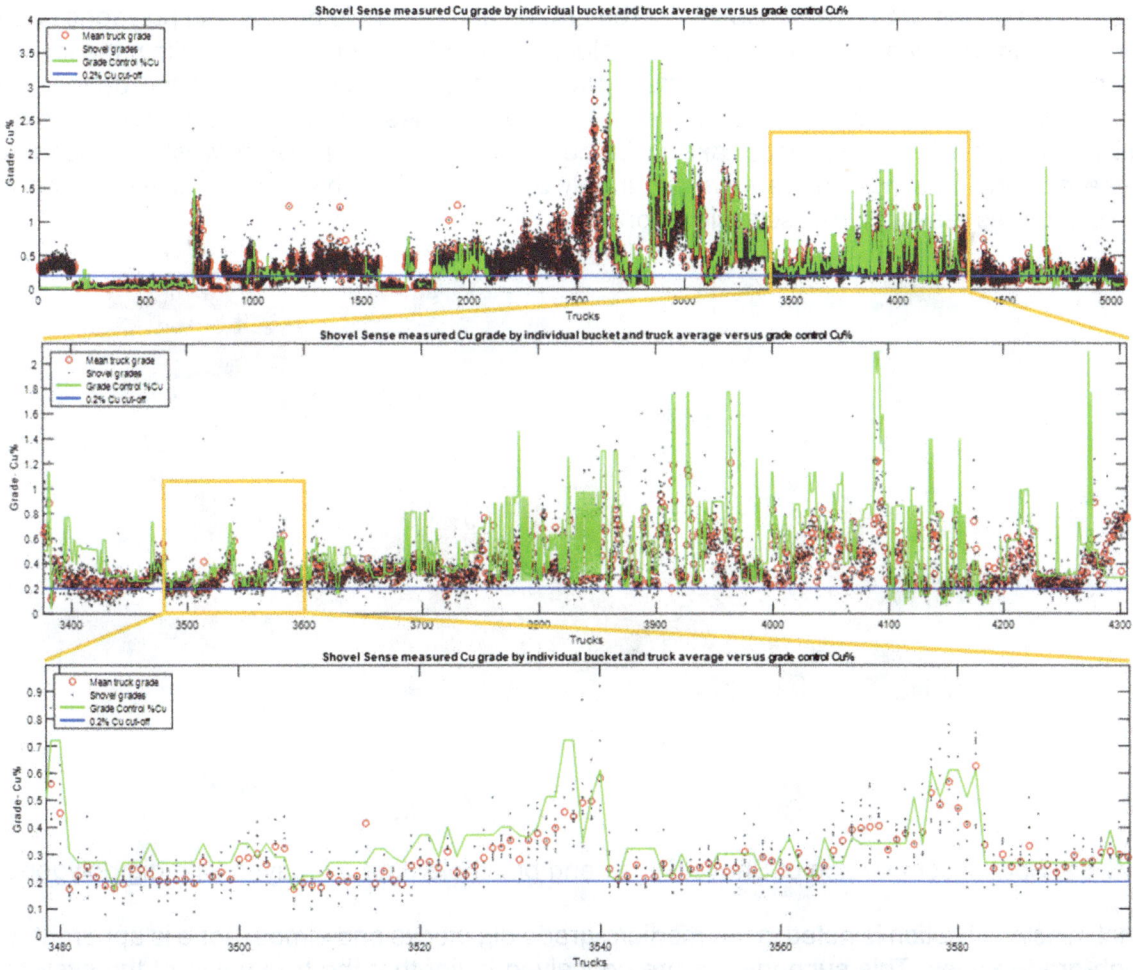

FIG 4 – ShovelSense measured copper grade by bucket and truck versus grade control predicted copper at three scales.

For ShovelSense, the potential for redirection is achieved by taking advantage of the finer resolution offered by the system. Each grade control assay characterises 800–1000 t of ore or waste and represents four to five truckloads of material. In contrast to this, each bucket that is mined with a ShovelSense equipped shovel produces a grade that is then averaged across the four to five buckets that make up the typical ~225 t load caried by a Cat 795 at Red Chris. This results in 16–25 times increase in measurement resolution and represents a significant step toward optimising grade delivery to the mill and the minimisation of metal loss to the waste dumps. Figure 5 graphically shows the resolution difference between individual bucket grades and grade control copper assays.

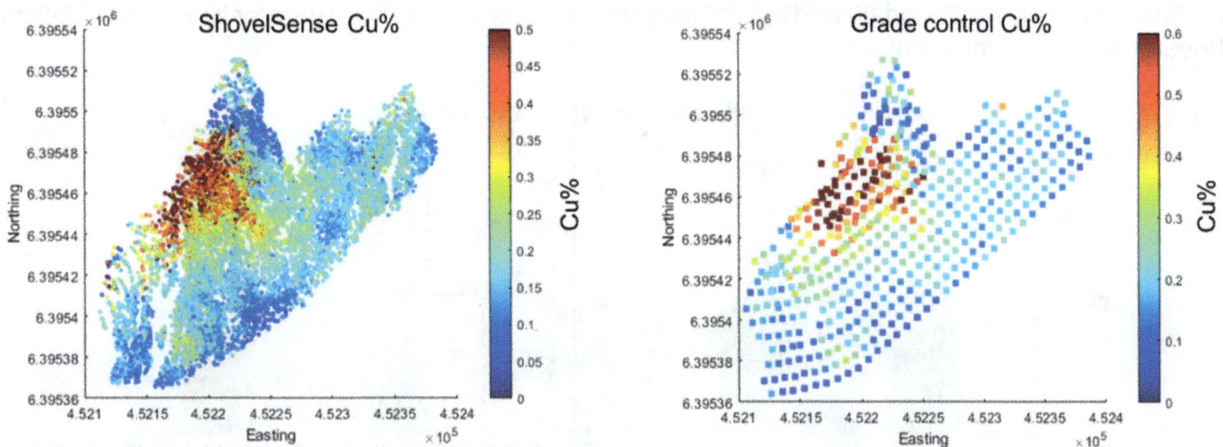

FIG 5 – Side by side comparison of ShovelSense and grade control measured copper in a blast pattern.

A spatial comparison of ShovelSense measured grades and blasthole defined dig polygons is shown in Figure 6. Misclassification is most apparent along ore/waste boundaries and within the low-grade dig blocks. This is partially due to the diffused nature of copper distribution in the deposit with a continuum existing between high-grade ore and waste. The designation of an ore/waste boundary in this continuum will inadvertently take in some waste in ore and ore in waste. Secondly, ore movement during blasting is not always accurately accounted for. This leads to some mixing along these boundaries resulting in misclassification.

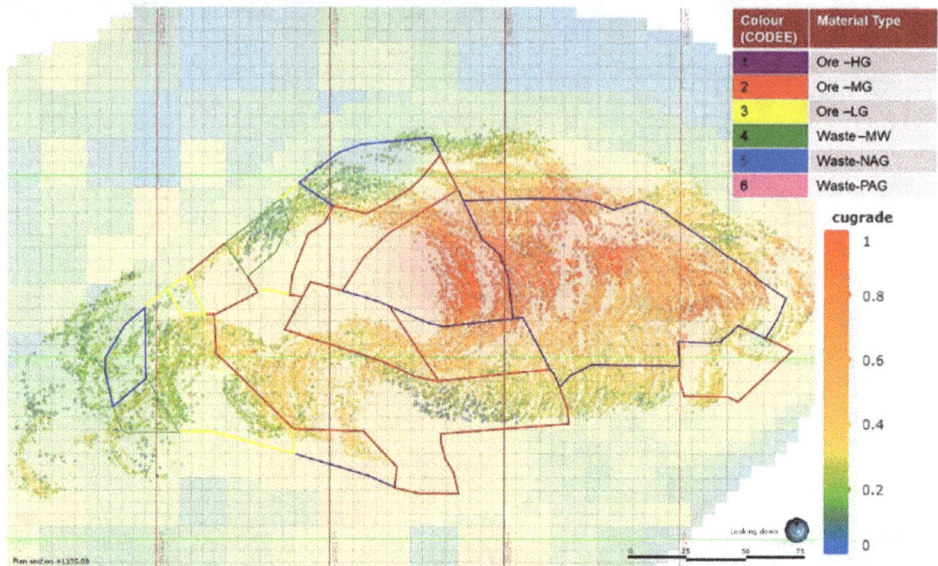

FIG 6 – Spatial comparison of ShovelSense and blasthole data overlayed with dig polygons.

Minimal misclassification is noted in the medium-grade dig blocks and almost none is apparent within the high-grade zones. This supports the pre-conceived belief that the true value of the system lies in bulk sorting material at the value margins of the deposit. Use of the system in homogenous high-grade, and to a slightly lesser extent, medium-grade zones will have a reduced benefit to Red Chris as the probability of diversion is less than when being used in low-grade ore or mineralised waste.

The ShovelSense trial showed that the grade control process is working correctly. ShovelSense measured grades correlated well with blasthole measured grade and the resulting dig blocks tightly bound the correct grade zones. The use of the ShovelSense system then is not to correct an existing problem but is to refine an already good grade control process.

Following successful completion of the trial, analysis of the data generated showed significant potential for re-claiming ore from waste and removing waste from mill feed. Figure 7 shows the potential for diversion from each of the major material classes with the low-grade and mineralised waste material types showing the greatest misclassification. This is expected as these material types straddle the ore waste boundary and any *in situ* grade heterogeneity has a greater likelihood of falling above or below the mine cut-off.

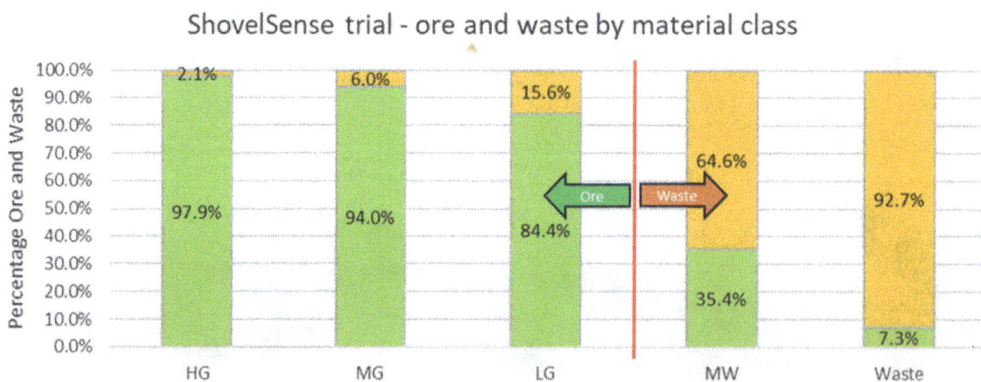

FIG 7 – Trial measured misclassification by material class.

For the duration of the trial, the system was recording data only, no connection back to dispatch was made to allow for truck diversions until the commercial deployment commenced. The reason for conducting the trial in this way was to ensure there was no impact on the current dispatch and control system until the value of the system had been confirmed.

Measured misclassification in each material type and resulting grade changes were overlayed across the existing mine plan, forming an understanding of potential for the operation. Based on this, commercial negotiations leading to the full-scale deployment were initiated with an agreement reached to install two systems on the large production shovels.

Commercial deployment of ShovelSense at Red Chris

The P&H 2800 (FS601) and the Komatsu PC7000 (FS620) were selected for commercial deployment as these two machines are the primary production shovels at Red Chris. ShovelSense was first installed on FS620, coming online on the 20th of March 2023 with FS601 following on the 28th of June 2023 (Figure 8).

FIG 8 – FS601 and FS620 at Red Chris currently using the ShovelSense system.

Both shovels use three XRF heads in the buckets to characterise the grade of material being mined. Figure 9 shows the armoured sensor heads in the shovel bucket. For the ShovelSense systems in use at Red Chris, copper is the only element currently being reported. Secondary elements of Fe, Mo, Zn, Sr and As are being tracked with the hope they can be used for discriminating other ore characteristics that can drive system performance in the future.

FIG 9 – Three head system on the shovels at Red Chris showing the outer armour and the inner laser, XRF emitter and X-ray receiver.

ShovelSense deployment outcomes

From installation to May 2024, ShovelSense equipped shovels have moved 11.3 Mt of material in ~51 000 trucks. Of the 11.3 Mt mined, 2.4 Mt of material has been redirected representing 21.6 per cent of all material mined using the ShovelSense systems:

- 1612 kt of ore has been re-directed from waste and at an average grade of 0.30 per cent Cu and 0.20 g/t Au. This material has a Net Smelter Return (NSR) of $24.58 and is classified as low-grade ore.

- 825 kt of waste has been removed from ore at an average grade of 0.17 per cent Cu and 0.13 g/t Au. This material has an NSR of $13.59 and is classified as waste.

- This has resulted in a 787 kt increase in ore available for processing at an average grade of 0.42 per cent Cu and 0.27 g/t Au and is on average classified as medium grade ore (Table 3).

TABLE 3

High-level summary of benefit derived by using the ShovelSense system.

ShovelSense benefit summary	Tonnes	Cu%	Au g/t	NSR	Material class
Diverted tonnes – waste to ore	1 612 300	0.30%	0.20	24.58	1 612 300
Diverted tonnes – ore to waste	825 294	0.17%	0.13	13.59	825 294
Net change in ore	**787 006**	**0.42%**	**0.27**	**36.21**	**787 006**

The generation of an additional ~787 kt of ore represents an 18.1 per cent increase in ore produced by the mine via the conversion of above cut-off grade waste material. When viewed as a strip ratio, the change is significant with the waste to ore ratio changing from 1.59 t of waste per tonne or ore to 1.19 t of waste per tonne of ore.

The percentage of misclassification since deployment (Figure 10) in each material class has exceeded that expected from the initial trial (Figure 7) with all material types showing an increase in the number of re-directable trucks. Of note is the MW where 50 per cent of all mined material is ore and the residual material now a true waste.

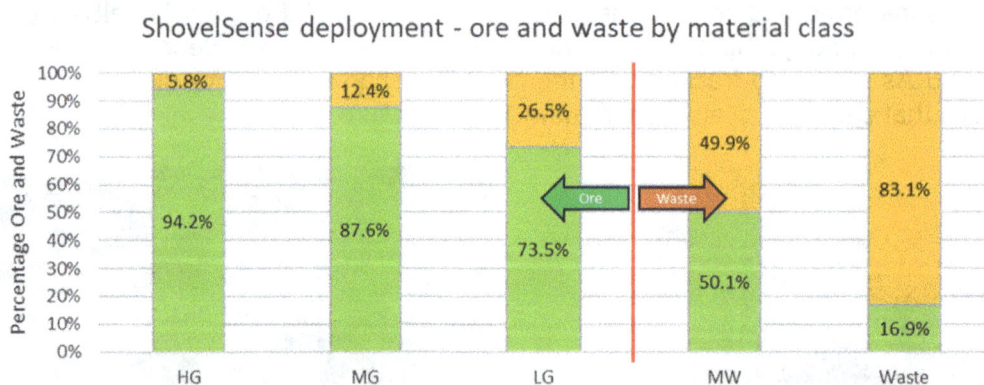

FIG 10 – Measured misclassification as characterised by ShovelSense during commercial deployment.

From installation to May 2024, 67 blast patterns across seven benches have been mined using the ShovelSense systems. Some of these patterns had very little material diverted while others had a significant percentage of the total blast redirected. Redirections occur in patterns where ShovelSense measured copper grades deviates from the grade control marked up grades. This can be due to mixing during basting or conservative standoffs used along ore waste boundaries to minimise dilution in plant feed.

One example of this occurred on bench 1376 with blast 33. ShovelSense was used to mine ~256 kt of material from this blast that was principally designated as PAG and mineralised waste. During mining of the location, ~155 kt of ore was reclaimed from waste via the redirection of 680 trucks. ~96 kt, or 62 per cent, of the redirected ~155 kt of material was classified as medium grade by ShovelSense. Table 4 shows the re-directed material tonnes due to ShovelSense and Figure 11 shows the grade control markup, ShovelSense measured Cu% and re-classified material in pattern 1376–33.

TABLE 4

Re-classified material from blast pattern 1376–33.

Pattern 1376–033	Grade control (t)	ShovelSense (t)	Change (t)
HG	0	459	459
MG	0	95 943	95 943
LG	27 631	85 712	58 081
MW	107 810	38 167	-69 643
Waste	120 104	35 265	-84 839
Total	**255 544**	**255 544**	**0**

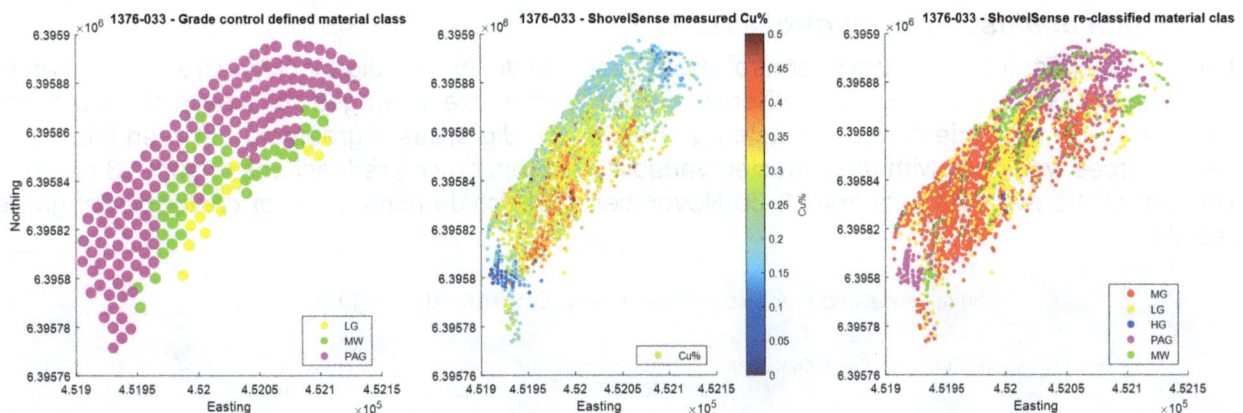

FIG 11 – Grade control markup, ShovelSense measured Cu% and re-classified material in pattern 1376–33.

A comparison of the percentage of misclassified material from the proof-of-concept assessment using downhole drill core assays data and the current commercial deployment shows a significant increase in actual redirections. The heterogeneity analysis assessed only vertical *in situ* orebody heterogeneity due to the constraints of the supplied data. This does not account for mixing during blasting or for conservative standoffs used along ore waste boundaries. As a result, actual waste from ore and ore from waste re-directions are higher than the theoretical assessment. This has ultimately led to a greater than anticipated number of truck re-directions, resulting in ShovelSense producing more than the predicted benefit for Red Chris. Figure 12 graphically illustrates the difference in the percentage of ore in each material class. Of note is the mineralised waste and waste classes that both show an increase in their total ore content by ~15 per cent each.

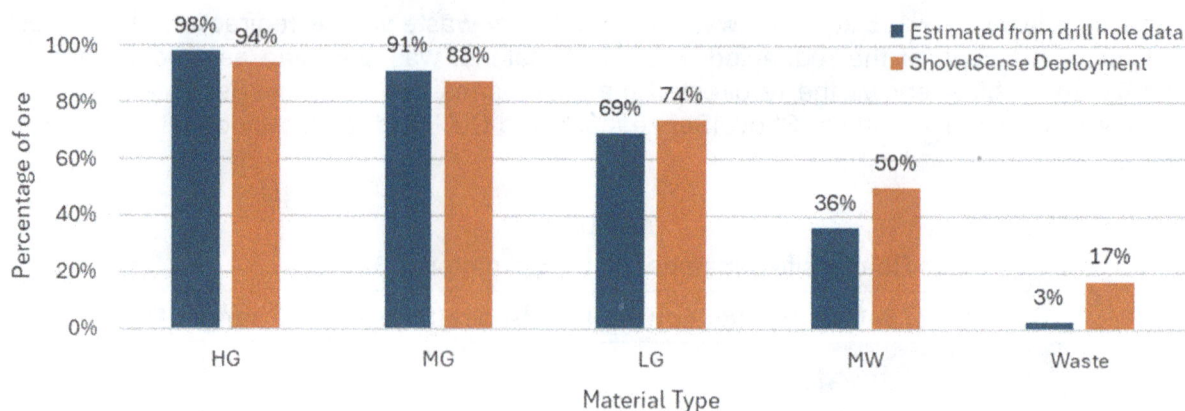

FIG 12 – comparison of the theoretical and actual ore content in individual material classes.

Validation of ShovelSense outputs

Verification of ShovelSense measured copper grades is important in developing trust in the outputs of the system. Poor measurement accuracy or low precision will result in the incorrect diversion of trucks and an erosion of value for Red Chris. For Red Chris, three methods for the validation of ShovelSense outputs were considered.

Validation against grade control assays

Initial checks against in-pit grade control assay, used to define ore dig blocks, were undertaken to verify ShovelSense outputs. ShovelSense bucket grades are averaged to individual trucks and assessed against grade control expected grades for the dig areas. Agreement between these two data sources was good with the expected variability present driving redirections. Figure 13 shows a line plot of the two data sets from 1–30 November 2023 for demonstration of conformity of grade results.

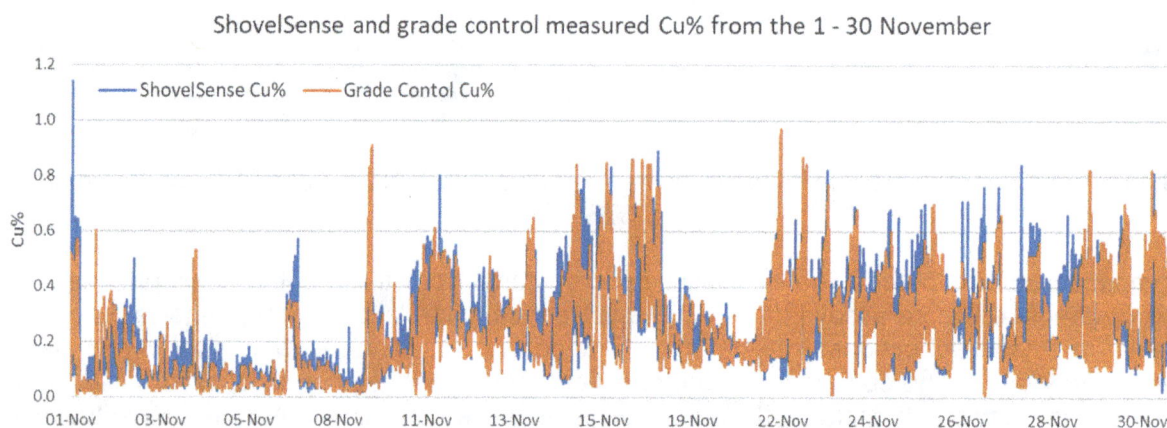

FIG 13 – ShovelSense and grade control measured Cu%.

Validation against belt sensor and mill OSA grades

Red Chris has two other independent measurement methods used to characterise the grade of the material being delivered to the mill from the pit. The first of these is the above mentioned PGNAA belt sensor located between the primary crusher and the COS.

The second meaurement point is the OSA measuring a range of elements in streams within the process plant including copper. The OSA is periodically calibrated and is checked against shift composites every day. Rock drawn in from the COS is milled in the grinding circuit and then directed to flotation via the primary cyclones. A measurement of the primary cyclone overflow is made by the OSA and represents feed to the flotation circuit. This material should, in a broad sense, have a similar grade to the rock delivered to the COS. Difference will be noted as there is mixing in the COS

Mill Operators Conference 2024 | Perth, Australia | 21–23 October 2024 and SAG mill with the re-circulating load in the ball mill further homogenising short-term grade variability.

For a comparison of these three independent data sets, only trucks filled using ShovelSense equipped shovels directed to the primary crusher were used. To compare ShovelSense grades against the belt sensor and OSA grades, time offsets were used to account for travel time on the belt and estimated residence time in the COS and grinding circuits. Table 5 defines the offsets used.

TABLE 5

Time offsets for ShovelSense, belt sensor and mill OSA grade alignment.

Time offsets for grade alignment	Minutes
Primary crusher to belt sensor	5.8
Belt sensor to COS	8.6
COS residence time	216
Grinding circuit residence time	6.5

ShovelSense, the PGNAA belt sensor and OSA measured copper grades show excellent argeement, providing further confidence in ShovelSense outputs. Figure 14 shows the grade data for the three data sets from 1 July – 30 September 2023.

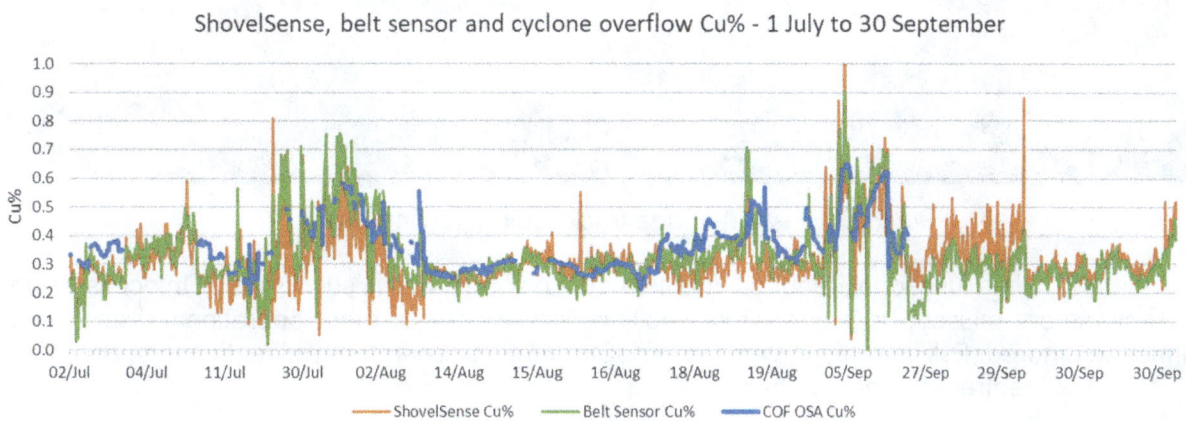

FIG 14 – ShovelSense, belt sensor and OSA measured copper grades.

Bulk sample verification of ShovelSense measured copper grades

To generate an additional layer of certainty on the grades being produced by ShovelSense, a bulk re-direct stockpile of ore reclaimed from waste was generated for batching through the belt sensor and mill to confirm the grade alignment. From 1–21 September 2023, ~48 kt of ShovelSense re-directed material from waste to ore was produced and stockpiled at an average measured grade of 0.25 per cent Cu.

The re-directed material was processed through the primary crusher in two campaigns. The first campaign involved 19 362 t of material fed between September 25th and 26th, following an extended crusher and mill shutdown. A mean grade of 0.28 per cent Cu was measured at by the PGNAA belt sensor while a composite of grab samples from the cyclone overflow produced a grade of 0.27 per cent Cu.

The second campaign processed 28 579 t of material, depleting the re-direct stockpile and completing the trial. For this second campaign, a mean grade of 0.27 per cent Cu was measured by the PGNAA belt sensor. The OSA for was offline during this period and no measurement of copper in cyclone overflow was available for comparison.

The overall average grade of the two campaigns was consistent at 0.27 per cent Cu and was supported by the cyclone overflow assay during the first campaign. These results were comparable to the ShovelSense stockpile average grade of 0.25 per cent Cu indicating that the system is successful in characterising the copper grade of mined material.

The grade validation steps used to determine ShovelSense's ability to discriminate between near cut-off ore and waste provided confidence in the continued use of the system and confirmed the belief that its continued use represents an important value contribution to production at Red Chris.

Fleet management integration

For the productive use of the ShovelSense system, some form of integration with dispatch and the fleet management system (FMS) is needed to facilitate truck diversion. At the time of installation, Red Chris used MineStar version 5.5 as their FMS and this had no ability to accept inputs from the ShovelSense system for truck diversion.

As a result, a third-party interface, MineMage, was used to access ShovelSense bucket data and MineStar truck routing data to produce a truck based average grade and assigned dump destination. When there was a mismatch between grade control assigned material class and ShovelSense characterised material class, a notification was provided to dispatch to reroute the truck. An example of the interface is shown in Figure 15, showing three trucks that need to be redirected from waste to ore.

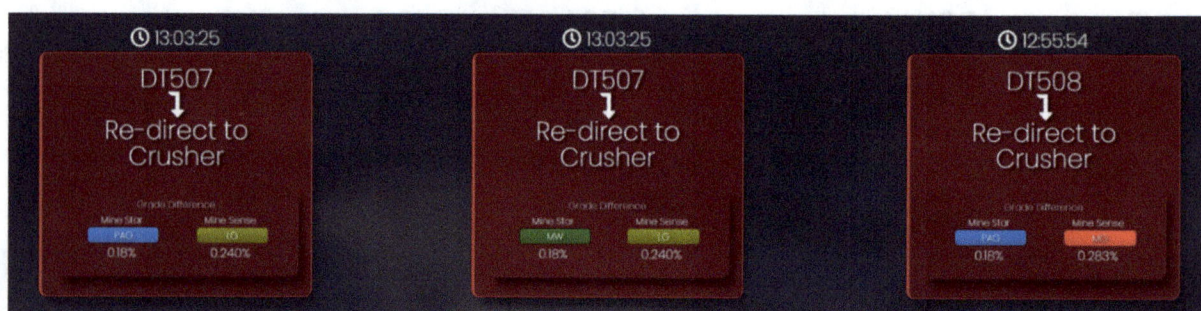

FIG 15 – MineMage truck redirection interface.

This method, while functional, requires attention from dispatch and is suboptimal. On occasion, truck redirections are missed, or they are incorrectly redirected due to other operational pressures that cannot be ignored. Full integration with the MineStar FMS for seamless redirection is needed as this will ensure all redirections are captured.

An upgrade of MineStar from version 5.5 to version 5.8 provided a mechanism for the integration of the two systems. Minestar Version 5.8 contains a ShovelSense add-in that can take in bucket grade data during loading to produce a truck average grade. Truck average grades are compared to a rule table and a material class assigned. If there is a mismatch between the originally assigned material class and the newly assigned material class, an auto redirection notification is sent to the truck. This method reduces the effort required by dispatch and maximise redirections.

Conversion to NSR based diversions

Material classification at Red Chris is defined using a calculated Net Smelter Return (NSR) with Cu and Au grades driving the outcome. NSR is the net revenue that Newmont receives from the sale of the copper and gold products produced by the operation less site costs and transportation and refining costs. The NSR calculation includes plant recovery in the equation with recovery being variable and dependant on both copper and gold head grade.

As ShovelSense is only measuring Cu in each bucket mined, material classification is linked to the mined Cu grade. A general relationship between copper and gold grade was established using grade control data to assist with material type classification. From this, a copper grade of 0.245 per cent Cu was established as the dividing point between waste and ore as at this copper grade, an average gold grade of 0.19 g/t will produce enough value that the material is classified as ore.

As a result of this, there will be occasions when material that has been classified as either ore or waste by grade control using both copper and gold grades, could be reclassified by ShovelSense due to the absence of a gold grade. To rectify this, gold grades, as measured by grade control, in the local vicinity of the shovel, are drawn into the material type classification so that an NSR for each shovel bucket, and then truck, can be calculated to better define material class. This aligns ShovelSense material classification with grade control material classification, reducing incorrect diversions due to incorrect material classification.

System reliability

Initial availability of the ShovelSense system was low, principally due to XRF head failure. Discussions with MineSense revealed a change in supplier of one of the components in the heads that was prematurely failing. This problem has now been mostly resolved with head failure significantly less than it was in the past. Current run hours on the heads have significantly improved with most heads now reaching or exceeding the expected 800 hrs of operational time. The current head 3 on FS620 has operated for over 2000 hrs.

For FS620, recent reliability issues are the result of damage to the main extension cable and the signal processing unit (SPU). The SPU is not located in a standard position for the system due to interference with the hydraulic lines to the bucket. FS620 is a Komatsu PC7000 shovel and one of only two in the world. The bucket, stick and hydraulic line routing is not common to any of the other Komatsu shovels and SPU placement during system installation caused some delays as a result of the fouling. Eventually the SPU was installed high on the outside of the bucket exposing the main cable to rock strikes during operation. This has caused some system downtime, however, cable replacement can be completed in a shift change.

Additional downtime for the system has been incurred as previously no maintenance on the ShovelSense systems was done outside of schedule PM windows. This has at times led to many unnecessary days of downtime for a relatively minor repair that could have brought the system back online and productively contribute to the success of Red Chris.

As an example, the system on FS620 was down for 11 days from 6–16 September 2023, due to extension cable and SPU damage. These issues were addressed in the PM from 14–16 September 2023, but could have been resolved earlier over one or more shift changes.

ShovelSense availability has been improving from September onwards with ShovelSense availability on FS601 exceeding that of the shovel. FS620 availability is also improving with 100 per cent availability in the month of November. Figure 16 shows the days in service for the two shovels and the corresponding ShovelSense systems.

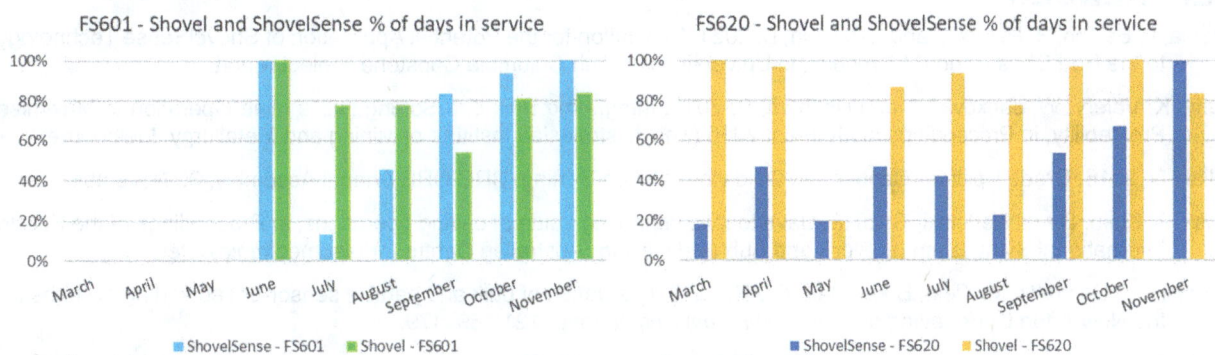

FIG 16 – Shovel and ShovelSense percentage of days in service.

PGNAA USAGE

Opportunities for greater integration of PGNAA outputs in plant operation

Full utilisation of the belt sensor as an input to the process control system is a work in progress. At present, copper and sulfur grades in feed to mill are monitored with intentions to begin to use these as inputs into the Met Accounting process. Additionally, sulfur grades in mill feed can be used as an input to reagent dosing for the flotation circuits.

NPR of the circuit tailings is an important criterion that needs monitoring as a measured value greater than two is required to avoid the requirement for sub-aqueous tailings deposition. The belt sensor is capable of measuring calcium, principally hosted in neutralising carbonates, and acid generating sulfur in plant feed. Using this data, a quick estimation of NPR in plant tailings can be generated. This will in turn minimise the need for sub-aqueous deposition and greater use of the rougher scavenging circuit for copper recovery rather than pyrite recovery as a PAG tailing stream.

A relationship using rock chemistry, measurable by the belt sensor, and Bond Work Index is being developed. This, when used in conjunction with recently installed SAG mill feed particle size cameras, will assist with isolating ore driven process deviations from operational upsets. When linked with pit location via ShovelSense, some forewarning of future process outcomes may be possible helping to continually maximise recovery.

CONCLUSIONS

Within the Red Chris deposit, there is significant *in situ* orebody heterogeneity at a scale much finer than that defined by grade control. Differences between truck scale and grade control scale heterogeneity allow for a waste rejection stage and additional ore recovery from waste between the mine and the mill. Estimation of heterogeneity conducted on downhole drill assay data was confirmed with actual measurements, and later truck diversions, providing confidence that a method for predicting bulk ore sorting potential from available *in situ* information is both feasible and realistic.

To date, ShovelSense has been responsible for the generation of an additional ~787 kt of ore, representing an 18.1 per cent increase in ore produced by the mine via the conversion of above cut-off grade waste material. When viewed as a strip ratio, the change is significant with the waste to ore ratio changing from 1.59 t of waste per tonne or ore to 1.19 t of waste per tonne of ore.

A combination of sensors has proven to be effective at measuring grades near cut-off to pre-process ore ahead of the mill with the potential for adaption to different mining techniques in development with field trials currently underway.

The use of multiple sensing system within an operation allows for ongoing cross-validation of measurements. This can then be used to detect individual system calibration drift and enhances operation wide grade data measurement confidence.

Belt sensing precision and measurement frequency can be fine-tuned depending on the application and the accepted measurement error by the end user.

REFERENCES

Acacia, L, Barton, B, Peters, J and Lapadula, B, 2021. Evaluation for the Potential Application of ShovelSense Technology to the Red Chris Open Pit Orebody, in University of British Columbia Capstone Project report.

Duffy, K, Valery, W, Jankovic, A and Holtham, P, 2015. Integrating Bulk Ore Sorting into a Mine Operation to Maximise Profitability, in Proceedings of MetPlant 2015 (The Australasian Institute of Mining and Metallurgy: Melbourne).

Miljak, D, 2016. Grade Uplift for Copper and Gold via Bulk Ore Sorting, CRC ORE Annual Assembly, 30 November.

Moss, A, Klein, B and Nadolski, S, 2018. Cave to mill: improving value of caving operations, in Proceedings of the Fourth International Symposium on Block and Sublevel Caving (Australian Centre for Geomechanics: Perth).

Nadolski, S, Samuels, M, Klein, B and Hart, C J R, 2018. Evaluation of bulk and particle sensor-based sorting systems for the New Afton block caving operation, *Minerals Engineering*, 121:169–179.

Noble, G, Strombotne, T and Gordon, K, 2024. Online Elemental Analysis of Copper Concentrator Feed Utilizing PGNAA Technology, in Proceedings 31st International Mineral Processing Congress.

Noble, G, Thompson, E and Sanhueza, I, 2022. Cross-belt elemental analysis, stockpile blending optimisation and waste bulk sorting in Copper using PGNAA, in *Proceedings of Copper 2022*, pp 171–182.

Redwood, N, 2018. Whittle Consulting - ShovelSense™ Economic Assessment, Whittle Consulting – Improving Mining Economics.

Rees, C, Brock Riedell, K, Proffett, J M, Macpherson, J and Robertson, S, 2015. The Red Chris Porphyry Copper-Gold Deposit, Northern British Columbia, Canada: Igneous Phases, Alteration and Controls of Mineralization, *Economic Geology*, 110:857–888.

Seaman, D R, Li, K, Lamson, G, Seaman, B A and Adams, M H, 2021. Overcoming rougher residence time limitations in the rougher flotation bank at Red Chris Mine, in *Proceedings Mill Operators Conference 2021,* pp 193–207 (The Australasian Institute of Mining and Metallurgy: Melbourne).

Stewart, R, Swanson, B, Sykes, M, Reemeyer, L, Wang, B and Stephenson, P, 2021. Red Chris Operations NI 43 101 Report.

Scats – what are they good for?

T McCredden[1] and C Fitzmaurice[2]

1. AAusIMM, Senior Metallurgist, IGO Ltd, South Perth WA 6151.
   Email: timothy.mccredden@igo.com.au
2. Group Manager Metallurgy, IGO Ltd, South Perth WA 6151.
   Email: craig.fitzmaurice@igo.com.au

ABSTRACT

Waste in mining operations is a function of economics and is entirely contingent on the treatment options considered and prevailing market conditions at any given time. As mined grades decline globally, waste generation (both total volume and as a fraction of total mined volume) increases. The resulting imperative is to improve primary treatment efficiencies and investigate options for economic retreatment of wastes. This is part of the drive toward a circular economy which focuses on reducing primary extraction through reusing or repurposing of existing material. At the Forrestania Nickel Operation, the Cosmic Boy Concentrator stockpiled ball mill scats over the course of 12 years. This resulted in the accumulation of >350 000 t of acid forming waste material with an average grade of 1.5 per cent nickel, representing a significant resource which was previously untreated due to potential grade displacement in the flotation plant, material hardness, and contamination with residual grinding media. A semi-mobile magnetic sorting circuit was designed and operated to treat this material by removing residual grinding media while upgrading the nickel. The circuit successful upgraded the nickel contents by 1.4 times, with nickel recovery to the accepts fraction exceeding 80 per cent while additionally reducing material hardness. This provided a relatively low-cost feed source for the flotation plant. The magnetic sorting circuit effectively reduced the waste stockpile mass by 60 per cent with the remaining waste material being earmarked for usage in mine closure plans. This has resulted in the complete utilisation of a material previously considered to be waste. This process has served two purposes in that it generated significant economic value while reducing the waste burden. It is a reminder to operating sites to revisit potential waste streams to explore value recovery and to forward plan the storage of waste for future use.

INTRODUCTION

Total mineral and metal production is growing exponentially to meet demand (Mudd, 2010). Simultaneously mined ore grades are declining, often with increasing mineralogical complexity (Mudd, 2010). To compensate, organisations are leveraging off technological developments to improve extraction efficiency and/or utilising larger resources to benefit from economies of scale. This, in part, has driven the lower mined ore grades as more deposits become economically viable but require increasing processing intensity and waste generation (Rötzer and Schmidt, 2018). This trend highlights the importance of moving away from a purely linear approach in regards to the generation of mining waste and tailings (Kinnunen and Kaksonen, 2019). With circular economics the aim is to reduce waste by utilising all processing steams (waste or otherwise) for economic gain. Within mining there has been instances of the reprocessing of historical tailings or waste stockpiles for net benefit (Talison Lithium, 2024; Tailings.info, 1989). This has usually been facilitated by high historical tailings grade, technological changes, and increased commodity prices, but long after its initial generation. The circular approach requires a change in mindset when approaching waste with a focus on minimisation or utilisation concurrent to the operation or at least planning for that end (Kinnunen and Kaksonen, 2019). By considering tailings or waste stockpiles as a potential future value source it allows operations to develop strategy around how this material should be stored, contained, or treated to maximise extractable value. This is incentivised by the significant sunk cost in terms of money, energy, and effort to produce these wastes (Kinnunen and Kaksonen, 2019; Curry, Ismay and Jameson, 2014).

At the Forrestania Nickel Operation (FNO), the Cosmic Boy Concentrator (CBC) stockpiled ball mill scats near the tailings storage facility (TSF) over the course of 12 years (Figure 1). Ball mill scats are generated during the milling process whereby harder ore material (usually lower grade) is not sufficiently ground at a given throughput rate to pass through the trommel discharge screen, and exit the circuit as a waste product. Similarly, as grinding media wears, this becomes mixed in with the discharging scat material. At FNO, this process produced a stockpile which contained >350 000 t. Due to the low-grade nature of the scats (~1.5 per cent nickel), reprocessing through CBC was not initially considered due to potential grade displacement. Flying Fox (FF) and Spotted Quoll (SQ) mined material was much higher nickel grade and while CBC was operating at full capacity, utilising any scats would reduce the overall nickel concentrate tonnes produced. Additionally, the scats posed other processing challenges due to its material hardness and the presence of the residual grinding media which could cause crusher damage.

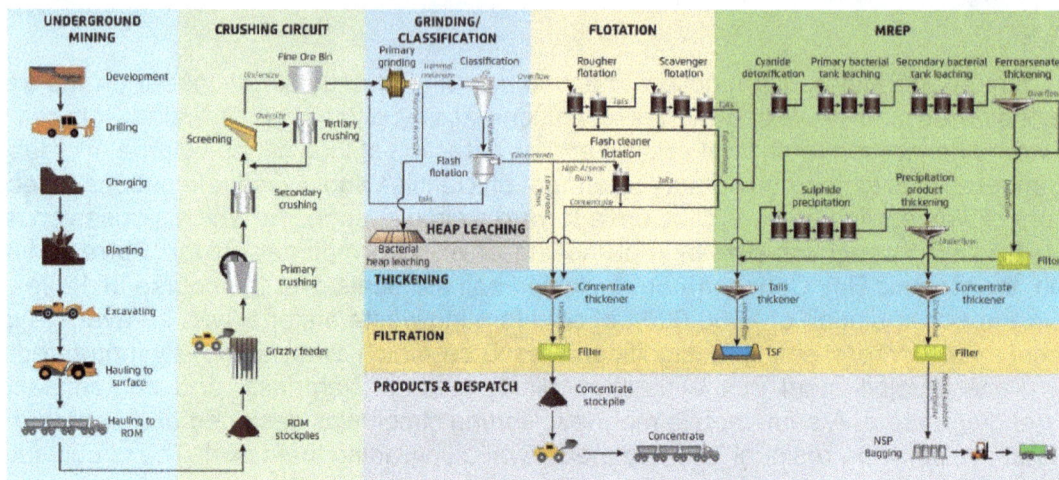

FIG 1 – FNO flow sheet.

In May 2021 a heap leaching pilot was commissioned using ball mill scats as feed material, with the intention of escalating to a full-scale operation. The timeline for the construction and operation of a heap leach is typically years and can have significant capital costs associated (Petersen, 2016). Early data from the pilot operation indicated that, although it is a viable processing method, the payback period for a full-scale scats heap leach would be too long. Additionally, forecast of declining production in terms of total nickel grade and tonnes of ore from the FF and SQ mines signalled a possible increase to CBC unit costs. This prompted an exploratory test program investigating beneficiation methods for ball mill scats to act as a feed to the flotation plant.

Proof of concept

Steinert Australia Pty Ltd were approach to conduct ore sorting tests, however this was excluded due to the size distribution of ball mill scats (P_{80} <10 mm) not being suited to sorting technology. The material was more suited to dry magnetic drum separation, with variability testing completed at Steinert Australia Pty Ltd from December 2021 to February 2022. This showed moderate nickel recoveries (66–70 per cent) to the accepts (magnetic) stream, with nickel upgrade ratios of >1.4 using a magnetic strength of 3320 Gauss (at drum surface). Testing at higher Gauss (8000) increased the nickel recovery to 90 per cent with similar upgrade ratios. The accepts were shown to have a lower Bond ball mill work index (2–3 kWh/t less than feed) due to the rejection of harder silicate gangue material. Laboratory scale flotation of the accepts achieved saleable concentrate grade specifications at nickel recoveries ranging from 70–75 per cent. This indicated that the scats material could be magnetically upgraded to increase nickel grade, improve grindability and be effectively floated/recovered. The residual grind media had to be removed due to possible damage it would cause to the crushers and the increased general wear to CBC. This grinding media removal was facilitated by bucket screening the feed and using a cross belt magnet within the sorting circuit. The option was available for the accepts to be fed via the emergency feeder bypassing the crushers. This was not pursued as it added significant additional processing costs (loader and labour), reduced the blending with other material and prevented any further particle size reduction prior to milling.

Design

The circuit design was a compromise between time, location, and cost. The primary constraint was time. Due to forecast reductions in the production from the FF and SQ mines, the scats became an important possible feed source to assist in reducing CBC unit costs. The secondary constraint was where the scats were located (adjacent to the TSF, Figure 2). To reduce the material transport costs component, the scats treatment circuit needed to be located near the stockpile where there were no existing services or other structures, necessitating an off-grid and semi-mobile (skid mounted) design. These constraints reduced the viable processing options and resulted in the project being divided into two stages.

FIG 2 – Top Left: Aerial image of scats stockpile; Top Right: Scats stockpile from TSF wall; Bottom Left: Bucket screening; Bottom Right: Oversized material removed via bucket screening.

Stage 1 was intended to allow for early material production to supply feed material to CBC, without significant focus on the sorting plant throughput rate or nickel recovery to accepts. To achieve this, stage 1 repurposed a second-hand magnetic drum from an eddy current separator. This consisted of a vibratory feeder with a rare earth magnetic drum 3320 Gauss mounted on a new basic skid structure. The drum was fed by a front-end loader via a mobile conveyor, with the products being discharged through chutes onto two mobile conveyors. A cross-belt magnet removed steel prior to the magnetic drum. The circuit was powered by a diesel genset in an area lit by mobile lighting towers. The circuit was centrally controlled though a purpose-built electrical skid switchboard with a PLC designating start/stop sequences and an integrated emergency stop circuit. The stage 1 circuit was designed with the intention of implementing stage 2, so all equipment was planned around being repurposed in the later upgrade. Most of the equipment was purchased outright to limit the impact of downtime on operating costs. This allowed the total timeline from capital approval to first production to be reduced to eight weeks.

Stage 2 was focused around improving throughput (t/h) and nickel recovery utilising an improved design and equipment, now possible as the time constraint of providing some accepts material to feed CBC was fulfilled by the stage 1 plant. This was achieved by adding an additional second-hand magnetic separator which was identical to the first albeit 0.5 m wider. To improve separation efficiency and nickel recovery, both magnetic drum arrays were replaced. The new arrays were arranged radially with a magnetic strength of 5500 Gauss. A two-part skid was designed to contain walkways, chutes, vibratory feeders, and the magnetic drums. This was designed and constructed with the ability to be transported to and operated at the remote stockpile location. Additional conveyors were added to facilitate the increased production to a nominal target of ~75 t/h of feed (Figures 3 and 4).

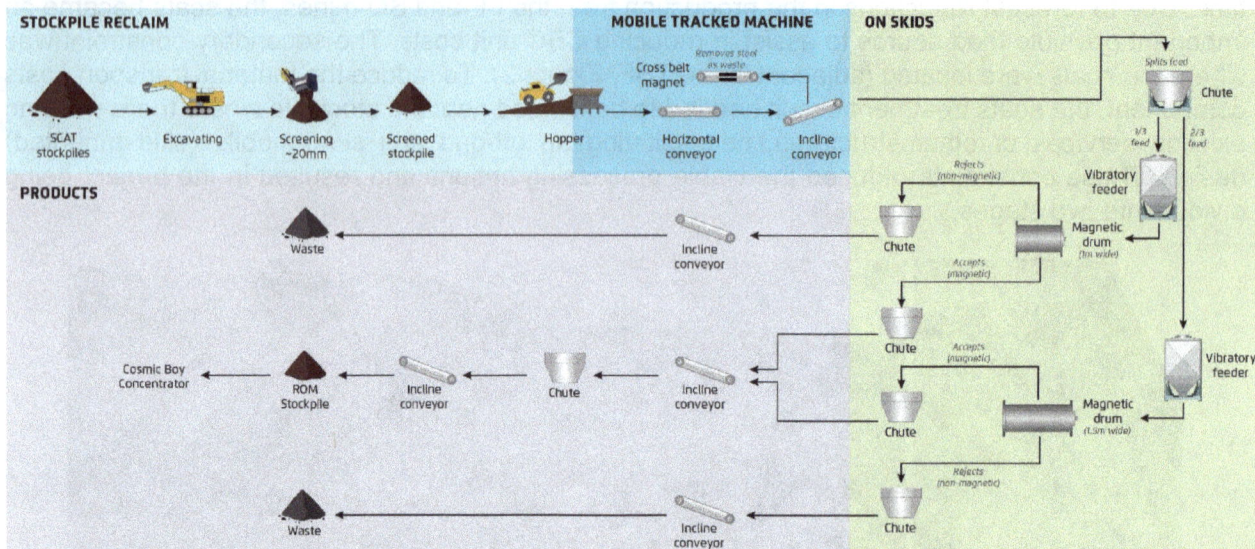

FIG 3 – Scats magnetic sorting project stage 2 flow sheet.

FIG 4 – Scat magnetic sorting project circuit.

Reclaiming material from the stockpile prior to treatment proved to be relatively technical, with several benches being constructed during de-stacking due to limited space and stockpile height. Once de-stacked, the material was bucket screened prior to feeding into the circuit. This was to breakup of any clumps of material formed due to oxidation and to remove the larger residual steel grinding media (>20 mm, Figure 2).

RESULTS

The total capital expenditure for the project was <A$2 M with a payback period of <30 days. The circuit had operated for 603 days (as at EOM January 2024) with planned production until October 2024. Rejects retreatment may be possible depending on the economics, which could further extend operations.

Stage 1

Stage 1 of the circuit operated for 17 weeks prior to being upgraded. This allowed the production of 33 000 t of accepts with an average nickel grade of 2.38 per cent, which were all fed into the CBC over the same period. The achieved nickel upgrade ratio ranged from 1.33–1.38 with nickel recoveries of 68–72 per cent. The low recoveries were expected due to the lower magnetic strength of the drum. Stage 1 rejects have been stockpiled for potential retreatment using the stronger magnetic drums.

Stage 2

Stage 2 production can be separated into the early and late-stage material. This is reflective of the mining source used at the time of production of the mill scats. FF material upgraded more effectively than SQ mine material and represented a greater proportion of the early-stage production. The upgrade difference was mineralogically driven as SQ material had more magnetically recoverable gangue. The production profile is shown in Figures 5 and 6 with cumulative nickel tonnes for different streams compared with average London Metal Exchange (LME) prices, nickel grade and recoveries. There is a clear reduction in the upgrade ratio and some nickel recovery loss in the in later stage of operation, in part due to mineralogical changes in feed. As at EOM January 2024, most of the nickel (54 per cent) originally contained in the stockpile was recovered to accepts early during stage 2 which benefited from high LME prices, with the average price over this period being ~A$37 000/t. An additional side benefit of the magnetic upgrade was the reduction in the nickel to arsenic ratio, as less arsenic was recovered relative to the nickel (Table 1).

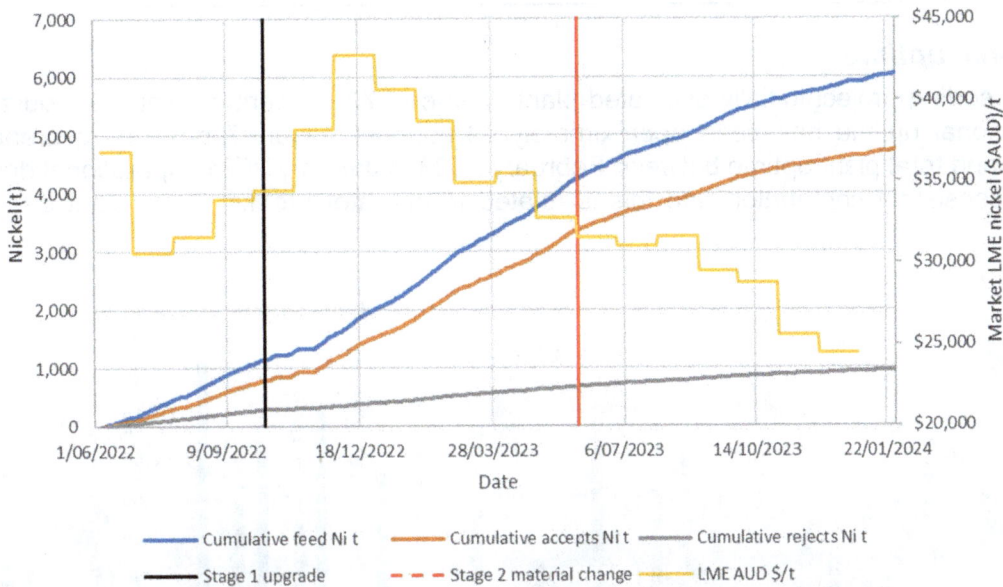

FIG 5 – Cumulative feed, accepts and reject nickel tonnes and market LME values over-time.

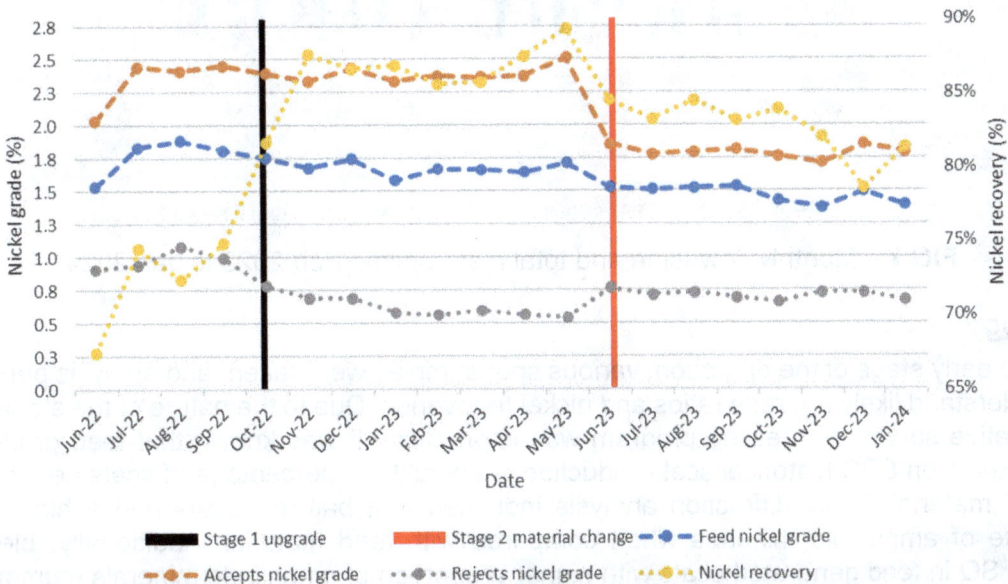

FIG 6 – Monthly data feed, accept and reject nickel grade and nickel recovery.

TABLE 1

Total accepts production to EOM January 2024.

| | Stage 1 | Stage 2 | | |
	Total	Early	Late	Total
Tonnes (t)	33 043	107 618	76 843	217 504
Ni (t)	787	2570	1384	4740
Ni (%)	2.38	2.39	1.80	2.18
Ni Recovery	72.92	87.02	83.04	83.19
As Recovery	62.86	74.64	75.07	75.07
Ni Upgrade Ratio	1.38	1.39	1.26	1.34

Operational uptime

The scats sorting project initially estimated plant uptime at 70 per cent of available hours. To-date the operational uptime has been approximately 71 per cent. Figure 7 provides a breakdown of downtime and total plant uptime between February 2023 to January 2024. Operational downtime is the most consistent contributor, and this is related to operator breaks, fly-in, plant clean up and handovers.

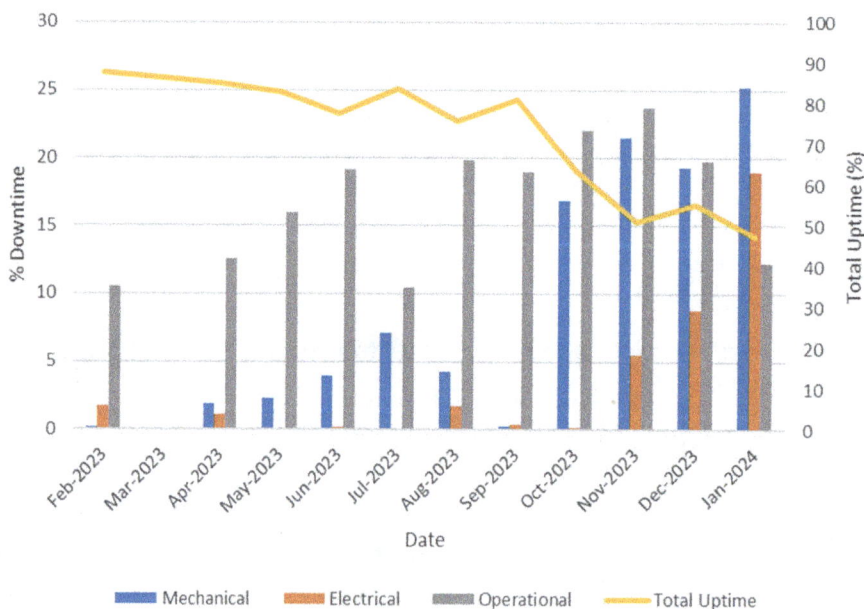

FIG 7 – Monthly downtime and total plant uptime Feb 2023 to Jan 2024.

Mineralogy

During the early stage of the operation, various spot samples were taken, and analysis performed to better understand likely upgrade ratios and nickel recoveries. Due to the nature of the stockpile a full representative survey and testing program was impractical. It was known that feed grades would decline (based on CBC historical scat production) as would the percentage of scats generated from FF mined material. X-ray diffraction analysis indicated that ball mill scats had a higher relative percentage of amphibole minerals when compared with feed material. Additionally, blends with increased SQ in feed generated scats with higher proportion of amphibole minerals (cummingtonite and actinolite) than FF blends. These amphibole minerals were the driver behind the declining grade upgrade in later part of stage 2 of the operation. Some scanning electron microscope work was conducted on as-received fractions to identify nickel deportment and distribution within the feed, accepts and rejects (Figure 8). The purpose of this was a qualitative assessment as full quantitative analysis on coarse particles would have require significant cost and sample numbers to meet statistical requirements. The imagery shows a clear upgrade in the accepts nickel content compared with the rejects, with the nickel containing minerals primarily represented by pentlandite, gersdorffite and nickeline (Figure 8).

FIG 8 – Accepts samples are in top row and reject bottom row; Left: Condense mineral list; Centre: Nickel heat map; Right: Accepts silicon, nickel, sulfur and arsenic distribution.

Impact on CBC plant

As at EOM January 2024, 214 587 t of accepts had been feed into the CBC. This represented 24 per cent of the total mill feed tonnage (20 per cent of total nickel metal in feed) over the period from June 2022 to January 2024 (Figure 9). Several minor operational changes were made to facilitate the addition of accepts as feed into CBC. The main concern was any residual grinding media damaging the crushing circuit. This was primarily in the form of small steel balls (~15 mm) as these could pass the bucket screen and be potentially missed by the cross-belt magnet. To address this, the maintenance department made minor adjustments to secondary and tertiary crusher gaps, changed out the cone crusher liner five to six days earlier than usual and made a modification to the liner attachment point. This minimised any additional damage or wear because of residual grinding within the accepts.

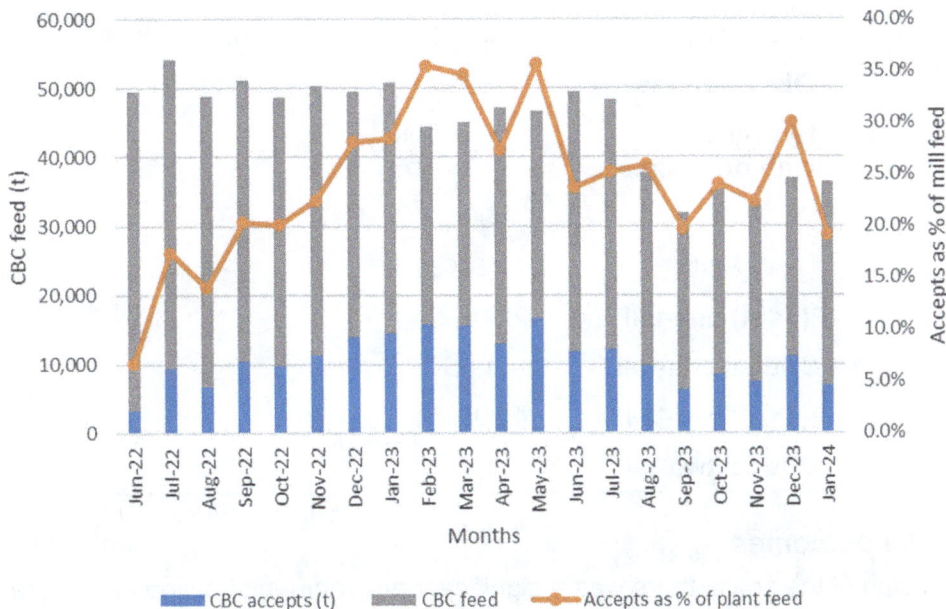

FIG 9 – Accepts as % of total CBC feed overtime.

The impact on plant operations was assessed using multiple linear regression techniques using daily plant data from the period January 2022 to December 2023. The analysis compared the impact of the accepts t/hr and nickel t/hr on several factors as shown in Table 2. The accepts as feed, had two unique impacts; increasing the t/hr resulted in an increase in the cyclone overflow (COF) P_{80} and increasing nickel t/hr reduced CBC nickel concentrate recovery. The COF P_{80} wasn't shown to be a significant determinant in CBC final concentrate nickel grade or recovery. In terms of the accepts impact to nickel recovery, a t-test showed that there is a significant nickel grade difference between the accepts and the remaining mill feed (Table 3). This indicates that the negative recovery effect can, in part, be explained by a typical grade/recovery relationship. As such, a lower nickel recovery rate was assigned to the accepts compared to fresh feed. This lower recovery was offset by the very low unit cost rate of the accepts relative to fresh feed (as mining and other historical costs were already 'sunk').

TABLE 2

Summary of multiple linear regression data: impact of accepts.

Change	Outcome	Factor	Comments
Increasing accepts t/hr	Increases	Scatting rate (Ni t/hr)	Equivalent impact as other feed sources.
	Increases	Scatting rate (t/hr)	Equivalent impact as other feed sources.
	Increases	COF P_{80}	Not a major determinant in concentrate recovery.
Increasing accepts Ni t/hr	Decreases	Scatting rate (t/hr)	Equivalent impact as other feed sources.
	Decreases	Concentrate nickel recovery (%)	See t-test regarding feed grade.

TABLE 3

Accepts and other feed grade.

T-Test – two-sample assuming unequal variances		
	Accepts nickel grade (%)	**Other feed nickel grade (%)**
Mean	2.10	2.56
Variance	0.14	0.27
Observations	521	541
Hypothesized mean difference	0	
df	981	
t Stat	-16.48	
P(T<=t) one-tail	2.50E-54	
t Critical one-tail	1.65	
P(T<=t) two-tail	4.99E-54	
t Critical two-tail	1.96	

Environmental outcomes

Prior to processing of the scats, there was a significant environmental obligation for the appropriate disposal or treatment of this material. The untreated scats had a stockpile of >350 000 t with a net acid production potential (NAPP) of 97.6 kg/t. As of EOM January 2024 there is a total of 134 000 t of reject material representing a 60 per cent mass reduction. Additionally, there is a 90 per cent reduction in its acid generating capacity (NAPP of 4–10 kg/t). These rejects have been earmarked for use as a capillary break or fill material for the TSF. This has significantly reduced FNO remediation costs.

Lesson learnt

The magnetic sorting circuit was effective in achieving its objectives; however, some modifications were required due to the rapid project implementation schedule. Many of these were implemented during the first six months of operations. Several key areas have been identified where improvement could have been made, given more time and planning or a more formal commissioning period.

In terms of circuit design, changes would have been made to the feed chute, feed hopper ramp and the inclusion of a vibrating screen. The feed chute didn't evenly distribute feed to the magnetic drums as planned and this reduced throughput. The feed hopper access ramp made maintenance difficult and prevented the feeder unit being tracked. The inclusion of a vibrating screen at the feed hopper would have removed any mid-size (12–15 mm) residual grinding media and reduced conveyor belt tearing from shattered waste rock (a result of the bucket screening process). In the initial design a static screen was included but proved unsuitable.

A significant difficulty in this project was the sourcing of ongoing and appropriate contract maintenance labour. This was in part due to market conditions (retention issues) and role requirements. The nature of the mobile equipment necessitates significant preventative maintenance to continually operate. The on-site spares were also critical. As such the project had significant downtime due to lead times for parts, breakdown situations (lack of preventative maintenance) or the availability of CBC maintenance personnel (Figure 7). As most equipment was purchased outright, this cost was lost contractor labour hours.

The cost and time required for stockpile reclamation was underestimated. This was a difficult undertaking due limitations in space and the stockpile height. The height of the heap meant several benches were created to safety de-stack. The requirement of bucket screening meant space needed to be allocated to 'throw down' material and another area allocated for screening. This increased the complexity of processing with the use of multiple temporary stockpiles and ever-changing traffic management plans.

The last area of improvement would have been in project operations. No additional internal staff were hired for the project, so once operational, managing and maintaining the circuit was spread across personnel with existing responsibilities. This inevitably reduced the productivity of the project. Several safety incidents resulted from insufficient equipment protections and poor contractor management. In retrospect, additional internal staff may have been merited to manage safety and maximise the value from the project.

REFLECTION

So, what are scats (or any other waste) good for? For FNO they represented a timely and economic feed source, the overall translation of waste into value. This project isn't necessarily an example of exceptional plant construction (in terms of efficiency or design) but was primarily about timing and the suitability of solving the immediate problem. This processing solution required very low capital investment and its timely implementation captured 54 per cent of the contained nickel in what was formerly considered a 'waste' stockpile at peak LME prices and was economically successful. The alterative options would have required significant capital investment and had long lead times, which in effect would have resulted in operating at a much lower LME nickel price. The magnetic sorting circuit has had the additional benefit of reducing FNO environmental remediation costs by removing 60 per cent of the stockpile mass and 90 per cent of its acid generating capacity. The reuse of the ball mill scat waste stockpile works within the circular mindset, by reusing waste to generate value. The original ball mill scat stockpilers didn't know the future processing outcome so barren waste rock among other contaminants were stockpiled within the heap and added unnecessary processing complexity. The challenge is for operators to continually consider waste streams a potential value source, for now or the future. The question to ask is can it be utilised concurrent with operations and, if not, how should you store, treat, and contain the material for future potential exploitation.

ACKNOWLEDGEMENTS

Those who assisted in the circuit design and supply: Steinert Australia Pty Ltd, Transeng, OPS Crushing and Screening and True North Renewables.

Contractors who operated the circuit: RAMS Goldfields.

IGO Ltd: CBC team: Production coordinators, maintenance teams, environmental department, metallurgists, and laboratory staff; BioHeap Ltd team; Paul Hudson and Shane Taylor.

REFERENCES

Curry, J A, Ismay, M J L and Jameson, G J, 2014. Mine operating costs and the potential impacts of energy and grinding, *Minerals Engineering*, 56:70–80.

Kinnunen, P H M and Kaksonen, A H, 2019. Towards circular economy in mining: Opportunities and bottlenecks for tailings valorization, *Journal of Cleaner Production*, 228:153–160.

Mudd, G M, 2010. The Environmental sustainability of mining in Australia: key mega-trends and looming constraints, *Resources Policy*, 35:98–115.

Petersen, J, 2016. Heap leaching as a key technology for recovery of values from low-grade ores – A brief overview, *Hydrometallurgy*, 165:206–212.

Rötzer, N and Schmidt, M, 2018. Decreasing Metal Ore Grades—Is the Fear of Resource Depletion Justified?, *Resources*, 7:88.

Tailings.info, 1989. Kaltails Project, *Tailings.info*. Available from: https://tailings.info/casestudies/kaltails.htm

Talison Lithium, 2024. Greenbushes-Project, *Talison Lithium*. Available from: https://www.talisonlithium.com/greenbushes -project [Accessed: 12/03/2024].

AUTHOR INDEX

Anghag, M	359	Gerold, C	573
Anis, F	59	Ghattas, R	195
Assmann, S	271	Gnoinski, G B	27
Avenido, N	359	Goode, I	159
Baaken, S	573	Greenhill, P G	95
Bai, G	159	Greet, C J	313
Baldock, J	385, 495	Griffin, P	463
Ballantyne, G	509	Guerney, P	17
Barsby, S	69	Gwynn-Jones, S	39
Bartholomew, K	359	Haines, C	195
Becker, M	211	Hanhiniemi, J J	323, 343
Begelhole, J	541	Heard, G	211
Bennett, D W	3	Heath, I	195
Bill, A	249	Heo, J	323
Bohorquez, J	39	Insalada, A	359
Brennan, L	525	Jain, A	59
Brown, L	81	Jankovic, A	463
Buaseng, S	343	Jessop, J	27
Buckman, J	299	Johannson, J	495
Burns, F	271, 475, 525	Kaartinen, J	121
Chandramohan, R	177, 249	Kanchibotla, S	427
Cornish, B	551	Kilcullen, A	323
Curtis, M D	107	Klein, B	587
David, D	573	Kramer, C	39
Downie, W	195	Lane, G	177, 509
Duffy, K	463, 525	Li, K	385
Ehrig, K	69	Li, Y	69
Everitt, M	573	Liang, H	81
Felipe, D	17	Liebezeit, V	69
Fiedler, K	147	Lipton, I T	95
Figueroa Salguero, G D	299	Lowes, C P	107
Fitzmaurice, C	605	Mainza, A N	427
Foggiatto, B	509	Malcolm, A	159
Futcher, W	587	Marcera, A J	359
Garside, M J	107	Martin, C	159

McCaffery, K M	27	Shean, B	313
McCredden, T	605	Silva, L	59
McIvor, R E	359	Simpson, M	231
Meinke, C	249	Sims, D	39
Mendoza, J	249	Siphanya, A	401
Moilanen, J	121	Smith, S	39
Mort, E	231	Smith, W	159
Munro, P D	3, 147, 413	Stanton, C	573
Newcombe, B	133	Stevens, J	211
Newcombe, G	133	Tavares, L M	427
Newell, A J H	147	Thanasekaran, H	323
Ntamuhanga, S	299	Thomson, J	509
Occena, E	359	Tlhobo, T	59
Ogden, T A	39	Toor, P	525
Olwagen, D	573	Torok, I A	541
Palmer, B	299	Torres, D	359
Paz, A	401	Torres, L	95
Peachey, G	121	Tsatouhas, G	271
Pease, J D	413	Turner, M	39
Powell, M S	427, 445	Valery, W	463, 525
Price, A	475	van Saarloos, B	231
Pyle, L	463	Van Sliedregt, J	475
Radford, R	211, 231	Vansana, K	401
Rajala, J	359	Villalvazo, A	249
Reinhold, J	271, 475, 525	Vollert, L	195
Remes, A	121	Wang, E	323
Rice, A	463	Weerasekara, N S	323, 343
Schmitz, C	573	Whittering, R	249
Seaman, B A	385, 463, 495	Wong, B	81
Seaman, D R	385, 495, 587	Zela, J	551
Seppelt, J	271, 475, 525